Essentials of Physical Chemistry

Essentials of Physical Chemistry

Editor: Finn Miller

NYRESEARCH
P R E S S

New York

Published by NY Research Press
118-35 Queens Blvd., Suite 400,
Forest Hills, NY 11375, USA
www.nyresearchpress.com

Essentials of Physical Chemistry
Edited by Finn Miller

© 2017 NY Research Press

International Standard Book Number: 978-1-63238-552-9 (Hardback)

Cataloging-in-Publication Data

Essentials of physical chemistry / edited by Finn Miller.
    p. cm.
Includes bibliographical references and index.
ISBN 978-1-63238-552-9
1. Chemistry, Physical and theoretical. 2. Chemistry. I. Miller, Finn.
QD453.3 .E87 2017
541--dc23

Printed in the United States of America.

Contents

Preface

The main aim of this book is to educate learners and enhance their research focus by presenting diverse topics covering this vast field. This is an advanced book which compiles significant studies by distinguished experts. This book addresses successive solutions to the challenges arising in the area of application, along with it; the book provides scope for future developments.

This book is an essential guide for both academicians and those who wish to pursue their interest in physical chemistry. Physical chemistry focuses on understanding chemical systems with respect to the diverse principles of physics. Different approaches, evaluations, methodologies and advanced studies on physical chemistry have been included herein. This book is compiled in such a manner that it will provide in-depth knowledge about the theory and practice of this field. It will serve as a reference to a broad spectrum of readers. With its detailed analyses and data, this book will prove immensely beneficial to professionals and students involved in physical chemistry at various levels.

It was a great honour to edit this book, though there were challenges, as it involved a lot of communication and networking between me and the editorial team. However, the end result was this all-inclusive book covering diverse themes in the field.

Finally, it is important to acknowledge the efforts of the contributors for their excellent chapters, through which a wide variety of issues have been addressed. I would also like to thank my colleagues for their valuable feedback during the making of this book.

Editor

A highly efficient polysulfide mediator for lithium–sulfur batteries

Xiao Liang[1], Connor Hart[1], Quan Pang[1], Arnd Garsuch[2], Thomas Weiss[2] & Linda F. Nazar[1]

The lithium–sulfur battery is receiving intense interest because its theoretical energy density exceeds that of lithium-ion batteries at much lower cost, but practical applications are still hindered by capacity decay caused by the polysulfide shuttle. Here we report a strategy to entrap polysulfides in the cathode that relies on a chemical process, whereby a host—manganese dioxide nanosheets serve as the prototype—reacts with initially formed lithium polysulfides to form surface-bound intermediates. These function as a redox shuttle to catenate and bind 'higher' polysulfides, and convert them on reduction to insoluble lithium sulfide via disproportionation. The sulfur/manganese dioxide nanosheet composite with 75 wt% sulfur exhibits a reversible capacity of 1,300 mA h g^{-1} at moderate rates and a fade rate over 2,000 cycles of 0.036%/cycle, among the best reported to date. We furthermore show that this mechanism extends to graphene oxide and suggest it can be employed more widely.

[1] Department of Chemistry, University of Waterloo, 200 University Avenue West, Waterloo, Ontario, Canada N2L 3G1. [2] BASF SE, 67056 Ludwigshafen, Germany. Correspondence and requests for materials should be addressed to L.F.N. (email: lfnazar@uwaterloo.ca).

As the development of portable electronics devices, electric vehicles and large-scale energy storage increases, so do the demands for energy storage batteries with high energy density and long service life. At the same time, it is recognized that traditional lithium-ion batteries are approaching their theoretical energy density limits[1–4]. Lithium–sulfur batteries are one of the most promising candidates to satisfy emerging market demands[5–7], as they possess a theoretical capacity and energy density of 1,675 mA h g^{-1} and 2,500 kW kg^{-1}, respectively, superior to current lithium-ion batteries[8,9]. In addition, they present an inherently low competitive cost due to the high natural abundance of sulfur. These advantages suggest that the lithium–sulfur battery should be capable of energy storage several times greater than conventional lithium-ion batteries, with reduced cost. However, practical applications are currently hindered by several obstacles. These predominantly relate to the insulating nature of sulfur and lithium sulfides, which require addition of conductive additives (hence lowering the active sulfur mass fraction), pronounced capacity fading on cycling and an internal redox shuttle that lowers Coulombic efficiency[10]. Over decades, much effort has been expended to try to solve these problems by trapping the polysulfides within the cathode structure. The most popular approaches are to encapsulate sulfur in the pores of carbon materials or a conductive polymer matrix. Meso/microporous carbons[11–14], carbon spheres[15–17], carbon nanotubes/fibres[18–20], polyaniline[21], polypyrrole[22,23] and poly(3,4-ethylenedioxythiophene)[24] sulfur-host cathodes have all had a positive effect on increasing the cycle life of lithium–sulfur cells. However, such architectures and polymer coatings can only partially retain polysulfides and are limited in their function over time, owing to structural changes that arise from the 80% volume change of the sulfur cathode on discharge. Two-dimensional materials such as graphene oxides (GOs) have also been very successfully employed as cathode hosts to increase the performance of the lithium–sulfur battery, where interaction between the oxygen/nitrogen functional groups and sulfur/polysulfides has been suggested[25–28]. Similarly, binding polysulfides onto hydrophilic metal oxide hosts was shown to significantly aid in maintaining high-capacity retention[29–31]. More recently, high-surface-area polar metallic oxides have been used as a two-in-one approach to provide both a 'sulfiphilic'

surface and supply electron transport to effect surface-enhanced redox chemistry[32]; or to spatially locate Li_2S deposition and enhance redox[33]. Organometallic redox mediators are shown to provide better utilization of Li_2S (ref. 34). Additives such as P_2S_5 also effectively control Li_2S deposition and aid in elimination of the polysulfide shuttle[35].

Herein, we present a quite different chemical approach to polysulfide retention in the sulfur cathode, which relies on mediating polysulfide redox through insoluble thiosulfate species in a two-step process. The thiosulfate groups are first created *in situ* by oxidation of initially formed soluble lithium polysulfide (LiPS) species on the surface of ultra-thin MnO_2 nanosheets. As reduction proceeds, the surface thiosulfate groups are proposed to anchor newly formed soluble 'higher' polysulfides by catenating them to form polythionates and converting them to insoluble 'lower' polysulfides. The polythionate complex formed on the surface is thus best described as a transfer mediator. This process curtails active mass loss during the discharge/charge process and supresses the polysulfide shuttle to result in high-performance cathodes with high sulfur loading, capable of high-capacity retention of 92% after 200 cycles at a C/5 rate and cycling over 2,000 cycles at 2C with a capacity decay of 0.036% per cycle.

Results

Synthesis and characterization. The MnO_2 nanosheets were prepared by reducing GO with $KMnO_4$ (ref. 36). Transmission electron microscopy (TEM) and scanning electron microscopy (SEM) images (Fig. 1a,b) reveal their two-dimensional lamellar structure. The selected area electron diffraction pattern of the MnO_2 nanosheets shows weak diffraction rings characteristic of δ-MnO_2, and the high-resolution TEM image in Fig. 1c exhibits a 0.25-nm lattice spacing corresponding to the (100) planes. X-ray diffraction (Fig. 1d) analysis confirms the MnO_2 nanosheets are monoclinic birnessite, δ-MnO_2 (JCPDS-01-080-1098). Birnessite is a phyllomanganate, a manganese oxide containing predominantly Mn^{4+} cations assembled in layers of edge-sharing octahedra. A layer charge deficit arises from the presence of Mn^{3+} cations and/or vacant-layer octahedra and is compensated by the presence of interlayer cations, which are hydrolysable K^+ in this case. Water molecules are also present between the

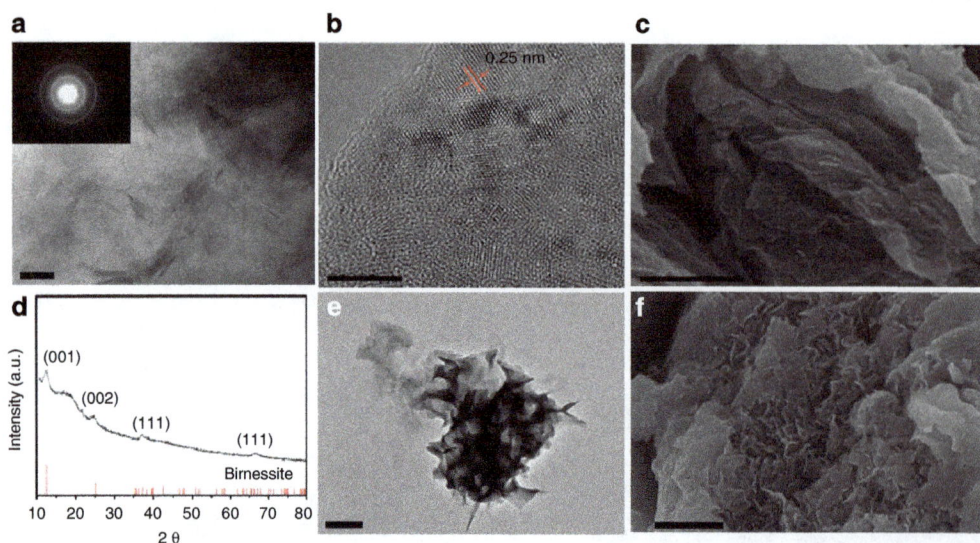

Figure 1 | Morphology and characteristics of the MnO₂ nanosheets and S/MnO₂ composite. (a) TEM image of MnO_2 nanosheets and its corresponding selected area electron diffraction pattern, **(b)** high-resolution TEM image of MnO_2 nanosheets, **(c)** SEM image of the MnO_2 nanosheets, **(d)** X-ray diffraction pattern of MnO_2 nanosheets, **(e)** TEM and **(f)** SEM images of the S/MnO_2 nanosheets composite. Scale bars, 50 nm **(a)**; 100 nm **(b,c,f)**; 200 nm **(e)**.

sheets of MnO_6 octahedra, coordinated both to the Mn and the K, with some fraction of the former being present as hydroxyl groups[37]. Thermogravimetric analysis (TGA, Supplementary Fig. 1a) reveals that the as-prepared MnO_2 nanosheets possess 8 wt% surface-absorbed water and 3 wt% water in the interlayers. Melt diffusion at $155\,^\circ C$ was used to uniformly disperse sulfur onto the surface of the MnO_2 nanosheets[28–30]. To maximize contact, the nanosheets were first pre-dispersed by sonication before being mixed with sulfur nanoparticles (see Methods) in a 25:75 weight ratio (Supplementary Fig. 2). The TEM and SEM images in Fig. 1e,f show a homogeneous sulfur coating on the nanosheet surface after melt diffusion, as confirmed by energy dispersive X-ray spectroscopy mapping (Supplementary Fig. 3). The targeted high sulfur loading of 75 wt% was achieved (TGA analysis of $75S/MnO_2$, Supplementary Fig. 1b).

Electrochemical properties. Electrochemical behaviour evaluated at different current densities for the $75S/MnO_2$ nanocomposite, showed characteristic Li-S voltage profiles (Fig. 2a)[12,14]. The cells exhibit a capacity of $\sim 1{,}300\,mA\,h\,g^{-1}$ at C/20 (discharge/charge of full theoretical capacity in 20 h). At a 20-fold higher current density (a C rate), only a modest drop in capacity to $950\,mA\,h\,g^{-1}$ was observed, suggestive of highly efficient kinetics. Most notable is the lack of overpotential in the initial charging process even on the first cycle, indicative of a good Li_2S interface with the MnO_2 nanosheets. Figure 2b shows the cycling stability of the cells at different rates. At C/5, the initial discharge capacity was $1{,}120\,mA\,h\,g^{-1}$, with $1{,}030\,mA\,h\,g^{-1}$ being sustained after more than 200 cycles and representing excellent capacity retention of 0.04% per cycle. At higher current densities (1C), the $75S/MnO_2$ composite still delivered exceptional reversible capacity of $800\,mA\,h\,g^{-1}$ after 200 cycles. This

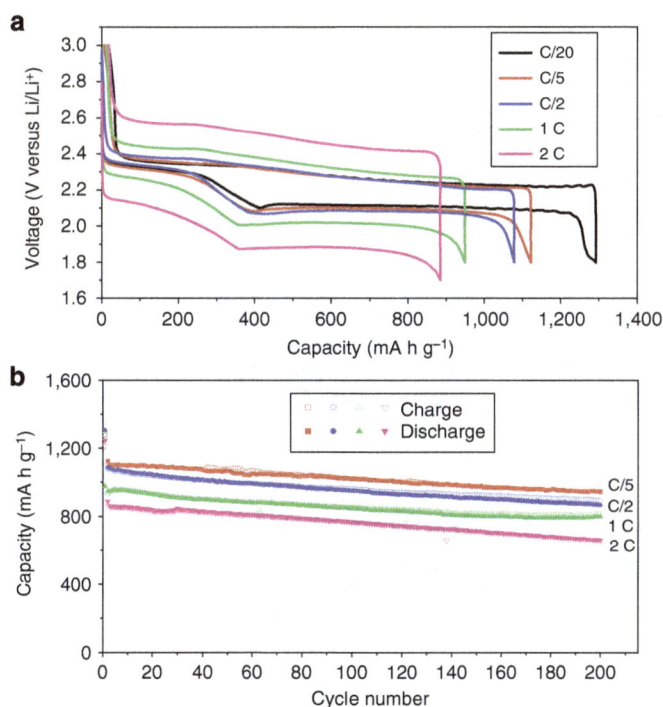

indicates that processes such as loss of electrical contact of Li_2S or sulfur with the conductive host caused by volume changes or the 'shuttle reaction' caused by the solubilized polysulfides are minimized.

***In-situ* visual–electrochemical study of $75S/MnO_2$.** To further illustrate the excellent properties discussed above, we examined electrochemical behaviour of $75S/MnO_2$ in an optically transparent Li–S cell, exploiting the very strong colour of LiPSs to probe their interaction with the cathode surface. Recent reports have visually demonstrated the binding of LiPSs with metallic metal oxide sulfur hosts such as Ti_4O_7 (ref. 32) and spatially localized deposition of Li_2S on conductive indium-tin oxide[33]. With δ-MnO_2—a poor semiconductor—the interaction is based on a very different principle, as detailed in the next section. The cell using $75S/MnO_2$ as the cathode clearly demonstrates the ability of the nanosheet host to control LiPS dissolution. For comparison, results are shown for a S/Ketjen Black (75S/KB) cathode prepared with the same 75% sulfur loading (Fig. 3). Figure 3a reveals that the electrolyte in the 75S/KB cell changed from colourless to bright yellow–green on partial discharge of the cell over 4 h. This point corresponds to the 'knee' in the discharge curve at a capacity of $400\,mA\,h\,g^{-1}$ (a potential of 2.1 V, see Fig. 2a), where the LiPS concentration is expected to be at a maximum. The characteristic colour of medium chain LiPSs (i.e, $\sim$, S_4^{2-}) indicates that the polysulfides diffuse out of the cathode and are solubilized in the electrolyte.

At the end of discharge (12 h), the electrolyte is still yellow, showing LiPSs remain in solution. In contrast, in the $75S/MnO_2$ cell the electrolyte exhibits only a faint yellow colour at 4 h. This provides striking visual evidence of very low LiPS content in the electrolyte and demonstrates effective trapping by the MnO_2. On full discharge, the electrolyte is rendered completely colourless, indicating effective conversion to insoluble reduced species, Li_2S_2 and Li_2S. A test of the polysulfide adsorption by the MnO_2 nanosheets was conducted by electrochemical titration to measure the degree of residual LiPS in solution after contact, confirming the strong binding (Supplementary Fig. 4)[38].

XPS study of the interaction of lithium (poly)sulfides and MnO_2 nanosheets. The nature of the interaction of the MnO_2

Figure 2 | Electrochemical performance of S/MnO₂. (a) Voltage profiles of S/MnO₂ nanosheets at current densities ranging from C/20 to 2C. The highlighted circle shows the lack of overpotential. (b) Cycling performance of S/MnO₂ nanosheets at C/5, C/2, 1C and 2C rates. The cells were subjected to an initial slow conditioning cycle at C/20 to allow complete access of the electrolyte to the active material.

Figure 3 | Visual confirmation of polysulfide entrapment at specific discharge depths. (a) 75S/KB and (b) 75S/MnO₂ cells. These were both discharged at C/20 between open circuit potential to 1.8 V under identical conditions.

nanosheets with LiPSs was determined from Mn $3p_{3/2}$ and S2p X-ray photoelectron spectroscopy (XPS) analysis. Only the peak position of the lower binding energy component of the $S2p_{3/2}/2p_{1/2}$ spin orbit doublet is described in the subsequent discussion following convention. The results are summarized in Figs 4 and 5.

Li$_2$S$_4$ was employed as the representative polysulfide species to probe the system at a partial state of discharge. Its XPS spectrum (Fig. 4a) shows a 1:1 ratio of the two $S2p_{3/2}$ contributions at 161.7 and 163.1 eV, ascribed to the terminal (S_T^{-1}) and bridging sulfur (S_B^0) atoms, respectively. This assignment is in accord with the chain structure of Li$_2$S$_4$, where the sulfur atoms at both ends have a formal charge of (-1), while those in the middle bear a formal charge of (0)[32,39]. Figure 4e displays the XPS Mn $2p_{3/2}$ spectra of δ-MnO$_2$, which is mainly composed of Mn^{4+}, with its near-surface enriched in Mn^{3+} relative to the bulk[40]. The MnO$_2$ nanosheet spectrum is typical of birnessite[41], showing a strong Mn $2p_{3/2}$ Mn^{4+} contribution that appears as a characteristic multiplet with a maximum at 642.5 eV, and three satellite peaks at higher binding energies of 643.7, 644.8 and 645.7 eV. The ratio of the multiplet peaks is in agreement with calculations[41]. The Mn $2p_{3/2}$ contribution at 641.4 eV corresponds to Mn^{3+}, which makes up ~4.6% of the MnO$_2$ nanosheets.

To examine the LiPS–Mn interaction, a suspension of vacuum-dried δ-MnO$_2$ in ether was added to a solution of yellow–green Li$_2$S$_4$ in dimethoxyethane, resulting in immediate discolouration. XPS data were collected on the recovered solid ('MnO$_2$–Li$_2$S$_4$'). Its S2p spectrum (Fig. 4b) reveals four sulfur environments. Two are the same terminal and bridging S environments as in Li$_2$S$_4$. The new significant contribution between 171 and 165 eV can be fit with two sulfur environments. The $S2p_{3/2}$ peak at 167.2 eV is in precise accord with the binding energy of thiosulfate[42], which

must arise from a surface redox reaction between Li$_2$S$_4$ and δ-MnO$_2$. This peak is from the 'central' or S=O sulfur in thiosulfate ([SSO$_3$]$^{2-}$), whereas its 'peripheral' sulfur lies at 161.5 eV, contributing to the S_T^{-1} from residual Li$_2$S$_4$ (thus accounting for the change in ratio of the (S_T^{-1}:S_B^0) $2p_{3/2}$ peaks from 1:1 to 2.4:1). The S2p XPS spectrum of sodium thiosulfate is provided for reference (Supplementary Fig. 5a). The other $S2p_{3/2}$ peak at 168.2 eV (ascribed to a polythionate complex in the MnO$_2$–Li$_2$S$_4$ material) will be discussed below. Oxidation of Li$_2$S$_4$ to thiosulfate is accompanied by a decrease in the Mn oxidation state, as evident in Fig. 4f. The contribution from Mn^{3+} at 641.4 eV in the Mn $2p_{3/2}$ XPS spectrum significantly increases (from ~5% to 35%) and two additional Mn $2p_{3/2}$ peaks arise at lower energy (640.4 and 639.4 eV), readily attributable to Mn^{2+}. The overall redox process is summarized in Fig. 4g. The presence of thiosulfate was also proven by Fourier transform infrared spectroscopy (FTIR). The FTIR spectrum of Li$_2$S$_4$ (Supplementary Fig. 6) shows an S–S band at 482 cm^{-1} (ref. 31), whereas MnO$_2$ has a strong antisymmetric MnO$_6$ stretching vibration at 519 cm^{-1} (ref. 43). The FTIR spectrum of MnO$_2$–Li$_2$S$_4$ exhibits a new peak at 670 cm^{-1}, characteristic of S$_2$O$_3^{2-}$ (ref. 44), and one very broad peak at 503 cm^{-1} midway between the contributions expected from S–S and MnO$_2$. We attribute this significant shift of both species to a strong interaction at the interface.

To further support the thiosulfate formation mechanism, we investigated the interaction of LiPS with other sulfur hosts—GO (ACS Material, USA) and graphene (ACS Material, single layer). The XPS S2p spectrum of GO-Li$_2$S$_4$ shows the same components as the MnO$_2$–Li$_2$S$_4$ (Fig. 4c), which suggests thiosulfate/polythionate surface species are formed here as well. The characteristic peaks are reduced in intensity compared with

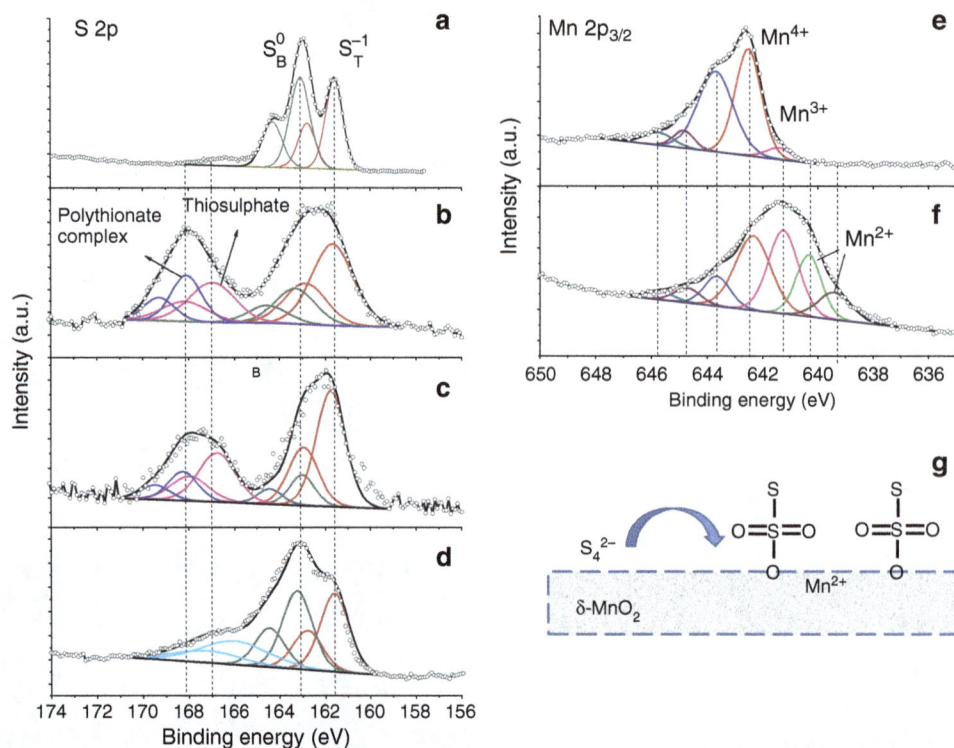

Figure 4 | XPS study of the interaction between sulfur species and MnO$_2$ nanosheets. S2p core spectra of (**a**) Li$_2$S$_4$ showing the terminal and bridging sulfur atoms in the expected 1:1 ratio; (**b**) MnO$_2$-Li$_2$S$_4$ and (**c**) GO–Li$_2$S$_4$ exhibiting the thiosulfate/polythionate active groups and (**d**) 'graphene'-Li$_2$S$_4$ showing their absence; the contribution in turquoise is due to sulfite (from the Li$_2$S$_4$)[32]. Mn $2p_{3/2}$ spectra of (**e**) MnO$_2$ nanosheets and (**f**) MnO$_2$-Li$_2$S$_4$. (**g**) Schematic showing the oxidation of initially formed polysulphide by δ-MnO$_2$ to form thiosulfate on the surface, concomitant with the reduction of Mn^{4+} to Mn^{2+}.

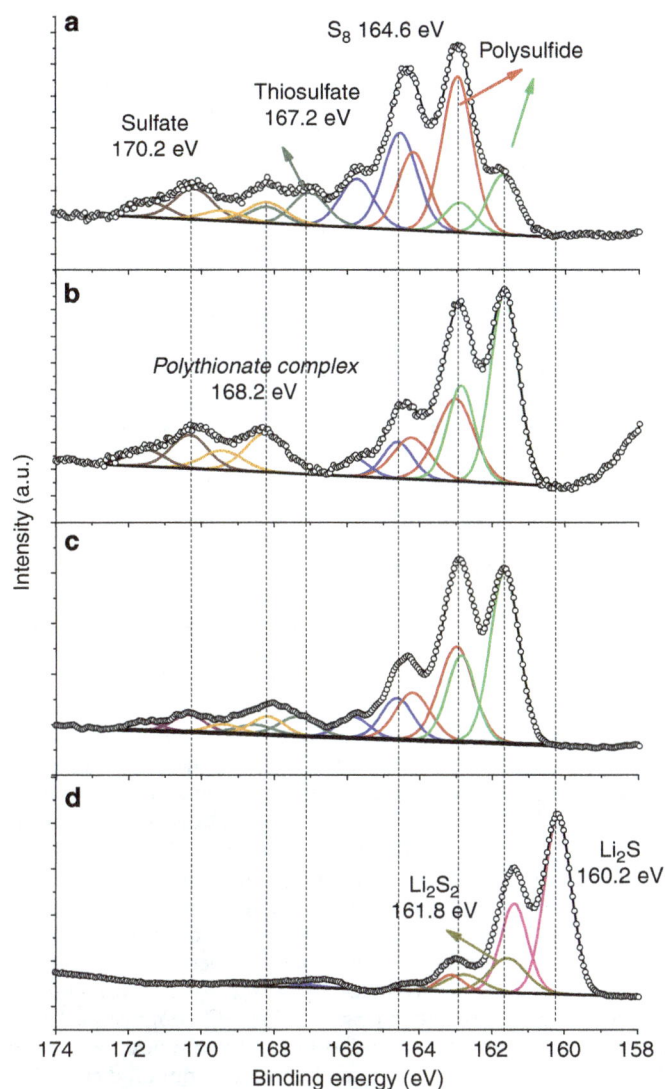

Figure 5 | Ex-situ XPS of S/MnO$_2$ nanosheets electrodes after discharge to specific states. (**a**) Discharged to 2.15 V, (**b**) discharged to 2.15 V and then aged in the cell for 20 h, (**c**) discharged to 800 mA h g^{-1} (the middle of the second plateau) and (**d**) discharged to 1.8 V. All cells were galvanostatically discharged at C/20 with 1 M LiClO$_4$ in DOL/DME (1:1 vol) as the electrolyte. The trace of sulfate at 170.2 eV is due to momentary air exposure during the XPS sample loading process.

δ-MnO$_2$, as expected. The XPS C 1s comparison of the GO and GO–Li$_2$S$_4$ confirms the chemical interaction of LiPS in GO–Li$_2$S$_4$, showing that 20% of the C-O(H) bonds were reduced to C–C (see Supplementary Fig. 7). Regarding the S2p spectrum of graphene–Li$_2$S$_4$, it shows two contributions at 161.7 and 163.1 eV, identical to that of Li$_2$S$_4$ itself (Fig. 4d). This suggests almost no interaction between graphene and polysulfides. The absence of thiosulfate in graphene–Li$_2$S$_4$ implies that the thiosulfate oxygen originates from surface oxy-groups on GO; for MnO$_2$, these arise from the OH groups that replace some of the O^{2-} in the birnessite structure for charge balance. This additional XPS analysis and insight into thiosulfate formation can explain the origin of the excellent cyclability of the MnO$_2$ nanosheets and that of GO materials[25,26], as discussed in detail below.

XPS analysis was further conducted on 75S/MnO$_2$ nanosheet electrodes extracted from cells at different discharge states (Fig. 5a–d). When the cell was discharged to 2.15 V, the presence of long-chain polysulfides is evidenced by strong S2p$_{3/2}$ contributions at

161.7 (S$_T^{-1}$) and 162.9 eV (S$_B^{-2}$) (Fig. 5a)[32,39], with their 1:2 ratio suggesting a chain length approximately equal to Li$_2$S$_6$ owing to the 2:4 ratio of terminal/bridging sulfur in this LiPS. Elemental sulfur was also detected at this and other intermediate states of discharge, consistent with incomplete reaction and/or the disproportionation of metastable polysulfides such as Li$_2$S$_8$ to form Li$_2$S$_6$ and sulfur as previously reported[45]. S2p peaks between 166 and 170 eV correspond to two sulfur environments: one at 167.2 eV (thiosulfate) and another at 168.2 eV, which is assigned to a polythionate complex as discussed above. These are the same as observed in the S2p spectrum of MnO$_2$–Li$_2$S$_4$ in Fig. 4b. The thiosulfate peak vanishes when the electrode is aged in the coin cell for 20 h after discharge to 2.15 V; only the peak due to the polythionate complex remains in this region. Thus, transformation of thiosulfate to the polythionate complex is accompanied by the conversion of longer-chain polysulfides to shorter-chain polysulfides ($\sim$Li$_2$S$_3$ based on the 2:1 ratio of the terminal:bridging sulfur; see Supplementary Fig. 8, for concentrations of each specific component). Exactly the same reaction is also observed between sodium thiosulfate and polysulfides (Li$_2$S$_4$), as shown in Supplementary Fig. 5b.

We conclude that S–S from longer-chain polysulfide species reacts with thiosulfate to yield the polythionate complex and shorter polysulfides. Prolonged contact time results in more complete conversion. XPS analysis of the electrode collected at the middle of the second plateau (Fig. 5c; 800 mA h g^{-1}) shows that the sulfur environments are the same as the electrode discharged to 2.15 V. Their different relative intensities represent the further consumption of longer-chain polysulfides accompanied by the formation of shorter-chain polysulfides. After the electrode was fully discharged to 1.8 V, this process is essentially complete (Fig. 5d). The contribution from thiosulfate and the polythionate complex is small, whereas a strong contribution from Li$_2$S (S2p$_{3/2}$ at 160.2 eV) along with $\sim$20% Li$_2$S$_2$ (161.7 eV) indicates almost full reduction of sulfur in the electrode. The minority contribution from Li$_2$S$_2$ was assigned based on its very similar binding energy as that of Na$_2$S$_2$ (161.8 eV[32], and see Supplementary Fig. 9). On charge, the XPS spectrum (Supplementary Fig. 10) is similar to Fig. 5a, revealing the re-appearance of the polythionate complex along with polysulfides and sulfur, and indicating that the process is reversible.

Discussion
We propose that the insoluble S$_2$O$_3^{2-}$ that forms on the surface serves as an internal mediator to anchor long-chain polysulfides from solution and trigger conversion to lower polysulfides via the reaction shown below (equation (1)):

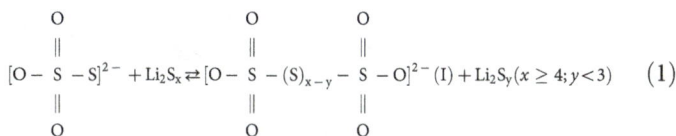

$$[O-\overset{\overset{O}{\|}}{\underset{\underset{O}{\|}}{S}}-S]^{2-} + Li_2S_x \rightleftarrows [O-\overset{\overset{O}{\|}}{\underset{\underset{O}{\|}}{S}}-(S)_{x-y}-\overset{\overset{O}{\|}}{\underset{\underset{O}{\|}}{S}}-O]^{2-} (I) + Li_2S_y (x \geq 4; y < 3) \quad (1)$$

In this mechanism, the polysulfide catenates to the thiosulfate by insertion in the S–S single bond to create the polythionate complex (I) and a lower polysulfide (that is, Li$_2$S$_2$ or Li$_2$S) by an internal disproportionation reaction. This can be explained by the attack of the highly nucleophilic thiosulfate on the bridging S(0) in the polysulfides followed by associative nucleophilic substitution. Sulfur is well known to undergo catenation reactions, such as recently proven for the reaction of sulfur with Li$_3$PS$_4$ to form polysulfidophosphates, Li$_3$PS$_{4+n}$ (ref. 46). Elemental sulfur is also susceptible to nucleophilic attack by HS$^-$ or SO$_3^{2-}$ in aqueous solution, resulting in the formation of chains of polysulfides or polythionate complexes, respectively. This is known as the 'Wackenroder reaction'[47]. The polythionate complex could also be formed with only one $-SO_3$ group (equation (2)):

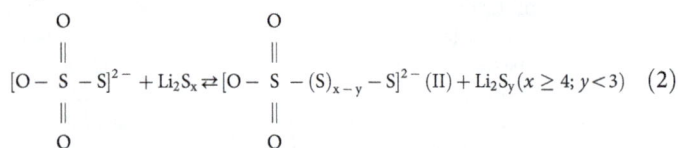

$$[O-\overset{\overset{O}{\|}}{\underset{\underset{O}{\|}}{S}}-S]^{2-} + Li_2S_x \rightleftarrows [O-\overset{\overset{O}{\|}}{\underset{\underset{O}{\|}}{S}}-(S)_{x-y}-S]^{2-} (II) + Li_2S_y (x \geq 4; y < 3) \quad (2)$$

However, the bi-polythionate complex (I) is preferred due to its known stability (albeit in aqueous media)[47]. Based on the above model, the electrochemistry of the Li–S system can be explained as follows: (1) initial formation of Li_2S_n ($4 \leq n \leq 8$) at the start of discharge gives rise to reaction with MnO_2, producing thiosulfate at the surface; (2) further reduction produces more polysulfides ($4 < n < 8$) that are immediately catenated to form an intermediate polythionate complex and shorter-chain polysulfides ($n < 3$). The polythionate complex is expected to be poorly soluble, curtailing the polysulfide shuttle. The process occurs progressively on discharge until full conversion to Li_2S_2/Li_2S is achieved. Deposition of lithium sulfide presumably overlays the signature of polythionate complex and thiosulfate, as XPS only probes <10 nm of the surface.

To evaluate the efficiency of the thiosulfate mediator, long-term cycling was performed at C/5, C/2 and 2C rates. The two former current densities are suitable for practical applications, while the latter is a benchmark for power utilization. For the C/5 cell (Fig. 6a), galvanostatic cycling for over half a year still results in a capacity of 380 mA h g^{-1} after 1,200 cycles. Similarly good performance was achieved at C/2 (Fig. 6b) over 1,500 cycles. After

2,000 cycles at 2C, 245 mA h g^{-1} useable capacity was still available with good Coulombic efficiency (>98.5%) and a very low decay rate of 0.036% per cycle. The rate was periodically changed to investigate the specific capacity that can be delivered at lower C-rates. After 1,000 cycles at 2C a discharge capacity of 715 mA h g^{-1} was obtained at C/20, and even after 2,000 cycles at 2C a reversible discharge capacity of 460 mA h g^{-1} was achieved at C/20 (Fig. 6c). The electrochemical robustness extends to cycling at very high C-rates. On switching the C-rate from 3C to C/5, the discharge capacity (985 mA h g^{-1}) was 96.3% of the original capacity (Supplementary Fig. 11). At the high rate of 4C, a stable capacity of 700 mA h g^{-1} was achieved and most of the original capacity was recovered on switching back to a C-rate. Moreover, the overall morphology and structure of the electrode after 1,000 cycles is very similar to that of the original material, as shown in Supplementary Fig. 12. This suggests that the MnO_2 nanosheet host acts to spatially locate and control both Li_2S/Li_2S_2 and sulfur deposition by providing an active interface via the thiosulfate intermediate and demonstrates the excellent stability of this cathode material.

In summary, by the in-situ generation of thiosulfate surface species, which react to form an active complex on the surface of MnO_2 nanosheets, we demonstrate that it is possible to cycle a Li–S cell with a capacity decay as low as 0.036% per cycle over 2,000 cycles, using a classic low viscosity dioxolane (DOL)–dimethyl ether (DME) electrolyte and polyvinylidene fluoride binder. We propose that a new, unique mechanism is responsible, whereby an active polythionate complex serves as an anchor and transfer mediator to inhibit active mass (polysulfide) dissolution into the electrolyte and control the deposition of Li_2S_2 or Li_2S. The resultant cells deliver high rechargeable capacity at practical current densities and high sulfur loading. We furthermore note that the thiosulfate mediator is not restricted to support on MnO_2 nanosheets, but is broadly applicable to other support materials and may be also responsible—in whole or in part—for the excellent cycling behaviour observed for GO composites. Unlike previous strategies to trap polysulfides by physical barriers or simple surface interactions, this chemistry is quite efficient. The discovery and understanding of a transfer mediator, which binds polysulfides and promotes stable redox activity, addresses one of the important challenges that face this chemistry. Along with future anticipated improvements in electrolytes and the lithium-negative electrode, this brings the Li–S battery a step closer to practical realization.

Methods

Preparation of the MnO_2 nanosheets. The layered MnO_2 nanosheets were synthesized by a one-step facile method using GO as template. Briefly, 20 mg of single-layer GO (ACS Materials) were dispersed in 100 ml deionized (DI) water by sonication. A well-mixed solution of 10 ml DI water and 160 mg of $KMnO_4$ (Sigma-Aldrich) was added into the GO suspension and stirred at room temperature for 30 min. The mixture was transferred into a thermostatic oven at 80 °C for 24 h. The resulting material was washed with DI water.

Preparation of the 75S/MnO_2 composite. For sulfur loading onto the MnO_2 nanosheets, the nano-sized sulfur was first synthesized by reacting 255 mg $Na_2S_2O_3$ (Sigma-Aldrich) with 278 μl concentrated HCl and 17 mg polyvinylpyrrolidone (Sigma-Aldrich) in 85 ml DI water. The MnO_2 nanosheets and nano-sized sulfur were dispersed in 40 ml DI water separately by sonication before being mixed to obtain a homogenous suspension. The suspension was filtered and then dried at 60 °C. The 75S/MnO_2 nanosheet composite was obtained by heating the mixture at 155 °C overnight.

Synthesis of Li_2S_4. Sulfur (Sigma-Aldrich) and Super-Hydride Solution (Aldrich, 1.0 M lithium triethylborohydride in tetrahydrofuran) were mixed together in a 2.75:1 molar ratio until the sulfur was fully dissolved. The resulting solution was dried under vacuum, resulting in precipitation of a yellow powder. A final wash with toluene was conducted, followed by centrifugation to separate the powder from the supernatant to isolate the Li_2S_4 powder.

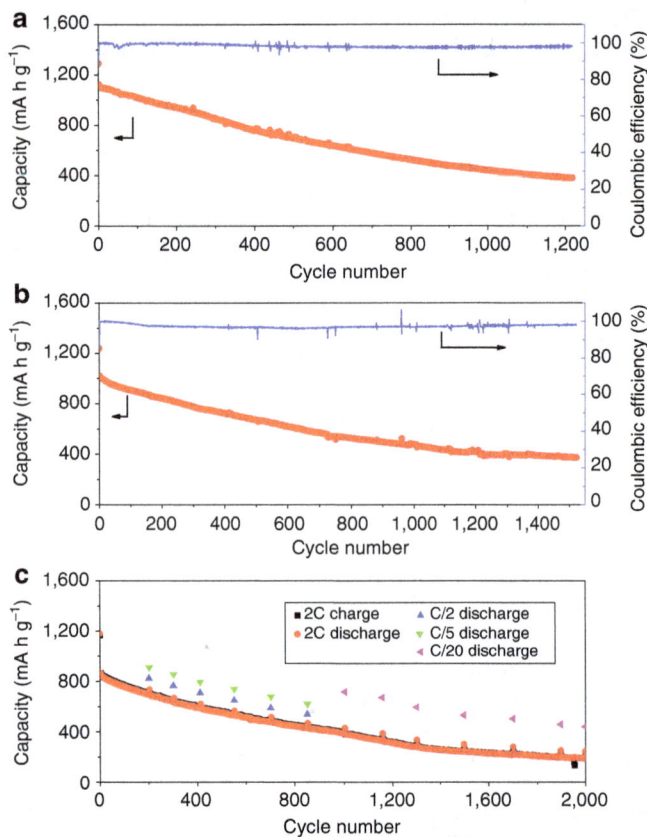

Figure 6 | Long-term cycling of 75S/MnO_2 at variable current densities. Capacity and the Coulombic efficiency at (**a**) C/5 over 1,200 cycles, (**b**) C/2 over 1,500 cycles, (**c**) 2C with periodic slow rate for over 2,000 cycles. The capacities of the C/2 and C/5 cells are quite similar to those shown in Fig. 2, demonstrating the reproducibility of the electrochemistry.

Preparation of MnO₂-Li₂S₄. The MnO_2 nanosheets (87 mg; 1 mmol) were dried at 90 °C under vacuum overnight and dispersed in 10 ml DME solution by sonication in a sealed vial. The mixture was transferred to a glovebox and combined with 10 ml of 0.1 M Li_2S_4 in DME. The mixture was stirred for 2 h and the product was centrifuged and dried under vacuum overnight. GO-Li_2S_4 and graphene-Li2S4 samples were prepared by the same procedure.

Preparation of MnO₂-Li₂S. 75S/MnO_2 was mixed with DMF and then cast as a slurry onto a carbon paper current collector. A 2325 coin cell was discharged to 1.8 V at a C/20 rate, using 1 M $LiClO_4$ in a 1:1 v/v ratio of DOL:DME as the electrolyte with lithium foil as the anode. The cathode was then removed from the coin cell in a glovebox and rinsed with acetonitrile three times before being dried.

Electrochemical measurements. Cathodes were prepared by casting the DMF slurry containing 75S/MnO_2 nanosheets, 15 wt% Super P carbon and 10 wt % polyvinylidene fluoride onto a carbon paper current collector. Electrochemical studies were carried out using an Arbin cycler in galvanostatic mode, employing electrodes with an average sulfur loading between 0.7 and 1.0 mg cm⁻² in 2325 coin cells with lithium foil as the anode and Celgard 3501 separator sheets. Cells were operated in a voltage window of 1.8–3.0 V, except for high-rate studies that were conducted between 1.7 and 3.0 V (2C and 3C rates), and between 1.6 and 3.0 V (4C rate), in an electrolyte comprising 1 M LiTFSI in a 1:1 volume of DME:DOL and 2 wt% LiNO₃. Specific capacity values were calculated with respect to the mass of sulfur. Electrodes for XPS analysis were prepared by discharging coin cells at C/20 in 1 M $LiClO_4$ (to avoid sulfur contributions from LiFTSI) in a 1:1 v/v ratio of DOL and DME.

Characterization. SEM studies were carried out on a Zeiss Ultra field emission SEM instrument and TEM was performed on a Jeol 2010F TEM/STEM operating at 200 KeV. TGA was used to determine the sulfur content of the material, on a TA Instruments SDT Q600 employing a heating rate of 10 °C/min from room temperature to 500 °C under an air flow. FTIR analysis was performed on a Bruker Tensor 37 spectrometer. For XPS, the samples were sealed in a vial before being quickly transferred to the chamber of a ultra-high vacuum Imaging XPS Microprobe system for analysis (Thermo VG Scientific ESCALab 250). All spectra were fitted with Gaussian–Lorentzian functions and a Shirley-type background using CasaXPS software. S2p peaks were fit using two equal full-width half maximum S2p doublets with 2:1 area ratios and splittings of 1.2 eV. The binding energy values were all calibrated using the C 1s peak at 285.0 eV.

References

1. Choi, N. S. et al. Challenges facing lithium batteries and electrical double-layer capacitors. *Angew. Chem. Int. Ed.* **51**, 9994–10024 (2012).
2. Yoo, H. D., Markevich, E., Salitra, G., Sharon, D. & Aurbach, D. On the challenge of developing advanced technologies for electrochemical energy storage. *Mater. Today* **17**, 110–121 (2014).
3. Manthiram, A. Materials challenges and opportunities of lithium ion batteries. *J. Phys. Chem. Lett.* **2**, 176–184 (2011).
4. Ellis, B., Lee, K. T. & Nazar, L. F. Positive electrode materials for Li-ion and Li-batteries. *Chem. Mater.* **22**, 691–714 (2010).
5. Bruce, P. G., Freunberger, S. A., Hardwick, L. J. & Tarascon, J. M. Li–O₂ and Li–S batteries with high energy storage. *Nat. Mater.* **11**, 19–29 (2012).
6. Yin, Y. X., Xin, S., Guo, Y. G. & Wan, L. J. Lithium–sulphur batteries: electrochemistry, materials, and prospects. *Angew. Chem. Int. Ed.* **52**, 2–18 (2013).
7. Evers, S. & Nazar, L. F. New approaches for high energy density lithium-sulfur batteries. *Acc. Chem. Res.* **46**, 1135–1143 (2013).
8. Peled, E., Gorenshtein, A., Segal, M. S. & Sternberg, Y. Rechargeable lithium sulfur battery. *J. Power Sources* **26**, 269–271 (1989).
9. Yang, Y., Zheng, G. Y. & Cui, Y. Nanostructured sulphur cathodes. *Chem. Soc. Rev.* **42**, 3018–3032 (2013).
10. Mikhaylik, Y. V. & Akridge, J. R. Polysulfide shuttle study in the Li/S battery system. *J. Electrochem. Soc.* **151**, A1969–A1976 (2004).
11. Ji, X. L., Lee, K. T. & Nazar, L. F. A highly ordered nanostructured carbon-sulphur cathode for lithium-sulphur batteries. *Nat. Mater.* **8**, 500–506 (2009).
12. Schuster, J. et al. Spherical ordered mesoporous carbon nanoparticles with high porosity for lithium-sulphur batteries. *Angew. Chem. Int. Ed.* **51**, 3591–3595 (2012).
13. He, G., Ji, X. L. & Nazar, F. L. High 'C' rate Li-S cathodes: sulfur imbibed bimodal porous carbons. *Energy Environ. Sci.* **4**, 2878–2883 (2011).
14. Liang, C. D., Dudney, N. J. & Howe, J. Y. Hierarchically structured sulphur/carbon nanocomposite material for high-energy lithium battery. *Chem. Mater.* **21**, 4724–4730 (2009).
15. He, G. et al. Tailoring porosity in carbon nanospheres for lithium sulphur battery cathodes. *ACS Nano* **7**, 10920–10930 (2013).
16. Jayaprakash, N., Shen, J., Moganty, S. S., Corona, A. & Archer, L. A. Porous hollow carbon@sulphur composites for high-power lithium-sulphur batteries. *Angew. Chem. Int. Ed.* **50**, 5904–5908 (2011).
17. Zhang, B., Qin, X., Li, G. R. & Gao, X. P. Enhancement of long stability of sulphur cathode by encapsulating sulphur into micropores of carbon spheres. *Energy Environ. Sci.* **3**, 1531–1537 (2010).
18. Elazari, R., Salitra, G., Garsuch, A., Panchenko, A. & Aurbach, D. Sulphur impregnated activated carbon fiber cloth as a binder-free cathode for rechargeable Li-S batteries. *Adv. Mater.* **23**, 5641–5644 (2011).
19. Zheng, G., Yang, Y., Cha, J. J., Hong, S. S. & Cui, Y. Hollow carbon nanofiber encapsulated sulphur cathodes for high specific capacity rechargeable lithium batteries. *Nano Lett.* **11**, 4462–4467 (2011).
20. Guo, J., Xu, Y. & Wang, C. Sulphur-impregnated disordered carbon nanotubes cathode for lithium-sulphur batteries. *Nano Lett.* **11**, 4288–4294 (2011).
21. Zhou, W. D., Yu, Y. C., Chen, H., DiSalvo, F. J. & Abruña, H. D. Yolk-shell structure of polyaniline coated sulphur for lithium-sulphur batteries. *J. Am. Chem. Soc.* **135**, 16736–16743 (2013).
22. Fu, Y. Z. & Manthiram, A. Orthorhombic bipyramidal sulphur coated with polypyrrole nanolayers as a cathode material for lithium–sulphur batteries. *J. Phys. Chem. C* **116**, 8910–8915 (2012).
23. Liang, X. et al. A nano-structured and highly ordered polypyrrole-sulphur cathode for lithium-sulphur batteries. *J. Power Sources* **196**, 6951–6955 (2011).
24. Yang, Y. et al. Improving the performance of lithium–sulphur batteries by conductive polymer coating. *ACS Nano* **5**, 9187–9193 (2011).
25. Ji, L. et al. Graphene oxide as a sulphur immobilizer in high performance lithium/sulphur cells. *J. Am. Chem. Soc.* **133**, 18522–18525 (2011).
26. Song, M. K., Zhang, Y. G. & Cairns, E. J. A long-life, high-rate lithium/sulphur cell: a multifaceted approach to enhancing cell performance. *Nano Lett.* **13**, 5891–5899 (2013).
27. Zhou, G. M. et al. Fibrous hybrid of graphene and sulphur nanocrystals for high-performance lithium–sulphur batteries. *ACS Nano* **7**, 5367–5375 (2013).
28. Qiu, Y. C. et al. High-rate, ultralong cycle-life lithium/sulfur batteries enabled by nitrogen-doped graphene. *Nano Lett.* **14**, 4821–4827 (2014).
29. Seh, Z. W. et al. Sulphur TiO₂ yolk-shell nanoarchitecture with internal void space for long-cycle lithium-sulphur batteries. *Nat. Commun.* **4**, 1331 (2013).
30. Ji, X., Evers, S., Black, R. & Nazar, L. F. Stabilizing lithium-sulphur cathodes using polysulphide reservoirs. *Nat. Commun.* **2**, 325 (2011).
31. Evers, S., Yim, T. & Nazar, L. F. Understanding the nature of absorption/adsorption in nanoporous polysulfide sorbents for the Li-S battery. *J. Phys. Chem. C.* **116**, 19653–19658 (2012).
32. Pang, Q., Kundu, D., Cuisinier, M. & Nazar, L. Surface-enhanced redox chemistry of polysulphides on a metallic and polar host for lithium-sulphur batteries. *Nat. Commun.* **5**, 4759 (2014).
33. Yao, H. B. et al. Improving lithium–sulphur batteries through spatial control of sulphur species deposition on a hybrid electrode surface. *Nat. Commun.* **5**, 3943 (2014).
34. Meini, S., Elazari, R., Rosenman, A. & Aurbach, D. The use of redox mediators for enhancing utilization of Li₂S cathodes for advanced Li-S battery systems. *J. Phys. Chem. Lett.* **5**, 915–918 (2014).
35. Lin, Z., Liu, Z. C., Fu, W. J., Dudney, N. J. & Liang, C. D. Phosphorous pentasulfide as a novel additive for high-performance lithium-sulfur batteries. *Adv. Funct. Mater.* **23**, 1064–1068 (2013).
36. Zhao, G. X. et al. synthesizing MnO₂ nanosheets from graphene oxide templates for high performance pseudosupercapacitors. *Chem. Sci.* **3**, 433–437 (2012).
37. Ma, R. Z., Bando, Y., Zhang, L. Q. & Sasaki, T. Layered MnO₂ nanobelts: hydrothermal synthesis and electrochemical measurements. *Adv. Mater.* **16**, 918–922 (2004).
38. Nazar, L. et al. Rational design of sulphur host materials for Li-S batteries: correlating lithium polysulphide adsorptivity and self-discharge capacity loss. *Chem. Commun.* doi: 10.1039/C4CC08980D (2014).
39. Kartio, I. J., Basilio, C. I. & Yoon, R. H. An XPS study of sphalerite activation by copper. *Langmuir* **14**, 5274–5278 (1998).
40. Najdoski, M., Koleva, V., Demiri, S. & Stojkovikj, S. A simple chemical method for deposition of electrochromic potassium manganese oxide hydrate thin films. *Mater. Res. Bull.* **47**, 2239–2244 (2012).
41. Nesbitt, H. W. & Banerjee, D. Interpretation of XPS Mn(2p) spectra of Mn oxyhydroxides and constraints on the mechanism of MnO₂ precipitation. *Amer. Miner.* **83**, 305–315 (1998).
42. Lindberg, B. J. et al. Molecular spectroscopy by means of ESCA II. Sulphur compounds. Correlation of electron binding energy with structure. *Phys. Scr.* **1**, 286–298 (1970).
43. Julien, C. M., Massot, M. & Poinsignon, C. Lattice vibrations of manganese oxides Part I. Periodic structures. *Spectrochim. Acta A* **60**, 689–700 (2004).
44. Miller, F. A. & Wilkins, C. H. Infrared spectra and characteristic frequencies of inorganic ions. *Anal. Chem.* **24**, 1253–1294 (1952).
45. Cuisinier, M. et al. Sulphur speciation in Li-S batteries determined by operando X-ray absorption spectroscopy. *J. Phys. Chem. Lett.* **4**, 3227–3232 (2013).

46. Lin, Z., Liu, Z. C., Fu, W. J., Dudney, N. J. & Liang, C. D. Lithium poly-sulfidophosphates: a family of lithium-conducting sulphur-rich compounds for lithium-sulphur batteries. *Angew. Chem. Int. Ed.* **125,** 1–5 (2013).
47. Holleman-Wiberg: *Inorganic Chemistry* (eds Wiberg, N.) 514 (Academic Press, 2001).

Acknowledgements

The research was supported by the BASF International Scientific Network for Electrochemistry and Batteries. We thank Dr Carmen Andrei, MacMaster University, and the Canadian Centre for Electron Microscopy for help with acquisition of the TEM images, and NSERC for platform funding through a Discovery Grant to LFN.

Author contributions

X.L. and L.F.N. designed this study. X.L. prepared materials and carried out the electrochemical experiments. X.L., C.H. and Q.P. carried out of the XPS investigation. X.L. carried out data processing and prepared figures. X.L. and L.F.N. wrote the manuscript. T.W. first suggested the proposed mechanism and all of the authors contributed to the scientific discussion.

Additional information

Competing financial interests: The authors declare no competing financial interests.

2

Anomalous dynamics of intruders in a crowded environment of mobile obstacles

Tatjana Sentjabrskaja[1,*], Emanuela Zaccarelli[2,3,*], Cristiano De Michele[2,3], Francesco Sciortino[2,3], Piero Tartaglia[3], Thomas Voigtmann[4,5], Stefan U. Egelhaaf[1] & Marco Laurati[1,6]

Many natural and industrial processes rely on constrained transport, such as proteins moving through cells, particles confined in nanocomposite materials or gels, individuals in highly dense collectives and vehicular traffic conditions. These are examples of motion through crowded environments, in which the host matrix may retain some glass-like dynamics. Here we investigate constrained transport in a colloidal model system, in which dilute small spheres move in a slowly rearranging, glassy matrix of large spheres. Using confocal differential dynamic microscopy and simulations, here we discover a critical size asymmetry, at which anomalous collective transport of the small particles appears, manifested as a logarithmic decay of the density autocorrelation functions. We demonstrate that the matrix mobility is central for the observed anomalous behaviour. These results, crucially depending on size-induced dynamic asymmetry, are of relevance for a wide range of phenomena ranging from glassy systems to cell biology.

[1] Condensed Matter Physics Laboratory, Heinrich Heine University, 40225 Düsseldorf, Germany. [2] CNR-ISC, Università di Roma 'La Sapienza', Piazzale A. Moro 2, Roma 00185, Italy. [3] Dipartimento di Fisica, Università di Roma 'La Sapienza', Piazzale A. Moro 2, Roma 00185, Italy. [4] Institut für Materialphysik im Weltraum, Deutsches Zentrum für Luft- und Raumfahrt (DLR), 51170 Köln, Germany. [5] Heinrich Heine University, Universitätsstraße 1, 40225 Düsseldorf, Germany. [6] División de Ciencias e Ingeniería, Universidad de Guanajuato, Loma del Bosque 103, León 37150, Mexico. * These authors contributed equally to this work. Correspondence and requests for materials should be addressed to E.Z. (email: emanuela.zaccarelli@cnr.it) or to M.L. (email: mlaurati@fisica.ugto.mx).

In the presence of a confining medium, the transport of objects deviates from normal diffusion. Anomalous behaviour, usually manifested by the presence of sub-diffusivity[1,2], emerges as a common feature of the dynamics. In the Lorentz gas[3,4], the prototype model for anomalous transport, point-like intruders move in voids between immobile, randomly-distributed particles. Their motion becomes sub-diffusive once the voids are barely interconnected. When a critical density of immobile particles is reached, they percolate and the intruder becomes localized[3]. Softness of the immobile particles or interactions among the intruders are known to modify this picture[5-10].

So far the slow movement of the host matrix has been largely ignored, despite representing realistic situations of biological[11-16] and industrial interest[17-23]. To address confined transport in slowly moving matrices, here we investigate a binary colloidal mixture of small and large hard spheres, of diameters σ_s and σ_l, which represent intruders and host matrix, respectively. Changing the size ratio $\delta = \sigma_s/\sigma_l$ we also modify the dynamic asymmetry of the system. We focus on volume fractions of large particles $\phi_l > 0.5$ approaching the glass transition, occurring at $\phi_l^g \approx 0.58$. In contrast the volume fraction of the intruders ϕ_s is very small with $x_s \equiv \phi_s/\phi = 0.01$. Such a system combines the confinement of a dilute fluid of mobile intruders with the slow dynamics of the matrix (Fig. 1a). It thus provides the simplest minimal model for the investigation of motion in crowded soft and biological matter.

Despite its conceptual simplicity, experimental investigations of the dynamics of small intruders in mixtures of Brownian particles with large size-asymmetry are scarce. This might be due to limitations in the spatial and temporal resolution of confocal microscopy which make it difficult to track particles that are significantly smaller than another species of Brownian, that is, at most micron-sized, particles. To overcome these limitations, we keep the selectivity of fluorescent labelling (Fig. 1b), which allows us to separately determine the small and large particles. However, instead of tracking we employ the recent Differential Dynamic Microscopy (DDM) technique[24-26]. This is based on the time correlation in Fourier space of the difference between images separated by a time delay Δt (Fig. 1c) and provides a measure of the (isotropic) collective intermediate scattering function or density autocorrelation function $f(q, \Delta t)$, where q is the modulus of the wave vector $\mathbf{q}$ (Fig. 1d). The decay of $f(q, \Delta t)$ as a function of time delay Δt corresponds to the loss of correlation of the particle density on a length scale determined by q^{-1} within the time delay Δt. The decay time is therefore related to the characteristic time of the particle motions on the length scale q^{-1}. Approaches similar to DDM, like fluorescence correlation spectroscopy, do not provide information on the probed length scale. This information is crucial to investigate the effect on the dynamics of the size of the voids in which the small particles move. The function $f(q, \Delta t)$ can also be obtained by dynamic light scattering, which, however, does not allow us to distinguish the two species by fluorescent labeling. We also study the same system by mode coupling theory (MCT) of the glass transition and, both in the case of mobile and immobile matrix particles, by numerical simulations, complementing the experimental results and providing insights on the underlying microscopic mechanisms. We observe anomalous dynamics of the small spheres at a critical size ratio δ_c and we show that this dynamical behavior is intimately connected to the slow dynamics of the matrix of large particles.

Results

Small particle dynamics. Figure 2a–d shows the measured collective intermediate scattering functions $f(q, \Delta t)$ of the small

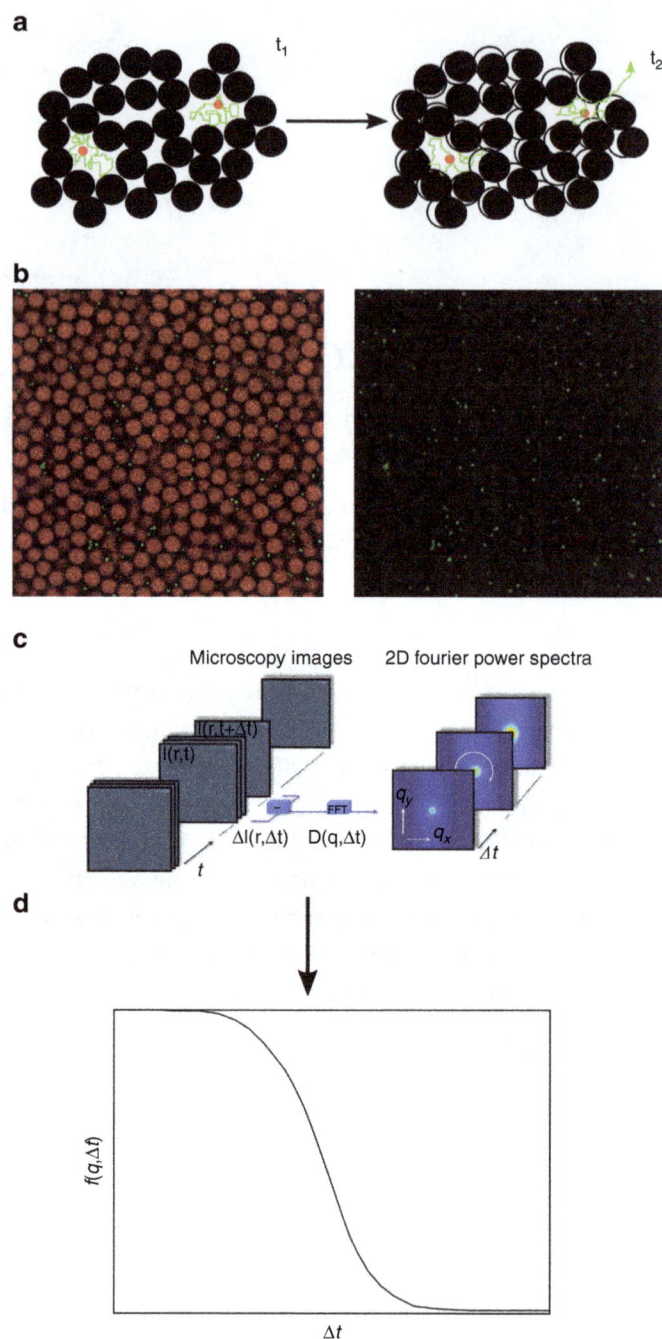

Figure 1 | Illustration of the system and measurement method.
(a) Schematic illustration of our system at two times t_1 and $t_2 > t_1$ highlighting the trajectories (green lines) of the intruders (red beads) in voids and between voids made possible due to the mobility of the matrix particles. (b) An exemplary confocal microscopy image of a mixture with $\delta = 0.18$ and $\phi = 0.58$ in which (left) both particles and (right) only the small particles are shown. (c) Image differences $\Delta I(r, \Delta t)$ at different delay times Δt are Fourier transformed to give 2D Fourier power spectra for different Δt. (d) After azimuthal averaging and additional treatment the intermediate scattering function $f(q, \Delta t)$ is obtained.

particles for size ratios $\delta = 0.18$ (Fig. 2a,c) and $\delta = 0.28$ (Fig. 2b,d) for different ϕ and q. For $\delta = 0.18$ and all ϕ and q, $f(q, \Delta t)$ versus Δt shows an initial decay, followed by a ϕ-dependent intermediate plateau, and eventually a decay to zero at longer times (Fig. 2a). The initial decay can be associated with the Brownian motion of small particles within the voids of the large particles

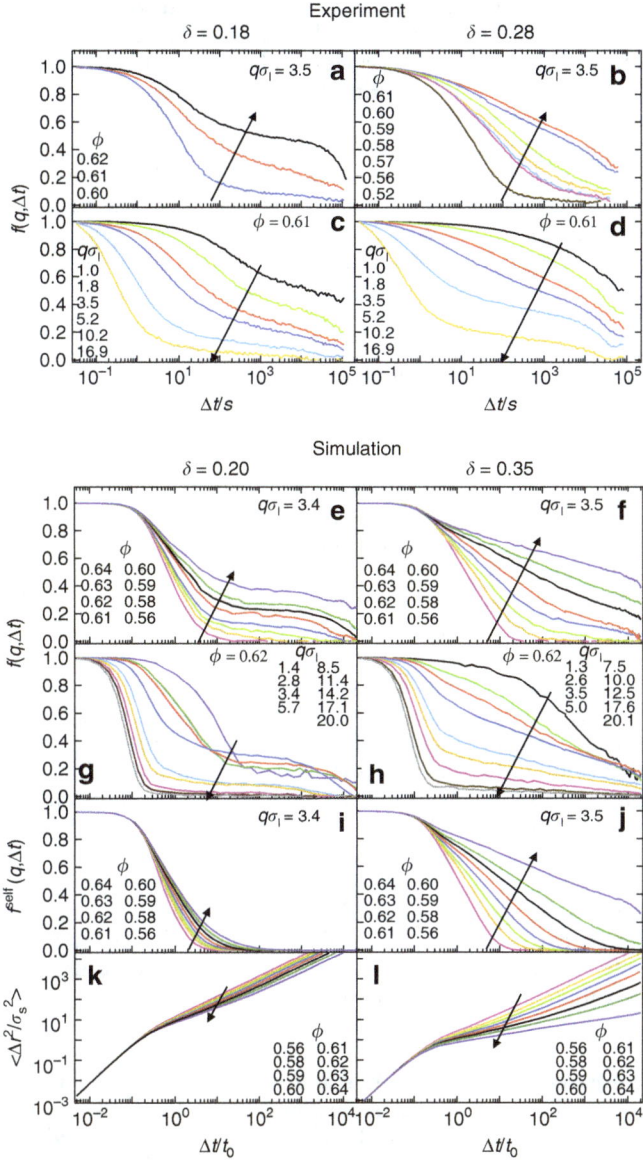

Figure 2 | Dynamics of the intruders as observed in experiments and simulations. Collective $f(q, \Delta t)$ (**a-h**) and self $f^{self}(q, \Delta t)$ (**i-j**) intermediate scattering functions and mean-squared displacements $\langle \Delta r^2 / \sigma_s^2 \rangle$ (**k-l**) as a function of delay time Δt, describing the dynamics of small spheres in binary mixtures with size ratios δ below (left) and around (right) the onset of anomalous dynamics, for different magnitudes of the scattering vector q and total volume fraction ϕ (as indicated). Arrows indicate increasing ϕ and increasing q accordingly.

matrix. It becomes increasingly slower for increasing ϕ (Fig. 2a) and decreasing q, which means increasing length scale (Fig. 2c). The intermediate plateau indicates the dynamical arrest of the collective dynamics, that is, of density fluctuations, and hence the absence of diffusion on the length scale determined by q^{-1}. The height of the plateau increases progressively with increasing ϕ, similarly to the scenario, in which a percolation-type transition is approached[5,27], and indicates that voids become smaller and particles are increasingly localised[28]. The final decay to zero of $f(q, \Delta t)$ shows that particles are still able to diffuse at long times. For a larger size ratio, $\delta = 0.28$, and comparable ϕ values, a completely different scenario appears. Beyond $\phi \approx 0.60$, $f(q, \Delta t)$ shows remarkable anomalous dynamics, manifested in an extended logarithmic decay over three decades in time. This intriguing behavior is mostly visible at $\phi \gtrsim 0.60$ and $q\sigma_l \approx 3.5$, that

is, when probing a length scale of about $2\sigma_l$ (Fig. 2b), which is comparable with the size of the matrix particles.

The experimental findings are confirmed by simulations. For $\delta = 0.20$ no anomalous behavior of the small particles is detected in the collective $f(q, \Delta t)$ (Fig. 2e,g) and in the self $f^{self}(q, \Delta t)$ correlation functions (Fig. 2i). Note that for $\delta = 0.20$, $f(q, \Delta t)$ displays a two step-relaxation and the presence of localisation (Fig. 2e,g), which is absent in $f^{self}(q, \Delta t)$ (Fig. 2i). Also the mean squared displacements (MSD) $\langle \Delta r^2 \rangle \equiv \langle |\vec{r}(t) - \vec{r}(0)|^2 \rangle$, with $\vec{r}(t)$ the position of a particle at time t, show almost no localisation at all ϕ (Fig. 2k). This decoupling between collective ($f(q, \Delta t)$) and self dynamics ($f^{self}(q, \Delta t)$, MSD) originates from the glassy environment in which the intruders move. Correlated motions of a group of intruders distributed within the matrix are more influenced by the slow dynamics of the matrix particles than uncorrelated single particle motions, which are mostly sensitive to the local structure of the voids[5,29]. For $\delta = 0.35$ we find the emergence of logarithmic anomalous relaxations of $f(q, \Delta t)$ (Fig. 2f,h) and $f^{self}(q, \Delta t)$ (Fig. 2j), for comparable q as in the experiments. Additional simulations for $\delta = 0.30$ and $\delta = 0.40$ also show a logarithmic decay over a smaller time window. Furthermore, for $\delta = 0.35$ and $\phi \gtrsim 0.60$ the MSD displays a clear sub-diffusive behavior, i.e., $\langle \Delta r^2 \rangle \sim t^\alpha$ with $\alpha < 1$ (Fig. 2l). Finally, for $\delta = 0.5$, $f(q, \Delta t)$ and $f^{self}(q, \Delta t)$ show a two-step decay and the MSD a localisation plateau at large ϕ, consistent with a standard glass transition of the small particles. At all investigated δ and for $\phi_l > 0.55$, the dynamics of the large particles are very slow and at intermediate times are indicating localisation and motion within nearest neighbour cages of approximate size $0.1\sigma_l$ (Supplementary Fig. 1 and Supplementary Note 1).

These results suggest the existence of a critical size ratio $\delta_c \simeq 0.35$ at which pronounced anomalous dynamics mark the transition from a diffusive to a glassy regime of the small particles moving in the large particles matrix. The δ_c and ϕ values where this transition is observed are slightly smaller in the experiments than in the simulations. This is attributed to the fact that in the experiments small particles are polydisperse, while in the simulations they are monodisperse. Polydispersity is expected to affect the transition since the average size particles might still be able to diffuse through the void spaces in the matrix, whereas the largest particles of the size distribution might no longer be able to diffuse through them. The crossover observed at δ_c is analogous to the transition from a diffusive to a localized state in models with fixed obstacles. However, the excluded volume of the intruder generates a coupling with the host matrix and, due to the mobility of the matrix, also between intruders in different voids, mutating localization into a glass transition due to the (slow) mobility of the matrix particles. Although this is apparently similar to intruders in a fixed matrix[5-7], the logarithmic decay of $f(q, \Delta t)$ stands out as a novel feature.

On the basis of MCT, the appearance of logarithmic decays in $f(q, \Delta t)$[30-32] is usually attributed to competing collective arrest mechanisms, like caging and bonding, and to higher-order glass transition singularities[29,33-35]. We solved MCT equations for a binary mixture of hard spheres and $x_s = 0.01$. The resulting correlators $f(q, \Delta t)$ for a range of packing fractions around the MCT glass transition, $\phi_c \approx 0.516$ and $\delta = 0.20$ and 0.35, are shown in Fig. 3. No clear sign of logarithmic decay of $f(q, \Delta t)$ is found for these states in MCT: while an approximate logarithmic dependence of the decay is observed at $\delta = 0.35$, $\phi = 0.51$ and $q\sigma_l = 3.4$, this extends over an interval of times much shorter than in experiments and simulations. In addition, upon further increasing ϕ the logarithmic dependence does not take over, but instead a two-step decay is found, followed by the arrest of the dynamics. Indeed higher-order singularities are not present in this region of ϕ and x_s values[29]. On the other hand, the MSD

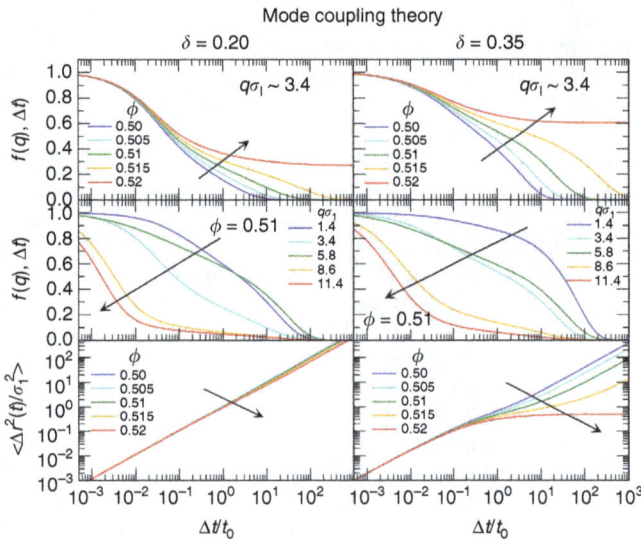

Figure 3 | Dynamics of the intruders as predicted by MCT. Intermediate scattering functions $f(q, \Delta t)$ (top, middle) and mean-squared displacements $\langle \Delta r^2 / \sigma_1^2 \rangle$ (bottom) describing the dynamics of small spheres in binary mixtures with size ratios delta below (left) and around (right) the onset of anomalous dynamics, for different magnitudes of the scattering vector q and total volume fraction ϕ (as indicated). Arrows indicate increasing ϕ or increasing q accordingly.

obtained from MCT shows the qualitative signatures found in simulations: for $\delta = 0.20 < \delta_c$, the long-time diffusion barely slows down with increasing ϕ, indicating a partially frozen glass in which the small particles are mobile. For $\delta = 0.35 \approx \delta_c$, anomalous sub-diffusion is observed, indicating that the glass-transition of the large particles and the localization transition of the small particles are close to each other. Thus, the appearance of approximately logarithmic decay in Fig. 3 could be a signal of the transition from coupled dynamics of the two species at large δ to decoupled dynamics at small δ.

Void space explored by small particles. A direct visualisation of small particle locations shows that the transition from diffusive dynamics at small δ to localised dynamics at large δ observed in experiments, simulations and theory is associated, similarly to models with immobile obstacles, with the transition from percolating to non-percolating voids within the matrix. However, a static picture of the void geometry cannot describe this transition, because the evolution of the void space involves a second timescale t_2 (Fig. 1a, right) associated with the mobility of the matrix. To analyse the dynamic rearrangements of the void structure, we monitor the evolution of the position of the small particles which explore this evolving structure. Accordingly, in Fig. 4a,b we show superpositions of small particle locations in 2D time series of confocal images over a long total observation time $t_f^{exp} = 297\,s$, at which $f(q, \Delta t)$ for $\delta = 0.18$ shows a decay of correlations, while $f(q, \Delta t)$ for $\delta = 0.28$ is in the logarithmic regime. For $\delta = 0.18$ we find that, within the observation time, small particles easily explore the whole space of the accessible voids which form a percolating network. In contrast, for $\delta = 0.28$ particles mostly explore their local environment, since voids only barely connect even at long times, allowing only a slow, partial exploration of the available void space. Simulations provide not only particle locations but also single-particle trajectories in three dimensions allowing a more quantitative determination of the percolation of the explored space. Visualisations of typical small particle trajectories for a fixed observation time $t_f^{sim} = 100 t_0$

(comparable with the experiments) and three different values of δ confirm the experimental features (Fig. 4c): within the observation time small particles explore a percolated space for small δ, while for the critical size ratio the space is barely connected, indicating that particles can rarely escape the local environment which is only possible due to the stochastic opening and closing of channels between neighbouring void spaces, associated with the matrix motion on the long time scale t_2. In addition, the simulations show that for even larger δ the explored space is disconnected. To quantify these observations we calculate the distribution $n(s)$ of the size s of the clusters in which the space explored by small particles within a certain time interval is organized, as explained in Methods. The results are shown in Fig. 4d for different δ values for an observation time equal to t_f^{sim}. This time corresponds to the interval over which the cluster size distribution of the explored space for δ_c is close to percolation, as indicated by the power-law dependence $n(s) \sim s^{-2.19}$, consistent with random percolation predictions[36]. Percolation at t_f^{sim} for δ_c is also indicated, in a finite-size system, by the maximum of the average size of finite-size clusters (excluding percolating clusters, calculated as explained in Methods) L_c as a function of time (Fig. 4e). For the other size ratios instead L_c is very small at t_f^{sim}. At small δ this is due to the fact that particles can easily move through channels connecting voids, and thus the explored space quickly associates into a percolating cluster. On the other hand, for large δ the creation of channels that allow the small particles to move between neighbouring void spaces is rare, and thus percolation of the explored space does not occur at t_f^{sim} and only voids corresponding to the size of monomers, dimers and few-mers are observed. This analysis reveals very different timescales at which the explored space percolates at different δ. These timescales depend, besides δ, on the timescale t_2 of the evolution of the void space, associated with the thermal motion of the matrix particles: yet this analysis is not offering substantial evidence that this mobility of the matrix is causing the logarithmic decays of the correlators observed at δ_c.

Comparison between mobile and immobile matrix. To go one step further and link the residual mobility of the matrix particles with the anomalous logarithmic decays, we perform additional simulations (for $\phi = 0.62$) for immobile matrix particles and compare the dynamics of the intruders with the case of a mobile matrix. When the large particles are immobile (Fig. 5a), the MSD shows a sub-diffusive regime (MSD $\sim t^\alpha$) followed by diffusion at long times (upward curvature) or localization (downward curvature), depending on δ. The crossover between these two long time behaviors takes place at a critical size ratio $\delta_c^{imm} \sim 0.275$ where the MSD remains subdiffusive also at long times[2]. The value of δ_c is smaller for the simulation with immobile large particles. This finding is consistent with the opening of channels as a consequence of the thermal motion of the matrix particles. In the case of mobile matrix particles localisation is *never* observed (Fig. 5b): even for large δ, the residual motion of the matrix allows the small particles to move and hence their MSD increases at long times. Furthermore, the subdiffusive regime observed in the case of an immobile matrix is only observed for $\delta < \delta_c^{imm}$ and thus in a smaller range than for mobile particles. This is consistent with the opening of channels as a consequence of the thermal motion of the matrix particles, which allows larger particles to move between voids. We also find that $f^{self}(q, \Delta t)$ calculated for the case of an immobile matrix displays a power-law dependence on time extending for several decades (Fig. 5c), as also observed in the Lorentz gas model[37], while the collective $f(q, \Delta t)$ displays neither a power-law nor a logarithmic dependence (Supplementary Fig. 2 and Supplementary Note 2). In the case of

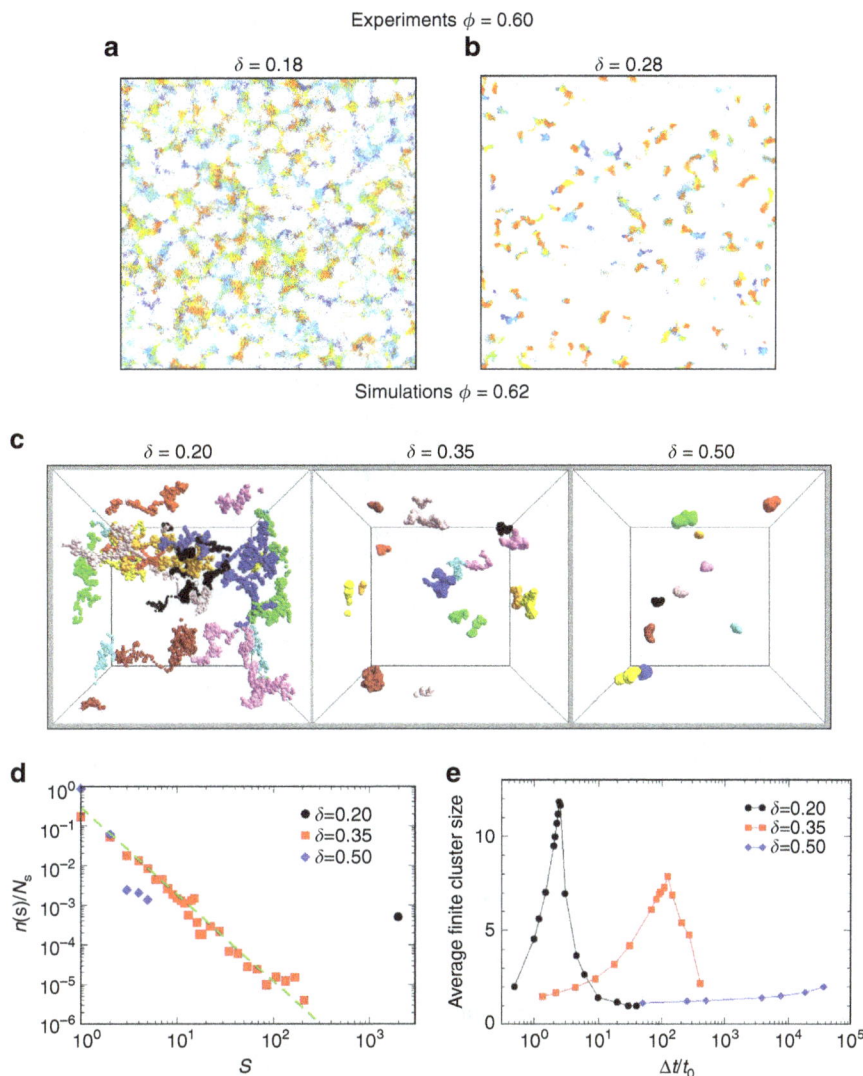

Figure 4 | Illustration of the space explored by small particles during their motions. (a–b) Overlay of small particle positions at different times (colour coded from blue, corresponding to $t_i^{exp}=0$ s to red, corresponding to $t_f^{exp}=297$ s with time steps of 33 s), obtained by particle tracking applied to 2D confocal microscopy images, for $\phi=0.60$ and (a) $\delta=0.18$, (b) $\delta=0.28$. (c) Positions of 10 small particles (distinguished by different colours) for (left) $\delta=0.2$, (middle) $\delta=0.35$ and (right) $\delta=0.5$, for a fixed total time of the trajectories $t_f^{sim}=100t_0$, comparable with the experiments (d). Distribution $n(s)$ (normalized by the number of small particles N_s) of the size s of clusters as defined in Methods, providing a measure of the space explored by small particles, evaluated within a fixed time interval $t_f^{sim}=100t_0$. For $\delta=0.35$ data follow a power-law dependence $n(s)\sim s^{-2.19}$, consistent with random percolation (dashed line), while for $\delta=0.20$ all particles belong to the same cluster. (e) Average size L_c of finite clusters as a function of time, for different δ, as indicated. The maximum in each curve signals the onset of percolation.

a mobile matrix, however, power law behaviour is not observed but, close to δ_c, a logarithmic dependence is found. Thus, thermal motion of the matrix particles gives rise to the logarithmic decay, a novel type of dynamics which does not occur in models with immobile obstacles.

Discussion

Our combined experimental, simulation and theoretical study shows that dynamics of intruders in a mobile crowded environment requires a description beyond that provided by models with a matrix of fixed obstacles. The novel application of the confocal DDM technique to concentrated binary colloidal mixtures allows us to investigate the collective dynamics of intruders in a mobile matrix, revealing extended anomalous dynamics for specific values of the size asymmetry and of the probed length scale. While the Lorentz model predicts

a power-law behavior, which is typical for systems close to a percolation transition, in the case of a mobile matrix we observe a logarithmic decay of the collective and self density fluctuations over at least three decades in time, at length scales comparable to the size of the matrix particles. This logarithmic decay marks the transition between a diffusive behaviour of intruders in a glassy medium for small size ratios $\delta<\delta_c$, where transient localization is due to the excluded volume of the mobile matrix, and glassy dynamics of the intruders at large size ratios $\delta>\delta_c$, due to crowding. Our results thus show that both percolation and glassy dynamics have to be considered. By comparing mobile and immobile matrix environments, we demonstrate that the dynamics of the small particles is profoundly altered, in a qualitative way, by the continuous evolution of channels in the mobile matrix, due to the thermal motion of large particles. A mobile matrix corresponds to an environment in which small intruders move in many real systems and applications, like in

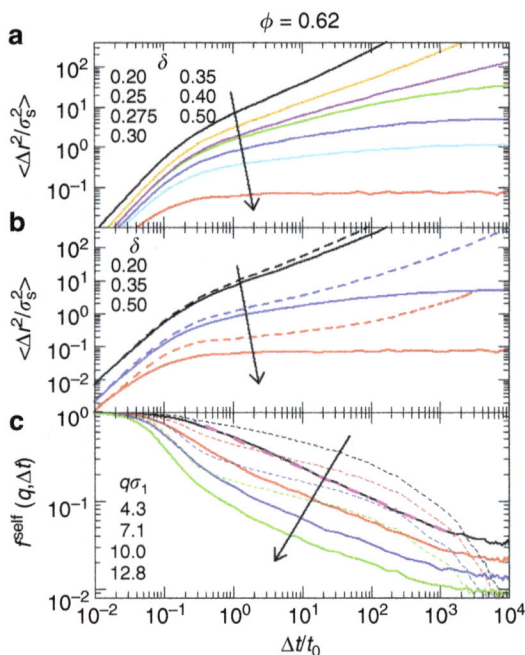

Figure 5 | Small particle dynamics in a mobile or immobile large particles matrix. (**a**) MSDs of the small particles for immobile large particles at $\phi = 0.62$ and various values of δ, as indicated. For $\delta_c^{imm} \sim 0.275$ a clear subdiffusive behavior is observed at all times. (**b**) Comparison of the MSDs of the small particles at $\phi = 0.62$ for mobile (dashed lines) and immobile (full lines) large particles, for increasing δ, as indicated. (**c**) Self intermediate scattering functions $f^{self}(q, \Delta t)$ at $\phi = 0.62$ and different wavevectors $q\sigma_l$, as indicated, for $\delta = 0.25$ (immobile, full lines) and $\delta = 0.35$ (mobile, dashed lines) highlighting the power-law dependence (dot-dashed line) in the immobile case. Arrows indicate increasing Δ or increasing q accordingly.

glasses, nanocomposite materials, chromatography, catalysis, oil recovery, drug delivery or cell signaling, cell interiors, human and animal crowds and vehicular traffic. We thus expect that our findings will inspire the development of a more realistic description of these situations and will stimulate theoretical studies to refine the MCT predictions.

Methods

Materials. We investigated dispersions of sterically stabilized polymethyl methacrylate (PMMA) spheres of diameters $\sigma_{1(1)} = 3.10 \, \mu m$ (polydispersity 0.07) or $\sigma_{1(2)} = 1.98 \, \mu m$ (polydispersity 0.07) mixed with spheres of diameter $\sigma_s = 0.56 \, \mu m$ (polydispersity 0.13) (fluorescently labeled with nitrobenzoxadiazole), in a cis-decalin/cycloheptyl-bromide mixture which closely matches their density and refractive index. The size ratio of the mixtures is $\sigma = 0.18(\sigma_{1(1)})$ and $\sigma = 0.28(\sigma_{1(2)})$, respectively. After adding salt (tetrabutylammoniumchloride), this system presents hard-sphere like interactions[38,39]. A sediment of the large spheres with $\phi = 0.65$ or of the small spheres with $\phi = 0.67$, as estimated from comparison with numerical simulations and experiments[40,41], is diluted to obtain one-component dispersions with desired volume fraction ϕ. Following a recent study[42], the uncertainty $\Delta\phi$ can be as large or above 3%. Using the nominal volume fraction ϕ of the large spheres as a reference, the volume fraction of the samples containing the small particles are adjusted in order to obtain comparable linear viscoelastic moduli in units of the energy density $3k_BT/4\pi R^3$, where k_B is the Boltzmann constant, T the temperature and R the particles' radius, while multiplying the frequency by the free-diffusion Brownian time $\tau_0 = 6\pi\eta R^3/k_B T$, where $\eta = 2.2 \, mPa \, s$ is the solvent viscosity. In this way we obtain samples with comparable rheological properties and, according to the generalised Stokes–Einstein relation[43], also dynamics and hence a similar location with respect to the glass transition. The comparable dynamics but different polydispersities of the one-component samples imply slightly different ϕ. Samples with different total volume fractions and a fixed composition, namely a fraction of small particles $x_s = \phi_s/\phi = 0.01$, where ϕ_s is the volume fraction of small particles, are prepared by mixing the one-component samples.

DDM measurements. Confocal microscopy images were acquired in a plane at a depth of approximately 30 µm from the coverslip. Images with 512×512 pixels,

corresponding to $107 \times 107 \, \mu m^2$, were acquired at a fast rate of 30 frames per second to follow the short-time dynamics and at a slow rate, between 0.07 and 0.33 frames per second, depending on sample, to follow the long-time dynamics. Image series were acquired using a Nikon A1R-MP confocal scanning unit mounted on a Nikon Ti-U inverted microscope, with a 60 × Nikon Plan Apo oil immersion objective (NA = 1.40). The pixel size at this magnification is $0.21 \times 0.21 \, \mu m$. The confocal images were acquired with the maximum pinhole size allowed by the microscope, corresponding to a pinhole diameter of 255 µm. Time series of 10^4 images were acquired for 2–5 different volumes, depending on sample.

DDM analysis. Particle movements induce fluctuations of the fluorescence intensity in the images, $i(x, y, t)$, with x, y the coordinates of a pixel in the image and t the time at which the image was recorded. To obtain additional information on the characteristic length scales of particle motions, $i(x, y, t)$ can be Fourier transformed, yielding $\hat{i}(\mathbf{q}, t)$, with $\mathbf{q}$ the wave vector in Fourier space, and then differences of the Fourier transformed image intensities can be correlated (Fig. 1c) to obtain the image structure function $D(\mathbf{q}, \Delta t)$:

$$D(\mathbf{q}, \Delta t) = \left\langle |\hat{i}(\mathbf{q}, t + \Delta t) - \hat{i}(\mathbf{q}, t)|^2 \right\rangle, \tag{1}$$

where $\langle \rangle$ represents an ensemble average. This analysis technique is named DDM[24]. The intermediate scattering function $f(\mathbf{q}, \Delta t)$ (Fig. 1d) can be extracted from the image structure function:

$$D(\mathbf{q}, \Delta t) = A(\mathbf{q})[1 - f(\mathbf{q}, \Delta t)] + B(\mathbf{q}), \tag{2}$$

with $A(q) = N|\hat{K}(q)|^2 S(q)$, where N is the number of particles in the observed volume, $\hat{K}(q)$ is the Fourier transform of the Point-Spread Function of the microscope, $S(q)$ is the static structure factor of the system, and $B(q)$ accounts for the camera noise. The inverse of the wave vector q determines the length scale over which the particle dynamics are probed. Thus $f(\mathbf{q}, \Delta t)$ is obtained, similarly to dynamic light scattering (DLS)[44], but for the present system the advantage of DDM over DLS is that fluctuations of the incoherent fluorescence signal can be correlated, a possibility which is excluded by the requirement of coherence of light in DLS. Furthermore, use of a confocal microscope drastically reduces the amount of background fluorescence of the measurements, significantly improving the determination of $f(\mathbf{q}, \Delta t)$. The effect of particles moving in and out of the observation plane on $f(\mathbf{q}, \Delta t)$ was found to be negligible for all samples, as determined by the q-dependence of the relaxation times of the initial decay of $f(\mathbf{q}, \Delta t)$, where no plateau at small q values was observed[25,45].

Particle localization. Coordinates of the small particles were extracted from time series of 2-dimensional images using standard particle localization routines based on the centroiding technique[46]. Only the particle positions at each time could be determined, not the full trajectories. Indeed the displacement of small particles during the time delay Δt between two successive frames is comparable or larger than their diameter, which implies that identifying particles after a Δt becomes too uncertain.

Simulations. We perform event-driven molecular dynamics simulations[47] in the *NVT* ensemble in a cubic box with periodic boundary conditions for binary mixtures of hard spheres, of which the large components are 7% polydisperse by a discrete Gaussian distribution[48] and the small ones are monodisperse. For each studied δ we vary the total number of particles in the range of a few thousands. The number of small particles thus varies from 1980 for $\delta = 0.2$ to 292 for $\delta = 0.5$. Mass and length are measured in units of particle mass m, average large particle diameter σ_l, whereas time is in units of $t_0 = \sqrt{m\sigma_l^2/\kappa_B T}$, where k_B is the Boltzmann constant and T the temperature. For the simulations with immobile hard spheres, after equilibration of the mixture, we freeze the large particles only. To roughly estimate the critical size ratio which demarcates the transition between diffusive and localized states, we averaged results over ten different matrix realizations.

Mode coupling theory. The equations determining $f(q, t)$ and $\langle \Delta r^2(t) \rangle$ within MCT were solved for a binary mixture of hard spheres within the Percus–Yevick approximation for the static structure; for details on the theory and the numerical procedure, see ref. 29. The $f(q, \Delta t)$ were obtained using a wave-number grid of equidistant steps $\Delta q = 0.4/\sigma_l$, with large-q cutoff $q\sigma_l = 400$. Brownian dynamics is assumed with the short-time diffusion coefficients following the Stokes–Einstein relation; the diffusion coefficient of the large particles sets the unit of time τ_0. In the calculations, the total packing fraction ϕ is varied, keeping $x_s = \phi_s/\phi = 0.01$ fixed.

Calculation of the size distribution of the explored space. To evaluate the distribution of space sampled by the small particles during time we employ the following procedure: First we generate a sequence of N_c configurations saved at equally spaced times t_i (with $i = 1 \dots N_c$) within a given time window t_{N_c}. The time interval Δt_c between two successive configurations, i.e., $\Delta t_c = t_{i+1} - t_i$ is chosen in such a way that $\langle \Delta r^2(\Delta t_c)\rangle/\sigma_s^2 = 0.5$. Second, we overlap all N_c configurations and perform a cluster size analysis according to the following criteria: (i) the same particle at different times t_i belong to the same cluster; (ii) if two particles overlap, they belong to the same cluster; (iii) the size s of a cluster is defined as the number of distinct particles belonging to the same cluster (running from one to the total number of small particles). To improve statistics we average the cluster size distribution $n(s)$ over a set of at least 10 independent groups of N_c configurations. The average size of finite clusters is calculated as $L_c = \sum s^2 n(s)/\sum sn(s)$, excluding percolating clusters.

References

1. Klafter, J. & Sokolov, I. M. Anomalous diffusion spreads its wings. *Phys. World* **18**, 29–32 (2005).
2. Höfling, F. & Franosch, T. Anomalous transport in the crowded world of biological cells. *Rep. Prog. Phys* **76**, 046602 (2013).
3. Lorentz, H. A. Le mouvement des électrons dans les métaux. *Arch. Neerl. Sci. Exact Natur* **10**, 336 (1905).
4. Höfling, F., Franosch, T. & Frey, E. Localization transition of the three-dimensional lorentz model and continuum percolation. *Phys. Rev. Lett.* **96**, 165901 (2006).
5. Krakoviack, V. Liquid-glass transition of a fluid confined in a disordered porous matrix: a mode-coupling theory. *Phys. Rev. Lett.* **94**, 065703 (2005).
6. Kurzidim, J., Coslovich, D. & Kahl, G. Single-particle and collective slow dynamics of colloids in porous confinement. *Phys. Rev. Lett.* **103**, 138303 (2009).
7. Kim, K., Miyazaki, K. & Saito, S. Slow dynamics in random media: crossover from glass to localization transition. *Europhys. Lett.* **88**, 36002 (2009).
8. Voigtmann, T. h. & Horbach, J. Double transition scenario for anomalous diffusion in glass-forming mixtures. *Phys. Rev. Lett.* **103**, 205901 (2009).
9. Schnyder, S. K., Spanner, M., Höfling, F., Franosch, T. & Horbach, J. Rounding of the localization transition in model porous media. *Soft Matter* **11**, 701–711 (2015).
10. Skinner, T. O. E., Schnyder, S. K., Aarts, D. G. A. L., Horbach, J. & Dullens, R. P. A. Localization dynamics of fluids in random confinement. *Phys. Rev. Lett.* **111**, 128301 (2013).
11. Ellis, R. J. & Minton, A. P. Cell biology: join the crowd. *Nature* **425**, 27–28 (2003).
12. Di Rienzo, C., Piazza, V., Gratton, E., Beltram, F. & Cardarelli, F. Probing short-range protein brownian motion in the cytoplasm of living cells. *Nat. Commun.* **5**, 5891 (2014).
13. Sadati, M., Nourhani, A., Fredberg, J. J. & Taheri Qazvini, N. Glass-like dynamics in the cell and in cellular collectives. *WIREs Syst. Biol. Med.* **6**, 137–149 (2014).
14. Angelini, T. E. *et al.* Glass-like dynamics of collective cell migration. *Proc. Natl. Acad. Sci* **108**, 4714–4719 (2011).
15. Trimble, W. S. & Grinstein, S. Barriers to the free diffusion of proteins and lipids in the plasma membrane. *J. Cell Biol.* **208**, 259–271 (2015).
16. Gravish, N., Gold, G., Zangwill, A., Goodisman, M. A. D. & Goldman, D. I. Glass-like dynamics in confined and congested ant traffic. *Soft Matter* **11**, 6552–6561 (2015).
17. Cherdhirankorn, T. *et al.* Fluorescence correlation spectroscopy study of molecular probe diffusion in polymer melts. *Macromolecules* **42**, 4858–4866 (2009).
18. Grabowski, C. A. & Mukhopadhyay, A. Size effect of nanoparticle diffusion in a polymer melt. *Macromolecules* **47**, 7238–7242 (2014).
19. Narayanan, S., Lee, D. R., Hagman, A., Li, X. & Wang, J. Particle dynamics in polymer-metal nanocomposite thin films on nanometer-length scales. *Phys. Rev. Lett.* **98**, 185506 (2007).
20. Kalathi, J. T., Yamamoto, U., Schweizer, K. S., Grest, G. S. & Kumar, S. K. Nanoparticle diffusion in polymer nanocomposites. *Phys. Rev. Lett.* **112**, 108301 (2014).
21. Babu, S., Gimel, J. C. & Nicolai, T. Tracer diffusion in colloidal gels. *J. Phys. Chem. B* **112**, 743–748 (2008).
22. Salami, S., Rondeau-Mouro, C., van Duynhoven, J. & Mariette, F. Probe mobility in native phosphocaseinate suspensions and in a concentrated rennet gel: effects of probe flexibility and size. *J. Agric. Food Chem.* **61**, 5870–5879 (2013).
23. Helbing, D. Traffic and related self-driven many-particle systems. *Rev. Mod. Phys.* **73**, 1067–1141 (2001).
24. Cerbino, R. & Trappe, V. Differential dynamic microscopy: probing wave vector dependent dynamics with a microscope. *Phys. Rev. Lett.* **100**, 188102 (2008).
25. Lu, P. J. *et al.* Characterizing concentrated, multiply scattering, and actively driven fluorescent systems with confocal differential dynamic microscopy. *Phys. Rev. Lett.* **108**, 218103 (2012).
26. Wilson, L. G. *et al.* Differential dynamic microscopy of bacterial motility. *Phys. Rev. Lett.* **106**, 018101 (2011).
27. Götze, W. & Hausmann, R. Further phase transition scenarios described by the self consistent current relaxation theory. *Z. Phys. B: Condens. Matter* **72**, 403–412 (1988).
28. Bosse, J. & Kaneko, Y. Self-diffusion in supercooled binary liquids. *Phys. Rev. Lett.* **74**, 4023–4026 (1995).
29. Voigtmann, T. Multiple glasses in asymmetric binary hard spheres. *Europhys. Lett.* **96**, 36006 (2011).
30. Moreno, A. J. & Colmenero, J. Logarithmic relaxation in a kinetically constrained model. *J. Chem. Phys.* **125**, 016101 (2006).
31. Moreno, A. J. & Colmenero, J. Relaxation scenarios in a mixture of large and small spheres: dependence on the size disparity. *J. Chem. Phys.* **125**, 164507 (2006).
32. Mayer, C. *et al.* Multiple glass transitions in star polymer mixtures: Insights from theory and simulations. *Macromolecules* **42**, 423–434 (2009).
33. Dawson, K. *et al.* Higher-order glass-transition singularities in colloidal systems with attractive interactions. *Phys. Rev. E* **63**, 011401 (2000).
34. Sciortino, F., Tartaglia, P. & Zaccarelli, E. Evidence of a higher-order singularity in dense short-ranged attractive colloids. *Phys. Rev. Lett.* **91**, 268301 (2003).
35. Gnan, N., Das, G., Sperl, M., Sciortino, F. & Zaccarelli, E. Multiple glass singularities and isodynamics in a core-softened model for glass-forming systems. *Phys. Rev. Lett.* **113**, 258302 (2014).
36. Stauffer, D. & Aharony, A. *Introduction to Percolation Theory* 2nd ed. (CRC Press, 1994).
37. Spanner, M., Schnyder, S. K., Höfling, F., Voigtmann, T. & Franosch, T. Dynamic arrest in model porous media-intermediate scattering functions. *Soft Matter* **9**, 1604–1611 (2013).
38. Yethiraj, A. & van Blaaderen, A. A colloidal model system with an interaction tunable from hard sphere to soft and dipolar. *Nature* **421**, 513–517 (2003).
39. Royall, C. P., Poon, W. C. K. & Weeks, E. R. In search of colloidal hard spheres. *Soft Matter* **9**, 17–27 (2013).
40. Schaertl, W. & Sillescu, H. Brownian dynamics of polydisperse colloidal hard spheres: equilibrium structures and random close packings. *J. Stat. Phys.* **77**, 1007–1025 (1994).
41. Desmond, K. W. & Weeks, E. R. Influence of particle size distribution on random close packing of spheres. *Phys. Rev. E* **90**, 022204 (2014).
42. Poon, W. C. K., Weeks, E. R. & Royall, C. P. On measuring colloidal volume fractions. *Soft Matter* **8**, 21–30 (2012).
43. Mason, T. G. Estimating the viscoelastic moduli of complex fluids using the generalized stokes-einstein equation. *Rheol. Acta* **39**, 371–378 (2000).
44. Berne, B. J. & Pecora, R. *Dynamic Light Scattering: With Applications to Chemistry, Biology, and Physics (Dover Books on Physics)* (Dover Publications, 2000).
45. Giavazzi, F. & Cerbino, R. Digital fourier microscopy for soft matter dynamics. *J. Opt.* **16**, 083001 (2014).
46. Crocker, J. C. & Grier, D. G. Methods of digital video microscopy for colloidal studies. *J. Coll. Interf. Sci* **179**, 298–310 (1996).
47. De Michele, C. Simulating hard rigid bodies. *J. Comput. Phys.* **229**, 3276–3294 (2010).
48. Zaccarelli, E. *et al.* Crystallization of hard-sphere glasses. *Phys. Rev. Lett.* **103**, 135704 (2009).

Acknowledgements

We thank Andrew Schofield (University of Edinburgh) for providing the PMMA particles, Vincent Martinez (University of Edinburgh), Wilson Poon (University of Edinburgh) and Matthias Reufer (LSI instruments) for providing routines for DDM analysis, and Thomas Franosch (ITP Innsbrück) and Manuel A. Escobedo-Sánchez (University of Düsseldorf) for discussions. T.S., S.U.E. and M.L. acknowledge funding by the Deutsche Forschungsgemeinschaft (DFG) through the research unit FOR1394, project P2, and funding of the confocal microscope through grant INST 208/617–1 FUGG. E.Z. and C.D.M. acknowledge support from MIUR through a Futuro in Ricerca grant FIRB ANISOFT (RBFR125H0M). E.Z., C.D.M. and F.S. acknowledge support from ERC-226207-PATCHYCOLLOIDS and ETN-COLLDENSE (H2020-MCSA-ITN-2014, grant no. 642774).

Author contributions

T.S., S.U.E. and M.L. planned, performed, analysed and interpreted the experiments. E.Z., P.T., C.D.M. and F.S. planned, ran and interpreted the simulations. T.V. obtained MCT predictions. All authors contributed to the interpretation and comparison of the data as well as the writing of the manuscript.

Additional information

3

Advanced intermediate temperature sodium–nickel chloride batteries with ultra-high energy density

Guosheng Li[1], Xiaochuan Lu[1], Jin Y. Kim[1], Kerry D. Meinhardt[1], Hee Jung Chang[1], Nathan L. Canfield[1] & Vincent L. Sprenkle[1]

Sodium-metal halide batteries have been considered as one of the more attractive technologies for stationary electrical energy storage, however, they are not used for broader applications despite their relatively well-known redox system. One of the roadblocks hindering market penetration is the high-operating temperature. Here we demonstrate that planar sodium–nickel chloride batteries can be operated at an intermediate temperature of 190 °C with ultra-high energy density. A specific energy density of 350 Wh kg^{-1}, higher than that of conventional tubular sodium–nickel chloride batteries (280 °C), is obtained for planar sodium–nickel chloride batteries operated at 190 °C over a long-term cell test (1,000 cycles), and it attributed to the slower particle growth of the cathode materials at the lower operating temperature. Results reported here demonstrate that planar sodium–nickel chloride batteries operated at an intermediate temperature could greatly benefit this traditional energy storage technology by improving battery energy density, cycle life and reducing material costs.

[1] Electrochemical Materials and Systems Group, Energy Processes and Materials Division, Pacific Northwest National Laboratory, Richland, 99352 Washington, USA. Correspondence and requests for materials should be addressed to G.L. (email: guosheng.li@pnnl.gov) or to V.L.S. (email: vincent.sprenkle@pnnl.gov).

Recently, molten-sodium (Na) beta-alumina batteries have been considered as one of the most attractive stationary electric energy storage systems, which are crucial to stimulate the growth of renewable energy resources and to improve the reliability of electric power grids[1–5]. Sodium–sulfur (Na–S)[6] and sodium-metal halide batteries (ZEBRA)[7] are two typical molten-Na beta-alumina batteries; however, recent fire incidents involving Na–S battery systems have increased general concern about the application of Na–S batteries as stationary energy storage devices. Although they share some features (for example, molten-sodium and β''-alumina solid electrolyte) in common with Na–S batteries, ZEBRA batteries can provide several advantages over Na–S batteries, including superior battery safety, high open-circuit voltage, lower operating temperature and ease of assembly in the discharged state without using metallic sodium in the anode[2,4,7].

Among various ZEBRA battery redox chemistries[8–12], the sodium–nickel chloride (Na–NiCl$_2$) battery has been most widely investigated in the past[13–15]. The overall redox reaction of a Na–NiCl$_2$ battery during charging and discharging processes is described as follows:

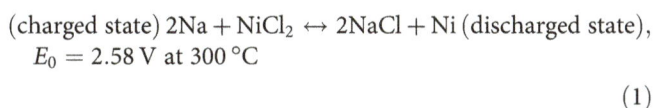

$$(\text{charged state}) \; 2Na + NiCl_2 \leftrightarrow 2NaCl + Ni \; (\text{discharged state}),$$
$$E_0 = 2.58\,\text{V at } 300\,°C$$

$$(1)$$

Despite the relatively simple redox reaction, cell degradation mechanisms of Na–NiCl$_2$ batteries have not been clearly understood in the past. In our recent studies[16], we have reported detailed correlations between NaCl/Ni particle growth (Ostwald ripening) and battery-operating conditions, such as C-rate, cathode formula and cycling capacity window. The main parameters that lead to faster Ni particle growth are higher current density, state of charge (SOC) at end of charge (EOC) and Ni/NaCl ratio. In the case of NaCl, significant growth has a close correlation with the cycling capacity window[16]. To achieve sustainable battery cycle life, the conventional tubular Na–NiCl$_2$ batteries are loaded with excessive Ni content in the cathode and also are operated with a shallow capacity window. The theoretical specific capacity and energy density of Na–NiCl$_2$ ZEBRA batteries obtained from reaction (1) are 305 mAh g^{-1} (without considering the melt) and 788 Wh kg^{-1} (open-circuit voltage at 2.58 V), see Supplementary Table 1. Despite the quite impressive theoretical energy density of Na–NiCl$_2$ batteries, general energy density obtained from a conventional tubular Na–NiCl$_2$ battery (operated at ca. 300 °C) is about 95–120 Wh kg^{-1} due to excessive Ni content and shallow capacity window[2,3,5]. Detailed plots of energy density versus Ni content with different cycling windows are shown in Fig. 1 (see Supplementary Fig. 1 as well). Resolving this shortcoming on a material and cost level requires creating new platforms based on innovative scientific and technical approaches. Excited by the magnitude and implications of revisiting Na–NiCl$_2$ ZEBRA battery technology, the research and industrial communities are seeking a revolutionary breakthrough that could enable substantially lower cost for materials and operations, as well as superior battery cycle life and safety.

In our previous work[17], we found that the operating temperature has significant influence on the cell chemistries during the battery cycling. The cell polarization, an important indicator of cell degradation, was found to increase faster at 280 °C than at 175 °C due to faster grain growth in the cathode ingredients. From a cell-operation point of view, lower temperature can potentially reduce costs associated with cell packing and reduce heat loss. In a recent report, Gerovasili et al.[18] concluded that lower heat transfer losses at 240 °C could result in

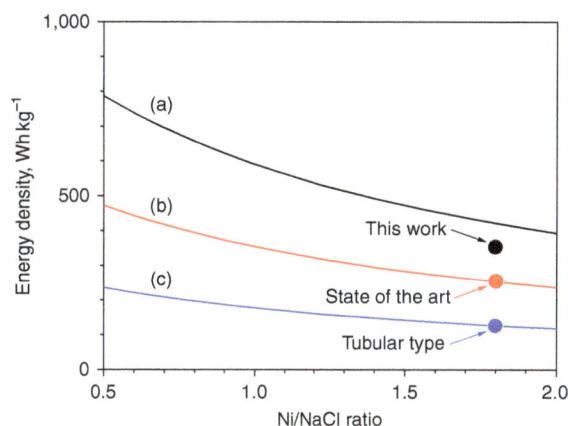

Figure 1 | Specific energy density. Specific energy densities were calculated on different cycling capacity windows for (a) 100, (b) 60 and (c) 30%. Energy density was calculated without considering the weight of melt required in the cathode. See Supplementary Fig. 1 for the specific energy density plots calculated with the weight of melt included.

up to 49% reduction in heating energy compared with operation at 275 °C. It should be noted that it is intrinsically difficult for a tubular ZEBRA battery to operate below 240 °C.

Drawing inspiration from the temperature-dependent particle growth, we construct a planar intermediate temperature (IT) Na–NiCl$_2$ ZEBRA battery technology, which allows the cells to be operated at an IT of 190 °C with considerable discharge power as high as 75 mW cm^{-2} (~ 0.6 C). Extensive investigations of cell performance and fundamental understanding of cathode degradation mechanisms at 190 °C are studied, and indicate that this novel planar IT Na–NiCl$_2$ ZEBRA battery technology could have a much higher specific energy density (350 Wh kg^{-1}) and much more stable cycle life than the state-of-the-art ZEBRA battery.

Results

Battery performances of planar IT ZEBRA battery. We constructed a planar IT ZEBRA battery, using a β''-alumina solid-state electrolyte (BASE; 3 cm^2 effective areas), a Ni/NaCl granule cathode (157 mAh, 52.3 mAh cm^{-2}) and NaAlCl$_4$ as a secondary electrolyte. Detailed information can be found in the experimental section. All batteries were charged and discharged between 2.8 V (EOC) and 2.0 V (end of discharge) to prevent side reactions occurring due to over-charging and over-discharging. Figure 2 shows battery performance for cells operated at two different temperatures (190 and 280 °C) with a constant discharge power of 25 mW cm^{-2} (~ 10 mA cm^{-2}, $\sim$C/5). An initial capacity of 106 mAh g^{-1} was observed for IT Na–NiCl$_2$ batteries tested at 190 °C. Interestingly, IT Na–NiCl$_2$ batteries (Fig. 2, black) showed a capacity increase for the first 100 cycles and then stabilized with a capacity of 137 mAh g^{-1} at the 200th cycle. The specific discharge energy density for IT Na–NiCl$_2$ batteries was 340 Wh kg^{-1} (200th cycle), which is by far the highest energy density demonstrated for Na–NiCl$_2$ batteries to the best of our knowledge. Identical batteries tested at 280 °C are also shown in Fig. 2 (blue). The initial capacity of 140 mAh g^{-1} was obtained for the battery operated at 280 °C. The higher initial capacity obtained at 280 °C compared with that at 190 °C is likely due to better sodium wetting on the BASE at 280 °C. In contrast with IT Na–NiCl$_2$ batteries at 190 °C, the capacity of batteries operated at 280 °C decreased drastically, to 107 mAh g^{-1} (76% retention) over 200 cycles. The more stable performance of the Na–NiCl$_2$ batteries at 190 °C than at 280 °C indicates that the lower operation temperature could be the most critical factor for

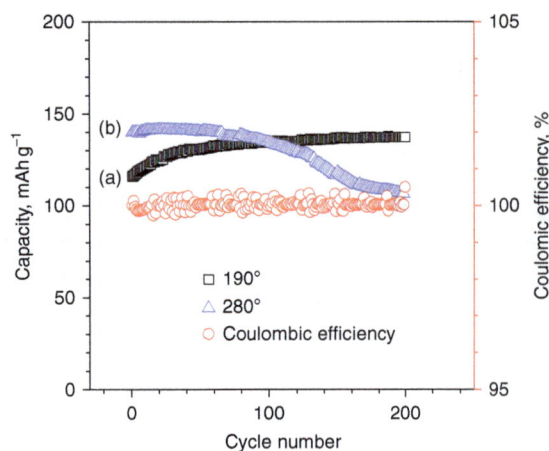

Figure 2 | Capacity retention and Coulombic efficiency plots. Na–NiCl$_2$ cells were operated at two different temperatures: (a) 190 (black) and (b) 280 °C (blue). Coulombic efficiency is shown in red. Cells were charged with a constant current (7 mA cm^{-2}, ~C/7) and were discharged with a constant power (25 mW cm^{-2}, ~10 mA cm^{-2}, ~C/5).

obtaining sustainable cell performance, which has been surprisingly understated in the past. As shown in Supplementary Fig. 2, cells have been also tested with a higher discharge power (75 mW cm^{-2}, ~30 mA cm^{-2}, ~0.6 C) at 190 and 280 °C. Similar to results shown in Fig. 2, the capacity was more stable for batteries operated at 190 °C than at 280 °C. The Coulombic efficiencies shown in Fig. 2 (red) and Supplementary Fig. 2 (red) of all tested batteries are nearly 100%, which is due to use of BASE as the sodium-ion conducting solid-state electrolyte.

To further understand the effect of temperature on cell performance, voltage profiles (1st, 100th and 200th cycles) versus SOC for cells tested at 190 and 280 °C are shown in Fig. 3 (25 mW cm^{-2}) and Supplementary Fig. 3 (75 mW cm^{-2}). For the cells operated at 190 °C (Fig. 3a), SOC at the EOC (SOC$_{EOC}$) and SOC at the end of discharge (SOC$_{EOD}$) for the 1st cycle were determined to be 90 and 17%, respectively. In further battery cycles, SOC$_{EOC}$ was gradually increased to 100% and SOC$_{EOD}$ was decreased to 12%. These adjustments are responsible for the progressively increasing capacity for the cells operated at 190 °C (Fig. 2), since the capacity of a battery can be calculated as follows:

$$\text{Capacity} = \text{SOC}_{EOC} - \text{SOC}_{EOD} \qquad (2)$$

For instance, capacities of the 1st and 200th cycles for the cells operated at 190 °C were 73 and 88%, respectively, as calculated using equation (2). In contrast with the lowest capacity having been observed in the early stage of tested cycles at 190 °C, the SOC$_{EOC}$ and SOC$_{EOD}$ for the 1st cycle at 280 °C were 100 and 14%, respectively, which resulted in the largest initial capacity of 86% as shown in Fig. 3b. The higher initial capacity observed at the operating temperature of 280 °C than at 190 °C is most likely due to better sodium wetting on the BASE at the higher operating temperature. However, the SOC$_{EOC}$ of cells operated at 280 °C rapidly decreased to 79% after the 200th cycle, which results in a capacity of 65% (76% capacity retention). On the other hand, a quite stable SOC$_{EOD}$ observed over 200 cycles indicates that degradation on the anode side (sodium wetting) is negligible for 280 °C as shown in Fig. 3b. The rapid degradation of capacity for the cells operated at 280 °C is more likely due to cathode degradation at the higher operating temperature.

Scanning electron microscopy of cathode materials. To investigate the correlation between morphology changes in Ni/NaCl

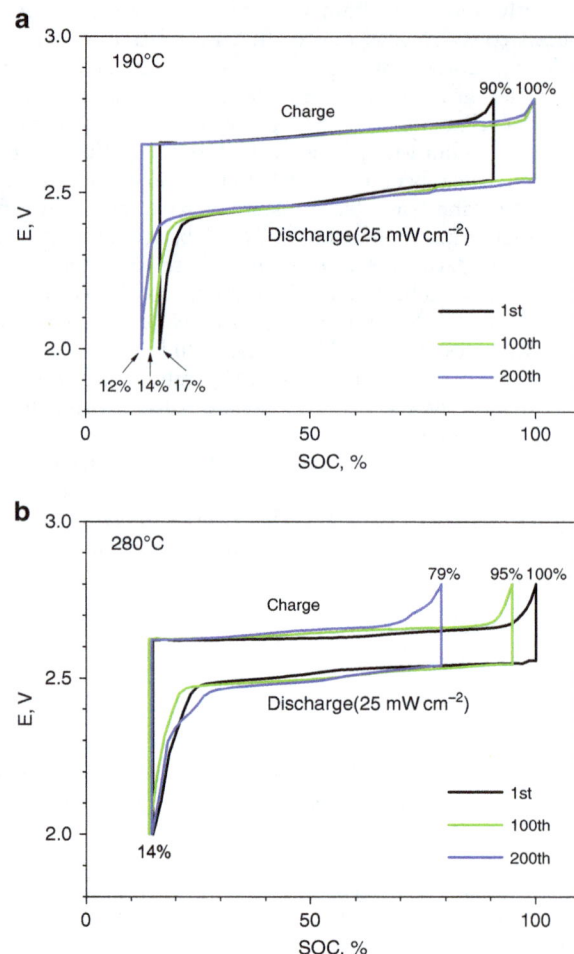

Figure 3 | Voltage profiles for planar IT Na–NiCl$_2$ batteries. Cells were operated with constant-current charge (7 mA cm^{-2}, ~C/7) and constant-power discharge (25 mW cm^{-2}, 10 mA cm^{-2}, ~C/5). Voltage profiles versus SOC are shown for 1st (black), 100th (green) and 200th (blue) cycles at two different temperatures (**a**) 190 and (**b**) 280 °C, respectively.

cathodes and cell performance, the cells operated with a constant discharge power of 25 mW cm^{-2} were disassembled after 200 cycles and the fracture surfaces were examined using scanning electron microscopy (SEM)/energy-dispersive x-ray spectrometry (EDS). For cathodes retrieved from cells tested at 190 °C, lighter spots shown in Fig. 4a (high resolution) and Fig. 4b represent Ni particles and corresponding images of Ni mapping are shown in Fig. 4c. Similarly, Nickel particles are shown in Fig. 4d (high resolution) and Fig. 4e,f (Ni mapping) for cells tested at 280 °C. Quite different sizes of Ni particles were observed for the cells operated at 190 °C compared with those at 280 °C. The typical particle size of Ni cathodes for the cells tested at 190 °C was 1–2 µm (Fig. 4a,b), which is similar to the initial particle size of raw Ni powders. However, significant Ni particle growth, up to 10 µm (Fig. 4d,e), was observed for the cells tested at 280 °C. Particle sizes of NaCl were also determined from SEM images (Fig. 4g,i) and corresponding images of Na mapping (Fig. 4h,j). NaCl particle sizes in the cells operated at 190 and 280 °C were ~5 µm and ~50 µm, respectively. SEM/EDS measurements were also performed for the cells operated with a higher discharge power of 75 mW cm^{-2} at 190 and 280 °C. Similar to the cells tested at a discharge power of 25 mW cm^{-2}, larger NaCl and Ni particles were observed in cells operated at 280 °C than in cells operated at 190 °C with a discharge power of 75 mW cm^{-2}, as shown in Supplementary Fig. 4.

Ni and NaCl particle growth. The sizes of Ni and NaCl particles from tested cells, cell cycling conditions and cell performances are summarized in Table 1 for a better comparison. It is quite clear that the particle sizes of Ni and NaCl for the tested cells show a strong dependence on the cell-operating temperature. For instance, the average Ni particle size at 280 °C is around 10 µm, which is significantly larger than the average particle size of 1–2 µm observed at 190 °C. Similarly, the NaCl particle size at 280 °C was 50 µm, which is much larger than an average particle size of 5–10 µm at 190 °C. Considering the particle size of raw Ni powders (1–2 µm) and NaCl powders (~5 µm), the particle growth at 190 °C is much slower than that at 280 °C.

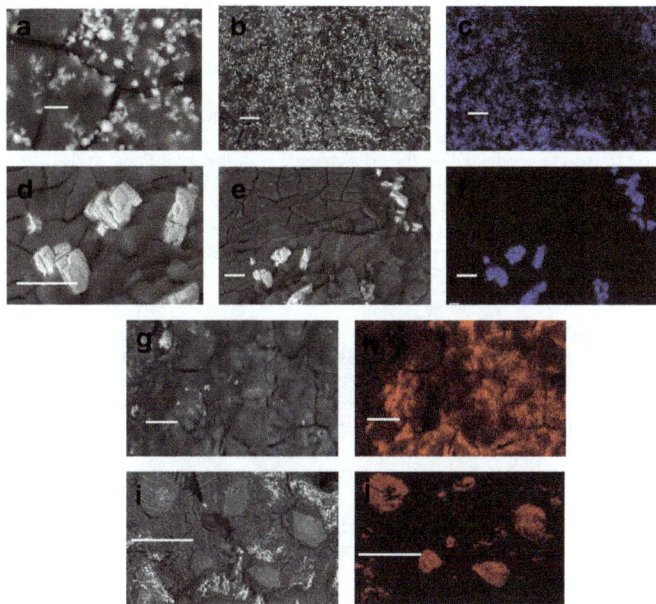

Figure 4 | SEM images for cathode materials. Cathode materials were retrieved from cells operated for 200 cycles. Ni particles are shown for cells operated at 190 °C (25 mW cm^{-2}) at (**a**) × 7,500 (scale bar, 2 µm), (**b**) × 1,000 (scale bar, 10 µm) and (**c**) Ni mapping × 1,000 (scale bar, 10 µm). Ni particles from cells operated at 280 °C (25 mW cm^{-2}) are shown at (**d**) × 3,000 (scale bar, 10 µm), (**e**) × 1,000 (scale bar, 10 µm) and (**f**) Ni mapping × 1,000 (scale bar, 10 µm). NaCl particles are shown for cells operated at 190 °C (25 mW cm^{-2}) at (**g**) × 2,000 (scale bar, 10 µm) and (**h**) Na mapping × 2,000 (scale bar, 10 µm); (**i**) × 300 (scale bar, 100 µm) and (**j**) Na mapping × 300 (scale bar, 100 µm) for 280 °C.

The morphology evolution of cathode materials has been considered as the most important cause of the degradation of Na–NiCl$_2$ batteries. For the charging process, the main reactions in the cathode side of Na–NiCl$_2$ batteries are the dissolution of NaCl particles into the melt and the formation of NiCl$_2$ layers on the surfaces of Ni particles. Larger particles of active ingredients existing in the cathode will lead to a sluggish dissolution of NaCl and less surface area of Ni particles, which will eventually cause a limited charging capacity. This is in good agreement with the observations shown in Table 1. For example, an IT Na–NiCl$_2$ battery operated at 190 °C can still be charged to 100% SOC after 200 cycles due to the minimal morphology changes in the cathode, but identical cells operated at 280 °C can be only charged up to 79% SOC after 200 cycles due to the accelerated particle growth in the cathode.

The mechanism of particle growth in Na–NiCl$_2$ batteries operated at 280 °C has been proposed in our previous study by attributing it to Ostwald ripening[16]. Here we would like to extend the particle growth mechanism by including the operating temperature as an important factor. In the literature[19,20], it has been generally understood that the influence of temperature on Ostwald ripening is through its effects on various parameters, such as the equilibrium solubility, the diffusion-influenced growth coefficient, the phase-transition energy and interfacial energy. Here, temperature effects on equilibrium solubility and diffusion-influenced growth coefficient are particularly important to understanding the enhanced particle growth in Na–NiCl$_2$ batteries at the higher temperature. For instance, the solubility of cathode materials (NaCl and NiCl$_2$) will drastically decrease with a decrease in the operating temperature[21]. The diffusion coefficients of dissolved NaCl and NiCl$_2$ will decrease as well, due to the increased viscosity of the melt. Since the equilibrium solubility and diffusion-influenced growth coefficient are proportional to the solubility and diffusion coefficient, respectively, particle growth by Ostwald ripening will be greatly suppressed at lower operating temperatures.

Discussion

For stationary energy storage, long-battery lifetime and lower materials cost are two critical factors. As shown in Fig. 5, long-term cycling of an IT Na–NiCl$_2$ battery showed an excellent stability over 1,000 cycles, which was a testing period of 1 year and 6 months. No battery degradation was observed until the

Table 1 | Grain sizes of Ni and NaCl particles and other battery performance data for cells cycled at different temperatures and discharge powers.

	190 °C		280 °C	
Discharge (mW cm^{-1})	25	75	25	75
C-rate	C/5	0.6 C	C/5	0.6 C
Ni (µm)	1–2	1–2	~10	~10
NaCl (µm)	<5	~10	~50	~50
EOC (%)†	100	100	79	78
EOD (%)†	12	17	14	20
Capacity window (%)	88	83	65	58
Energy density (Wh g^{-1})	340	300	258	220
Degradation (%)	19*	13*	− 24	− 30

EOC, end of charge; EOD, end of discharge.
*Positive degradation is due to the increased capacity of the 200th versus the 1st cycle at 190 °C.
†EOC and EOD of 200th cycle are shown in the table.

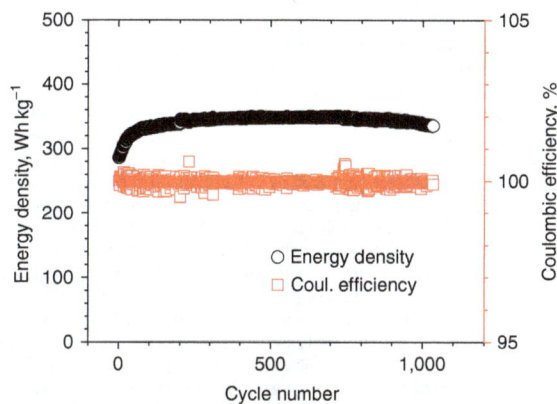

Figure 5 | Energy density and Coulombic efficiency. Planar IT Na–NiCl$_2$ cells were operated at 190 °C over 1,000 cycles in a period of 1 year and 6 months. No degradation was observed for the first 700 cycles. A degradation rate of <0.01% per cycle was obtained for the test after the 700th cycle.

700th cycle, and the degradation rate thereafter was $<0.01\%$ per cycle. The energy density of an IT Na–NiCl$_2$ battery shown in Fig. 5 was generally over 330 Wh kg^{-1}. Because of the ultra-high energy density of the IT Na–NiCl$_2$ batteries demonstrated in this work, the materials cost of cathodes (notably the cost of Ni) can be greatly reduced compared with the conventional tubular Na–NiCl$_2$ batteries, which would be considered as a significant saving. One concern with decreasing the operational temperature from 280 °C to an IT of 190 °C is the inevitable loss in energy efficiency of a Na–NiCl$_2$ battery. Cells operated at the lower temperature showed generally larger overpotentials than cells operated at the higher temperature as shown in Fig. 3 and Supplementary Fig. 3. Overall energy efficiency plots are shown in Supplementary Fig. 5. Typical overall energy efficiencies at the 200th cycle are 91.6% (190 °C, 25 mW cm^{-2}), 86.5% (190 °C, 75 mW cm^{-2}), 92.3% (280 °C, 25 mW cm^{-2}) and 88.4% (280 °C, 75 mW cm^{-2}). The energy efficiencies of cells operated at 190 °C show about 1% decrease in overall efficiencies versus those operated at 280 °C; this is due to the close relationship between the operating temperature, interfacial resistance and sodium-ion conductivity in the cell. However, the decrease in energy efficiency is trivial compared with the advantages of the lower operating temperature.

In conclusion, we have developed a novel planar Na–NiCl$_2$ battery that can be operated at an IT of 190 °C. This planar IT Na–NiCl$_2$ technology was able to deliver an ultra-high energy density (350 Wh kg^{-1}) with very long cycle life (over 1,000 cycles) and excellent capacity retention (no decay until the 700th cycle, 0.01% per cycle thereafter). This work accomplished a breakthrough towards making Na–NiCl$_2$ battery technology more competitive for stationary energy storage applications.

Methods

Material preparation and battery assembly. Detailed methods for preparing the Ni/NaCl cathodes and planar cell assembly procedures have been reported previously[16]. Cathode materials in the discharged state consisted of 62.2 wt% Ni (Novamet, Type 255), 34.2 wt% NaCl (Alfa Aesar, 99.99%) and 3.6 wt% additives (1.6 wt% of FeS, 0.6 wt% of aluminium powder, 0.5 wt% of NaI and 0.9 wt% NaF). The cathode materials were thoroughly mixed using a low-energy ball-milling method. Then, well-mixed cathode powders were transferred into a granulator (Freund TF-Labo) to make cathode granules. NaAlCl$_4$, the secondary electrolyte, was synthesized following our previously reported method[22]. The ratio of NaAlCl$_4$ and Ni/NaCl cathode is 0.5:1. The specific capacity of the button cell in this work was 157 mAh g^{-1} for without considering and 105 mAh g^{-1} with including the mass of the melt (NaAlCl$_4$). The button cell consisted of battery cases (stainless steel) for cathode (attached with a thin molybdenum film liner) and anode sides, a molybdenum cathode current collector, a stainless steel anode current collector, an α-alumina (99.5% purity) fixture and a planar composite yttria-stabilized zirconia (YSZ)/ BASE (3 cm^2 active area) disc. A schematic view of the planar design of the cell used in this work is shown in Supplementary Fig. 7. BASE discs with 500-μm thickness were fabricated using the vapour–phase-conversion process[7,11], which is briefly described as follows, 70% of α-alumina (α-Al$_2$O$_3$, Almatis A16-SG, >99%) powder and YSZ powder (8YSZ, Imerys Fused Minerals, 5.4 wt% Y$_2$O$_3$) were first mixed to the desired volume ratios (7:3), and followed by attrition milling in iso-propanol to achieve a median particle size (d$_{50}$) <0.5 μm. This not only ensures nearly uniform particle size, but also eliminates regions of compositional non-uniformity. Slurries were prepared by mixing attrition milled powders with solvents made of methyl ethyl ketone/ethanol (4:1), dispersant (PS-131, AkzoNobel, 0.5 wt%), a binder (polyvinylbutyral, B-79, 4.38 wt%) and a plasticizer (benzyl butyl phathalate, Alfa Aesar, 98%, 4.73 wt%) and then tape cast to a thin sheet (125 μm). Four layers of thin sheets were stacked and run through a hot roll laminator at 135 °C to create a monolithic laminate. The laminated sheet was laser cut to a desired size (3 cm^2) and sintered under a programed thermal process (0.5 °C min^{-1} to 190 °C for 2 h, 0.5 °C min^{-1} to 400 °C for 1 h and 3 °C min^{-1} to 1,500 °C for 2 h) to allow removing solvents, burn-out of the binder and sintering. After sintering, the α-Al$_2$O$_3$/YSZ parts were placed in 'packing powder' of β″-Al$_2$O$_3$ powder with 10 wt% NaAlO$_2$ and heat treated at 1,400 °C for 10 h. The packing powder acts as the sodium source to convert the sintered α-Al$_2$O$_3$ to β″-Al$_2$O$_3$. The results of ionic conductivity measurement and SEM cross-section for BASEs used in this work are shown in Supplementary Fig. 6. The activation energy obtained from Arrhenius plot is about 0.24 ev. The cell assembly process can be briefly described as follows. First, a BASE disc was glass-sealed to an α-alumina fixture, and the anode side of the BASE disc was pretreated with an aqueous lead acetate (Pb[CH$_3$COO]$_2$) solution. Then, the fixture was heat treated at 400 °C in an inert environment to form a thin layer of Pb/PbO on the BASE surface facing the anode. It has been reported that this thin layer of Pb/PbO improves initial sodium wetting on the surface of the BASEs[16,23]. The fixture was then transferred into a nitrogen-purged glove box for the cell assembly. In all, 1 g of cathode granules was added to cathode side of the fixture and 0.5 g of melt was vacuum infiltrated into the granules at an elevated temperature of 200 °C. A small amount of metallic sodium (<5 mg) was added to the anode shim (an initial contact) before the fixture was enclosed with cell cases by the compression sealing using metal O-rings.

Battery tests. Assembled cells were transferred from the glove box into furnaces for battery cycle tests in air. The cells were tested at two different temperatures, 190 and 280 °C, and the temperatures of the cells were closely monitored by K-type thermocouple wire gauges attached to the cell cases. Galvanostatic tests were performed using an Arbin potentiostat (MSTAT 8000). The cell test consisted of conditional and regular cycles. During the conditional cycles, the first charging process of the cells is conducted in three stages with different charging currents (0.6 mA for 2 h, 1 mA for 10 h and 10 mA thereafter) until the cell voltage reaches 2.8 V. For the first discharge, the cells are discharged at a constant current (10 mA, C/15) until the cell voltage decreases to 1.8 V. After the first cycle is completed, the cells are cycled between 1.8 V (discharge limit) and 2.8 V (charge limit) with a constant current (10 mA, C/15) for two additional cycles. Regular cycle tests were conducted on the cells charged with the conditional cycle. During the regular cycle test, the cells were cycled between 2.0 V (discharge limit) and 2.8 V (charge limit). A constant charge current of 20 mA (~C/7) and a constant discharge power of 25 mW cm^{-2} (~C/5, 10 mA cm^{-2}) or 75 mW cm^{-2} (~0.6 C, 30 mA cm^{-2}) were applied for the charging and discharging steps, respectively.

Scanning electron microscopy. SEM characterization on the fracture surfaces of the cathodes was done on JEOL JSM-5900LV equipped with an Oxford EDS with lithium drift detector and JEOL JSM-7001F (field emission) equipped with an Oxford EDS system with silicon drift detector. Batteries were disassembled in a glove box and the desired cathode materials were delivered to the SEM chamber with minimum exposure to the air. The particle sizes of Ni in the cathodes were determined from SEM backscattering images combined with elemental mapping of Ni. Similar to Ni analysis, the particle sizes of NaCl in the cathodes were determined from SEM backscattering images combined with elemental mapping of Na.

References

1. Dunn, B., Kamath, H. & Tarascon, J. M. Electrical energy storage for the grid: a battery of choices. *Science* **334**, 928–935 (2011).
2. Yang, Z. G. et al. Electrochemical energy storage for green grid. *Chem. Rev.* **111**, 3577–3613 (2011).
3. Hueso, K. B., Armand, M. & Rojo, T. High temperature sodium batteries: status, challenges and future trends. *Energy Environ. Sci.* **6**, 734–749 (2013).
4. Ha, S., Kim, J. K., Choi, A., Kim, Y. & Lee, K. T. Sodium-metal halide and sodium-air batteries. *ChemPhysChem* **15**, 1971–1982 (2014).
5. Soloveichik, G. L. Battery technologies for large-scale stationary energy storage. *Annu. Rev. Chem. Biomol.* **2**, 503–527 (2011).
6. Wen, Z. Y., Hu, Y. Y., Wu, X. W., Han, J. D. & Gu, Z. H. Main challenges for high performance NAS battery: materials and interfaces. *Adv. Funct. Mater.* **23**, 1005–1018 (2013).
7. Lu, X. C., Lemmon, J. P., Sprenkle, V. & Yang, Z. G. Sodium-beta alumina batteries: status and challenges. *JOM* **62**, 31–36 (2010).
8. Distefano, S., Ratnakumar, B. V. & Bankston, C. P. Advanced rechargeable sodium batteries with novel cathodes. *J. Power Sources* **29**, 301–309 (1990).
9. Ratnakumar, B. V., Attia, A. I. & Halpert, G. Alternate cathodes for sodium-metal chloride batteries. *J. Electrochem. Soc.* **138**, 883–884 (1991).
10. Li, G. S. et al. An advanced Na-FeCl2 ZEBRA battery for stationary energy storage application. *Adv. Energy Mater.* **5**, 1500357 (2015).
11. Parthasarathy, G., Weber, N. & Virkar, A. V. High temperature sodium–zinc chloride batteries with sodium Beta"-alumina solid electrolyte. *ECS Trans.* **6**, 67–76 (2007).
12. Lu, X. C. et al. A novel low-cost sodium-zinc chloride battery. *Energy Environ. Sci.* **6**, 1837–1843 (2013).
13. Sudworth, J. L. Sodium/nickel chloride (ZEBRA) battery. *J. Power Sources* **100**, 149–163 (2001).
14. Galloway, R. C. & Dustmann, C. H. ZEBRA battery–material cost availability and recycling. *EVS* **19**, 15 (2003).
15. Dustmann, C. H. Advances in ZEBRA batteries. *J. Power Sources* **127**, 85–92 (2004).
16. Li, G. S., Lu, X. C., Kim, J. Y., Lemmon, J. P. & Sprenkle, V. L. Cell degradation of a Na-NiCl2 (ZEBRA) battery. *J. Mater. Chem. A* **1**, 14935–14942 (2013).
17. Lu, X. C. et al. The effects of temperature on the electrochemical performance of sodium-nickel chloride batteries. *J. Power Sources* **215**, 288–295 (2012).

18. Gerovasili, E., May, J. F. & Sauer, D. U. Experimental evaluation of the performance of the sodium metal chloride battery below usual operating temperatures. *J. Power Sources* **251,** 137–144 (2014).

19. Voorhees, P. W. The theory of Ostwald Ripening. *J. Stat. Phys.* **38,** 231–252 (1985).

20. Madras, G. & McCoy, B. J. Temperature effects on the transition from nucleation and growth to Ostwald ripening. *Chem. Eng. Sci.* **59,** 2753–2765 (2004).

21. Macmillan, M. G. & Cleaver, B. Solubility of nickel chloride in molten sodium tetrachloroaluminate saturated with sodium-chloride over the temperature-range 200-degrees-c-400-degrees-C. *J. Chem. Soc. Faraday T.* **89,** 3817–3819 (1993).

22. Li, G. S. *et al.* Novel ternary molten salt electrolytes for intermediate-temperature sodium/nickel chloride batteries. *J. Power Sources* **220,** 193–198 (2012).

23. Sudworth, J. L. & Tilley, A. R. *The Sodium Sulfur Battery* (Kluwer, 1985).

Acknowledgements

This work was supported by the U.S. Department of Energy (DOE) Office of Electricity Delivery and Energy Reliability under the Contract No. 57558. PNNL is a multiprogram laboratory operated by Battelle Memorial Institute for the DOE under Contract DE-AC05-76RL01830. G.L. and V.L.S. are grateful for the financial support from the International Collaborative Energy Technology, R&D Program of the Korea Institute of Energy Technology Evaluation and Planning (KETEP), grated financial resource from POSCO and the Ministry of Trade, Industry and Energy, and Republic of Korea (No. 20158510050010).

Author contributions

G.L., J.Y.K. and V.L.S. conceived and designed the work. G.L. performed experimental works and wrote the manuscript. G.L., X.L. and K.D.M. were involved in the cell design. G.L. and N.L.C. performed SEM measurements. All the authors participated in revising the manuscript and approved the final version.

Additional information

Competing financial interests: The authors declare no competing financial interests.

4

Tracking the shape-dependent sintering of platinum–rhodium model catalysts under operando conditions

Uta Hejral[1,2,3], Patrick Müller[1,2,3], Olivier Balmes[4,5], Diego Pontoni[5] & Andreas Stierle[1,2,3]

Nanoparticle sintering during catalytic reactions is a major cause for catalyst deactivation. Understanding its atomic-scale processes and finding strategies to reduce it is of paramount scientific and economic interest. Here, we report on the composition-dependent three-dimensional restructuring of epitaxial platinum–rhodium alloy nanoparticles on alumina during carbon monoxide oxidation at 550 K and near-atmospheric pressures employing *in situ* high-energy grazing incidence x-ray diffraction, online mass spectrometry and a combinatorial sample design. For platinum-rich particles our results disclose a dramatic reaction-induced height increase, accompanied by a corresponding reduction of the total particle surface coverage. We find this restructuring to be progressively reduced for particles with increasing rhodium composition. We explain our observations by a carbon monoxide oxidation promoted non-classical Ostwald ripening process during which smaller particles are destabilized by the heat of reaction. Its driving force lies in the initial particle shape which features for platinum-rich particles a kinetically stabilized, low aspect ratio.

[1] Deutsches Elektronen-Synchrotron (DESY), NanoLab, Notkestrasse 85, D-22607 Hamburg, Germany. [2] Universität Hamburg, Fachbereich Physik, Jungiusstraße 9, 20355 Hamburg, Germany. [3] Universität Siegen, Fachbereich Physik, Walter-Flex-Straße 3, 57072 Siegen, Germany. [4] MAX IV Laboratory, Fotongatan 2, 225 94 Lund, Sweden. [5] ESRF - The European Synchrotron, Radiation Facility, 71 Avenue des Martyrs, 38043 Grenoble, France. Correspondence and requests for materials should be addressed to A.S. (email: andreas.stierle@desy.de).

Heterogeneous, nanoparticle based catalysts are widely used in chemical industry as well as automotive exhaust converters or fuel cells for energy conversion. A major challenge are the harsh conditions during catalytic reactions since they often lead to nanoparticle sintering, which results in catalyst deactivation primarily due to the loss of active surface area[1–4]. Thus, research increasingly focuses on unravelling opportunities to avoid sintering mainly by particle encapsulation or size distribution tuning[5–9]. Alloy and bimetallic nanoparticles offer promising routes to mitigate particle sintering, in addition to tailored selectivity and activity due to ligand effects between neighbouring atoms[10–15]. The beneficial effect of alloying for sinter prevention under reaction conditions is however not well understood. Although much effort has been put into the study of sintering reversal via redispersion, its application is limited to certain systems and often requires the presence of undesired chlorine[16,17], emphasizing the strong need to develop strategies for the production of more sinter resistant catalysts.

To reduce or avoid nanoparticle sintering an atomic scale understanding of the underlying mechanisms is mandatory. Two scenarios are proposed in the literature for thermally induced sintering processes[18–22]: (1) Ostwald ripening, in which larger nanoparticles grow at the expense of smaller ones by diffusion of single atoms or complexes driven by a gradient in chemical potential; (2) thermally activated complete particle migration and coalescence. Both processes get more complex when surrounding gas atmospheres are involved. Previous *ex situ* and *in situ* studies focused mainly on the sintering of nanoparticles induced by elevated temperatures and exposure to individual gases such as O_2, H_2, CO and H_2O vapour or air[23–26]. In the case of Au nanoparticles on TiO_2 a synergistic effect of CO and O_2 for enhanced sintering under CO oxidation reaction conditions was reported[25]. Under pure CO atmospheres at 10^{-4} mbar platinum (Pt) nanoparticles on MgO(100) with diameters above 3.6 nm are observed to be stable up to 700 K (ref. 27). Recent density functional theory calculations demonstrated that CO can disintegrate 2 nm rhodium (Rh) nanoparticles via complex formation[28].

Here we report a systematic study of the composition-dependent sintering, shape and size changes of epitaxial Pt–Rh alloy nanoparticles on Al_2O_3(0001) under catalytic CO oxidation reaction conditions close to atmospheric pressures. We record large area reciprocal space maps using high-energy grazing incidence x-ray diffraction as an *in situ* probe, and we use a combinatorial sample with composition gradient, assuring identical experimental conditions for different Pt–Rh compositions. Switching from reducing conditions to catalytic activity for CO oxidation is followed by *in situ* mass spectrometry. We find a composition-dependent nanoparticle sintering behaviour under near-atmospheric pressure CO oxidation conditions, which we attribute to a kinetically stabilized initial shape of more Pt-rich particles after their growth. Such a shape-dependent sintering mechanism has to our knowledge not yet been reported in literature.

Result

Measurement principle and sample design. The experiment was realized using a dedicated *in situ* x-ray diffraction chamber for catalysis experiments under flow conditions, as illustrated in Fig. 1 (ref. 29). The combinatorial sample contained stripes of Pt–Rh nanoparticles with varying composition from pure platinum to pure Rh and a constant height of ~20 Å. The focused high-energy x-ray beam facilitated studying one stripe at a time and by translating the sample perpendicular to the incident beam nanoparticles of different Pt–Rh compositions could be probed subsequently under identical reaction conditions[30,31]. Monitoring the partial gas pressures inside the reactor allowed a

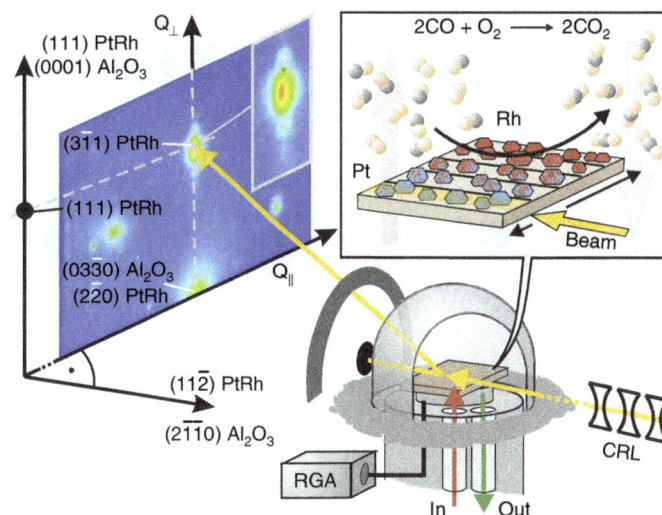

Figure 1 | Measurement principle. The high-energy x-ray beam is focused by compound refractive lenses (CRLs) on the sample surface containing pure Pt, $Pt_{0.85}Rh_{0.15}$, $Pt_{0.7}Rh_{0.3}$, $Pt_{0.5}Rh_{0.5}$ and pure Rh nanoparticle stripes. The diffraction pattern is collected by a 2D detector while the heated sample is exposed to a computer-controlled gas flow mixture. The composition of the gas phase is controlled by leaking into a residual gas analyser (RGA). In the inset of the diffraction pattern a close-up of the $(3,\bar{1},1)$ nanoparticle Bragg reflection is displayed.

direct correlation of particle structural changes to the catalytic activity[32].

For all Pt–Rh compositions we find that the nanoparticles grow on the Al_2O_3(0001) substrate in (111) direction with distinct epitaxy as shown in Fig. 1. Note that the in-plane misfit between the alloy nanoparticles and the Al_2O_3(0001) substrate varies between $+0.9\%$ for pure Pt to -2.3% for pure Rh. Epitaxial nanoparticles, as employed in our study, give rise to well-defined x-ray diffraction Bragg peaks in reciprocal space with the absence of powder diffraction rings, see Fig. 1. While the particles' in- and out-of-plane lattice parameters are given by the Bragg peak position, the intensity distribution in the vicinity of the Bragg peaks holds quantitative information on the nanoparticle size and shape at the atomic scale[33,34]. To systematically unravel structural changes during CO oxidation we measured for each Pt–Rh composition and reactivity condition a two-dimensional (2D) reciprocal space map centred around the particle $(3\bar{1}1)$ Bragg peak (Fig. 1). At high-photon energies the in-plane particle mosaicity of 1–2° probes a significant section of an azimuthal plane in reciprocal space and, therefore, gives rise to a diffraction pattern similar to transmission electron microscopy. The correlation between the wide-range detectability in reciprocal space and the particle mosaicity are illustrated and elucidated in Supplementary Fig. 1 and Supplementary Note 1. This allows a fast collection of large area reciprocal space maps without scanning sample or detector angles, which is crucial under reaction conditions[32].

Figure 2a shows an overview of the 2D diffraction patterns for different Pt–Rh compositions (vertical panels) under the various conditions applied during the CO oxidation experiment (horizontal panels). The position of the Bragg peak maxima is given by Vegard's law according to their nominal composition, indicating that the as grown particles are present in the state of randomly mixed face centered cubic alloys. The intensity distribution around the particle Bragg reflections is smeared out in the L-direction perpendicular to the surface because of the finite particle height. It is characterized by periodical fringes above and below the Bragg peaks, especially pronounced for

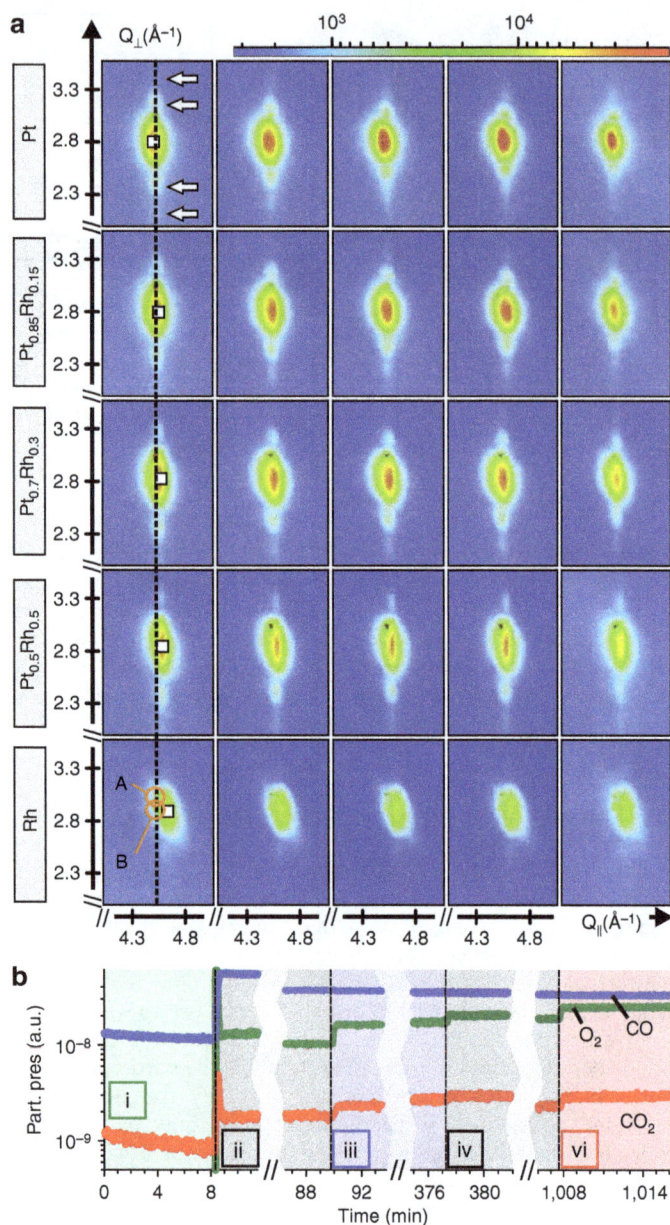

Figure 2 | *In situ* monitoring of composition-dependent sintering during CO oxidation. (a) Particle (3$\bar{1}$1) Bragg peak maps collected for different Pt-Rh compositions at 550 K and $p_{tot} = 200$ mbar when successively changing towards conditions of higher activity (i → vi; exposure time: 50 s). Set partial flows: i: flow $f_{O2} = 0$ ml min^{-1}, $f_{CO} = 2$ ml min^{-1}; ii: $f_{O2} = 2$ ml min^{-1}, $f_{CO} = 10$ ml min^{-1}; iii: $f_{O2} = 4$ ml min^{-1}, $f_{CO} = 10$ ml min^{-1}; iv: $f_{O2} = 5$ ml min^{-1}, $f_{CO} = 10$ ml min^{-1}; vi: $f_{O2} = 7$ ml min^{-1}, $f_{CO} = 10$ ml min^{-1}. Between iv and vi (not shown here): 2 cycles of gas switching between $f_{O2} = 5$ ml min^{-1}, $f_{CO} = 10$ ml min^{-1} and $f_{O2} = 0$ ml min^{-1}, $f_{CO} = 2$ ml min^{-1} (condition v, Fig. 3). Orange circles: position of substrate (30$\bar{3}$6) Bragg peak (A) and the intersection point of the substrate (30$\bar{3}$L) crystal truncation rod with the Ewald sphere (B). Vertical dashed line: in-plane substrate (30$\bar{3}$0) reference; white squares: positions of particle Bragg peak maxima. **(b)** Partial pressures of CO, O$_2$ and CO$_2$ as measured by the RGA under different conditons (i-vi) and as a function of time.

Pt-rich particles, indicated by arrows for the Pt particles in Fig. 2a. These Laue oscillations imply the presence of plate-like nanoparticles which are characterized by a low-defect internal structure and a small-vertical size distribution[35].

***In situ* monitoring of the composition-dependent sintering.** In a first step the clean particles were probed under reducing atmosphere (condition i: $f_{CO} = 2$ ml min^{-1}; no O$_2$, $T = 550$ K). For all subsequent settings (conditions ii-vi) the CO flow was kept constant at 10 ml min^{-1}, as was the nominal temperature at 550 K, while the O$_2$ flow was increased in steps up to 7 ml min^{-1}. Ar was used as carrier gas adding up to a constant total flow of 100 ml min^{-1} at a constant total pressure of 200 mbar. The period of the Laue oscillations is very similar for all Pt-Rh compositions under condition i, implying that the particles of all stripes exhibited a very similar initial height. Figure 2b displays the residual gas analyser signal of CO, O$_2$ and CO$_2$ proportional to the respective average partial pressures inside the reactor for the aforementioned conditions. When switching to higher oxygen partial flows the CO$_2$ production increased accordingly. Low-index Pt, Rh as well as Pt-Rh alloy surfaces and nanoparticles are reported to be catalytically active for CO oxidation at 550 K (refs 16,36,37), but since we measure the integral CO$_2$ production over all stripes, we can not directly conclude which composition is the more active one. At 550 K and 20 mbar CO partial pressure (corresponding to our conditions) CO is expected to still partially cover the different Pt-Rh surfaces, in line with a Langmuir-Hinshelwood reaction mechanism[36,37].

As key observation we note that the finite height fringes for Pt-rich particles progressively shift towards the Bragg peak position, when increasing the oxygen pressure and switching to higher catalytic activity (Fig. 2a). This implies that their height increases significantly when CO oxidation sets in because the L-spacing of their minima is inversely proportional to the particle height. We conclude that at same time the epitaxial arrangement of the particles is maintained since Debye-Scherrer powder diffraction rings are absent. It is striking that the shift of the fringes is progressively suppressed with increasing Rh compositions, which implies that sintering is strongly reduced for Rh-rich particles.

Sinter-induced quantitative particle shape changes. To obtain a deeper insight into the shape and size changes the particles underwent throughout the varying environmental conditions, we extracted the respective scattering profiles along L through the nanoparticle Bragg reflections from the 2D maps of Fig. 2a. The extracted rods were intensity corrected for the detector background, the missing θ-rotation of the stationary sample and the Lorentz factor, as is thoroughly described in Supplementary Note 2. It gets evident, that the Laue oscillations are damped in a characteristic way, which we attribute to a variation in occupancy of the different atomic layers directly related to the particle shape. The generic shape of a (111) oriented nanoparticle consists of (111) top and bottom facets as well as (111) and (100) side facets (see Fig. 3g,h, neglecting higher index facets for the moment). The equilibrium shape of the particles can be constructed for a given ratio $g = \frac{\gamma_{(100)}}{\gamma_{(111)}}$ of the (100) and (111) surface energies γ (Gibbs free energy per surface unit, depending also on the surrounding gas atmosphere) by the Wulff-construction[38]. Geometrically meaningful values for g in the range from $g = 0.9$ to $g = \sqrt{3} = 1.73$ were tested as starting models to fit the data, they are depicted in the top parts of Supplementary Figs 2 and 3 for the case of Pt$_{0.7}$Rh$_{0.3}$ particles under condition i. The truncation of the particles was performed in such a way that the observed finite height Laue oscillations were best reproduced, providing atomic layer sensitivity for the height determination. For a given particle shape, the different atomic layers exhibit a predetermined atomic occupancy, which leads to a characteristic damping of the Laue oscillations. In total, eight different starting models were tested for each condition and composition which were further

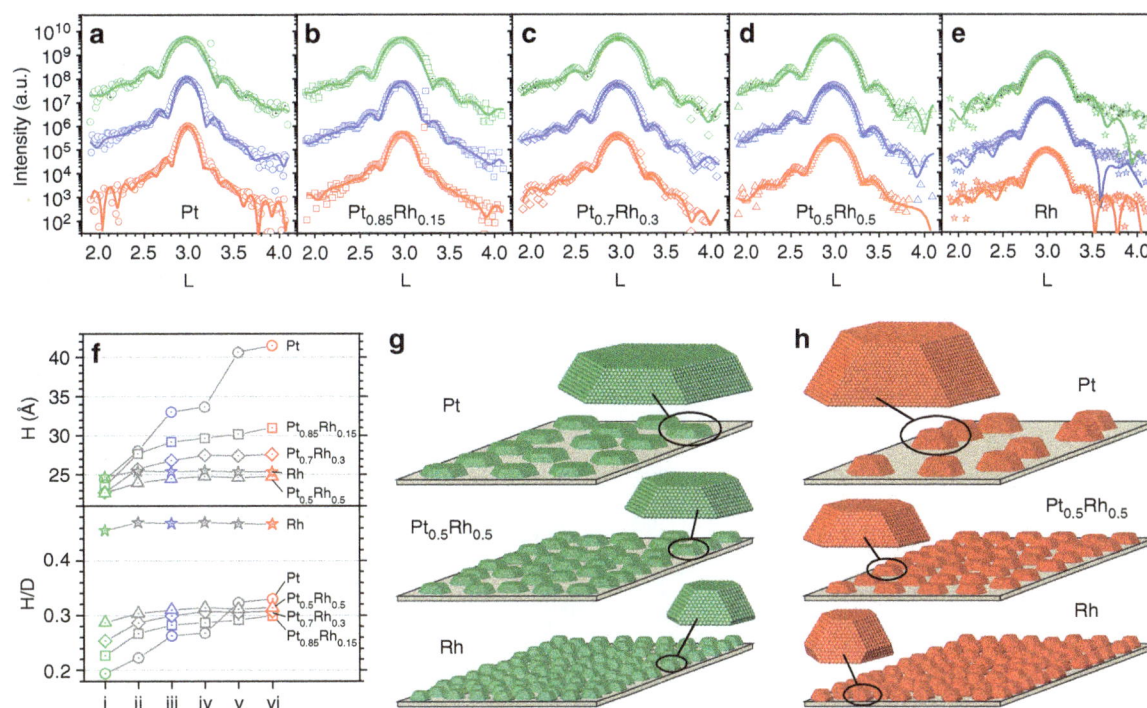

Figure 3 | Sintering-induced particle shape changes for different Pt-Rh compositions. (**a–e**) Linescans in (111) direction (open symbols: data; solid lines: fit for underlying model particle shape with $\frac{\gamma_{(100)}}{\gamma_{(111)}}=1.1$) through the respective particle ($3\bar{1}1$) Bragg peaks extracted from the 2D maps of Fig. 2a for different Pt-Rh compositions ((**a**) Pt; (**b**) $Pt_{0.85}Rh_{0.15}$; (**c**) $Pt_{0.7}Rh_{0.3}$; (**d**) $Pt_{0.5}Rh_{0.5}$; (**e**) Rh) and various conditions (green: i; blue: iii; and red: vi). (**f**) composition-dependent particle heights and aspect ratios $\frac{H}{D}$ for different CO oxidation conditions (i–vi) as obtained from fits to the data, taking into account the average particle diameter as obtained by AFM after the experiment. The error bars are on the order of the symbol sizes. Average particle shapes and sample morphology before (**g**) and after (**h**) sintering as deduced from the Bragg peak and reflectivity fits are shown for some selected Pt-Rh compositions.

refined during the fit in the following way: the occupancy numbers of the atomic layers at the substrate-particle interface and up to five layers on the top part of the particle accounting for particle height fluctuations were allowed to vary. Further fit parameters comprised the displacement parameters of these atomic layers. The fit parameters are depicted in Supplementary Fig. 4, the fitting procedure is elucidated in the Methods part at the end of this article.

As a general trend for all conditions and compositions, we obtain good fit results when starting with shapes of truncated (111)-oriented particles with ratios of $g=\frac{\gamma_{(100)}}{\gamma_{(111)}}$ ranging approximately from 0.9 to 1.3, compatible with particles possessing distinct (100)-type facets (Supplementary Fig. 2). Particles with highly reduced (100)-type facets and nearly triangular shaped particles are unlikely to occur ($g=1.5$). Figure 3a–e shows the fits for $g=1.1$ to the experimental data for the different Pt–Rh compositions and conditions i, iii and vi. The three-dimensional (3D) models for particles for a starting model with $g=1.1$ are represented in Fig. 3g,h for some selected Pt–Rh compositions. Supplementary Note 3 contains a more thorough explanation of the fit results.

Under vacuum conditions, theoretical g ratios vary from 1.05 for Pt to 1.16 for Rh (refs 39,40). In the case of CO and CO/O mixtures, no systematic investigation of the surface energy on Pt, Rh as well as Pt–Rh alloy surfaces as a function of coverage and chemical potentials is reported, which would allow a direct prediction of the particle shape using the Wulff construction. The fitted occupancies for the topmost particle layers (Supplementary Fig. 3 and Supplementary Note 3) show a reduction compared with the values for ideal particle shapes. This can be explained by a moderate particle height distribution and/or the presence of higher indexed facets, in-line with a slightly rounder particle shape[39]. Experimentally, the stabilization of low-index facets was

observed during CO oxidation at 470 K by transmission electron microscopy[41], supporting a moderate particle height distribution in our case.

From the fits we find that the occupancy parameters of the interface layer for nearly all compositions and conditions range between 0.4 and 0.7 (Supplementary Table 1 and Supplementary Note 3), respectively. This points to misfit induced defects at the interface to the substrate, which locally disturb the lattice. The interfacial occupancy values decrease after reaction-promoted sintering, pointing to a growing number of interfacial defects such as misfit dislocations in line with previous findings[42]. The first and second metal layer distance at the interface is found to increase by 0.1–0.2 Å, likely because of the interaction with the Al_2O_3 substrate (Supplementary Table 1).

Figure 3f summarizes the average particle height in the course of the CO oxidation experiment as obtained from the fit. While for pure Pt particles the height almost doubles changing from 24 to 42 Å; the height increase is reduced for the $Pt_{0.85}Rh_{0.15}$ particles changing from 23 to 31 Å, whereas the height of pure Rh particles stays nearly constant (26 Å compared with 24 Å). This composition-dependent discrepancy in height of the sintered particles is also confirmed by our atomic force microscopy (AFM) and x-ray reflectivity data (Supplementary Table 2, Supplementary Figs 5c and 6). At the same time, we estimate from rocking curve measurements that the average particle diameter D stays approximately constant within the error bars ($D=12$ nm for pure Pt, decreasing to 6 nm for pure Rh, our approach for deducing the particle diameter from rocking scans is illustrated and explained in Supplementary Figs 7,8 and Supplementary Note 4, respectively). The values at the end of the experiment are confirmed by AFM performed after the in situ experiments (Fig. 4g–i, Supplementary Fig. 5 and Supplementary Note 5), which in addition disclose a large size distribution (Supplementary Fig. 5b). For a second

Intensity (a.u.)   Coverage θ

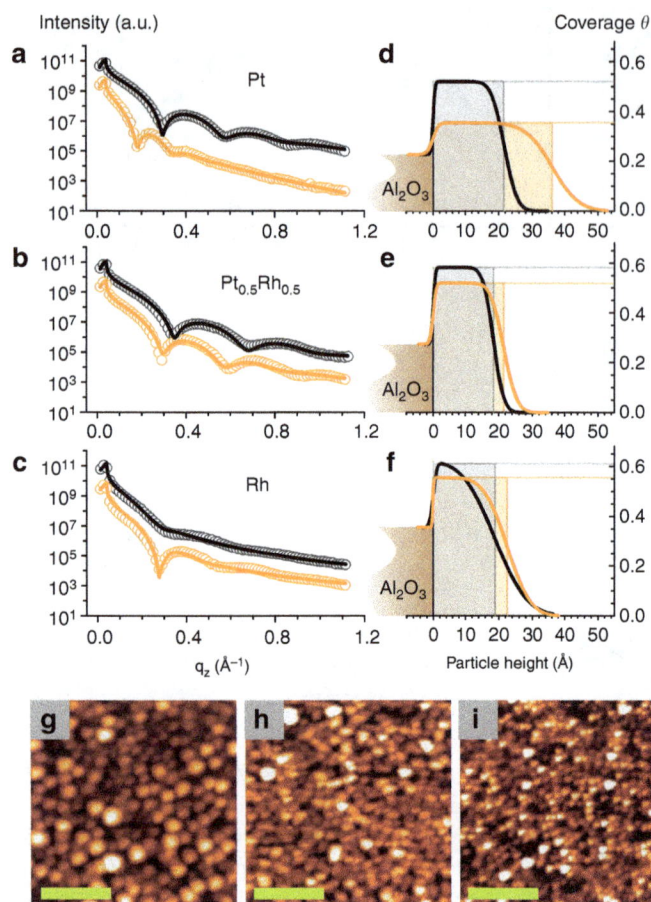

Figure 4 | Composition-dependent particle height and coverage and diameter before and during CO oxidation. (**a-c**) x-ray reflectivity measurements (open circles: data, solid lines: fit) performed on particles with different Pt-Rh compositions and (**d-f**) electron density profiles obtained from the fit results which yield information on the particle height and percental particle coverage (black, before CO oxidation; orange, during CO oxidation at flows of 7 ml min^{-1} O$_2$ and 10 ml min^{-1} CO). The shaded boxes represent the average particle height and coverage. (**g-i**) AFM measurements performed under UHV after the CO oxidation experiments: (**g**) Pt; (**h**) Pt$_{0.5}$Rh$_{0.5}$; (**i**) Rh. The scale bars correspond to a length of 100 nm.

composition gradient sample which was investigated in the $(111)/(11\bar{2})$ azimuthal plane (Supplementary Fig. 9) a similar behaviour was observed (Supplementary Fig. 10 and Supplementary Note 6). The results allow us to plot the nanoparticle average aspect ratio as a function of the reaction conditions (Fig. 3f), exemplifying a strong increase in aspect ratio for Pt-rich particles.

Adhesion energy determination. The height of the particles depends under equilibrium conditions on the interfacial energy according to the Wulff–Kashiew theorem[43]. The determined particle shape, therefore, also holds information on the adhesion energy W_{adh}, which can be calculated for strongly truncated nanoparticles based on the truncation of the particles and reference values for (111) and (100) facet surface energies[39,44]. A sketch of such a strongly truncated particle resembling the ones on our sample after sintering is depicted in Supplementary Fig. 11. Assuming the sintered particles at the end of the experiment have adopted equilibrium shape, we find the following adhesion energies taking a surface energy ratio of $g = \frac{\gamma_{(100)}}{\gamma_{(111)}} = 1.1$ and the theoretical values for the clean surface energies $\gamma_{(111)}$ into account:

2 J m^{-2} for pure Pt, 2.2 J m^{-2} for Pt$_{0.85}$Rh$_{0.15}$, 2.24 J m^{-2} for Pt$_{0.7}$Rh$_{0.3}$, increasing to a maximum of 2.6 J m^{-2} for Pt$_{0.5}$Rh$_{0.5}$ and decreasing again to 1.9 J m^{-2} for pure Rh (the systematic error bar of ± 0.2 J m^{-2} takes the experimental uncertainty of g from 0.9 to 1.3 into account, see Supplementary Table 3 and the Methods section part at the end of this article for further explanations). The values presented here represent an upper limit, since $\gamma_{(111)}$ may be slightly reduced in the presence of adsorbates under reaction conditions[39]. The adhesion energy values are compatible to literature values for Pd nanoparticles on ultrathin Al$_2$O$_3$ films[44]. The observed trend reflects the competing influence of the increasing chemical interaction from Pt to Rh for stoichiometric Al$_2$O$_3$ surfaces[45], and the variation in misfit to the Al$_2$O$_3$ substrate. A perfect match to the substrate is expected for a composition of Pt$_{0.7}$Rh$_{0.3}$. For an increasing misfit to the Rh-rich side the adhesion energy is expected to decrease[42,46]. At the same time the chemical interaction is increasing, resulting in a shift of the maximum of the adhesion energy to a Rh composition of 50%. Such a competition between chemical interaction and misfit-induced deformation energy is expected also for other oxide supported metal nanoparticles, and allows to tune the adhesion of the nanoparticles to the oxide support.

Mass transport on the sample surface during reaction. To further corroborate the observed sintering behaviour, *in situ* x-ray reflectivity experiments were performed before CO oxidation and under reaction conditions at flows of 7 ml min^{-1} O$_2$ and 10 ml min^{-1} CO. X-ray reflectivity provides information on the average electron density profile perpendicular to the surface, related to the average particle height and the surface coverage on an absolute scale[47]. The x-ray reflectivity measurements performed before CO oxidation (black curves in Fig. 4a) confirm the initially similar particle heights of 18–22 Å and result in relative particle coverages on the different stripes of 50–60%, respectively. The qualitative inspection of the reflectivity curves in Fig. 4 (and also Supplementary Fig. 6) reveals that the period of the particle height-dependent interference fringes decreases much more strongly for pure Pt and Pt-rich particles, as compared with pure Rh. The fit reveals that the average height of the pure Pt particles increases from 22 to 36 Å, whereas the height increase is less pronounced for Pt$_{0.85}$Rh$_{0.15}$ particles (from 19–29 Å) and almost not present for particles with higher Rh composition (Pt$_{0.5}$Rh$_{0.5}$: from 18 to 22 Å, pure Rh: from 19 to 23 Å, Supplementary Table 2). The particle heights obtained from the reflectivity data are in general smaller than the ones deduced from the Bragg peak fits. This is due to the fact that the simple box model used for fitting the reflectivity data systematically underestimates the particle height, because it does not take into account the details of the particle shape[48].

Interestingly, the height increase is found to be proportional to the reduction of the particle surface coverage, keeping the integral of the electron density profile constant within the error bars. The reduction in coverage is most pronounced for pure Pt particles for which it decreases from 52 to 35%, as can be inferred from the electron density profiles (Fig. 4d–f and Supplementary Fig. 6). The reflectivity results strongly suggest that a huge mass transport takes place on the sample surface during the activity-induced restructuring of the Pt-rich nanoparticles. In addition, they prove the conservation of catalyst material which is represented by the area of the shaded boxes in Fig. 4d–f (Supplementary Fig. 6) and which remains constant within error bars for all Pt–Rh compositions. Despite the tremendous material transport the sample features separated and well-defined nanoparticles, as can be inferred from the AFM images measured after the CO oxidation experiment (Fig. 4g–i and Supplementary Fig. 5a). The height histograms obtained from the AFM images (Supplementary Fig. 5b) correspond well to the roughness values determined by x-ray reflectivity, see Supplementary Table 2.

Discussion

We now turn to the key question concerning the driving force for the observed 3D sintering behaviour of Pt-rich particles. According to our observations, Rh and Rh-rich particles essentially maintain their original size and shape under reaction conditions, which they had initially obtained after deposition at 800 K. From this we conclude that they are close to equilibrium shape. For Pt-rich particles the total surface coverage decreases under CO oxidation conditions, which implies a tremendous lateral inter-particle mass transport, as well as intra-particle mass transport. The Pt-rich particles undergo a significant change in aspect ratio, while keeping their average diameter approximately constant. As a realistic scenario, we propose that within the inherently large lateral size distribution (see AFM images in Fig. 4g–i and Supplementary Fig. 5a,b) smaller particles are present which are less stable. The energy released during the CO oxidation reaction on Pt is 3 eV per produced CO_2 molecule[49], which when locally thermalized may easily promote the destabilization of smaller particles. The observed phenomenon can thus be viewed as a CO oxidation promoted non-classical Ostwald ripening process during which as a net result larger Pt-rich particles essentially grow vertically. The net growth of larger Pt particles takes place predominantly perpendicular to the surface, as a result of the balance between the tendency of the particles to obtain a more 3D equilibrium shape and the lateral growth during the non-classical Ostwald ripening process. The particles overcome the kinetically stabilized particle shape with low aspect ratio towards a shape closer to thermal equilibrium. The CO oxidation thus locally releases energy opening new pathways for thermal atomic diffusion rendering this tremendous catalyst restructuring possible.

In summary, we monitored, operando, the composition-dependent sintering of alloy nanoparticles during catalytic CO oxidation, which we found to be more pronounced for Pt-rich particles. Rh-rich particles resist sintering under identical reaction conditions. We employed high-energy x-ray diffraction as in situ tool together with a combinatorial sample geometry, beneficial for efficient data acquisition. The restructuring is characterized by a strong particle height increase accompanied by a decrease of the particle surface coverage. For Pt nanoparticles, the addition of Rh gives rise to an advantageous higher dispersion and the increased mismatch to the substrate promotes a more 3D shape, closer to equilibrium. Our findings have implications for the preparation of more sinter resistant particles: to avoid shape transformations, particles with equilibrium shape are preferable, which are sufficiently large to sustain the reaction-induced local heat dissipation. A reduction in size distribution will be beneficial to reduce sintering, since it is expected to minimize the reaction-induced destabilisation of small particles. The reaction heat itself sets, however, a lower limit to the useful particle size.

Methods

Sample preparation. The Pt–Rh alloy nanoparticles were prepared by means of physical vapour deposition under ultrahigh vacuum (UHV) conditions (base pressure below 3×10^{-11} mbar) at a substrate temperature of 800 K. For the simultaneous evaporation of the two materials an Omicron EFM 3T evaporator was used. The particle stripes of varying composition were prepared successively by using a Ta mask with a slit in front of the sample. The calibrated Pt and Rh fluxes were in case of each stripe tuned according to the desired Pt–Rh composition. The stripes were grown perpendicular to the reciprocal space plane of interest (sample I: stripes along the $(2\bar{1}\bar{1}0)_{Al_2O_3}$–direction to monitor the particle $(3\bar{1}1)$ peak in the $(01\bar{1}0)/(0001)$ Al_2O_3-plane; sample II: stripes along the $(\bar{1}100)_{Al_2O_3}$–direction to measure the particle $(1\bar{1}1)$ peak in the $(11\bar{2}0)/(0001)$ Al_2O_3-plane). Before deposition the $Al_2O_3(0001)$ substrates had been chemically treated in an ultrasonic bath (15 min in acetone and 15 min in ethanol). Thereafter the substrates underwent cracking with atomic oxygen ($T_{substrate} = 570$ K; $p_{O2} = 1 \times 10^{-7}$ mbar; and $P_{cracker} = 65$ W) to remove residual carbon. The correct substrate orientation with respect to the slit position and the cleanliness of the surface before deposition were confirmed by low-energy electron diffraction and Auger electron spectroscopy, respectively.

Experimental set-up and sample environment. The x-ray measurements were performed in a dedicated in situ catalysis chamber[29] (base pressure: 2×10^{-9} mbar). Inside the chamber we annealed the samples before the CO oxidation experiments under hydrogen ($p_{H2} = 2 \times 10^{-5}$ mbar, $T = 490$–540 K) to remove possible oxides since the samples had been exposed to air after growth. Apart from the use of CO and O_2 we employed Ar as carrier gas to keep the total pressure and total flow inside the reactor at 200 mbar and at 100 ml min^{-1}, respectively. The cleanliness of the CO gas was ensured by the use of a carbonyl trap inserted in the CO line before the reactor.

The x-ray measurements were carried out at the high-energy beamline ID15A (ESRF) at a photon energy of 78.7 keV. The 2D reciprocal space maps were measured with a FReLoN 14 bit CCD camera, x-ray reflectivity was performed using a NaI scintillation point detector. For imaging the 2D maps we used a shallow incident angle ($\alpha_{in} = 0.0333°$) below the substrate critical angle to suppress scattering from the bulk. The beam was focussed by 200 compound refractive lenses which resulted in a vertical and horizontal beam size of 8 and 25 µm at the sample position. At $\alpha_{in} = 0.0333°$ this results in a beam footprint of 13.8 mm which covers the whole length of the particle stripes (sample dimensions: $10 \times 10 \times 1$ mm^3).

Fitting the particle rods extracted from the 2D maps. The particle diffraction rods were extracted from the 2D maps parallel to the (111) direction. The obtained intensity distribution was corrected for the detector background, the missing θ-rotation of the stationary sample, and the Lorentz factor (Supplementary Note 2). They were simulated using the programme package of ROD (refs 50,51), which allows to refine surface structure models, where the fitting parameters include among others occupancy parameters describing the percentaged atomic density of the different layers, as well as displacement parameters accounting for the relaxation of the atomic layers in vertical direction. In our approach we used the underlying surface structure model to mimic the nanoparticle structure: as the generic nanoparticle shape is determined by the atomic layer density profile in the vertical direction it was in our particle model shapes accounted for by adjusting the occupancy parameters for each atomic layer accordingly. Based on the Wulff-construction, different underlying particle shape models were put to the test in the particle rod fitting procedure (see next paragraphs as well as Supplementary Note 3, and Supplementary Figs. 2–4).

Particle model shapes. The different underlying particle shape models used in the fitting procedure differed in the particle shape itself, but adhered for a certain condition and composition to the particle height H and diameter D deduced from the x-ray data on the basis of the finite thickness oscillations and/or the AFM measurements. The model shapes comprised eight truncated (111)-oriented particle models of varying surface energy ratios $g = \frac{\gamma_{100}}{\gamma_{111}}$ based on the Wulff-construction[38], a truncated sphere model and a simple model of a continuous layer. A juxtaposition of the different model particle shapes is presented at the top of Supplementary Figs. 2 and 3 for the case of $Pt_{0.7}Rh_{0.3}$ particles under condition i (no O_2, 2 ml min^{-1} CO, $p_{tot} = 200$ mbar and $T = 550$ K).

Fitting procedure. To illustrate the fitting procedure and the parameters varied therein Supplementary Fig. 4 shows as an example the fit results obtained for $Pt_{0.7}Rh_{0.3}$ particles under condition i. The particle height and diameter in this case comprised $H = 18$ and $D = 90$ Å, respectively, and the particle was accordingly concluded to consist of nine atomic layers. Supplementary Figure 4a shows the corresponding particle model for a surface energy ratio of $g = 1.1$, Supplementary Fig. 4b the respective layer model. Each layer has an occupancy value determined from the corresponding atomic density, where the ninth and thus the topmost layer (occupancy value t_0) is indicated by black dashes. The occupancy values according to the (unfitted) model shape of Supplementary Fig. 4a are represented by the red line in Supplementary Fig. 4c.

To compensate for the particle height distribution, up to two layers were added to the layer model. During the fit the occupancy values of these two layers (t_1, t_2), the ones of layers t_{-1}, t_{-2}, t_0 and of the two layers at the interface (b_0, b_1) were allowed to vary. If the fit suggested unphysical values the number of parameters was reduced accordingly (for example, only one layer was added on top and only the occupancy values t_1, t_0, t_{-1}, b_0 and b_1 were used as fit parameters, or even no layer was added on top and only the occupancy values t_0, b_0 and b_1 were used as fit parameters). The values of the occupancy parameters were considered as physically meaningful as long as with increasing distance of the atomic layer from the particle bulk the corresponding occupancy values decreased (that is, $t_{-2} \geq t_{-1} \geq t_0 \geq t_1 \geq t_2$ and $b_1 \geq b_0$).

Apart from the occupancy parameters also displacement parameters, which consider the layer relaxation, were included in the fit. Supplementary Figure 4c shows the fitted occupancy values for the case of the $Pt_{0.7}Rh_{0.3}$ particles under condition i (black diamonds). The fitted displacement parameters can be extracted from Supplementry Table 1.

Adhesion energy determination. Assuming that after sintering (condition vi) the particles have adopted their equilibrium shape, the adhesion energy W_{adh} was deduced for the different Pt–Rh compositions.

Supplementary Figure 11 shows the side view of a strongly truncated particle, similar to the ones present on our sample, where the cross-section of the untruncated particle was obtained according to the Wulff-construction[38]. The height of the truncated particle is less than half the height of the unsupported particle and the effective surface energy γ^* can be written as $\gamma^* = \gamma_{interface} - \gamma_{substrate}$ (refs 43,44). Knowing the particle height H, the length w of the top facet, the surface energy ratio $g = \frac{\gamma_{100}}{\gamma_{111}}$ and the surface energy of the (111)-type facet this effective surface energy can be expressed as[44]:

$$\gamma^* = \gamma_{111} \cdot \left(\sqrt{\frac{3}{2}} \cdot \frac{H}{w} \cdot g - 1 \right), \tag{1}$$

from which the adhesion energy W_{adh} can be determined:

$$W_{adh} = \gamma_{111} - \gamma^*. \tag{2}$$

This formula was used to estimate the adhesion energy of the Pt-containing particles after sintering. In the case of the pure Rh particles formulas (1) and (2) could not be applied since the height of the truncated particles was higher than half of the height of the unsupported particles. Instead, the Wulff–Kaishew theorem, which can be expressed as

$$W_{adh} = \gamma_{111} \cdot \frac{2h}{H+h}, \tag{3}$$

was employed, where h is the height of the 'buried' part of the particle, as can be seen in Supplementary Fig. 11. The results for the adhesion energies W_{adh} for the particle shapes of the different compositions with a surface energy ratio of $g = \frac{\gamma_{100}}{\gamma_{111}} = 1.1$ are written in bold and are summarized in Supplementary Table 3. The results for the values of $g = 0.9$ and $g = 1.3$ are included and can be considered as the maximum and minimum values of the respective error bars.

Employed coordinate systems. For the metal nanoparticles face-centred cubic bulk reciprocal lattice indices are used (cubic room temperature lattice constants $a_{Pt} = 3.924$ Å and $a_{Rh} = 3.801$ Å) and for the Al_2O_3 substrate hexagonal reciprocal lattice indices (room temperature lattice constants $a = b = 4.763$ Å, $c = 13.003$ Å, $\alpha = \beta = 90°$ and $\gamma = 120°$).

Fitting of the x-ray reflectivity data. The x-ray reflectivity data were corrected for the absorbers and the dead time of the point detector. In the fit model the layer of nanoparticles is represented by a closed box whose height is obtained from the fit. The interface roughnesses are in the underlying modified Parratt formalism taken into account as Gaussian fluctuations around the interface layers. Comparison of the fit parameter for the electron density with the corresponding density literature value for a closed layer yields the fraction of surface covered by nanoparticles. The fitting programme accounts for the angle-dependent beam footprint on the sample.

AFM measurements. The AFM measurements were performed after the CO oxidation experiments under UHV conditions in non-contact mode using an Omicron VT AFM XA instrument.

References

1. Bartholomew, C. H. Mechanisms of catalyst deactivation. *Appl. Catal. A Gen.* **212**, 17–60 (2001).
2. Forzatti, P. & Lietti, L. Catalyst deactivation. *Catal. Today* **52**, 165–181 (1999).
3. Moulijn, J. A., van Diepen, A. E. & Kapteijn, F. Catalyst deactivation: is it predictable? What to do? *Appl. Catal. A Gen.* **212**, 3–16 (2001).
4. Campbell, C. T., Parker, S. C. & Starr, D. E. The effect of size-dependent nanoparticle energetics on catalyst sintering. *Science* **298**, 811–814 (2002).
5. Lee, I. *et al.* New nanostructured heterogeneous catalysts with increased selectivity and stability. *Phys. Chem. Chem. Phys.* **13**, 2449–2456 (2011).
6. Lu, J. *et al.* Coking- and sintering-resistant palladium catalysts achieved through atomic layer deposition. *Science* **335**, 1205–1208 (2012).
7. Wettergren, K. *et al.* High sintering resistance of size-selected platinum cluster catalysts by suppressed Ostwald ripening. *Nano Lett.* **14**, 5803–5809 (2014).
8. Shinjoh, H. *et al.* Suppression of noble metal sintering based on the support anchoring effect and its application in automotive three-way catalysis. *Top. Catal.* **52**, 1967–1971 (2009).
9. Lu, P., Campbell, C. T. & Xia, Y. A. Sinter-resistant catalytic system fabricated by maneuvering the selectivity of SiO_2 deposition onto the TiO_2 surface versus the Pt nanoparticle surface. *Nano Lett.* **13**, 4957–4962 (2013).
10. Tao, F., Zhang, S., Nguyen, L. & Zhang, X. Action of bimetallic nanocatalysts under reaction conditions and during catalysis: evolution of chemistry from high vacuum conditions to reaction conditions. *Chem. Soc. Rev.* **41**, 7980–7993 (2012).
11. Ponec, V. Alloy catalysts: the concepts. *Appl. Catal. A Gen.* **222**, 31–45 (2001).
12. Park, J. Y., Zhang, Y., Grass, M., Zhang, T. & Somorjai, G. A. Tuning of catalytic CO oxidation by changing composition of Rh-Pt bimetallic nanoparticles. *Nano Lett.* **8**, 673–677 (2008).
13. Cao, A., Lu, R. & Veser, G. Stabilizing metal nanoparticles for heterogeneous catalysis. *Phys. Chem. Chem. Phys.* **12**, 13499–13510 (2010).
14. Cao, A. & Veser, G. Exceptional high-temperature stability through distillation-like self-stabilization in bimetallic nanoparticles. *Nat. Mater.* **9**, 75–81 (2010).
15. Greeley, J. & Mavrikakis, M. Alloy catalysts designed from first principles. *Nat. Mater.* **3**, 810–815 (2004).
16. Newton, M. A., Belver-Coldeira, C., Martínez-Arias, A. & Fernández-García, M. Dynamic in situ observation of rapid size and shape change of supported Pd nanoparticles during CO/NO cycling. *Nat. Mater.* **6**, 528–532 (2007).
17. Birgersson, H., Eriksson, L., Boutonnet, M. & Järås, S. G. Thermal gas treatment to regenerate spent automotive three-way exhaust gas catalysts (TWC). *Appl. Catal. B Environ.* **54**, 193–200 (2004).
18. Thiel, P. A., Shen, M., Liu, D.-J. & Evans, J. W. Coarsening of two-dimensional nanoclusters on metal surfaces. *J. Phys. Chem. C* **113**, 5047–5067 (2009).
19. Parker, S. C. & Campbell, C. T. Kinetic model for sintering of supported metal particles with improved size-dependent energetics and applications to Au on $TiO_2(110)$. *Phys. Rev. B* **75**, 035430-1–035430-15 (2007).
20. Jak, M. J. J., Konstapel, C., van Kreuningen, A., Verhoeven, J. & Frenken, J. W. M. Scanning tunnelling microscopy study of the growth of small palladium particles on $TiO_2(110)$. *Surf. Sci.* **457**, 295–310 (2000).
21. Challa, S. R. *et al.* Relating rates of catalyst sintering to the disappearance of individual nanoparticles during Ostwald ripening. *J. Am. Chem. Soc.* **133**, 20672–20675 (2011).
22. Datye, A. K., Xu, Q., Kharas, K. C. & McCarty, J. M. Particle size distributions in heterogeneous catalysts: what do they tell us about the sintering mechanism? *Catal. Today* **111**, 59–67 (2006).
23. Simonsen, S. B. *et al.* Direct observations of oxygen-induced platinum nanoparticle ripening studied by in situ TEM. *J. Am. Chem. Soc.* **132**, 7968–7975 (2010).
24. Behafarid, F. & Roldan Cuenya, B. Towards the understanding of sintering phenomena at the nanoscale: geometric and environmental effects. *Top. Catal.* **56**, 1542–1559 (2013).
25. Yang, F., Chen, M. S. & Goodman, D. W. Sintering of Au particles supported on $TiO_2(110)$ during CO oxidation. *J. Phys. Chem. C* **113**, 254–260 (2009).
26. Parkinson, G. S. *et al.* Carbon monoxide-induced adatom sintering in a Pd-Fe_3O_4 model catalyst. *Nat. Mater.* **12**, 724–728 (2013).
27. Chaâbane, N., Lazzari, R., Jupille, J., Renaud, G. & Avellar Soares, E. CO-induced scavenging of supported Pt nanoclusters: a GISAXS study. *J. Phys. Chem. C* **116**, 23362–23370 (2012).
28. Ouyang, R., Liu, J.-X. & Li, W.-X. Atomistic theory of Ostwald ripening and disintegration of supported metal particles under reaction conditions. *J. Am. Chem. Soc.* **135**, 1760–1771 (2013).
29. van Rijn, R. *et al.* Ultrahigh vacuum/high-pressure flow reactor for surface x-ray diffraction and grazing incidence small angle x-ray scattering studies close to conditions for industrial catalysis. *Rev. Sci. Instrum.* **81**, 014101-1–014101-8 (2010).
30. Nolte, P. *et al.* Combinatorial high energy x-ray microbeam study of the size-dependent oxidation of Pd nanoparticles on MgO(100). *Phys. Rev. B* **77**, 115444-1–115444-7 (2008).
31. Müller, P., Hejral, U., Rütt, U. & Stierle, A. In situ oxidation study of Pd-Rh nanoparticles on $MgAl_2O_4(001)$. *Phys. Chem. Chem. Phys.* **16**, 13866–13874 (2014).
32. Gustafson, J. *et al.* High-energy surface x-ray diffraction for fast surface structure determination. *Science* **343**, 758–761 (2014).
33. Nolte, P. *et al.* Shape changes of supported Rh nanoparticles during oxidation and reduction cycles. *Science* **321**, 1654–1658 (2008).
34. Nolte, P. *et al.* Reversible shape changes of Pd nanoparticles on MgO(100). *Nano Lett.* **11**, 4697–4700 (2011).
35. Pietsch, U., Holy, V. & Baumbach, T. *High-Resolution X-Ray Scattering* (Springer, 2004).
36. Peden, C. H. F. *et al.* Kinetics of CO oxidation by O_2 or NO on Rh(111) and Rh(100) single crystals. *J. Phys. Chem.* **92**, 1563–1567 (1988).
37. Su, X., Cremer, P. S., Ron Shen, Y. & Somorjai, G. A. High-pressure CO oxidation on Pt(111) monitored with infrared-visible sum frequency generation (SFG). *J. Am. Chem. Soc.* **119**, 3994–4000 (1997).
38. Wulff, G. Zur Frage der Geschwindigkeit des Wachsthums und der Auflösung der Krystallflächen. *Z. Kristallogr.* **34**, 449–530 (1901).
39. Seriani, N. & Mittendorfer, F. Platinum-group and noble metals under oxidizing conditions. *J. Phys. Condens. Matter* **20**, 184023–184033 (2008).
40. Mittendorfer, F., Seriani, N., Dubay, O. & Kresse, G. Morphology of mesoscopic Rh and Pd nanoparticles under oxidizing conditions. *Phys. Rev. B* **76**, 233413–233416 (2007).
41. Yoshida, H. *et al.* Temperature-dependent change in shape of platinum nanoparticles supported on CeO_2 during catalytic reactions. *Appl. Phys. Express* **4**, 065001–065004 (2011).

42. Graoui, H., Giorgio, S. & Henry, C. R. Effect of the interface structure on the high-temperature morphology of supported metal clusters. *Philos. Mag.* **81**, 1649–1658 (2001).
43. Winterbottom, W. L. Equilibrium shape of a small particle in contact with a foreign substrate. *Acta Metall.* **15**, 303–310 (1967).
44. Hansen, H. *et al.* Palladium nanocrystals on Al_2O_3: structure and adhesion energy. *Phys. Rev. Lett.* **83**, 4120–4123 (1999).
45. Li, H.-T. *et al.* Interfacial stoichiometry and adhesion at metal/α-Al_2O_3 interfaces. *J. Am. Ceram. Soc.* **94**, S154–S159 (2011).
46. Vervisch, W., Mottet, C. & Goniakowski, J. Effect of epitaxial strain on the atomic structure of Pd clusters on MgO(100) substrate - a numerical simulation study. *Eur. Phys. J. D* **24**, 311–314 (2003).
47. Kasper, N. *et al. In situ* oxidation study of MgO(100) supported Pd nanoparticles. *Surf. Sci.* **600**, 2860–2867 (2006).
48. Hejral, U., Vlad, A., Nolte, P. & Stierle, A. *In situ* oxidation study of Pt nanoparticles on MgO(001). *J. Phys. Chem. C* **117**, 19955–19966 (2013).
49. Ertl, G. in *Catalysis: Science and Technology* (eds Anderson, J. R. & Boudart, M.) vol. 4, 245 (Springer-Verlag, 1983).
50. Vlieg, E. Integrated intensities using a six-cirlce surface x-ray diffractometer. *J. Appl. Crystallogr.* **30**, 532–543 (1997).
51. Vlieg, E. *ROD*: a program for surface x-ray crystallography. *J. Appl. Crystallogr.* **33**, 401–405 (2000).

Acknowledgements

We thank the ID15 team for the excellent support during the synchrotron beamtimes at the ESRF. We thank the ID03 team and R. van Rijn for their help during a preliminary experiment. The Omicron application laboratory is acknowledged for conduction of the AFM experiments. Financial support by the Bundesministerium für Bildung und Forschung (BMBF), Project No. 05K10PS1 (NanoXcat) is gratefully acknowledged.

Author contributions

U.H. has written the manuscript together with A.S.; O.B. and D.P. set-up the experiment, U.H. and P.M. grew the combinatorial samples, all authors contributed to the conduction of the experiment and A.S. developed the experimental idea.

Additional information

Competing financial interests: The authors declare no competing financial interests.

5

Calcium-based multi-element chemistry for grid-scale electrochemical energy storage

Takanari Ouchi[1], Hojong Kim[2], Brian L. Spatocco[1] & Donald R. Sadoway[1]

Calcium is an attractive material for the negative electrode in a rechargeable battery due to its low electronegativity (high cell voltage), double valence, earth abundance and low cost; however, the use of calcium has historically eluded researchers due to its high melting temperature, high reactivity and unfavorably high solubility in molten salts. Here we demonstrate a long-cycle-life calcium-metal-based rechargeable battery for grid-scale energy storage. By deploying a multi-cation binary electrolyte in concert with an alloyed negative electrode, calcium solubility in the electrolyte is suppressed and operating temperature is reduced. These chemical mitigation strategies also engage another element in energy storage reactions resulting in a multi-element battery. These initial results demonstrate how the synergistic effects of deploying multiple chemical mitigation strategies coupled with the relaxation of the requirement of a single itinerant ion can unlock calcium-based chemistries and produce a battery with enhanced performance.

[1] Department of Materials Science and Engineering, Massachusetts Institute of Technology 77 Massachusetts Avenue, Cambridge, Massachusetts 02139-4307, USA. [2] Department of Materials Science and Engineering, The Pennsylvania State University, 320 Forest Resources Laboratory, University Park, Pennsylvania 16802-4705, USA. Correspondence and requests for materials should be addressed to D.R.S. (email: dsadoway@mit.edu).

A liquid metal battery (LMB) consists entirely of liquid active components: a low-density liquid metal negative electrode, an intermediate-density molten salt electrolyte and a high-density liquid metal positive electrode. Due to their mutual immiscibility these active components further self-segregate into three distinct layers according to their densities. On discharge, the negative electrode is oxidized to form an itinerant ion, which migrates across the molten salt electrolyte to the positive electrode, where the itinerant ion is electrochemically reduced to neutral metal, alloying with the positive electrode. This process is reversed upon charging. The LMB is well-positioned to satisfy the demands of grid-scale energy storage due to its ability to vitiate capacity fade mechanisms present in other battery chemistries and to do so with earth abundant materials and easily scalable means of construction[1,2].

Owing to its high solubility in molten salts calcium is impractical as an electrode[1,3,4]. Metal solubility renders the molten salt electronically conductive[5], which leads to loss of coulombic efficiency in electrolysis and loss of stored energy in a battery, that is, so-called self-discharge. In addition, the strong reducing capability of this electropositive element dispersed in the molten salt makes containment problematic, as most commonly used materials are susceptible to calciothermic reduction[6]. Herein we have made calcium the negative electrode of the LMB by devising parallel mitigation strategies to dramatically decrease its chemical potential so as to suppress both solubility and reactivity while advantageously lowering the melting temperature of the metal–salt couple.

The detrimental dissolution reaction of calcium metal in calcium halides can be represented by the following[4,5,7]:

$$Ca + Ca^{2+} \leftrightarrow 2Ca^{+} \quad \text{or} \quad Ca + Ca^{2+} \leftrightarrow Ca_2^{2+} \quad (1)$$

where calcium metal (Ca) reacts with calcium cations (Ca^{2+}) to form subvalent ions (Ca^{+} or Ca_2^{2+}). Using the latter case as an example, the equilibrium constant of the dissolution reaction is therefore given as:

$$K_{eq} \propto a_{Ca_2^{2+}} \Big/ [a_{Ca} \cdot a_{Ca^{2+}}] \quad (2)$$

where $a_{Ca_2^{2+}}$ is the activity of dissolved subvalent calcium, a_{Ca} the activity of calcium metal in negative electrode, and $a_{Ca^{2+}}$ the activity of calcium cation in the electrolyte. Focusing on the contributions of the reactants, we reason that suppressing the activity of calcium metal in the negative electrode, a_{Ca}, by alloying with more electronegative metals acting as diluents and that lowering the activity of Ca^{2+} in the electrolyte, $a_{Ca^{2+}}$, by the introduction of other cations, should result in attendant reductions in the concentration of subvalent Ca_2^{2+} while simultaneously decreasing the reactivity and melting temperature of the negative electrode.

It has been shown in our previous studies that alloying calcium with various positive electrodes drops the activity of calcium to values as low as 10^{-9} (for bismuth) and 10^{-10} (for antimony)[8,9], indicative of strong chemical interactions. The low activity of calcium in these electrodes contributes to the high-cell voltage of calcium-based cells as well as to the suppression of calcium metal dissolution from the positive electrodes[3]. In previous work, we have confirmed the bi-directionality of these positive electrodes with coulombic efficiencies exceeding 99% for both calcium–bismuth (Ca–Bi) and calcium–antimony (Ca–Sb)[10,11]. Clearly, to obtain high-cell voltage, the activity of calcium in the negative electrode should be as high as possible. The challenge is how to suppress the solubility of calcium metal from the negative electrode without making it denser than the electrolyte or raising the melting point, while minimally reducing cell voltage. The present study shows that magnesium (Mg) is an effective diluent

as it lowers the liquidus temperature of calcium–magnesium (Ca–Mg) alloy ($T_{eutectic} = 443\,°C$ and $517\,°C$ (ref. 12)) while supporting adequate calcium activity over a wide range of composition[12]. Specifically, an electromotive force study of Ca–Mg alloys suggests that the cell voltage reduction can be minimized to $<0.1\,V$ as long as the calcium concentration remains above 30 mol% (ref. 12).

Choosing the proper composition of multi-cation salts lowers the liquidus temperature of the electrolyte and decreases calcium solubility. The halides of the other cations in the multi-cation salt also work as supporting electrolytes, which desirably results in a lower ohmic drop under current flow. The choice of lithium chloride (LiCl) is advantageous due to its high ionic conductivity ($\sim 4\,S\,cm^{-1}$)[13] and its ability to form a low-melting eutectic mixture of lithium chloride and calcium chloride (LiCl–CaCl$_2$, 65–35 mol%, $T_{eutectic} = 485\,°C$ (ref. 14)). Furthermore, from the perspective of minimizing the self-discharge current, LiCl–CaCl$_2$ exhibited the best performance in the Ca–Mg||Bi cell. A liquid metal battery cell utilizing the Ca–Mg alloy as a negative electrode, the LiCl–CaCl$_2$ multi-cation mixture as an electrolyte, and a Bi positive electrode shows at 550 °C both high coulombic and energy efficiencies ($>99\%$ and $>70\%$, respectively) and demonstrates no capacity fade in excess of 1,400 cycles. In addition, due to the similarity in the deposition potentials of lithium and calcium and their mutual solubility we have found that these two metals can jointly engage in charge-transfer reactions at both electrodes to establish a co-deposition energy storage device.

Results

Effects of electrode composition on cell performance. A Ca–Mg (20–80 mol%)||Bi cell whose design is depicted in Fig. 1a was cycled at $200\,mA\,cm^{-2}$ current density and 650 °C operating temperature. As shown in Fig. 1b the cell achieved 98% coulombic and 62% energy efficiencies with a discharge voltage of 0.52 V. The self-discharge current density of the cell, $\sim 1\,mA\,cm^{-2}$, is extremely low (see Supplementary Table 1 and Supplementary Note 1), comparable with that of the best-performing Li||Sb–Pb cells[2]. Even though the concentration of calcium in the negative electrode was quite low (20 mol%) the open-circuit voltage decreased only 0.18 V from that of a cell fitted with a pure calcium negative electrode[1,8,12]. The volumetric energy density (electrodes basis) of this cell was 197 Wh L^{-1} (see Supplementary Table 2). In contrast, the higher calcium concentration Ca–Mg (90–10 mol%)||Bi cell maintained a cell-discharge voltage above 0.71 V and energy efficiency exceeding 70% (Fig. 1b). The volumetric energy density of the cell increased to 329 Wh L^{-1} (see Supplementary Table 3). The high concentration of calcium in the negative electrode caused a slight reduction in coulombic efficiency to 96% as a result of a somewhat elevated self-discharge current density ($\approx 4\,mA\,cm^{-2}$ as shown in Supplementary Table 1 and Supplementary Note 1). This is well explained by the fact that at 600 °C the activity of calcium in magnesium is 2.9×10^{-2} for 20–80 mol% Ca–Mg and ≈ 1 for 90–10 mol% Ca–Mg (ref. 12). Upon addition of 10 mol% of magnesium, the self-discharge current density was halved with almost no voltage penalty ($\approx 0.01\,V$ (refs 8,12)) when compared with that of the pure Ca||Bi cell ($\approx 10\,mA\,cm^{-2}$). In sum, the Ca–Mg||Bi cells were designed to suppress the self-discharge current density while maintaining high cell voltage at a reduced operating temperature. In addition, a Ca–Mg negative electrode with a LiCl–CaCl$_2$ electrolyte can be operated with an antimony (Sb) positive electrode. With higher-cell voltage (0.88 V), higher energy efficiency (74%), and higher volumetric energy density (384 Wh L^{-1}, see Supplementary Table 4) the cell with an Sb

a

- Ceramic insulator
- Negative current collector
- Negative electrode
- Salt
- Positive electrode
- Positive current collector

b

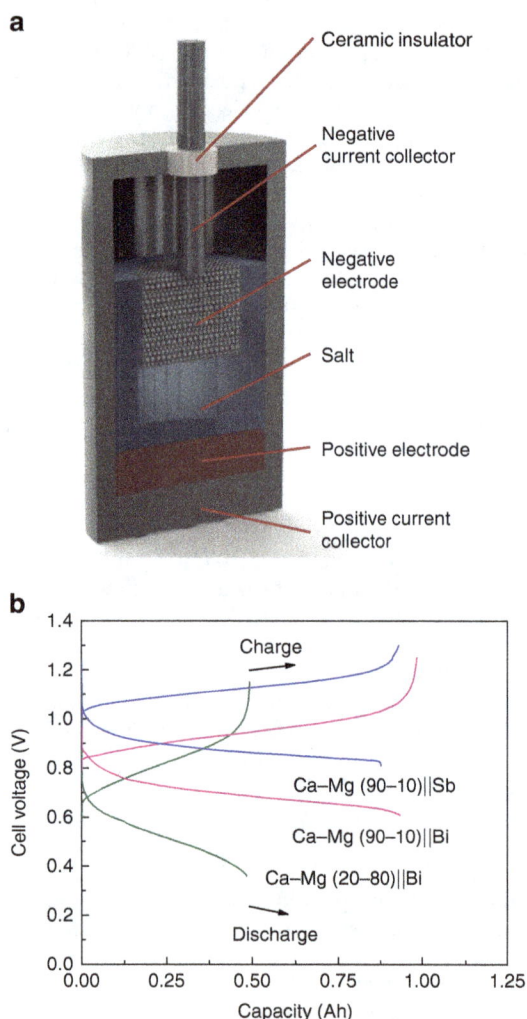

Figure 1 | Schematic and performance of Ca-Mg|LiCl-CaCl₂|Bi and Sb cells. (a) Schematic of cell with the negative current collector consisting of a stainless steel rod and Fe-Ni foam and the positive current collector made of mild steel or graphite. The foam contains the negative electrode. Current collectors are electrically isolated by means of an alumina insulator. **(b)** Charge-discharge voltage time traces of Ca-Mg (20–80 mol%)||Bi, Ca-Mg (90–10 mol%)||Bi, and Ca-Mg (90–10 mol%)||Sb operated at current density 200 mA cm^{-2} and temperature 650 °C. The theoretical capacities of Ca-Mg (20–80 mol%)||Bi, Ca-Mg (90–10 mol%)||Bi, and Ca-Mg (90–10 mol%)||Sb cells were 0.569, 1.33 and 1.08 Ah, respectively. The results of measurements were replicated more than five times, four times and twice, respectively.

positive electrode outperforms one with a Bi positive electrode as shown in Fig. 1b.

Cell performance at lower operating temperature. Exploiting the low eutectic temperatures of both the Ca–Mg alloy (90–10 mol%) and the binary LiCl–CaCl₂ electrolyte, we were able to decrease the cell operating temperature to 550 °C and obtain smooth voltage time traces at current densities ranging from 100 to 955 mA cm^{-2} (Fig. 2a). In parallel, the self-discharge current density of Ca–Mg (90–10 mol%)||Bi cells was found to decrease to <1 mA cm^{-2} (Supplementary Fig. 1), which we attribute to the decrease in solubility of Ca in the electrolyte[4,7,15]. The fall in discharge capacity with increasing current density is a consequence of the fact that at high current densities the deposition rate of Ca at the electrode-electrolyte interface exceeds

the rate of diffusion of Ca metal away from the electrode-electrolyte interface in the Bi positive electrode[10,11]. Because the theoretical capacity of the cell was defined as the mole fraction of calcium, at which the nucleation of the Ca₁₁Bi₁₀ phase occurs (25 mol% at 550 °C (ref. 8)), at high-current densities it is possible to reach this stoichiometry locally at the interface before the bulk metal electrode reaches this composition. The dependence of cell voltage on current density at several discharged states (9.3, 22.6 and 41.0% of theoretical capacity) is shown in Fig. 2b. The linearity in the I–V characteristic of the cell is evidence of the facile kinetics of the charge/discharge processes. This cell cycled at 200 mA cm^{-2} for over 1,400 cycles with 99% coulombic efficiency and 70% round-trip energy efficiency as depicted in Fig. 2c. The average volumetric energy density of this cell was 228 Wh L^{-1} (see Supplementary Table 5). Additionally noteworthy was that these metrics were achieved without observable voltage loss, side reactions or capacity fade as shown in Fig. 2c,d. Long service life-time translates to low cost of ownership, which is the most important parameter for grid-scale energy storage. No calcium metal battery has ever exhibited such stability.

Analysis of charge–discharge reactions. Figure 3 shows the cross-section of a Ca–Mg (90–10 mol%)|| Bi cell in a partially discharged state. After 400 cycles at 550 °C, electrodes were subjected to chemical analysis by direct current plasma emission spectrometry. No corrosion was observed on either the negative or positive current collector. Significantly, both negative and positive electrodes in the partially discharged cells contain Li (Supplementary Table 6). This means that both the Ca and Li ions in the multi-cation salt co-alloy and co-dealloy with both the negative and positive electrodes during charge and discharge. Furthermore, the absence of Mg in the positive electrode proves that there is no participation on the part of Mg in the cell electrochemistry, that is, Mg acts purely as a solvent to reduce the melting point of the negative electrode. The cell reactions can be represented as

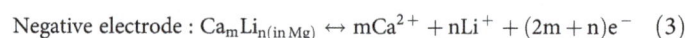

$$\text{Negative electrode}: \text{Ca}_m\text{Li}_{n(\text{in Mg})} \leftrightarrow m\text{Ca}^{2+} + n\text{Li}^+ + (2m+n)e^- \quad (3)$$

$$\text{Positive electrode}: m\text{Ca}^{2+} + n\text{Li}^+ + (2m+n)e^- \leftrightarrow \text{Ca}_m\text{Li}_{n(\text{in Bi})} \quad (4)$$

Discussion

This unique multi-cation salt gives rise to the possibility of a multi-element LMB, in which a plurality of active metals participates in faradaic reactions. The high coulombic efficiency (>99%) and excellent retention of discharge capacity observed during lithium co-deposition indicate that the electrode reactions are fully bi-directional. Recognition that co-deposition in a battery can be managed to advantage broadens the selection of salt composition and extends the prediction of the capacity limit. This is in sharp contrast to the practice in calcium electrorefining which considers only those salts, for example, CaCl₂–KCl, that obviate or suppress significant co-deposition to maintain the purity of the metal product[16]. The fact that this cell operates without need for an ion-selective membrane between the electrodes as is the case with sodium–sulfur and sodium–nickel chloride (ZEBRA) batteries, which rely on a β″-alumina separator, for example, sets this battery chemistry apart from all others. This membrane-free cell designed with a plurality of mitigation strategies opens up multi-element participation, which suppresses Ca reactivity, lowers operating temperature, and finally realizes a Ca-metal-based rechargeable battery that has the potential to meet the performance requirements of grid-scale energy storage applications.

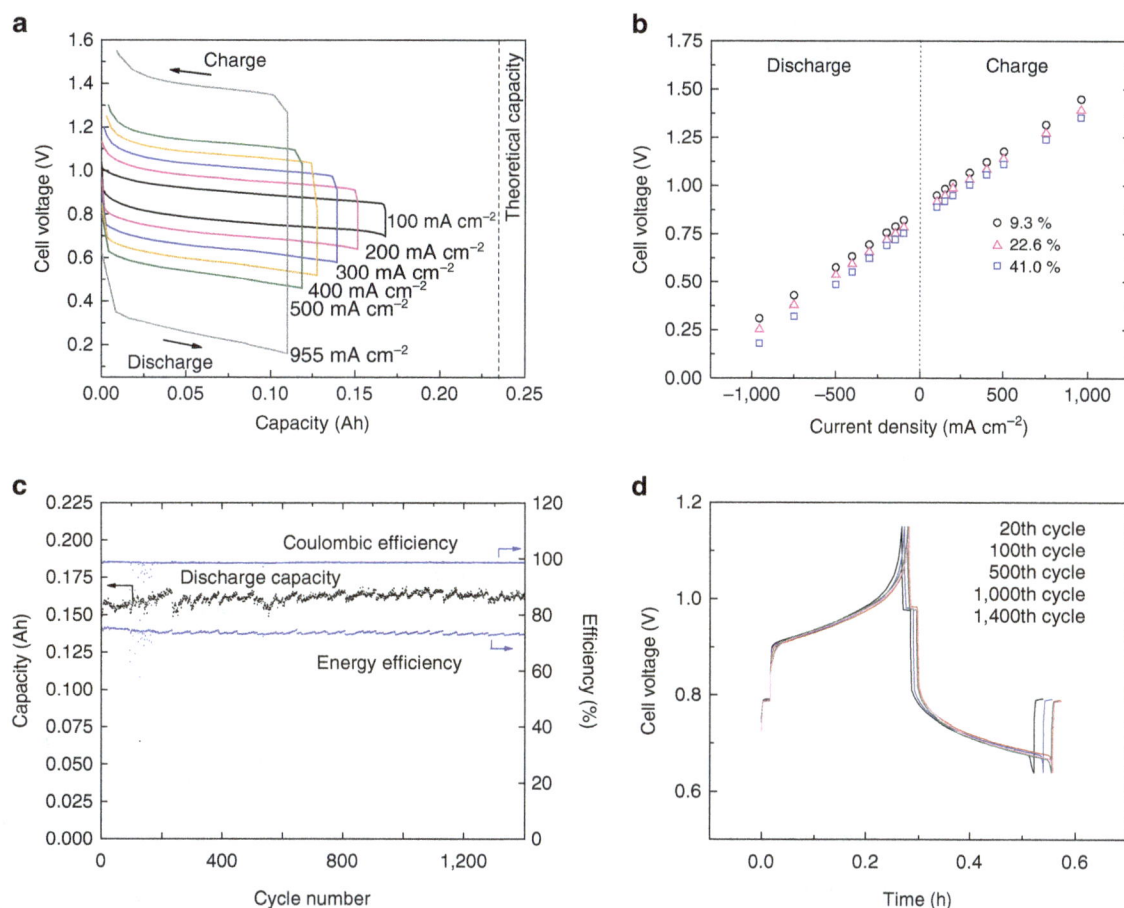

Figure 2 | Cell performance of Ca–Mg (90–10 mol%)||Bi cell. (**a**) Charge–discharge voltage time trace at current densities 100–955 mA cm^{-2}. (**b**) Cell voltage with varying current density at 9.3, 22.6 and 41% of theoretical capacity. (**c**) Discharge capacity, coulombic efficiency and energy efficiency with cycling. (**d**) Representative charge–discharge voltage time traces at different cycle numbers. Theoretical capacity of this cell was 0.239 Ah. Operating temperature was 550 °C. Current density of (**c,d**) was 200 mA cm^{-2}. The results derive from measurements on more than two cells.

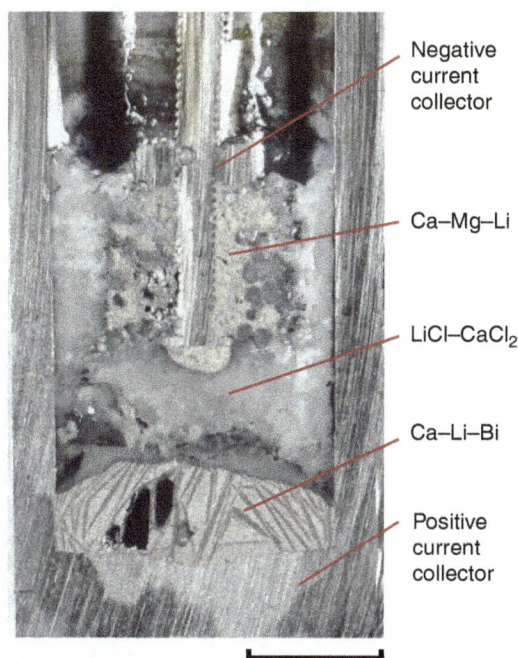

Figure 3 | Cross-section of Ca–Mg (90–10 mol%)||Bi cell at partial state of charge. The scale bar is 10 millimeter (mm). The results derive from measurements on more than three cells.

Methods

Cell configuration. The negative current collector (NCC) consists of an 18–8 stainless steel threaded rod and a column-shaped Fe–Ni foam. The Fe–Ni foam was 13 mm in diameter, 10 mm in height, and 60 ppi in porosity. A mild steel (AISI 1018) crucible served as the positive current collector (PCC) for Bi. A graphite crucible served as the PCC for Sb. The dimensions of the PCCs were 25 mm in outer diameter, 83 mm in height, 77 mm in depth and 20 mm in inner diameter. The foam NCC was positioned 15 mm above the PCC. By surface tension the foam NCC holds the metal of the negative electrode and thus prevents keeps it from shorting to the positive current collector. The surface area of positive electrode was estimated at 3.14 cm^2 by assuming the interface between it and the electrolyte to be perfectly flat. This value was used to derive the current density.

Materials preparation. Pure Ca (99.99%, Aldrich), Mg (99.95%, Alfa Aesar), Bi (99.999%, Aldrich) and Sb (99.9999%, Alfa Aesar) were used for electrodes. These metals were melted in the PCCs by an induction heater (MTI Corporation, EQ-SP-15A), installed inside the glovebox with an inert argon atmosphere (O$_2$ < 0.1 p.p.m., H$_2$O < 0.1 p.p.m.). Pure Ca and Mg were melted in a mild steel cup by the induction heater, and then the NCCs were immersed in the liquid Ca–Mg alloy for 10 min.

High-purity anhydrous salts of LiCl (99.995%, Alfa Aesar) and CaCl$_2$ (99.99%, Alfa Aesar) were used. The salts were weighed out in appropriate quantities in ~500 g batches, mixed, and transferred to an alumina crucible which was then placed in a stainless steel vacuum chamber, the chamber sealed, and loaded into a furnace. To prepare a dry, homogenous molten salt solution, the test chamber was (1) evacuated to ~1 Pa, (2) heated at 80 °C for 12 h under vacuum, (3) heated at 230 °C for 12 h under vacuum, (4) purged with ultra-high-purity argon gas (99.999%, O$_2$ < 1 p.p.m., H$_2$O < 1 p.p.m., Airgas Inc) and (5) heated at 700 °C for 3 h under argon gas flowing at 0.2 cm^3 s^{-1}. After cooling to room temperature, the pre-melted electrolyte was transferred back to the glove box and ground into powder (≈10 μm in diameter). Supplementary Table 7 reports the masses of metal and salt in each cell. To accelerate the measurement of long-term capacity fade at deep discharge (within several months instead of years), we reduced the size of the electrodes of the cell in Fig. 2. It is important to restrict the depletion of LiCl so as to avoid solidification of the molten salt. According to the phase diagram of LiCl–CaCl$_2$ (ref. 14), at 550 °C for example, the liquid range of composition in the LiCl–CaCl$_2$ molten salt is 58–77 mol% of LiCl. To operate the cells at $\sim$1 Ah capacity we used $\sim$12 g of electrolyte thereby preventing solidification of the LiCl–CaCl$_2$ melt caused by depletion of LiCl via co-deposition of Li.

Cell assembly. Cells were loaded in the stainless steel vacuum chamber and subjected to the same drying process as were the salts (80 °C for 12 h and 230 °C for 12 h under vacuum). Then, the cells were connected to a battery tester (Model 4300, Maccor). The cells were heated under open circuit condition. After the temperature reached the liquidus of the electrolyte (485 °C), the cell voltage was held at 1.25 V until the current became steady.

Analysis. The measurement of self-discharge current density was performed after the cell reached the operating temperature and the current became steady at a constant applied voltage of 1.1 V reported in Supplementary Fig. 1 and of 1.25 V reported in Supplementary Table 1. The charging and discharging voltage time traces were measured at constant current density. Data were acquired by battery testing instrumentation (Model 4300, Maccor). Coulombic efficiency (η_Q) was calculated from charging capacity (Q_C) and discharging capacity (Q_D), ($\eta_Q = 100 \cdot Q_D/Q_C$). Energy efficiency ($\eta_E$) was calculated from charging energy (E_C) and discharging energy (E_D), ($\eta_E = 100 \cdot E_D/E_C$). Discharge voltage ($V_D$) was calculated from discharge capacity and discharge energy ($V_D = E_D/Q_D$). Theoretical discharge capacity of the cell was defined as the mole fraction of calcium, at which the nucleation of the Ca$_{11}$Bi$_{10}$ phase occurs (25 mol% at 550 °C and 27 mol% at 650 °C (ref. 8)) or the Ca$_{11}$Sb$_{10}$ phase (23 mol% at 650 °C (ref. 11)) assuming that only calcium participates in charge/discharge reaction. The volumetric energy density was calculated using the volumes of electrodes as shown in Supplementary Tables 2–5. Direct-current plasma emission spectrometry was carried out by Luvak Inc., following ASTM E 1097–12.

References

1. Kim, H. *et al.* Liquid metal batteries: past, present, and future. *Chem. Rev.* **113**, 2075–2099 (2013).
2. Wang, K. *et al.* Lithium-antimony-lead liquid metal battery for grid-level energy storage. *Nature* **514**, 348–350 (2014).
3. Sharma, R. A. Solubilities of calcium in liquid calcium chloride in equilibrium with calcium–copper alloys. *J. Phys. Chem.* **74**, 3896–3900 (1970).
4. Bredig, M. A. *Mixtures of Metals with Molten Salts, ORNL-3391* (Oak Ridge National Laboratory, 1963).
5. Dworkin, A. S., Bronstein, H. R. & Bredig, M. A. Ionic melts as solvents for electronic conductors. Discuss. *Faraday Soc.* **32**, 188–196 (1961).
6. Ono, K. & Suzuki, R. O. A new concept for producing ti sponge: calciothermic reduction. *JOM* **54**, 59–61 (2002).
7. Dworkin, A. S., Bronstein, H. R. & Bredig, M. A. Electrical Conductivity of Solutions of Metals in their Molten Halides.VIII. Alkaline earth metal systems. *J. Phys. Chem.* **70**, 2384–2388 (1966).
8. Kim, H. *et al.* Thermodynamic properties of calcium-bismuth alloys determined by emf measurements. *Electrochim. Acta* **60**, 154–162 (2012).
9. Poizeau, S., Kim, H., Newhouse, J. M., Spatocco, B. L. & Sadoway, D. R. Determination and modeling of the thermodynamic properties of liquid calcium–antimony alloys. *Electrochim. Acta* **76**, 8–15 (2012).
10. Kim, H., Boysen, D. A., Ouchi, T. & Sadoway, D. R. Calcium–bismuth electrodes for large-scale energy storage (liquid metal batteries). *J. Power Sources* **241**, 239–248 (2013).
11. Ouchi, T., Kim, H., Ning, X. & Sadoway, D. R. Calcium-antimony alloys as electrodes for liquid metal batteries. *J. Electrochem. Soc.* **161**, A1898–A1904 (2014).
12. Newhouse, J. M., Poizeau, S., Kim, H., Spatocco, B. L. & Sadoway, D. R. Thermodynamic properties of calcium–magnesium alloys determined by emf measurements. *Electrochim. Acta* **91**, 293–301 (2013).
13. Janz, G. J. *et al.* Molten salts: Volume 4, part 2, chlorides and mixtures—electrical conductance, density, viscosity and surface tension data. *J. Phys. Chem. Ref. Data* **4**, 871–1178 (1975).
14. Mahendran, K. H., Nagaraj, S., Sridharan, R. & Gnanasekaran, T. Differential scanning calorimetric studies on the phase diagram of the binary LiCl–CaCl$_2$ system. *J. Alloys Compd.* **325**, 78–83 (2001).
15. Dosaj, V., Aksaranan, C. & Morris, D. R. Thermodynamic properties of the calcium – calcium chloride system measured by an electrochemical technique. *J. Chem. Soc., Faraday Trans. 1.* **71**, 1083–1098 (1975).
16. Zaikov, Y. P., Shurov, N. I., Batukhtin, V. P. & Molostov, O. G. Calcium production by the electrolysis of molten CaCl$_2$—part II. Development of the electrolysis devices and process technology approval. *Metall. Mater. Trans. B* **45B**, 968–974 (2014).

Acknowledgements

Financial support from the Advanced Research Projects Agency-Energy (US Department of Energy) and Total S.A. is gratefully acknowledged.

Author contributions

T.O. and D.R.S. conceived the idea for the project. T.O. performed all experiments. T.O., H.K., B.L.S. and D.R.S. contributed in discussing and drafting the manuscript.

Additional information

6

Broadband single-molecule excitation spectroscopy

Lukasz Piatkowski[1], Esther Gellings[1] & Niek F. van Hulst[1,2]

Over the past 25 years, single-molecule spectroscopy has developed into a widely used tool in multiple disciplines of science. The diversity of routinely recorded emission spectra does underpin the strength of the single-molecule approach in resolving the heterogeneity and dynamics, otherwise hidden in the ensemble. In early cryogenic studies single molecules were identified by their distinct excitation spectra, yet measuring excitation spectra at room temperature remains challenging. Here we present a broadband Fourier approach that allows rapid recording of excitation spectra of individual molecules under ambient conditions and that is robust against blinking and bleaching. Applying the method we show that the excitation spectra of individual molecules exhibit an extreme distribution of solvatochromic shifts and distinct spectral shapes. Importantly, we demonstrate that the sensitivity and speed of the broadband technique is comparable to that of emission spectroscopy putting both techniques side-by-side in single-molecule spectroscopy.

[1]ICFO—Institut de Ciencies Fotoniques, The Barcelona Institute of Science and Technology, 08860 Castelldefels (Barcelona), Spain. [2]ICREA—Institució Catalana de Recerca i Estudis Avançats, 08010 Barcelona, Spain. Correspondence and requests for materials should be addressed to L.P. (email: Lukasz.Piatkowski@icfo.eu) or to N.F.v.H. (email: Niek.vanHulst@icfo.eu).

Optical spectroscopy is a primary analysis tool underlying almost any field of science; absorption, emission and excitation spectra are routinely recorded on bulk samples. The advent of single-molecule detection pushed spectroscopy to a next level: probing molecules one by one unravels inhomogeneities and dynamical processes that are otherwise hidden in the ensemble average. Intrinsic molecular diversity and distinct interactions of molecules with their nanoenvironment lead to wide distributions of their spectral properties. Pursuing a single molecule in time reveals discrete dynamics such as blinking[1] and spectral diffusion[2-6]. To date single molecules, quantum dots, nanoparticles and proteins are detected and tracked with wide-ranging applications in molecular biology, polymer chemistry, nanoscopy and so on.

Historically, the first single molecules were detected through absorption[7] and fluorescence excitation[8] at cryogenic temperatures—conditions leading to high photostability and large absorption cross-section have allowed the recording of both excitation and emission spectra of individual molecules[5,6,9-12]. Far superior signal-to-noise ratios, however, are obtainable with the background-free fluorescence-based detection as opposed to absorption-based spectroscopy; as a result detection of single molecules through their fluorescence has established as the predominant method of choice.

The vast majority of applications in biology and chemistry involve the study of molecules in their natural state that is under ambient conditions. At room temperature, however, the significantly broadened absorption lines combined with a substantial reduction of the photostability and absorption cross-section demands more sensitive fluorescence detection methods and spatial separation of individual molecules. The introduction of near-field imaging[13] and non-invasive confocal microscopy[14,15] have guaranteed the necessary sensitivity and spatial resolution, however, the limited number of photons emitted before photobleaching still puts a major constraint on all room temperature single-molecule spectroscopy techniques. Therefore, only time and detection efficient spectroscopic schemes are feasible, limiting room temperature experiments largely to the recording of emission spectra[15-18].

In recent efforts to go beyond single-molecule emission spectroscopy, individual molecules have been detected using photothermal contrast[19], scattered light[20,21], plasmonic structures[22] and even direct detection of a single molecule through absorption has been demonstrated at room temperature[23-25]. Although very interesting, these approaches are far from routine and their main drawback is that they typically lack spectral information.

Previously, Stopel et al.[26] determined the Stokes shift of individual molecules recording both excitation and emission spectrum at room temperature. Against the general believe, they found that the Stokes shift varies between individual molecules. In this study, the excitation spectra were recorded by serial scanning of the narrowband excitation wavelength derived from a white-light continuum. A disadvantage of this approach, however, is that blinking and bleaching of the molecule compete with the sequential scanning, which may result in an incomplete excitation spectrum. Recently Weigel et al.[27] were the first to close a coherent control loop on a single molecule, showing that essentially the excitation spectrum is probed.

The ability to routinely perform single-molecule excitation spectroscopy under ambient conditions side-by-side to the already well-established single-molecule emission spectroscopy would be highly valuable. The crucial difference between the two spectroscopies is that emission spectra probe the final spontaneous, nanosecond decay to the ground state and its vibrational progression, while excitation spectra explore the excited state, its vibrational manifold and any intermediate short-lived (ps) states towards the decay. The excitation spectra are particularly sensitive to coupling between molecules or the nanoenvironment of the single molecules in general. Having both spectra at our disposal, a complete picture of the spectral characteristics and excited state dynamics of the molecule emerges.

Beyond the fundamental properties of individual molecules, such as assymetry in absorption–emission spectra, origins of blinking–bleaching dynamics and intra/intermolecular inter-system-crossings, excitation spectra provide valuable information in applied fields such as environmental sciences, medicine and material sciences, just to name a few. For example excitation spectra are used to determine the nature, relative quantities and composition of chromophores in plants, photosynthetic units and coral reef matter[28,29]. Even in the studies on photodynamic therapy for cancer treatment, excitation spectra are used to non-invasively determine in vivo the penetration depth of drugs in skin[30]. However, in such applications the multichromophoric composition leads to complex photophysics including energy transfer, self-quenching, reabsorption and reemission, complicating the multicomponent analysis of excitation spectra. Clearly the ability to measure excitation spectra of molecules in extremely small volumes and at interfaces is desirable. A novel approach to single-molecule excitation spectroscopy at ambient conditions would provide sufficient sensitivity to probe even monolayers of such samples with diffraction-limited spatial resolution.

Here we demonstrate a fast and efficient broadband excitation spectroscopy method to record single-molecule excitation spectra at room temperature. Our results uncover heterogeneities in the excitation spectra with unexpectedly large spectral shifts over 100 nm between individual molecules, which have remained beyond observation so far. Exploiting an ultra-broadband laser, we probe the entire excitation spectrum at once while recording the resulting fluorescence response. Different excitation wavelengths are sampled when scanning a time delay between two interfering broadband (over 100 nm) laser pulses. The time-dependent fluorescence response then yields the excitation spectrum through a Fourier transformation. Conceptually, our method is a pulsed single-molecule approach, in contrast to earlier continuous-wave Fourier excitation spectroscopy on bulk samples[31,32]. An important advantage of interferometric excitation over scanned narrowband excitation is that in the presence of blinking and even bleaching (provided that the main part of the interferogram around $\Delta t = 0$ is largely intact) full spectral information is still contained in the measured (though incomplete) interferogram, with no significant effect on the quality of the measured excitation spectrum. The insensitivity of the interferometric approach to fluorescence intensity fluctuations and spectral jumps has also been shown for interferometric detection of the emission spectra of individual molecules at cryogenic temperatures[33]. Moreover, the coherence and fs-resolution of the interfering broadband pulse-pair can be exploited to control the excited state of individual molecules before fluorescence emission[34].

Results

Inhomogeneity of single molecules. All our measurements of the excitation and emission spectra were performed on single quaterrylene diimide (QDI) molecules derived from a rylene dye family[35,36]. Rylene dyes have been extensively used in single-molecule experiments because of their extraordinary brightness and photostability[23,33,37-39]. The absorption spectrum of a solution of QDI in toluene peaks around 750 nm, as shown in Fig. 1b (solid, black line), whereas the emission

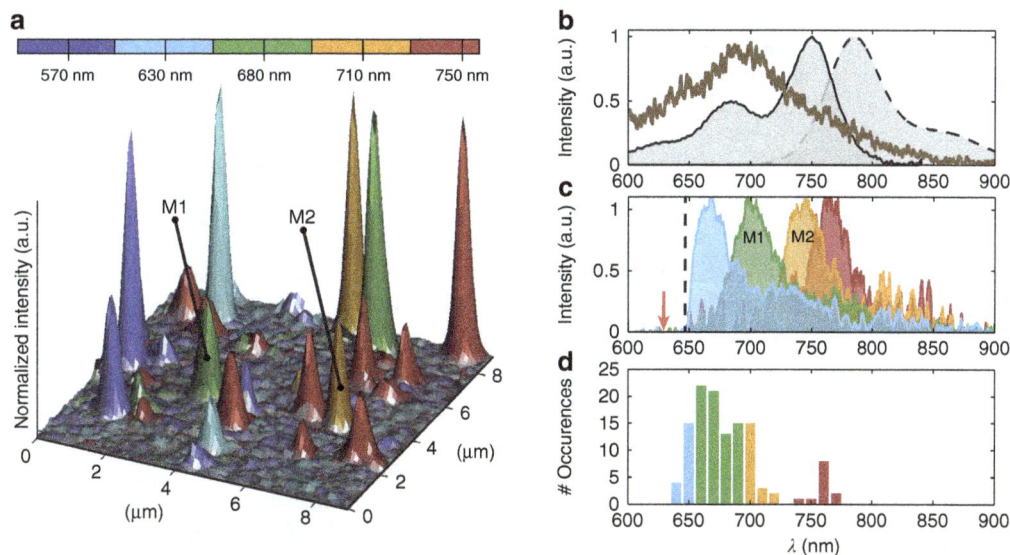

Figure 1 | Heterogeneity of single molecule emission. (a) A composite confocal image of dispersed QDI molecules in a PMMA matrix constructed from a series of five normalized images recorded at different excitation wavelengths. The assigned colour corresponds to appearance of the molecules in the image for a specific excitation wavelength. The height of the peaks reflects the relative fluorescence intensity between the molecules. **(b)** Absorption (black solid line) and emission (black dashed line) spectra of a solution of QDI in PMMA/toluene mixture. The brown solid line indicates the absorption spectrum of QDI molecules embedded in a solidified PMMA matrix. **(c)** Exemplary emission spectra associated with molecules absorbing at different wavelengths. Molecules M1 and M2 correspond to the two molecules from the composite confocal image in **a**. The red arrow indicates the excitation wavelength (633 nm), the black dashed line indicates the cutoff wavelength of the long-pass filter. **(d)** A histogram representing the positions of the maxima of the emission spectra for all measured molecules. The assigned colours correspond to the excitation wavelengths indicated in panel **(a)**.

spectrum of the same QDI solution (upon excitation at 633 nm) is Stokes shifted to 780 nm (black dashed line).

We first turn our attention to the emission spectra of individual QDI molecules, which exhibit intriguing properties[40]. In Fig. 1a we show a composite confocal fluorescence image of QDI molecules embedded in a polymethyl methacrylate (PMMA) matrix. The image is constructed by superimposing five individually recorded images of the same area using different excitation (detection lower limit) wavelengths: 570 nm (610 nm); 630 nm (648 nm); 680 nm (730 nm); 710 nm (740 nm); and 750 nm (778 nm). Before the composition, images were normalized to unity. The height of the peaks reflects the relative fluorescence intensity between the molecules. Each molecule was labelled with a different colour according to the spectral range, in which it starts absorbing and emitting light. Molecules excited at the longest wavelength (750 nm) are labelled red; molecules that only start emitting when excited at 710 nm are labelled orange and so on. The efficient excitation of QDI molecules at short excitation wavelengths (around and below 650 nm) is surprising, as the absorption of QDI solution in this spectral range is very low (below 20% of the maximum). We have found, however, that the QDI absorption spectrum undergoes a significant blueshift (hypsochromic shift) and modification of the spectral shape, upon embedding the molecules in a PMMA matrix (see brown spectrum in Fig. 1b and Supplementary Note 1). The QDI/PMMA absorption spectrum is broadened and lacks the vibrational progression feature, which is clearly resolved in the solution absorption spectrum. The strong spectral shifts imply that to understand the spectral properties of QDI in PMMA, it is necessary to disentangle the excitation spectra of individual molecules.

To assess the spectral variation, we measured a total of 122 single-molecule emission spectra at excitation (detection) wavelength of 633 nm (above 655 nm). The emission spectra of four typical molecules representing the light blue, green, orange and red type molecules in the composite confocal image (Fig. 1a) are

shown in Fig. 1c. The main emission peak position of the emission spectra was determined with a Gaussian fit, and the distribution of peak positions is plotted in Fig. 1d.

The presented data evidently shows that the emission spectra of individual QDI molecules in the PMMA matrix exhibit a remarkably large distribution of spectral shifts towards lower wavelengths with respect to the emission spectrum of the QDI ensemble solution spectrum. Moreover, it underlines that the maximum fluorescence peak can shift by more than 100 nm compared with the ensemble solution which has a maximum peak at 780 nm (see Fig. 1b, black, dashed line). The confocal image (Fig. 1a) illustrates that a significant amount of molecules only becomes visible when changing the wavelength by more than 150 from 750 to 570 nm. Conversely, the majority of molecules, emitting upon 750 nm excitation are not visible anymore when excited at 570 nm. The absorption of individual molecules shift spectrally more than the width of the QDI solution absorption spectrum (roughly 160 nm) and, thus, will remain unnoticed when the excitation wavelength is chosen solely based on the solution spectrum. This risk of selective detection is clearly observed in Fig. 1d. Both shape and position of the detected fluorescence emission distribution depend on the choice of excitation wavelength. These findings clearly indicate that a broad range of excitation wavelength is necessary to probe the entire distribution of emission spectra. The majority of single-molecule detection methods rely on the acquisition of fluorescence upon narrowband excitation with just a single wavelength, which limits and biases the quantitative analysis of spectral distributions in single-molecule experiments.

Although quite large, the observed spectral shifts and spectral variability are not uncommon in single-molecule emission spectra. In fact, analogous effects have been reported for a number of other molecules, including perylene diimide (molecular analogue of QDI) embedded in polyvinyl alcohol (PVA), for which ~40-nm wide distribution of emission spectra has been found because of the heterogeneity of the polymer matrix sites as well as twisted conformations of the core[41,42]. It has also been shown that changes in the photophysical form of the green fluorescent protein in polyvinyl alcohol lead to a spectral variations of nearly 100 nm due to π-stacking interactions, extended chromophoric π-systems and photoactivation[43]. Comparable spectral shifts have been measured in single carbocyanine dye molecules adsorbed on bare glass along with strong alterations in the shape of the vibronic bands[44.]

Clearly, the dramatic variability of the observed emission spectra should be reflected in the excitation spectra, while differences give clues on the nature of the heterogeneity. In the following, we, thus, focus on the challenge to measure the excitation spectra of individual molecules to address their large spectral variability and directly compare excitation and emission spectra.

Broadband excitation spectroscopy. The experimental set-up is sketched in Fig. 2a (for details see Methods). The broadband laser excitation spectrum is derived from a Ti:sapphire oscillator and covers the wavelength range 655–770 nm, with 12 fs pulse duration. After propagating through a Mach–Zehnder interferometer including a delay line with a sub-fs interpulse delay precision, the interfering laser pulses with interpulse delay Δt excite a single molecule whose fluorescence response is detected. The total fluorescence intensity $F(\omega)$ is directly proportional to both the molecular excitation probability (absorption cross-section) $F_{QDI}(\omega)$ and the laser spectral power $F_{laser}(\omega)$ at excitation frequency ω. Each spectrum $F(\omega)$ has a Fourier-related time-dependent interferogram $H(\Delta t)$. A different combination of wavelengths is sampled at each delay line position and information on the excitation spectrum (that is, absorption cross-section) is contained in Δt dependent fluorescence intensity.

When exciting a molecule in its linear optical response regime, the measured fluorescence response $H(\Delta t)$ is the convolution of the laser interferometric autocorrelation function $H_{laser}(\Delta t)$ and the interferogram of the excitation spectrum $H_{QDI}(\Delta t)$ of the molecule: $H(\Delta t) \propto (H_{laser} \otimes H_{QDI})(\Delta t)$. Using the convolution theorem, the Fourier transformation of $H(\Delta t)$ turns into the product of the laser spectrum and QDI excitation spectrum: $F(\omega) \propto F_{laser}(\omega) \cdot F_{QDI}(\omega)$. Dividing $F(\omega)$ by the laser spectrum

$F_{laser}(\omega)$, we directly obtain the excitation spectrum of the molecule.

In Fig. 2b we show three examples of the time-dependent fluorescence response of single QDI molecules. Differences in the extent of the interference and beating in the interferogram recordings are a clear sign that each of the molecules interacts differently with the broadband excitation laser. Molecule M4 (red) shows a beating pattern on a time scale of ~ 25 fs, a clear signature of the presence of a superposition of (at least) two distinct frequency bands. The other two interferograms (molecule M1 and M2) clearly lack this feature. The corresponding Fourier transforms are shown in Fig. 2c. As expected, the red spectrum shows two bands whereas the other two (green, blue) show only one. Since the laser spectrum was identical in all three cases, the clear disparities between these spectra indicate differences in the excitation spectra of the molecules. The narrow spectra of M1 and M2 are either caused by a very narrow excitation spectrum of the molecule or the molecular excitation spectrum being shifted towards lower wavelengths, out of our laser spectral window.

Single-molecule excitation spectra. We have measured the fluorescence response of 25 molecules along with interferograms of the excitation laser. The excellent photostability of the QDI molecules enabled the recording of several consecutive interferograms for most molecules (a total of 95 individual interferograms).

In Fig. 3 we show distinct excitation spectra for five single QDI molecules (M1:M5). In Fig. 3a1–a5, we present the Fourier transformations of the fluorescence response (that is the product spectrum $F(\omega)$) measured on each individual molecule (solid line, green) and the laser spectrum measured aside the molecule (shaded area, blue). Dividing the green product spectra by the blue-laser spectra directly yields the excitation spectra of the molecules, which are presented in Fig. 3b1–b5 (solid line, red). For comparison, we also show the ensemble solution absorption spectrum (shaded area, grey), which has been spectrally offset from the ensemble position to match the measured single-molecule excitation spectra. Here we assume that only the absolute spectral position, and not the vibrational progression of individual QDI molecules undergoes significant change under encapsulation of the molecules in the PMMA matrix. The

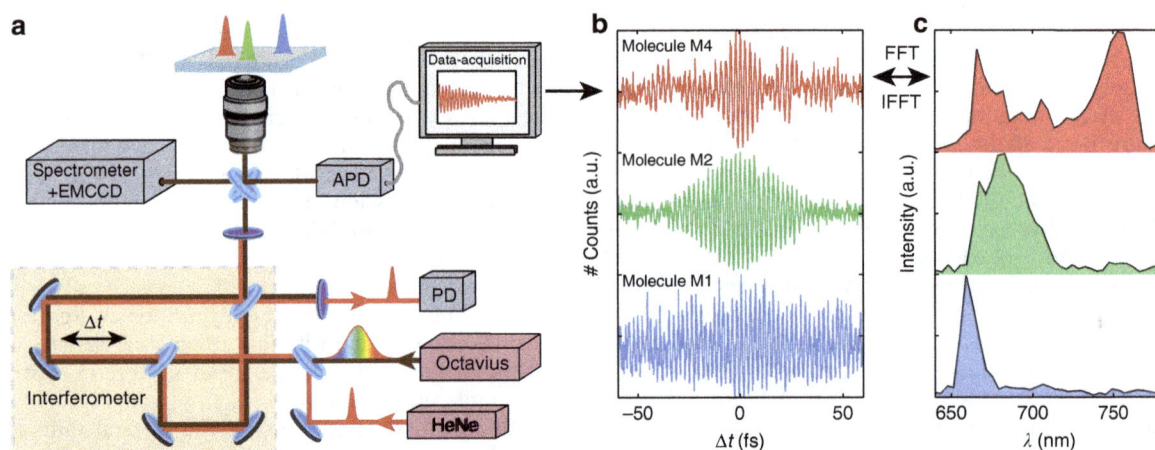

Figure 2 | The concept of broadband single-molecule excitation spectroscopy. (**a**) Schematic representations of the experimental set-up: a broadband fs laser (Octavius) is split in two branches, and a molecule of choice excited by the pulse-pair, while scanning the time delay Δt. (**b**) Three typical fluorescence interferograms of individual QDI molecules (M1, M2 and M4), each exhibiting a specific delay time-dependent fluorescence response and (**c**) their corresponding Fourier transformations. For full fluorescence interferograms see Supplementary Fig. 1. The typical contrast between the background count level and the fluorescence signal is 1/5–1/8 (see Supplementary Fig. 1). FFT, fast Fourier transform; IFFT, inverse fast Fourier transform

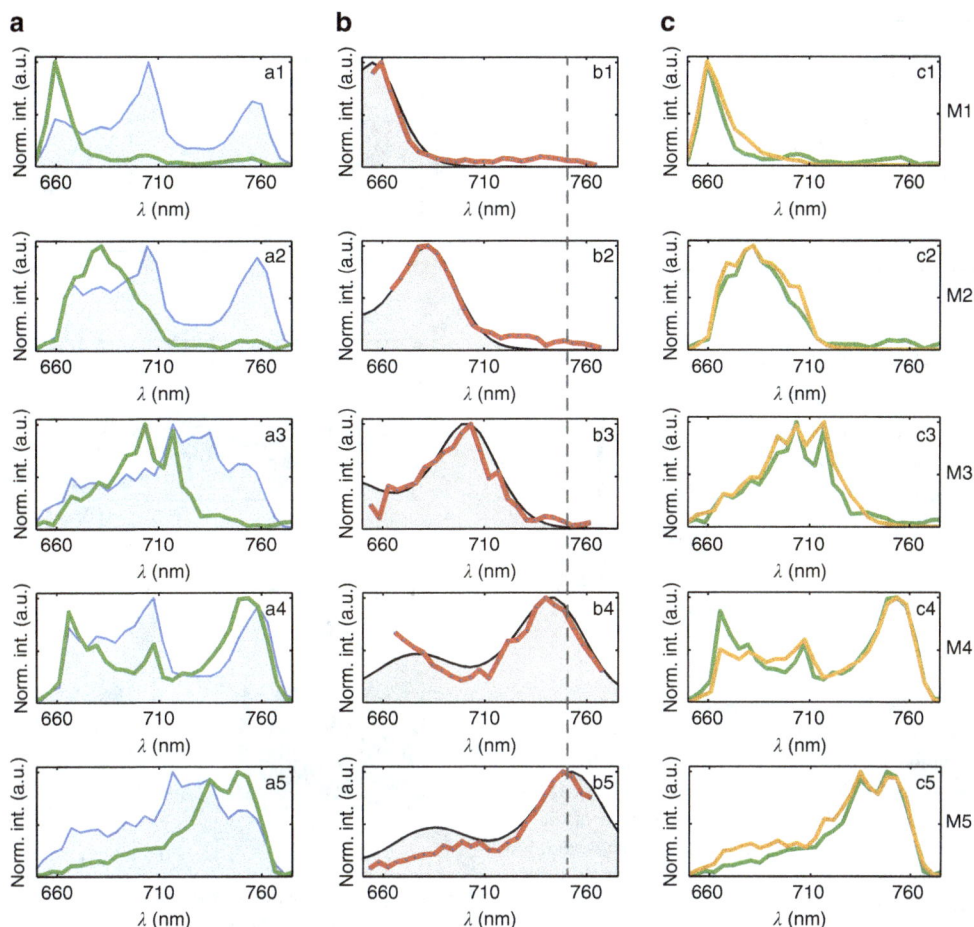

Figure 3 | Excitation spectra of single molecules at room temperature. (**a**) Experimental spectra of single molecules (green lines) for a series of five molecules (M1:M5), together with the laser spectrum (blue, shaded). (**b**) Single-molecule excitation spectra (red lines) obtained by dividing the experimental single-molecule product spectra by the corresponding laser spectra. Shaded grey spectra in **b** represent the ensemble absorption spectra shifted accordingly for direct comparison. The dashed line indicates the position of a maximum of the QDI ensemble absorption spectrum at 750 nm. (**c**) Single-molecule product spectra (green lines) compared with reconstructed theoretical spectra (orange lines) obtained by multiplying the appropriately shifted ensemble QDI absorption spectrum (grey, shaded) by the experimental laser spectra (blue, shaded).

excitation spectra of another 6 individual QDI molecules (M6:M11) are shown in Supplementary Fig. 2.

We find that the measured single-molecule excitation spectra are strongly shifted compared with the ensemble absorption spectrum. Interestingly, we can, however, reproduce the experimental product spectrum by multiplying the measured laser spectrum by the appropriately spectrally shifted QDI ensemble solution spectrum. The comparison between the measured (solid line, green) and reproduced (solid line, orange) single-molecule product spectrum is shown in Fig. 3c. The two curves agree quite well for all the measured molecules.

As can be seen in Fig. 3a, we have used two different shapes of laser spectra for the measurements (M1, M2 and M4—first shape; M3 and M5—second shape). We thus confirm that our approach and its sensitivity do not depend on the spectral shape of the laser. Each of the 25 molecules we measured exhibited a different excitation spectrum. While we observed hypsochromic shifts of up to 100 nm (see M1), we did not observe any molecules red-shifted by more than a few nm (see M5) with respect to the ensemble solution spectrum. The use of broadband excitation allows us to directly sample the distribution of single-molecule spectra over a 100 nm broad band, which with narrowband excitation would be difficult to probe. Because the measured excitation spectra of the molecules with small hypsochromic shifts (that is, large overlap with our laser spectrum)

indeed resemble the ensemble solution absorption spectrum, we expect that the more (above 100 nm) blue-shifted molecules exhibit similar excitation spectrum containing the vibrational progression. For M1, M2 and M3, however, the vibrational bands are outside the spectral window of our broadband excitation laser.

Variations in the shape of the excitation spectra. Upon closer inspection, our results do display more spectral variability than just the spectral shifts shown in Fig. 3. The single-molecule excitation spectra exhibit a variety of inter-vibronic band distances, spectral band widths and relative peak intensities. In Fig. 4a, we demonstrate three typical excitation spectra for which the main electronic transition band ($0'$–$0''$, number indicating vibrational level, single/double apostrophe noting ground/excited electronic state) overlap the best. Differences in the relative peak intensities are clearly visible. Moreover, changes in the distance between the two vibronic transitions ($0'$–$0''$ and $0'$–$1''$) are noticeable in the spectra. Similar spectral variations can also be found in the emission spectra, which for comparison we show in Fig. 4b. These observations illustrate that we have achieved sufficient sensitivity in the measured excitation spectra to probe the effect of the nanoenvironment on the vibrational modes of individual molecules.

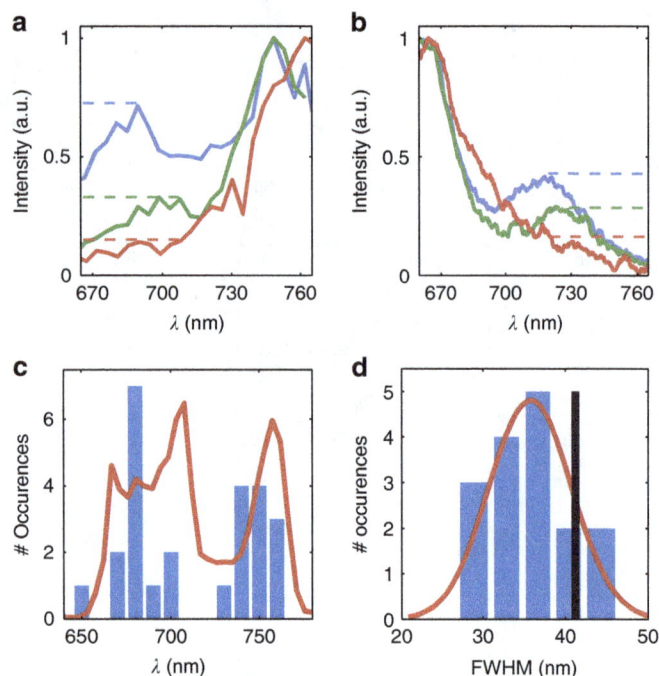

Figure 4 | Heterogeneity of the excitation and emission spectra.
Comparison of three single-molecule excitation (**a**) and emission (**b**) spectra showed with red, green and blue lines. Both types of spectra show a similar spread of relative peak intensities and inter-vibronic band distances (see corresponding colours). (**c**) A histogram presenting spread of the 0′-0″ transition positions of all measured excitation spectra along with the laser spectrum (red). (**d**) Distribution of the 0′-0″ transition peak widths along with a fitted Gaussian spread function. The vertical black line indicates the width of the transition in the bulk solution spectrum.

We determined a spread of the main transition positions for all the measured excitation spectra in Fig. 4c. We find that the 25 excitation spectra are rather homogeneously distributed across the laser spectrum (shown in red). As the fluorescence is detected above 785 nm, less emission is collected for molecules that are very blue-shifted compared with those absorbing at longer wavelengths, making the former underrepresented in our choice of excitation range. Furthermore, the shape of the laser spectrum may also have an indirect effect on the measured distribution—in case the laser spectrum would be more intense on the red side than on the blue side, the molecules absorbing on the blue side would be even further underrepresented as they would appear even dimmer in the confocal image. This once again pinpoints the importance of the right choice of excitation wavelength, detection range and even intensity distribution of the laser spectrum. We did not find any correlation between the position of the main electronic transition band and its spectral width. In forthcoming experiments, utilizing a broader laser excitation spectrum, our approach should allow us to study possible correlations between the spectral position of the excitation spectrum and the spectral properties of the vibrational progression (separation, intensity and widths).

For completeness, we have analysed the width of the main transition band (0′-0″) for 16 out of 25 measured excitation spectra (those for which the main transition band was completely overlapping with the spectral window of the laser) by fitting a Gaussian profile. The result is shown in a form of a histogram in Fig. 4d. We found that the average width of the main electronic transition band for the measured single molecules is 35.5 nm, which is nearly 20% narrower (roughly 6 nm) than that of the ensemble absorption band (41.5 nm, indicated with vertical

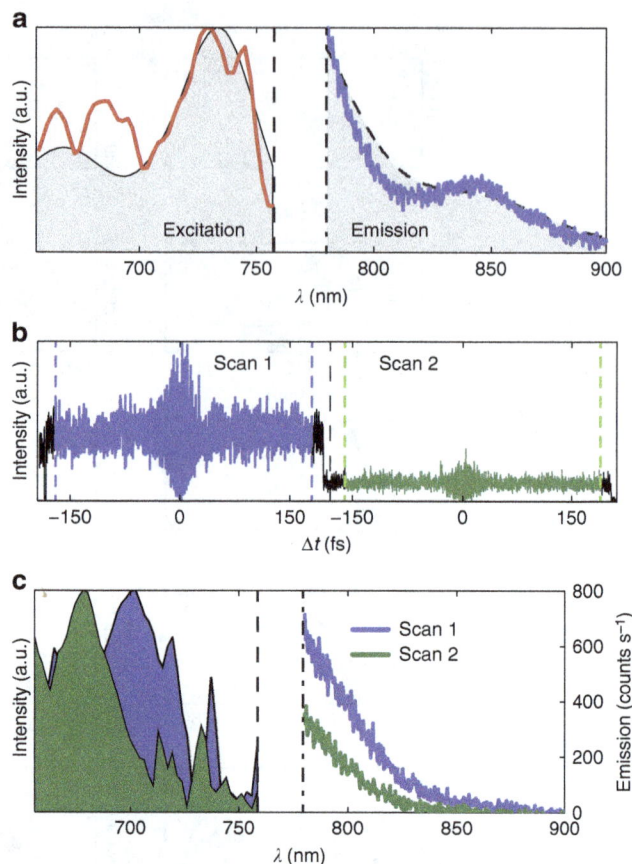

Figure 5 | Simultaneous detection of emission and excitation spectra of individual molecules. (**a**) Exemplary excitation (red) and emission (blue) spectra measured simultaneously on the same molecule. For comparison we plot the QDI solution absorption and emission spectra (shaded grey) separated by the solution Stokes shift. (**b**) Two (blue and green) consecutive fluorescence interferogram scans measured on the same molecule. Interferogram sections marked with a black line are the parts ommited in the Fourier transformation analysis. (**c**) The corresponding excitation spectra alongside the simultaneously recorded emission spectra for each (blue and green) of the two scans. Vertical dashed and dash-dotted lines in **a** and **c** indicate cutoff wavelengths of the laser and long-pass filter, respectively.

black line). The reduced width of the main electronic transition band for the single molecules clearly indicates that we are resolving the inhomogeneous broadening of the ensemble. The observed widths for individual molecules, however, might still be broadened due to spectral diffusion occurring during the acquisition time of the interferogram.

The observed variations in the excitation spectra are not because of random fluctuations in the experiment, or low signal-to-noise ratio as the variations between successively acquired excitation spectra on the same molecule are typically much smaller than differences in excitation spectra between different molecules.

Simultaneous detection of excitation and emission spectra. Finally, we modified the experimental set-up to allow simultaneous detection of excitation and emission spectra of the same molecule at room temperature. To this end we placed a 50/50 beamsplitter in the fluorescence detection path and used half of the fluorescence signal for the detection of excitation spectra with the APD and the other half for the detection of emission spectra with the electron multiplying charge-coupled device (EMCCD) camera. In Fig. 5a we show the simultaneously recorded excitation (red) and emission (blue) spectrum of an individual QDI molecule. For comparison we also plot the QDI solution absorption and emission spectra (shaded grey). We spectrally offsetted both solution spectra to match the single-molecule spectra without changing the Stokes shift separation of the solution spectra. Both solution and single-molecule spectra match each other quite well. A part of the main peak of the emission spectrum is missing due to the cutoff wavelength of the long-pass filter used to filter out the laser light. For future experiments it might be advantageous to use a long-pass filter with a sharper cutoff slope and to move the cutoff point towards the middle of the Stokes separation.

QDI molecules are generally stable in their fluorescence emission; however, on a few occasions we observed jumps in the fluorescence intensity while recording fluorescence interferograms. On even fewer occassions such intensity fluctuations occurred towards the start or end of the recorded interferogram. A nice example of such discrete jump is shown in Fig. 5b. The investigated molecule clearly emits photons at distinct fluorescence levels. When analysing the two scans separately, the excitation spectra on the left side of Fig. 5c are obtained. They clearly have the main excitation peak at distinct spectral positions, separated by ∼20 nm. The corresponding, simultaneously acquired emission spectra are plotted alongside for both scans, showing similar shape but different intensity.

As this particular molecule absorbed and emitted more towards the blue side, we could not draw any direct conclusions on the spectral shift or position of the emission spectra, other than that we only detected the slope of the second emission band. However, we were able to compare the ratio in measured total fluorescence intensity to the ratio in excitation efficiency due to the spectral shift of the molecules' absorption spectrum. Taking the integrated product spectrum (which is the molecule's excitation spectrum times the laser spectrum), we found a ratio of 1/1.4 in excitation efficiency between the two scans. The ratio of fluorescence intensity between the two scans, however, is much larger and amounts to 1/4. It is thus clear that the emission spectrum must shift along with the excitation spectrum towards the blue, this way moving further out of the detection window (determined by the long-pass filter) and accounting for the lower detected fluorescence in the interferogram scans.

Discussion
We have demonstrated that our broadband Fourier excitation approach effectively captures excitation spectra of individual molecules at room temperature. We, thus, put forward yet another way of detecting and investigating single molecules, right along with absorption, emission and scattering. The single-molecule excitation spectra allow us to resolve the intrinsic and environmentally induced inhomogeneities in the excited state potential of single molecules. These are inaccessible through emission spectroscopy (or Raman scattering techniques), which only probe the ground state potential. Furthermore, we have shown that the interplay between the excitation wavelength and detection spectral window may obscure the real extent of spectral inhomogeneity among single molecules, as indicated recently in the single-molecule experiment at cryogenic temperature[45]. One might fail to detect a significant fraction of the spectrally distributed molecules. This further corroborates the need of broadband excitation spectroscopy for the study of the single-molecule static and dynamic properties and their quantitative analysis of both emission and excitation spectroscopy.

The presented broadband excitation spectroscopy technique has the potential to contribute to the understanding of many dynamical processes such as intersystem crossing which typically

affects the excitation spectrum but not the emission and absorption spectra. Consequently, we can probe the mechanisms and spectral dependence of intra- and intermolecular intersystem crossing, which so far has mainly been investigated at cryogenic conditions through statistical analysis of the blinking dynamics[46–50] and through fluorescence-detected magnetic resonance spectroscopy[51–53]. Similarly, our approach offers new insights into single-molecule blinking dynamics. For single-molecule emission spectra acquired with a narrow excitation bandwidth, it is very challenging to differentiate between different blinking mechanisms like spectral jumps outside the excitation window and transitions to dark states. As we demonstrated, the broadband excitation spectroscopy approach allows us to follow the excitation spectrum in time and thus to verify correlations between blinking and spectral changes on the excitation side. We presented a proof-of-principal experiment, which demonstrates the feasibility of simultaneous acquisition of both the emission and excitation spectra and ambient conditions. We believe that this advancement in the single-molecule spectroscopy will allow for detailed studies on the interaction between individual molecules and their environment as well as for characterization of molecules in more complex systems. Finally, the technique will prove useful to follow slow-occurring chemical reactions in time through changes in the molecule's excitation spectrum both in solution and on the single-molecule level.

We note that our experimental approach is based on the concept of Fourier transform spectroscopy, however, with two important differences with respect to already established techniques. Firstly, commercially available FT spectrometers do not offer single-molecule sensitivity and do not allow for active control over excitation processes. As we use a coherent broadband light source we can manipulate the phase, time delay and chirp of the pulses. Transform limited pulses (in case of our set-up 12 fs) offer a time resolution that can be used to coherently probe femtosecond dynamics and (de)coherence of single molecules[54,55]. Using a laser with a sufficiently broad spectrum (exceeding the width of the absorption spectrum) it is possible to extract the dephasing time directly from the measured interferograms. Secondly, using two interfering excitation beams it should be possible to measure the excitation spectrum of non-fluorescent (single) molecules by detecting stimulated emission photons, a detection scheme which has already shown to reach nearly single-molecule sensitivity[56]. This would give access to the information normally obtained through fluorescence, but from non-fluorescent molecules.

On a technical note, the presented technique requires a broadband laser source and a simple (Mach–Zehnder) interferometer. However, it does not require any other modifications to a fluorescence detection scheme and thus is compatible with any confocal single-molecule detection set-up. It is worth noting that acquiring an excitation spectrum or an emission spectrum of a single molecule typically requires a similar number of photocounts and thus can be acquired in a comparable time. To obtain the presented excitation spectra of single molecules (with 4 nm resolution), a single interferogram was acquired for ∼120 s (see Methods). By reducing the sampling frequency and/or the data range (which decreases spectral resolution and spectral range) it is possible to record an interferogram within roughly 10 s without significantly affecting the shape of the measured excitation spectrum. The effect of reducing the resolution of the measured interferogram or its temporal range on the shape and quality of the excitation spectrum for molecule M4 (see Fig. 3), is shown in Supplementary Fig. 3. The acquisition time to obtain a reasonable fluorescence emission spectrum with the same excitation power is typically 2–10 s, depending on the brightness of the molecule. The few tens of seconds needed to measure an excitation spectrum of single molecules is comparable with the survival time of many biologically relevant fluorescing molecules. Therefore excitation spectra of molecules with low quantum effciency and photostability, like light harvesting complexes, is within reach.

Methods

Broadband excitation spectroscopy. The experimental set-up is schematically depicted in Fig. 2a. We used a broadband titanium-sapphire laser (Octavius-85M, Thorlabs) operating at 85 MHz and tuned to a central wavelength of 710 nm with a bandwidth of about 120 nm. The laser pulses were split into two parts in a Mach–Zehnder-type interferometer consisting of two identical 50/50 beamsplitters (Semrock) and a mechanical delay line (NRT 100/M, Thorlabs), which we used to control the path difference between the two arms of the interferometer. The interfering pulse pairs were propagated collinearly into an inverted microscope (Observer D1, Zeiss). A reference HeNe laser beam was propagated through the same interferometer, separated from the Ti:sapphire light using two band-pass filters (617 ± 36, 632 ± 11 nm, Semrock) and detected with a photodiode (PDA36A, Thorlabs). The derived reference interferogram was used to precisely determine the optical path difference between the interfering broadband pulse-pair. Before entering the microscope, the HeNe light was filtered out using a long-pass filter (635 nm LP, Semrock). In the microscope, the broadband Octavius pulses were reflected from a 50/50 beamsplitter and focused to a diffraction-limited spot on the sample with a high numerical aperture objective (1.3NA, $\times 100$, Zeiss Fluar). The sample was placed on a piezo-controlled stage (Mad City Labs) allowing precise positioning of the molecules in the focal spot. The fluorescence from the sample was collected in reflection through the same objective and beamsplitter and sent either to a spectrometer equipped with an EMCCD camera for spectral detection (Newton, Andor) or to an avalanche photodiode (Perkin-Elmer) that allowed confocal optical imaging of the sample. The fluorescence was separated from the laser light using two long-pass filters (780 nm LP and 785 nm LP, Semrock). The laser interferograms were detected with a diffuser placed in front of an APD.

The experiment was performed as follows: first a confocal image of the sample was recorded. A molecule was placed in the focal spot, its position was optimized based on the intensity of the fluorescence signal and the interferogram scans were recorded until the molecule photobleached. Finally, the long-pass filter was replaced with the diffuser, the molecule was translated out of the focal spot and a series of reference laser interferograms was recorded.

Detection efficiency. The laser power was set to roughly 1 µW for a single beam at the sample position, which corresponds to around 450 W cm^{-2}. Considering transmission of optical elements in the detection path we were able to detect approximately 35% of the emitted fluorescence photons. For a typical QDI molecule this yielded 2–3 kcounts s^{-1}. The delay line was scanned over 60 µm with an interpulse delay ranging from -200 to $+200$ fs. The delay line velocity was set to 0.5 µm s^{-1}, which yielded a total single interferogram acquisition time of 120 s. The acquisition time was set to 10 ms, which including software data handling time, resulted in 12–13 measurement points per oscillation period of the broadband laser (corresponding to sampling rate of $\sim 5 \times 10^{15}$ Hz or a measurement point every 0.15 fs). The resulting typical resolution in the frequency domain was 4.5 nm.

Emission spectra. For the emission spectra, single molecules were excited by the HeNe laser (at 633 nm) using excitation intensity of 1–2 µW (450–900 W cm^{-2}). The fluorescence was separated from the excitation light using a notch filter (633/25 nm) for the HeNe laser or a long-pass filter (690 nm cutoff) for the Ti:sapphire laser. Typically 20 emission spectra were recorded in a series with integration time of 11 s/spectrum. The gain of the charge-coupled device camera was set to 200. The emission spectrum of the QDI solution (10^{-6} M in 1% w/v toluene/PMMA) shown in Fig. 1b (shaded grey) was measured using HeNe excitation in combination with a 635 nm LP filter.

Absorption spectra. The absorption spectrum of QDI (10^{-6} M) in toluene/PMMA (1% w/v) solution (Fig. 1b, shaded grey) was measured using a commercial spectrophotometer (NanoDrop 2000, Thermo Scientific). The absorption spectra as a function of evaporation (shown in Fig. 1b, also see Supplementary Fig. 4) were measured using the microscope's built-in halogen lamp. Light transmitted through the sample was collected through the objective and detected with the EMCCD camera. Typically 40 µl of solution (10^{-5} M) was placed on top of a microscope cover slip and a kinetic series of 8,000 transmitted light spectra was acquired, with integration time of 1 s per spectrum.

Multicolour excitation. Confocal fluorescence images at different excitation wavelengths were recorded using a 5 nm broad excitation bands (in combination with a long-pass filter for detection): 570 nm (610 nm); 630 nm (648 nm); 680 nm (730 nm); 710 nm (740 nm); and 750 nm (778 nm); derived from an ultra-broadband laser (SuperK, NKT).

Sample preparation. The QDI molecules were obtained from the Müllen group (Max Planck Institute for Polymer Research, Mainz, Germany). The samples were prepared by spin-coating a solution of QDI molecules at a roughly nM concentration in PMMA/toluene mixture (1% w/v). Approximately 50 µl of the solution was spin-coated on a #1 microscope cover slip for 60 s, at a spinning rate of 2,000 r.p.m. Before sample deposition, microscope cover slips were cleaned by leaving them in a piranha solution (1:2 ratio hydrogen peroxide to sulfuric acid) for about 30 min, then rinsing with deionized water and blow drying with nitrogen. We found this procedure to yield no or very little contamination on the cover slips.

References

1. Dickson, R. M., Cubitt, A. B., Tsien, R. Y. & Moerner, W. E. On/off blinking and switching behaviour of single molecules of green fluorescent protein. *Nature* **388**, 355–358 (1997).
2. Basché, T. & Moerner, W. E. Optical modification of a single impurity molecule in a solid. *Nature* **355**, 335–337 (1992).
3. Zumbusch, A., Fleury, L., Brown, R., Bernard, J. & Orrit, M. Probing individual two-level systems in a polymer by correlation of single molecule fluorescence. *Phys. Rev. Lett.* **70**, 3584–3587 (1993).
4. Ambrose, W. P. & Moerner, W. E. Fluorescence spectroscopy and spectral diffusion of single impurity molecules in a crystal. *Nature* **349**, 225–227 (1991).
5. Myers, A. B., Tchénio, P., Zgierski, M. Z. & Moerner, W. E. Vibronic spectroscopy of individual molecules in solids. *J. Phys. Chem.* **98**, 10377–10390 (1994).
6. Kiraz, A., Ehrl, M., Bräuchle, C. & Zumbusch, A. Low temperature single molecule spectroscopy using vibronic excitation and dispersed fluorescence detection. *J. Chem. Phys.* **118**, 10821–10824 (2003).
7. Moerner, W. E. & Kador, L. Optical detection and spectroscopy of single molecules in a solid. *Phys. Rev. Lett.* **62**, 2535–2538 (1989).
8. Orrit, M. & Bernard, J. Single pentacene molecules detected by fluorescence excitation in a p-terphenyl crystal. *Phys. Rev. Lett.* **65**, 2716–2719 (1990).
9. Feist, F. A., Tommaseo, G. & Basché, T. Observation of very narrow linewidths in the fluorescence excitation spectra of single conjugated polymer chains at 1.2 K. *Phys. Rev. Lett.* **98**, 208301–208305 (2007).
10. Feist, F. A. & Basché, T. Fluorescence excitation and emission spectroscopy on single MEH-PPV chains at low temperature. *J. Phys. Chem. B* **112**, 9700–9708 (2008).
11. Tchénio, P., Myers, A. B. & Moerner, W. E. Vibrational analysis of the dispersed fluorescence from single molecules of terrylene in polyethylene. *Chem. Phys. Lett.* **213**, 325–332 (1993).
12. Tchénio, P., Myers, A. B. & Moerner, W. E. Dispersed fluorescence spectra of single molecules of pentacene in p-terphenyl. *J. Phys. Chem.* **97**, 2491–2493 (1993).
13. Betzig, E. & Chichester, R. J. Single molecules observed by near-field scanning optical microscopy. *Science* **262**, 1422–1425 (1993).
14. Nie, S., Chiu, D. T. & Zare, R. N. Probing individual molecules with confocal fluorescence microscopy. *Science* **266**, 1018–1021 (1994).
15. Trautman, J. K. & Macklin, J. J. Time-resolved spectroscopy of single molecules using near-field and far-field optics. *Chem. Phys.* **205**, 221–229 (1996).
16. Xie, X. S. & Trautman, J. K. Optical studies of single molecules at room temperature. *Annu. Rev. Phys. Chem.* **49**, 441–480 (1998).
17. Trautman, J. K., Macklin, J. J., Brus, L. E. & Betzig, E. Near-field spectroscopy of single molecules at room temperature. *Nature* **369**, 40–42 (1994).
18. Xie, X. S. Single molecule spectroscopy and dynamics. *Acc. Chem. Res.* **29**, 598–606 (1996).
19. Gaiduk, A., Yorulmaz, M., Ruijgrok, P. V. & Orrit, M. Room-temperature detection of a single molecule's absorption by photothermal contrast. *Science* **330**, 353–356 (2010).
20. Piliarik, M. & Sandoghdar, V. Direct optical sensing of single unlabelled proteins and super-resolution imaging of their binding sites. *Nat. Commun.* **5**, 4495 (2014).
21. Ortega-Arroyo, J. & Kukura, P. Interferometric scattering microscopy (iSCAT): new frontiers in ultrafast and ultrasensitive optical microscopy. *Phys. Chem. Chem. Phys.* **14**, 15625–15636 (2012).
22. Zijlstra, P., Paulo, P. M. R. & Orrit, M. Optical detection of single non-absorbing molecules using the surface plasmon resonance of a gold nanorod. *Nat. Nanotechnol.* **7**, 379–382 (2012).
23. Celebrano, M., Kukura, P., Renn, A. & Sandoghdar, V. Single-molecule imaging by optical absorption. *Nat. Photon.* **5**, 95–98 (2011).
24. Kukura, P., Celebrano, M., Renn, A. & Sandoghdar, V. Single-molecule sensitivity in optical absorption at room temperature. *J. Phys. Chem. Lett.* **1**, 3323–3327 (2010).
25. Chong, S., Min, W. & Xie, X. S. Ground-state depletion microscopy: detection sensitivity of single-molecule optical absorption at room temperature. *J. Phys. Chem. Lett.* **1**, 3316–3322 (2010).
26. Stopel, M. H. W., Blum, C. & Subramaniam, V. Excitation spectra and stokes shift measurements of single organic dyes at room temperature. *J. Phys. Chem. Lett.* **5**, 3259–3264 (2014).

27. Weigel, A., Sebesta, A. & Kukura, P. Shaped and feedback-controlled excitation of single molecules in the weak-field limit. *J. Phys. Chem. Lett.* **6,** 4032–4037 (2015).

28. Pfündel, E. & Baake, E. A quantitive description of fluorescence excitation spectra in intact bean leaves greened under intermittent light. *Photosynth. Res.* **26,** 19–28 (1990).

29. Matthews, B. J. H., Jones, A. C., Theodoru, N. K. & Tudhope, A. W. Excitation-emission-matrix fluorescence spectroscopy applied to humic acid bands in coral reefs. *Mar. Chem.* **55,** 317–332 (1996).

30. Juzenas, P., Juzeniene, A., Kaalhus, O., Iani, V. & Moan, J. Noninvasive fluorescence excitation spectroscopy during application of 5-aminolevulinic acid in vivo. *Photochem. Photobiol. Sci.* **1,** 745–748 (2002).

31. Hirschberg, J. G. *et al.* Interferometric measurement of fluorescence excitation spectra. *Appl. Opt.* **37,** 1953–1957 (1998).

32. Paul, R., Steiner, A. & Gemperlein, R. Spectral sensitivity of *Calliphora erythrocephala* and other insect species studied with Fourier interferometric stimulation (FIS). *J. Comp. Physiol. A* **158,** 669–680 (1986).

33. Korlacki, R., Steiner, M., Qian, H., Hartschuh, A. & Meixner, A. J. Optical fourier transform spectroscopy of a single-walled carbon nanotube and single molecules. *Chem. Phys. Chem.* **8,** 1049–1055 (2007).

34. Dantus, M. & Lozovoy, V. V. Experimental coherent laser control of physicochemical processes. *Chem. Rev.* **104,** 1813–1859 (2004).

35. Chen, L., Li, C. & Müllen, K. Beyond perylene diimides: synthesis, assembly and function of higher rylene chromophores. *J. Mater. Chem. C* **2,** 1938–1956 (2014).

36. Geerts, Y. *et al.* Quaterrylenebis(dicarboximide)s: near infrared absorbing and emitting dyes. *J. Mater. Chem.* **8,** 2357–2369 (1998).

37. Hildner, R., Brinks, D. & van Hulst, N. F. Femtosecond coherence and quantum control of single molecules at room temperature. *Nature Phys.* **7,** 172–177 (2011).

38. Hwang, J., Fejer, M. M. & Moerner, W. E. Scanning interferometric microscopy for the detection of ultrasmall phase shifts in condensed matter. *Phys. Rev. A* **73,** 021802 (2006).

39. Brinks, D. *et al.* Visualizing and controlling vibrational wave packets of single molecules. *Nature* **465,** 905–908 (2010).

40. Piatkowski, L., Gellings, E. & van Hulst, N. F. Multicolour single molecule emission and excitation spectroscopy reveals extensive spectral shifts. *Faraday Discuss.* **184,** 207–220 (2015).

41. Margineanu, A. *et al.* Photophysics of a water – soluble rylene dye: comparison with other fluorescent molecules for biological applications. *J. Phys. Chem. B* **108,** 12242–12251 (2004).

42. Hofkens, J. *et al.* Conformational rearrangements in and twisting of a single molecule. *Chem. Phys. Lett.* **333,** 255–263 (2001).

43. Blum, C., Meixner, A. J. & Subramaniam, V. Room temperature spectrally resolved single-molecule spectroscopy reveals new spectral forms and photophysical versatility of *Aequorea* green fluorescent protein variants. *Biophysical J.* **87,** 4172–4179 (2004).

44. Weston, K. D., Carson, P. J., Metiu, H. & Buratto, S. K. Room-temperature fluorescence characteristics of single dye molecules adsorbed on a glass surface. *J. Chem. Phys.* **109,** 7474–7485 (1998).

45. Kunz, R. *et al.* Single-molecule spectroscopy unmasks the lowest exciton state of the B850 assembly in LH2 from Rps. acidophila. *Biophys. J.* **106,** 2008–2016 (2014).

46. Kol'chenko, M. A., Kozankiewicz, B., Nicolet, A. & Orrit, M. Intersystem crossing mechanisms and single molecule fluorescence: terrylene in anthracene crystals. *Opt. Spectrosc.* **98,** 681–686 (2005).

47. Kozankiewicz, B. & Orrit, M. Single-molecule photophysics, from cryogenic to ambient conditions. *Chem. Soc. Rev.* **43,** 1029–1043 (2014).

48. Mais, S. *et al.* Terrylenediimide: A novel fluorophore for single-molecule spectroscopy and microscopy from 1.4 K to room temperature. *J. Phys. Chem. A* **101,** 8435–8440 (1997).

49. Kummer, S., Basché, T. & Bräuchle, C. Terrylene in p-terphenyl: a novel single crystalline system for single molecule spectroscopy at low temperatures. *Chem. Phys. Lett.* **229,** 309–316 (1994).

50. Lang, E. *et al.* Comparison of the photophysical parameters for three perylene bisimide derivatives by single-molecule spectroscopy. *Chem. Phys. Chem.* **8,** 1487–1496 (2007).

51. Brouwer, A. C. J., Groenen, E. J. J. & Schmidt, J. Detecting magnetic resonance through quantum jumps of single molecules. *Phys. Rev. Lett.* **80,** 3944–3947 (1998).

52. Brouwer, A. C. J., Köhler, J., van Oijen, A. M., Groenen, E. J. J. & Schmidt, J. Single-molecules fluorescence autocorrelation experiments on pentacene: the dependence of intersystem crossing on isotopic composition. *J. Chem. Phys.* **110,** 9151–9159 (1999).

53. Brown, R., Wrachtrup, J., Orrit, M., Bernard, J. & von Borczyskowski, C. Kinetics of optically detected magnetic resonance of single molecules. *J. Chem. Phys.* **100,** 7182–7191 (1994).

54. Scherer, N. F. *et al.* Fluorescence-detected wave packet interferometry: time resolved molecular spectroscopy with sequences of femtosecond phase-locked pulses. *J. Chem. Phys.* **95,** 1487–1512 (1991).

55. Milota, F., Sperling, J., Szöcs, V., Tortschanoff, A. & Kauffmann, H. F. Correlation of femtosecond wave packets and fluorescence interference in a conjugated polymer: towards the measurement of site homogeneous dephasing. *J. Chem. Phys.* **120,** 9870–9884 (2004).

56. Min, W. *et al.* Imaging chromophores with undetectable fluorescence by stimulated emission microscopy. *Nature* **461,** 1105–1109 (2009).

Acknowledgements

We thank Yves L. A. Rezus and Daan Brinks for critical reading of the manuscript. L.P. acknowledges financial support from the Marie-Curie International Fellowship COFUND and the ICFOnest program. E.G. acknowledges financial support from the Erasmus + program. This research was supported by the European Commission (ERC Advanced Grant 247330-NanoAntennas), Spanish MINECO (PlanNacional project FIS2012-35527, network FIS2014-55563-REDC and Severo Ochoa grant SEV2015-0522), the Catalan AGAUR (2014 SGR01540) and Fundació CELLEX (Barcelona).

Author contributions

L.P. and N.F.v.H. designed the experiment. E.G. and L.P. performed the experiments and data analysis. E.G., L.P. and N.F.v.H. wrote the manuscript.

Additional information

7

Coupled molybdenum carbide and reduced graphene oxide electrocatalysts for efficient hydrogen evolution

Ji-Sen Li[1,2,*], Yu Wang[1,*], Chun-Hui Liu[1], Shun-Li Li[1], Yu-Guang Wang[2], Long-Zhang Dong[1], Zhi-Hui Dai[1], Ya-Fei Li[1] & Ya-Qian Lan[1]

Electrochemical water splitting is one of the most economical and sustainable methods for large-scale hydrogen production. However, the development of low-cost and earth-abundant non-noble-metal catalysts for the hydrogen evolution reaction remains a challenge. Here we report a two-dimensional coupled hybrid of molybdenum carbide and reduced graphene oxide with a ternary polyoxometalate-polypyrrole/reduced graphene oxide nanocomposite as a precursor. The hybrid exhibits outstanding electrocatalytic activity for the hydrogen evolution reaction and excellent stability in acidic media, which is, to the best of our knowledge, the best among these reported non-noble-metal catalysts. Theoretical calculations on the basis of density functional theory reveal that the active sites for hydrogen evolution stem from the pyridinic nitrogens, as well as the carbon atoms, in the graphene. In a proof-of-concept trial, an electrocatalyst for hydrogen evolution is fabricated, which may open new avenues for the design of nanomaterials utilizing POMs/conducting polymer/reduced-graphene oxide nanocomposites.

[1] Jiangsu Key Laboratory of Biofunctional Materials, College of Chemistry and Materials Science, Nanjing Normal University, Nanjing 210023, China.
[2] Key Laboratory of Inorganic Chemistry in Universities of Shandong, Department of Chemistry and Chemical Engineering, Jining University, Qufu, Shandong 273155, China. * These authors contributed equally to this work. Correspondence and requests for materials should be addressed to Z.-H.D. (email: daizhihuii@njnu.edu.cn) or to Y.-Q.L. (email: yqlan@njnu.edu.cn).

To address the energy crisis and ameliorate environmental contamination, researchers have devoted considerable attention to hydrogen as promising alternative to fossil fuels. Electrochemical water splitting to produce hydrogen, or the hydrogen evolution reaction (HER), is the most economical and sustainable method for large-scale hydrogen production. Achieving this goal requires inexpensive electrocatalysts with high efficiency for the HER[1,2]. Although the best electrocatalysts are Pt or Pt-based materials, their high cost and low abundance substantially hamper their large-scale utilization[3-5]. Thus, the development of low-cost and earth-abundant non-noble-metal catalysts to replace Pt is an important and urgently needed for practical applications.

Because of their Pt-like catalytic behaviours[6], Mo-based compounds, such as Mo_2C[7-10], MoN[11-13], MoS_2 (refs 14–17), and others[18-20] have attracted substantial interest as a new class of electrocatalysts. To further enhance the HER activity, Mo-based compounds have been anchored onto conductive supports, such as carbon nanosheets (NSs)[21-23] and carbon nanotubes (CNTs)[11,24,25], which not only prevent Mo-based compounds from aggregating but also increase the dispersion of active sites. Among these conductive supports, reduced graphene oxide (RGO), particularly nitrogen (N)-doped RGO, has garnered much attention because of its excellent electron transport properties and chemical stability[26,27]. Therefore, RGO-supported Mo-based compounds appear to be highly active and stable electrocatalysts[11,25,28-30]. However, carbonization at high-reaction temperature during synthesis procedures leads to the sintering and aggregation of Mo-based-compound nanoparticles (NPs), thus reducing their number of exposed active sites and their specific surface area[8,19]. In addition, due to its strong π-stacking and hydrophobic interactions, RGO NSs usually aggregate, which hinders their practical application[31,32]. Preventing the RGO from re-stacking and the Mo-based compound NPs from aggregating during the synthesis of a porous uniform thin layer RGO-supported Mo-based electrocatalysts is critical to enhancing their catalytic performance.

We developed a new approach to integrate polyoxometalates (POMs) and pyrrole (Py) on graphene substrates via a "one-pot" method to obtain ternary POMs–polypyrrole/RGO (POMs–PPy/RGO) nanohybrid sheets with a uniform distribution. As an important family of transition-metal oxide clusters with excellent redox features[33,34], POMs provided an essential oxidizing medium for the oxidative polymerization of Py[35], and the POMs finally were converted into "heteropoly blue"[36]. Heteropoly blue can be used as a highly localized reducing agent and can further react with graphene oxide (GO) to restore the original POMs. With the polymerization of the Py monomers, POMs were dispersed into the PPy framework. Meanwhile, RGO was homogeneously dispersed and segregated by both the POMs and PPy during the synthesis of POMs–PPy/RGO. Thus, RGO-supported Mo-based

catalysts prepared with POMs–PPy/RGO as a precursor may efficiently hinder the Mo sources and graphene from aggregating during the process of forming the RGO-supported NPs. To the best of our knowledge, reports on POMs, PPy and RGO ternary hybrids by a green and one-pot redox relay reaction are rare. More importantly, the coupled hybrid with both Mo_2C and RGO has not been previously prepared with a ternary hybrid as the precursor.

In this work, we carefully design and fabricate a two-dimensional (2D) coupled hybrid consisting of Mo_2C encapsulated by N, phosphorus (P)-codoped carbon shells and N, P-codoped RGO (denoted as $Mo_2C@NPC/NPRGO$) using a PMo_{12} ($H_3PMo_{12}O_{40}$)–PPy/RGO nanocomposite as the precursor. Notably, the entire polymerization and the reductive reactions are triggered by PMo_{12} without any additional oxidants or reductants, leading to a synthetic process that is green, efficient and economical. PPy was used as both the carbon and nitrogen sources as well as the reducing agent for GO. Three main advantages of this method are attributed to the $Mo_2C@NPC/NPRGO$ hybrid: (1) due to the unique structure of PMo_{12}–PPy/RGO, the Mo_2C NPs are nanosized and uniformly embedded in the carbon matrix without aggregation; (2) the Mo_2C NPs are coated with carbon shells, which effectively prevent Mo_2C NPs from aggregating or oxidizing and impart them with fast electron transfer ability; and (3) owing to the heteroatom dopants (N, P), a large number of active sites are exposed. Overall, taking advantage of the synergistic catalytic effects, the $Mo_2C@NPC/NPRGO$ catalyst exhibits excellent electrocatalytic activity for the HER, with a low onset overpotential of 0 mV (vs reversible hydrogen electrode (RHE)), a small Tafel slope of 33.6 mV dec^{-1}, and excellent stability in acidic media. Its HER catalytic activity, which is comparable to that of commercial Pt–C catalyst, even superior to those of the best reported non-noble-metal catalysts. In addition, we further investigate the nature of catalytically active sites for the HER using density functional theory (DFT). This approach provides a perspective for designing 2D nanohybrids with transition-metal carbides and RGO as HER catalysts.

Results

Catalyst synthesis and characterization. $Mo_2C@NPC/NPRGO$ was synthesized as follows: (1) the PMo_{12}–PPy/RGO nanocomposite was synthesized via a green one-pot redox relay reaction. The nanocomposite was then carbonized under a flow of ultrapure N_2 at 900 °C for 2 h at a heating rate of 5 °C min^{-1}. Finally, the obtained samples were acid etched in 0.5 M H_2SO_4 for 24 h with continuous agitation at 80 °C to remove unstable and inactive species. The etched samples were thoroughly washed with de-ionized water until the pH of the wash water was neutral (Fig. 1).

Figure 2a shows a scanning electron microscopy (SEM) image of PMo_{12}–PPy/RGO. The rough surfaces and wrinkled edges on the sheet-like structures were due to the intercalation and

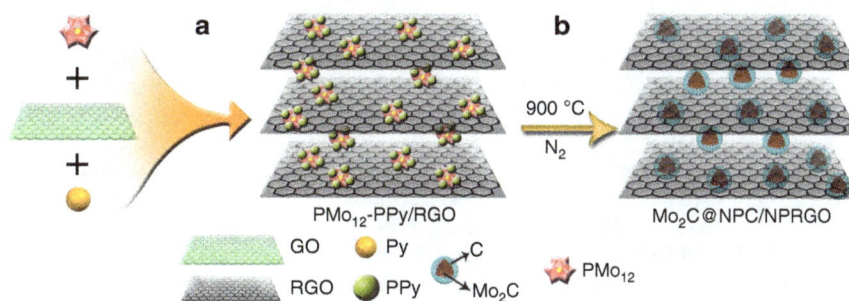

Figure 1 | Schematic illustration of the synthetic process of Mo₂C@NPC/NPRGO. (a) Synthesis of PMo_{12}–PPy/RGO via a green one-pot redox relay reaction. (b) Formation of $Mo_2C@NPC/NPRGO$ after carbonizing at 900 °C.

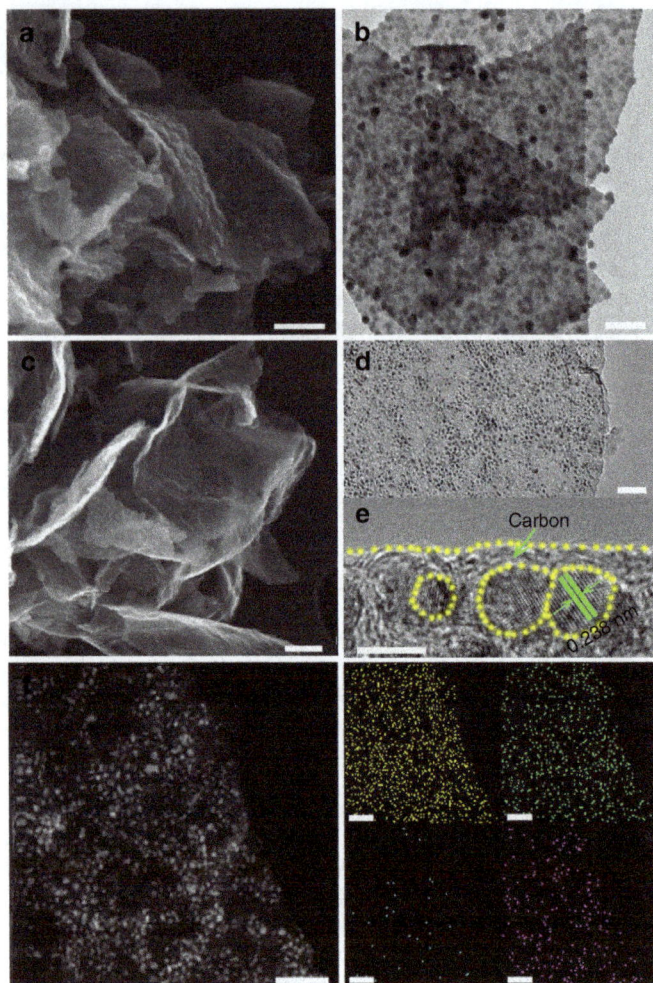

**Figure 2 | Characterization of the PMo₁₂–PPy/RGO and Mo₂C@NPC/
NPRGO hybrids. (a)** SEM and **(b)** TEM images of PMo₁₂–PPy/RGO.
(c) SEM, **(d)** TEM, **(e)** HRTEM and **(f)** STEM images and EDX elemental
mapping of C, N, P and Mo of Mo₂C@NPC/NPRGO. Scale bar: **a,b,c**
(200 nm); **d** (100 nm); **e** (5 nm) and **f** (50 nm).

polymerization of Py. A transmission electron microscopy (TEM)
image of PMo₁₂–PPy/RGO revealed that a large amount of
PPy/PMo₁₂ NPs were homogeneously coated onto the RGO NSs
and that voids were present (Fig. 2b). As evident in Fig. 2c and d,
the morphologies of Mo₂C@NPC/NPRGO were similar to that of
PMo₁₂–PPy/RGO after carbonization. The nanosized Mo₂C NPs
with diameters of ∼2–5 nm were uniformly decorated on the
RGO sheets at a high density, which was attributed to the distinct
porous structure of PMo₁₂–PPy/RGO. The high-resolution TEM
(HRTEM) image exhibited clear lattice fringes with an interplanar
distance of 0.238 nm, corresponding to the (111) planes of
Mo₂C (Fig. 2e)[37]. Notably, the Mo₂C NPs were embedded in
the carbon shells, which can efficiently prevent the aggregation
and/or excessive growth of Mo₂C NPs[22]. Figure 2f shows the
scanning TEM (STEM) and corresponding energy dispersive
X-ray spectroscopy (EDX) elemental mapping images, which
confirmed that C, Mo, N and P were distributed on the
Mo₂C@NPC/NPRGO surface, consistent with the EDX
spectrum (Supplementary Fig. 1). These results confirm the
successful synthesis of the Mo₂C@NPC/NPRGO nanocomposite.

For comparison, the nanohybrid of Mo₂C encapsulated by N,
P-codoped carbon (defined as Mo₂C@NPC) was also synthesized
through a similar preparation procedure without GO.
Supplementary Fig. 2a shows aggregation of PPy/PMo₁₂ NPs.

Supplementary Fig. 2b and c reveals that Mo₂C NPs tended to
agglomerate during the heat treatment to form large NPs, which
decreased the exposed active surface. Supplementary Fig. 2d
demonstrates the STEM and EDX elemental mapping images of
Mo₂C@NPC. These data verified that the Mo₂C@NPC material
contained C, N, P and Mo elements, consistent with the EDX
results (Supplementary Fig. 1b). Hence, these results sufficiently
confirm that the presence of GO plays an important role in the
generation of highly dispersed and nanosized Mo₂C NPs.

Supplementary Fig. 3 shows the powder X-ray diffraction
patterns of Mo₂C@NPC and Mo₂C@NPC/NPRGO. The broad
peak at ∼25° was ascribed to carbon[38,39]. The other peaks located
at 37.9, 43.7, 61.6 and 75.6° were indexed to the (111), (200), (220)
and (311) planes of Mo₂C (JCPDS, No. 15-0457), respectively;
these peaks were broad and exhibited low intensity because of the
smaller crystallites of Mo₂C or Mo₂C coated with amorphous
carbon[21,40,41]. Beside, the degrees of graphitization of the two
catalysts were analyzed by Raman spectra (Supplementary Fig. 4).
As is well-known, the ratio between the D (1,350 cm⁻¹) and G
band (1,580 cm⁻¹) intensities (I_D/I_G) is an important criterion to
judge the degree of the graphitization[9,28]. Compared to
Mo₂C@NPC, the I_D/I_G of Mo₂C@NPC/NPRGO is higher,
implying that more defects formed on the RGO sheets, thus
favoring the accessibility of more active sites and enhancing the
electrocatalytic performance. The Brunauer–Emmett–Teller (BET)
surface areas of Mo₂C@NPC and Mo₂C@NPC/NPRGO calculated
by the N₂ sorption isotherms were 55 and 190 m² g⁻¹, respectively
(Supplementary Fig. 5a). Mo₂C@NPC showed a microporous
structure, with pore sizes mainly in the range from 1 to 2 nm
(Supplementary Fig. 5b), whereas the corresponding pore size
distribution of Mo₂C@NPC/NPRGO was mainly concentrated in
the range from 1 to 10 nm, which was characteristic of a
microporous and mesoporous structure (Supplementary Fig. 5c).
Overall, the large surface area and enriched porous structures
efficiently facilitate electrolyte penetration and charge transfer[9].

X-ray photoelectron spectroscopy (XPS) analyses of
Mo₂C@NPC/NPRGO catalysts were carried out to elucidate their
valence states and compositions. As observed, the XPS spectrum
of Mo₂C@NPC/NPRGO (Supplementary Fig. 6) indicated the
presence of C, N, O, P and Mo in the catalyst. The deconvoluted
C1s spectrum is shown in Fig. 3a, and the main peak at 284.6 eV
implies that the graphite carbon is the majority species[22]. The
deconvolution of N1s energy level signals for Mo₂C@NPC/
NPRGO revealed the peaks at 398.6 and 401.3 eV, which were
assigned to pyridinic and graphitic N (Fig. 3b), respectively[21,27].
From Fig. 3c, it can be seen that the P2p peaks at about 133.5, and
134.8 eV were attributed to P–C and P–O bonding,
respectively[18,28]. Besides, the high-resolution Mo 3d XPS
revealed that the peak at 228.8 eV was attributable to Mo²⁺,
stemming from Mo₂C. In parallel, as a consequence of surface
oxidation, the peaks at 232.05 and 235.2 were attributable to
MoO₃ and that at 232.7 eV was assignable to MoO₂ (refs 8,21);
both of these species are inactive toward the HER (Fig. 3d). For
comparison, Mo₂C@NPC is shown in Supplementary Fig. 7. All
of these data were similar to those for Mo₂C@NPC/NPRGO. The
corresponding atomic percentages of the different catalysts
measured by XPS are listed in Supplementary Table 1.

Electrocatalytic HER performance. A three-electrode system was
adopted to evaluate the electrocatalytic activities of Mo₂C@NPC/
NPRGO toward the HER in 0.5 M H₂SO₄ at 100 mV s⁻¹. For
comparison, Mo₂C@NPC and commercial Pt–C (20 wt% Pt on
carbon black from Johnson Matthey) were also assessed. The
corresponding polarization curves without IR compensation are
shown in Fig. 4a. All potentials in this work are reported vs RHE.
As expected, the commercial Pt–C displayed the highest

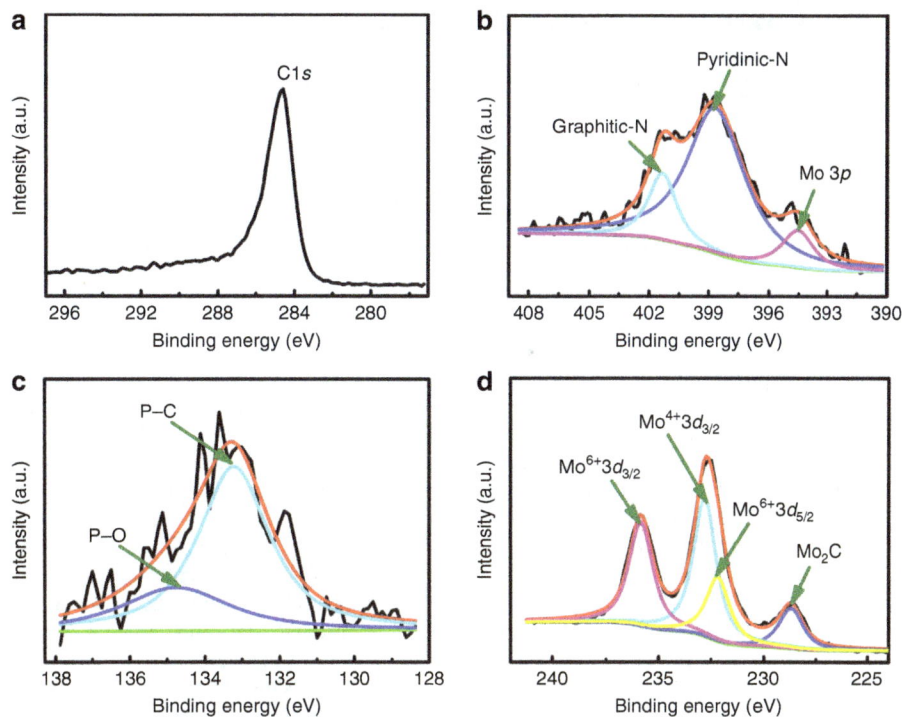

Figure 3 | Compositional characterization of the Mo₂C@NPC/NPRGO. XPS high-resolution scans of (**a**) C 1s, (**b**) N 1s, (**c**) P 2p and (**d**) Mo 3d electrons of Mo₂C@NPC/NPRGO.

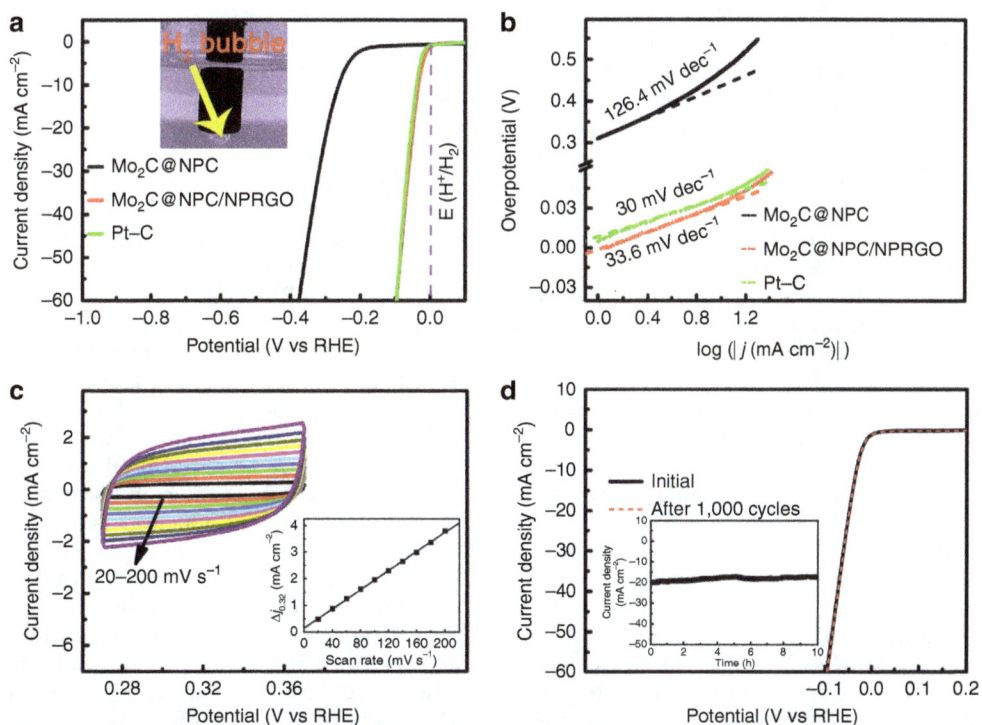

Figure 4 | HER activity characterization. (**a,b**) Polarization curves and Tafe plots of Mo₂C@NPC, Mo₂C@NPC/NPRGO and Pt–C. (inset: the production of H₂ bubbles on the surface of Mo₂C@NPC/NPRGO). (**c**) CVs of Mo₂C@NPC/NPRGO with different rates from 20 to 200 mV s⁻¹. Inset: The capacitive current at 0.32 V as a function of scan rate for Mo₂C@NPC/NPRGO. (**d**) Polarization curves of Mo₂C@NPC/NPRGO initially and after 1,000 CV cycles. Inset: Time-dependent current density curve of Mo₂C@NPC/NPRGO under a static overpotential of 48 mV for 10 h.

electrocatalytic activity, with an onset overpotential of nearly zero[30]. The Mo₂C@NPC catalyst exhibited far inferior HER activity. Impressively, Mo₂C@NPC/NPRGO exhibited the lowest onset overpotential of 0 mV, approaching the performance of

commercial Pt–C. Moreover, it was clearly observed that the cathodic current rose sharply with more negative potentials. Generally, the potential value for a current density of 10 mA cm⁻² is an important reference because solar-light-

Table 1 | Comparison of catalytic parameters of different HER catalysts.

Catalyst	Onset potential (mV vs RHE)	Overpotential at 10 mA cm^{-2} (mV vs RHE)	j_0 (mA cm^{-2})	Tafel slope (mV dec^{-1})
Mo$_2$C@NPC	137	260	3.16×10^{-3}	126.4
Mo$_2$C@NPC/NPRGO	0	34	1.09	33.6
Pt–C	0	40	0.39	30

HER, H$_2$ evolution reaction; Pt–C, 20 wt% Pt on carbon lack from Johnson-Matthey; RHE, reversible hydrogen electrode.
j_0 represents exchange current density that was calculated from Tafel curves using extrapolation method.

coupled HER apparatuses usually operate at 10–20 mA cm^{-2} under standard conditions (1 sun, AM 1.5)[4]. To achieve this current density, Mo$_2$C@NPC requires an overpotential of 260 mV. Strikingly, Mo$_2$C@NPC/NPRGO required only ~34 mV to achieve a 10 mA cm^{-2} current density, even superior to commercial Pt–C (40 mV) (Table 1). To the best of our knowledge, this overpotential is superior to those of all previously reported non-noble-metal electrocatalysts for the HER, such as MoS$_2$/CoSe$_2$ (ref. 15), MoO$_2$ (ref. 18), Mo$_2$C/CNT[24] and CoNi@NC[40] (Supplementary Table 2).

To elucidate the HER mechanism, Tafel Plots were fitted to Tafel equation (that is, $\eta = b \log(j) + a$, where b is the Tafel slope, and j is the current density), as shown Fig. 4b. The Tafel slope of commercial Pt–C was ~30 mV dec^{-1}, which was in agreement with the reported value, thus supporting the validity of our electrochemical measurements[30]. The Tafel slope of Mo$_2$C@NPC/NPRGO was 33.6 mV dec^{-1}, which indicated higher performance than that of Mo$_2$C@NPC (126.4 mV dec^{-1}). Meanwhile, the Tafel slope of Mo$_2$C@NPC/NPRGO suggested that hydrogen evolution on the Mo$_2$C@NPC/NPRGO electrode probably proceeds via a Volmer–Tafel mechanism, where the recombination is the rate-limiting step[17]. The exchange current density (j_0) was extrapolated from the Tafel plots. Notably, Mo$_2$C@NPC/NPRGO displayed the largest j_0 of 1.9×10^{-3} A cm^{-2}, which was nearly three times larger than the j_0 of Pt–C (0.39×10^{-3} A cm^{-2}) (Table 1) and was substantially greater than those of other recently reported non-noble-metal catalysts (Supplementary Table 2). This performance of Mo$_2$C@NPC/NPRGO demonstrates favourable HER kinetics at the Mo$_2$C@NPC/NPRGO/electrolyte interface.

The electrochemical double-layer capacitance (EDLC, C_{dll}) was measured to investigate the electrochemically active surface area. Cyclic voltammetry (CV) was performed in the region from 0.27 to 0.37 V at rates varying from 20 to 200 mV s^{-1} (Fig. 4c and Supplementary Fig. 8). The C_{dll} of Mo$_2$C@NPC/NPRGO (17.9 mF cm^{-2}) was ~195 times larger than that of Mo$_2$C@NPC (0.092 mF cm^{-2}). Thus, the large j_0 value of Mo$_2$C@NPC/NPRGO may benefit from both its large BET surface area and its large EDLC.

To gain further insight into the electrocatalytic activity of Mo$_2$C@NPC/NPRGO for the HER, we performed electrochemical impedance spectroscopy (EIS). The Nyquist plots of the EIS responses are shown in Supplementary Fig. 9. Compared with the Nyquist plot of Mo$_2$C@NPC, that of Mo$_2$C@NPC/NPRGO showed a much smaller semicircle, suggesting that Mo$_2$C@NPC/NPRGO has lower impedance. This result proves that the catalyst affords markedly faster HER kinetics due to the presence of the RGO support.

Long-term stability is also critical for HER catalysts. To probe the durability of the Mo$_2$C@NPC/NPRGO catalyst, continuous CV was performed between −0.2 and 0.2 V at a 100 mV s^{-1} scan rate in 0.5 M H$_2$SO$_4$ solution (Fig. 4d). As observed, the polarization curve for Mo$_2$C@NPC/NPRGO remained almost the same after 1,000 cycles. In addition, the durability of Mo$_2$C@NPC/NPRGO was also examined by electrolysis at a static overpotential of 48 mV. The inset of Fig. 4d shows that the

current density experienced a negligible loss at ~20 mA cm^{-2} for 10 h. For comparison, the durability of the Mo$_2$C@NPC catalyst was examined by the same methods (Supplementary Fig. 10). This is reconfirming that Mo$_2$C@NPC and Mo$_2$C@NPC/NPRGO are stable electrocatalysts in acidic solutions.

In control experiments, we investigated the effect of the PMo$_{12}$ content on electrocatalytic performance. Two other catalysts with different PMo$_{12}$ contents (1.1 and 3.3 g) were synthesized (denoted as Mo$_2$C@NPC/NPRGO-1.1 and Mo$_2$C@NPC/NPRGO-3.3). The morphology, structure and composition of these two catalysts were studied by SEM, TEM, HRTEM, STEM, EDX, elemental mapping, powder X-ray diffraction patterns and XPS in detail (Supplementary Figs 11–16). The HER activities of Mo$_2$C@NPC/NPRGO-1.1 and -3.3 were also evaluated using the same measurements. As seen from Fig. 5a,b, Mo$_2$C@NPC/NPRGO showed the lowest onset overpotential and the smallest Tafel slope among the three samples. We speculate that these results are likely related to the amount and distribution of active sites. Because of the lower amount of Mo$_2$C NPs in Mo$_2$C@NPC/NPRGO-1.1, the corresponding electrocatalytic activity was poorer than that of Mo$_2$C@NPC/NPRGO. In contrast, a larger number of Mo$_2$C NPs in Mo$_2$C@NPC/NPRGO-3.3 aggregated together, which is also unfavourable for the HER. These results demonstrate that the amount of PMo$_{12}$ substantially influences the HER performance.

We subsequently studied the influence of carbonization temperature under the given conditions. Supplementary Figs 17–22 show the morphology, structure and composition of the two samples carbonized at 700 and 1,100 °C (defined as PMo$_{12}$–PPy/RGO-700 and Mo$_2$C@NPC/NPRGO-1100), respectively. The onset overpotentials of PMo$_{12}$–PPy/RGO-700 and Mo$_2$C@NPC/NPRGO-1100 were 20 and 27 mV, respectively, and the Tafel slopes were 48.4 and 70.1 mV dec^{-1}, respectively (Fig. 5c and d). Among these catalysts, the Mo$_2$C@NPC/NPRGO catalyst exhibited the optimal HER activity, possibly because active sites of Mo$_2$C were not produced when PMo$_{12}$–PPy/RGO is carbonized at 700 °C; the high–carbonization temperature led to substantial sintering and aggregation of Mo$_2$C NPs, which further reduced the density of highly active sites. Meanwhile, the N content decreased with increasing carbonization temperature (Supplementary Table 1). All of these results were consistent with the SEM, TEM, XRD, thermogravimetric analysis and XPS results (Supplementary Figs 17–22). Therefore, in this work, the selection of the correct PMo$_{12}$ content and carbonization temperature was critical to forming high-HER active sites.

Theoretical investigation. The aforementioned experimental results demonstrated that the Mo$_2$C@NPC/NPRGO composite exhibits excellent electrocatalytic activity toward the HER because of the synergistic effects of Mo$_2$C and NPC/NPRGO. To elucidate the mechanism underlying the superior HER activity of the Mo$_2$C@NPC/NPRGO composite, we performed a series of DFT calculations (Supplementary Fig. 23 and Supplementary Table 3). Theoretically, the HER pathway can be depicted as a three-state

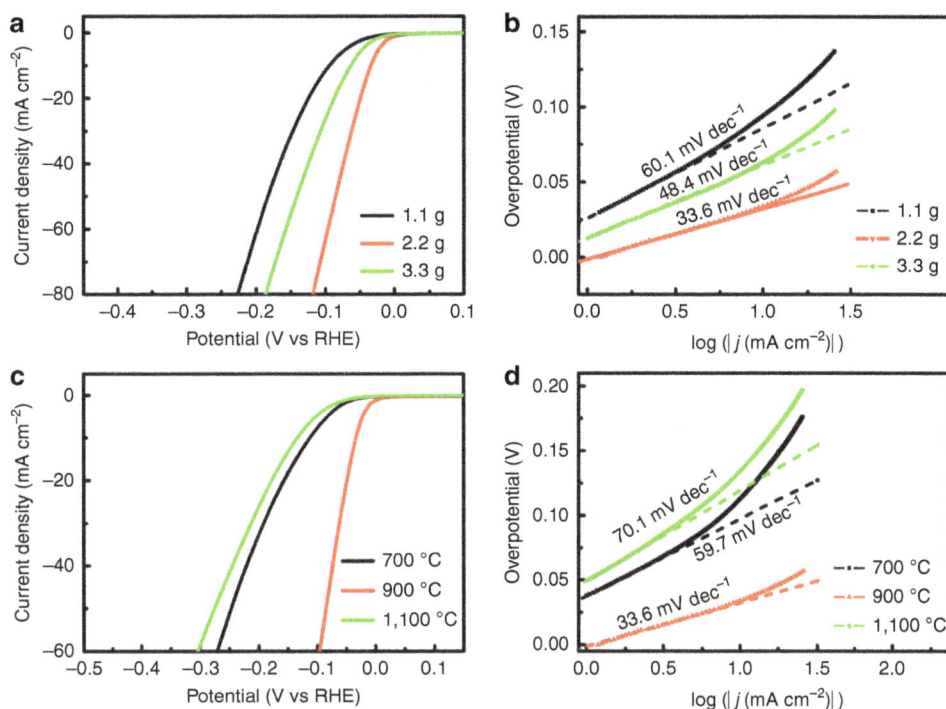

Figure 5 | Comparison of the HER performance of different electrocatalysts. (**a,b**) Polarization curves and Tafe plots of Mo$_2$C@NPC/NPRGO with different mass of PMo$_{12}$ (1.1, 2.2 and 3.3 g). (**c,d**) Polarization curves and Tafe plots of Mo$_2$C@NPC/NPRGO (2.2 g) at different carbonization temperature.

diagram containing an initial state of $H^+ + e^-$, an intermediate state of adsorbed H (H*, where * denotes an adsorption site), and a final state of 1/2 the H$_2$ product[5,22]. Generally, a good hydrogen evolution catalyst should have a free energy of adsorbed H of approximately zero ($\Delta G_{H*} \approx 0$), which can provide a fast proton/electron-transfer step as well as a fast hydrogen release process[42]. Because only trace amounts of P were present in the Mo$_2$C@NPC/NPRGO hybrid compared to the N content, we investigated only the effect of N doping (graphitic N and pyridinic N) on the catalytic effect of the hybrids. Figure 6 shows the calculated free energy diagram for the HER in various studied systems.

According to our computational results, pristine graphene had an endothermic ΔG_{H*} of 1.82 eV, implying an energetically unfavourable interaction with hydrogen. Therefore, the HER can barely proceed on pristine graphene because of the slow proton/electron transfer. On the other hand, the (001) surface of Mo$_2$C had a strong interaction with H, as indicated by the exothermic ΔG_{H*} of − 0.82 eV, which would subsequently lead to poor HER performance because of the foreseeable difficulty of hydrogen release. Moreover, N-doped graphene exhibited low catalytic activity toward the HER. Specifically, the ΔG_{H*} values for graphitic-N- and pyridinic-N-doped graphene were 0.89 and − 2.04 eV, respectively.

However, the catalytic activity of graphene and N-doped graphene were substantially improved when they were anchored to the surface of Mo$_2$C. For example, the ΔG_{H*} values for Mo$_2$C@C and Mo$_2$C@C-graphitic N were 0.41 and 0.69 eV, respectively, which were much lower than those of suspended graphene (1.82 eV) and N-doped graphene (0.89 eV). The ΔG_{H*} of Mo$_2$C@C (0.41 eV) indicated that the graphene C atoms in the hybrid also play an important role in the HER activity. In particular, due to the synergistic effect between Mo$_2$C and C-pyridinic N, Mo$_2$C@C-pyridinic N had a favourable ΔG_{H*} (− 0.22 eV) for the adsorption and desorption of hydrogen. Therefore, the active sites for the HER should be composed mainly of pyridinic N atoms and C atoms of graphene rather than

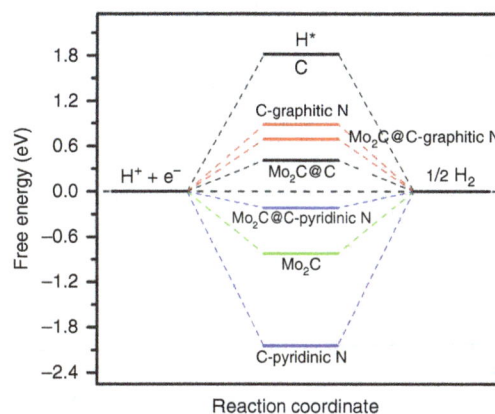

Figure 6 | DFT-calculated HER activities. Calculated free energy diagram for HER on various studied system.

graphitic N atoms. We note here that, according to the results of XPS analysis, the major type of N in Mo$_2$C@NPC/NPRGO was pyridinic N, which means that Mo$_2$C@NPC/NPRGO would manifest a high density of active sites and would consequently present a high-current density at a low overpotential for the HER. Overall, the experimental and theoretical results verified that as-synthesized Mo$_2$C@NPC/NPRGO is an unexpected and highly efficient HER electrocatalyst.

Discussion

In view of the aforementioned considerations, the amazing HER activities of the Mo$_2$C@NPC/NPRGO are postulated to originate from the following reasons: (1) the small size of Mo$_2$C NPs favors the exposure of an abundance of available active sites, which may enhance the catalytic activity for the HER[7,21,28]; (2) the introduction of heteroatoms (N, P) into the carbon structure results in charge density distribution and asymmetry spin, thus enhancing the interaction with H^+ (refs 18,27). Especially, pyridinic N is favourable for highly efficient catalytic performance[43,44]; (3) as an advanced support, RGO can increase the dispersion of PMo_{12} to further obtain highly dispersed Mo_2C during the carbonization process. Meanwhile, the outstanding electrical conductivity of RGO facilitates charge transfer in the catalyst[11,25]; (4) the robust conjugation between Mo_2C and NPC/NPRGO provides a resistance-less path favourable for fast electron transfer. The carbon shells may hamper the aggregation of Mo_2C NPs[21] and promote electron penetration from Mo_2C to RGO[22]. Furthermore, the geometric confinement of Mo_2C inside the carbon shells can also enhance the catalytic activity for the HER[40] and (5) the unique structure of $Mo_2C@NPC/NPRGO$ is favourable for the fast mass transport of reactants and facilitates electron transfer[26,39]. Because of the synergistic catalytic effects of the aforementioned factors, the $Mo_2C@NPC/NPRGO$ catalyst exhibits potent HER activity.

In summary, we designed and developed a novel architecture that is composed of Mo_2C NPs, NPC and NPRGO by simply carbonizing a ternary PMo_{12}–PPy/RGO nanocomposite. The effect of the PMo_{12} content and carbonization temperature on the HER activity was investigated in detail. The RGO-supported Mo-based catalysts prepared with PMo_{12}–PPy/RGO as the precursor may efficiently hinder Mo sources and graphene from aggregating during the formation of RGO-supported Mo_2C NPs. The obtained $Mo_2C@NPC/NPRGO$ nanocomposite exhibits the best HER performance and high stability as an electrocatalyst in an acidic electrolyte reported to date. Theoretical studies demonstrated that the synergistic effect between Mo_2C and C-pyridinic N contributes to the excellent HER activity of the $Mo_2C@NPC/NPRGO$ nanocomposite, in accordance with the experimental results. This proof-of-concept study not only offers novel hydrogen-evolving electrocatalysts with excellent activity but also opens new avenues for the development of other 2D coupled nanohybrids with transition-metal carbides and RGO using POMs/conducting polymer/RGO as a precursor. These catalysts can also be explored as highly efficient electrocatalysts for oxygen reduction reaction (ORR), HER and lithium batteries.

Methods

Synthesis of PMo$_{12}$–PPy/RGO and Mo$_2$C@NPC/NPRGO hybrids. In a typical synthesis, GO NSs were pre-synthesized by chemical oxidation exfoliation of natural graphite flakes using a modified Hummers method[45]. The obtained GO NSs were dispersed in de-ionized water by ultrasonication to form a suspension with the concentration of 1 mg ml^{-1}. Around 12.5 ml of such GO suspension and 150 ml of 2 mM PMo_{12} solution were added into a clean three-necked flask, respectively, and mixed uniformly under a strong ultrasonication bath. Subsequently, Py monomer solution by dispersing 230 μl of Py in 15 ml de-ionized water, was slowly dropped into the above mixed PMo_{12}/GO suspension. With the addition of Py monomer solution, the reaction system gradually turned from yellow-brown to deep blue and a black precipitate began to generate after about 5 min. Finally, the reactor was transferred to an oil bath and allowed to react for 30 h at 50 °C under vigorously magnetic stirring. After separated by centrifugation and washed with deionized water and anhydrous ethanol for several times, the black PMo_{12}–PPy/RGO ternary nanohybrids were obtained, which were dried in vacuum at 50 °C. In control experiments, PMo_{12}–PPy/RGO (1.1) and PMo_{12}–PPy/RGO (3.3) were synthesized by identical condition except that the amount of PMo_{12} is 1.1 and 3.3 g, respectively.

To prepare the $Mo_2C@NPC/NPRGO$ nanocomposite, 2 g of PMo_{12}–PPy/RGO was carbonized in a flow of ultrapure N_2 at 900 °C for 2 h with the heating rate of 5 °C min^{-1}. The obtained samples were acid etched in H_2SO_4 (0.5 M) for 24 h with continuous agitation at 80 °C to remove unstable and inactive species. The etched samples were then thoroughly washed with de-ionized water until reaching a neutral pH, and defined as $Mo_2C@NPC/NPRGO$, $Mo_2C@NPC/NPRGO$-1.1 and -3.3, respectively.

Synthesis of PMo$_{12}$–PPy and Mo$_2$C@NPC composites. The synthetic procedure is very similar to PMo_{12}–PPy/RGO without GO. Likewise, the preparation of $Mo_2C@NPC$ composite is identical with that of $Mo_2C@NPC/NPRGO$.

Characterization. The TEM and HRTEM images were recorded on JEOL-2100F apparatus at an accelerating voltage of 200 kV. Surface morphologies of the carbon materials were examined by a SEM (JSM-7600F) at an acceleration voltage of 10 kV. The EDX was taken on JSM-5160LV-Vantage-typed energy spectrometer. The XRD patterns were recorded on a D/max 2500VL/PC diffractometer (Japan) equipped with graphite monochromatized Cu Kα radiation ($\lambda = 1.54060$ Å). Corresponding work voltage and current is 40 kV and 100 mA, respectively. XPS was recorded by a scanning X-ray microprobe (PHI 5000 Verasa, ULAC-PHI, Inc.) using Al Kα radiation and the C1s peak at 284.8 eV as internal standard. The Raman spectra of dried samples were obtained on Lab-RAM HR800 with excitation by an argon ion laser (514.5 nm). The nitrogen adsorption–desorption experiments were operated at 77 K on a Micromeritics ASAP 2050 system. BET surface areas were determined over a relative pressure range of 0.05–0.3, during which the BET plot is linear. The pore size distributions were measured by using the nonlocalized density functional theory method. Before the measurement, the samples were degassed at 150 °C for 10 h.

Electrochemical measurements. All electrochemical experiments were conducted on a CHI 760D electrochemical station (Shanghai Chenhua Co., China) in a standard three electrode cell in 0.5 M H_2SO_4 at room temperature. A glassy carbon electrode (3 mm in diameter), an Ag/AgCl with saturated KCl, and a Pt wire were used as the working electrode, reference and counter electrode, respectively. A total of 4 mg of the catalysts were dispersed in 2 ml of 9:1 v/v water/Nafion by sonication to form a homogeneous ink. Typically, 5 μl well-dispersed catalysts were covered on the glassy carbon electrode and then dried in an ambient environment for measurements. The electrocatalyst was prepared with a catalyst loading of 0.14 mg cm^{-2}. Commercial 20% Pt–C catalyst was also used as a reference sample. Linear sweep voltammetry was tested with a scan rate of 5 mV s^{-1}. EIS measurements were carried out from 1,000 kHz to 100 mHz with an amplitude of 10 mV at the open-circuit voltage. The electrochemical stability of the catalyst was conducted by cycling the potential between -0.3 and 0.3 V vs RHE at a scan rate of 100 mV s^{-1}. The Chronoamperometry were tested at an overpotential of -0.12 V vs RHE after equilibrium. To estimate the electrochemical active surface areas of the catalysts, CV was tested by measuring EDLC under the potential window of 0.19–0.39 vs RHE with various scan rate (20, 40, 60, 80, 100, 120, 140, 160, 180 and 200 mV s^{-1}). A flow of N_2 was maintained over the electrolyte during the experiment to eliminate dissolved oxygen. The potential vs RHE was converted to RHE via the Nernst equation: $E_{RHE} = E_{Ag/AgCl} + 0.059pH + E^\theta_{Ag/AgCl}$. In 0.5 M H_2SO_4, $E_{RHE} = 0.21$ V $+ E_{Ag/AgCl}$.

Computational details. DFT calculations were performed using the plane-wave technique implemented in the Vienna ab initio Simulation package[46]. The ion–electron interaction was treated within the projector-augmented plane wave pseudopotentials[47,48]. The generalized gradient approximation expressed by Perdew − Burke − Ernzerhof functional[49] and a plane-wave cutoff energy of 360 eV were used in all computations. The electronic structure calculations were employed with a Fermi-level smearing of 0.1 eV for all surface calculations and 0.01 eV for all gas-phase species. The Brillouin zone was sampled with $3 \times 3 \times 1$ k-points. The convergence of energy and forces were set to 1×10^{-5} eV and 0.02 eV Å^{-1}, respectively. A vacuum region of around 12 Å was set along the z direction to avoid the interaction between periodic images. More computational details are provided in Supplementary Note 1.

References

1. Turner, J. A. Sustainable hydrogen production. *Science* **305**, 972–974 (2004).
2. Wang, J. *et al.* Recent progress in cobalt-based heterogeneous catalysts for electrochemical water splitting. *Adv. Mater.* **28**, 215–230 (2016).
3. Stamenkovic, V. R. *et al.* Trends in electrocatalysis on extended and nanoscale Pt-bimetallic alloy surfaces. *Nat. Mater.* **6**, 241–247 (2007).
4. Walter, M. G. *et al.* Solar water splitting cells. *Chem. Rev.* **110**, 6446–6473 (2010).
5. Zheng, Y. *et al.* Hydrogen evolution by a metal-free electrocatalyst. *Nat. Commun.* **5**, 3783 (2014).
6. Levy, R. B. & Boudart, M. Platinum-like behavior of tungsten carbide in surface catalysis. *Science* **181**, 547–549 (1973).
7. Wu, H. B., Xia, B. Y., Yu, L., Yu, X. Y. & Lou, X. W. Porous molybdenum carbide nano-octahedrons synthesized via confined carburization in metal-organic frameworks for efficient hydrogen production. *Nat. Commun.* **6**, 6512 (2015).
8. Zhao, Y., Kamiya, K., Hashimoto, K. & Nakanishi, S. *In situ* CO$_2$-emission assisted synthesis of molybdenum carbonitride nanomaterial as hydrogen evolution electrocatalyst. *J. Am. Chem. Soc.* **137**, 110–113 (2015).
9. Ma, F. X., Wu, H. B., Xia, B. Y., Xu, C.Y. & Lou, X. W. Hierarchical β-Mo$_2$C nanotubes organized by ultrathin nanosheets as a highly efficient electrocatalyst for hydrogen production. *Angew. Chem. Int. Ed.* **54**, 15395–15399 (2015).
10. Liao, L. *et al.* A nanoporous molybdenum carbide nanowire as an electrocatalyst for hydrogen evolution reaction. *Energy Environ. Sci.* **7**, 387–392 (2014).

11. Youn, D. H. *et al.* Highly active and stable hydrogen evolution electrocatalysts based on molybdenum compounds on carbon nanotube–graphene hybrid support. *ACS Nano* **8,** 5164–5173 (2014).

12. Chen, W. F. *et al.* Biomass-derived electrocatalytic composites for hydrogen evolution. *Energy Environ. Sci.* **6,** 1818–1826 (2013).

13. Ma, L., Ting, L. R. L., Molinari, V., Giordano, C. & Yeo, B. S. Efficient hydrogen evolution reaction catalyzed by molybdenum carbide and molybdenum nitride nanocatalysts synthesized via the urea glass route. *J. Mater. Chem. A* **3,** 8361–8368 (2015).

14. Wang, H. *et al.* Transition-metal doped edge sites in vertically aligned MoS$_2$ catalysts for enhanced hydrogen evolution. *Nano Res.* **8,** 566–575 (2015).

15. Gao, M. R. *et al.* An efficient molybdenum disulfide/cobalt diselenide hybrid catalyst for electrochemical hydrogen generation. *Nat. Commun.* **6,** 5982 (2015).

16. Merki, D. & Hu, X. Recent developments of molybdenum and tungsten sulfides as hydrogen evolution catalysts. *Energy Environ. Sci.* **4,** 3878–3888 (2011).

17. Jaramillo, T. F. *et al.* Identification of active edge sites for electrochemical H$_2$ evolution from MoS$_2$ nanocatalysts. *Science* **317,** 100–102 (2007).

18. Tang, Y. J. *et al.* Porous molybdenum-based hybrid catalysts for highly efficient hydrogen evolution. *Angew. Chem. Int. Ed.* **54,** 12928–12932 (2015).

19. Vrubel, H. & Hu, X. Molybdenum boride and carbide catalyze hydrogen evolution in both acidic and basic solutions. *Angew. Chem. Int. Ed.* **51,** 12703–12706 (2012).

20. Faber, M. S. & Jin, S. Earth-abundant inorganic electrocatalysts and their nanostructures for energy conversion applications. *Energy Environ. Sci.* **7,** 3519–3542 (2014).

21. Ma, R. *et al.* Ultrafine molybdenum carbide nanoparticles composited with carbon as a highly active hydrogen-evolution electrocatalyst. *Angew. Chem. Int. Ed.* **54,** 14723–14727 (2015).

22. Liu, Y. *et al.* Coupling Mo$_2$C with nitrogen-rich nanocarbon leads to efficient hydrogen-evolution electrocatalytic sites. *Angew. Chem. Int. Ed.* **54,** 10752–10757 (2015).

23. Cui, W. *et al.* Mo$_2$C nanoparticles decorated graphitic carbon sheets: biopolymer-derived solid-state synthesis and application as an efficient electrocatalyst for hydrogen generation. *ACS Catal.* **4,** 2658–2661 (2014).

24. Chen, W. F. *et al.* Highly active and durable nanostructured molybdenum carbide electrocatalysts for hydrogen production. *Energy Environ. Sci.* **6,** 943–951 (2013).

25. Seol, M. *et al.* Mo-compound/CNT-graphene composites as efficient catalytic electrodes for quantum-dot-sensitized solar cells. *Adv. Energy Mater.* **4,** 1300775 (2014).

26. Duan, J., Chen, S., Chambers, B. A., Andersson, G. G. & Qiao, S. Z. 3D WS$_2$ nanolayers@heteroatom-doped graphene films as hydrogen evolution catalyst electrodes. *Adv. Mater.* **27,** 4234–4241 (2015).

27. Duan, J., Chen, S., Jaroniec, M. & Qiao, S. Z. Porous C$_3$N$_4$ nanolayers@N-graphene films as catalyst electrodes for highly efficient hydrogen evolution. *ACS Nano* **9,** 931–940 (2015).

28. Yan, H. *et al.* Phosphorus-modified tungsten nitride/reduced graphene oxide as a high-performance, non-noble-metal electrocatalyst for the hydrogen evolution reaction. *Angew. Chem. Int. Ed.* **54,** 6325–6329 (2015).

29. He, C. & Tao, J. Synthesis of nanostructured clean surface molybdenum carbides on graphene sheets as efficient and stable hydrogen evolution reaction catalysts. *Chem. Commun.* **51,** 8323–8325 (2015).

30. Li, Y. *et al.* MoS$_2$ nanoparticles grown on graphene: an advanced catalyst for the hydrogen evolution reaction. *J. Am. Chem. Soc.* **133,** 7296–7299 (2011).

31. Kamat, P. V. Graphene-based nanoarchitectures. anchoring semiconductor and metal nanoparticles on a two-dimensional carbon support. *J. Phys. Chem. Lett.* **1,** 520–527 (2009).

32. Huang, C., Li, C. & Shi, G. Graphene based catalysts. *Energy Environ. Sci.* **5,** 8848–8868 (2012).

33. Du, D. Y., Qin, J. S., Li, S. L., Su, Z. M. & Lan, Y. Q. Recent advances in porous polyoxometalate-based metal-organic framework materials. *Chem. Soc. Rev.* **43,** 4615–4632 (2014).

34. Cronin, L. & Müller, A. From serendipity to design of polyoxometalates at the nanoscale, aesthetic beauty and applications. *Chem. Soc. Rev.* **41,** 7333–7334 (2012).

35. Wang, T. *et al.* Electrochemically fabricated polypyrrole and MoS$_x$ copolymer films as a highly active hydrogen evolution electrocatalyst. *Adv. Mater.* **26,** 3761–3766 (2014).

36. Zhou, D. & Han, B. H. Graphene-based nanoporous materials assembled by mediation of polyoxometalate nanoparticles. *Adv. Funct. Mater.* **20,** 2717–2722 (2010).

37. Ma, R., Hao, W., Ma, X., Tian, Y. & Li, Y. Catalytic ethanolysis of kraft lignin into high-value small-molecular chemicals over a nanostructured α-molybdenum carbide catalyst. *Angew. Chem. Int. Ed.* **53,** 7310–7315 (2014).

38. Zhou, W. *et al.* N-doped carbon-wrapped cobalt nanoparticles on *N*-doped graphene nanosheets for high-efficiency hydrogen production. *Chem. Mater.* **27,** 2026–2032 (2015).

39. Wu, R., Zhang, J., Shi, Y., Liu, D. & Zhang, B. Metallic WO$_2$–carbon mesoporous nanowires as highly efficient electrocatalysts for hydrogen evolution reaction. *J. Am. Chem. Soc.* **137,** 6983–6986 (2015).

40. Deng, J., Ren, P., Deng, D. & Bao, X. Enhanced electron penetration through an ultrathin graphene layer for highly efficient catalysis of the hydrogen evolution reaction. *Angew. Chem. Int. Ed.* **54,** 2100–2104 (2015).

41. Zheng, W. *et al.* Experimental and theoretical investigation of molybdenum carbide and nitride as catalysts for ammonia decomposition. *J. Am. Chem. Soc.* **135,** 3458–3464 (2013).

42. Hinnemann, B. *et al.* Biomimetic hydrogen evolution: MoS$_2$ nanoparticles as catalyst for hydrogen evolution. *J. Am. Chem. Soc.* **127,** 5308–5309 (2005).

43. Lai, L. *et al.* Exploration of the active center structure of nitrogen-doped graphene-based catalysts for oxygen reduction reaction. *Energy Environ. Sci.* **5,** 7936–7942 (2012).

44. Rao, C. V., Cabrera, C. R. & Ishikawa, Y. In search of the active site in nitrogen-doped carbon nanotube electrodes for the oxygen reduction reaction. *J. Phys. Chem. Lett.* **1,** 2622–2627 (2010).

45. Lee, J. H. *et al.* Restacking-inhibited 3D reduced graphene oxide for high performance supercapacitor electrodes. *ACS Nano* **7,** 9366–9374 (2013).

46. Kresse, G. & Hafner, J. *Ab initio* molecular dynamics for liquid metals. *Phys. Rev. B* **47,** 558–561 (1993).

47. Blöchl, P. E. Projector augmented-wave method. *Phys. Rev. B* **50,** 17953–17979 (1994).

48. Kresse, G. & Joubert, D. From ultrasoft pseudopotentials to the projector augmented-wave method. *Phys. Rev. B* **59,** 1758–1775 (1999).

49. Perdew, J. P., Burke, K. & Ernzerhof, M. Generalized gradient approximation made simple. *Phys. Rev. Lett.* **77,** 3865–3868 (1996).

Acknowledgements

This work was financially supported by the National Natural Science Foundation of China (No. 21371099, 21522305 and 21471080), the NSF of Jiangsu Province of China (No. BK20130043 and BK20141445), the Natural Science Foundation of Shandong Province (No. ZR2014BQ037), the Youths Science Foundation of Jining University (No. 2014QNKJ08), the Priority Academic Program Development of Jiangsu Higher Education Institutions and the Foundation of Jiangsu Collaborative Innovation Center of Biomedical Functional Materials.

Author contributions

Y.-Q.L. and J.-S.L. conceived the idea. J.-S.L., C.-H.L., Y.-G.W. and L.-Z.D. designed the experiments, collected and analysed the data. Y.-F.L. and Y.W. performed the DFT calculations. S.-L.L. and. Z.-H.D. assisted with the experiments and characterizations. J.-S.L and Y.-Q.L. co-wrote the manuscript. All authors discussed the results and commented on the manuscript.

Additional information

8

Making the hydrogen evolution reaction in polymer electrolyte membrane electrolysers even faster

Jakub Tymoczko[1,2], Federico Calle-Vallejo[3], Wolfgang Schuhmann[1,2] & Aliaksandr S. Bandarenka[1,4,5]

Although the hydrogen evolution reaction (HER) is one of the fastest electrocatalytic reactions, modern polymer electrolyte membrane (PEM) electrolysers require larger platinum loadings (~ 0.5-$1.0\,\mathrm{mg\,cm^{-2}}$) than those in PEM fuel cell anodes and cathodes altogether ($\sim 0.5\,\mathrm{mg\,cm^{-2}}$). Thus, catalyst optimization would help in substantially reducing the costs for hydrogen production using this technology. Here we show that the activity of platinum(111) electrodes towards HER is significantly enhanced with just monolayer amounts of copper. Positioning copper atoms into the subsurface layer of platinum weakens the surface binding of adsorbed H-intermediates and provides a twofold activity increase, surpassing the highest specific HER activities reported for acidic media under similar conditions, to the best of our knowledge. These improvements are rationalized using a simple model based on structure-sensitive hydrogen adsorption at platinum and copper-modified platinum surfaces. This model also solves a long-lasting puzzle in electrocatalysis, namely why polycrystalline platinum electrodes are more active than platinum(111) for the HER.

[1] Center for Electrochemical Sciences—CES, Ruhr-Universität Bochum, Universitätsstrasse 150, D-44780 Bochum, Germany. [2] Analytische Chemie—Elektroanalytik & Sensorik, Ruhr-Universität Bochum, Universitätsstrasse 150, D-44780 Bochum, Germany. [3] Leiden Institute of Chemistry, Leiden University, PO-Box 9502, 2300 RA Leiden, The Netherlands. [4] Physik-Department ECS, Technische Universität München, James-Franck-Strasse 1, 85748 Garching, Germany. [5] Nanosystems Initiative Munich (NIM), Schellingstrassee 4, 80799 Munich, Germany. Correspondence and requests for materials should be addressed to A.B. (email: bandarenka@ph.tum.de).

Heterogeneous redox reactions at electrified interfaces are of growing importance in contemporary science and technology[1-3], as they determine the performance of several electrochemical devices for future sustainable provision, storage and redistribution of renewable energy[4-7]. In particular, the efficiency of electrolysers and fuel cells largely depends on their electrode/electrolyte interfaces and catalytic properties[8-11].

In this context, hydrogen (H_2) production from water is an important electrocatalytic process due to its dual impact: it is a good model catalytic system[12-15] for the evaluation of new material design methodologies and it is significant for future energy provision and storage[16-19]. In spite of numerous achievements[18], only ~4% of H_2 produced comes from water electrolysis[20]. The main impediments to a wider utilization of water electrolysis are the high energy losses in electrolysers due to the insufficient activity of state-of-the-art electrodes. Considering the global hydrogen production of ~15 trillion moles per year (2011)[21] and average prices (2016) in the United States and Europe of ~0.1 Euro per kWh (refs 22,23), the electricity costs to produce just 4% of H_2 using polymer electrolyte membrane (PEM) electrolysers would exceed ~6 billion Euros. Compared with these expenses for electricity, the material costs (noble-metal catalysts, supports and so on) are relatively small. For instance, decreasing the operating voltage of PEM electrolysers from presently ~2.0 (ref. 18) by 0.1 V using improved hydrogen and oxygen evolution electrocatalysts could decrease the electricity expenses for electrolysis by ~0.3 billion Euros. Assuming a current density in state-of-the-art PEM electrolysers of 1 A cm^{-2} (ref. 18) and catalyst loadings of 1 mg cm^{-2} for anodes and cathodes[18], only the reduction in electricity expenses exceeds ~10 times the whole price of platinum (with the amounts, which are equivalent to ~0.5% of its annual production) or iridium (~30% of its annual production, correspondingly) catalysts necessary to electrochemically produce the above-mentioned amount of hydrogen annually. In other words, a ~20 mV decrease in the operating voltage of PEM electrolysers corresponds to the price of noble-metal catalysts needed to produce 4% of H_2 electrochemically. Although the long-term goal is to replace scarce electrocatalysts with more abundant and similarly active analogues, fundamental and application-related issues require further optimization of state-of-the-art hydrogen evolution reaction (HER) and oxygen evolution reaction electrocatalysts[18].

Here we show that incorporating (sub)monolayer amounts of copper (Cu) to platinum (Pt) enhances the catalytic activity ~2 times at low overpotentials, surpassing the highest HER-specific activities reported under similar conditions. These results are rationalized in terms of a structure-sensitive analysis of hydrogen adsorption on Pt- and Cu-modified Pt surfaces that also explains why polycrystalline Pt is more active than Pt(111) towards the HER.

Results

General considerations. According to the current understanding, the HER (as well as hydrogen oxidation reaction, HOR) mechanisms involve adsorbed hydrogen (denoted *H) at the electrode surface. As stated by the Sabatier principle[24], the optimal catalytic surface should bind reaction intermediates neither too weak nor too strong. This qualitative rule can be converted to a quantitative tool using calculated or measured adsorption energies for the relevant reaction intermediates at specific active sites on the surface. Figure 1 shows theoretical adsorption energies and experimental activity data for HER at pure metal surfaces[25]. As can be seen from Fig. 1, the trends in the measured HER can be fairly explained using the hydrogen

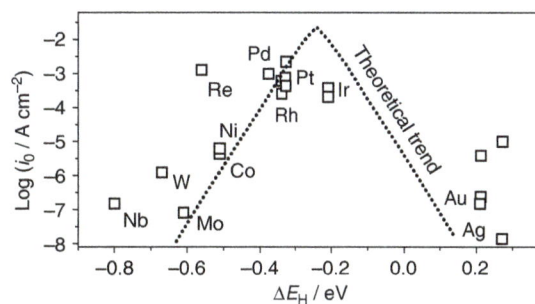

Figure 1 | Trends in hydrogen evolution reaction activity. Experimental HER activity expressed as the exchange current density, $\log(i_0)$, for different metal surfaces as a function of the calculated *H chemisorption energy, ΔE_H. The result of a simple theoretical kinetic model is also shown as a dotted line. Original data are taken from ref. 25.

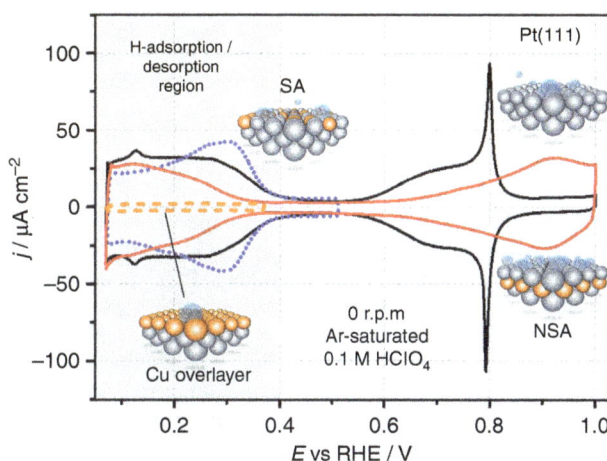

Figure 2 | Cyclic voltammetry in Ar-saturated 0.1 M HClO$_4$. The CVs of the Pt(111), Cu overlayer on Pt(111), Cu-Pt(111) SA and NSA (1 ML Cu initially deposited) demonstrate how the relative position of the Cu atomic layer governs the adsorption/desorption of hydrogen (grey area) at the electrode surface. $dE/dt = 50$ mV s^{-1}.

binding energy as a 'descriptor', ΔE_H, estimated via density functional theory (DFT) calculations, as reported by Nørskov et al.[25]

Although exact ΔE_H values depend on the surface coverage of hydrogen[12], a straightforward outcome of this approach is that the optimum electrocatalytic sites for the HER should bind *H slightly weaker (~0.09 eV) than Pd, Rh or Pt. In principle, the electronic properties of metal surfaces can be modulated by different means. One of the common ways to do this is to prepare bulk alloys, where the bulk crystal composition and structure influence the properties of the surfaces and, hence, their catalytic activity through strain and ligand effects[26-28]. An alternative way is to modify the properties of the topmost layer at the surface by selectively positioning atomic layers of solute metals directly at the surface to form either overlayers[29], surface or subsurface alloys. An example of the latter approach is shown in Fig. 2.

Electrochemical performance. Figure 2 shows cyclic voltammograms (CVs) taken in Ar-saturated 0.1 M HClO$_4$ electrolytes for the unmodified Pt(111), the Cu overlayer on Pt(111), the Cu-Pt(111) near-surface alloys (NSAs) and surface alloys (SAs) within the regions of their electrochemical stability. In the CVs, the potential region between ~0.4 and 0.07 V corresponds to hydrogen adsorption/desorption before the formation of H_2 at more negative potentials. Notably, the position of the Cu atomic layer significantly changes the hydrogen binding energy. For example, the SA apparently binds *H stronger than Pt(111), as revealed by the corresponding CVs between ~0.2 and ~0.4 V (Fig. 2). In contrast, positioning the Cu atoms into the second layer (1 ML Cu initially deposited) weakens *H binding compared with unmodified Pt(111). In addition, the Cu pseudomorphic overlayer (POL) does not adsorb hydrogen species at these potentials (Fig. 2).

Experimentally observable changes in ΔE_H are mainly due to ligand effects, as Cu and Pt have dissimilar valence configurations ($s^1 d^{10}$ versus $s^1 d^9$) and the differences in the lattice constants are not negligible (3.61 versus 3.92 Å). These differences have a direct impact on the kinetics of reactions that involve adsorbed hydrogen species as reaction intermediates, in particular for the HER. For example, the SA, which binds hydrogen species stronger than Pt(111), would also probably be less active for both HER and HOR: this corresponds to the left part of the volcano plot in Fig. 1 and more negative ΔE_H values, relative to Pt. On the other hand, the Cu overlayer is 'too noble' for the hydrogen species to be active towards HER. This corresponds to the right part of the volcano in Fig. 1, far from the optimum towards more positive ΔE_H values. In contrast, one can expect that the NSA would probably be more active than Pt(111): its surface binds hydrogen species slightly weaker than Pt, which corresponds to the direction towards the theoretical maximum in Fig. 1. In Fig. 3, we confirm all these expectations.

At low overpotentials, a Cu-Pt(111) POL does not show noticeable HER activities (Fig. 3). The SA is less active than Pt(111), as expected. Finally, the voltammogram for the NSA (1 ML Cu initially deposited) in Fig. 3 reveals a substantially higher hydrogen evolution activity as compared with that for Pt(111). It is noteworthy that the results presented in Fig. 3 correspond to measurements in Ar-saturated electrolytes, as these are the simplest tests to derive activity trends with minimal influence of complex experimental factors (those are especially important for overlayers, when the electrolyte is saturated with electroactive species such as H_2 or CO)[30]. In the following, we focus on a more detailed electrochemical characterization of the active NSA electrodes.

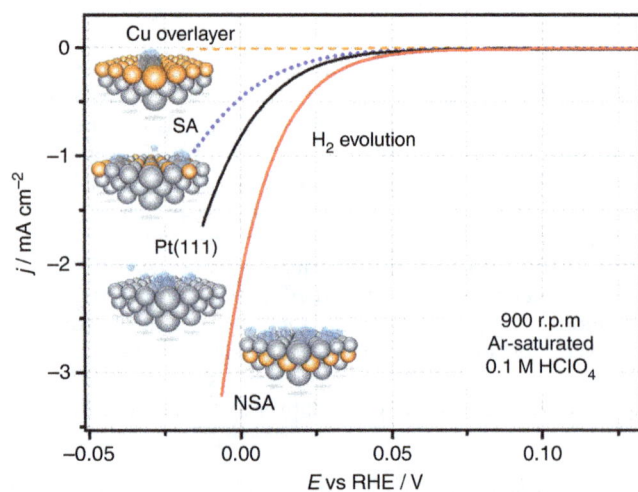

Figure 4 shows typical rotating disk electrodes (RDE) voltammograms recorded in H_2-saturated 0.1 M $HClO_4$ for Pt(111) and NSA electrodes. The NSA surface is more active than unmodified Pt for both HER and HOR. As those reactions involve the same intermediates, the same ΔE_H descriptor can be used to explain this fact. Although we do not use iR correction to avoid additional errors in this particular case (see ref. 30) and rather compare the model surfaces under the same conditions, it is still possible to approximately estimate the 'apparent' exchange current density, i_0, at very low overpotentials close to 0.0 V reversible hydrogen electrode (RHE)[15]. This value reflects the intrinsic activity of materials and can be used to compare different electrocatalysts reported by different research groups. The estimated i_0 values are at least ~1.5 mA cm^{-2} for Pt(111) and ~3.0 mA cm^{-2} for the NSA. Notably, the apparent exchange current density for Pt(111) is higher than that reported in a very detailed investigation performed by Markovic et al.[15] for low-index Pt(hkl) single-crystal surfaces measured in H_2SO_4 at the same pH value. We hypothesize that this is due to a difference in the exact experimental protocols, as discussed recently in detail in ref. 30. Nevertheless, this fact additionally prevents misinterpretation of the NSA activity results, as those are compared with already very active reference Pt surfaces.

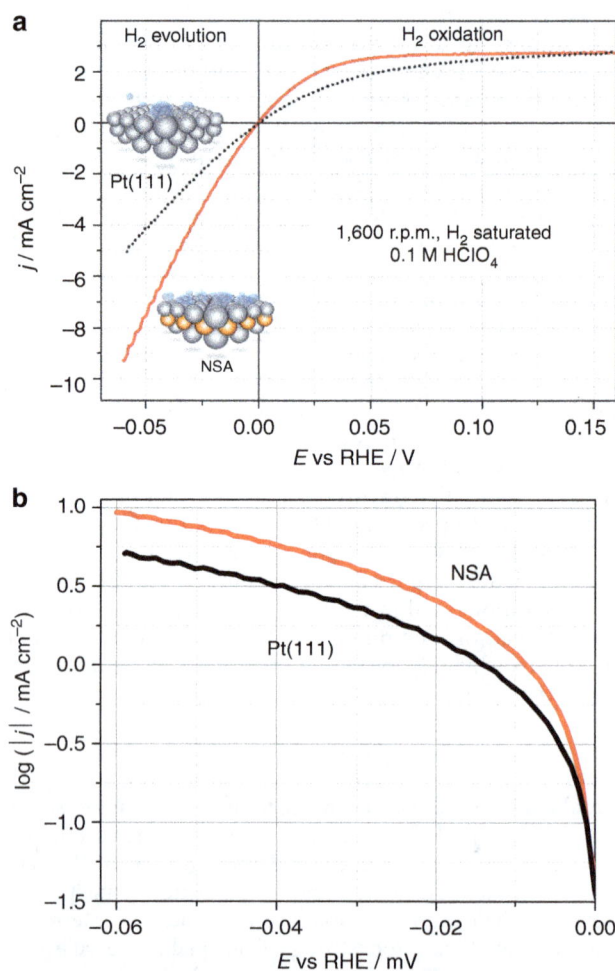

Figure 3 | RDE voltammetry in Ar-saturated 0.1 M HClO₄. The RDE voltammograms for Cu overlayer, Pt(111), SA and NSA (1 ML Cu initially deposited) electrodes show the correlation between the HER activity of the electrodes and the position of the Cu atomic layers relative to the topmost Pt layer. The negative currents start before 0.0 V RHE, because the electrolyte is saturated with the inert gas. $dE/dt = 10$ mV s^{-1}.

Figure 4 | RDE voltammetry in H₂-saturated HClO₄. (a) RDE voltammograms of the Cu-Pt(111) NSA (1 ML Cu initially deposited) compared with the unmodified Pt(111) electrode. $dE/dt = 10$ mV s^{-1}. **(b)** Logarithmic plot of the currents related to hydrogen evolution for the NSA and Pt(111).

Table 1 | Activities for the HER/HOR at room temperature.

Electrode	$i_{0,apparent}$ (mA cm^{-2})*	Source
Pt(111)	~0.45	ref. 15
Pt(100)	~0.6	
Pt(110)	~0.98	
Pd$_{OL}$/PtRu(111)	~2.0	refs 29,31
Pd$_{OL}$/Pt(111)	~2.0	
Pt(111)	~1.5	This work
Cu-Pt(111) NSA (1 ML Cu initially deposited)	~3.0	This work
Pt (nanoparticles)	~1.0 (at 80 °C)	ref. 18

HER, hydrogen evolution reaction; HOR, hydrogen oxidation reaction.
*At pH 1, without iR correction, as reported in the literature.

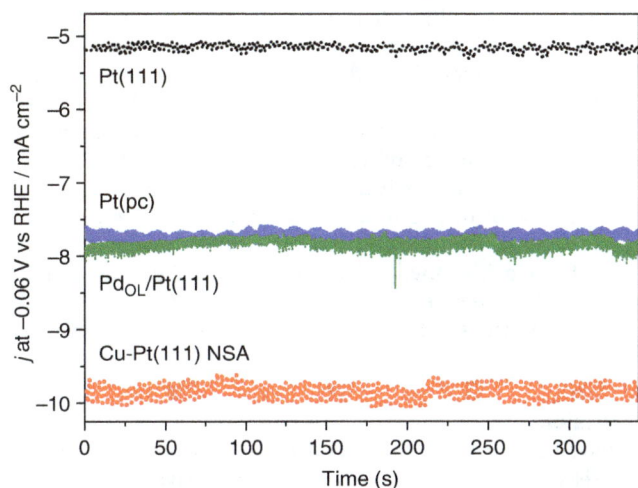

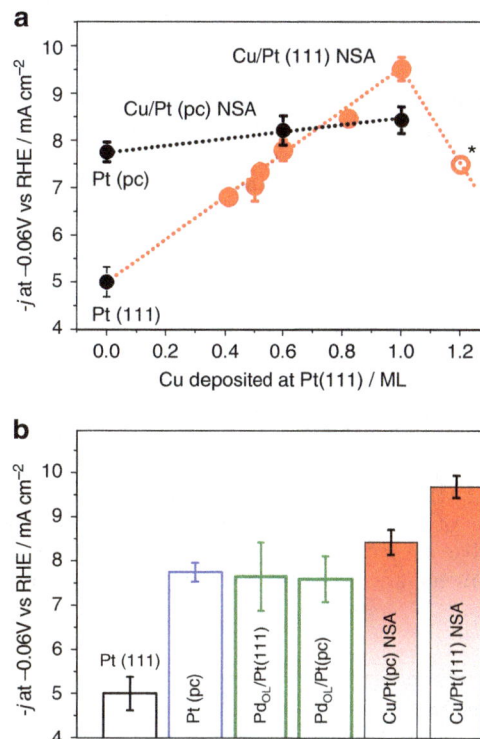

Figure 5 | Hydrogen evolution activity as a function of time. The figure contains the current–time curves for hydrogen evolution reaction under potentiostatic conditions ($E = -0.06$ V) for the Cu-Pt(111) NSA, polycrystalline Pt, Pd$_{OL}$-modified Pt and unmodified Pt(111) electrodes (RDE measurements at 1,600 r.p.m.).

Figure 6 | Effect of Cu and Pd addition on the hydrogen evolution reaction activity. (**a**) Effect of the Cu subsurface concentration on the HER activity measured at a potential of -0.06 V versus RHE for Pt(111) and polycrystalline Pt electrodes (Pt(pc), RDE measurements at 1,600 r.p.m., scan rate: 50 mV s^{-1}), given with corresponding error bars (standard deviation) estimated using the results of at least five independent measurements. Open symbol with '*' represents the situation when more than 1 ML of copper was introduced intentionally. (**b**) Current densities measured at -0.06 V versus RHE for Pt(111), Pt(pc), Pd overlayer on Pt(111) (Pd$_{OL}$/Pt(111), Pd overlayer on Pt(pc) (Pd$_{OL}$/Pt(pc)), Cu/Pt(pc) NSA and Cu/Pt(111) NSA with corresponding error bars (standard deviation) estimated using the results of at least five independent measurements. The values are displayed without iR correction.

If the so-called Tafel plot is used (Fig. 4b), the slopes of the curves for the NSA and Pt(111) samples at each electrode potential are rather similar, suggesting that there are no significant changes in the HER mechanism among these two surfaces.

Table 1 compares the activities for the HER/HOR at room temperature for active model surfaces, as summarized in refs 15,18,31 and measured in this work. In addition, as can be seen from Table 1, typical values for the apparent exchange current densities of Pt nanoparticles (averaged among values used by different groups), even at elevated temperatures and in real devices, are approximately three times lower than that for the NSA sample.

To further evaluate the activity of the Cu-Pt(111) NSA with respect to the best known catalysts and additionally account for possible artefacts caused by the formation of the non-conducting H$_2$ gas phase at the electrode surface during the cathodic/anodic scans, we compare chronoamperograms (current versus time curves taken at a certain potential) for the most active surfaces reported up to date in Fig. 5.

Figure 5 compares the activity of the Cu-Pt(111) NSA (1 ML Cu initially deposited) with polycrystalline Pt and Pd$_{OL}$ deposited on Pt(111). First, the activity for all samples remains practically unchanged, as well as their basic CVs, indicating that the differences in activities are not due to artefacts caused by

generation of the non-conducting gas phase. Notably, the activity of the Cu-Pt(111) NSA (1 ML Cu initially deposited) is reproducibly better than any other reported state-of-the-art electrocatalysts including polycrystalline Pt[32], which has been suggested as one of the most active surface towards HER.

We performed additional benchmark measurements using polycrystalline Pt samples including iR correction. The activity results show that our polycrystalline samples possess exactly the same activity towards HER/HOR as reported by Sheng *et al.*[32] in their detailed study of the activities of polycrystalline Pt[30,33].

Furthermore, the influence of the subsurface concentration of Cu in Cu-Pt(111) and Cu-Pt(pc) NSAs has been tested by varying the amount of Cu initially deposited (Fig. 6a). Interestingly, the activity of the Cu-Pt(111) NSA sample with 1 ML Cu deposited initially remains the most active one. Attempts to introduce even more Cu into the subsurface region through a two-stage deposition/annealing procedure leads to a decrease in the activity (marked with '*' in Fig. 6a). All active samples evaluated in this work are compared in Fig. 6b.

Notably, the Cu-Pt(111) NSA samples demonstrated good stabilities towards H-induced segregation and anodic corrosion after 5,000 cycles between 0.05 and 1.0 V (versus RHE), as reported recently[34]. This additionally suggests that modification of just the subsurface region of HER electrocatalysts is a promising approach to enhance not only their catalytic activity but also their stability.

Computational. Finally, we rationalize our most important experimental findings based on the computational results as shown in Fig. 7. Figure 7a contains the trends in hydrogen adsorption for pure Pt(111) and CuPt(111) NSAs and SAs. Clearly, subsurface Cu at all concentrations has the same net effect of weakening the adsorption energies of atomic hydrogen, whereas surface Cu has the opposite effect in line with previous results[35]. DFT calculations confirm that hydrogen atoms are indeed bound more weakly at NSAs and more strongly at SAs.

To assess the structural sensitivity of the HER, we have tested the adsorption of *H at various sites apart from the hollow sites at (111) terraces usually considered in computational models[25]. Figure 7b shows the differences in adsorption energies of *H with respect to Pt(111) on numerous sites at Pt surfaces and NSAs with 1 ML Cu in the subsurface. The trends are described as a function of the generalized coordination numbers $(\overline{CN})$ of the active sites[36,37]. In simple terms, generalized coordination numbers are a weighted average of the conventional

coordination numbers. The weights are the coordination numbers of the nearest neighbours of the active sites. The various sites considered in this study and the way of estimating their generalized coordination numbers are provided in Supplementary Table 1.

Two noteworthy features of hydrogen atom adsorption on Pt and Cu-Pt NSAs are captured in Fig. 7b. First, sites with coordination lower than Pt(111) bind *H more strongly, whereas those with larger coordination bind more weakly. Second, *H adsorption on Cu-Pt NSAs is systematically weaker than on their counterparts at pure Pt, regardless of surface coordination.

Figure 7c contains the HER volcano-type activity plot built following the model by Nørskov et al.[25] (see also Fig. 1 and Supplementary Methods). The plot reflects simultaneously the effect of geometric coordination and Cu content on the HER activity. First of all, highly coordinated defects on pure Pt are substantially more active than sites at Pt(111), which justifies the fact that polycrystalline Pt is more active than Pt(111) for the HER (see Supplementary Methods for further experimental evidence). Undercoordinated defects, however, are less active than (111) terraces. On the other hand, Cu-Pt(111) NSAs are highly active and both overcoordinated and undercoordinated defects decrease their activity, which explains why (111) NSAs are more active than the polycrystalline ones. Finally, SAs are not active in view of their strong *H adsorption energies. Therefore, structure- and composition-sensitive experimental trends for the HER in acidic media are well captured by the trends in *H adsorption energies. In turn, these energies are substantially influenced by the surface coordination of the active sites and the presence of Cu.

Discussion

We have provided experimental and theoretical evidence to claim that selective positioning of Cu atomic layers modifies the adsorption properties of platinum electrodes for the electrochemical hydrogen evolution, accelerating one of the fastest electrocatalytic reactions known to date. Using predominantly the ligand effect, submonolayer amounts of Cu atoms in the second atomic layer induce a twofold increase in the electrocatalytic activity of Pt(111). This makes them the most active electrocatalysts ever reported for the HER in acidic media under comparable conditions, to the best of our knowledge. Further efforts to improve the performance of nanoparticle materials for the cathodes in PEM electrolysers may use this rationale based on the purposeful and delicate location of submonolayer amounts of foreign metals at surfaces.

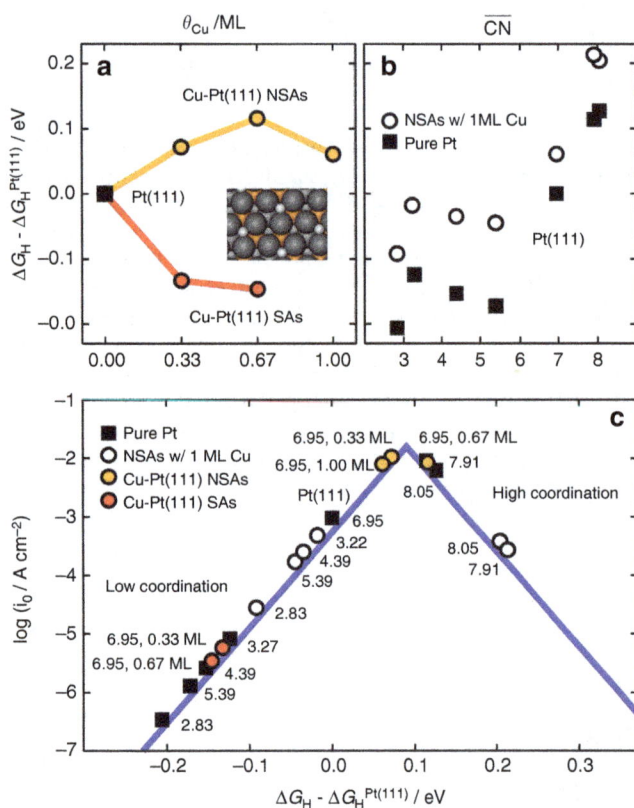

Figure 7 | Trends in H₂ adsorption and evolution. (a) *H adsorption energies with surface Cu (SAs, red) and subsurface Cu (NSAs, orange) on Pt(111). *H binds more strongly to SAs compared with Pt(111), whereas the opposite is observed for the NSAs. The inset shows *H at a NSA with 0.67 ML Cu (grey balls, Pt; red, Cu atoms; white, adsorbed *H). **(b)** Adsorption energies of *H on Pt and NSAs with 1ML Cu as a function of the generalized coordination numbers of the active sites. In all cases, *H adsorption energies are weaker on the NSAs than on the corresponding sites on Pt. Moreover, more coordinated sites than in the case of (111) terraces bind *H weaker, whereas the less coordinated ones bind it stronger. **(c)** Volcano plot showing the individual HER activity, coordination and composition of all studied sites. Sites at Cu-Pt(111) NSAs and overcoordinated Pt sites possess the highest activities, whereas SAs and undercoordinated sites are not active.

Methods

Electrode preparation. Details relating to the electrode surface preparation and characterization are given in Supplementary Figs 1–13. The relative position of Cu atomic layers at the surface was controlled as described elsewhere[35,38,39]. Briefly, to form a copper POL or deposit submonolayer amounts of it, underpotential deposition was performed from a solution containing 2 mM Cu^{2+} in 0.1 M $HClO_4$. The Cu-Pt(111) NSAs, where Cu atoms are preferentially located in the subsurface layer, were obtained by short annealing of the overlayer ($\sim$2 min) at 400 °C in Ar/H₂ atmosphere containing 5% of H₂ in Ar (6.0, AirLiquide, Germany). Cu-Pt(111) SAs, where Cu atoms are located in the first atomic layer of the Pt host, were subsequently obtained by annealing the NSAs in Ar/CO atmosphere (0.1% CO in Ar, $\sim$2 min at 400 °C). The preparation procedures result in single-crystalline samples of Pt(111), Cu-Pt(111) NSAs and SAs, which were atomically smooth.

Activity measurements. Electrolytes containing 0.1 M $HClO_4$ (Merck Suprapur, Germany) were used for activity measurements. A mercury–mercury sulfate reference electrode was kept in a separate compartment and separated from the working electrolyte with an ionically conducting ceramic insert. A polycrystalline Pt wire was used as counter electrode. All potentials are referred to the RHE scale. A SP-300 potentiostat (Bio-Logic, France) was used to control the electrochemical measurements. Electrochemical experiments including the activity measurements were performed using a specifically designed electrochemical cell for the preparation and *in-situ* electrochemical characterization of single-crystal alloy electrodes, previously described in ref. 40. Measurements with RDE were performed using a Pine RDE 710 instrument (USA).

DFT calculations. Full details of the DFT calculations, the assessment of adsorption energies, the model for estimating the current densities and the computation of generalized coordination numbers are provided in Supplementary Methods section with additional explanations illustrated in Supplementary Fig. 14 and Supplementary Table 1.

References

1. Nørskov, J. K., Bligaard, T., Rossmeisl, J. & Christensen, C. H. Towards the computational design of solid catalysts. *Nat. Chem.* **1**, 37–46 (2009).
2. Koper, M. T. M. Structure sensitivity and nanoscale effects in electrocatalysis. *Nanoscale* **3**, 2054–2073 (2011).
3. Stephens, I. E. L., Bondarenko, A. S., Grønbjerg, U., Rossmeisl, J. & Chorkendorff, I. Understanding the electrocatalysis of oxygen reduction on platinum and its alloys. *Energy Environ. Sci.* **5**, 6744–6762 (2012).
4. Bard, A. J. Inner-sphere heterogeneous electrode reactions. Electrocatalysis and photocatalysis: the challenge. *J. Am. Chem. Soc.* **132**, 7559–7567 (2010).
5. Gasteiger, H. A. & Markovic, N. M. Just a dream—or future reality? *Science* **324**, 48–49 (2009).
6. Vesborg, P. C. K. & Jaramillo, T. F. Addressing the terawatt challenge: scalability in the supply of chemical elements for renewable energy. *RSC Adv.* **2**, 7933–7947 (2012).
7. Rabis, A., Rodriguez, P. & Schmidt, T. J. Electrocatalysis for polymer electrolyte fuel cells: recent achievements and future challenges. *ACS Catal.* **2**, 864–890 (2012).
8. Symes, M. D. & Cronin, L. Decoupling hydrogen and oxygen evolution during electrolytic water splitting using an electron-coupled-proton buffer. *Nat. Chem.* **5**, 403–409 (2013).
9. Bandarenka, A. S. & Koper, M. T. M. Structural and electronic effects in heterogeneous electrocatalysis: Toward a rational design of electrocatalysts. *J. Catal.* **308**, 11–24 (2013).
10. Holewinski, A., Xin, H., Nikolla, E. & Linic, S. Identifying optimal active sites for heterogeneous catalysis by metal alloys based on molecular descriptors and electronic structure engineering. *Curr. Opin. Chem. Eng.* **2**, 312–319 (2013).
11. Gasteiger, H. A., Baker, D. R. & Carter, R. N. in *Hydrogen Fuel Cells: Fundamentals and Applications* (Wiley-CPH, 2010).
12. Skulason, E. *et al.* Modeling the electrochemical hydrogen oxidation and evolution reactions on the basis of density functional theory calculations. *J. Phys. Chem. C* **114**, 18182–18197 (2010).
13. Greeley, J., Jaramillo, T. F., Bonde, J., Chorkendorff, I. & Nørskov, J. K. Computational high-throughput screening of electrocatalytic materials for hydrogen evolution. *Nat. Mater.* **5**, 909–913 (2006).
14. Björketun, M. E., Bondarenko, A. S., Abrams, B. L., Chorkendorff, I. & Rossmeisl, J. Screening of electrocatalytic materials for hydrogen evolution. *Phys. Chem. Chem. Phys.* **12**, 10536–10541 (2010).
15. Markovic, N. M., Grgur, B. N. & Ross, P. N. Temperature-dependent hydrogen electrochemistry on platinum low-index single-crystal surfaces in acid solutions. *J. Phys. Chem. B* **101**, 5405–5413 (1997).
16. Züttel, A., Borgschulte, A. & Schlapback, L. *Hydrogen as a Future Energy Carrier* (Wiley-VCH, 2008).
17. Wang, M., Wang, Z., Gong, X. & Guo, Z. The intensification technologies to water electrolysis for hydrogen production—a review. *Renew. Sust. Energ. Rev.* **29**, 573–588 (2014).
18. Carmo, M., Fritz, D. L., Mergel, J. & Stolten, D. A comprehensive review on PEM water electrolysis. *Int. J. Hydrogen Energy* **38**, 4901–4934 (2013).
19. Jaramillo, T. F. *et al.* Identification of active edge sites for electrochemical H_2 evolution from MoS_2 nanocatalysts. *Science* **317**, 100–102 (2007).
20. Bicakova, O. & Straka, P. Production of hydrogen from renewable resources and its effectiveness. *Int. J. Hydrogen Energy* **37**, 11563–11578 (2012).
21. US Department of Energy Hydrogen Analysis Resource Center, Hydrogen Production, Worldwide and US Total Hydrogen Production http://hydrogen.pnl.gov/cocoon/morf/hydrogen/article/706 (2012).
22. European Commission (Eurostat) Energy Price Statistics, http://ec.europa.eu/eurostat/statistics-explained/index.php/Energy_price_statistics (accessed on December 2015).
23. US Energy Information Administration, http://www.eia.gov/electricity/data/browser/#/topic/7?agg=2,0,1&geo=g&freq=M (accessed on December 2015).
24. Sabatier, P. Hydrogentions et deshydrogenations par catalyse. *Ber. Deut. Chem. Gesell.* **44**, 1984–2001 (1911).
25. Nørskov, J. K. *et al.* Trends in the exchange current for hydrogen evolution. *J. Electrochem. Soc.* **152**, J23–J26 (2005).
26. Mavrikakis, M., Hammer, B. & Nørskov, J. K. Effect of strain on the reactivity of metal surfaces. *Phys. Rev. Lett.* **81**, 2819–2822 (1998).
27. Kitchin, J. R., Nørskov, J. K., Barteau, M. A. & Chen, J. G. Role of strain and ligand effects in the modification of the electronic and chemical properties of bimetallic surfaces. *Phys. Rev. Lett.* **93**, 156801–156804 (2004).
28. Kitchin, J. R., Nørskov, J. K., Barteau, M. A. & Chen, J. G. Modification of the surface electronic and chemical properties of Pt(111) by subsurface 3d transition metals. *J. Chem. Phys.* **120**, 10240–10246 (2004).
29. Greeley, J., Nørskov, J. K., Kibler, L. A., El-Aziz, A. M. & Kolb, D. M. Hydrogen evolution over bimetallic systems: understanding the trends. *ChemPhysChem* **7**, 1032–1035 (2006).
30. Čolić, V. *et al.* Experimental aspects in benchmarking of the electrocatalytic activity. *ChemElectroChem* **2**, 143–149 (2015).
31. Kibler, L. A. Hydrogen electrocatalysis. *ChemPhysChem* **7**, 985–991 (2006).
32. Sheng, W. C., Gasteiger, H. A. & Shao-Horn, Y. Hydrogen oxidation and evolution reaction kinetics on platinum: acid versus alkaline electrolytes. *J. Electrochem. Soc.* **157**, B1529–B1536 (2010).
33. Ganassin, A., Colic, V., Tymoczko, J., Bandarenka, A. S. & Schuhmann, W. Non-covalent interactions in water electrolysis: influence on the activity of Pt(111) and iridium oxide catalysts in acidic media. *Phys. Chem. Chem. Phys.* **17**, 8349–8355 (2015).
34. Tymoczko, J., Calle-Vallejo, F., Colic, V., Schuhmann, W. & Bandarenka, A. S. Evaluation of the electrochemical stability of model Cu-Pt(111) near-surface alloy catalysts. *Electrochim. Acta* **179**, 469–474 (2015).
35. Bandarenka, A. S. *et al.* Design of an active site towards optimal electrocatalysis: overlayers, surface alloys and near-surface alloys of Cu/Pt(111). *Angew. Chem. Int. Ed.* **51**, 11845–11848 (2012).
36. Calle-Vallejo, F., Martínez, J. I., García-Lastra, J. M., Sautet, P. & Loffreda, D. *Angew. Chem. Int. Ed.* **53**, 8316–8319 (2014).
37. Calle-Vallejo, F. *et al.* Finding optimal surface sites on heterogeneous catalysts by counting nearest neighbors. *Science* **350**, 185–189 (2015).
38. Stephens, I. E. L. *et al.* Tuning the activity of Pt(111) for oxygen electroreduction by subsurface alloying. *J. Am. Chem. Soc.* **133**, 5485–5491 (2011).
39. Bondarenko, A. S., Stephens, I. E. L. & Chorkendorff, I. A cell for the controllable thermal treatment and electrochemical characterisation of single crystal alloy electrodes. *Electrochem. Commun.* **23**, 33–36 (2012).
40. Tymoczko, J., Schuhmann, W. & Bandarenka, A. S. A versatile electrochemical cell for the preparation and characterisation of model electrocatalytic systems. *Phys. Chem. Chem. Phys.* **15**, 12998–13004 (2013).

Acknowledgements

Financial support from SFB 749, the cluster of excellence Nanosystems Initiative Munich (NIM), Cluster of Excellence RESOLV (EXC 1069) funded by the DFG (Deutsche Forschungsgemeinschaft) and in the framework of Helmholtz-Energie-Allianz 'Stationäre elektrochemische Speicher und Wandler' (HA-E-0002) is gratefully acknowledged. F.C.V. acknowledges funding by the Netherlands Organization for Scientific Research (NWO), Veni project number 722.014.009. The Stichting Nationale Computerfaciliteiten (NCF) is acknowledged for the use of their supercomputer facilities, with financial support from NWO.

Author contributions

A.B. and W.S. conceived and supervised the experiments, which were carried out by J.T. F.C.V. made the DFT calculations and the theoretical modelling. The manuscript was written through contributions of all authors. All authors have given approval to the final version of the manuscript.

Additional information

9

Explanation of efficient quenching of molecular ion vibrational motion by ultracold atoms

Thierry Stoecklin[1], Philippe Halvick[1], Mohamed Achref Gannouni[2], Majdi Hochlaf[2], Svetlana Kotochigova[3] & Eric R. Hudson[4]

Buffer gas cooling of molecules to cold and ultracold temperatures is a promising technique for realizing a host of scientific and technological opportunities. Unfortunately, experiments using cryogenic buffer gases have found that although the molecular motion and rotation are quickly cooled, the molecular vibration relaxes at impractically long timescales. Here, we theoretically explain the recently observed exception to this rule: efficient vibrational cooling of $BaCl^+$ by a laser-cooled Ca buffer gas. We perform intense close-coupling calculations that agree with the experimental result, and use both quantum defect theory and a statistical capture model to provide an intuitive understanding of the system. This result establishes that, in contrast to the commonly held opinion, there exists a large class of systems that exhibit efficient vibrational cooling and therefore supports a new route to realize the long-sought opportunities offered by molecular structure.

[1] Université de Bordeaux, Institut des Sciences Moléculaires, UMR 5255 CNRS, 33405 Talence, France. [2] Université Paris-Est, Laboratoire Modélisation et Simulation Multi Echelle, MSME UMR 8208 CNRS, 5 bd Descartes, 77454 Marne-la-Vallée, France. [3] Department of Physics, Temple University, 1925 N 12th Street, Philadelphia, Pennsylvania 19122, USA. [4] Department of Physics and Astronomy, University of California, 475 Portola Plaza, Los Angeles, California 90095, USA. Correspondence and requests for materials should be addressed to T.S. (email: thierry.stoecklin@u-bordeaux.fr).

The internal structure of molecules offers a host of scientific and technological opportunities[1], including the manipulation of quantum information, critical insight into quantum chemistry and improved tests of the Standard Model. To utilize this potential of molecules typically requires the preparation of molecular samples at very low temperatures, where only a single quantum state is occupied. Unfortunately, experiments attempting to reach these temperatures by buffer gas cooling have found that although the molecular motion and rotation are quickly cooled to the cryogenic temperature[2,3], the molecular vibration relaxes at impractically long timescales[4]. However, in a recent study[5], sympathetic cooling through collisional interaction with *laser-cooled atoms* was demonstrated to be an alternative and efficient approach to quench molecular ion vibrational motion. Although this result was predicted based on a semi-classical argument[6], it had not been verified by detailed quantum mechanical calculations. Further, it was not known how widely this technique could be applied, nor how to estimate its efficiency for other systems.

Here, we perform a detailed theoretical study of the Ca-BaCl$^+$ system and compare to the recent experimental results, as well determine efficiency criteria to predict vibrational quenching rates in similar systems. Specifically, we first build an analytical model of the potential energy surface (PES) of the Ca-BaCl$^+$ collision using a large grid of *ab initio* points, taking special care to accurately describe its long-range behaviour, which has an important role at very low collision energy. We then perform close-coupling calculations of the vibrational quenching of BaCl$^+$ by collisions with Ca atoms. Although such calculations are nowadays straightforward and relatively fast, it is highly computationally demanding in the case of Ca-BaCl$^+$ owing to the large mass and bonding energy. We therefore give a brief account of the method used to make the calculations feasible in a reasonable amount of computer time and compare the results with experiment. In addition, we use a scattering model based on quantum defect theory (QDT) with generalized short-range boundary conditions to gain an insight into the vibrational cooling of the molecular ions by the ultracold atoms. Finally, we compare the close-coupling vibrational quenching results with those obtained for four other similar systems: He-N$_2^+$, He-NO$^+$, He-CH$^+$ and Ar-NO$^+$ and propose a very simple statistical capture model, which reproduces the close-coupling results and provides a simple means to estimate the efficiency of vibrational quenching for a given system. This result establishes that, in contrast to the commonly held opinion, there exists a large class of systems that exhibit efficient vibrational cooling and therefore supports a new route to realize the long-sought opportunities offered by molecular structure.

Results

Ca-BaCl$^+$ PES calculation. To calculate the Ca-BaCl$^+$ vibrational quenching rate, we first built an analytical model of the PES of the Ca-BaCl$^+$ collision using a large grid of *ab initio* points. The electronic ground state of the CaBaCl$^+$ complex, a $^1A'$ state, was calculated with the MOLPRO programme suite[7] (see the Methods for details) on a three-dimensional (3D) grid of points in the Jacobi coordinates space r, R and θ. Here, r represents the BaCl$^+$ bond length, R the distance between Ca and the centre of mass of BaCl$^+$, and θ the angle between $\mathbf{r}$ and $\mathbf{R}$, with the linear structure Ba-Cl-Ca corresponding to $\theta = 0°$. The functional form $V(r,R,\theta)$ of this PES is defined as the sum of the interaction energy V_I between Ca and BaCl$^+$ and the potential of the isolated diatomic BaCl$^+$, V_{BaCl}:

$$V(r,R,\theta) = V_I(r,R,\theta) + V_{BaCl}(r). \quad (1)$$

These terms are found by interpolation over the *ab initio* calculations (see the Methods for details).

The results of these calculations are shown in Fig. 1 along two dimensions in the Jacobi space. Figure 1a shows the existence of a relatively deep potential well in good agreement with the charge-transfer nature of the bonding within this ionic complex. Figure 1b reveals the existence of two minimal structures and two saddle points connecting these equilibrium structures. It also shows that the potential is strongly anisotropic. Although the Ba–Cl bond length is only slightly extended by the interaction with the calcium, we observe in Fig. 1a that the vibrational potential of BaCl$^+$ is significantly modified by the latter interaction. This indicates there is a significant coupling between the vibration of BaCl$^+$ and the other modes of motion. This coupling is expected to promote vibrationally inelastic collisions.

Close-coupling calculation of the vibrational quenching rate. Calculation of the vibrational quenching rate on such a complex PES is normally done through the use of approximation

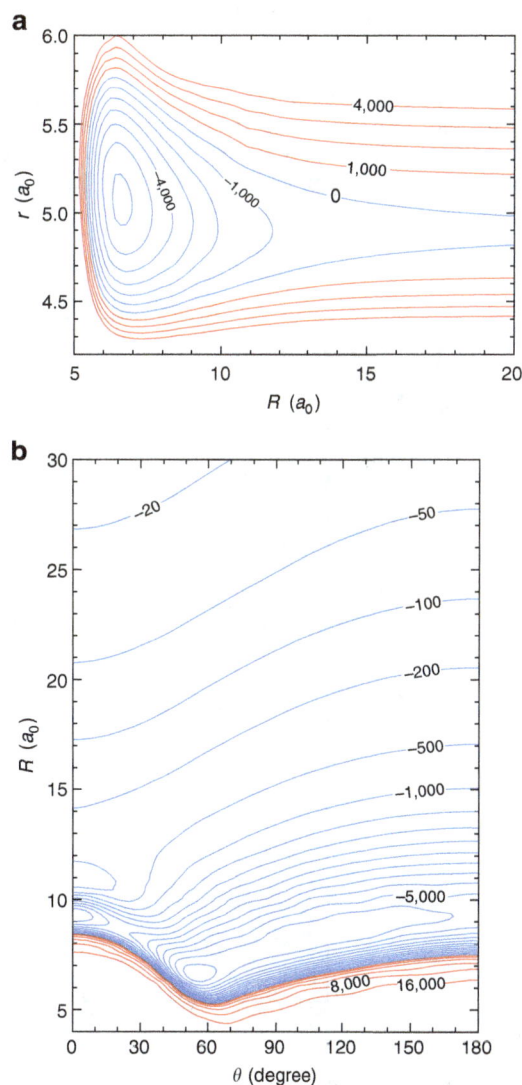

Figure 1 | Potential energy. (a) Contour plot of the total PES for $\theta = 55°$. Contour energies are regularly spaced by $1,000\,cm^{-1}$. **(b)** Contour plot of the three-body interaction energy for $r = 5\,a_0$. Below $-500\,cm^{-1}$, contour energies are regularly spaced by $500\,cm^{-1}$. Red contours correspond to positive energies, and blue to zero and negative energies.

techniques (see the Methods for details), which are not applicable at ultracold temperatures. Therefore, the use of the close-coupling method[8] is compulsory, despite the fact that several other features of this system make the calculation tremendously intense—in fact, to our knowledge, no similar calculation has ever been performed before. Owing to the strong long-range and anisotropic potential, the small value of the rotational constant of $BaCl^+$, and the large relative mass, the calculation must be performed to extraordinarily large distances (up to 2,000 a_o) with matrices for a given value of J and parity that are as large as $10^4 \times 10^4$. Therefore, several theoretical and numerical improvements were required to make the calculations possible (see the Methods for details), including the development of a new scattering code using asynchronous task parallelization to calculate the vibrational quenching rates.

Figure 2b shows the calculated vibrational quenching rate coefficients for a selected set of rovibrational states of $BaCl^+$. A thick horizontal line represents the experimental measurement[4] of the population averaged vibrational relaxation rate for $v=1$ and $v=2$, where the length of the line is representative of the energy range in the experiment. The ion-induced-dipole Langevin law[9] is shown as a dashed line. The calculated values compare very well with the experiment, whereas the Langevin law is roughly double the experimental value. The close-coupling rate follows the Langevin law in the temperature domain of the experiment and departs from it at lower temperature—as discussed later, this departure is not due to quantum suppression[10] as one might expect. In Figures 2a, 2c, 2d, the vibrational and rotational quenching are compared for several initial rotational levels belonging, respectively, to the vibrational levels $v=1$, 2 and 3. The vibrational quenching is always larger than the rotational quenching. This very unusual result is due to the low value of the vibrational frequency of $BaCl^+$ and the deep potential well, together yielding a strong coupling between many vibrational levels. This is in contrast with previously studied atom-diatom van der Waal neutral or ionic complexes[11–13], where the potential well is usually not deep enough to couple even two different vibrational levels of the diatom. The only other possibility to obtain vibrational quenching comparable to rotational quenching is when the bond length of the complex is smaller than expected with a pure Van der Waals interaction, indicating the rise of chemical bonding induced by electron sharing between monomers. This is, for example, the case of the He-CH^+ complex[14].

QDT calculation of the vibrational quenching rate. Despite its obvious utility, the close-coupling calculation does not lend itself to an intuitive understanding of the collision physics. Therefore, we have also performed a QDT calculation[15–17] of $BaCl^+$-Ca vibrational quenching, where the radial Schrodinger equation is solved for the long-range, isotropic R^{-4} induction term from large separation and matched with boundary conditions at short range describing the amplitude and phase of flux returning from the chemical bonding region $\eta_{\ell m}(E)$ and $\delta_{\ell m}(E)$, respectively (see the Methods for details).

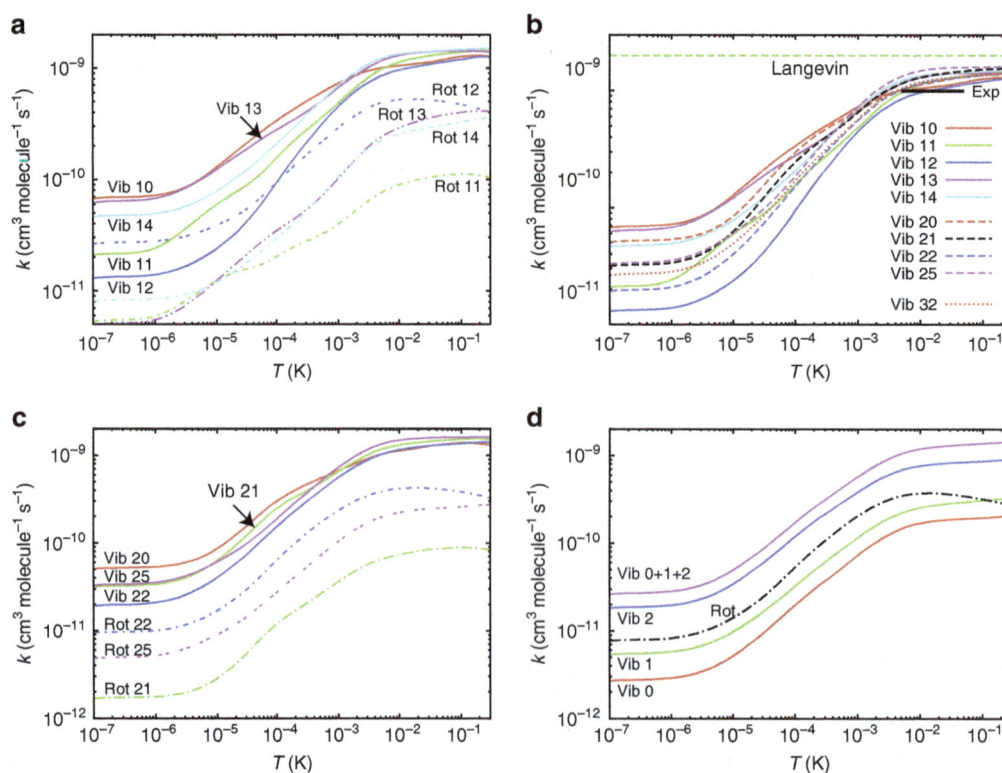

Figure 2 | Calculated quenching rates. (a) Comparison between the vibrational and rotational quenching rate coefficients for $BaCl^+$ in the initial states ($v=1, j=0,1,2,3,4$). The first and second numbers designate, respectively, the initial vibrational and rotational quantum number of $BaCl^+$. **(b)** Comparison between the vibrational quenching rate coefficients of several excited rovibrational levels (v, j) of $BaCl^+$ resulting from collisions with Ca with the experimental results and with the Langevin law. The first and second numbers designate, respectively, the initial vibrational and rotational quantum number of $BaCl^+$. **(c)** Comparison between the vibrational and rotational quenching rate coefficients for $BaCl^+$ in the initial states ($v=2, j=0,1,2,5$). The first and second numbers designate, respectively, the initial vibrational and rotational quantum number of $BaCl^+$. **(d)** Comparison between the vibrational and rotational quenching rate coefficients for $BaCl^+$ in the initial state ($v=3, j=2$). The label of each curve designates the final vibrational level.

Figure 3a shows the quenching rate coefficient in the so-called universal limit, where all collisions reaching short range lead to quenching, that is, $\eta_{\ell m}(E) = 0$. This rate agrees reasonably well with experiment and theory at the experimentally relevant energies, but dramatically overestimates the quenching rate at low energies. We thus conclude that the reduction in quenching rate at low energy is not due to quantum suppression as would be expected in systems with shorter ranged potentials[8]. Therefore, we match the QDT result to the close-coupling calculation by parameterizing the short-range boundary condition as $\delta_{\ell m}(E) = \delta_0$ and $\eta_{\ell m}(E) = \eta_0 + \eta_1 l(l+1)$ when $0 \leq \eta_{min} \leq \eta_{\ell m}(E) \leq 1$ and $\eta_{\ell m}(E) = \eta_{min}$ or 1 otherwise, where $\eta_0, \eta_{min}, \eta_1$ and δ_0 are determined by least-squares fitting (see the Methods for details). Figure 3b shows the good agreement between this fit and the close-coupling calculation and indicates that the suppression of the rate coefficient at low energies comes from lower quenching probabilities for small partial waves, as might be expected since the vibrational quenching is driven by asymmetry in the interaction potential.

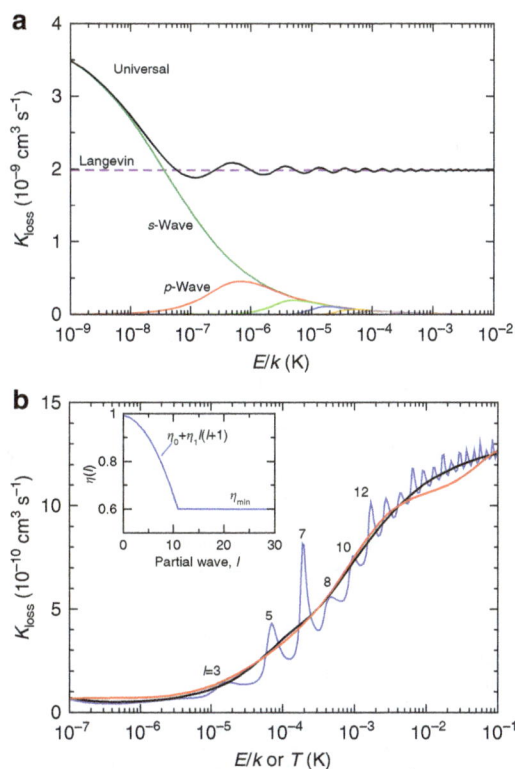

Statistical capture model for the vibrational quenching rate. In addition to these calculations, it is desirable to develop a model for vibrational quenching, which provides useful rules-of-thumb to both better understand the collision physics and guide future experiments. Unfortunately, the available models of vibrational quenching[18] are not appropriate in this regime. The Landau–Teller model[19], which is the most common model, is restricted to high collision energy. The only model available at low collision energy, owing to Dashevskaia and Nikitin[20], cannot be applied to ionic systems as it makes the hypothesis that the long-range interaction potential is weak. Therefore, in order to rationalize our results, we introduce a statistical capture model following the recent work of Lara et al. on the ultra-low-temperature reactivity of $D^+ + H_2$ reaction[21]. This model divides the quenching process into two steps: the formation of the collision complex through the long-range interaction potential followed by its fragmentation to produce a vibrationally quenched diatomic cation. This scheme relies on the statistical ansatz, which is valid for collisions involving deep intermediate wells and/or ultra low temperature. The capture process is described by the Langevin rate, whereas the probability of vibrational quenching is proportional to the number of accessible vibrational channels, which is roughly measured by the ratio D_e/ω_e, where D_e is the triatomic molecular ion dissociation energy and ω_e the diatomic vibrational frequency. Therefore, we define the statistical capture rate constant k_{10}^{SC}, independent of the temperature, as:

$$k_{10}^{SC} \sim \frac{D_e}{\omega_e} k_{\text{Langevin}} \sim \frac{D_e}{\omega_e} \left[2\pi \sqrt{\frac{\alpha}{\mu}} \right] \sim \frac{D_e}{\omega_e} \sqrt{\frac{\alpha}{\mu}} \quad (2)$$

The validity of this model can be tested by comparing to known atom-ion quenching rates. Figure 4 shows a comparison of the statistical capture model to the quenching rate calculated from the Wigner threshold law for Ca-BaCl$^+$ as well as several other species, which have been analysed by some of the authors[9–12]. Given the strong correlation with statistical capture model, we can create rules-of-thumb to aid future experiments realize efficient vibrational cooling. Namely, the vibrational quenching efficiency depends both on the strength of the long-range interaction potential, given by α, and on the density of states of the complex, given by D_e/ω_e. The increase of the state density increases the lifetime of the complex and facilitates vibrational quenching.

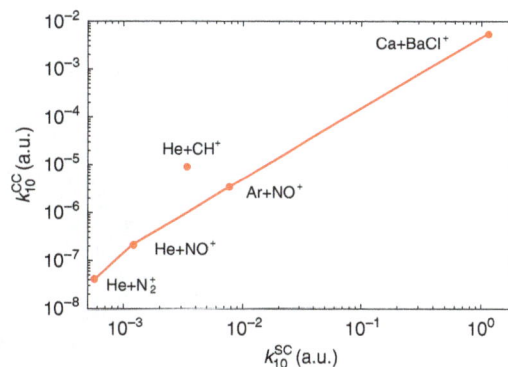

Figure 3 | Quantum defect theory (QDT) of the Ca + BaCl$^+$ collision. (a) Loss rate coefficients based on the QDT as functions of collision energy. The black solid line corresponds to the loss rate coefficient assuming universal or completely absorbing short-range boundary conditions. Coloured lines correspond to loss rates contributions from individual partial waves and its projections with $l \leq 3$. Finally, the dashed line is the loss rate coefficient found by the Langevin capture theory. **(b)** Loss rate coefficients K_{loss} as functions of collision energy or temperature for a partial-wave dependent QDT optimized to agree with the coupled-channels BaCl$^+(v=0, j=1)$ + Ca vibrational quenching rate coefficient. The blue curve shows $K_{loss}(E)$ as a function of collision energy, whereas the black curve shows the corresponding thermally averaged rate coefficient. The red curve corresponds to the close-coupling results 'Vib 1 0' shown in Fig. 2. Shape resonances in $K_{loss}(E)$ are assigned by their partial wave l. The inset shows the short-range amplitude with parameters $\eta_0 = 0.9916$, $\eta_1 = -0.003$ and $\eta_{min} = 0.6$. The short-range phase is $\delta_{\ell m}(E) = 0.34\pi$.

Figure 4 | Close-coupling and statistical capture comparison. Comparison between the close-coupling vibrational quenching rate coefficients $k_{10}^{CC}(T)$ with the statistical capture rate $k_{10}^{SC}(T)$ for five different colliding systems with the diatomic cation in the initial state $(v=1, j=0)$.

Discussion

Among the possible explanations of the very large rate coefficients for the vibrational quenching, which is even larger than those for rotational quenching, one could be tempted to think about quasi resonant vibration-rotation energy transfer. This type of transfer is well documented for neutral systems like the collisions of Li_2 and HF with neutral noble gases at high temperatures[22] and has also been predicted theoretically to occur for the neutral He-H_2 collision at very low temperatures[23]. Quasi resonant vibration-rotation energy transfer, however, is expected for molecules in highly excited initial rotational states and takes place when the rapidly rotating diatom is stretched to its outer turning point and collinear with the atom[24]. In the present case, the calculated vibrational quenching rate coefficients are already unusually large for the lowest $j = 0$-5 rotational quantum numbers of $BaCl^+$. This mechanism then cannot be at play.

The calculations presented here confirm the observation of efficient vibrational cooling in the Ca-$BaCl^+$ system and provide a simple means to predict new systems that will exhibit efficient cooling. As all laser coolable atoms exhibit large polarizabilities, the only requirement on future experiments is to choose molecular ions that are strongly bound. This requirement is not particularly restrictive as other requirements of the effort, for example, non-reactivity with the ultracold atom and the existence of a large dipole moment, are typically only satisfied for strongly bound molecular ions. Therefore, sympathetic cooling of molecular ions with ultracold atoms appears to be poised to provide a generic and robust route to harness the potential of molecular structure for science and technology.

Methods

PES calculation. The electronic ground state of the CaBaCl$^+$ complex is a $^1A'$ state. In preliminary computations, we found that the wavefunction is dominantly described by a single electronic configuration. Therefore, the interaction energy between Ca and BaCl$^+$ entities was computed using the coupled cluster method based on single and double electronic excitations and a perturbative treatment of triple excitations, CCSD(T). The counterpoise procedure was used to correct the interaction energy for the basis set superposition error. All electronic calculations were performed with the MOLPRO suite of programmes[7]. We used the def2-QZVPPD basis set[25], which was augmented by diffuse functions. For chlorine, this results on (21s,15p,5d,2f,1g) primitives, which were contracted into (10s,7p,5d,2f,1g) and for calcium (24s,18p,6d,3f) contracted into (11s,6p,4d,3f) basis sets. For the barium atom, only the valence electrons were described by the (8s,8p,5d,3f) primitives contracted into (7s,5p,3d,3f), whereas we used also the quasi-relativistic 10-valence-electron pseudopotential ECP46MWB[26] for the 46 inner electrons.

In this manner, the interaction energy was calculated on a 3D grid of points in the Jacobi coordinates space. The 3D grid is a direct product of the three 1D grids spanning the Jacobi coordinates r, R and θ. Here, r represents the BaCl$^+$ bond length, R the distance between Ca and the centre of mass of BaCl$^+$, and θ the angle between $\mathbf{r}$ and $\mathbf{R}$, with the linear structure Ba-Cl-Ca corresponding to $\theta = 0°$. For the r bond length, we selected 15 values ranging from 3.85 to 6.05 a_0. The distance R took 50 values varying from 4 to 50 a_0, with a variable step increasing from 0.2 to 10 a_0. The θ angle was uniformly distributed from 0° to 180° by step of 10°.

The evolution of the potential energy curve (PEC) and the dipole moment of the isolated BaCl$^+$ diatomic along the r coordinate were calculated with the Davidson corrected multi-reference configuration interaction (MRCI + Q) method[27] using the molecular orbitals calculated by the complete active space self consistent field (CASSCF) technique[28]. State-averaged CASSCF calculations (over 8 states with equal weight) were performed, along with 12 active orbitals and 6 active electrons. The basis set and the core pseudopotential defined above were used. A grid of 26 points from 3.4 to 20 a_0 was calculated.

The parallel and perpendicular static polarizations of BaCl$^+$ were calculated using the finite-field procedure. For that purpose, we computed the ground-state CASSCF/MRCI + Q energy, the CASSCF calculations being performed with the ground-state orbitals only. The polarizabilities were calculated on a grid of 17 points from 3.7 to 8 a_0. The ionization energy of BaCl$^+$ was obtained as the energy difference between the energy of BaCl$^+$ and of BaCl^{++}, both taken in their respective electronic ground state. The potential of BaCl^{++} was calculated at the CASSCF/MRCI + Q level where the CASSCF was state-averaged on two states and using 5 active electrons distributed in 12 active orbitals.

The functional form $V(r,R,\theta)$ of the PES is defined as the sum of the interaction energy V_I between Ca and BaCl$^+$ and the potential of the isolated diatomic

BaCl$^+$, V_{BaCl}:

$$V(r, R, \theta) = V_I(r, R, \theta) + V_{BaCl}(r). \tag{3}$$

The potential V_{BaCl} is obtained by cubic splines interpolation of the *ab initio* data. The functional form of the interaction energy $V_I(r,R,\theta)$ is defined as the sum of the short-range V^S and long-range V^L contributions, combined with the switching function $S(R)$:

$$V_I(r, R, \theta) = S(R) V^S(r, R, \theta) + (1 - S(R)) V^L(r, R, \theta) \tag{4}$$

with the short-range function defined by

$$V^S(r, R, \theta) = \sum_{k=0}^{k_{max}} (r - r_0)^k \sum_{k=0}^{l_{max}} P_l(\cos\theta) C_{kl}(R) \tag{5}$$

where $P_l(\cos\theta)$ are normalized Legendre polynomials. The coefficients $C_{kl}(R)$ have been first calculated at the points of the R-grid by the linear least squares method, using $k_{max} = 10$, $l_{max} = 10$ and $r_0 = 5$ a_0. Then, a cubic splines interpolation is used to obtain the coefficients for any value of R inside the limits of the R-grid.

The switching function, which ensures a smooth connection between the long-range and the short-range functions, is defined by

$$S(R) = \frac{1}{2}[1 - \tanh(A_0(R - R_0))] \tag{6}$$

with $R_0 = 20$ a_0, and $A_0 = 0.14$ a_0^{-1}.

The long-range part is defined as a the sum of the leading terms of induction and dispersion energies[29],

$$V^L(r, R, \theta) = U_{ind}(r, R, \theta) + U_{disp}(r, R, \theta) \tag{7}$$

$$U_{ind}(r, R, \theta) = -\frac{\alpha_{Ca}}{2R^4} + \frac{2\alpha_{Ca}\mu(r)\cos\theta}{R^5} - \frac{1}{2}\frac{\alpha_{Ca}\mu^2(r)}{R^6}(3\cos^2\theta - 1) \tag{8}$$

$$U_{disp}(r, R, \theta) = -\frac{1}{2}\frac{I_{Ca}I_{BaCl}}{I_{Ca}+I_{BaCl}}\frac{\alpha_{Ca}}{R^6}\left[\alpha_\parallel(r) + 2\alpha_\perp(r) + \frac{1}{2}[\alpha_\parallel(r) - \alpha_\perp(r)](3\cos^2\theta - 1)\right] \tag{9}$$

where the dipole moment μ, the parallel $\alpha_\parallel$ and perpendicular $\alpha_\perp$ polarizabilities of BaCl$^+$ were obtained by polynomial interpolation of the *ab initio* data

$$\mu(r) = \sum_{k=0}^{4} a_k(r - r_0)^k \tag{10}$$

$$\alpha_\parallel(r) = \sum_{k=0}^{5} b_k(r - r_0)^k \tag{11}$$

$$\alpha_\perp(r) = \sum_{k=0}^{5} c_k(r - r_0)^k \tag{12}$$

and where the coefficients a_k, b_k and c_k have been determined by the linear least square method. The polarizability of calcium[30] is $\alpha_{Ca} = 168.71$ a_0^3, the ionization energy of calcium[28] $I_{Ca} = 6.113$ eV, and the ionization energy of BaCl$^+$ $I_{BaCl} = 13.9$ eV. Let us note that, although using the long-range interaction model without any dependence in the internal coordinates is a customary approximation, we have chosen here to take in account the vibrational coordinate because this work focuses on vibrational quenching.

The spectroscopic properties of the fitted diatomic potential for BaCl$^+$(X$^1\Sigma^+$) are $r_e = 4.899$ a_0, $D_e = 39,278$ cm^{-1}, $\omega_e = 333.8$ cm^{-1} and $\omega_e x_e = 0.837$ cm^{-1}. These values are close to the theoretical data previously reported[31]. The dipole moment calculated at the equilibrium bond length is $\mu_e = 8.927$ Debye. This calculation was done with the origin of coordinates fixed at the centre of mass, using the average atomic masses.

We display in Fig. 1a the contour plot of the total potential energy along the r and R coordinates, and in Fig. 1b the contour plot of the interaction energy along the R and θ coordinates. The plot in Fig. 1a shows the existence of a relatively deep potential well in good agreement with the charge-transfer nature of the bonding within this ionic complex. Figure 1b reveals the existence of two minimal structures and two saddle points connecting these equilibrium structures. It also shows that the potential is strongly anisotropic. The existence of these stationary points on the ground potential of the Ca-BaCl$^+$ system was checked by *ab initio* geometry optimizations where all coordinates were kept free. The Jacobi coordinates of the global minimum of $V(r,R,\theta)$ are $r = 5.08$ a_0, $R = 6.54$ a_0 and $\theta = 54.7°$. This means that the calcium atom is bonded to the chlorine end of BaCl$^+$, with bond length $r_{CaCl} = 5.34$ a_0 and angle $\theta_{CaClBa} = 87.1°$. The electronic dissociation energy is $D_e^I = 6,263.8$ cm^{-1}. If we consider the interaction potential alone, the coordinates of the global minimum are then $r = 5.41$ a_0, $R = 6.49$ a_0 and $\theta = 52.8°$, and the corresponding dissociation energy is 7,441.9 cm^{-1}.

Although the Ba-Cl bond length is only slightly extended by the interaction with the calcium, we observe in Fig. 1a that the vibrational potential of BaCl$^+$ is significantly modified by the latter interaction. This indicates there is a significant coupling between the vibration of BaCl$^+$ and the other modes of motion. This coupling is expected to promote vibrationally inelastic collisions.

At long range, the induction term that scales as R^{-5} is the leading *anisotropic* interaction. The angular dependence (see equation 6) of this long-range interaction potential tends to align the three atoms Ca-Ba-Cl. Thus, at low collision energies, we expect a propensity for the alignment of the reactants, resulting in pendular states of BaCl$^+$ around the linear structure. It is only for $R < 10$ a_0 that the propensity for the linear structure begins to disappear gradually with decreasing R. The minimum of the potential energy of the linear structure is found at $r = 4.94$ a_0

and $R = 9.50\,a_0$, with a dissociation energy of $5{,}387.6\,\text{cm}^{-1}$. This minimum is a saddle point, as bending the Ca-Ba-Cl structure leads to the global minimum.

Close-coupling calculations. The use of the long established close-coupling method[6] to study the dynamics of atom-diatom rovibrational inelastic collisions seems at first to be a simple task. However, the deep potential well of the system and both the very small vibrational and rotational quanta[29] of $BaCl^+$ make the size of such calculations tremendous and one could be tempted to consider the use of more approximate alternative approaches like the infinite order sudden approximation[32] or the coupled states approximation[33,34] methods. Unfortunately, a prerequisite for use of the infinite order sudden approximation is that the rotational spacing of the diatom has to be negligible compared with the collision energy, which would become true above $10\,\text{cm}^{-1}$, but is not satisfied in the energy domain of the ultracold experiment. Further, this approximation requires that the well depth be small compared with the collision energy, such that couplings with the closed channels remain small. With a well depth of $7{,}442\,\text{cm}^{-1}$, the use of this approximation is then restricted to very high collision energies for this system. Similarly, the coupled states approximation method is expected to be valid for rotor states whose relative kinetic energy is large compared with the well depth; it is therefore not applicable to Ca-BaCl$^+$ at very low collision energy.

The use of the close-coupling method then appears to be compulsory for this collision, despite the fact that several other features of this system that make such calculations especially intense. First, the strong angular anisotropy of the potential and the small value of the rotational constant of $BaCl^+$ result in the coupling of many rotational levels. Second, the strong long-range potential requires propagation of the calculation to very long intermolecular distances. Third, the large relative mass necessitates many values of the total angular momentum quantum number J to reach convergence. The size of the matrices that have to be propagated for a given value of J and parity can then be as big as $10^4 \times 10^4$ (above a collision energy of $1\,\text{cm}^{-1}$), which make the calculations unusually heavy, even at very low collision energy.

Therefore, several theoretical and numerical improvements were required to make the calculations possible. First, the vibrational coupling matrices $\langle \varphi_{vj} | (r - r_0)^k | \varphi_{v'f} \rangle$, where φ_{vj} denotes the asymptotic rovibrational eigenfunctions, were calculated once and stored before the close coupling calculation. Thus, the large vibrational quadrature (about 100 points) necessary to compute these coupling terms was reduced to a summation over 11 terms (k runs from 0 up to 10) in the close-coupling calculation. Second, the calculation of the cross-sections for a single collision energy, even the lowest one, converged as a function of the total angular momentum, takes several hundred of hours of CPU time and reaches quickly (around a collision energy of $1\,\text{cm}^{-1}$) a thousand of hours. Therefore, we developed an MPI version of our Newmat code[35] using asynchronous task parallelization. The elementary task is the propagation of the wavefunction at one particular collision energy. The MPI code distributes N tasks over M processors. Because the tasks are independent, this parallelization scheme requires no overhead and feeds efficiently all the processors.

We included 25 rotational levels in each of the 19 vibrational levels used to perform the calculations. The propagation step size was taken to be $0.015\,a_0$ and the maximum distance of propagation was $400\,a_0$. The calculations were performed in the collision energy interval $(10^{-6}, 10^0)\,\text{cm}^{-1}$. The relative convergence criterion of the inelastic cross-sections as a function of the total angular momentum was taken to be 0.1% for the lowest energies, 1% for intermediate energies and 5% for the highest energies around $1\,\text{cm}^{-1}$.

Quantum defect theory. Despite its obvious utility, the close-coupling calculation does not readily lend itself to an intuitive understanding of the collision physics.

Therefore, we have also developed a QDT for scattering between $BaCl^+$ and Ca based on solutions of a single radial Schrödinger equation where the potential between the molecular ion and neutral atom is dominated by its longest-ranged and isotropic induction potential, $V(R) = -C_4/R^4$. At $R = R_{\text{short}}$, where R_{short} is a characteristic short-range separation defined more precisely below, these solutions are uniquely specified by a boundary condition with a limited number of parameters that summarize the reflections (elastic collisions) and absorption (quenching collisions) in the complex, anisotropic and possibly chaotic evolution at short separations. By carefully choosing the collision-energy and partial-wave dependence of the boundary condition, we reproduce the temperature dependence of the rate coefficients for rovibrational relaxation as obtained by close-coupling calculations. The behaviour of these boundary conditions gives a simple, intuitive picture of the collision processes.

The long-range induction potential is much longer ranged than the van der Waals potential between two neutral atoms. As a result, many partial waves contribute to the rates; in fact only for temperatures of $100\,\text{nK}$ are collisions s-wave dominated. Nevertheless, as we will show below, even for mK collision energies quantum mechanical effects profoundly affect the scattering of $BaCl^+$ and Ca and play an important role in the description of the interplay between inelastic and elastic collisional processes.

In our QDT, we numerically solve the single-channel radial Schrödinger equation for collision energy E, partial wave l and projection m from separation $R = R_{\text{short}}$ to ∞ with boundary condition $\psi_{lm}(R = R_{\text{short}}) \sim e^{i(y - \pi/4)} + \eta_{lm}(E)e^{2i\delta_{lm}(E)}e^{-i(y - \pi/4)}$, where $y = R_4/R$ and $R_4 = \sqrt{2\mu C_4/\hbar^2}$. The functions $\exp(\pm i[y - \pi/4])$ can be recognized as Wentzel–Kramers–Brillouin solutions of a $-C_4/R^4$ potential at zero collision energy and partial wave. The short-range amplitude $\eta_{lm}(E)e^{2i\delta_{lm}(E)}$ with real-valued functions $\eta_{\ell m}(E)$ and $\delta_{\ell m}(E)$ determines the flux returning from the chemical bonding region where all three atoms are separated by no more than $R_{\text{short}} << R_4$; we use $R_{\text{short}} \approx 30a_0$, just outside the separations where the electron wave functions have significant overlap. The short-range phase, in principle, are determined by the close coupling simulations. Flux conservation requires that $0 \leq \eta_{\ell m}(E) \leq 1$, where $\eta_{\ell m}(E) = 0$ corresponds to the case where no flux is returned from short range and $\eta_{\ell m}(E) = 1$ corresponds to the case where everything is reflected back. The phase $\delta_{\ell m}(E)$ describes the relative phase shift of the returning flux. In the limit $R \to \infty$, the wavefunction approaches $\psi_{lm}(R) \to e^{-i(kR - l\pi/2)} - S_{ii}(E, lm)e^{i(kR - l\pi/2)}$, where $E = \hbar^2 k^2/(2\mu)$, k is the relative collision wavevector, and $S_{ii}(E, lm)$ is the diagonal S-matrix element from which elastic and total inelastic rate coefficients can be determined. In fact, the total inelastic rate coefficient is $K_{\text{loss}}(E) = \sum_{lm} K_{\text{loss}}^{lm}(E)$ with $K_{\text{loss}}^{lm}(E) = v_{\text{r}} \frac{\pi}{k^2}(1 - |S_{ii}(E, lm)|^2)$, where v_{r} is the relative velocity. Here, we have used flux conservation or the unitarity of the S-matrix to rewrite the loss rate coefficient solely in terms of the diagonal S-matrix element, $S_{ii}(E, lm)$. At ultracold temperatures, only a few partial waves l contribute as for higher l the centrifugal barrier prevents the atom and molecule from approaching each other and $K_{\text{loss}}^{lm}(E)$ rapidly goes to zero with increasing l.

The $-C_4/R^4$ potential is the largest energy scale at $R = R_{\text{short}}$ and, consequently, we initially assume that $\eta_{\ell m}(E)$ and $\delta_{\ell m}(E)$ are independent of collision energy E, partial wave l and projection m. In fact, ultracold reactions between neutral KRb and K, with a long-range van der Waals dispersion potential, forming K_2 have been successfully modelled[36] with $\eta_{\ell m}(E) << 1$. It is then natural to first study the universal limit of $\eta_{\ell m}(E) = 0$, that is, all short-range collisions lead to quenching, for our induction potential. Figure 3a shows the universal loss rate coefficient as a function of E from $E/k = 1\,\text{nK}$ to $0.01\,\text{K}$. At very low energy, the rate is dominated by s-wave scattering and is a decreasing function of E. At higher energy, other

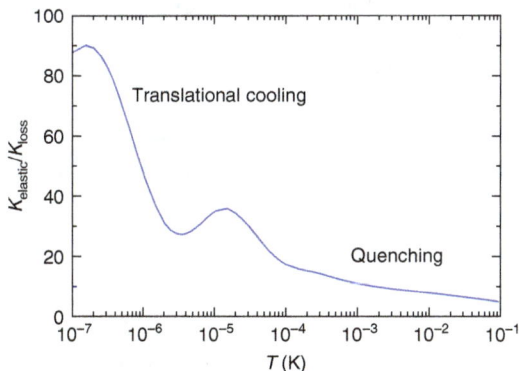

Figure 5 | Relative rate dependence on partial wave. Ratio of the thermalized elastic and loss rate coefficients, $K_{\text{elastic}}/K_{\text{loss}}$, as function of temperature for a partial-wave-dependent optimized QDT of $BaCl^+ + Ca$. Parameters as in Fig. 3b.

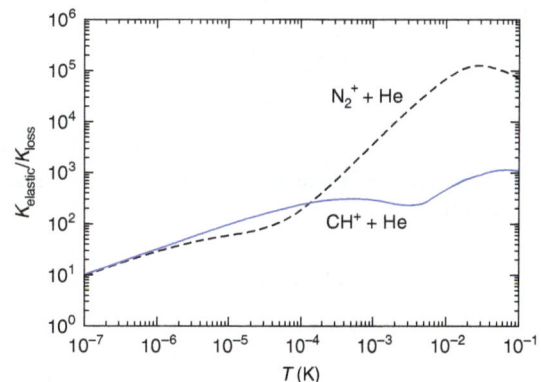

Figure 6 | Quantum defect theory (QDT) for previously studied systems. Ratio of the thermalized elastic and loss rate coefficients, $K_{\text{elastic}}/K_{\text{loss}}$, as function of temperature for the $N_2^+ + He$ (dashed line) and $CH^+ + He$ (solid line) systems using optimized QDT.

partial waves contribute creating a rate coefficient that weakly oscillates around the rate coefficient predicted by Langevin capture theory. Although the QDT universal approximation roughly agrees with both the experimental determination and the close-coupling calculation for energies above 1 mK, the universal result dramatically over estimates the quenching rate for temperatures below 1 mK.

Therefore, in order to reproduce the temperature dependence of the coupled-channel calculation with QDT, we must use a collision-energy- and partial-wave-dependent short-range amplitude $\eta_{lm}(E)e^{2i\delta_{lm}(E)}$. For practical purposes, the number of parameters that describe $\eta_{\ell m}(E)$ and $\delta_{\ell m}(E)$ must be limited and, here, we use the fact that QDT aims to only describe cross-sections over a small range of collision energies and limited number of partial waves. In our case, we need to represent collision energies below 0.1 K and partial waves up to 35. We then observe that at $R = R_{short}$ the rotational energy $\hbar^2 l(l+1)/(2\mu R_{short}^2)$ for $l = 35$, although much smaller than the $-C_4/R_{short}^4$ potential energy, is much larger than the relevant range of E. Hence, we can assume that the short-range parameters only depend on l. In fact, we further limit the parameterization to $\delta_{\ell m}(E) = \delta_0$ and $\eta_{\ell m}(E) = \eta_0 + \eta_1 l(l+1)$ when $0 \leq \eta_{min} \leq \eta_{\ell m}(E) \leq 1$ and $\eta_{\ell m}(E) = \eta_{min}$ or 1 otherwise. There are four free variables η_0, η_{min}, η_1 and δ_0.

Figure 3b shows the best fit of QDT to the $BaCl^+$ ($v = 1, j = 0$) + Ca vibrational quenching rate coefficient. The figure shows both its energy dependence and the corresponding thermally averaged value. The partial wave dependence of $\eta_{\ell m}(E)$ is shown in the inset. The energy-dependent rate coefficient is less than 10^{-10} cm^3 s^{-1} for $E/k \approx 1$ µK. For these energies only $l = 0$ and 1 contribute, $\eta_{\ell m}(E)$ is close to one, and losses are suppressed. The rate coefficient then increases, interrupted by multiple resonances, to a rate coefficient of about one half of the Langevin rate coefficient near $E/k \approx 0.1$ K, where nearly 35 partial waves contribute and $\eta_{\ell m}(E) = \eta_{min}$. In the thermally averaged rate coefficient, the resonances, except for those below $T = 100$ µK, have washed out. Rate coefficients for other rovibrational levels (v, j) can be found in a similar manner and lead to slightly different parameters η_0, η_{min}, η_1 and δ_0.

The location and origin of the series of resonances explains much of our successful fit. They are shape resonances behind the even partial-wave centrifugal barriers. Analysis of the analytic $E = 0$ solution of the $-C_4/R^4$ potential with short-range amplitude $\eta_{\ell m}(E = 0) = 1$ and l-independent phase $\delta_{\ell m}(E = 0)$ has shown that if an $E = 0$ bound state occurs for partial wave l then such bound state also exist for the ..., l-4, l-2, l+2, l+4, ... partial waves. This condition is only satisfied for two values of the short-range phase δ_0, one corresponding to all even partial waves, one to the odd waves. For different values of δ_0 as well as $\eta_{\ell m}(E) < 1$, these bound states become shape resonances where the inelastic loss is resonantly enhanced. For $BaCl^+ + Ca$ collisions with its $\eta_{\ell m}(E) \approx \eta_0 \approx 1$ for small partial waves, we find that near the optimal δ_0 resonances occur for odd partial waves and, in particular, near $E/k = 10$ µK the rate coefficient is sensitive to the location of the $l = 3$, f-wave resonance. For $E/k > 0.5$ mK, where $l \geq 8$ wave collisions become prominent, the amplitude $\eta_{\ell m}(E)$ deviates sufficiently from one, such that resonances from even partial wave collisions are observed. Finally, for the optimal conditions, the absolute value of the elastic scattering length a for s-wave $BaCl^+ + Ca$ collisions is larger than the natural length scale of the C_4 potential. In this case, the finite Wigner-threshold prediction for the loss rate coefficient is only reached for collision energies below $E/k = 1$ nK not shown in Fig. 3b. Figure 5 predicts the ratio of the QDT elastic and loss rate coefficient as a function of temperature. Ratios much larger than one indicate that the kinetic energy in the centre-of-mass motion of the molecular ion is more efficiently relaxed than its vibrational energy. For example, for $BaCl^+$ ions at $T = 0.1$ K, the ratio is close to one. We anticipate two cooling stages for $BaCl^+$ ions in collisions with ultracold and highly polarizable Ca atoms. In the first stage cooling or relaxation of the rovibrational states occurs, as the small ratio $K_{elastic}/K_{loss}$ implies that translational cooling in this region will be inefficient. Only when the molecules are in the lowest rovibrational state, the second stage, will elastic collisions cool the external motion of the molecules.

Traditional cooling schemes often involve collisions of ionic molecules with He gas at cryogenic temperatures. Here, we analyse the ratio $K_{elastic}/K_{loss}$ for two ionic systems $He-N_2^+$ and $He-CH^+$ to demonstrate a different cooling mechanism than for cooling with Ca. First, we fit the $He-N_2^+$ and $He-CH^+$ vibrational quenching rate coefficients obtained in the coupled-channel calculations[10,12] to our QDT

theory with partial wave-dependent short-range parameters. The parameters of our best fit are $\eta_0 = 1-5.7{*}10^{-6}$, $\eta_1 = -2{*}10^{-6}$, $\delta_{\ell m}(E) = 0.76\pi$ for $He-N_2^+$ and $\eta_0 = 1-7{*}10^{-3}$, $\eta_1 = 0$ and $\delta_{\ell m}(E) = 0.06\pi$ for $He-CH^+$. Figure 6 shows the temperature-dependent ratio $K_{elastic}/K_{loss}$ for these two systems. It is evident that for both systems the elastic rate coefficient is much larger than that of the inelastic processes. Hence, for a molecular ion in a given excited rovibrational level its translational motion will be cooled first. Occasionally, an inelastic process relaxes this internal state at the cost of rapid increase of the translational temperature. Elastic collisions will then start the cooling all over again.

Statistical capture model. To rationalize our results, we compare in Table 1 five collisions involving a diatomic cation and a rare gas. We report in this table the dissociation energy of the complex, the vibrational frequency of the diatomic cation, the relative mass of the system, the polarizability of the rare gas atom and the imaginary part of the scattering length β_{10} associated with the collisional vibrational quenching of the lowest rotational level of the first excited vibrational level of the diatomic cation. In the Wigner regime, the quenching rate coefficient is independent of the temperature and directly proportional to β_{10}:

$$k_{10} = \frac{4\pi\beta_{10}}{\mu}. \tag{13}$$

The values reported in the table were obtained from our close-coupling calculations for the collision energy of 10^{-6} cm^{-1}, where only the s-wave contributes to the collision. The propagation at such low collision energy has to be performed up to very large values of the intermolecular separation coordinate. For $Ca-BaCl^+$, this maximum distance had to be extended to 2,000 a_0. Because of the long-range ion induced dipole potential, which furthermore requires applying modified effective range theory[37], the use of the scattering length approximation is limited to even lower collision energies. The usual formulas linking the real and imaginary parts of the scattering length with the scattering S matrix are in this case valid only for collision energies lower that 10^{-10} cm^{-1}. The properties of the $Ca-BaCl^+$ system, namely a very low vibrational frequency, a deep potential well, a large value of the three quantities—equilibrium diatomic distance, relative mass and atomic rare gas polarizability—are seen to differ strongly from those of the other systems. But what makes $BaCl^+$ especially singular is the unusually small value of its vibrational frequency (330 cm^{-1})—compared with ~2,000 cm^{-1} for the other diatoms considered. As seen in Table 1, the vibrational quenching scattering length β_{10} monotonously increases as a function of the triatomic ion well depth D_e and also as a function of D_e divided by the diatom vibrational frequency ω_e. This correlation was noticed long ago by many authors[18] and the

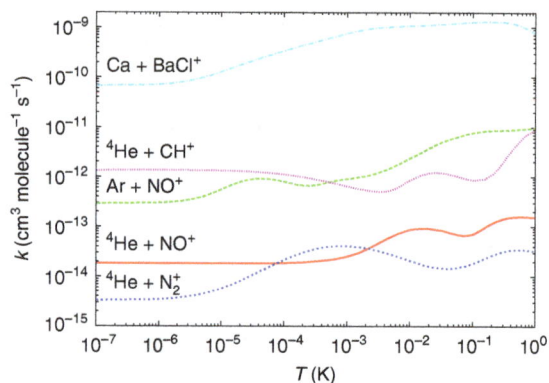

Figure 7 | Comparison of quenching rates for present and previously studied systems. Close coupling vibrational quenching rate coefficients $k_{10}^{CC}(T)$ for five different colliding systems with the diatomic cation in the initial state ($v = 1, j = 0$).

Table 1 | Comparison of some of the main features of several atom-cationic diatom vibrationally inelastic collisions in the limit of zero temperature.

System A-BC$^+$	ω_e(cm^{-1})	B_{rot}(cm^{-1})	R_e(a_0)	D_e(cm^{-1})	μ (g mol^{-1})	β_{10} (a_0)	α (Å^3)	Reference
^{4}He-N$_2^+$	2,207	1.93	6.08	-84.5	3.50	$2.5\ 10^{-4}$	0.2	ref. 12
^{4}He-NO$^+$	2,376	1.997	5.26	-195.4	3.53	$1.38\ 10^{-3}$	0.2	ref. 13
^{4}He-CH$^+$	2,380	14.24	4.10	-513.6	3.06	$5.3\ 10^{-2}$	0.2	ref. 14
Ar-NO$^+$	2,376	1.997	5.86	-980.4	17.1	$1.1\ 10^{-1}$	1.64	ref. 11
Ca-BaCl$^+$	334	0.09	6.49	$-7,442$	40	$3.26\ 10^2$	22.8	

β_{10} is the value of the Close Coupling imaginary part of the scattering length for the collisional vibrational quenching rate coefficient of the ($v = 1, j = 0$) state of the cationic diatom, computed at the collision energy of 10^{-6} cm^{-1}. For each system, the relative mass μ, the equilibrium intermonomers distance R_e, the well depth D_e of the triatomic ABC$^+$ complex, the diatom vibrational frequency ω_e, the diatom rotational constant B_{rot} and the polarizability of the impinging atom are reported.

Landau–Lifshitz probability for vibrational quenching of neutral molecules used by Dashevskaya and Nikitin[18], for example, is mainly dependent on both parameters ω_e and D_e. At very low collision energy, the long-range part of the interaction potential plays a major role, whereas the ratio D_e/ω_e offers a crude measurement of the efficiency of the vibrational energy redistribution. Therefore, we define the statistical capture rate constant k_{10}^{SC}, independent of the temperature, by simply multiplying the Langevin rate coefficient by this ratio D_e/ω_e:

$$k_{10}^{SC} \sim \frac{D_e}{\omega_e} k_{\text{Langevin}} \sim \frac{D_e}{\omega_e}\left[2\pi\sqrt{\frac{\alpha}{\mu}}\right] \sim \frac{D_e}{\omega_e}\sqrt{\frac{\alpha}{\mu}} \quad (14)$$

This relation can be formally introduced through a capture statistical approach, which was presented in the body of this manuscript. This simple relation makes the vibrational quenching proportional not only to the D_e/ω_e ratio but also to the polarizability of the ultra cold atom and to the relative mass of the system. These two last results were expected as they express the dependence of the quenching efficiency on the strength of the long-range interaction potential and on the density of states of the complex. The increase of the state density increases the lifetime of the complex and facilitates vibrational quenching. We notice, however, that the $^4\text{He-CH}^+$ system seems to behave differently than the other systems. This is effectively what can be seen on Fig. 7 where the Close coupling rate coefficients were reported for these five systems in the $[10^{-7}, 1]$ Kelvin interval. The $^4\text{He-CH}^+$ rate coefficient is the only one for which the rate coefficient decreases instead of increasing above the Wigner regime. This behaviour was shown in our paper dedicated to this system[12] to be due to virtual state scattering[38]. This is indeed the only one of these five systems for which the real part of the scattering length is negative. Virtual state scattering increases the close coupling value of the quenching rate coefficient and our simple model then allows predicting a lower bound for the zero temperature limit for all the studied systems.

References

1. Carr, L. D. et al. Cold and ultracold molecules: science, technology and applications. N. J. Phys. **11**, 055049 (2009).
2. Hutzler, N. R., Lu, H. I. & Doyle, J. M. The Buffer Gas Beam: an intense, cold, and slow source for atoms and molecules. Chem. Rev. **112**, 4803–4827 (2012).
3. Hansen, A. K. et al. Efficient rotational cooling of coulomb-crystallized molecular ions by a helium buffer gas. Nature **508**, 76–79 (2014).
4. Campbell, W. C. et al. Time-domain measurement of spontaneous vibrational decay of magnetically trapped NH. Phys. Rev. Lett. **100**, 083003 (2008).
5. Rellergert, W. G. et al. Evidence for sympathetic vibrational cooling of translationally cold molecules. Nature **495**, 490–495 (2013).
6. Hudson, E. R. Method for producing ultracold molecular ions. Phys. Rev. A **79**, 032716 (2009).
7. Werner, H.-J. et al. MOLPRO http://www.molpro.net (2012).
8. Arthurs, A. & Dalgarno, A. The theory of scattering by a rigid rotator. Proc. R. Soc. Lond. Ser. A **256**, 540–551 (1960).
9. Levine, R. D. & Bernstein, R. B. Molecular Reaction Dynamics and Chemical Reactivity (Oxford Univ., 1987).
10. Côté, R., Heller, E. J. & Dalgarno, A. Quantum suppression of cold atom collisions. Phys. Rev. A **53**, 234–241 (1996).
11. Halvick, P., Stoecklin, T., Lique, F. & Hochlaf, M. Explicitly correlated treatment of the Ar-NO$^+$ cation. J. Chem. Phys. **135**, 044312 (2011).
12. Stoecklin, T. & Voronin, A. Strong isotope effect in ultracold collison of N$_2^+$ ($v=1$, $J=0$) with He: a case study of virtual-state scattering. Phys. Rev. A **72**, 042714 (2005).
13. Stoecklin, T. & Voronin, A. Vibrational and rotational cooling of NO$^+$ in collisions with He. J. Chem. Phys. **134**, 204312 (2011).
14. Stoecklin, T. & Voronin, A. Vibrational and rotational energy transfer of CH$^+$ in collisions with ^{4}He and ^{3}He. Eur. Phys. J. D **46**, 259–265 (2008).
15. Idziaszek, Z. & Julienne, P.S. Universal rate constants for reactive collisions of ultracold molecules. Phys. Rev. Lett. **104**, 113202 (2010).
16. Kotochigova, S. Dispersion interactions and reactive collisions of ultracold polar molecules. N. J. Phys. **12**, 073041 (2010).
17. Gao, B. Quantum defect theory for -1/r^4-type interactions. Phys. Rev. A **88**, 022701 (2013).
18. Smith, I. W. M. Collisional energy transfer, intramolecular vibrational relaxation and unimolecular reactions. J. Chem. Soc. Faraday. Trans. **93**, 3741–3750 (1997).
19. Ferguson, E. E. Vibrational quenching of small molecular ions in neutral collisions. J. Phys. Chem. **90**, 731–738 (1986).
20. Dashevskaya, E. I. & Nikitin, E. E. Quantum Suppression and enhancement of the quasiclassical Landau-Lifshitz matrix elements: Application to the inelastic H$_2$-H scattering at ultralow energies. Phys. Rev. A. **63**, 012711 (2000).
21. Lara, M. et al. Cold and ultracold dynamics of the barrierless D$^+$ + H2 reaction: Quantum reactive calculations for ~R^{-4} long range interaction potentials. J. Chem. Phys. **143**, 204305 (2015).
22. Miklavc, A., Markovic, N., Nyman, G., Harb, V. & Nordholm, S. Mechanism of quasiresonant vibration–rotation energy transfer in atom–diatom encounters. J. Chem. Phys. **97**, 3348–3356 (1992).
23. Forrey, R. C., Balakrishnan, N., Dalgarno, A., Haggerty, M. R. & Heller, E. Quasiresonant energy transfer in ultracold atom-diatom collisions. Phys. Rev. Lett. **82**, 2657–2660 (1999).
24. Magill, P. D., Stewart, B., Smith, N. & Pritchard, D. E. Dynamics of Quasiresonant Vibration-Rotation Transfer in Atom-Diatom Scattering. Phys. Rev. Lett. **60**, 1943–1946 (1988).
25. Rappoport, D. & Furche, F. Property-optimized Gaussian basis sets for molecular response calculation. J. Chem. Phys. **133**, 134105 (2010).
26. Kaupp, M. et al. Pseudopotential approaches to Ca, Sr, an Ba hydrides. Why are some alkaline earth MX$_2$ Compounds Bent? J. Chem. Phys. **94**, 1360–1366 (1991).
27. Werner, H.-J. & Knowles, P. J. An efficient internally contracted multiconfiguration–reference configuration interaction method. J. Chem. Phys. **89**, 5803 (1988).
28. Werner, H.-J. & Knowles, P. J. A second order multiconfiguration SCF procedure with optimum convergence. J. Chem. Phys. **82**, 5053 (1985).
29. Buckingham, A. D. Permanent and induced molecular moments and long-range intermolecular forces. Adv. Chem. Phys. **12**, 107–142 (1967).
30. CRC Handbook of Chemistry and Physics 75 edn (ed Lide, D. R.) CRC Press, 1995).
31. Chen, K. et al. Molecular ion trap-depletion spectroscopy. Phys. Rev. A **83**, 030501 (2011).
32. Goldflam, R., Green, S. & Kouri, D. J. Infinite order sudden approximation for rotational energy transfer in gaseous mixtures. J. Chem. Phys. **67**, 4149–4161 (1977).
33. Kouri, D. J., Heil, T. G. & Shimoni, Y. Sufficiency conditions for the validity of the j_z-conserving coupled states approximation. J. Chem. Phys. **65**, 1462–1473 (1976).
34. McGuire, P. Validity of the coupled states approximation for molecular collisions. Chem. Phys. **13**, 81–94 (1976).
35. Stoecklin, T., Voronin, A. & Rayez, J. C. Vibrational quenching of N$_2$ ($v=1$,$J_{\text{rot}}=j$) by ^{3}He: Surface and Close-Coupling Calculations at Very Low Energy. Phys. Rev. A **66**, 042703 (2002).
36. Ospelkaus, S. et al. Quantum-state controlled chemical rections of ultracold potassium-rubidium molecules. Science **327**, 853–857 (2010).
37. O'Malley, T. F., Spruch, L. & Rosenberg, L. Modification of Effective-Range Theory in the Presence of a Long-Range (r^{-4}) Potential. J. Math. Phys **2**, 491–498 (1961).
38. Joachain, C. J. Quantum Collision Theory third edition (North-Holland Publishing Company, 1983).

Acknowledgements

M.A.G. and M.H. acknowledge Marie Curie International Research Staff Exchange Scheme Fellowship within the 7th European Community Framework Program under Grant No. PIRSES-GA-2012-31754 and COST ACTION CM1405 MOLIM. Computer time for this study was provided by the Mésocentre de Calcul Intensif Aquitain computing facilities of Université de Bordeaux and Université de Pau et des Pays de l'Adour. E.R.H. and S.K. acknowledge support from the ARO, ARO-MURI (grant Nos. W911NF-15-1-0121, W911NF-14-1-0378) and NSF (grant Nos. PHY-1255526, PHY-1308573).

Author contributions

M.A.G. and M.H. did the CCSD(T) calculations. P.H. did the CASSCF/MRCI calculations and the PES interpolation. T.S. worked out the statistical capture model. M.A.G. did the calculation of the cross-sections, using the quantum collision software written by T.S. S.K. developed the quantum defect theory, contributed to the interpretation of the results and wrote sections of the manuscript. E.R.H. contributed to the interpretation of the results and the writing of the manuscript.

Additional information

10

Thermodynamics of deposition flux-dependent intrinsic film stress

Amirmehdi Saedi[1,†] & Marcel J. Rost[1]

Vapour deposition on polycrystalline films can lead to extremely high levels of compressive stress, exceeding even the yield strength of the films. A significant part of this stress has a reversible nature: it disappears when the deposition is stopped and re-emerges on resumption. Although the debate on the underlying mechanism still continues, insertion of atoms into grain boundaries seems to be the most likely one. However, the required driving force has not been identified. To address the problem we analyse, here, the entire film system using thermodynamic arguments. We find that the observed, tremendous stress levels can be explained by the flux-induced entropic effects in the extremely dilute adatom gas on the surface. Our analysis justifies any adatom incorporation model, as it delivers the underlying thermodynamic driving force. Counterintuitively, we also show that the stress levels decrease, if the barrier(s) for adatoms to reach the grain boundaries are decreased.

[1] Huygens-Kamerlingh Onnes Laboratory, Leiden University, Niels Bohrweg 2, Leiden 2333 CA, The Netherlands. † Present address: ARCNL, Science Park 102, Amsterdam 1098 XG, The Netherlands. Correspondence and requests for materials should be addressed to M.J.R. (email: rost@physics.leidenuniv.nl).

During the growth of a polycrystalline film on a substrate, the film usually develops a significant amount of internal stress. If the film temperature is high enough to reach Volmer–Weber-type growth conditions[1,2], the film stress during deposition follows a compressive–tensile–compressive evolution, as is indicated with stages I, II and III in Fig. 1. During stage I, the nucleated islands develop a compressive stress due to surface tension effects[3,4]. Stage II occurs during film closure when the three-dimensional (3D) islands coalesce and form grain boundaries (GBs). At this stage, the film free energy can be further lowered by GB zipping, which in turn delivers tensile stress[5,6]. Without sufficient mobility (low temperature or high deposition flux) the film remains tensile on further growth. In contrast, a maximum tensile stress develops for Volmer–Weber-type growth of high-mobility materials, which occurrence coincides approximately with the moment the film closes. From this moment on, the stress turns once again towards compressive values (stage III)[7]. Surprisingly, a significant part of the compressive stress has a reversible nature: on interrupting the deposition flux (Fig. 1), the film stress jumps to less compressive values and the original compressive stress state before interruption is almost fully restored when the flux is switched on again. These stress jumps can be as large as ∼ 150 MPa and the time constant of the stress variation on resuming the deposition is in the order of 20 s (refs 8,9).

In the last 20 years several mechanisms have been proposed aiming to explain the observed effects: (1) pre-coalescence surface tension continuation combined with ongoing grain growth[10]; (2) surface roughness development during deposition combined with step–step interactions[2]; (3) adatom insertion into GBs[11–13]; (4) interaction of adatoms with surface and each other[9]; (5) inside bundling–outside grooving of GBs[14]; (6) depth changes in the GB grooves[15]; and so on. Whereas several of these mechanisms rely on kinetically limited processes, the GB adatom insertion model suggests that the compressive stress is generated via adatoms that are forced into the GBs by the enhanced chemical potential (CP) of the surface that is set-up by the deposition flux. By switching off this flux, the CP should drop, which should lead to an outflow of the excess atoms from the GBs and, thereby, to a relaxation of the compressive stress[11]. Recent experiments confirmed that GBs are prerequisite for the existence of the reversible stress jumps[16]. However, more questions arise, as the time constant of the stress relaxation on interruption seems to be temperature-independent[17]. On the other hand, surface stress effects[9] are expected to be too low in magnitude[18] to explain the reversible stress jumps. While the discussion on the mechanism(s) still continues, at a more fundamental level, the underlying driving force behind the effect has never been addressed.

In this paper we derive the magnitudes and the changes of the CP on the surface next to the position of the GBs and show that this indeed forms the driving force for any adatom insertion model. From a thermodynamic point of view, the most fundamental question has never been addressed, probably due to conceptual difficulties in calculating the CP of the surface during the growth: 'how can a flux (change) as low as ∼ 0.1 monolayer per second (ML per s) lead to stress jumps as high as ∼ 150 MPa?' Our study focuses exactly on this question and we show not only that these low fluxes can generate such huge stresses but also that the driving force for the stress jumps is decreased, if it is easier for the adatoms to diffuse towards and into the GBs.

Results

Basic thermodynamic description. To derive our model, it is important to realize that the film is under growth conditions and therefore naturally not in equilibrium. However, as long as the growth conditions do not change, it can be treated in steady state, like the famous Growth–Wulff construction[19]. The enhanced surface CP with respect to equilibrium sets up an adatom current to steps, which finally leads to the film growth. This also means that the CP on the surface varies locally and that positions connected to each other will try to balance their difference. If atom transport is sufficiently active on the timescale of consideration, one can approximate adjacent positions to be in equilibrium. Therefore, for constant small deposition fluxes and the absence of kinetic limitations, thermodynamic equilibrium can be assumed between the positions on the surface immediately next to grain boundaries (s/GB), the GBs and the grain interiors (g). This assumption is further underpinned by the small number of total additional atom that have to be incorporated in the GBs. For the surface we solve rate equations to determine the CP immediately next to the GBs and we further treat this position, the GBs, and the grain interior to be in equilibrium. Thermodynamic equilibrium certainly does not hold for the transition between the flux on and off states, but is justified a few tens of seconds after the flux change (see above). Therefore, at constant or zero flux, a change of the CP of the surface next to the GBs, will finally change the CP of the GBs with the same amount, which in turn will change the CP of the grains:

$$\Delta\mu_{s/GB} \approx \Delta\mu_{GB} \approx \Delta\mu_g \Rightarrow \Delta\mu_{s/GB} \approx \Delta\mu_g \quad (1)$$

This core equation enables us to bypass the determination of the CP of the GBs, as well as the absolute CP values on the surface and within the grains.

Chemical potential of the surface. The free energy of a surface depends on the formation and interaction energies of a myriad of surface features such as terraces, steps, kinks, step adatoms, adatoms and so on. As the surface morphology evolves during deposition, the population of these features changes accordingly. However, it is known that the reversible, compressive stress can develop within seconds after starting the flux even with rates as low as 0.1 ML per s. Obviously, the population of point-like features, like adatoms, step adatoms and kinks, can (and will) change abruptly on the arrival of flux on the surface, but extended surface features, like terraces and step edges, do not change significantly within such short timescales, as they consist of a large number of atoms[20]. For example, the surface roughness is directly linked to changes in the appearance, distribution and amount of steps and terraces. The compressive–tensile–compressive behaviour is usually observed under step-flow growth mode conditions, where changes in the surface roughness are known to happen very slowly[21,22]. This means that the gradual increase in

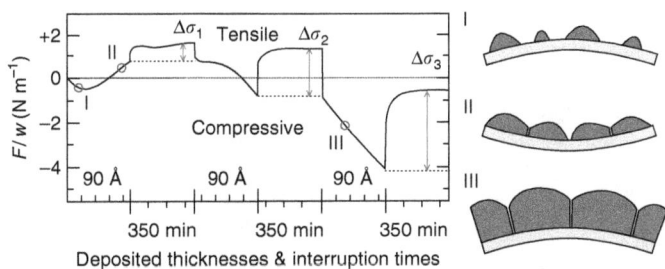

Figure 1 | Stress evolution during Volmer–Weber-type film growth for copper deposition with a flux of 0.1 Å s⁻¹ onto silicon oxide at room temperature. It consists of three main stages: nucleation (I); coalescence (II); and thickening (III). The deposition was interrupted three times for 350 min. The reversible stress jumps are indicated by $\Delta\sigma_i$. Part of graph reprinted with permission from ref. 8. Copyright *Journal of Applied Physics*, 1996, AIP Publishing LLC.

surface roughness (and hence the extended surface features) during deposition will only have a long time effect on the surface CP via the Gibbs–Thomson relation. Since the reversible stress jumps occur in a matter of seconds, we safely can ignore these long-term changes in our analysis. Moreover, in contrast to the adatoms that live in a two-dimensional (2D)-space on the terraces, the step adatoms and step kinks are confined to the one-dimensional space on the step edges. This causes the rate, at which the step adatoms and kinks meet and annihilate each other, to be significantly higher than the adatoms on the terraces. As a result, the increase in adatom population on the terraces, on starting the deposition, is orders of magnitude higher than that of kinks or step adatoms[23]. The conclusion is that the surface CP change, which is responsible for the almost instantaneous stress jumps, is mainly dominated by a change in the adatom density. Consequently, we can neglect all other contributions, as they would lead only to higher-order correction terms in determining the surface CP variations:

$$\Delta\mu_s = \Delta\mu_{adatom} + O\left(\Delta\mu_{step\,adatom},\ \Delta\mu_{kink},\ \Delta\mu_{step},\ ..\right)$$
$$\approx \Delta\left[\frac{\partial U_{adatom}}{\partial N} + \frac{\partial U_{adatom\,int.}}{\partial N} - T\frac{\partial S_{adatom}}{\partial N}\right] \quad (2)$$

The first term in the brackets, $\partial U_{adatom}/\partial N$, accounts for the surface temperature-dependent change in average energy (potential and kinetic) of individual adatoms. Given by the radiation of the evaporator and the kinetic energy of the arriving atoms, the increase in film temperature is $<10\,$K for Cu, Ag and Au (ref. 24), which corresponds to $\sim2.6\,$meV per film atom according to the classical Dulong–Petit limit of the heat capacity in solids. As these materials all have an excellent thermal conductivity, the surface and the bulk temperatures are virtually identical. Since we finally have to compare only the CP variation of the surface and the grains, we can safely neglect this term, as we would have to add the same value to both sides of equation (1).

The second term $\partial U_{adatom\,int.}/\partial N$ corresponds to the interaction energy between the individual adatoms given by (combinations of) van der Waals, electrostatic (dipole), elastic and electronic (substrate-mediated) effects[25]. We safely can ignore this term, as scanning tunneling microscopy (STM) experiments at $\sim15\,$K have shown that the absolute value of the interaction energy drops below $0.1\,$meV for two Cu adatoms separated more than $60\,$Å on a Cu(111) surface[26]. This is equivalent to an adatom density (fractional coverage) of $<6.0\times10^{-4}\,$ML, and as it will be shown in the following, we never reach such densities during the deposition.

The third term $-T\times\partial S_{adatom}/\partial N$ involves the entropic effects of the 2D adatom gas. In general, depending on the adatom mobility, adatoms can be assumed to be confined on discreet lattice sites (adatom lattice gas) or to be delocalized behaving as a 2D van der Waals surface gas (2D adatom gas)[27] (Supplementary Note 1). As these two models naturally set the lower and upper limits for the adatom gas entropy, we calculated the boundary values of the CP for copper in Fig. 2a (Supplementary Fig. 1). Although the absolute values differ more than $0.3\,$eV, both models show a linear behaviour in this logarithmic plot below $0.01\,$ML such that the following approximation holds for the CP variations:

$$\Delta\left[-T\frac{\partial S}{\partial N}\right] \approx k_B T\ln(\theta_2/\theta_1) \quad (3)$$

Adatom density on terraces. To calculate the adatom densities during deposition and interruption, Fig. 2b shows a simplified model of the film surface, in which we define the position of the first lattice row next to the ascending step edge as the origin of a terrace with width w in lattice units. By solving the differential equation for mass conservation, the adatom density at site n on the terrace, θ_n, can be derived as a function of deposition flux F (Supplementary Note 2 and Supplementary Figs 2, 3 and 4):

$$\theta_n = \theta_{eq} + \frac{Fw(an+1)(sw+2)}{2v_d(asw+a+s)} - \frac{Fn^2}{2v_d} \quad (4)$$

where $\theta_{eq} = \exp(-E_{form}/k_B T)$, $v_d = v_0\exp(-E_{diff}/k_B T)$, $a = \exp(-\Delta E_{att}/k_B T)$ and $s = s_0\exp(-\Delta E_{ES}/k_B T)$, in which v_0, E_{diff}, E_{form}, ΔE_{att}, s_0 and ΔE_{ES} are the diffusion rate prefactor, diffusion barrier, adatom formation energy from a kink site of the step, the attachment barrier, correction prefactor for hop over the step and the Ehrlich–Schwoebel barrier, respectively (Fig. 2b)[27,28]. Note that at zero deposition flux, the adatom density at each position n of the terrace is equal to the equilibrium density θ_{eq}. For constant deposition with $\Delta E_{ES} \gg \Delta E_{att}$, $s \ll a$, the adatom density is highest close to the end of the terrace (Fig. 2b), whereas the maximum shifts to the middle of the terrace for low values of ΔE_{ES} (Fig. 2c). Note that the maximum of the adatom density is only exactly at the end of the terrace for $E_{ES} = \infty$.

Combining equations (2)–(4), one can calculate $\Delta\mu_s$ as a function of deposition flux for any site n on the terrace.

Chemical potential of the grains. Considering an in-plane isotropic biaxial film ($\sigma_x = \sigma_y$ and $\sigma_z = 0$), it can be proven for the right-hand side of equation (1) that the CP within the grains is proportional to its total internal stress level σ_g (Supplementary

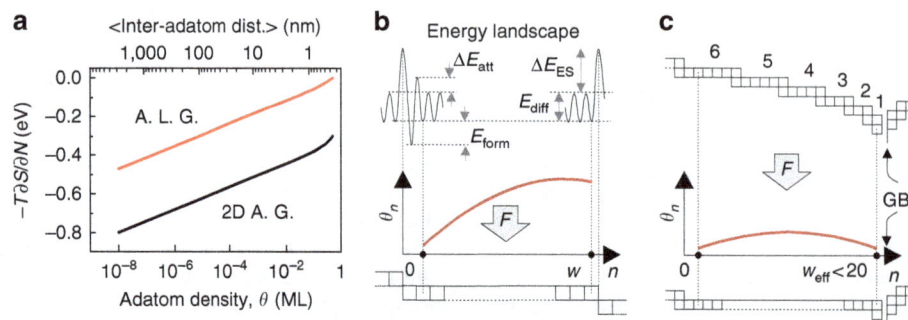

Figure 2 | Surface CP on terraces and next to GBs. (**a**) Entropic component of the CP on Cu(111), estimated by the adatom lattice gas and the 2D adatom gas model. (**b**) Adatom density profile, θ_n, on a terrace; the corresponding energy landscape is indicated on the top. (**c**) Typical step configuration in the vicinity of a GB caused by the Zeno effect (top). As ΔE_{ES} vanishes for terrace widths with $w < 6$, the combination of these terraces can be approximated with one effective terrace (bottom). This leads to a lower adatom density near the GB and, therefore, to lower stresses. dist., distance.

Note 3 and Supplementary Fig. 5):

$$\Delta\mu_g = -\int_{\sigma_{rs}^0}^{\sigma_{rs}^0 + \Delta\sigma_{rs}^{rev}} \Omega_{ij} d\sigma_{ij} = -\Omega_x \Delta\sigma_x - \Omega_y \Delta\sigma_y \approx -\frac{2}{3}\Omega\Delta\sigma_g \quad (5)$$

where Ω is the atomic volume[29].

Derivation of the stress jumps. Motivated by the fact that stress emergence has to be GB-related[16], it is crucial to derive the stress jumps using the $\Delta\mu_s$ immediately next to the GBs. Combining the above equations at position $n = w$, the predicted upper limit for the reversible stress jumps based on pure thermodynamics is given by

$$|\Delta\sigma_{rev}| \approx \frac{3}{2}\frac{k_B T}{\Omega}\ln\left(1 + \frac{Fw(aw+2)}{2v_d(asw+s+a)\theta_{eq}}\right) \quad (6)$$

It has been shown for gold at room temperature that step-flow growth is the underlying atomic process for polycrystalline film growth in the Volmer–Weber regime[21]. On the basis of the so-called Zeno effect, terraces closer to the GB get progressively decreased in their width during deposition, which leads to an enhanced surface curvature at the GB vicinity, as sketched in Fig. 2c, top[21,30]. This results in a deviation from the macroscopic equilibrium surface of an annealed polycrystalline film[31]. Although not mentioned in ref. 21, this deviation changes only slightly back over 1 h on stopping the deposition while keeping the film at room temperature. The same effect, which is due to the existence of a significant Ehrlich–Schwoebel barrier, has also been observed on Cu(111)[20]. For our study, we, therefore, safely omit Gibbs–Thompson correction terms associated with macroscopic surface curvature variations.

The red dashed lines in Fig. 3 show the predicted stress jumps derived via equation (6) for surface terrace widths between 1 and 500 atomic spacings and the existence of an Ehrlich–Schwoebel barrier. It is striking that the obtained stress values exceed even the experimental ones (crosses), and that we receive a rather good agreement already for a terrace width with $w = 1$ atomic spacing. Please note also that the experimentally observed stress values are often above the compressive yield strength of copper, which is ~ 63 MPa (ref. 32). To get a feeling for the numbers, the top horizontal axis shows the adatom density at the end point ($w = n$) for a terrace with a width of $w = 500$ (74 nm). Note that the adatom density is $< 10^{-4}$ ML even for fluxes as high as 100 ML per s. This validates our dilute adatom gas assumption while calculating the surface CP and justifies our steady-state thermodynamic approach. Note also that it is surprising that a dilute adatom gas of $< 10^{-4}$ ML has the potential to induce ~ 1 GPa stress in the film, which is more than both the yield and the ultimate strength of copper. Please note that this stress can be realized in the bulk only, if there exists a kinetically not limited atomic mechanism that transfers the CP variation of the surface to the grain interior. The curves are calculated for Cu(111) and we have used $T = 298$ K, $\Omega = \frac{1}{4}(361.49\,\text{pm})^3$, $v_0 = 10^{12}$ Hz, $E_{diff} = 0.040$ eV (ref. 33), $s_0 = 15$ (ref. 27), $\Delta E_{ES} = 0.224$ eV (ref. 34), $\Delta E_{att} = 0$ eV ($\Delta E_{att} \approx 0$ for most metals at room temperature) and $E_{form} = 0.714$ eV (ref. 35), for our calculations, which implies an adatom equilibrium density of $\theta_{eq} = 8.6 \times 10^{-13}$ ML at zero deposition flux.

It is known that stress relaxation mechanisms are active both during the growth and after stopping the deposition, which reduce the absolute intrinsic film stress[36–38]. Indeed, the experimental stress values (black crosses) are in general lower than the red dashed stress lines. However, if one considers that the time constant of the stress relaxation processes is distinctively larger than the time constant of the reversible jumps[39], the discrepancy between the observed and the predicted values for

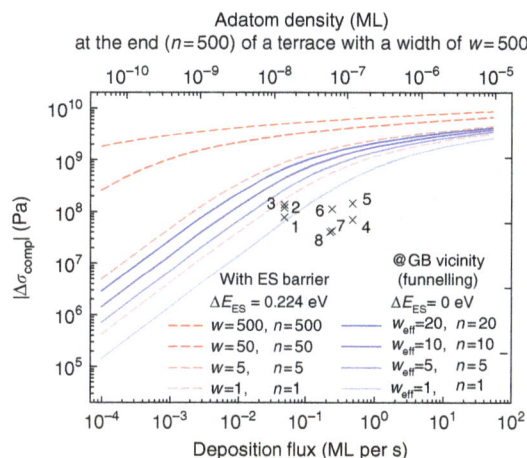

Figure 3 | Reversible stress jumps for a (111) textured copper film as a function of flux. The dashed red curves are calculated for typical terraces ($1 < w < 500$ and $\Delta E_{ES} = 0.224$ eV), whereas the solid blue curves describe terraces in the vicinity of GBs ($w_{eff} \leq 20$ and $\Delta E_{ES} = 0$ eV), where funnelling takes place. The top horizontal axis shows the adatom density for a typical surface terrace ($w = 500$ and $\Delta E_{ES} = 0.224$ eV). Crosses show experimentally reported literature values: 1, 2, 3 (ref. 8), 4 (ref. 43), 5, 6 (ref. 9), 7 (ref. 36) and 8 (ref. 44). Note that the stresses are lower, if it is easier for atoms to diffuse to and into GBs.

typical surface terraces is too large to be explained by the stress relaxation effects alone.

Stress jumps considering funnelling. As our derivation of the stress jumps exceeds in general the experimental values, we turn our attention to experiments on both Cu(111) and Ag(111) at room temperature: these experiments revealed that the Ehrlich–Schwoebel barrier of the lower step vanishes, if the distance between two neighbouring steps becomes less than six atomic spacings[40,41]. This effect opens a fast mass transfer channel for terraces with $w \leq 5$ called funnelling. The fast mass transport region can extend up to 21 atomic sites (over five steps) away from the GB. We account for this by setting the critical terrace width to $w = 5$ and by assuming that the Zeno effect reduces the width of the subsequent terraces by one atomic spacing. With this approximations, the last six terraces next to a GB can be treated as one single terrace with an effective width of $w_{eff} \simeq 20$, with $\Delta E_{ES} = 0$ ($s = 1$) (Fig. 2c, bottom). However, as intermediate step edges given by the five steps can potentially act as adatom sinks, we evaluate the stress jumps for effective terrace widths, w_{eff}, ranging 1–20 atomic spacings (see blue lines in Fig. 3). As our calculations define upper limits for the stress jumps, a funnelling width of three spacings represents the best fit. This means that is it enough, if only the last step (and not five) before the GB shows funnelling. The fact that our result with the inclusion of funnelling delivers a rather good fit with the experimental values is a strong indication for the validity of the GB insertion model, especially, as we calculate the CP on the surface exactly next to the GBs. Note, however, that we do not address the exact atomic pathway (for example, diffusion, exchange and so on), as we evaluate only the thermodynamics.

Discussion

Although we lowered the barriers for the atoms to diffuse towards and into the GB, we receive lower stresses than without funnelling. This seemingly counterintuitive behaviour demonstrates that we are not addressing a certain atomic diffusion/incorporation model, but calculate the thermodynamic driving force on the basis of the CP. Lowering the barriers for atoms to diffuse towards and into the GB, decreases the adatom density near the GB and results in a reduction of the driving force for the atoms to diffuse into the GB. Evidently, the funnelling curves predict the correct order of magnitude for the stress levels. By setting $(s = a = 1)$ in equation (4), one gets an estimate of the adatom density at the GB vicinity $n = w_{eff}$: for a deposition rate of 1 ML per s the adatom density is predicted to be lower than 10^{-10} ML.

Finally, we address another hypothetical mechanism in combination with the equilibrium situation between the surface and the bulk. For materials with a low enough Ehrlich–Schwoebel barrier and funnelling terraces, the CP distribution shows a maximum around the middle of the terrace (Fig. 2c). In addition, the CP at the end position of, for example, a large terrace in the middle of the grain (with Ehrlich–Schwoebel barrier) is significantly larger than the CPs of funnelling terraces that connect to the grain boundaries. Pure thermodynamic considerations imply that also these maxima tend to establish equilibrium with the interior of the grain underneath. As the grain is comparable to a single crystal, dislocation nucleation would be an imaginable pathway to balance the CP differences. However, on epitaxial films and single crystals the reversible stress jumps are not observed[16]. The reason for this is a high nucleation barrier: a critical stress of 1.3 GPa has been determined to nucleate dislocations in Cu at 300 K (ref. 42). Such high absolute stress values are neither observed experimentally nor does our model predict an equivalent rise of the surface CP (except for large terraces in combination with high deposition fluxes). Dislocation nucleation is, therefore, kinetically limited. The overall picture is that during the growth all points on the surface are in parallel trying to balance their CPs with adjacent positions, as well as with the grain underneath. However, in which way and by which rate the CP of the grains will change clearly depends on the rates of the pathways between all subsystems: surface, GB and bulk. With a significant dislocation nucleation barrier and active GB diffusion, as well as atom incorporation, the whole system quickly evolves towards equilibrium via atom insertion in GBs. As a result the CP difference between the grain and the surface CP maxima is reduced, which effectively lowers the driving force for dislocation nucleation making this latter process even less favourable.

If one intends to compare our results with experiments, it is important to realize that we determine only the pure reversible equilibrium jumps. For a proper comparison the experiments should have no kinetic limitation of atoms going in/out of the grain boundaries, should be performed long enough such that equilibrium has reached (all GBs show the equilibrium density of additional atoms), and no stress relaxation mechanisms should occur. In this limit, we expect the stress jumps to be GB density-independent. Kinetic limitations would immediately result in a GB density dependence, as equilibrium will not be reached and the rate towards equilibrium scales with the number of the pathways and hence the GB density.

The deposition flux and temperature dependence is more complex, as the growth mode (layer-by-layer, step-flow and 3D/rough growth) that determines the size of the terrace next to the GBs also changes with deposition rate and mobility. If one, for example, lowers the rate for a film that grows in 3D growth mode, one might enter step-flow growth conditions in which effective larger terraces (with higher adatom density) might communicate with the GBs, such that the stress jumps are even higher instead of lower.

Our analysis shows that entropic effects in the extremely dilute adatom gas on the surface of a polycrystalline film during vapour deposition are strong enough to cause plastic deformation in the film. The predicted film stresses are even higher than the observed

ones. If we lower the barriers for atoms to diffuse towards and into the GB by funnelling, the stresses decrease and the predicted values perfectly match the experimental ones. With this we deliver the, until now missing, thermodynamic driving force for any GB atom insertion model. Further experimental research, similar to[15,21], is needed to clarify the exact atomistic mechanisms and pathways behind this effect.

References

1. Floro, J. A., Chason, E., Cammarata, R. C. & Srolovitz, D. J. Physical origins of intrinsic stresses in VolmerWeber thin films. *MRS Bull.* **27**, 19–25 (2002).
2. Spaepen, F. Interfaces and stresses in thin films. *Acta Mater.* **48**, 31–42 (2000).
3. Laugier, M. Intrinsic stress in thin films of vacuum evaporated LiF and ZnS using an improved cantilevered plate technique. *Vacuum* **31**, 155–157 (1981).
4. Cammarata, R. C., Trimble, T. M. & Srolovitz, D. J. Surface stress model for intrinsic stresses in thin films. *J. Mater. Res.* **15**, 2468–2474 (2000).
5. Hoffman, R. W. Stresses in thin films: the relevance of grain boundaries and impurities. *Thin Solid Films* **34**, 185–190 (1976).
6. Freund, L. B. & Chason, E. Model for stress generated upon contact of neighbouring islands on the surface of a substrate. *J. Appl. Phys.* **89**, 4866–4873 (2001).
7. Abermann, R. Measurements of the intrinsic stress in thin metal films. *Vacuum* **41**, 1279–1282 (1990).
8. Shull, A. L. & Spaepen, F. Measurements of stress during vapor deposition of copper and silver thin films and multilayers. *J. Appl. Phys.* **80**, 6243–6256 (1996).
9. Friesen, C., Seel, S. C. & Thompson, C. V. Reversible stress changes at all stages of VolmerWeber film growth. *J. Appl. Phys.* **95**, 1011–1020 (2004).
10. Abermann, R., Koch, R. & Kramer, R. Electron microscope structure and internal stress in thin silver and gold films deposited onto MgF2 and SiO substrates. *Thin Solid Films* **58**, 365–370 (1979).
11. Chason, E., Sheldon, B. W., Freund, L. B., Floro, J. A. & Hearne, S. J. Origin of compressive residual stress in polycrystalline thin films. *Phys. Rev. Lett.* **88**, 156103 (2002).
12. Tello, J. S., Bower, A. F., Chason, E. & Sheldon, W. Kinetic model of stress evolution during coalescence and growth of polycrystalline thin films. *Phys. Rev. Lett.* **98**, 216104 (2007).
13. Pao, C.-W., Foiles, S. M., Webb, III E. B., Srolovitz, D. J. & Floro, J. A. Thin film compressive stresses due to adatom insertion into grain boundaries. *Phys. Rev. Lett.* **99**, 036102 (2007).
14. González-González, A. et al. E. Postcoalescence evolution of growth stress in polycrystalline films. *Phys. Rev. Lett.* **110**, 056101 (2013).
15. Yu, H. Z. & Thompson, C. V. Correlation of shape changes of grain surfaces and reversible stress evolution during interruptions of polycrystalline film growth. *Appl. Phys. Lett.* **104**, 141913 (2014).
16. Leib, J., Mönig, R. & Thompson, C. V. Direct evidence for effects of grain structure on reversible compressive deposition stresses in polycrystalline gold films. *Phys. Rev. Lett.* **102**, 256101 (2009).
17. Leib, J. & Thompson, C. V. Weak temperature dependence of stress relaxation in as-deposited polycrystalline gold films. *Phys. Rev. B* **82**, 121402 (R) (2010).
18. Pao, C. W., Srolovitz, D. J. & Thompson, C. V. Effects of surface defects on surface stress of Cu(001) and Cu(111). *Phys. Rev. B* **74**, 155437 (2006).
19. Sekerka, R. F. in *Crystal Growth—From Fundamentals to Technology.* (eds Müller, G., Mètois, J. J. & Rudolph, P.) Ch. 1 (Elsevier, 2004).
20. Giesen, M. & Ibach, H. Step edge barrier controlled decay of multilayer islands on Cu(111). *Surf. Sci.* **431**, 109–115 (1999).
21. Rost, M. J. In situ real-time observation of thin film deposition: roughening, Zeno effect, grain boundary crossing barrier, and steering. *Phys. Rev. Lett.* **99**, 266101 (2007).
22. Michely, T. & Krug, J. *Islands, Mounds and Atoms.* Ch. 4.3 (Springer, 2004).
23. Zhang, J. & Nancollas, G. H. Kink densities along a crystal surface step at low temperatures and under nonequilibrium conditions. *J. Cryst. Growth* **106**, 181–190 (1990).
24. Abermann, R., Martinz, H. P. & Kramer, R. Thermal effects during the deposition of thin silver, gold and copper films and their influence on internal stress measurements. *Thin Solid Films* **70**, 127–137 (1980).
25. Naumovets, A. G. Collective surface diffusion: an experimentalists view. *Phys. A* **357**, 189–215 (2005).
26. Repp, J. et al. Substrate mediated long-range oscillatory interaction between adatoms: Cu(111). *Phys. Rev. Lett.* **85**, 2981–2984 (2000).
27. Ibach, H. *Physics of Surfaces and Interfaces.* Chs. 4.3, 5.4, 10.1, 10.4 and 11.4 (Springer, 2006).
28. Kürpick, U. Self-diffusion on (100), (110), and (111) surfaces of Ni and Cu: a detailed study of prefactors and activation energies. *Phys. Rev. B* **64**, 075418 (2001).
29. Eliaz, N. & Banks-Sills, L. Chemical potential, diffusion and stress—common confusions in nomenclature and units. *Corros. Rev.* **26**, 87–103 (2008).

30. Elkinani, I. & Villain, J. Le paradoxe de Zenon d'Elee. *Solid State Commun.* **87,** 105–108 (1993).
31. Rost, M. J., Quist, D. & Frenken, J. W. M. Grains, growth, and grooving. *Phys. Rev. Lett.* **91,** 026101 (2003).
32. Sandström, R. & Hallgren, J. The role of creep in stress strain curves for copper. *J. Nucl. Mater* **422,** 51–57 (2012).
33. Knorr, N. *et al.* Long-range adsorbate interactions mediated by a two-dimensional electron gas *K. Phys. Rev. B* **65,** 115420 (2002).
34. Giesen, M., Schulze Icking-Konert, G. & Ibach, H. Interlayer mass transport and quantum confinement of electronic states. *Phys. Rev. Lett.* **82,** 3101–3104 (1999).
35. Stoltze, P. Simulation of surface defects. *J. Phys. Condens. Matter* **6,** 9495–9517 (1994).
36. Chocyk, D., Proszynski, A., Gladyszewski, G., Pienkos, T. & Gladyszewski, L. Post-deposition stress evolution in Cu and Ag thin films. *Opt. Appl.* **35,** 419–424 (2005).
37. Koch, R. & Abermann, R. Microstructural changes in vapour-deposited silver, copper and gold films investigated by internal stress measurements. *Thin Solid Films* **140,** 217–226 (1986).
38. Thompson, C. V. & Carel, R. Stress and grain growth in thin films. *J. Mech. Phys. Solids* **44,** 657–673 (1996).
39. Yu, H. Z., Leib, J. S., Boles, S. T. & Thompson, C. V. Fast and slow stress evolution mechanisms during interruptions of Volmer-Weber growth. *J. Appl. Phys.* **115,** 043521 (2014).
40. Giesen, M., Schulze Icking-Konert, G. & Ibach, H. Fast decay of adatom islands and mounds on Cu(111): a new effective channel for interlayer mass transport. *Phys. Rev. Lett.* **80,** 552–555 (1998).
41. Giesen, M. & Ibach, H. On the mechanism of rapid mound decay. *Surf. Sci* **464,** L697–L702 (2000).
42. Zhu, T., Li, J., Samanta, A., Leach, A. & Gall, K. Temperature and strain-rate dependence of surface dislocation nucleation. *Phys. Rev. Lett.* **100,** 025502 (2008).
43. Abermann, R. & Koch, R. The internal stress in thin silver, copper and gold films. *Thin Solid Films* **129,** 71–78 (1985).
44. Gladyszewski, G., Chocyk, D., Proszynski, A. & Pienkos, T. Stress evolution during intermittent deposition of metallic thin films. *Microelec. Eng.* **83,** 2351–2354 (2006).

Acknowledgements

We acknowledge R.V. Mom for discussions about thermodynamics and chemical potentials. The research described in this paper has been solely performed at the University of Leiden within the 'vidi' project of M.J.R. that was financed via by the Dutch Technology Foundation STW (Project No. 10779), which is the Applied Science Division of NWO, and the Technology Program of the Ministry of Economic Affairs.

Author contributions

The project was initiated and conceptualized by M.J.R. In his discussions with A.S., he pointed out the deficiencies of stress generation models at the time. Inspired by the discussions, A.S. envisioned a possible path to calculate the reversible stress analytically. Encouraged by this, the authors developed the thermodynamical description together, for which A.S. worked out the equations. Together they discussed the approach, interpreted the results and wrote the manuscript.

Additional information

11

Crystallographic and spectroscopic snapshots reveal a dehydrogenase in action

Lu Huo[1,2,*,†], Ian Davis[1,2,*], Fange Liu[1,†], Babak Andi[3], Shingo Esaki[1,2], Hiroaki Iwaki[4], Yoshie Hasegawa[4], Allen M. Orville[3,5] & Aimin Liu[1,2]

Aldehydes are ubiquitous intermediates in metabolic pathways and their innate reactivity can often make them quite unstable. There are several aldehydic intermediates in the metabolic pathway for tryptophan degradation that can decay into neuroactive compounds that have been associated with numerous neurological diseases. An enzyme of this pathway, 2-aminomuconate-6-semialdehyde dehydrogenase, is responsible for 'disarming' the final aldehydic intermediate. Here we show the crystal structures of a bacterial analogue enzyme in five catalytically relevant forms: resting state, one binary and two ternary complexes, and a covalent, thioacyl intermediate. We also report the crystal structures of a tetrahedral, thiohemiacetal intermediate, a thioacyl intermediate and an NAD^+-bound complex from an active site mutant. These covalent intermediates are characterized by single-crystal and solution-state electronic absorption spectroscopy. The crystal structures reveal that the substrate undergoes an E/Z isomerization at the enzyme active site before an sp^3-to-sp^2 transition during enzyme-mediated oxidation.

[1] Department of Chemistry, Georgia State University, Atlanta, Georgia 30303, USA. [2] Molecular Basis of Disease Area of Focus Program, Georgia State University, Atlanta, Georgia 30303, USA. [3] Photon Sciences Directorate, Brookhaven National Laboratory, Upton, New York 11973, USA. [4] Department of Life Science and Biotechnology and ORDIST, Kansai University, Suita, Osaka 564-8680, Japan. [5] Biosciences Department, Brookhaven National Laboratory, Upton, New York 11973, USA. * These authors contributed equally to this work. † Present addresses: Department of Pharmaceutical Sciences, University of Connecticut (L.H.); Department of Chemistry, University of Chicago (F.L.). Correspondence and requests for materials should be addressed to A.L. (email: Feradical@gsu.edu).

The dominant route of tryptophan catabolism, the kynurenine pathway, has recently garnered increased attention given its apparent association with numerous inflammatory and neurological conditions, for example, gastrointestinal disorders, depression, Parkinson's disease, Alzheimer's disease, Huntington's disease and AIDS dementia complex[1–6]. Though the precise mechanism by which the kynurenine pathway influences these diseases has not yet been fully elucidated, it has been determined that several metabolites of this pathway are neuroactive. Notably, the concentration of quinolinic acid, a non-enzymatically derived decay product of an intermediate of the kynurenine pathway used for NAD^+ biosynthesis, is elevated over 20-fold in patients' cerebrospinal fluid with AIDS dementia complex, aseptic meningitis, opportunistic infections or neoplasms[7], and more than 300-fold in the brain of human immunodeficiency virus-infected patients[8]. This NAD^+ precursor has also been shown to be an agonist of N-methyl-D-aspartate receptors, and an increase of its concentration may lead to over-excitation and death of neuronal cells[9,10].

The apparent medical potential of the kynurenine pathway warrants detailed study and characterization of its component enzymes and their regulation. One enzyme in particular, 2-aminomuconate-6-semialdehyde dehydrogenase (AMSDH), is responsible for oxidizing the unstable metabolic intermediate 2-aminomuconate-6-semialdehyde (2-AMS) to 2-aminomuconate (2-AM) (Fig. 1a). On the basis of sequence alignment, AMSDH is a member of the hydroxymuconic-semialdehyde dehydrogenase (HMSDH) family under the aldehyde dehydrogenase (ALDH) superfamily[11]. ALDHs are prevalent in both prokaryotic and eukaryotic organisms and are responsible for oxidizing aldehydes to their corresponding carboxylic acids. They use $NAD(P)^+$ as a hydride acceptor to harvest energy from their primary substrate and generate $NAD(P)H$, which provides the major reducing power to maintain cellular redox balance[12,13]. In addition to being commonly occurring metabolic intermediates, aldehydes are reactive electrophiles, making many of them toxic. Enzymes of the ALDH superfamily are typically promiscuous with regards to their substrates; however, in recent years, this superfamily has had several new members identified with greater substrate fidelity, especially when the substrate is identified as a semialdehyde[14].

The putative native substrate of AMSDH, 2-AMS, is a proposed metabolic intermediate in both the 2-nitrobenzoic acid degradation pathway of *Pseudomonas fluorescens* KU-7 (ref. 15) and the kynurenine pathway for L-tryptophan catabolism in mammals[9,10,16]. In the presence of NAD^+ and AMSDH, 2-AMS is oxidized to 2-AM (Fig. 1a); however, it can also spontaneously decay to picolinic acid and water with a half-life of 35 s at neutral pH[17]. Due to its instability, 2-AMS has not yet been isolated, leaving its identity as the substrate of AMSDH an inference based on decay products and further metabolic reactions. There are several reasons for the poor understanding of this pathway: it is complex with many branches, some of the intermediates are unstable and difficult to characterize, and several enzymes of the pathway, including AMSDH, are not well understood. Hence, the structure of AMSDH will help to address questions such as what contributes to substrate specificity for the semialdehyde dehydrogenase and how 2-AMS is bound and activated during catalysis.

In the present study, we have cloned AMSDH from *Pseudomonas fluorescens*, generated an *E. coli* overexpression system and purified the target protein for molecular study. We also constructed several mutant expression systems to characterize the role of specific active site residues. Enzymatic assays were performed for all forms of the enzyme, and crystal structures were solved for the wild type and one mutant. We were able to capture several catalytic intermediates *in crystallo* by soaking protein crystals in mother liquor containing either the primary organic substrate or a substrate analogue and discovered that in addition to dehydrogenation, the substrate undergoes isomerization at the active site.

Results

Catalytic activity of wild-type AMSDH. Due to the unstable nature of its substrate, 2-AMS, the activity of AMSDH was detected using a coupled-enzyme assay that employed its upstream partner, α-amino β-carboxymuconate ε-semialdehyde decarboxylase (ACMSD), to generate 2-AMS *in situ*. ACMSD transforms α-amino β-carboxymuconate ε-semialdehyde (ACMS) (λ_{max} at 360 nm) to 2-AMS (λ_{max} at 380 nm)[16,17]. As seen in Fig. 1b, in an assay that uses only ACMSD, the absorbance peak of its substrate, ACMS, red-shifts to 380 nm as 2-AMS is formed. The absorbance at 380 nm then quickly decreases as 2-AMS decays to picolinic acid, a compound with no absorbance features above 200 nm. In a coupled-enzyme assay, ACMSD, AMSDH and NAD^+ are included in the reaction system. As shown in Fig. 1c, ACMS is still consumed; however, there is no red shift observed because 2-AMS is enzymatically converted to 2-AM (λ_{max} at 325 nm) rather than accumulating and decaying to picolinic acid. The production of 2-AM requires that an equimolar amount of NAD^+ be reduced to NADH (λ_{max} at 339 nm). A stable alternative substrate, 2-hydroxymuconate-6-semialdehyde (2-HMS), was used to pursue kinetic parameters (Fig. 1d), when using saturating NAD^+ concentrations (≥ 1 mM), the k_{cat} and K_m of AMSDH for 2-HMS were $1.30 \pm 0.01 \, s^{-1}$ and $10.4 \pm 0.2 \, \mu M$, respectively (Fig. 1e).

Structural snapshots of the dehydrogenase catalytic cycle. We solved five crystal structures of wild-type AMSDH, including the ligand-free (2.20 Å resolution), NAD^+-bound binary complex (2.00 Å), ternary complex with NAD^+ and substrate 2-AMS (2.00 Å) or 2-HMS (2.20 Å) and a thioacyl intermediate (1.95 Å). All five structures belong to space group $P2_12_12_1$. Data collection and refinement statistics are listed in Supplementary Table 1. The complete AMSDH model includes four polypeptides per asymmetric unit describing one homotetramer (Supplementary Fig. 1a). Each monomer of AMSDH contains three domains: a subunit interaction domain, a catalytic domain and an NAD^+ binding domain (Supplementary Fig. 1b). For details of the secondary structure, see Supplementary Discussion.

In the structure of the co-crystallized binary complex, an NAD^+ molecule is present in an extended, anti-conformation in the amino-terminal, co-substrate-binding domain of each monomer (Fig. 2a). The electron density map of NAD^+ is well defined, and the interactions between the protein and NAD^+ are equivalent in all four subunits as shown in Fig. 2e. The NAD^+-bound AMSDH structure is similar to the ligand-free structure with an aligned r.m.s.d. of 0.239 Å. Residues that belong to the NAD^+-binding pocket are also well aligned with the exception of Cys302, Arg108 and Leu116 (Supplementary Fig. 2). On binding NAD^+, the thiol moiety of Cys302 rotates so that the sulfur is 2.3 Å closer to the substrate-binding pocket and away from the nicotinamide head of NAD^+.

Crystal structures of enzyme–substrate ternary complexes. Structures of AMSDH in ternary complex with co-substrate NAD^+ and its primary substrates were obtained by soaking co-crystallized AMSDH-NAD^+ crystals with 2-AMS and 2-HMS, respectively. Extra density that fits with the corresponding substrate molecule was observed in the active site of each subunit. The co-substrate NAD^+ in the ternary complex structures is

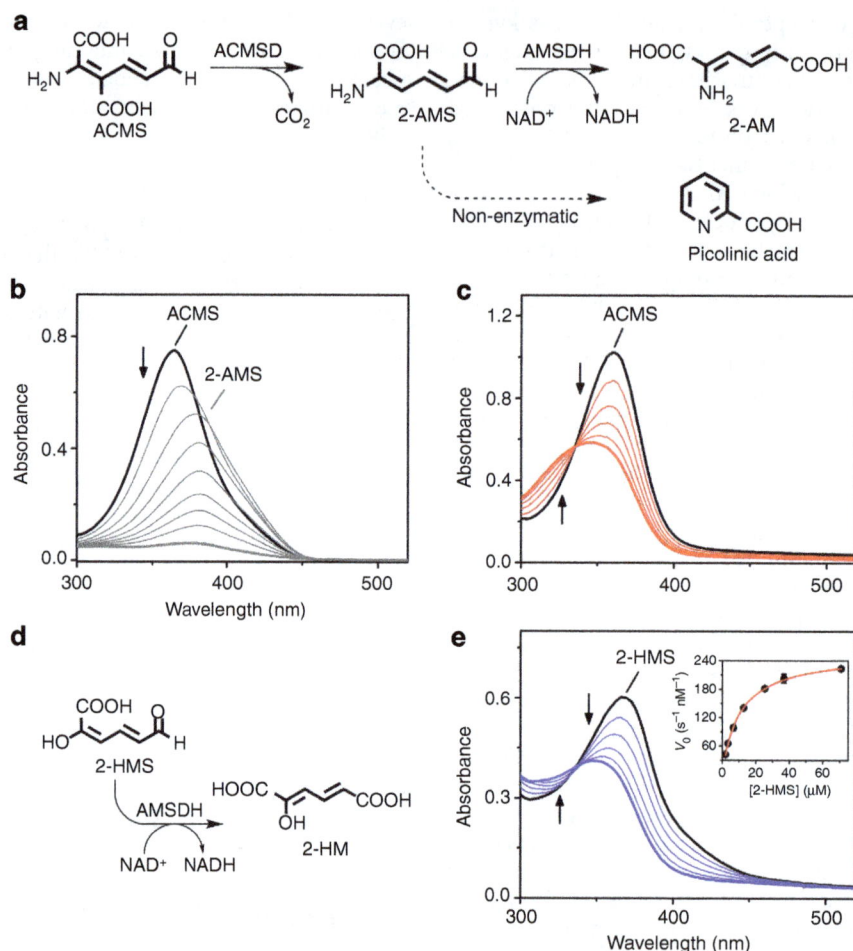

Figure 1 | Activity of AMSDH. (**a**) Reaction scheme showing the enzymatic generation of 2-AMS, the reaction catalysed by AMSDH, and the competing non-enzymatic decay of 2-AMS to picolinic acid. (**b**) Representative assay showing the ACMSD (1 μM)-catalysed conversion of ACMS (λ_{max} 360 nm) to 2-AMS (λ_{max} 380 nm), which decays to picolinic acid (transparent). (**c**) Coupled-enzyme assay in which AMSDH (200 nM) oxidizes 2-AMS, produced *in situ* as shown in **b** in 50 s, to 2-AM (λ_{max} 325 nm). (**d**) Reaction scheme showing 2-HMS oxidation by AMSDH. (**e**) Representative assay showing the activity of AMSDH (200 nM) on 2-HMS (λ_{max} 375 nm) in 50 s. The inset is a Michaelis–Menten plot.

bound in the same manner as in the binary complex. Binding of the primary substrates introduced minimal change to the protein structure; the r.m.s.d. for the superimposed structures of substrate-free with 2-AMS- and 2-HMS-bound ternary complex structures are 0.170 and 0.276 Å, respectively. These two primary substrates bind to AMSDH in an identical fashion, with two arginine residues, Arg120 and Arg464, playing an important role in stabilizing the substrate by forming two sets of bifurcated hydrogen bonds with one of the carboxyl oxygens and the 2-amino or hydroxyl group of 2-AMS (Fig. 2b) or 2-HMS (Fig. 2c), respectively. The observation of two hydrogen bonds being donated by the active site arginines to the 2-amino group of 2-AMS indicates that in the substrate-bound form, 2-AMS may be in its 2-imine rather than 2-enamine tautomer, as an amino group unlikely to accept two hydrogen bonds. Mutation of Arg120 to alanine causes a moderate decrease of the k_{cat} to $0.7 \pm 0.2\,s^{-1}$ from $1.30 \pm 0.01\,s^{-1}$ and a dramatic increase of the K_m with a lower bound of $446.3 \pm 195.9\,\mu M$ (an accurate determination of the K_m is hindered by insufficient 2-HMS concentrations) compared with $10.4 \pm 0.2\,\mu M$ in the wild type (Supplementary Fig. 3a). Mutation of Arg464 to alanine decreased the k_{cat} to $\sim 0.3\,s^{-1}$, and not only increased the K_m to $\sim 170\,\mu M$, but also leads to a significant substrate inhibition effect with a K_i of $\sim 6\,\mu M$ (Supplementary Fig. 3b). This substrate inhibition is likely caused by the unproductive binding of a second substrate molecule in the space created by the deletion of

Arg464 or by a failure of the enzyme to properly bind and stabilize the imine form of the substrate.

Catalytic intermediates trapped after ternary complex formation. Enzyme–NAD$^+$ binary complex crystals were soaked in mother liquor containing 2-HMS for a range of time points from 5 min to more than 3 h before flash cooling in liquid nitrogen. In a crystal that was soaked for 40 min, an intermediate was trapped and refined to a resolution of 1.95 Å (Fig. 2d). Crystals soaked for longer time points gave a similar intermediate with poorer resolution. In this structure, 2-HMS is observed in the 2Z, 4E isomer rather than the 2E, 4E isomer as seen in the substrate-bound ternary structure. Also, the substrate interacts with Arg120 and Arg464 with both of its carboxyl oxygens rather than one carboxyl oxygen and the 2-hydroxy oxygen as shown in the 2-HMS ternary complex structure. Fitting this density with the 2E, 4E conformation resulted in unsatisfactory $2F_o - F_c$ and $F_o - F_c$ density maps as shown in Supplementary Fig. 4a. Likewise, attempting to fit the 2Z, 4E isomer to the ternary complex structure did not produce satisfactory results (Supplementary Fig. 4b). On E to Z isomerization, the carbon chain of the substrate extends, and the distance between its sixth carbon and Cys302's sulfur is now at 1.8 Å, which is within covalent bond distance for a carbon–sulfur bond. Also, the continuous electron density

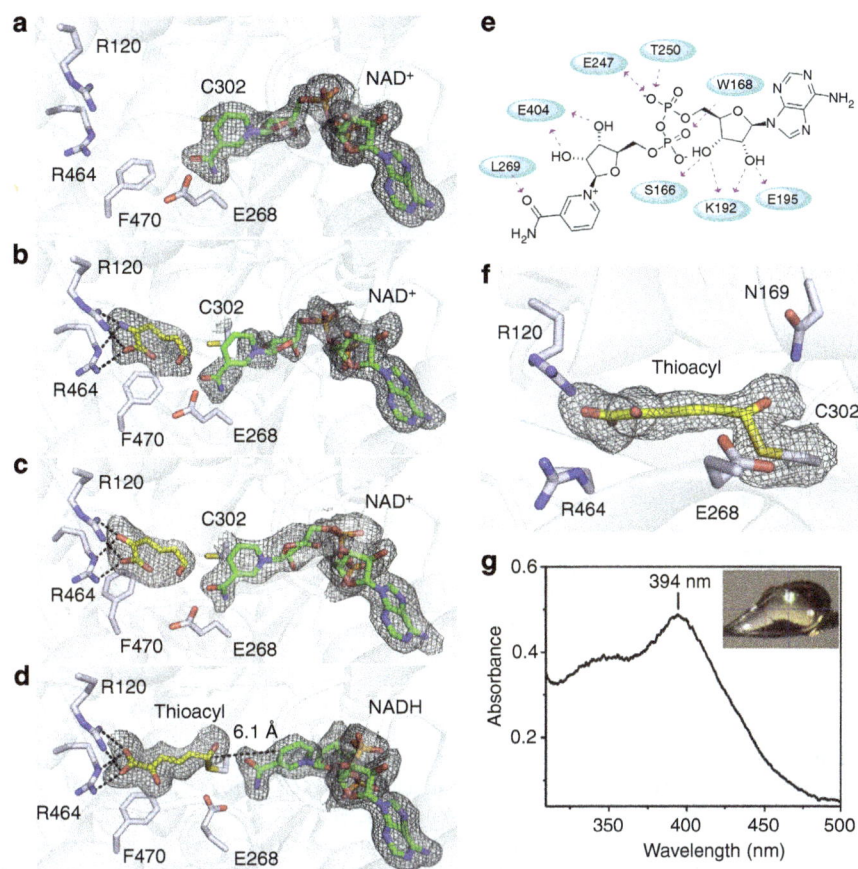

Figure 2 | Crystal structures of wild-type AMSDH and single-crystal electronic absorption spectrum of a catalytic intermediate. AMSDH was co-crystallized with NAD$^+$ to give AMSDH-NAD$^+$ binary complex crystals that were used for soaking experiments. (**a**) Active site structure of the binary AMSDH-NAD$^+$ complex, (**b**) the ternary complex of AMSDH-NAD$^+$ crystals soaked with 2-AMS for 5 min before flash cooling, (**c**) the ternary complex of AMSDH-NAD$^+$ soaked with 2-HMS for 10 min before flash cooling, (**d**) the trapped thioacyl, NADH-bound intermediate obtained by soaking AMSDH-NAD$^+$ crystals with 2-HMS for 40 min before flash cooling. (**e**) Two-dimensional interaction diagram for NAD$^+$ binding. (**f**) Close-up of the thioacyl intermediate in **d**. (**g**) Single-crystal electronic absorption spectrum of **d**. Protein backbone and residues are shown as light blue cartoons and sticks, respectively. The substrates and intermediate are shown as yellow sticks, and NAD$^+$ and NADH are shown as green sticks. The omit map for ligands is contoured to 2.0 σ and shown as a grey mesh.

between Cys302-SG and 2-HMS-C6 indicates the presence of a covalent bond (Fig. 2f). Another feature of this intermediate is that the nicotinamide ring of NAD$^+$ has moved 4.6 Å away from the active site and adopted a bent conformation (Fig. 2d) compared with the position in the binary or ternary complex structures (Fig. 2a–c). The structural changes of NAD$^+$ associated with reduction has been observed and well documented[18,19]. In the oxidized state, NAD(P)$^+$ lies in the Rossmann fold in an extended conformation, allowing for hydride transfer from the substrate to its nicotinamide carbon during the first half of the reaction. Reduced NAD(P)H then adopts a bent conformation in which the nicotinamide head moves back towards the protein surface. This movement provides more space in the active site for the second half of the reaction, acyl-enzyme adduct hydrolysis, to take place. Thus, the coenzyme in this intermediate structure is likely to have been reduced to NADH and, as such, the structure is assigned as a thioacyl-enzyme–substrate adduct. The single-crystal electronic absorption spectrum of the sample has an absorbance maximum at 394 nm (Fig. 2g). The same absorbance band was observed in crystals soaked with 2-HMS from 30 min to 2 h (Supplementary Fig. 5a). However, this long-lived intermediate in the crystal was not observed in solution with millisecond-to-second time resolution in stopped-flow experiments (Supplementary Fig. 6a). Thus, it is either present in an earlier time domain

(sub-milliseconds), or alternatively, it may not accumulate in solution because NADH can readily dissociate in solution, whereas it may be trapped in the active site when in the crystalline state.

Another notable change in the intermediate structure is the movement of the side chain of Glu268, which rotates 73° towards the active site (Fig. 2c,d). To probe the function of Glu268, we constructed an alanine mutant and found that it exhibited no detectable activity in steady-state kinetic assays. Interestingly, E268A exhibits completely different pre-steady state activity than the wild-type enzyme. As shown in Supplementary Fig. 6b, an absorbance band at 422 nm was formed concomitant with the decay of the 2-HMS peak within 1 s of the reaction. This new species is generated stoichiometrically on titration of 2-HMS with E268A (Fig. 3d). The moiety that gives rise to this new absorbance band is stable for minutes at room temperature and cannot be separated from the protein by membrane filtration-based methods[20], suggesting that it is covalently bound to the protein. The formation of an enzyme–substrate adduct in the E268A mutant was investigated by mass spectrometry. For the as-isolated E268A, the resultant multiply charged states (Supplementary Fig. 7) were deconvoluted to obtain a molecular weight (MW) of 56,252 Da (Fig. 4a). This value is in good agreement with the predicted MW of E268A AMSDH plus an amino-terminal His-tag and linking residues, 56,251 Da. The

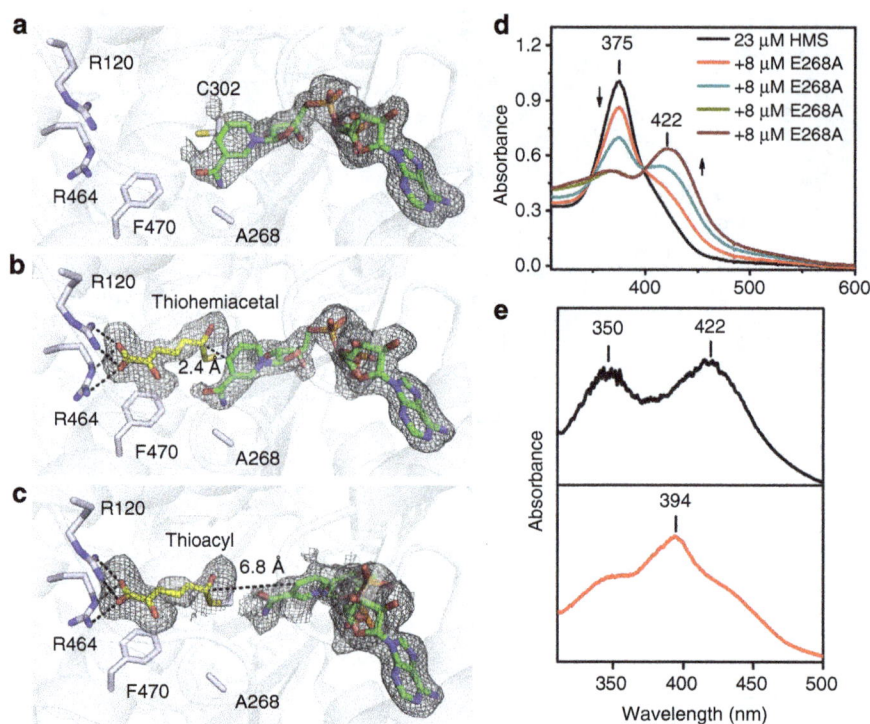

Figure 3 | Crystal structures of the E268A mutant and its solution and single-crystal electronic absorption spectra. (**a**) Structure of the active site of the co-crystallized E268A-NAD$^+$ binary complex, (**b**) a thiohemiacetal intermediate obtained by soaking the E268A-NAD$^+$ crystals with 2-HMS for 30 min before flash cooling and (**c**) a thioacyl intermediate obtained by soaking the E268A-NAD$^+$ crystals with 2-HMS for 180 min before flash cooling. (**d**) Solution electronic absorption spectra of a titration of 2-HMS with E268A. (**e**) Single-crystal electronic absorption spectrum of the intermediate in **b** (top panel) and single-crystal electronic absorption spectrum of the intermediate in **c** (bottom panel). Protein backbone and residues are shown as light blue cartoons and sticks, respectively. The substrate and intermediate are shown as yellow sticks, and NAD$^+$ and NADH are shown as green sticks. The omit map for ligands is contoured to 2.0 σ and shown as a grey mesh.

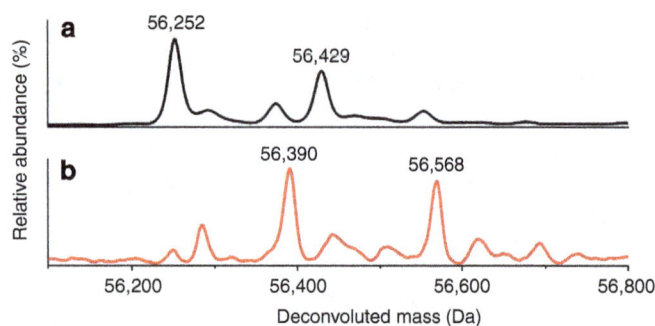

Figure 4 | Deconvoluted positive-mode electrospray ionization mass spectra of as-isolated E268A (a) and 2-HMS treated-E268A (b). The two major components are labelled with their respective molecular weights.

second largest peak in the deconvoluted spectrum has a MW 177 Da greater than that of the most abundant signal. This is likely due to post-translational modification of the His-tag; α-N-Gluconoylation of His-tags has been observed in *E. coli*-expressed proteins, which cause 178 Da extra mass[21]. The mutant protein was then treated with the alternate substrate, 2-HMS, and the mass spectrum shows a new major peak at 56,390 Da (Fig. 4b), 138 Da heavier than the as-isolated mutant. Similarly, the second most abundant peak corresponds to a His-tag modified mutant plus 139 Da. In this spectrum, the peaks arising from the as-isolated mutant are substantially reduced, indicating that 2-HMS, 141 Da, is bound to the E268A mutant enzyme.

We determined the crystal structure of E268A co-crystallized with NAD$^+$ and refined it to 2.00 Å resolution (Fig. 3a). The overall structure aligns very well with the wild-type binary complex structure with an r.m.s.d. of 0.139 Å. The active site of E268A also resembles the native AMSDH structure (Supplementary Fig. 8). The nature of the absorbing species at 422 nm was further investigated by soaking co-crystallized E268A-NAD$^+$ crystals in mother liquor containing 2-HMS. By doing so, two temporally, structurally and spectroscopically distinct intermediates were identified.

When E268A-NAD$^+$ crystals are soaked with 2-HMS for 40 min or less, their single-crystal electronic absorption spectra show an absorbance maximum at 422 nm (Supplementary Fig. 5b), as was observed in the solution-state titration and the stopped-flow assays. An individual electronic absorption spectrum for an E268A-NAD$^+$ crystal soaked with 2-HMS for 15 min can be found in Fig. 3e (top). The structure of E268A-NAD$^+$ soaked with 2-HMS for 30 min before flash cooling was solved and refined to 2.15 Å resolution (Fig. 3b). In this structure, a continuous electron density between Cys302-SG and 2-HMS-C6 is observed, similar to the thioacyl intermediate observed in the wild-type enzyme. However, in contrast to the thioacyl intermediate, the density around C6 is less flat, indicating an sp^3- rather than sp^2-hybridized carbon (Fig. 5a). The angle between the plane of the carbon backbone of the substrate and the formerly aldehydic oxygen is $55 \pm 9°$, compared with the angle of the wild-type thioacyl intermediate at $26 \pm 4°$ (Supplementary Table 2). More importantly, the C6 of 2-HMS and the C4N of NAD$^+$ are very close (2.4–2.8 Å), making it unlikely that the hydride has been transferred from the substrate. Taken together, these data allow us to assign this intermediate to a thiohemiacetal enzyme adduct (Fig. 3b). A similar intermediate has only been trapped once previously

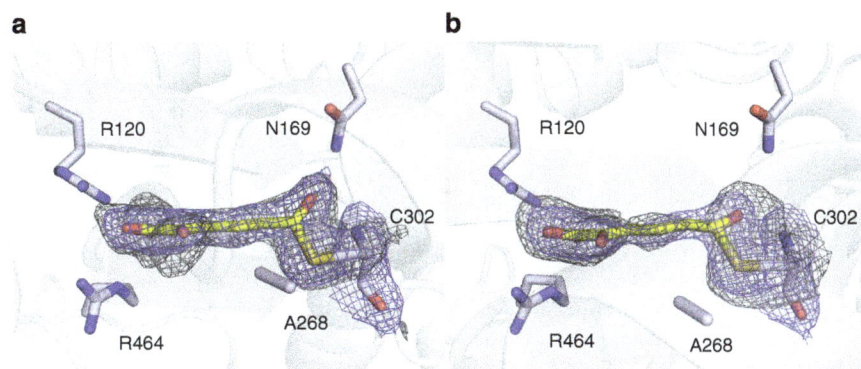

Figure 5 | Crystal structures of two distinct catalytic intermediates. (a) Electron density map of the thiohemiacetal intermediate obtained from E268A-NAD$^+$ crystal soaked with 2-HMS for 30 min. **(b)** Electron density map of the thioacyl intermediate obtained from E268A-NAD$^+$ crystal soaked with 2-HMS for 180 min. The $2F_o − F_c$ electron density map for ligands and Cys302 is contoured to 1.0 σ and shown as a blue mesh. The omit map for ligands and Cys302 is contoured to 2.0 σ and shown as a gray mesh.

in a crystal that contains no co-substrate[22]. Hence, this is the first time for this intermediate to be trapped in the presence of NAD$^+$.

If the E268A-NAD$^+$ crystals are soaked with 2-HMS for longer than 1 h, their single-crystal electronic absorption spectra begin to resemble that of the wild-type, thioacyl intermediate with a corresponding absorbance maximum at 394 nm (Supplementary Fig. 5b), as seen in wild-type, thioacyl intermediate crystals. An individual electronic absorption spectrum for an E268A-NAD$^+$ crystal soaked with 2-HMS for 120 min can be found in Fig. 3e (bottom). The structure of an E268A-NAD$^+$ crystal soaked with 2-HMS for 180 min was solved and refined to 2.20 Å (Fig. 3c). The structure of this intermediate is also similar to the wild-type, thioacyl-enzyme adduct with NADH, rather than NAD$^+$ found at the active site. The distance between the C4N of NADH and C6 of 2-HMS is longer than 6.1 Å (Fig. 3c). The electron density around C6 is flatter (Fig. 5b) compared with the thiohemiacetal intermediate and similar to the thioacyl intermediate trapped in the wild-type AMSDH structure (Fig. 2f), and the angle between the plane of the carbon backbone of the substrate and the carbonyl oxygen is 20 ± 5°, which is statistically indistinguishable from that of the wild-type, thioacyl intermediate, 26 ± 4° (Supplementary Table 2). On the basis of the similarities in their absorbance and structures, we conclude that this latter intermediate is equivalent to the wild-type, thioacyl intermediate. It is also worth noting that the strictly conserved asparagine 169 (Fig. 5) is seen to stabilize both the thiohemiacetal and thioacyl intermediates through hydrogen-bonding interactions.

Investigation of isomerization by computational modelling. The isomerization of 2-AMS from the 2E to 2Z isomer implied by the solved crystal structures was probed with density functional theory calculations. The free energy profiles obtained were used to help illuminate the nature of 2-AMS and gain insight into how the active site of AMSDH may facilitate the isomerization. The total energies of different isomers and rotamers of 2-AMS in its enamine/aldehyde and imine/eneol tautomers and the rotational barriers about their respective 2–3 bond were compared. For the imine/eneol tautomer, additional computations were performed with the side groups from Arg120 and Arg464 to investigate what effect, if any, they will have on the free energy profile for rotation about the 2–3 bond of 2-AMS.

First, 2-AMS was constructed and optimized in its 2-enamine, 6-aldehyde, 2E isomer with a negatively charged 2-carboxylate group (Fig. 6a). To estimate the energy barrier for an uncatalysed

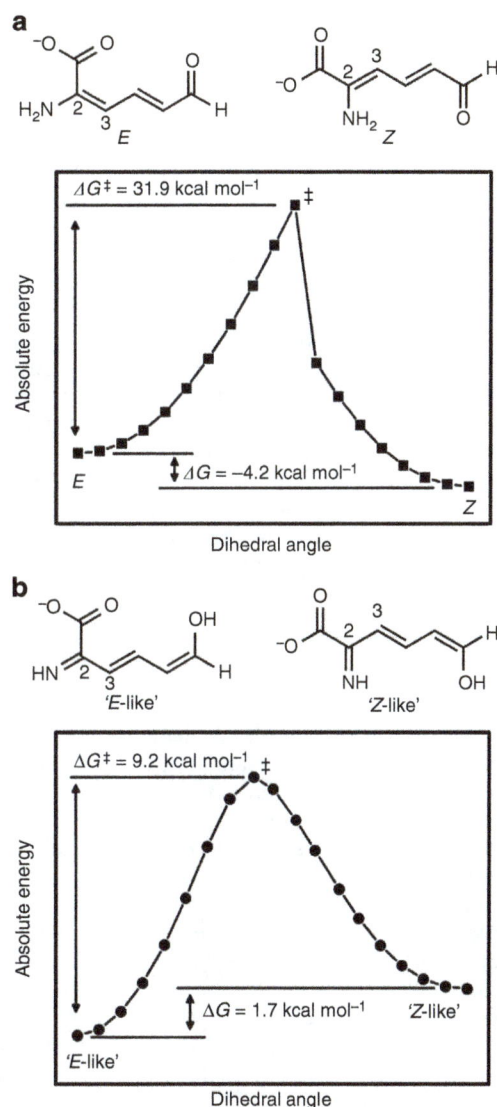

Figure 6 | Free energy profiles for the rotation about the 2–3 bond of 2-AMS in its (a) enamine and (b) imine form, respectively. DFT calculations were performed at the B3LYP/6-31G* + level of theory. The dihedral angle about the 2–3 bond was restrained in 10° increments and the structures were optimized at each point.

isomerization from the 2*E* to the 2*Z* isomer, the 2–3 double bond was then restrained at 10° intervals from 180 to 0°, and the structure was optimized at each point. On the basis of the free energy profile (Fig. 6a), the uncatalysed isomerization barrier is 31.9 kcal mol^{-1}. The profile also shows that the 2*Z* isomer, as might be expected, is lower in energy than the 2*E* isomer by 4.2 kcal mol^{-1}. Next, the rotational barrier about the 2–3 bond of 2-AMS when in its 2-imine, 6-enol tautomer, as is suggested by the ternary complex structure, was calculated in the same manner. The barrier was found to be 9.2 kcal mol^{-1}, and opposite to the enamine tautomer, the '2*Z*-like' rotamer is higher in energy than the '2*E*-like' rotamer by 1.7 kcal mol^{-1} (Fig. 6b). Unsurprisingly, the rotational barrier about the 2–3 bond is much lower in the imine tautomer; however, the '2*Z*-like' rotamer of the imine tautomer is 21.8 kcal mol^{-1} higher in energy than the 2*Z* isomer of the enamine tautomer.

Possible influences of the two active site arginines on the free energy profile for rotation were also considered. To mimic the conditions of the enzyme active site, similar calculations as those above were performed, which included the guanidinium heads of Arg120 and Arg464. The starting model was built using the active site geometry of the ternary complex crystal structure (PDB entry: 4I25), and on inspection, it is immediately apparent that with two arginines in such close proximity to the substrate, there is insufficient space for two hydrogen atoms on the nitrogen at the 2-position of 2-AMS, and attempts to optimize an enamine tautomer with the hydrogen-bonding pattern of the ternary complex produced structures within which the entire 2-AMS molecule rotates so that only the carboxylate group interacts with the guanidinium moieties. The absolute positions of the guanidinium groups were fixed and the structure of 2-AMS in the imine tautomer was optimized. The dihedral angle of the 2–3 bond of 2-AMS was then increased in 45° increments and the structure optimized while restraining the position of the guanidinium groups and the 2–3 bond to build a rough free energy profile to estimate the rotational barrier. In the presence of the active site arginines, the barrier about the 2–3 bond of 2-AMS is further reduced to 8.5 kcal mol^{-1} (Supplementary Table 3). Another interesting finding is that in the presence of the guanidinium groups, the '2*E*-like' and '2*Z*-like' rotamers of 2-AMS are nearly isoenergetic, with a free energy difference of 0.2 kcal mol^{-1} (Supplementary Table 3).

Discussion

The substrate of AMSDH, 2-AMS, contains an unstable aldehyde in conjugation with an enamine and can decay to picolinic acid and water, presumably through an electrocyclization reaction similar to its metabolic precursor, ACMS[23]. To assay the enzymatic activity, the upstream enzyme was utilized in the reaction mixture to generate substrate, and it was shown that AMSDH is catalytically active. Unfortunately, no kinetic parameters can be reliably determined because the concentration of 2-AMS is not well defined in the coupled-enzyme assay. To circumvent this issue, a previously-identified, stable alternative substrate, 2-HMS[24,25], in which a hydroxyl group replaces the amino group in 2-AMS to prevent cyclization, was used to characterize the activity of AMSDH and to examine the activity of the mutants.

Substrate-bound, ternary complex structures were obtained by soaking co-crystallized protein and NAD$^+$ with 2-AMS or 2-HMS. 2-AMS is an unstable compound which decays with a $t_{1/2}$ of about 9 s at pH 7.5 and 37 °C or 35 s at pH 7.0 and 20 °C. Notably, this is its first reported structure. It appears to be stabilized in the enzyme active site in its imine tautomer by forming two sets of bifurcated hydrogen bonds with Arg120 and

Arg464 so that the electrocyclization reaction cannot occur. Both arginine residues are close to the protein surface and in good positions to serve as gatekeepers, bringing the substrate into the active site. As a residue residing on a loop, Arg464 should be relatively flexible. The electron density for the side chain of Arg120 is partially missing in the binary complex structure but very well resolved in both ternary complex structures. This observation indicates that the presence of substrate can stabilize what may be a flexible residue. It becomes evident from the coordinates that Arg120 and Arg464 play an important role in substrate recognition, stabilization and possibly product release. Two arginine residues are rarely observed in such close proximity, stabilizing one end of the same molecule with multiple hydrogen bonds. With the exception of the hydrogen bonds provided by Arg120 and 464, the substrate-binding pocket is mostly composed of hydrophobic residues. On the basis of sequence alignment (Supplementary Fig. 9), these two arginine residues are strictly conserved throughout the HMSDH family but are not found in other members of the ALDH superfamily. We propose that these dual arginines combined with the size restrictions provided by the hydrophobic pocket endow this enzyme with its specificity towards small α-substituted carboxylic acids with an aldehyde moiety, such as 2-AMS and 2-HMS. Furthermore, our computational work suggests that these arginines are crucial for stabilizing the imine tautomer of 2-AMS to allow for rotation about its 2–3 bond.

Two strictly conserved catalytic residues, Cys302 and Glu268, are present at the interior of the substrate-binding pocket. General features regarding these residues in the ALDH super-family are (1) that the cysteine serves as a catalytic nucleophile, which is anticipated to form a covalent-adduct intermediate with the substrate by a nucleophilic addition during catalysis[26-28] and (2) that the glutamate serves as a base to activate water for hydrolysis of the thioacyl-enzyme adduct[29-32]. Previous studies indicate that the catalytic cysteine can adopt two conformations, resting and attacking[19]. In the ligand-free structure, Cys302 is far from where the carbonyl carbon of the substrate should be and is in the resting state. In the ternary complex structures, Cys302 is located at an ideal position to initiate catalysis, which is the attacking state. It is proposed to attack the aldehydic carbon (C6) of the substrate. In the two ternary complex structures, the distance between the sulfur of Cys302 and the C6 of the substrate is ∼3.3 Å. Cys302 and the aldehydic carbon form a covalent bond in both thioacyl and thiohemiacetal intermediates. Mutation of Cys302 to serine led to enzyme with no detectable dehydrogenase activity, further confirming its catalytic significance.

Examining the wild-type AMSDH structures shows that in the NAD$^+$-bound binary complex, Glu268 adopts a 'passive' conformation, pointing away from the substrate-binding pocket, and forms hydrogen bonds with both NE of Trp177 (3.2 Å distance) and the backbone oxygen of Phe470 (3.6 Å) to leave space for the reduction of NAD$^+$. Its electron density is very well resolved and the side chain B-factor is close to average: 28.2 Å^2/28.5 Å^2. The thiol moiety of Cys302 is 7.14 Å from Glu268 and is unlikely to form interactions. Interestingly, in both substrate-bound structures, Glu268 becomes more flexible and exhibits much weaker electron density and increased side chain B-factors compared with average protein B-factors: 37.8 Å^2/28.5 Å^2 and 66.37 Å^2/39.7 Å^2. In the thioacyl intermediate structure, the electron density of Glu268 becomes very well defined again, but its side chain rotates 73° towards the bound substrate and seems to be in an 'active' position to abstract a proton from a deacylating water (Fig. 2d). At this point in the reaction cycle, the NADH molecule needs to leave the active site to make room for the catalytic water molecule. Movement of the nicotinamide ring of

NAD$^+$ coupled with the rotation of an active site glutamate has previously been observed in other ALDHs during catalysis[30–32].

Mutation of Glu268 to alanine led to the accumulation of the thiohemiacetal intermediate in both solution and crystalline states. The strictly conserved glutamate residue in the active site of ALDH enzymes has been proposed to play up to three possible roles during catalysis. It is strictly required to activate the deacylating water that allows for product release, it is in a 'passive' conformation during NAD(P)$^+$ reduction, and in some cases, it may serve to activate cysteine for nucleophilic attack[33]. On the basis of these roles, mutation to alanine would be expected to decrease the rate of hydrolysis of the thioacyl adduct, have no effect on the rate of reduction of NADH and possibly decrease the rate of nucleophilic attack by cysteine. With this understanding, deletion of the active site glutamate should cause an accumulation of the thioacyl intermediate. However, in this work, the E268A mutant is shown to accumulate the preceding thiohemiacetal intermediate both in crystal and in solution. This finding suggests an additional catalytic role for this residue: rotation of Glu268 towards the active site facilitates the hydride transfer from the tetrahedral thiohemiacetal adduct to NAD$^+$. The rapid formation of the intermediate in solution indicates that Glu268 of AMSDH does not play a role in activating cysteine. However, it does appear necessary to complete hydride transfer from the substrate to NAD$^+$, and its removal turns the native, primary substrate into a suicide inhibitor.

On the basis of previous studies of the ALDH mechanism, the eight high-resolution crystal structures solved (Supplementary Table 1) as well as our biochemical and computational studies, we propose a catalytic mechanism for AMSDH. As shown in Fig. 7, NAD$^+$ binds to the enzyme, **1**, to form an NAD$^+$-bound AMSDH complex, **2**. The substrate, 2-AMS, is then recognized by Arg120 and Arg464 through multiple hydrogen-bonding inter-actions, and its imine tautomer is stabilized in the active site, **3**. At this point, the order of the rotation, tautomerization and nucleophilic attack by C302 on the aldehydic carbon to produce

the tetrahedral, thiohemiacetal intermediate, **4**, is not yet clear. The isomerization and nucleophilic attack drive a translation of the substrate away from Arg120 and Arg464 so that they are only able to interact with the carboxylate group of the substrate. Next, NAD$^+$ is reduced to NADH by abstraction of a hydride from **4**, forming a thioacyl intermediate, **5**, a process which involves an sp^3-to-sp^2 transition during oxidation of the organic substrate by NAD$^+$. On reduction, the nicotinamide portion of NADH moves away from the substrate as Glu268 rotates into position to activate a water molecule to perform a nucleophilic attack on the same carbon that was previously attacked by Cys302, forming a second tetrahedral intermediate, **6**. Finally, the second tetrahedral intermediate collapses, breaking the C–S bond and releasing the final products, 2-AM and NADH. Species **1–5** are spectro-scopically and structurally characterized, while intermediate **6** was not seen to accumulate.

In this work, five catalytically relevant structures of the wild-type AMSDH and three mutant structures yield a comprehensive understanding of the protein's overall structure, co-substrate-binding mode and elucidate the primary residues responsible for substrate specificity among the HMSDH family of the ALDH superfamily. The structural and spectroscopic snapshots capture the crystal structure of an unstable kynurenine metabolite, 2-AMS, and two catalytic intermediates, including stabilizing a tetrahedral intermediate in a mutant protein, which was further verified by mass spectrometry. Capture of a thiohemiacetal intermediate upon deletion of E268 also points to a new role for this well-established active site base in hydride transfer from the substrate to NAD$^+$. Another interesting finding revealed through solving the ternary complex and intermediate crystal structures and supported by computational studies is that an E to Z isomerization of the substrate occurs in this dehydrogenase before hydride transfer. To the best of our knowledge, this is the first piece of structural evidence illustrating an ALDH that proceeds via an E/Z isomerization of its substrate during catalysis.

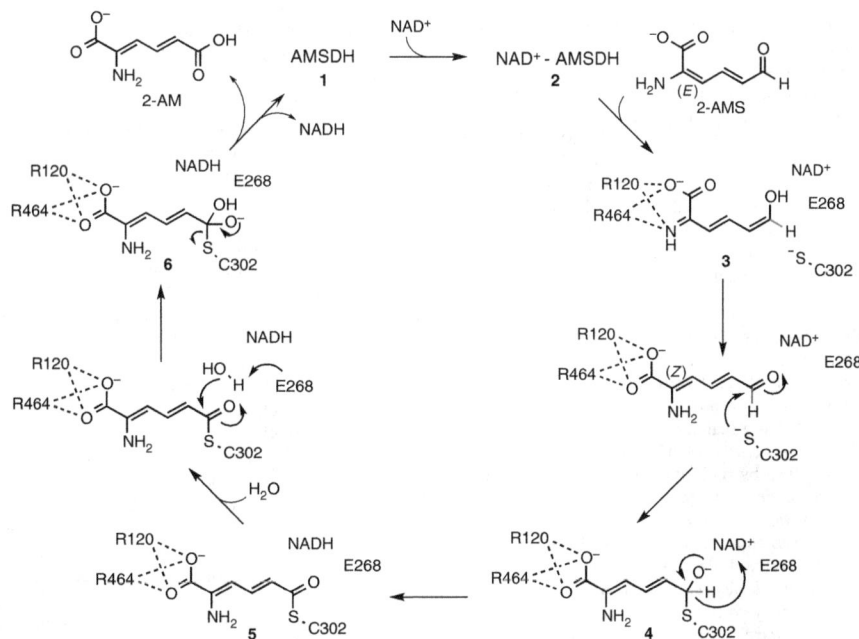

Figure 7 | Proposed catalytic mechanism for the oxidation of 2-AMS by AMSDH. The primary substrate (2E, 4E)-2-aminomuconate-semialdehyde binds to the enzyme in its imine tautomer to form the ternary complex (**3**). An isomerization and attack by cysteine on the aldehydic carbon form the (2Z, 4E)-2-aminomuconate-thiohemiacetal adduct (**4**). AMSDH-mediated oxidation of **4** concomitant with reduction of NAD$^+$ to NADH follows, generating a thioacyl-enzyme intermediate (**5**). Both **4** and **5** are the catalytic intermediates covalently attached to the enzyme. Hydrolysis of **5** then allows the release of the products 2-AM and NADH, restoring the ligand-free enzyme for the next catalytic cycle.

Methods

General methods. The cloning, expression, purification and site-directed mutagenesis of AMSDH are described in the Supplementary Methods.

Preparation of ACMS and 2-HMS. ACMS was generated by catalysing the insertion of molecular oxygen to 3-hydroxyanthranilic acid by purified, Fe^{2+} reconstituted 3-hydroxyanthranilate 3,4-dioxygenase as described previously[16,20]. 2-HMS is generated non-enzymatically from ACMS following a previously established method[24]. The pH of solutions containing ACMS was adjusted to ~ 2 by the addition of hydrochloric acid. 2-HMS formation was monitored on an Agilent 8453 diode-array spectrophotometer at 315 nm. The solutions were then neutralized with sodium hydroxide once the absorbance at 315 nm stopped increasing. 2-HMS at neutral pH has a maximum absorbance at 375 nm (ref. 24).

Enzyme activity assay using 2-HMS as substrate. Steady-state kinetics analyses were carried out at room temperature on an Agilent 8453 diode-array spectrophotometer. Reaction buffer contains 25 mM HEPES and 1 mM NAD^+, pH 7.5. Consumption of 2-HMS by 200 nM AMSDH was detected by monitoring the decrease of its absorbance at 375 nm with a molar extinction coefficient of $43,000 \, M^{-1} cm^{-1}$ (ref. 24) for 15 s with a 0.5 s integration time. For mutants, 700 nM protein and a wavelength of 420 nm, $\varepsilon_{420} \, 11,180 \, M^{-1} cm^{-1}$, was used. Absorbance at 375 nm decreased and blue shifted to 295 nm, the maximum ultraviolet absorbance for the product, 2-hydroxymuconic acid. This is consistent with previous reports in which the ending compound was purified and verified as the correct product[24]. The pre-steady state spectra were obtained with an Applied Photophysics Stopped-Flow Spectrometer SX20 (UK) with the mixing unit hosted inside an anaerobic chamber made by Coy Laboratory Products (MI, USA). Pre-steady state activity used the same reaction buffer but with 23 μM AMSDH or E268A and 25 μM 2-HMS and were carried out at 10 °C. The change in absorbance was monitored for 1.0 s.

X-ray crystallographic data collection and refinement. Purified AMSDH samples at a final concentration of 10 mg ml^{-1} containing no NAD^+ or 10 equiv. of NAD^+ were used to set up sitting-drop vapour diffusion crystal screening trays in Art Robbins 96-well Intelli-Plates using an ARI Gryphon crystallization robot. The initial crystallization conditions were obtained from PEG-Ion 1/2 (Hampton Research) screening kits at room temperature. The screened conditions were optimized by increasing protein concentration to 40 mg ml^{-1} and lowering crystallization temperature to 18 °C. NAD^+-bound AMSDH crystals were obtained from drops assembled with 1.5 μl of protein (preincubated for 10 min with 10 equiv. of NAD^+) mixed with 1.5 μl of a reservoir solution containing 20% polyethylene glycol 3350 and 0.2 M sodium phosphate dibasic monohydrate, pH 9.1, by hanging drop diffusion in VDX plates (Hampton Research). Pyramid shaped crystals that diffract up to ~ 1.9 Å appeared overnight. The reservoir solution for crystallizing the cofactor-free AMSDH crystals contains 12% polyethylene glycol 3350, 0.1 M sodium formate, pH 7.0. Crystals belonging to the same space group formed within 2–3 days with an irregular plate shape and diffracted up to ~ 2.2 Å. NAD^+-AMSDH crystals were used for substrate-soaking experiments. Crystals were transferred to mother liquor solution containing ~ 1 mM 2-HMS and incubated for 10–180 min before flash cooling in liquid nitrogen. Soaking 2-AMS as a substrate is more complicated because of its instability. Crystallization solution containing ~ 1.5 mM ACMS were used for soaking. After transferring several crystals to the soaking solution (8 μl), 2 μl of 1 mM purified ACMSD was included to catalyse the conversion of ACMS to 2-AMS. Crystals were flash frozen after a 5 min-incubation. Crystallization solution containing 20% glycerol or ethylene glycol was used as cryoprotectant. X-ray diffraction data were collected on SER-CAT beamline 22-ID or 22-BM of the Advanced Photon Source, Argonne National Laboratory.

Ligand refinement and molecular modelling. The first AMSDH structure, the cofactor NAD^+ bound structure, was solved by the molecular replacement method with the Advanced Molecular Replacement coupled with Auto Model Building programs from the PHENIX software using 5-carboxymethyl-2-hydroxymuconate semialdehyde dehydrogenase (PDB: 2D4E) as a search model, which shares 39% of amino-acid sequence identity with *P. fluorescens* AMSDH. The ligand-free, mutant and ternary complex structures were solved by molecular replacement using the refined NAD^+-AMSDH as the search model. Refinement was conducted using PHENIX software[34]. The program Coot was used for electron density map analysis and model building[35]. NAD$^+$/NADH, substrates 2-AMS and 2-HMS and Cys-substrate covalent-adduct intermediate were well defined and added to the model based on the $2F_o - F_c$ and $F_o - F_c$ electron density maps. Refinement was assessed as complete when the $F_o - F_c$ electron density contained only noise. The structural figures were generated using PyMOL software (http://www.pymol.org/).

Single-crystal spectroscopy. Electronic absorption spectra from single crystals held at 100 K were collected at beamline X26-C of the National Synchrotron Light Source (NSLS)[36]. The electronic absorption data were typically obtained between 200 and 1,000 nm with a Hamamatsu (Bridgewater, N.J.) L10290 high-power ultraviolet–visible light source. The lamp was connected to one of several 3-m long solarization-resistant optical fibres with an internal diameter of 115, 230, 400 or 600 μm (Ocean Optics, Dunedin, FL). The other end was connected to a 40-mm diameter, 35 mm working distance 4 ×, Schwardchild reflective microscope objective (Optique Peter, Lentilly France). The spectroscopy spot size is a convolution of the optical fibre diameter and the magnification of the objective, which in this case produced 28, 50, 100 or 150 μm diameter spots, respectively. Photons that passed through the crystal were collected with a second, aligned objective that was connected to a similar optical fibre or one with a slightly larger internal diameter. The spectrum was then recorded with either an Ocean Optics USB4000 or QE65000 spectrometer. Anisotropic spectra and an image of the crystal/loop were collected as a function of rotation angle in 5° increments. These were analysed by XREC[37] to determine the flat face and optimum orientation.

Mass spectrometry. To prepare samples for ESI mass spectrometry, as-isolated E268A AMSDH was buffer-exchanged to 10 mM Tris (pH 8.0) by running through a desalting column (GE Healthcare). Intermediate bound E268A was obtained by mixing E268A with 3 equiv. of 2-HMS. Excess 2-HMS was removed by desalting chromatography using the same buffer. Desalted proteins were concentrated to a final concentration of 20 μM. Freshly prepared samples were rinsed by acetonitrile and 0.1% formic acid (1:1 ratio) before injection. Mass spectrometry experiments were conducted using a Waters (Milford, MA) Micromass Q-TOF micro (ESI-Q-TOF) instrument operating in positive mode. The capillary voltage was set to 3,500 V, the sample cone voltage to 35 V and the extraction cone voltage to 2 V. The source block temperature and the desolvation temperature were set to 100 and 120 °C, respectively. The samples were introduced into the ion source by direct injection at a flow rate of 5 μl min^{-1}. The raw data containing multiple positively charged protein peaks were deconvoluted and smoothed using MassLynx 4.1.

Computational studies. All ground-state density functional theory calculations were performed with Gaussian 03 Revision-E.01 at the B3LYP/6-31G* + level of theory[38]. The chemical structures were optimized using the ternary complex crystal structure (PDB entry: 4I25) as a starting model. To calculate the isomerization barrier, the dihedral angle about the 2–3 bond was restrained and the rest of the molecule was optimized. For the calculations that included the guanidinium heads of Arg120 and Arg464, the geometry was obtained from the crystal structure, and their positions were fixed while the substrate was optimized.

References

1. Keszthelyi, D., Troost, F. J. & Masclee, A. A. Understanding the role of tryptophan and serotonin metabolism in gastrointestinal function. *Neurogastroenterol. Motil.* **21**, 1239–1249 (2009).
2. Myint, A. M. *et al.* Kynurenine pathway in major depression: evidence of impaired neuroprotection. *J. Affect. Disord.* **98**, 143–151 (2007).
3. Ogawa, T. *et al.* Kynurenine pathway abnormalities in Parkinson's disease. *Neurology* **42**, 1702–1706 (1992).
4. Guillemin, G. J. *et al.* Quinolinic acid in the pathogenesis of Alzheimer's disease. *Adv. Exp. Med. Biol.* **527**, 167–176 (2003).
5. Guidetti, P. & Schwarcz, R. 3-Hydroxykynurenine and quinolinate: pathogenic synergism in early grade Huntington's disease? *Adv. Exp. Med. Biol.* **527**, 137–145 (2003).
6. Kerr, S. J., Armati, P. J., Guillemin, G. J. & Brew, B. J. Chronic exposure of human neurons to quinolinic acid results in neuronal changes consistent with AIDS dementia complex. *AIDS* **12**, 355–363 (1998).
7. Heyes, M. P. *et al.* Quinolinic acid in cerebrospinal fluid and serum in HIV-1 infection: relationship to clinical and neurological status. *Ann. Neurol.* **29**, 202–209 (1991).
8. Heyes, M. P. *et al.* Sources of the neurotoxin quinolinic acid in the brain of HIV-1-infected patients and retrovirus-infected macaques. *FASEB J.* **12**, 881–896 (1998).
9. Schwarcz, R. The kynurenine pathway of tryptophan degradation as a drug target. *Curr. Opin. Pharmacol.* **4**, 12–17 (2004).
10. Stone, T. W. & Darlington, L. G. Endogenous kynurenines as targets for drug discovery and development. *Nat. Rev. Drug Discov.* **1**, 609–620 (2002).
11. Perozich, J., Nicholas, H., Wang, B. C., Lindahl, R. & Hempel, J. Relationships within the aldehyde dehydrogenase extended family. *Protein Sci.* **8**, 137–146 (1999).
12. Unden, G. & Bongaerts, J. Alternative respiratory pathways of *Escherichia coli*: energetics and transcriptional regulation in response to electron acceptors. *Biochim. Biophys. Acta* **1320**, 217–234 (1997).
13. Nicholls, D. G. Mitochondrial function and dysfunction in the cell: its relevance to aging and aging-related disease. *Int. J. Biochem. Cell Biol.* **34**, 1372–1381 (2002).
14. Hempel, J., Nicholas, H. & Lindahl, R. Aldehyde dehydrogenases: widespread structural and functional diversity within a shared framework. *Protein Sci.* **2**, 1890–1900 (1993).

15. Hasegawa, Y. *et al.* A novel degradative pathway of 2-nitrobenzoate via 3-hydroxyanthranilate in *Pseudomonas fluorescens* strain KU-7. *FEMS Microbiol. Lett.* **190,** 185–190 (2000).

16. Li, T., Walker, A. L., Iwaki, H., Hasegawa, Y. & Liu, A. Kinetic and spectroscopic characterization of ACMSD from *Pseudomonas fluorescens* reveals a pentacoordinate mononuclear metallocofactor. *J. Am. Chem. Soc.* **127,** 12282–12290 (2005).

17. Li, T., Ma, J., Hosler, J. P., Davidson, V. L. & Liu, A. Detection of transient Intermediates in the metal-dependent non-oxidative decarboxylation catalyzed by α-amino-β-carboxymuconate-ε-semialdehyde decarboxylase. *J. Am. Chem. Soc.* **129,** 9278–9279 (2007).

18. Perez-Miller, S. J. & Hurley, T. D. Coenzyme isomerization is integral to catalysis in aldehyde dehydrogenase. *Biochemistry* **42,** 7100–7109 (2003).

19. Muñoz-Clares, R. A., González-Segura, L. & Díaz-Sánchez, A. G. Crystallographic evidence for active-site dynamics in the hydrolytic aldehyde dehydrogenases. Implications for the deacylation step of the catalyzed reaction. *Chem. Biol. Interact.* **191,** 137–146 (2011).

20. Huo, L., Davis, I., Chen, L. & Liu, A. The power of two: arginine 51 and arginine 239* from a neighboring subunit are essential for catalysis in α-amino-β-carboxymuconate-ε-semialdehyde decarboxylase. *J. Biol. Chem.* **288,** 30862–30871 (2013).

21. Geoghegan, K. F. *et al.* Spontaneous α-N-6-phosphogluconoylation of a 'His tag' in *Escherichia coli*: the cause of extra mass of 258 or 178 Da in fusion proteins. *Anal. Biochem.* **267,** 169–184 (1999).

22. Blanco, J., Moore, R. A. & Viola, R. E. Capture of an intermediate in the catalytic cycle of L-aspartate-β-semialdehyde dehydrogenase. *Proc. Natl Acad. Sci. USA* **100,** 12613–12617 (2003).

23. Colabroy, K. L. & Begley, T. P. The pyridine ring of NAD is formed by a nonenzymatic pericyclic reaction. *J. Am. Chem. Soc.* **127,** 840–841 (2005).

24. Ichiyama, A. *et al.* Studies on the metabolism of the benzene ring of tryptophan in mammalian tissues. II. Enzymic formation of α-aminomuconic acid from 3-hydroxyanthranilic acid. *J. Biol. Chem.* **240,** 740–749 (1965).

25. He, Z., Davis, J. K. & Spain, J. C. Purification, characterization, and sequence analysis of 2-aminomuconic 6-semialdehyde dehydrogenase from *Pseudomonas pseudoalcaligenes* JS45. *J. Bacteriol.* **180,** 4591–4595 (1998).

26. Abriola, D. P., Fields, R., Stein, S., MacKerell, Jr. A. D. & Pietruszko, R. Active site of human liver aldehyde dehydrogenase. *Biochemistry* **26,** 5679–5684 (1987).

27. Kitson, T. M., Hill, J. P. & Midwinter, G. G. Identification of a catalytically essential nucleophilic residue in sheep liver cytoplasmic aldehyde dehydrogenase. *Biochem. J.* **275,** 207–210 (1991).

28. Farres, J., Wang, T. T., Cunningham, S. J. & Weiner, H. Investigation of the active site cysteine residue of rat liver mitochondrial aldehyde dehydrogenase by site-directed mutagenesis. *Biochemistry* **34,** 2592–2598 (1995).

29. Steinmetz, C. G., Xie, P., Weiner, H. & Hurley, T. D. Structure of mitochondrial aldehyde dehydrogenase: the genetic component of ethanol aversion. *Structure* **5,** 701–711 (1997).

30. Moore, S. A. *et al.* Sheep liver cytosolic aldehyde dehydrogenase: the structure reveals the basis for the retinal specificity of class 1 aldehyde dehydrogenases. *Structure* **6,** 1541–1551 (1998).

31. D'Ambrosio, K. *et al.* The first crystal structure of a thioacylenzyme intermediate in the ALDH family: new coenzyme conformation and relevance to catalysis. *Biochemistry* **45,** 2978–2986 (2006).

32. Park, J. & Rhee, S. Structural basis for a cofactor-dependent oxidation protection and catalysis of cyanobacterial succinic semialdehyde dehydrogenase. *J. Biol. Chem.* **288,** 15760–15770 (2013).

33. Wang, X. & Weiner, H. Involvement of glutamate 268 in the active site of human liver mitochondrial (class 2) aldehyde dehydrogenase as probed by site-directed mutagenesis. *Biochemistry* **34,** 237–243 (1995).

34. Adams, P. D. *et al.* PHENIX: a comprehensive Python-based system for macromolecular structure solution. *Acta Crystallogr. D Biol. Crystallogr.* **66,** 213–221 (2010).

35. Emsley, P. & Cowtan, K. Coot: model-building tools for molecular graphics. *Acta Crystallogr. D Biol. Crystallogr.* **60,** 2126–2132 (2004).

36. Orville, A. M. *et al.* Correlated single-crystal electronic absorption spectroscopy and X-ray crystallography at NSLS beamline X26-C. *J. Synchrotron Radiat.* **18,** 358–366 (2011).

37. Pothineni, S. B., Strutz, T. & Lamzin, V. S. Automated detection and centring of cryocooled protein crystals. *Acta Crystallogr. D Biol. Crystallogr.* **62,** 1358–1368 (2006).

38. Frisch, M. J. *et al.* Gaussian 03, Revision E.01 (Gaussian, Inc., 2004).

Acknowledgements

This work was supported, in whole or in part, by the National Science Foundation grant CHE-0843537, National Institutes of Health grants GM108988 and GM107529 and Georgia Research Alliance Distinguished Scientist Program (A.L.), Molecular Basis of Disease Area of Focus graduate fellowship (L.H., I.D. and S.E.), Center for Diagnostics and Therapeutics (F.L.), Georgia State University Dissertation Award (L.H.) and funds from Mext Haiteku (Y.H.), Offices of Biological and Environmental Research award FWP BO-70 of the U.S. Department of Energy and NIH grant P41GM103473 (B.A. & A.M.O.). We thank Dr. Siming Wang for assistance with the mass spectrometry analysis and Dr. Donald Hamelberg for valuable discussions. X-ray data were collected at the Southeast Regional Collaborative Access Team (SER-CAT) 22-ID and 22-BM beamlines at the Advanced Photon Source, Argonne National Laboratory. Use of the Advanced Photon Source was supported by the U.S. Department of Energy, Office of Science, Office of Basic Energy Sciences, under Contract No. W-31-109-Eng-38. Single-crystal spectroscopy data were obtained at beamline X26-C of the National Synchrotron Light Source (NSLS), Brookhaven National Laboratory with the support of the U.S. Department of Energy under Contract No. DE-AC02-98CH10886.

Author contributions

A.L. conceived of and led the study. H.I. and Y.H constructed the initial expression system. F.L. optimized the expression and established protein isolation and activation procedures. I.D. solved the apo-AMSDH structure. L.H. solved all complex and intermediate structures. L.H. and I.D. performed the kinetic assays. L.H. performed the mass spectrometry experiment. S.E. constructed the mutant expression systems. B.A. and A.M.O. collected the single-crystal electronic absorption spectra. I.D. performed the quantum chemical calculations. The manuscript was written by L.H. I.D. and A.L. All authors approved the final submitted manuscript.

Additional information

Accession codes: Coordinates and structure factors for apo-AMSDH, NAD$^+$-bound AMSDH, NAD$^+$- and 2-AMS-bound AMSDH, NAD$^+$- and 2-HMS-bound AMSDH, thioacyl intermediate AMSDH, E268A AMSDH, E268A thiohemiacetal intermediate, and E268A thioacyl intermediate have been deposited in the RCSB Protein Data Bank under accession codes 4I26, 4I1W, 4I25, 4I2R, 4NPI, 4OE2, 4OU2, and 4OUB, respectively.

Competing financial interests: The authors declare no competing financial interests.

12

Bridge helix bending promotes RNA polymerase II backtracking through a critical and conserved threonine residue

Lin-Tai Da[1,*], Fátima Pardo-Avila[1,*], Liang Xu[2,*], Daniel-Adriano Silva[1,3], Lu Zhang[1], Xin Gao[4], Dong Wang[2] & Xuhui Huang[1,5,6]

The dynamics of the RNA polymerase II (Pol II) backtracking process is poorly understood. We built a Markov State Model from extensive molecular dynamics simulations to identify metastable intermediate states and the dynamics of backtracking at atomistic detail. Our results reveal that Pol II backtracking occurs in a stepwise mode where two intermediate states are involved. We find that the continuous bending motion of the Bridge helix (BH) serves as a critical checkpoint, using the highly conserved BH residue T831 as a sensing probe for the 3′-terminal base paring of RNA:DNA hybrid. If the base pair is mismatched, BH bending can promote the RNA 3′-end nucleotide into a frayed state that further leads to the backtracked state. These computational observations are validated by site-directed mutagenesis and transcript cleavage assays, and provide insights into the key factors that regulate the preferences of the backward translocation.

[1] Department of Chemistry, School of Science and Institute for Advance Study, Hong Kong University of Science and Technology, Clear Water Bay, Kowloon, Hong Kong. [2] Department of Cellular and Molecular Medicine, School of Medicine; Skaggs School of Pharmacy and Pharmaceutical Sciences, University of California San Diego, La Jolla, California 92093, USA. [3] Department of Biochemistry, University of Washington, Seattle, Washington 98195, USA. [4] King Abdullah University of Science and Technology, Computational Bioscience Research Center, Computer, Electrical and Mathematical Sciences and Engineering Division, Thuwal 23955-6900, Saudi Arabia. [5] Division of Biomedical Engineering, School of Science and Institute for Advance Study, Hong Kong University of Science and Technology, Clear Water Bay, Kowloon, Hong Kong. [6] Center of Systems Biology and Human Health, School of Science and Institute for Advance Study, Hong Kong University of Science and Technology, Clear Water Bay, Kowloon, Hong Kong. * These authors contributed equally to this work. Correspondence and requests for materials should be addressed to D.W. (email: dongwang@ucsd.edu) or to X.H. (email: xuhuihuang@ust.hk).

RNA polymerase is the key enzyme in gene expression responsible for synthesizing messenger RNA (mRNA) based on the DNA template[1-4]. RNA polymerase II (Pol II) can efficiently detect and cleave mis-incorporated nucleotides via its proofreading mechanism that contributes its high transcription fidelity[5,6]. The transcription proofreading is enabled by Pol II backtracking (see Fig. 1a for details): Pol II moves in a reverse direction from the pre-translocation state to a 'backtracked' state where the RNA 3'-end nucleotide dislodges from the active site, and extrudes through the pore region of the secondary channel[7,8]. This backtracking motion can be greatly favoured by damaged DNA template, mismatched base-pairing or a nucleosomal barrier[9-12] (see Fig. 1b for details). Finally, 1 or 2 backtracked RNA nucleotides can be removed via an intrinsic activity of Pol II, whereas the cleavage of more than two backtracked RNA nucleotides requires the recruitment of the transcription factor IIS (TFIIS). The backtracked Pol II Elongation Complex (EC) has been implicated in many critical biological processes, such as control of promoter-proximal pausing, transcription elongation dynamics (pausing and arrest), termination and so on (refs 10,13–16).

Recently, the X-ray structures of Pol II EC arrested in backtracked states have been solved[7,8]. Wang et al.[8] revealed atomistic details of a backtracked Pol II EC with one or two RNA 3'-end nucleotides retreated to the backtracked form. This allows identifying the critical interactions between the backtracked RNA nucleotides and the Pol II residues. In particular, a single backtracked RNA 3'-end nucleotide was found to directly contact the Rpb2 residue Y769, Trigger Loop (TL) residues Q1078, N1082 and Bridge helix (BH) residues T827 and so on. More recently, Cheung et al.[7] employed CTP extension from a tailed DNA template and captured the X-ray structure of the Pol II EC with nine backtracked poly(C) RNA transcript, providing the details of the extended interaction network between the further

backtracked RNA nucleotides and the Pol II residues. In addition, Sydow et al.[17] reported a structure of Pol II EC in a frayed state (not yet backtracked), in which the RNA 3'-end frayed nucleobase stacks directly with the Rpb2 residue Y769, this structure was thus suggested as an intermediate state before the complete backtracking[17] (see the fraying-dependent stepwise model in Supplementary Fig. 1). However, the frayed state was not crystallized with a 3'-end mismatched ribonucleotide in the active site, instead, the mismatched nucleotides were located at an upstream position. Therefore, there still exists the possibility that this frayed state is an off-pathway metastable state, and 3'-end RNA nucleotide can backtrack concurrently with backtracking of its complementary template DNA nucleotide (see the fraying-independent concerted model in Supplementary Fig. 1). Although the above X-ray structures have revolutionized our understandings of the backtracked Pol II ECs at an atomic detail, several critical questions remain largely unsolved. For example, Does backtracking take place in a stepwise or a concerted mode? (that is, Whether the terminal base pair is disrupted before or co-currently in comparison with pol II reverse translocation?) What is the driving force for the backtracking? How does Pol II modulate the energetics of backtracking when mismatch or DNA damage is present?

The experimental approaches to address the above issues are limited and challenging, as most of these approaches cannot directly provide dynamical information for backtracking at the molecular level. Molecular dynamics (MD) simulations can greatly complement experimental observations on elucidating the dynamics at the atomic level. MD simulations have been widely applied to study proteins and nucleic acids, including systems as large as Pol II (refs 18–24). One major challenge faced by MD simulations of a system as large as the Pol II complex in an explicit solvent box ($\sim 370,000$ atoms) is to reach biologically relevant timescales[25]. Most all-atom MD simulations of RNA polymerases are performed at timescales from tens to hundreds of nanoseconds, whereas relevant conformational changes such as backtracking typically occur at timescales on the order of hundreds of microseconds or even longer[26].

The above-mentioned timescale gap can be bridged by Markov State Models (MSMs), which allow us to model long timescale dynamics from many parallel short MD simulations[27-37]. MSMs have been successfully employed to investigate conformational changes that are difficult to be studied using straightforward MD simulations[37,38]. These studies cover a wide aspect of areas, including protein[39-43] or RNA[32] conformational changes, ligand-receptor binding[44-46], allostery[47] and so on. Recently, we have constructed a MSM to elucidate the forward translocation mechanism of Pol II EC at atomic resolution, and found that the translocation in Pol II EC takes place at the timescale of tens of microseconds in the absence of incoming nucleoside triphosphate (NTP)[48].

In this study, we built a MSM to investigate the mechanisms of Pol II backtracking upon nucleotide mis-incorporation, a critical step for proofreading. We identified a stepwise backtracking process in which the first step is the fast fraying motion (at sub-microsecond) of the RNA 3'-end nucleotide. This step is followed by the rotation of the DNA transition nucleotide (TN) to stack with BH residues, which results in the formation of an additional intermediate state. During the last step, the reverse translocation of the upstream RNA:DNA hybrid takes place at around 100 μs to reach the final backtracked register. Interestingly, we found that the bending of the BH can serve as a checkpoint that examines the stability of the base pair (bp) of RNA:DNA hybrid at $i + 1$ site, and determine the direction of translocation (whether the RNA:DNA hybrid moves forward or backward by fraying and backtracking). In particular, the

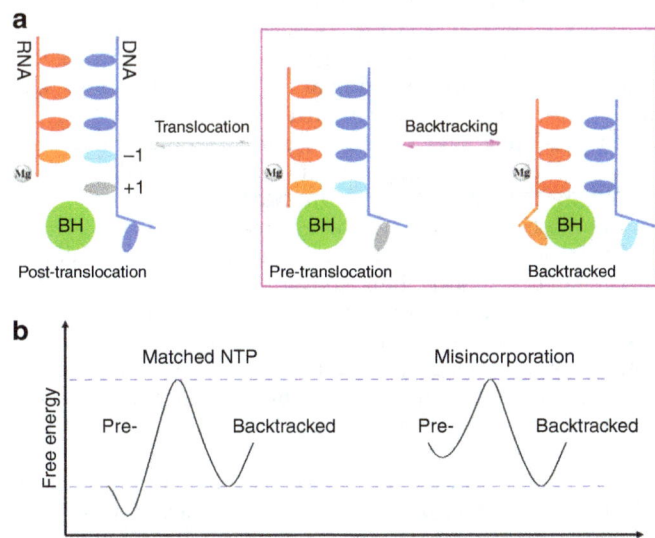

Figure 1 | Translocation and backtracking for RNA Pol II Elongation Complex (EC). (a) Cartoon models of the translocation and backtracking for RNA Pol II Elongation Complex (EC) during transcription elongation. Upstream DNA (in blue), RNA (in red), Bridge helix (BH, in green), Mg^{2+} A (in grey) are shown. In particular, the DNA transition nucleotide (TN) in translocation and backtracking models is highlighted in grey and cyan colour, respectively, and the RNA 3'-end nucleotide in backtracking model is highlighted in orange. (b) Schematic free energy landscape for pre-translocation and backtracked states. The transition barrier between these two states will vary under different conditions: matched NTP (left) and mis-incorporation that can be captured by proofreading (right).

conserved BH residue T831, which directly points at the $i+1$ site, can facilitate the separation of the mismatched base pair and drive the Pol II complex into the backtracked state. The critical role of T831 predicted by simulations was further validated by site-directed mutagenesis experiments, which show that the T831A substitution can substantially decrease the backtracking-dependent transcript cleavage rate.

Results

Backtracking in RNA Pol II EC follows a stepwise model. We first set up the frayed and one-nucleotide backtracked states based on the crystal structures with PDB ID: 3HOZ[17] and 3GTG[8], respectively (Supplementary Fig. 1). The pre-translocation state was modelled by placing an rG:dG mismatched base pair in the active site and removing the backtracked nucleotides from the crystal structure (PDB ID: 3GTG). Next, we obtained the initial backtracking pathways using the Climber algorithm[49], which is designed to search for the low-energy pathways connecting two conformational states. To take into account the possibility of both stepwise and concerted mechanisms, we generated initial pathways following both models, that is, pre-translocation $\leftrightarrow$ backtracked and pre-translocation $\leftrightarrow$ frayed $\leftrightarrow$ backtracked states. Finally, to eliminate the bias introduced by the initial pathways, three rounds of MD simulations were performed and only the last two rounds of MD simulations (with an aggregated simulation time of $\sim 48\,\mu s$) were used to construct the final MSM (see the Methods section for details).

From the MSM, we found that backtracking follows a stepwise mechanism rather than a concerted one. In particular, we used Transition Path Theory to compute the probability of each of the two competing mechanisms of backtracking from the flux through each pathway[41,50]. This flux analysis shows that the entire flux goes through the stepwise pathway (see Fig. 2 and Supplementary Movie 1 for details). The projection of the free energy landscapes on a pair of reaction coordinates (root-mean-square deviations (RMSDs) of the two RNA 3'-end nucleotides and their corresponding base-paired DNA

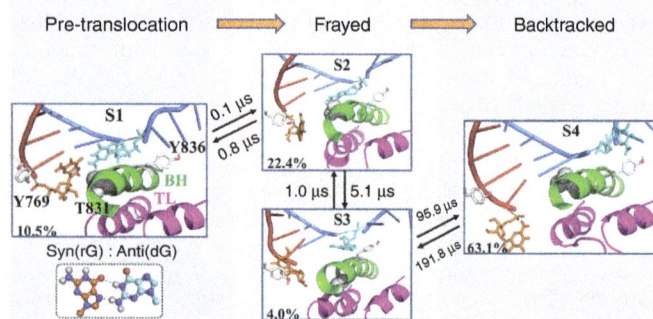

Figure 2 | Backtracking in RNA Pol II EC follows a stepwise model. Representative structures for each of the four metastable states identified by our Markov State Model (MSM; S1–S4) and connected by the top one pathway from the pre-translocation state (S1) to the backtracked state (S4): S1→S2→S3→S4. The equilibrium population for each state (left bottom corner) and the Mean First Passage Time (MFPT) between each transition (beside arrow, unit in μs) are presented, with their corresponding errors given as followings: $10.5 \pm 1.5\%$, $22.4 \pm 2.1\%$, $4.0 \pm 0.7\%$ and $63.1 \pm 2.8\%$ (equilibrium population); $0.1 \pm 0.0\,\mu s$, $0.8 \pm 0.1\,\mu s$, $5.1 \pm 3.0\,\mu s$, $1.0 \pm 0.7\,\mu s$, $95.9 \pm 42.3\,\mu s$ and $191.8 \pm 69.0\,\mu s$ (MFPT). The hybrid RNA/DNA chains (red/blue), the Trigger Loop (purple), Bridge helix (green), Rpb1 residue Y836 and Rpb2 residue Y769 (both in grey) are shown. The DNA TN (dG) and its mismatched RNA nucleotide (rG) are highlighted with cyan and orange stick model, respectively.

nucleotides) obtained from our MSM also clearly shows that the RNA fraying and DNA template translocation occur in a stepwise mode (see Supplementary Fig. 2a for details). Furthermore, we observe that one of the initial pathways corresponding to the concerted mechanism (the black line in Supplementary Fig. 2a) clearly deviates from the low free energy regions of the MSM projection (contour lines in Supplementary Fig. 2a). To further prove that the MSM is capable of eliminating the bias introduced by the initial pathways, we projected the conformations from our MSM sampling onto three reaction coordinates (three top eigenvectors identified by the Isomap dimensionality reduction technique)[51]. As shown in Supplementary Fig. 2b, the MSM sampling largely overlaps with the stepwise initial pathways (in yellow), even though it covers a much larger region of the conformational space. However, the MSM sampling clearly deviates from the concerted initial pathways (in cyan). In summary, our MSM built upon the wild-type (WT) MD simulations has clearly indicated the existence of the frayed intermediate state and supported the stepwise model, whereas strongly disfavouring the concerted mechanism.

Our MSM for backtracking contains four metastable states (S1–S4 states; Fig. 2). As observed in the crystal structure[5], favourable interactions between the RNA 3'-end nucleotide and a number of Pol II residues (for example, BH residues T827, and TL residues Q1078 and N1082) are formed to stabilize the backtracked state (S4) for the rG:dG mismatched system (see Supplementary Fig. 1 for details). In addition to the pre-translocation (S1) and backtracked states (S4), two additional intermediate states are identified along the major pathway (Fig. 2). One of them (S2) is consistent with the frayed crystal structure[17]. The other intermediate also contains a frayed RNA 3'-end nucleotide, but two additional interactions are formed: the BH residue Y836 interacts with the DNA TN, and Rpb2 residue Y769 stacks with the RNA 3'-end nucleotide (Fig. 2). These two residues play a critical role in stabilizing the frayed RNA nucleotide in the S3 state. In particular, the stable stacking interactions between the base group of the frayed RNA nucleotide and the aromatic ring of the Y769 exist in most of the MD conformations ($\sim 80\%$) from the S3 state. However, only limited conformations ($<10\%$) from the other three states contain the above stacking interactions. The transition between the intermediate state S3 and the backtracked state (S4) that requires the translocations of the upstream RNA:DNA hybrid is the rate-limiting step (Fig. 2), at an order of a $100\,\mu s$. In addition, the system can quickly inter-convert between the other three states (S1–S3) at a timescale of a few microseconds. Finally, the backtracked state is the most stable state ($\sim 63.1\%$ of the conformations), whereas the pre-translocation state is much less stable with a population around an order of magnitude smaller, consistent with the kinetic model derived from a recent single-molecule experimental study[12]. These results strongly suggest that the rG:dG mismatched base pair can significantly shift the translocation bias towards the backtracked state.

RNA 3'-end nucleotide frays due to mismatch and BH bending. As shown in Fig. 2, for the rG:dG mismatched base pair, the frayed state (S2, 22.4%) is energetically more favourable than the pre-translocation state (S1, 10.5%). To evaluate the effects of the fully matched base pair on the stability of the frayed state, we designed an additional Pol II EC with rG:dC bp at $i+1$ site for further MD simulation studies (Fig. 3). To obtain the starting structures, we chose seven representative conformations near the transition state (TS) region that separates the S1 and S2 metastable states of the rG:dG system (corresponding to the

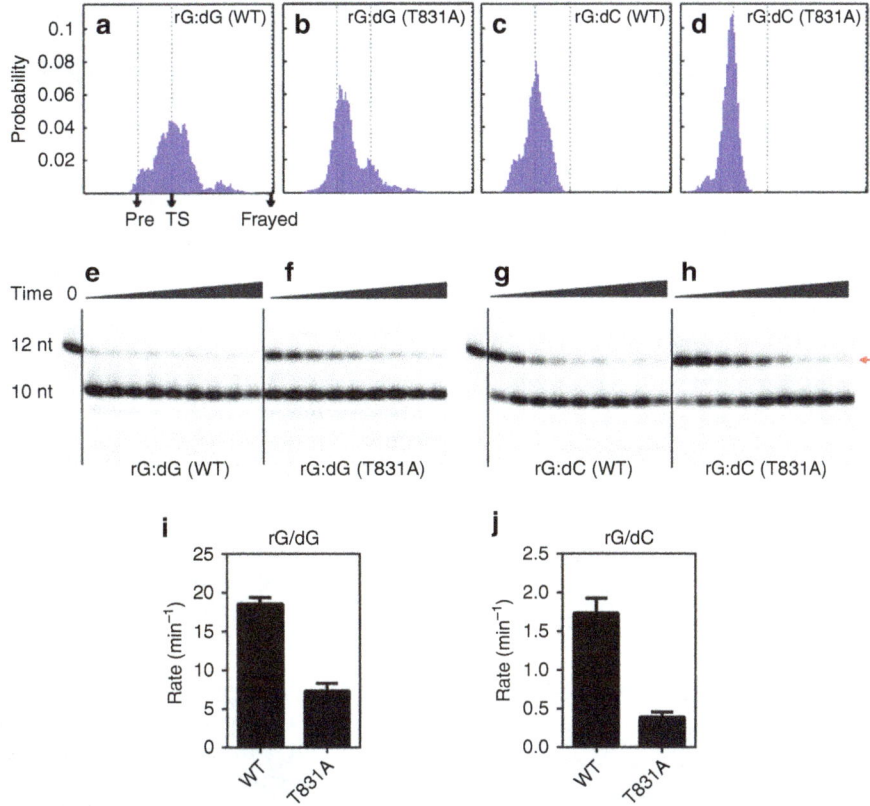

Figure 3 | Fraying motion of the RNA 3′-end nucleotide is dictated by the mismatched base pair in the active site and promoted by BH residue T831 serving as a sensing probe. As a comparison, one additional matched base pair (rG:dC) was designed for the mutant MD simulations starting from the regions near to the transition state (TS) between S1 and S2 states. (**a–d**) Histogram for the movements of the RNA 3′-end nucleotide for the rG:dG (**a**) and rG:dC (**c**) base pairs starting from the above TS, and their corresponding T831A mutants on the Pol II BH domain (**b** and **d**, respectively). Three critical states: Pre-translocation state (S1), TS and Frayed state (S2), are mapped with dashed lines and black arrows in the plots. The x coordinate represents a transition index, defined as $x = \frac{d_i}{d_{ref}}\left(\frac{\mathbf{d}_i \mathbf{d}_{ref}}{d_i d_{ref}}\right)$, where $\mathbf{d}_i = \mathbf{r}_{MD} - \mathbf{r}_{pre}$ is the vector connecting the pre-translocation structure and a certain MD conformation (each conformation $\mathbf{r}$ is represented using the c.o.m coordinates of its DNA TN). $\mathbf{d}_{ref} = \mathbf{r}_{frayed} - \mathbf{r}_{pre}$ is the reference vector connecting the pre-translocation and frayed state. The transition index indicates where a MD conformation locates between the pre-translocation and frayed state. In particular, $x = 0$ corresponds to the pre-translocation (pre) state while $x = 1$ corresponds to the frayed state. (**e–h**) Evaluation of the functional roles of the BH residue T831 in backtracking by site-directed mutagenesis studies and TFIIS cleavage assays. Comparing to the WTs of the rG:dG and rG:dC systems (**e** and **g**, respectively), the corresponding T831A mutant has weaker TFIIS-facilitated cleavage activities (**f** and **h**, respectively). The upper bands refer to the initial RNA transcript (12 nt) as indicated by red arrow; the lower bands refer to the cleaved product (10 nt). Quantitative analysis of TFIIS-stimulated cleavage rates is shown in **i** and **j**. Error bars represent standard deviations derived from three independent experiments.

region around the RMSD coordinate (3, 2.5) in the free energy profile in Supplementary Fig. 2a), and then mutated the rG:dG base pair to the rG:dC bp. Next, we performed 10-ns MD simulations for each of the above newly generated conformations with both rG:dC and rG:dG base pair systems.

The results clearly show that for the rG:dG base pair, nearly half of the conformations (46%) move back to the S1 state (Fig. 3a), indicating that the selected conformations indeed locate near the TS between the S1 and S2 states involved in the backtracking process for the rG:dG system. In sharp contrast, the rG:dC matched base pair completely shifts the transition equilibrium towards the S1 state (~100%; Fig. 3c), suggesting that the hydrogen bonds of the rG:dC base pair significantly favour the pre-translocation state and prevent the RNA 3′-end nucleotide from fraying. Our mutant MD simulation results strongly support the idea that the strength of the hydrogen bonds within the base pair located in the active site can significantly modulate the relative stabilities of S1 and S2 states, which in turn can shift the backtracking bias. Consistently, the experiments conducted by Wang and co-workers showed that the RNA cleavage efficiencies for the rG:dG pair was significantly higher

than that for the rG:dC base pair[8]. In addition to the rG:dG mismatch, we also performed MD simulations from the TS between S1 and S2 for three other mismatched base pairs in the active site: rA:dG, rC:dT and rU:dT. For all these three systems, we also observed a significant fraction of MD simulations that transit towards the frayed state (S2), which indicates that the frayed state could be a metastable intermediate for different mismatched systems (see Supplementary Fig. 3 and Supplementary Note 6 for details).

To reveal the functional roles of the BH in the fraying motion of the RNA 3′-end nucleotide, we carried out a cross-correlation analysis for both the RNA 3′-end nucleotide and the DNA TN (Fig. 4). We found that in the S1 state, the BH residue T831 is tightly coupled with both RNA and DNA nucleotides in the active site (Fig. 4a,b). Interestingly, the residue T831 was found to locate right on the helical turn of the BH that bends the most (Supplementary Fig. 4). Therefore, we suggest that the BH serves as a checkpoint to examine the stability of the base pair in the active site through its bending motion, using residue T831 as a sensing probe. In other words, if the base pair in the active site is unstable (for example, mismatched), the RNA 3′-end nucleotide

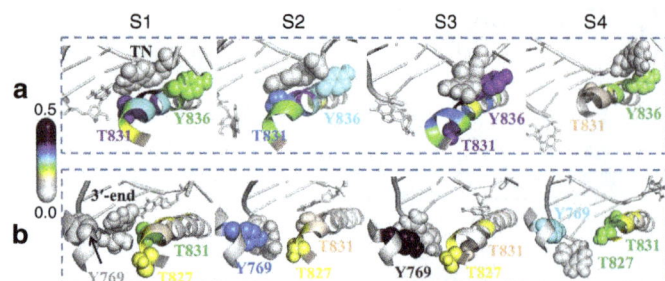

Figure 4 | Functional roles of the Pol II BH residues during the backtracking revealed by cross-correlation analysis. In each metastable state, the correlations were calculated between the Pol II residues and the DNA TN (**a**) or RNA 3′-end nucleotide (**b**), respectively. For DNA TN (**a**), two critical BH residues, T831 and Y836, are highlighted with sphere models. In both S1 and S2 states, residue T831 is tightly correlated with the TN. Compared with S1 state, the DNA TN enhances its correlations with the residue Y836 in S2 state because of increased flexibility. In S3 state, the TN further increases its correlations with residue Y836 by forming direct packing interactions, which is critical for the TN to cross over the BH. In S4 state, the tight correlation between the Y836 and TN remains. For the RNA 3′-end nucleotide (**b**), two BH residues, T827 and T831, and one Rpb2 residue Y769, are highlighted as sphere models. In S1 state, the T831 shows the strongest correlation with the RNA 3′-end nucleotide, and weakly correlated with T827 and Y769. In S2 state, the RNA 3′-end nucleotide loses the coupling with residue T831 and strengthens the correlations with residues T827 and Y769. Strikingly, in S3 state, the RNA 3′-end nucleotide is tightly coupled to the residue Y769 by forming stable packing interactions, and it is barely coupled with BH residues. In the final state, RNA 3′-end nucleotide shows the correlation with all of these three residues. For each representation, the RNA:DNA hybrid chains are shown in grey, the DNA TN in **a** and RNA 3′-end nucleotide in **b** are shown as grey sphere models. The colours of the Pol II residues, including the BH residues and the Rpb2 residue Y769, correspond to the colours in the spectrum bar on the left.

tends to fray, which can be further promoted by the continuous bending motion of the BH. Next, in the frayed state, the motion of the RNA 3′-end nucleotide becomes less correlated with that of T831, whereas the motion of RNA 3′-end nucleotide becomes more correlated with that of the BH residue T827 and Rpb2 residue Y769 (Fig. 4a,b). On the other hand, the dynamics of the DNA TN is still highly correlated with T831, and establishes new correlations with the dynamics of the BH residue Y836 (Fig. 4a). This may be due to the increased flexibility of the DNA TN after breaking its interactions with the RNA 3′-end nucleotide.

Site-directed mutagenesis validates role of T831. To further examine the specific roles of the BH residue T831 in backtracking, we first performed additional MD simulations for the rG:dG system with the T831A mutation on the BH. We started the simulations from the same set of conformations near the TS between S1 and S2 states, as used in the above MD simulations of different base pairs at the $i+1$ site. The results indicate a significant difference between the WT and the T831A mutant. As shown in Fig. 3b, the T831A substitution on the BH can greatly prevent the fraying of the RNA 3′-end nucleotide with nearly 86% conformations moving towards the S1 state, sharply contrasting to the corresponding WT results (Fig. 3a), suggesting that the BH residue T831 can substantially facilitate the fraying motion of the RNA 3′-end nucleotide.

To validate the above prediction, we conducted mutagenesis studies to evaluate the role of the BH residue T831 by measuring the transcript cleavage rate, which is dependent on the extent of

the backtracking. Based on our model, the T831A substitution on Pol II can diminish the ability of the BH domain to promote the fraying of the RNA 3′-end nucleotide, which, in turn, would slow down the cleavage reaction. We designed the same RNA:DNA scaffold for the rG:dG system as we used in our theoretical studies and measured the cleavage rate with both WT and T831A mutant of Pol II EC. Here, we employed both intrinsic and TFIIS-stimulated transcript cleavage assays to probe the impact of T831A on backtracking behaviour of RNA pol II. Indeed, as expected, we found reduced transcript cleavage activity with T831A pol II mutant in comparison with WT pol II (see Fig. 3e,f and Supplementary Fig. 5 for details). Intriguingly, we observed much more striking reduction of cleavage activities for T831A pol II mutant in the TFIIS-mediated cleavage. Quantitative analysis indicates the substitution T831A can significantly reduce the TFIIS-aided transcript cleavage rate by 2.5-fold compared with the WT (Fig. 3i), which provides experimental verification for our theoretical prediction.

As a control, we further evaluated the effects of the T831A mutation on the rG:dC system by performing both MD simulations and experimental studies, following the same procedure as we did for the rG:dG system. The MD simulations show that, for the rG:dC system, the T831A substitution slightly favours the forward translocation over the fraying motions (the population ratio between the conformations that translocate forward and the conformations that tend to fray is ∼69% versus 31%; Fig. 3d), whereas the above distribution is reversed in the WT where the population ratio is 46% versus 54% (see Fig. 3c). Consistent with the above theoretical observation, the intrinsic and TFIIS-stimulated cleavage experiments (Fig. 3g,h, j and Supplementary Fig. 5) also support that, for the rG:dC system, the presence of T831 in WT could substantially promote the backtracking process and thus increase cleavage rates (Fig. 3g), compared with the T831A mutant (Fig. 3h). Notably, we found the effect of substitution of T831A on TFIIS-stimulated cleavage (experiment data) is much more significant than intrinsic cleavage (both experiment and simulation). This result suggests that a tighter coupling of tri-helical bundle (composed of BH and helical portions of TL) interaction upon TFIIS insertion within the secondary channel, thus the BH bending and T831 may have much more significant effect on backtracked translocation.

Backtracking of the upstream RNA:DNA hybrid. We found that the fraying of the RNA 3′ nucleotide and the entry of the DNA TN into the backtracking pathway are not accompanied by translocation of the RNA:DNA hybrid. As shown in Fig. 5 and Supplementary Fig. 6, the upstream RNA:DNA hybrid backtracks in a single step during the S3 to S4 transition, and this motion occurs after the fraying of the RNA 3′-end nucleotide and the formation of the stable base stacking between DNA TN and BH residue Y836 (Fig. 4). In our model, the backtracking of the upstream RNA:DNA hybrid is the rate-limiting step of the whole backtracking process (at ∼100 μs), as it requires the breaking and re-forming of numerous interactions between Pol II and the RNA:DNA hybrid.

This asynchronous movement of the hybrid backbones and DNA TN is also observed in the forward translocation process for the matched nucleotide system[48]. In the absence of incoming NTP, the Pol II EC with matched nucleotides can oscillate between the pre- and post-translocation state. For the forward translocation, the upstream RNA:DNA hybrid movement is also the rate-limiting step, but its timescale is ∼5-fold faster than what we observed for the backtracking. The slower dynamics of backtracking is likely due to the breaking of the stacking interactions between the frayed RNA nucleotide and the Rpb2

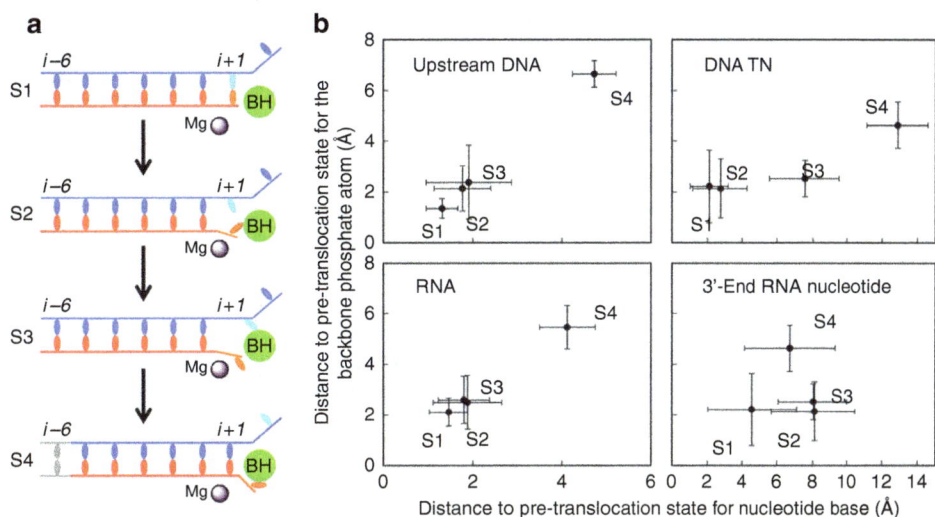

Figure 5 | Fraying of the RNA 3′-end nucleotide and the entry of the DNA TN into the backtracking pathway are not accompanied by translocation of the RNA:DNA hybrid. (a) Cartoon models illustrating the movements of the RNA 3′-end nucleotide (orange), DNA TN (cyan) and upstream RNA:DNA hybrids (red/blue) during the backtracking. From S1 to S2, the RNA 3′-end nucleotide frays from the $i+1$ site, but the DNA TN remains at $i+1$ site. From S2 to S3, the DNA TN starts to cross over the BH and stacks with the BH residues. From S3 to S4, the backbone translocation of the upstream RNA:DNA hybrids occurs at a single-step and facilitates the DNA TN to transfer to the complete backtracked state. **(b)** Plots showing the distance to the pre-translocation state for the backbone phosphate and the nucleotide base. For the upstream DNA chain (upper left panel) and RNA chain (lower left panel), the distance values were averaged over 5 upstream DNA and RNA nucleotides (From i-2 to i-6 sites, see Supplementary Fig. 3). The panels on the right show only the RNA 3′-end nucleotide and DNA TN (lower and upper rows, respectively). The fluctuations of the distance within each state are indicated with the error bars.

residue Y769 during the transition from S3 to S4, as the Y769 is tightly correlated and forms direct contacts with the frayed RNA 3′-end nucleotide in the S3 state, but the correlation is largely lost in the S4 state (Fig. 4). In addition, the aromatic side chain of the BH residue Y836 is found to form direct stacking interactions with the base of DNA TN in S3 (Figs 2 and 4).

To reveal the specific role of the above two residues Y836 and Y769 in the transition of the DNA TN from S3 to S4 states, we chose four representative conformations near the TS between the states S3 and S4 (the region with the RMSD value of ~6.3 Å along the Y axis in the free energy profile in Supplementary Fig. 2a). For those four selected conformations, we then replaced the Y836 and Y769 individually with alanine, and performed 10 ns MD simulations for each of the mutant systems as well as the WT (see the Methods section for details). Our results show that the Y836A mutation can prevent the transition of the DNA TN to the S4 state (~14% MD conformations move towards S4 state) compared with the WT system (Supplementary Fig. 7a,b). In a previous forward translocation study[48], Y836 has been suggested to stabilize the DNA TN over the BH and further facilitate the translocation of the upstream RNA:DNA hybrid. We thus suggest that Y836 plays a similar role in facilitating the transition from the S3 to S4 state during the backtracking. However, the substitution Y769A can conversely promote the backtracking with ~71% MD conformations moving towards S4 state (Supplementary Fig. 7c), supporting our previous conclusion that Y769 can substantially prevent the backtracking process.

Discussion

We revealed atomic-level details of the molecular mechanism of the backtracking in Pol II EC by constructing a MSM from extensive MD simulations, combined with site-directed mutagenesis and transcript cleavage studies. We employed the rG:dG mismatched base pair in our study and found that the backtracking occurred in a stepwise mode comprising four

well-defined metastable states. At the beginning, the pre-translocation state (S1) can quickly move to the frayed state (S2), which is promoted by the continuous bending motions of BH (Fig. 6). We suggest that the bending motion of the BH works as a checkpoint that can examine the stability of the base pair in the active site using the residue T831 as a probe that interacts with the base groups of both DNA TN and RNA 3′-end nucleotide (Fig. 6), which is well consistent with our mutagenesis results. Next, the S2 state can quickly equilibrate with another metastable state (S3) where two tyrosine residues, BH residue Y836 and Rpb2 residue Y769, play an important role in stabilizing the DNA TN and RNA 3′-end nucleotide, respectively. Finally, the last transition (from S3 to S4 states) in which the upstream RNA:DNA hybrid backtracks in a single-step, occurring at around hundreds of microseconds, was found to be the rate-limiting step (Fig. 6). Interestingly, one recent study conducted by Imashimizu et al.[52] suggested that the binding of the double-mutant elongation factor TFIIS to the yeast Pol II can result in a paused state of the Pol II complex where rearrangements of the 3′-end RNA nucleotide and the Pol II residues in the active site may take place, and this paused state may finally lead to the backtracking. Related to this experimental finding, we speculate that the frayed state may also play a role in the above-mentioned pausing process. In bacterial RNA polymerase, similar connection has also been suggested by the Landick group[16].

It has been well documented that the backtracking is sequence-dependent and particularly on the sequence near the RNA 3′-end nucleotide[10,53,54]. Our mutant MD simulations provided the molecular basis underlying these experimental observations. Our findings indicated that decreasing the strength of the hydrogen bonds within the base pair in the active site could potentially shift the backtracking bias from backtracking-resistant to backtracking-prone by favouring the frayed state. This emphasizes the dominant role of the hydrogen bonds in regulating the backtracking. More intriguingly, the BH bending

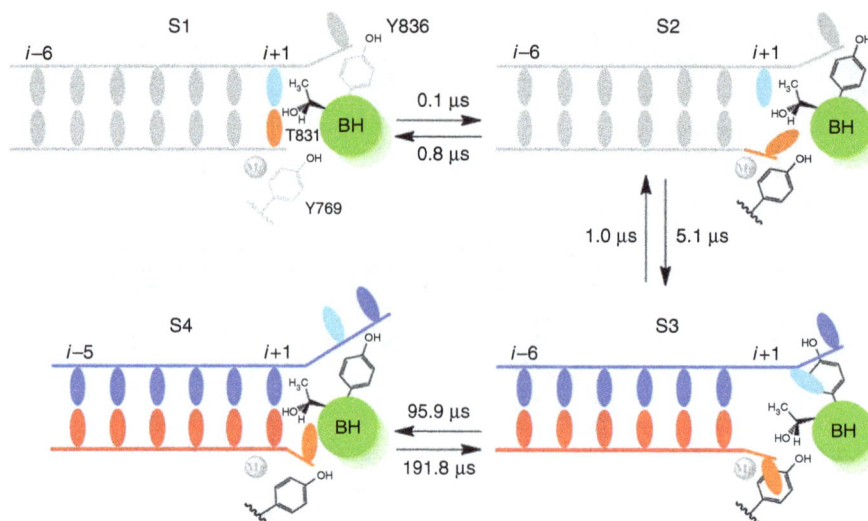

Figure 6 | Cartoon model of the backtracking mechanism in Pol II EC. Bending of the BH towards the $i+1$ site serves as a checkpoint to examine the stabilities of the base pair in the active site using residue T831 as a sensing probe. If the base pairing at $i+1$ is unstable (for example, mismatched), BH bending can trigger the motion of the RNA 3'-end nucleotide to the frayed state (S2 state). Next, the Pol II EC moves to S3 state where BH residue Y836 stacks with DNA TN and Rpb2 residue Y769 stacks with RNA 3'-end nucleotide through their aromatic rings. Finally, the movement of the RNA:DNA hybrid takes place. Refer to the Fig. 4a for details of the representations. The MFPTs for each transition calculated from the MSM are also provided.

can continuously examine the base groups of the RNA and DNA nucleotides at the $i+1$ site, which in turn, depending on the stabilities of the base pairs, results in either facilitating the upstream RNA:DNA hybrid to move one register forward (translocation) or the RNA 3'-end nucleotide to fray and consequently lead to backtracking.

Quantitative comparisons of dynamics between our model's predictions and experimental observations are challenging as it is difficult for existing experimental techniques to directly monitor the dynamics of backtracking conformational changes alone. However, recent optical tweezers experiments monitored the dynamics of complete Pol II elongation cycles upon external force, and then they obtained rates of individual conformational changes such as translocation and backtracking by fitting to a kinetic model of the elongation cycle[12]. In this way, they observed that the backtracking rate is significantly slower than the forward translocation ($\sim$13-fold)[12], which is qualitatively consistent with our simulation predictions (backtracking is $\sim$5-fold slower than forward translocation[55]). Our model suggests that the slower kinetics of the backtracking is due to the stacking interaction between the RNA 3'-end nucleotide and the Rpb2 residue Y769, which can block the extended backtracking motion. It is also worth noting that their kinetic model was derived based on a completely matched system and the rates they obtained for translocation/backtracking include the TL opening/closing motion, which is not simulated here. Recently, we also fitted a kinetic model with TL opening/closing motion separated from translocation based on both single-molecular and simulation data[56]. Our results suggest that the TL motion is the rate-limiting step of transcription, which explains apparently faster kinetics of simulated translocation and backtracking ($\sim$10–100 µs) compared with single-molecular[12] and fluorescence studies ($\sim$10 ms)[26]. We also note that the partial DNA scaffold used in our MD simulations may accelerate the dynamics of backtracking. With the recently available structures of full transcription bubble[57,58], we will further investigate this issue in future studies.

In the future, it will be interesting to extend our study to the mismatched pyrimidine base pair, for example, U:T wobble pair, as it may have different dynamic behaviour during the back-tracking process compared with the purine base pair because of

its smaller molecular size[59]. Second, we can go beyond the one-nucleotide backtracking event and study the backtracking mechanisms for two or even more RNA 3'-end nucleotides, which may give insight into the molecular mechanisms of the transcriptional pausing and arrest[10,15,55,60]. Third, it will also be interesting to investigate the backtracking mechanisms for different mutagenic DNA lesions, such as 8-oxoG and O^6-methylguanine (O^6-mdG) and so on, and how some of these lesions can escape Pol II proofreading[61].

Methods

We constructed the MSM to investigate the backtracking mechanism in Pol II EC, and our algorithm consists of the following steps: (i) Model the Pol II EC in pre-, frayed and backtracked states; (ii) generate initial low-energy pathways using a modified version of the Climber[49] algorithm for two proposed backtracking pathways (concerted and stepwise); (ii) seed three rounds of unbiased MD simulations from these initial backtracking pathways and (iv) construct and validate the MSM and further use the MSM to identify metastable intermediate states and obtain both thermodynamics and kinetics.

System setup and MD simulations. The backtracked and frayed states of Pol II EC were modelled from the X-ray structures of PDB id: 3GTG[8] and 3HOZ[17], respectively. The pre-translocation state was built from the above backtracked model by placing an rG:dG mismatched base pair in the active site and removing the backtracked RNA 3'-end nucleotide. The system was first energy-minimized, and then solvated in explicit SPC waters[62]. 77 Na$^+$ ions were added to neutralize the system. The final system contains 369,947 atoms. The AMBER force field was used to describe the system (99SB[63] for protein and metal ions, 99χ (refs 64,65) for DNA and RNA nucleotides), and all MD simulations were performed at 310K using GROMACS 4.5 (ref. 66; see Supplementary Note 1 for more details of the model construction).

Generating initial low-energy backtracking pathways. We proposed two models of backtracking: a concerted and a stepwise model. For the concerted model, we applied the Climber[49] algorithm to generate the pathways along pre-translocation → backtracked states, as well as its reverse route: backtracked → pre-translocation states. For the stepwise model, the initial pathway was produced to follow the route of pre-translocation → frayed → backtracked, and its reverse transition: backtracked → frayed → pre-translocation states. Then, for each backtracking pathway, we geometrically grouped the snapshots from the corresponding Climber simulations into 40 clusters, and 2 random conformations from each cluster were chosen (80 conformations for each model, 160 conformations in total) and were employed for the first round of MD simulations (see Supplementary Note 2 for more details).

Seeding unbiased MD simulations. In total, we performed four rounds of unbiased MD simulations. In the first round, the above 160 conformations from the initial backtracking pathways were employed as the starting structures for the 10 ns NVT MD simulations. We then selected 80 representative conformations from the first round of MD simulations as the starting structures for the second round of 100 ns MD simulations. Next, we selected another 80 representative conformations from the second round simulations as the starting structures for our third round of 100 ns simulations. To further enhance the sampling, we performed the fourth round of 320×100 ns MD simulations by initiating two additional independent simulations with different initial velocities from each of the initial conformations used in our second and third round. Finally, we collected a total of 480×100 ns MD simulation trajectories for further analysis (see Supplementary Note 2 for more details).

Constructing and validating MSMs. To construct the MSM, we followed a two-step procedure[29]: (i) we grouped the MD conformations with similar structures together to generate a set of 800 microstates using the K-centre clustering algorithm[29]; (ii) to visualize the mechanism of backtracking, we further grouped the 800 microstates into 4 macrostates using the Robust Perron Cluster Cluster Analysis (PCCA+) algorithm that is implemented in the MSMbuilder package[31] (see Supplementary Note 3 for additional details of the MSM construction).

All the reported quantative properties are computed based on the 800-state MSM constructed using the RMSD metrics (Supplementary Figs 8 and 9). To validate our 800-state MSM, we plotted the implied timescale as a function of lag time. The implied timescale plots reach the plateau at the lag time of ~8 ns (Supplementary Fig. 8b), suggesting our MSM is Markovian at 8 ns or longer lag times. To further validate the model[41], we predicted the residence probability for a given microstate using our MSM, and these predicted values are in good agreement with those directly obtained from the original MD simulations (Supplementary Fig. 8c).

To ensure that our MSM is independent of the choice of distance metric used for clustering, we built our model based on RMSD as well as the time-structure-based Independent Component Analysis (tICA)[67,68] (see Supplementary Figs 10 and 11, 13–15, and Supplementary Note 3 for details). For each method, we grouped the MD conformations (~4,800,000) into different number of microstates to verify that constructed MSM is robust to the number of clusters chosen. Notably, both methods render very similar dynamic properties. In particular, the predicted timescale for the slowest transition for the backtracking process was the same for all the MSMs built (Supplementary Figs 9 and 10). Moreover, we observed that the metastable macrostates we identified have consistent structural features regardless of the choice of metrics (see Supplementary Figs 11 and 12 and Supplementary Note 4 for details). We also tested the convergence of our conformational sampling by projecting the sampling from data sets with different sizes (6.4, 19.2, 28.8, 38.4 and 48 μs) onto two slowest tICs. In particular, for each data set, we projected the MD conformations in each microstate separately, and then performed a weighted sum according to their equilibrium populations obtained from the corresponding MSM. As shown in Supplementary Fig. 13, we can clearly see that the projections of the free energy landscape reach reasonable convergence, and particularly no new metastable regions in the phase space was discovered with the increase of conformational sampling. In addition, we also show that MSMs constructed from data sets with different sizes predicted consistent slowest implied timescales (Supplementary Fig. 16). These observations suggest that our MD data set is fairly sufficient to address the backtracking mechanism. We note that the recently developed TRAM method may further improve the quality of MSMs by using thermodynamic data obtained from other enhanced sampling methods to help the construction of MSMs[69]. Moreover, we show that the 4-macrostate MSM is not perfectly Markovian (see Supplementary Figs 17 and 18, and Supplementary Note 3 for details). In this paper, we finally adopted the 800-state MSM, constructed using the RMSD metrics, to report the key metastable states and their associated thermodynamic and kinetic properties.

MD simulations of matched/mismatched base pair or mutants. We performed additional MD simulations to investigate the backtracking propensity for one matched base pair: rG:dC. In addition, mutant MD simulations were designed to evaluate the functional roles of several critical residues in the backtracking. Finally, to test if the frayed intermediate state is also required for other mismatched base pair systems during the backtracking process, we designed three more Pol II systems with the following base pairs: rA:dG, rC:dT and rU:dT (see Supplementary Note 2 for additional details).

Generating millisecond trajectories based on our MSM. We used the transition probability matrix at microstate level to generate one 10-ms parallel trajectory, from which we quantitatively calculated several properties, including the RMSD for the backbone movements, cross-correlation values and the RMSF for the BH bending motion. In addition, the average values and the standard deviations of the populations and MFPTs were calculated by bootstrapping the MD trajectories for 100 times. During each bootstrapping, we repeatedly selected a MD trajectory by random from our data set for 480 times with replacement. Then, for each bootstrapped sample, we constructed the MSM and calculated the transition probability matrix at microstate level using a lag time of 8 ns (Markovian time) to generate one 10-ms long trajectory, from which we calculated the stationary distributions of the 4 macrostates and the MFPT for each transition. Finally, the

average values and corresponding standard deviations were calculated by averaging the results from all 100 samples.

Cross-correlation calculations. To study the correlated motions between the DNA TN/RNA 3′-end nucleotide and the Pol II residues, we calculated the cross-correlations for the pairs we are interested in (see Supplementary Note 3 for details).

TFIIS-dependent transcript cleavage assays. TFIIS cleavage reactions were performed by pre-incubating purified Pol II (WT or Rpb1 T831A mutant) with the RNA:DNA scaffolds as used in the theoretical studies with either rG:dC or rG:dG at $i+1$ site. Reactions were quenched at various time points and the products were separated by denaturing PAGE (see Supplementary Note 5 for more experimental details).

Intrinsic transcript cleavage assays. Cleavage reactions were performed by pre-incubating purified Pol II (WT or Rpb1 T831A mutant) with the RNA:DNA scaffolds containing either rG:dC or rG:dG bp. The Pol II EC was assembled with the following scaffold in a 20 mM Tris-HCl (pH = 7.5) without Mg^{2+}.

5′   CGTCTGCTTATCGGTAG NTS
3′ GTAGCTCTCCTAXGCAGACGAATAGCCATC Template
5′ AUCGAGAGGAUG 12G
            X = dG or dC

Intrinsic cleavage was initiated after addition of Mg^{2+}. Final concentrations for intrinsic cleavage were 20 mM Tris-HCl (pH = 9), 100 nM Pol II, 25 nM scaffold and 50 mM MgCl$_2$. The reaction was quenched in 0.5 mM EDTA at various time points and analysed by denatured PAGE. Time points in this assay were 1 min, 5 min, 20 min, 1 h, 3 h, 8 h and 24 h. (see Supplementary Note 5 for more experimental details).

References

1. Kornberg, R. D. The molecular basis of eukaryotic transcription. *Proc. Natl Acad. Sci. USA* **104,** 12955–12961 (2007).
2. Nudler, E. RNA polymerase active center: the molecular engine of transcription. *Annu. Rev. Biochem.* **78,** 335 (2009).
3. Hirata, A., Klein, B. J. & Murakami, K. S. The X-ray crystal structure of RNA polymerase from Archaea. *Nature* **451,** 851–854 (2008).
4. Jun, S.-H. *et al.* The X-ray crystal structure of the euryarchaeal RNA polymerase in an open-clamp configuration. *Nat. Commun.* **5,** 5132 (2014).
5. Saxowsky, T. T. & Doetsch, P. W. RNA polymerase encounters with DNA damage: transcription-coupled repair or transcriptional mutagenesis? *Chem. Rev.* **106,** 474–488 (2006).
6. Xu, L. *et al.* Molecular basis of transcriptional fidelity and DNA lesion-induced transcriptional mutagenesis. *DNA Repair* **19,** 71–83 (2014).
7. Cheung, A. C. M. & Cramer, P. Structural basis of RNA polymerase II backtracking, arrest and reactivation. *Nature* **471,** 249–253 (2011).
8. Wang, D. *et al.* Structural basis of transcription: backtracked RNA polymerase II at 3.4 angstrom resolution. *Science* **324,** 1203–1206 (2009).
9. Jin, J. *et al.* Synergistic action of RNA polymerases in overcoming the nucleosomal barrier. *Nat. Struct. Mol. Biol.* **17,** 745–752 (2010).
10. Nudler, E. RNA polymerase backtracking in gene regulation and genome instability. *Cell* **149,** 1438–1445 (2012).
11. Sydow, J. F. & Cramer, P. RNA polymerase fidelity and transcriptional proofreading. *Curr. Opin. Struct. Biol.* **19,** 732–739 (2009).
12. Dangkulwanich, M. *et al.* Complete dissection of transcription elongation reveals slow translocation of RNA polymerase II in a linear ratchet mechanism. *Elife* **2,** e00971 (2013).
13. Dutta, D., Shatalin, K., Epshtein, V., Gottesman, M. E. & Nudler, E. Linking RNA polymerase backtracking to genome instability in E. coli. *Cell* **146,** 533–543 (2011).
14. Hein, P. P. *et al.* RNA polymerase pausing and nascent-RNA structure formation are linked through clamp-domain movement. *Nat. Struct. Mol. Biol.* **21,** 794–802 (2014).
15. Sigurdsson, S., Dirac-Svejstrup, A. B. & Svejstrup, J. Q. Evidence that transcript cleavage is essential for RNA polymerase II transcription and cell viability. *Mol. Cell* **38,** 202–210 (2010).
16. Zhang, J., Palangat, M. & Landick, R. Role of the RNA polymerase trigger loop in catalysis and pausing. *Nat. Struct. Mol. Biol.* **17,** 99–104 (2010).
17. Sydow, J. F. *et al.* Structural basis of transcription: mismatch-specific fidelity mechanisms and paused RNA polymerase II with frayed RNA. *Mol. Cell* **34,** 710–721 (2009).
18. Kireeva, M. L. *et al.* Molecular dynamics and mutational analysis of the catalytic and translocation cycle of RNA polymerase. *BMC Biophys.* **5,** 11 (2012).
19. Da, L.-T., Wang, D. & Huang, X. Dynamics of pyrophosphate ion release and its coupled trigger loop motion from closed to open state in RNA polymerase II. *J. Am. Chem. Soc.* **134,** 2399–2406 (2012).

20. Huang, X. *et al.* RNA polymerase II trigger loop residues stabilize and position the incoming nucleotide triphosphate in transcription. *Proc. Natl Acad. Sci. USA* **107**, 15745–15750 (2010).

21. Feig, M. & Burton, Z. F. RNA polymerase II with open and closed trigger loops: active site dynamics and nucleic acid translocation. *Biophys. J.* **99**, 2577–2586 (2010).

22. Batada, N. N., Westover, K. D., Bushnell, D. A., Levitt, M. & Kornberg, R. D. Diffusion of nucleoside triphosphates and role of the entry site to the RNA polymerase II active center. *Proc. Natl Acad. Sci. USA* **101**, 17361–17364 (2004).

23. Wang, B., Predeus, A. V., Burton, Z. F. & Feig, M. Energetic and structural details of the trigger-loop closing transition in RNA polymerase II. *Biophys. J.* **105**, 767–775 (2013).

24. Wang, B. B., Opron, K., Burton, Z. F., Cukier, R. I. & Feig, M. Five checkpoints maintaining the fidelity of transcription by RNA polymerases in structural and energetic details. *Nucleic Acids Res.* **43**, 1133–1146 (2015).

25. Lane, T. J., Shukla, D., Beauchamp, K. A. & Pande, V. S. To milliseconds and beyond: challenges in the simulation of protein folding. *Curr. Opin. Struct. Biol.* **23**, 58–65 (2013).

26. Malinen, A. M. *et al.* Active site opening and closure control translocation of multisubunit RNA polymerase. *Nucleic Acids Res.* **40**, 7442–7451 (2012).

27. Noé, F. & Fischer, S. Transition networks for modeling the kinetics of conformational change in macromolecules. *Curr. Opin. Struct. Biol.* **18**, 154–162 (2008).

28. Buchete, N.-V. & Hummer, G. Coarse master equations for peptide folding dynamics. *J. Phys. Chem. B* **112**, 6057–6069 (2008).

29. Bowman, G. R., Huang, X. & Pande, V. S. Using generalized ensemble simulations and Markov state models to identify conformational states. *Methods* **49**, 197–201 (2009).

30. Chodera, J. D., Singhal, N., Pande, V. S., Dill, K. A. & Swope, W. C. Automatic discovery of metastable states for the construction of Markov models of macromolecular conformational dynamics. *J. Chem. Phys.* **126**, 155101–155117 (2007).

31. Deuflhard, P. & Weber, M. Robust Perron cluster analysis in conformation dynamics. *Linear Algebra Its Applicat.* **398**, 161–184 (2005).

32. Huang, X., Bowman, G. R., Bacallado, S. & Pande, V. S. Rapid equilibrium sampling initiated from nonequilibrium data. *Proc. Natl Acad. Sci. USA* **106**, 19765–19769 (2009).

33. Pan, A. C. & Roux, B. Building Markov state models along pathways to determine free energies and rates of transitions. *J. Chem. Phys.* **129**, 064107 (2008).

34. Chodera, J. D. & Noé, F. Markov state models of biomolecular conformational dynamics. *Curr. Opin. Struct. Biol.* **25**, 135–144 (2014).

35. Prinz, J.-H., Keller, B. & Noé, F. Probing molecular kinetics with Markov models: metastable states, transition pathways and spectroscopic observables. *Phys. Chem. Chem. Phys.* **13**, 16912–16927 (2011).

36. Swope, W. C., Pitera, J. W. & Suits, F. Describing protein folding kinetics by molecular dynamics simulations. 1. Theory. *J. Phys. Chem. B* **108**, 6571–6581 (2004).

37. Malmstrom, R. D., Lee, C. T., Van Wart, A. T. & Amaro, R. E. Application of molecular-dynamics based Markov state models to functional proteins. *J. Chem. Theory Comput* **10**, 2648–2657 (2014).

38. Bowman, G. R., Beauchamp, K. A., Boxer, G. & Pande, V. S. Progress and challenges in the automated construction of Markov state models for full protein systems. *J. Chem. Phys.* **131**, 124101 (2009).

39. Zhuang, W., Cui, R. Z., Silva, D. A. & Huang, X. Simulating the T-jump-triggered unfolding dynamics of trpzip2 peptide and its time-resolved IR and two-dimensional IR signals using the Markov state model approach. *J. Phys. Chem. B* **115**, 5415–5424 (2011).

40. Bowman, G. R., Voelz, V. A. & Pande, V. S. Atomistic folding simulations of the five-helix bundle protein λ6 − 85. *J. Am. Chem. Soc.* **133**, 664–667 (2010).

41. Noé, F., Schütte, C., Vanden-Eijnden, E., Reich, L. & Weikl, T. R. Constructing the equilibrium ensemble of folding pathways from short off-equilibrium simulations. *Proc. Natl Acad. Sci. USA* **106**, 19011–19016 (2009).

42. Kohlhoff, K. J. *et al.* Cloud-based simulations on Google Exacycle reveal ligand modulation of GPCR activation pathways. *Nat. Chem.* **6**, 15–21 (2014).

43. Qiao, Q., Bowman, G. R. & Huang, X. Dynamics of an intrinsically disordered protein reveal metastable conformations that potentially seed aggregation. *J. Am. Chem. Soc.* **135**, 16092–16101 (2013).

44. Silva, D. A., Bowman, G. R., Sosa-Peinado, A. & Huang, X. A role for both conformational selection and induced fit in ligand binding by the LAO protein. *PLoS Comput. Biol.* **7**, e1002054 (2011).

45. Choudhary, O. P. *et al.* Structure-guided simulations illuminate the mechanism of ATP transport through VDAC1. *Nat. Struct. Mol. Biol.* **21**, 626–632 (2014).

46. Plattner, N. & Noé, F. Protein conformational plasticity and complex ligand-binding kinetics explored by atomistic simulations and Markov models. *Nat. Commun.* **6**, 7653 (2015).

47. Malmstrom, R. D., Kornev, A. P., Taylor, S. S. & Amaro, R. E. Allostery through the computational microscope: cAMP activation of a canonical signalling domain. *Nat. Commun.* **6**, 7588 (2015).

48. Silva, D.-A. *et al.* Millisecond dynamics of RNA polymerase II translocation at atomic resolution. *Proc. Natl Acad. Sci. USA* **111**, 7665–7670 (2014).

49. Weiss, D. R. & Levitt, M. Can morphing methods predict intermediate structures? *J. Mol. Biol.* **385**, 665–674 (2009).

50. Weinan, E. & Vanden-Eijnden, E. Transition-path theory and path-finding algorithms for the study of rare events. *Phys. Chem.* **61**, 391–420 (2010).

51. Tenenbaum, J. B., De Silva, V. & Langford, J. C. A global geometric framework for nonlinear dimensionality reduction. *Science* **290**, 2319–2323 (2000).

52. Imashimizu, M. *et al.* Intrinsic translocation barrier as an initial step in pausing by RNA polymerase II. *J. Mol. Biol.* **425**, 697–712 (2013).

53. Nudler, E., Mustaev, A., Goldfarb, A. & Lukhtanov, E. The RNA–DNA hybrid maintains the register of transcription by preventing backtracking of RNA polymerase. *Cell* **89**, 33–41 (1997).

54. Hein, P. P. *et al.* RNA polymerase pausing and nascent-RNA structure formation are linked through clamp-domain movement. *Nat. Struct. Mol. Biol.* **21**, 794–802 (2014).

55. Landick, R. Transcriptional pausing without backtracking. *Proc. Natl Acad. Sci. USA* **106**, 8797–8798 (2009).

56. Yu, J., Da, L.-T. & Huang, X. Constructing kinetic models to elucidate structural dynamics of a complete RNA polymerase II elongation cycle. *Phys. Biol.* **12**, 016004–016004 (2015).

57. Bernecky, C., Herzog, F., Baumeister, W., Plitzko, J. M. & Cramer, P. Structure of transcribing mammalian RNA polymerase II. *Nature* **529**, 551–554 (2016).

58. Barnes, C. O. *et al.* Crystal structure of a transcribing RNA polymerase II complex reveals a complete transcription bubble. *Mol. Cell* **59**, 258–269 (2015).

59. Hawryluk, P. J., Újvári, A. & Luse, D. S. Characterization of a novel RNA polymerase II arrest site which lacks a weak 3′ RNA–DNA hybrid. *Nucleic Acids Res.* **32**, 1904–1916 (2004).

60. Weixlbaumer, A., Leon, K., Landick, R. & Darst, S. A. Structural basis of transcriptional pausing in bacteria. *Cell* **152**, 431–441 (2013).

61. Kellinger, M. W. *et al.* 5-Formylcytosine and 5-carboxylcytosine reduce the rate and substrate specificity of RNA polymerase II transcription. *Nat. Struct. Mol. Biol.* **19**, 831–833 (2012).

62. Berendsen, H. J. C., Postma, J. P. M., van Gunsteren, W. F. & Hermans, J. in *Intermolecular Forces.* (ed. Pullman, B) 331–342 (Reidel Publishing Company, 1981).

63. Hornak, V. *et al.* Comparison of multiple Amber force fields and development of improved protein backbone parameters. *Proteins* **65**, 712–725 (2006).

64. Yildirim, I., Stern, H. A., Kennedy, S. D., Tubbs, J. D. & Turner, D. H. Reparameterization of RNA χ Torsion parameters for the AMBER force field and comparison to NMR spectra for cytidine and uridine. *J. Chem. Theory Comput.* **6**, 1520–1531 (2010).

65. Zgarbová, M. *et al.* Refinement of the Cornell *et al.* nucleic acids force field based on reference quantum chemical calculations of glycosidic torsion profiles. *J. Chem. Theory Comput.* **7**, 2886–2902 (2011).

66. Van Der Spoel, D. *et al.* GROMACS: fast, flexible, and free. *J. Comput. Chem.* **26**, 1701–1718 (2005).

67. Pérez-Hernández, G., Paul, F., Giorgino, T., De Fabritiis, G. & Noé, F. Identification of slow molecular order parameters for Markov model construction. *J. Chem. Phys.* **139**, 015102 (2013).

68. Schwantes, C. R. & Pande, V. S. Improvements in Markov state model construction reveal many non-native interactions in the folding of NTL9. *J. Chem. Theory Comput.* **9**, 2000–2009 (2013).

69. Hao, W., Mey, A. S. J. S., Edina, R. & Frank, N. Statistically optimal analysis of state-discretized trajectory data from multiple thermodynamic states. *J. Chem. Phys.* **141**, 214106–214106 (2014).

Acknowledgements

X.H. acknowledges the Hong Kong Research Grants Council (16302214, 609813, HKUST C6009-15G, AoE/M-09/12, T13-607/12R, and M-HKUST601/13) and National Science Foundation of China 21273188. D.W. acknowledges the NIH (GM102362), Kimmel Scholars award from the Sidney Kimmel Foundation for Cancer Research, start-up funds from Skaggs School of Pharmacy and Pharmaceutical Sciences, UCSD and Academic Senate Research Award from UCSD. X.G. was supported by funding from the King Abdullah University of Science and Technology. F.P. acknowledges the support from Hong Kong PhD Fellowship Scheme (2011/12) and the partial support for PhD studies from the CONACYT. This research made use of the resources of the Supercomputing Laboratory at King Abdullah University of Science and Technology. We thank Dr Jeffery Strathern (NCI) for providing yeast strain containing Pol II Rpb1 T831A mutant and Dr Mikhail Kashlev for providing purified Pol II Rpb1 T831A mutant.

Author contributions

Lin-Tai Da, Fátima Pardo-Avila, Lu Zhang and Xin Gao performed MD simulations. Liang Xu performed *in vitro* transcription assays. All authors contributed the ideas, designed the MD simulations, discussed the results and wrote the manuscript.

Additional information

Competing financial interests: The authors declare no competing financial interests.

13

Directed block copolymer self-assembly implemented via surface-embedded electrets

Mei-Ling Wu[1,2,3], Dong Wang[1,2] & Li-Jun Wan[1,2]

Block copolymer (BCP) nanolithography is widely recognized as a promising complementary approach to circumvent the feature size limits of conventional photolithography. The directed self-assembly of BCP thin film to form ordered nanostructures with controlled orientation and localized pattern has been the key challenge for practical nanolithography applications. Here we show that BCP nanopatterns can be directed on localized surface electrets defined by electron-beam irradiation to realize diverse features in a simple, effective and non-destructive manner. Charged electrets can generate a built-in electric field in BCP thin film and induce the formation of perpendicularly oriented microdomain of BCP film. The electret-directed orientation control of BCP film can be either integrated with mask-based patterning technique or realized by electron-beam direct-writing method to fabricate microscale arbitrary lateral patterns down to single BCP cylinder nanopattern. The electret-directed BCP self-assembly could provide an alternative means for BCP-based nanolithography, with high resolution.

[1] Key Laboratory of Molecular Nanostructure and Nanotechnology, Institute of Chemistry, Chinese Academy of Sciences (CAS), Beijing 100190, China. [2] Beijing National Laboratory for Molecular Sciences, Beijing 100190, China. [3] University of CAS, Beijing 100049, China. Correspondence and requests for materials should be addressed to D.W. (email: wangd@iccas.ac.cn).

Block copolymer (BCP) self-assembly can spontaneously generate periodic arrays of microdomains with versatile morphology and nanoscale feature size in the range of ca. 10–100 nm (ref. 1). The well-defined nanostructures of BCP have been utilized as templates to control spatial order of other functional materials for a variety of applications[2,3] such as microelectronic devices[4,5], photovoltaic devices[6], magnetic storage devices[7], microreactors[8] and porous filtration membranes[9]. In particular, BCP-based nanolithography[10,11], which utilizes the self-assembled BCP thin film to pattern the substrates, has been widely recognized as a viable alternative or complementary approach to conventional photolithography[12,13]. It is promised to address the conflict between the continuous demanding to shrink the feature size of electronic devices and the technology and cost limits of photolithography. Owing to the favourable advantages such as high throughput, low cost and most notably, compatibility with current nanolithography streamline, BCP-based nanolithography has been targeted as one of the most important next-generation lithography techniques in the International Technology Roadmap for Semiconductors[12].

For a typical BCP-based nanolithography, orientation of domains normal to the substrates is essential to facilitate robust pattern transfer[13,14]. The most common way to control the orientation of BCP microdomains is to fabricate a neutral surface through surface energy modification to balance its interfacial energy with both BCP components[15,16]. Taking advantage of the dielectric constant difference of each block, orientation control of the BCP self-assembly can also be achieved by applying an external electric field, although it is generally limited to BCP films with micrometre-scale thickness[17,18]. In addition, surface topology modulation[19,20], solvent annealing[21,22] and other methods[23,24] have also been proposed to control BCP microdomain orientation. Besides the perpendicular orientation, controlled BCP self-assembly with lateral order and localized nanostructures[25–28] is also necessary in practical lithography processes to realize pattern registration and addressable device-oriented patterns with low defects[29,30]. Directed self-assembly, which integrates the BCP self-assembly with the traditional lithography processes, has been developed to achieve oriented and lateral ordering of BCP films[31,32] utilizing chemical epitaxy and graphoepitaxy methods[33–36]. In addition, electrohydrodynamic jet printing was utilized to fabricate complex and hierarchical patterns of BCP films[37,38]. While these methods have improved the ordering of BCP films and advanced the process of registration and addressing, they are generally subject to complex lithographic steps, limited substrates and/or physical destruction of the substrates. It would be rewarding to develop a simple, effective and non-destructive method to realize device-oriented features with versatile patterns and high nanopatterning resolution.

Herein, we introduce a novel method to achieve simultaneously orientation control and localization of BCP self-assembly implemented by surface-embedded electrets, which is fabricated by electron-beam (e-beam) irradiation. Electrets are materials that retain trapped charges or polarization, forming a quasi-permanent, macroscopic electric field around the perimeter[39–41]. Similar to external electric filed, which can align the BCP blocks, the electric field around the electrets is also found to be able to control the orientation of BCP film. The surface electrets-mediated BCP self-assembly provides a facile approach to BCP nanopatterning, and can yield customized nanopatterns with arbitrary geometry and high fidelity by employing an e-beam direct-writing technique. More attractively, by tuning the parameters of e-beam irradiation carefully, we can realize the formation of individual cylinder nanopattern with extremely high

accuracy, which represents the utmost resolution of BCP-based nanolithography and is critical for the formation of contact and via holes in integrated circuit.

Results

Fabrication of nanopatterns. The overall fabricating process of electrets-induced polystyrene-b-poly(methyl methacrylate) (PS-b-PMMA) self-assembly is shown in Fig. 1a. Briefly, a SiO$_2$/Si wafer was first irradiated by e-beam, with proper dose and then spin-coated with BCP film. After thermal annealing, the BCP film underwent microphase separation and self-assembled into stable cylindrical or lamellar nanostructures, depending on the molecular weight ratio of the blocks. Figure 1b shows the plain view scanning electron microscopy (SEM) image of a cylinder-forming BCP thin film on the SiO$_2$/Si wafer, of which the central region has been irradiated by the e-beam. The as-obtained hexagonal arrays (Fig. 1b and Supplementary Fig. 1) in the irradiated region prove that microdomains of PS-b-PMMAs orient normal to the substrate surface. Similar results are found in the lamellar BCP film. Fingerprint arrays are formed in the irradiated area, with the lamella microdomains oriented normal to the substrate (Fig. 1c and Supplementary Fig. 2). In contrast, in the area where SiO$_2$/Si wafers have not been treated by e-beam irradiation, parallel orientations are observed (Supplementary Fig. 3). To probe the internal perpendicular orientation of the entire thin films, we

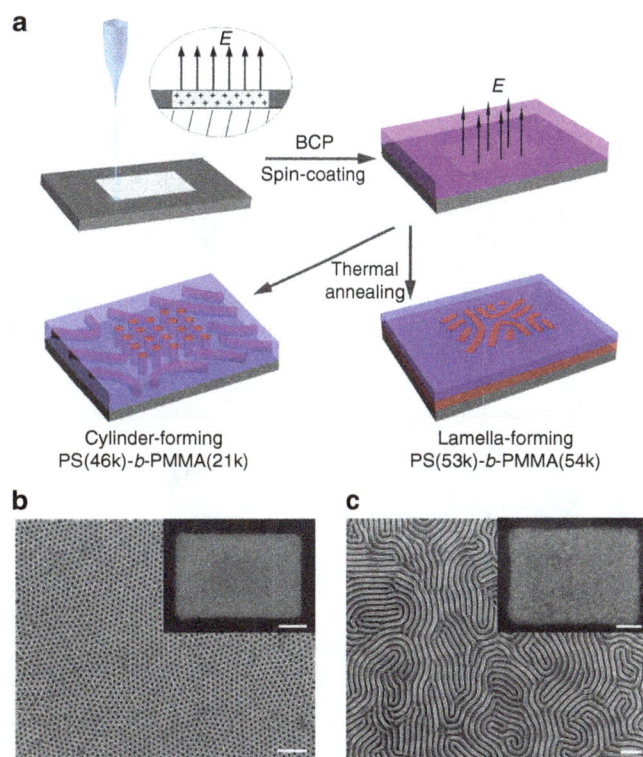

Cylinder-forming
PS(46k)-b-PMMA(21k)

Lamella-forming
PS(53k)-b-PMMA(54k)

Figure 1 | PS-b-PMMA self-assembly on e-beam-irradiated SiO$_2$/Si substrate. (a) Schematic description for fabricating PS-b-PMMA films with perpendicular orientation on e-beam-irradiated SiO$_2$/Si substrates. SEM images of PS-b-PMMA films self-assembled on the SiO$_2$/Si wafers whose central regions were irradiated by e-beam: (b) Cylinder-forming PS-b-PMMA (46k-b-21k) with thickness of 37 nm and (c) lamella-forming PS-b-PMMA (53k-b-54k) with thickness of 56 nm (scale bars, 200 nm). The insets in b and c are the corresponding low-magnification images (scale bars, 1 μm), in which the bright contrast indicates a uniform formation of the perpendicular orientation. The samples were treated by Ar plasma to selectively remove PMMA to improve the SEM image contrast.

performed grazing-incidence small-angle X-ray scattering (GISAXS) experiment. In the two-dimensional GISAXS pattern of cylinder PS-b-PMMA film self-assembled on the irradiated SiO$_2$/Si of ~ 1 cm^2 (Supplementary Fig. 4a), the symmetric scattering peaks are confined in the Q_y direction and the first-order scattering peaks at $Q_y = 0.196$ nm^{-1} reflect the perpendicularly oriented microdomains of the entire film with a period (domain space) $L_0 = 32.1$ nm ($2\pi/Q_y$). Similarly, perpendicular orientation of lamellae is supported by the GISAXS pattern in Supplementary Fig. 4b ($Q_y = 0.119$ nm^{-1} and $L_0 = 52.8$ nm). In addition, the e-beam-induced BCP self-assembly is found widely feasible; all the tested PS-b-PMMAs of different molecular weights and block ratios show vertically oriented patterns (Supplementary Fig. 5). The above results indicate that with the aid of e-beam irradiation, we can readily achieve perpendicularly oriented microdomain of PS-b-PMMA films.

The e-beam irradiation is considered critical to induce BCP orientation. To ascertain this point, we studied the influence of e-beam parameters on the self-assembly of BCP film. Here we use the area ratio of perpendicularly oriented cylinder microdomains ($C_\perp$) in the whole irradiated region, that is, coverage of $C_\perp$, to represent the degree of orientation. As shown in Fig. 2, the coverage of $C_\perp$ increases gradually with e-beam dose (dose = beam current × irradiation time/area) and then stabilizes to $\sim 100\%$ at an optimized e-beam dose (~ 200 mC cm^{-2}). A threshold e-beam dose is required to achieve 100% coverage of $C_\perp$, which is affected by the e-beam parameters, such as beam current and accelerating voltage (Supplementary Fig. 6). The dependence of the BCP orientation on the e-beam parameters indicates that the BCP self-assembly by this method hinges largely on the surface charges induced by e-beam irradiation.

Mechanism study. The orientation dependence of PS-b-PMMA self-assembly on irradiation dosage provides an important cue to understand the mechanism. In this work, native layer of SiO$_2$ exists on Si wafer after standard cleaning procedure (thickness: 2.3 nm). It has been well documented that SiO$_2$ is an electret material that can trap charges[41,42]. We preformed Kelvin probe force microscope (KPFM) to trace the charges of the SiO$_2$ surface, as it can measure relative surface potential and is an effective technique to uncover the charge trapping of substrates[43]. Figure 3a shows a schematic diagram of KPFM measurement and a typical KPFM mapping image of a SiO$_2$/Si wafer, of which the central region has been irradiated by e-beam. According to

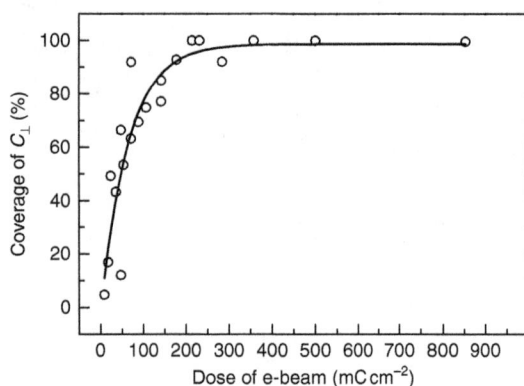

Figure 2 | Evolution of coverage of C$_\perp$ with e-beam dose. E-beam dose is controlled by changing irradiation area and time. Coverage of $C_\perp$ is calculated by the area ratio of $C_\perp$ in the whole irradiated region. The line is a fitted curve based on the scattered points. All experiments were conducted using a 16.9-pA beam current, a 5-kV accelerating voltage and PS(46k)-b-PMMA(21k) films with thickness of 37 nm.

the KPFM image, the contact potential difference (CPD, between the substrate and the tip) in the irradiated region increases significantly compared with the pristine region. The positive CPD reflects that positive charges are trapped in the SiO$_2$ layer after e-beam irradiation, forming a surface-embedded electret. The CPD increment (ΔCPD) increases gradually with the e-beam dose and finally stabilizes to 300–350 mV (Fig. 3b). The evolution tendency of ΔCPD is similar to that of nanopattern coverage (Fig. 2), indicating that orientation of BCP film is closely correlated to the charge state of the surface.

In addition to KPFM measurement, SEM and X-ray photo-electronic spectroscopy (XPS) can also reflect the charging behaviour of the SiO$_2$ layer. The SEM image of electret irradiated by e-beam shows higher contrast difference compared with the non-irradiated region (Supplementary Fig. 7a). The voltage contrast image is attributed to the charging of the electret[44]. XPS was performed to examine the surface constitution and chemical states of the SiO$_2$/Si wafers before and after irradiation (Supplementary Fig. 7b). Binding energies of Si–Si and Si–O bonds increase gradually with the e-beam irradiation, and 0.3 eV increment of the Si–Si binding energy is found in the SiO$_2$/Si substrate after irradiation (3.9 mC cm^{-2}). The small positive shift of binding energy indicates the accumulation of positive charges (holes). The positive charges result from the net outcome of electron injection and secondary electron emission during irradiation. As the secondary electron emitting from the wafer occurs mainly on the surface and cannot be counteracted by the electron injection, accumulated holes are formed on the SiO$_2$ layer, thereby charging the SiO$_2$ electret[42].

Earlier reports have revealed that external electric field can align the blocks of PS-b-PMMA along the field and control its orientation due to the dielectric permittivity difference of blocks[18,45]. The e-beam irradiation introduces positive charges into the surface-embedded electret and generates a static electric field E at the SiO$_2$/Si surface (inset in Fig. 1a). For a planar electret, whose lateral dimension is much larger than its thickness D and the film thickness L, one has

$$E = \frac{\varphi}{L + D\varepsilon_{AB}/\varepsilon_s} \tag{1}$$

where φ is the surface potential at the electret (determined by the charges trapped in the electret), ε_{AB} is the mean (arithmetic) dielectric constant of diblock copolymer A–B (Supplementary Discussion) and ε_s is the dielectric constant of the electret substrate[41]. Therefore, the electric field strength is 1.5–9.5 V μm^{-1} through BCP films with thicknesses of 30–200 nm (Supplementary Discussion). These values are comparable with typical external fields (1.0–40 V μm^{-1}, Supplementary Table 1) used previously in the electric field alignment of the BCP films. As the electric field is strong enough to compensate the energetic penalty from the orientation change[46–48], it would induce PS-b-PMMA films to orient normal to the electret substrate.

It is noted that the surface energy of substrate could also be changed after e-beam irradiation. To clarify whether electric field or the surface-wetting property underlies the orientation control, a series of control experiments was conducted. A thin self-assembled monolayer of phenyltrichlorosilane was grafted onto the e-beam-irradiated SiO$_2$/Si wafer, which would alter the surface energy but should not shield the electric field. As a result, on the irradiated region, PS-b-PMMA microdomains can still form perpendicular orientation similar to the uncoated samples, whereas parallel orientation is obtained in the non-irradiated area (Supplementary Fig. 8). Thus, the electric field is considered the driving force for BCP alignment; if the orientation control were caused by the surface energy change during e-beam irradiation,

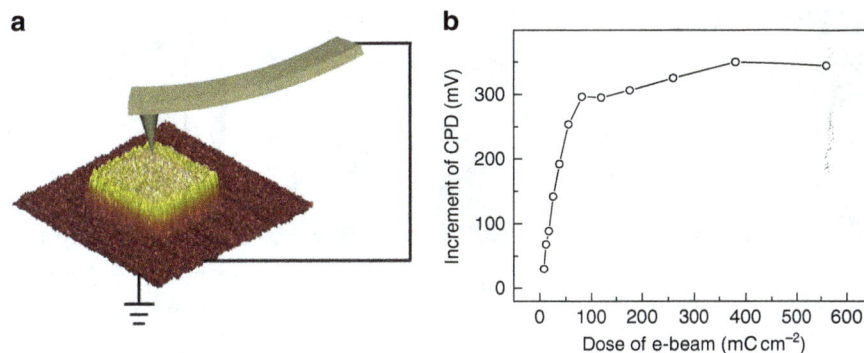

Figure 3 | Surface potential analysis of electrets. (**a**) Schematic diagram of KPFM measurement and a representative surface potential distribution of a SiO$_2$/Si surface with central region irradiated by e-beam. (**b**) Evolution of corresponding incremental CPD (ΔCPD) with e-beam dose.

the self-assembled monolayer modification should have removed the surface energy difference at the irradiated/non-irradiated regions and the morphology of BCP film should be the same. Furthermore, we removed the charges in the SiO$_2$/Si electrets deliberately by thermal annealing ($>500\,°C$; electrets are relatively stable at $\sim 250\,°C$) or by piranha solution cleaning before PS-b-PMMA self-assembly. Parallel orientation is obtained on the SiO$_2$/Si without the electric field alignment (Supplementary Fig. 9). Third, we have extended the present method to achieve perpendicular orientation of BCP film on other electret substrates. Uniform vertical cylindrical PS-b-PMMA films are attained on various e-beam-irradiated electrets, including SiO$_2$/Si, PI and Si$_3$N$_4$/SiO$_2$/Si (nitride–oxide–silicon (NOS); Supplementary Fig. 10). Intriguingly, wettability experiment shows that charged SiO$_2$/Si is preferential to PMMA block, charged PI is relatively preferential to PS block and charged NOS is non-preferential for both blocks (see Supplementary Table 2 and Supplementary Discussion for more discussion). Perpendicular orientation of PS-b-PMMA film is successfully achieved regardless of the preferential wetting properties (SiO$_2$/Si, PI). These results further prove that the electric field from the charged surface electret is the critical driving force for the perpendicular orientation of PS-b-PMMA films.

Features and merits. As the directed self-assembly of PS-b-PMMA film is implemented with surface-embedded electrets, this method exhibits several pronounced advantages. First, this method is feasible, requiring virtually no chemical pretreatments, and is applicable to many electret materials. As discussed above, other electret substrates besides SiO$_2$ are suitable for this method. In addition, this method is applicable to SiO$_2$ layer with a wide thickness range (Supplementary Fig. 11).

In the conventional approaches to control the BCP orientation by surface energy modification, the BCP thickness is generally limited to below $\sim 3\,L_0$ in PS-b-PMMA. For thicker films, driving force from neutral interfaces declines and the vertical micro-domains would not penetrate through the entire film[10,49]. External electric field-based alignment method is applicable for film thickness of micrometres, but it is difficult to be applied to BCP with nanometre scale, as BCP film with sufficient thickness is required for electrode separation[31]. In contrast, the built-in electric field originated from the surface electret eliminates the limitation of external electric field to apply to film thickness down to tens of nanometres. As shown in Supplementary Fig. 12, the PS-b-PMMA films of different film thicknesses up to 161 nm can orient normal to surface of the electret and the degree of long-range lateral order slightly increases with the film thickness. We can further fabricate PS(53k)-b-PMMA(54k) film with thickness

of 208 nm ($\sim 3.9\,L_0$), with vertical orientation penetrating the entire film thickness (Supplementary Fig. 13a). Moreover, in the GISAXS pattern (Supplementary Fig. 13b), the pronounced first-order peak at $Q_y = 0.12\,nm^{-1}$ and the high-order peaks with relative Q_y ratios of 1:2:3:4 also evidence that the lamella BCP microdomains orient normal to the substrate throughout the entire film.

The electrets-mediated BCP orientation can yield nanopatterns in millimetre scale (mm scale) and produce customized patterns. By increasing e-beam diameter and beam current, the process of e-beam irradiation can be significantly accelerated. For example, when e-beam of 20 μm in diameter (see Mode 3 in the Method) is employed, the irradiation speed increases from 0.4 to $2.5 \times 10^3\,\mu m^2\,s^{-1}$ and it takes $\sim 30\,min$ to irradiate a region of $3.0 \times 2.2\,mm^2$. As shown in Supplementary Fig. 14, perpendicu-larly oriented cylinder morphology is realized in several square millimetres with high uniformity. The orientation quality of the mm-scale BCP film is comparable to those fabricated by surface energy modification, surface topology modulation and external electric alignment[15,18,50]. In addition, the electret implantation can be guided by customized masks to realize patterned BCP orientation in mm scale (see Supplementary Fig. 15 for the process and pattern images).

Finally, the electret charging can be realized by the technique of e-beam direct writing (EBD) for patterns with precise size and customized geometry. EBD technique has been extensively applied in fabricating patterned templates as a mask-free method. By exploiting EBD to manipulate the e-beam moving path, we can fabricate e-beam-irradiated region with arbitrary geometries and sizes. The fine and arbitrary features of EBD endow this method with splendid abilities to produce versatile user-defined layouts for localized and self-registered BCP nanopatterns. We elucidate this combined technique with a pattern of Tai Chi diagram. The Tai Chi-patterned electret was first fabricated by EBD, which was demonstrated by the KPFM image (Fig. 4a). After BCP self-assembly on the customized electret pattern, Tai Chi patterns can be easily built with perpendicularly oriented lamella microdomains in high fidelity (Fig. 4b,c and Supplementary Fig. 16). Using BCP with different molecular weight ratios, we can form nanopattern of fingerprints and dots in Yin and Yang deliberately (see Supplementary Fig. 16 for more nanopatterns with Tai Chi diagram). Furthermore, patterns of stars and word 'NANO' with perpendicular orientation of cylinder-forming BCP can be also fabricated successfully by a similar method (Fig. 4d–f).

Intriguingly, we further demonstrate orientation control over individual cylinders, which represents the ultimate resolution limit for directed BCP self-assembly, as a most attractive feature by controlling the size of the surface-embedded electrets with a

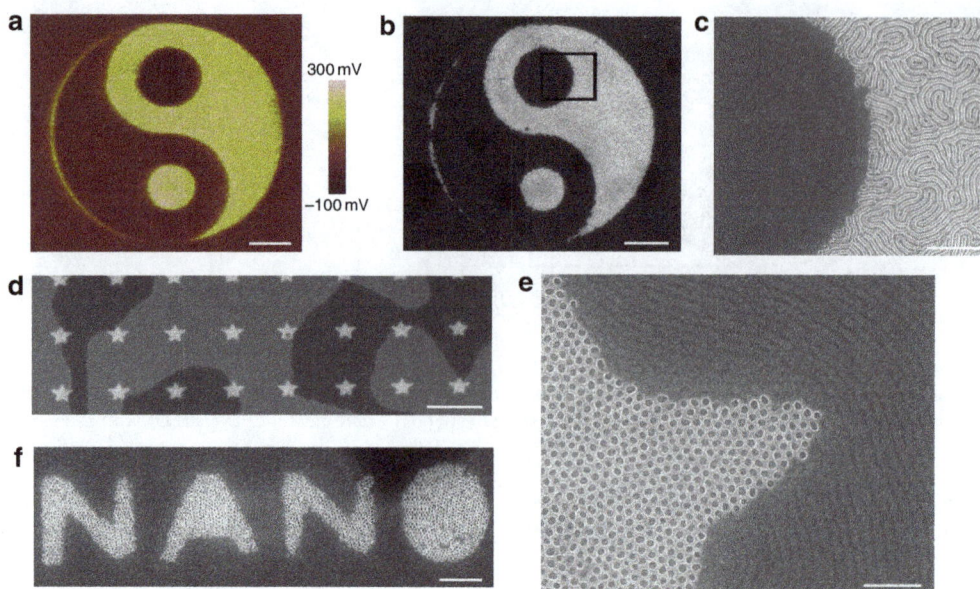

Figure 4 | BCP nanopatterns with user-defined diagrams. (a) Surface potential image of a Tai Chi diagram fabricated by e-beam irradiation with EBD. **(b,c)** SEM images of the corresponding self-assembled morphology of PS(53k)-*b*-PMMA(54k) film (thickness: 56 nm). Spatially oriented PS(46k)-*b*-PMMA(21k) film with various geometries: **(d)** patterned array of stars; **(e)** magnified view of **d**; **(f)** pattern of letters NANO. The film thickness in **d,e** and **f** is 37 nm. Scale bars, 2 μm **(a,b)**; 500 nm **(c)**; 5 μm **(d)**; 200 nm **(e)**; 500 nm **(f)**.

highly focused e-beam. Patterns of isolated hole or several holes can be formed with an irradiated circle region by careful e-beam manipulation (Fig. 5). Highly focused e-beam can precisely control the spatial distribution of the electret, at surface of which local electric field is strong enough to control the orientation of PS-*b*-PMMA for nucleation in a minimum volume[51]. As shown in Fig. 5b, individual cylinders can be realized on the SiO$_2$ surface electret obtained from superfine e-beam irradiation. The control over individual microdomain of BCP self-assembly, as the resolution limit of the BCP nanopattern, is of paramount importance for application as contact holes in integrated circuit[12]. Moreover, the high-resolution nanopattern implemented on the localized electrets is free from small guiding templates or physical destruction of the substrate, and thus remarkably facilitates the process of subsequent high-resolution nanopattern transfer to other substrates[35]. Increasing the irradiated size or e-beam dosage, several cylinders are achieved in a controllable manner (Fig. 5c–e). An array composed of ordered vertical cylinder can also be successfully fabricated on an irradiated linear region (Supplementary Fig. 17a). When applying a lamella-forming polymer to such linear electret, vertical lamellae orient normal to the domain boundary (Supplementary Fig. 17b) for minimizing the free energy[52]. We can envision that the approach of surface electret-directed BCP self-assembly holds feasibility and opportunity to form accurate registering patterns provided with ultra-high-resolution e-beam[53].

Discussion

We have experimentally demonstrated the mechanism and appealing merits of the electret-mediated BCP self-assembly. The charged electret introduces an electric field that aligns the BCP film. In the electric field E over the electret, diblock copolymer A–B consisting of blocks with different dielectric constants (ε_A and ε_B) is aligned by the driving force[47,54]

$$f_e \propto \frac{(\Delta\varepsilon)^2}{\varepsilon_{AB}} E^2 \qquad (2)$$

where $\Delta\varepsilon = |\varepsilon_A - \varepsilon_B|$ is the dielectric constant contrast of blocks A

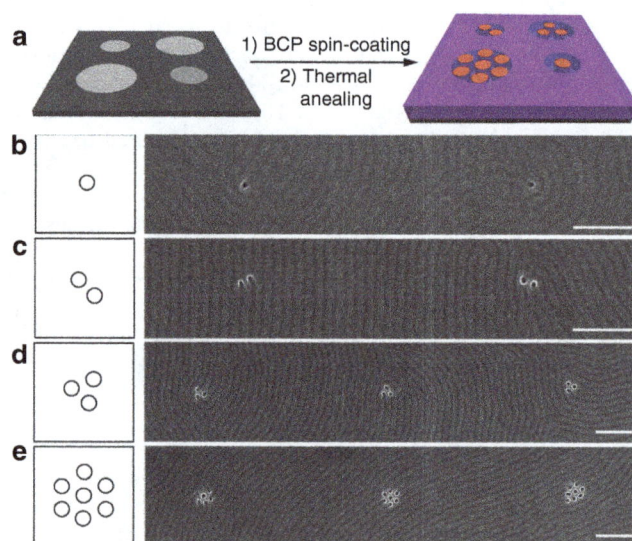

Figure 5 | High-resolution BCP patterns realized by e-beam control.
(a) Schematic process of high-resolution perpendicular cylindrical BCP nanopattern self-assembled on charged electrets with controlled circular size and e-beam dose. **(b-e)** SEM images of individual hole and patterned holes of PS-*b*-PMMA film after PMMA-block removal (scale bars, 200 nm). Each hole corresponds to the position of a vertically oriented PMMA microdomain of PS-*b*-PMMA film. The film thickness is 37 nm.

and B, and ε_{AB} is the dielectric constant of A–B. The microdomains would be aligned parallel to the electric field (that is, perpendicular to the substrate) as long as the electrostatic energy (F_e), which favours the perpendicular orientation of BCP, overcomes the energy penalty from the BCP orientation transformation (F_p, mainly the elastic free energy and interfacial energy). Under a critical condition, the change of free energy (ΔF_e and ΔF_p) of BCP film from parallel to perpendicular orientation equals 0, that is,

$$\Delta F = \Delta F_e + \Delta F_p = 0 \qquad (3)$$

The critical surface potential (φ_c) can be derived from equation (3), which is the threshold of surface potential to compensate the surface energy penalty and provides a theoretical assessment of possibility to align the BCP via the charged electret. Key parameters influencing φ_c include dielectric constants (ε_A and ε_B), BCP film thickness L and interfacial energy mismatch δ. On the basis of the lamellar diblock copolymer A–B (in a strong-segregation regime) and a planar electret model, the dependence of φ_c on these parameters is theoretically calculated and plotted in Supplementary Fig. 18 (see Supplementary Discussion for details). According the theoretical estimation, the influence of dielectric constant contrast $\Delta\varepsilon$ (electrically relative) is significant, whereas the effect of interfacial energy is minor. The interfacial energy influences the BCP self-assembly by changing the threshold of the surface potential. The dependence of the electret-directed self-assembly on the film thickness is also insignificant.

In this study, the surface electrets were charged via e-beam irradiation to produce BCP nanopatterns with sizes ranging from nanoscale to mm scale. In fact, electrets can also be charged via corona discharge or a metal-coated polydimethylsiloxane stamp, to yield inch-scale built-in electric field, which is promising to make the electrets-aided patterning practically scalable. It is noteworthy that though this BCP self-assembly method applying local electric field can effectively achieve spatial orientation control of PS-b-PMMA film, further work is necessary to control the defects and achieve highly ordered BCP assembly.

In summary, we have presented a novel approach to achieve orientation control of the self-assembled PS-b-PMMA film with the aid of surface electret. The surface-embedded electret, which is charged by e-beam irradiation, generates static electric field in polymer film and control the orientation of BCP film perpendicular on the surface. This method is marked with simplicity and compatibility. Various customized patterns are achievable with size ranging from millimetre with a mask-based patterning technique, to nanometre or micrometre integrated with EBD. By precisely tuning the e-beam parameters, we demonstrate the formation of individual cylinder nanopattern, which represents the resolution limit of BCP nanopattern and provides promising prospect in achieving accurate registering.

Methods

Cleaning of substrates. Silicon wafers were cleaned in piranha solution at boiling water bath for 30 min and then rinsed with super-pure water. Polyamide film was cleaned in ethanol and acetone solution with ultrasonic cleaner. Si substrate with Si_3N_4/SiO_2 (2.3 nm/2.3 nm) double layer (NOS) was fabricated by magnetron sputtering and was directly used. All other substrates were blow-dried with nitrogen flow before use.

E-beam irradiation and post treatment. Three e-beam irradiation modes were used in this work for different experimental purposes.

Mode 1: E-beam irradiated in repeated raster scan mode, producing numerous frames. The scan time for each frame was 472 ms. Each scan consisted of 768×512 pixels, and the irradiation time was 1 µs for each pixel. Diameter of the e-beam was ~ 5 nm. E-beam dose was determined by beam current, irradiation time and irradiation area.

Mode 2: E-beam irradiated in single-frame mode (the same as EBD technique). E-beam dose was user-defined. The point-to-point spacing was 4.6 nm. Diameter of the e-beam was ~ 5 nm.

Mode 3: E-beam irradiation is the same as that in Mode 1, except that the e-beam diameter was ~ 20 µm.

Irradiation process in Modes 1 and 2 were performed on Helios NanoLab 600i (FEI company, Germany). The chamber pressure was $\sim 2.0 \times 10^{-4}$ Pa and the working distance was 4 mm. Irradiation process in Mode 3 was performed on the electron microprobe (Shimadzu Corp., Japan). The chamber pressure was $\sim 2.0 \times 10^{-4}$ Pa and the working distance was 5 mm.

Modes 1 and 2 were used to fabricate small-size or patterned irradiation region. Beam current and accelerating voltage can be adjusted as required. Unless otherwise stated, the beam current was 16.9 pA and the accelerating voltage was 5 kV. Beam current was measured accurately by a Faraday cup. To form the isolated or superfine patterns, e-beam dose was ~ 50 mC cm^{-2} and the irradiated region was set as a circle with a diameter of 50–100 nm.

Mode 3 was used to form large-area irradiation region (mm scale). The beam current was 100 nA and the accelerating voltage was 15 kV. The SiO_2/Si wafer was irradiated for 30 min to form an irradiated region of 3×2.25 mm^2. To achieve patterned irradiation, Cu grid was used as a mask to block partial e-beam irradiation.

To study the post-treatment effects on the BCP self-assembly, SiO_2/Si substrates irradiated by Mode 1 were annealed at 250 or 500 °C for 2 h in Ar flow or immersed in piranha solution (H_2SO_4/H_2O_2 mixture) at 100 °C for 0.5 h. To alter the surface energy of SiO_2/Si, SiO_2/Si substrate irradiated by Mode 1 was immersed in phenylteichlorosilane/toluene mixture (1 vol%) for 0.5 h and was then washed in toluene to remove the unreacted silane.

Formation of PS-b-PMMA film. Cylinder-forming PS-b-PMMAs (46k-b-21k, 68k-b-33.5k and 35k-b-12.5k g mol^{-1}) and lamella-forming PS-b-PMMAs (53k-b-54k and 23k-b-22k g mol^{-1}) diblock copolymers were purchased from Polymer Source. The BCPs were dissolved in toluene and spin-coated onto the irradiated SiO_2/Si wafers. Unless otherwise stated, the film thickness of PS-b-PMMA (46k-b-21k) and PS-b-PMMA (53k-b-54k) was 37 and 56 nm, respectively, which was determined by a spectroscopic ellipsometer (SE 850 DUV, SENTECH Instruments). The BCP films were annealed at 250 °C for 2 h in Ar flow for microphase separation.

KPFM measurement. The KPFM measurement was operated on Multimode 8 (Bruker Corp.). The samples were prepared with the Helios 600i system, and a marker was added near the irradiated zone to locate the irradiated zone quickly under the charge-coupled device of KPFM system. We used Si probe with 1.5 N m^{-1} force constant and 250 kHz resonance frequency. Scan velocity was 0.6 Hz and number of scan lines was 512. All KPFM images were analysed using Nanoscope Analysis software.

Characterization. The surface morphology of the BCP films was observed by Helios 600i SEM system with an accelerating voltage of 5 kV. To increase the contrast of SEM image, all the annealed samples were treated with Ar plasma (power, 50 W; flow rate, 20 s.c.c.m.) to remove the PMMA domains selectively. GISAXS experiments were conducted at beamline 16B of the Shanghai Synchrotron Radiation Facility. The synchrotron X-ray energy was 10 keV and sample-to-detector distance was 1,820 mm. The incident angle of X-ray beam was 0.185°. A vacuum guide tube in which the scattered beam passed through was used to minimize air scattering. The two-dimensional GISAXS patterns were recorded on a Mar 165 charge-coupled device detector ($2,048 \times 2,048$ pixels, 80 µm per pixel). The surface elemental information was analysed by the X-ray photoelectron spectroscopy on the Thermo Scientific ESCALab 250Xi using 200 W Al Kα radiation. The irradiated SiO_2/Si wafers for GISAXS and XPS measurement were prepared by Mode 3.

References

1. Tseng, Y. C. & Darling, S. B. Block copolymer nanostructures for technology. *Polymers* **2**, 470–489 (2010).
2. Chai, J., Wang, D., Fan, X. & Buriak, J. M. Assembly of aligned linear metallic patterns on silicon. *Nat. Nanotechnol.* **2**, 500–506 (2007).
3. Jeong, S.-J. et al. Universal block copolymer lithography for metals, semiconductors, ceramics, and polymers. *Adv. Mater.* **20**, 1898–1904 (2008).
4. Black, C. T. et al. Polymer self assembly in semiconductor microelectronics. *IBM J. Res. Dev.* **51**, 605–633 (2007).
5. Black, C. T. Self-aligned self assembly of multi-nanowire silicon field effect transistors. *Appl. Phys. Lett.* **87**, 163116 (2005).
6. Hua, B., Lin, Q., Zhang, Q. & Fan, Z. Efficient photon management with nanostructures for photovoltaics. *Nanoscale* **5**, 6627–6640 (2013).
7. Naito, K., Hieda, H., Sakurai, M., Kamata, Y. & Asakawa, K. 2.5-inch disk patterned media prepared by an artificially assisted self-assembling method. *IEEE Trans. Magn.* **38**, 1949–1951 (2002).
8. Deng, X., Buriak, J. M., Dai, P. X., Wan, L. J. & Wang, D. Block copolymer-templated chemical nanopatterning on pyrolyzed photoresist carbon films. *Chem. Commun.* **48**, 9741–9743 (2012).
9. Yang, S. Y. et al. Nanoporous membranes with ultrahigh selectivity and flux for the filtration of viruses. *Adv. Mater.* **18**, 709–712 (2006).
10. Bang, J., Jeong, U., Ryu du, Y., Russell, T. P. & Hawker, C. J. Block copolymer nanolithography: translation of molecular level control to nanoscale patterns. *Adv. Mater.* **21**, 4769–4792 (2009).
11. Bates, C. M., Maher, M. J., Janes, D. W., Ellison, C. J. & Willson, C. G. Block copolymer lithography. *Macromolecules* **47**, 2–12 (2014).

12. The international technology roadmap for semiconductors: lithography. http://www.itrs2.net/itrs-reports.html (2015).

13. Kim, S. Y., Gwyther, J., Manners, I., Chaikin, P. M. & Register, R. A. Metal-containing block copolymer thin films yield wire grid polarizers with high aspect ratio. *Adv. Mater.* **26**, 791–795 (2014).

14. Park, O. H. *et al.* High aspect-ratio cylindrical nanopore arrays and their use for templating titania nanoposts. *Adv. Mater.* **20**, 738–742 (2008).

15. Mansky, P., Liu, Y., Huang, E., Russell, T. P. & Hawker, C. J. Controlling polymer-surface interactions with random copolymer brushes. *Science* **275**, 1458–1460 (1997).

16. Bates, C. M. *et al.* Polarity-switching top coats enable orientation of sub-10-nm block copolymer domains. *Science* **338**, 775–779 (2012).

17. Violetta, O. *et al.* Electric field alignment of a block copolymer nanopattern: direct observation of the microscopic mechanism. *ACS Nano* **3**, 1091–1096 (2009).

18. Thurn-Albrecht, T. *et al.* Ultrahigh-density nanowire arrays grown in self-assembled diblock copolymer templates. *Science* **290**, 2126–2129 (2000).

19. Man, X., Tang, J., Zhou, P., Yan, D. & Andelman, D. Lamellar diblock copolymers on rough substrates: self-consistent field theory studies. *Macromolecules* **48**, 7689–7697 (2015).

20. Hong, S. W. *et al.* Unidirectionally aligned line patterns driven by entropic effects on faceted surfaces. *Proc. Natl Acad. Sci. USA* **109**, 1402–1406 (2012).

21. Gu, X., Gunkel, I., Hexemer, A., Gu, W. & Russell, T. P. An in situ grazing incidence X-ray scattering study of block copolymer thin films during solvent vapor annealing. *Adv. Mater.* **26**, 273–281 (2014).

22. Lo, T. Y. *et al.* Phase transitions of polystyrene-b-poly(dimethylsiloxane) in solvents of varying selectivity. *Macromolecules* **46**, 7513–7524 (2013).

23. Thebault, P. *et al.* Tailoring nanostructures using copolymer nanoimprint lithography. *Adv. Mater.* **24**, 1952–1955 (2012).

24. Yu, B. *et al.* Confinement-induced novel morphologies of block copolymers. *Phys. Rev. Lett.* **96**, 138306 (2006).

25. Park, S. *et al.* Macroscopic 10-terabit-per-square-inch arrays from block copolymers with lateral order. *Science* **323**, 1030–1033 (2009).

26. Tang, C., Lennon, E. M., Fredrickson, G. H., Kramer, E. J. & Hawker, C. J. Evolution of block copolymer lithography to highly ordered square arrays. *Science* **322**, 429–432 (2008).

27. Majewski, P. W., Rahman, A., Black, C. T. & Yager, K. G. Arbitrary lattice symmetries via block copolymer nanomeshes. *Nat. Commun.* **6**, 7448 (2015).

28. Maher, M. J. *et al.* Photopatternable interfaces for block copolymer lithography. *ACS Macro Lett.* **3**, 824–828 (2014).

29. Tsai, H. *et al.* Two-dimensional pattern formation using graphoepitaxy of PS-b-PMMA block copolymers for advanced FinFET device and circuit fabrication. *ACS Nano* **8**, 5227–5232 (2014).

30. Doerk, G. S. *et al.* Pattern placement accuracy in block copolymer directed self-assembly based on chemical epitaxy. *ACS Nano* **7**, 276–285 (2013).

31. Hu, H., Gopinadhan, M. & Osuji, C. O. Directed self-assembly of block copolymers: a tutorial review of strategies for enabling nanotechnology with soft matter. *Soft Matter* **10**, 3867–3889 (2014).

32. Cheng, J. Y., Ross, C. A., Smith, H. I. & Thomas, E. L. Templated self-assembly of block copolymers: top-down helps bottom-up. *Adv. Mater.* **18**, 2505–2521 (2006).

33. Mark, P. S. *et al.* Directed assembly of block copolymer blends into nonregular device-oriented structures. *Science* **308**, 1442–1446 (2005).

34. Ion, B. *et al.* Graphoepitaxy of self-assembled block copolymers on two-dimensional periodic patterned templates. *Science* **321**, 939–943 (2008).

35. Yi, H. *et al.* Flexible control of block copolymer directed self-assembly using small, topographical templates: potential lithography solution for integrated circuit contact hole patterning. *Adv. Mater.* **24**, 3107–3114 (2012).

36. Doerk, G. S. *et al.* Enabling complex nanoscale pattern customization using directed self-assembly. *Nat. Commun.* **5**, 5805 (2014).

37. Onses, M. S. *et al.* Hierarchical patterns of three-dimensional block-copolymer films formed by electrohydrodynamic jet printing and self-assembly. *Nat. Nanotechnol.* **8**, 667–675 (2013).

38. Onses, M. S. *et al.* Block copolymer assembly on nanoscale patterns of polymer brushes formed by electrohydrodynamic jet printing. *ACS Nano* **8**, 6606–6613 (2014).

39. Jacobs, H. O. & Whitesides, G. M. Submicrometer patterning of charge in thin-film electrets. *Science* **291**, 1763–1766 (2001).

40. Zhao, D. *et al.* Self-organization of thin polymer films guided by electrostatic charges on the substrate. *Small* **7**, 2326–2333 (2011).

41. Sessler, G. M. *Electrets* 2nd edn (Springer, 1987).

42. Li, W.-Q. & Zhang, H.-B. The positive charging effect of dielectric films irradiated by a focused electron beam. *Appl. Surf. Sci.* **256**, 3482–3492 (2010).

43. Tsui, B. Y., Hsieh, C. M., Su, P. C., Tzeng, S. D. & Gwo, S. Two-dimensional carrier profiling by Kelvin-probe force microscopy. *Jpn J. Appl. Phys.* **47**, 4448–4453 (2008).

44. Nakasugi, T., Ando, A., Sugihara, K., Miyoshi, M. & Okumura, K. New registration technique using voltage-contrast images for low-energy electron-beam lithography. *Proc. SPIE* **4343**, 334–341, (2001).

45. Liedel, C., Pester, C. W., Ruppel, M., Urban, V. S. & Boker, A. Beyond orientation: the impact of electric fields on block copolymers. *Macromol. Chem. Phys.* **213**, 259–269 (2012).

46. Thurn-Albrecht, T., DeRouchey, J., Russell, T. P. & Jaeger, H. M. Overcoming interfacial interactions with electric fields. *Macromolecules* **33**, 3250–3253 (2000).

47. Kyrylyuk, A. V. & Fraaije, J. G. Electric field versus surface alignment in confined films of a diblock copolymer melt. *J. Chem. Phys.* **125**, 164716 (2006).

48. Kyrylyuk, A. V., Sevink, G. J. A., Zvelindovsky, A. V. & Fraaije, J. G. E. M. Simulations of electric field induced lamellar alignment in block copolymers in the presence of selective electrodes. *Macromol. Theory Simul.* **12**, 508–511 (2003).

49. Maher, M. J. *et al.* Interfacial design for block copolymer thin films. *Chem. Mater.* **26**, 1471–1479 (2014).

50. Kulkarni, M. M., Yager, K. G., Sharma, A. & Karim, A. Combinatorial block copolymer ordering on tunable rough substrates. *Macromolecules* **45**, 4303–4314 (2012).

51. Gopinadhan, M., Majewski, P. W. & Osuji, C. O. Facile alignment of amorphous poly(ethylene oxide) microdomains in a liquid crystalline block copolymer using magnetic fields: toward odered electrolyte membranes. *Macromolecules* **43**, 3286–3293 (2010).

52. Shin, D. O. *et al.* One-dimensional nanoassembly of block copolymers tailored by chemically patterned surfaces. *Macromolecules* **42**, 1189–1193 (2009).

53. van Dorp, W. F. *et al.* Molecule-by-molecule writing using a focused electron beam. *ACS Nano* **6**, 10076–10081 (2012).

54. Ashok, B., Muthukumar, M. & Russell, T. P. Confined thin film diblock copolymer in the presence of an electric field. *J. Chem. Phys.* **115**, 1559–1564 (2001).

Acknowledgements

This work was supported by National Natural Science Foundation of China (21433011, 21127901 and 91527303) and the Strategic Priority Research Program of the Chinese Academy of Sciences (Grant No. XDB12020100). We thank Shanghai Synchrotron Radiation Facility and Dr Yuzhu Wang for the help in the GISAXS experiments and Prof. Shien-Der Tzeng in National Tsing Hua University for providing the NOS samples.

Author contributions

D.W. and M.-L.W. conceived the idea; M.-L.W. conducted the experiments; all the authors participated in discussing the results; M.-L.W. and D.W. drafted the manuscript.

Additional information

14

Evidence and implications of direct charge excitation as the dominant mechanism in plasmon-mediated photocatalysis

Calvin Boerigter[1], Robert Campana[1], Matthew Morabito[1] & Suljo Linic[1]

Plasmonic metal nanoparticles enhance chemical reactions on their surface when illuminated with light of particular frequencies. It has been shown that these processes are driven by excitation of localized surface plasmon resonance (LSPR). The interaction of LSPR with adsorbate orbitals can lead to the injection of energized charge carriers into the adsorbate, which can result in chemical transformations. The mechanism of the charge injection process (and role of LSPR) is not well understood. Here we shed light on the specifics of this mechanism by coupling optical characterization methods, mainly wavelength-dependent Stokes and anti-Stokes SERS, with kinetic analysis of photocatalytic reactions in an Ag nanocube–methylene blue plasmonic system. We propose that localized LSPR-induced electric fields result in a direct charge transfer within the molecule–adsorbate system. These observations provide a foundation for the development of plasmonic catalysts that can selectively activate targeted chemical bonds, since the mechanism allows for tuning plasmonic nanomaterials in such a way that illumination can selectively enhance desired chemical pathways.

[1] Department of Chemical Engineering, University of Michigan, Arbor, Michigan 48109, USA. Correspondence and requests for materials should be addressed to S.L. (email: linic@umich.edu).

Numerous studies have shown that plasmonic metal nanoparticles can drive photochemical reactions directly on the surface of the nanoparticles when illuminated with relatively low-intensity light of particular frequencies[1-4]. It has been proposed that these processes are driven by the excitation of localized surface plasmon resonance (LSPR) on the metal nanoparticles[4,5]. LSPR is the resonant collective oscillation of valence electrons, established when the frequency of an incident electromagnetic field matches the natural frequency of surface electrons oscillating against the restoring force of positive nuclei[6]. It has been proposed that the interaction of LSPR with energetically accessible adsorbate orbitals can yield a charge injection process, where energized charge carriers (electrons or holes, formed by the plasmon decay) transiently populate otherwise unpopulated electronic states (orbitals), centred on the adsorbate molecule[7-9]. In this process, the adsorbate (more specifically, the entire adsorbate–nanoparticle system) is moved to a different potential energy surface and forces are induced on atoms in the adsorbate. These forces lead to nuclear motion of atoms in the adsorbate, which can result in the activation of chemical bonds and chemical transformations[10]. This mechanism of charge-driven chemical transformations is known by its original name 'desorption (reaction) induced by electronic transitions'.

The mechanism of the charge injection process on excited plasmonic metal nanoparticles and the role of LSPR in the process are not well understood. The illustration in Fig. 1 outlines the two potential microscopic mechanisms by which charge injection can occur: indirect and direct. The indirect mechanism assumes that charge carriers are excited from occupied states to higher-energy unoccupied states within the metal nanoparticle yielding an excited electron distribution. Subsequently, charge carriers of appropriate energy from this distribution can scatter through the adsorbate states (orbitals) forming transiently

charged adsorbates. The role of LSPR in this process is to increase the rates of charge excitation, and the energetic electron distribution associated with this mechanism is very similar to the one obtained if a high-intensity laser was illuminated on an extended metal structure. On the other hand, the direct transfer mechanism assumes the direct LSPR-induced electron excitation from occupied to unoccupied orbitals of the molecule–nanoparticle complex (that is, the excitation process) is not mediated by the formation of an excited electron distribution within the metal nanoparticle. Instead, the decay of an oscillating surface plasmon results in the excitation of an electron directly into an unoccupied adsorbate orbital of matching energy. The direct and indirect charge excitation processes, as well as nanoparticle heating due to photon absorption can all theoretically result in photochemistry, and it is not clear which of these processes is dominant[11].

The identification of the dominant mechanism of charge excitation and the impact of LSPR is highly consequential in answering the question whether plasmonic nanostructures can be tuned in a way that under illumination they selectively enhance particular chemical pathways, while suppressing other chemical pathways. As illustrated in Fig. 1b, if indirect charge transfer plays the dominant role then the steady-state concentration of energetic electrons is the highest close to the Fermi level (that is, similar to Fermi–Dirac electron distribution) and therefore the electron-driven chemical transformation would preferentially proceed through the adsorbate orbitals closest to the Fermi level. This mechanism does not offer much opportunity to tune the chemical pathways by tuning the plasmonic properties of nanostructures since LSPR would only enhance the rates of electron excitation without impacting the electron distribution. On the other hand, the direct mechanism assumes that the LSPR-mediated photon absorption occurs through the creation of electron–hole pairs of specific energy that are localized on the adsorbate as illustrated in

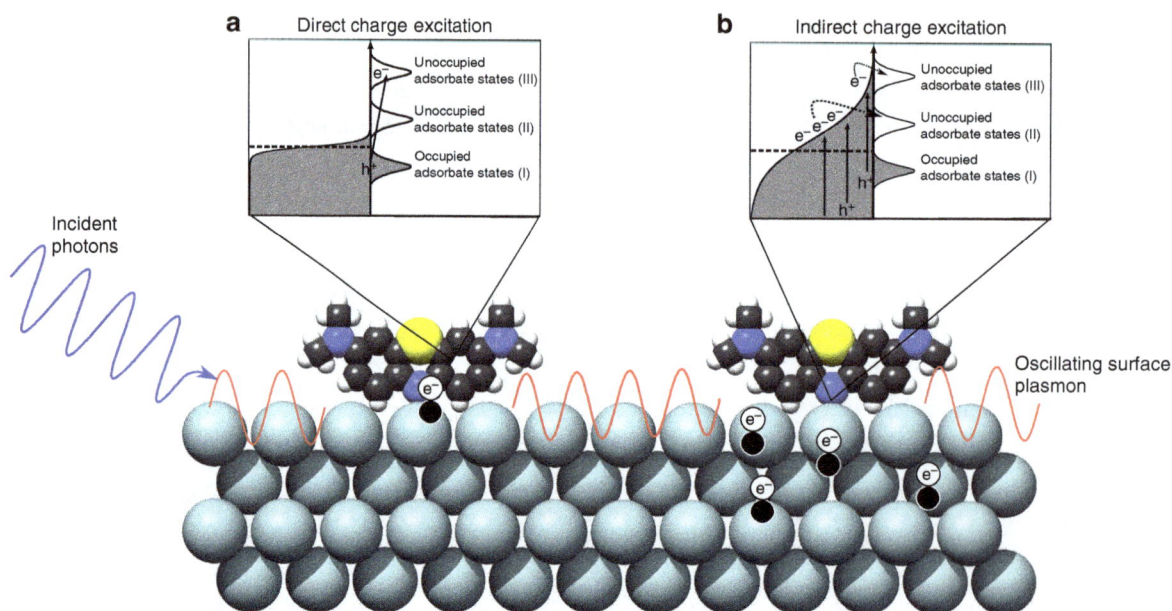

Figure 1 | Illustration of LSPR-mediated charge excitation mechanisms. Incident photons excite oscillating surface plasmons on an adsorbate-covered Ag nanoparticle surface. These surface plasmon oscillations decay through the formation of energetic electron–hole pairs. In the direct process (**a**), the electron is excited directly into an unoccupied orbital of matching energy within the adsorbate. In the indirect process (**b**), the energetic electrons formed by decaying plasmons form a distribution within the metal nanoparticle. Electrons with proper energy can then scatter into available adsorbate orbitals. Because of the nature of the electron distribution formed in the indirect mechanism, more electrons will scatter into lower energy orbitals (II) and chemical transformation will preferentially proceed through that lower energy activated pathway. In the direct mechanism, however, the electrons can potentially excite into higher energy orbitals (III) when that energy matches the incident photon energy. This opens the possibility for selective chemical pathway targeting that impossible in the indirect mechanism.

Fig. 1a. The direct mechanism offers an opportunity to affect chemical selectivity by creating plasmonic nanostructures with the LSPR characteristics that enhance the rates of targeted electronic transitions and therefore the rates of chemical transformations that proceed through only those specific transitions.

Herein, we shed light on the underlying physical mechanisms involved in the LSPR-induced charge transfer and chemical transformations on optically excited plasmonic metal nanoparticles. To accomplish this, we have coupled optical characterization methods, mainly wavelength-dependent Stokes and anti-Stokes surface-enhanced Raman spectroscopy (SERS) of probe molecules chemisorbed on plasmonic Ag nanostructures, with the kinetic analysis of the rates of photocatalytic reactions of these molecules in otherwise inert environment. We demonstrate that on optically excited plasmonic nanoparticles, the direct charge transfer mechanism plays an important role and that it results in accelerated rates of chemical decomposition (or desorption) of a probe molecule in an inert atmosphere. We propose that strong localized LSPR-induced electric fields result in photon absorption through the direct excitation of an electron from specific occupied to unoccupied orbitals of the molecule–nanoparticle complex (that is, at the surface of the nanoparticle). This can induce photochemical transformation of the chemisorbed molecules via a desorption (or reaction) induced by electronic transitions mechanism described above[7]. We also discuss the potential impact of these findings and suggest ways to develop plasmonic catalysts that can potentially selectively activate targeted chemical bonds.

It is important to put our results in the context of numerous previous SERS studies of probe molecules on plasmonic nanoparticles. These previous studies have given us a wealth of information about various mechanisms and consequences of surface enhancement in Raman spectroscopy, even demonstrating that laser-induced electron-driven reactions on plasmonic metals are possible[9,12,13]. The results presented herein build upon these findings by focusing on the missing link between LSPR and charge excitation. We distinguish between the above-discussed two LSPR-mediated mechanisms of charge excitation by analysing the unique anti-Stokes spectral behaviour that is interpreted in terms of LSPR-mediated vibroelectronic excitation.

Results

Optical properties of the Ag nanocube SERS platform. The SERS platforms (that is, the model system) used in our studies contained small aggregate clusters of silver (Ag) nanocubes with the cube length of ~ 75 nm dispersed on a silicon substrate as shown in the optical micrograph in the inset of Fig. 2. We perform SERS measurements of probe molecules using two different lasers with the wavelengths of 532 and 785 nm. The optical excitation of LSPR at these wavelengths results in high-intensity electric fields at the surface of the Ag nanoparticles. Since these local fields facilitate the photophysical and photochemical processes studied herein, we first employed classical finite difference time domain (FDTD) electrodynamics simulations to compute the local near field electric field intensities of Ag systems that modelled our SERS platform. These simulations serve to provide part of the framework with which to interpret the key Raman results in this study.

In our FDTD studies we, used model systems consisting of two and three Ag nanocubes positioned at different configurations at 1 and 2 nm separations from each other. Not surprisingly, the FDTD simulations show that the highest plasmonic electric field intensities are generated in the gaps between the nanoparticles (Fig. 2). We find that for both laser source wavelengths, the field intensity enhancements in the 1-nm gap between the two nanoparticles are similar to each other (($|\mathbf{E}|/|\mathbf{E_0}|)^2 \sim 4 \times 10^4$). Furthermore, we have also integrated the field intensity enhancement across the entire surface of the two nanoparticle system to obtain the average surface field intensity enhancements (Supplementary Table 1). We find that the average field intensity enhancements are within an order of magnitude of each other for the 532- and 785-nm sources. Similar analysis for the Ag nanoparticles separated by 2 nm from each other, as well as for the systems containing aggregates of three Ag nanoparticles, showed very similar results (Supplementary Figures 1–6, Supplementary Table 2, Supplementary Note 1).

It should be noted that the geometries simulated in the FDTD studies are idealized representation of our SERS platforms, which in practice contain aggregates of multiple nanoparticles with varying orientations and separation distances. The effect of particle orientations and separation distances on the optical properties of collections of plasmonic nanoparticles has been studied extensively previously. These studies showed that the resonant wavelength of the coupled dimer plasmon resonance peak is dependent on the particle separation, moving to longer wavelengths and higher maximum field intensities for smaller separations[14,15]. Therefore, due to the random nature of the Ag nanoparticle clustering in our system, we expect that there are different spots that support plasmon resonance peaks over a broad range of wavelengths. The data in Fig. 2a, which show the ultraviolet–visible extinction spectrum of a sample of silver nanocube aggregates used in the SERS studies, show this precise signature with a very broad extinction peak ranging between 600 and 900 nm. The slightly stronger and more confined extinction peak ranging from 450 to 550 nm is the result of dipolar LSPR within single particles, while still higher-energy peaks come from higher-order multipole resonances within individual particles[16,17]. The wavelengths of the Raman lasers used in our studies are overlaid on the spectrum.

It is worth noting that in addition to the above described photophysical properties, which can be captured by classical electrodynamics, there have been reports of quantum charge tunnelling between optically excited plasmonic nanoparticles[18,19]. This charge tunnelling process has been reported for the particle separations smaller than ~ 0.3–0.5 nm (refs 20–22). It is a consequence of very large electric fields confined in the nanogaps that force the movement of charge between the particles. At these extremely small gap distances, the tunnelling process broadens and slightly blue shifts the LSPR resonance peaks while decreasing the maximum field intensity from the classical prediction.

To summarize, our analysis of the photophysical properties of the SERS platforms suggests that (i) there are sites in our system that can support LSPR with significantly elevated electric field intensities at both Raman laser wavelengths, (ii) while we know the field intensities are significantly elevated, we cannot know the exact intensity at each site due to the disordered nature of Ag aggregates in our SERS platforms and (iii) potential quantum tunnelling could be taking place in aggregates with interparticle gaps of 0.5 nm or less. With regards to the final two points, our analysis of Raman results and the conclusions drawn are not affected by the lack of exact knowledge of field intensities at every site in the system. By focusing our analysis on the ratios of anti-Stokes to Stokes signals in a given measurement and probing the dependence of only these ratios on the incident photon intensity, we find the exact value of the electric field intensity to be irrelevant so long as it is strong enough to support SERS.

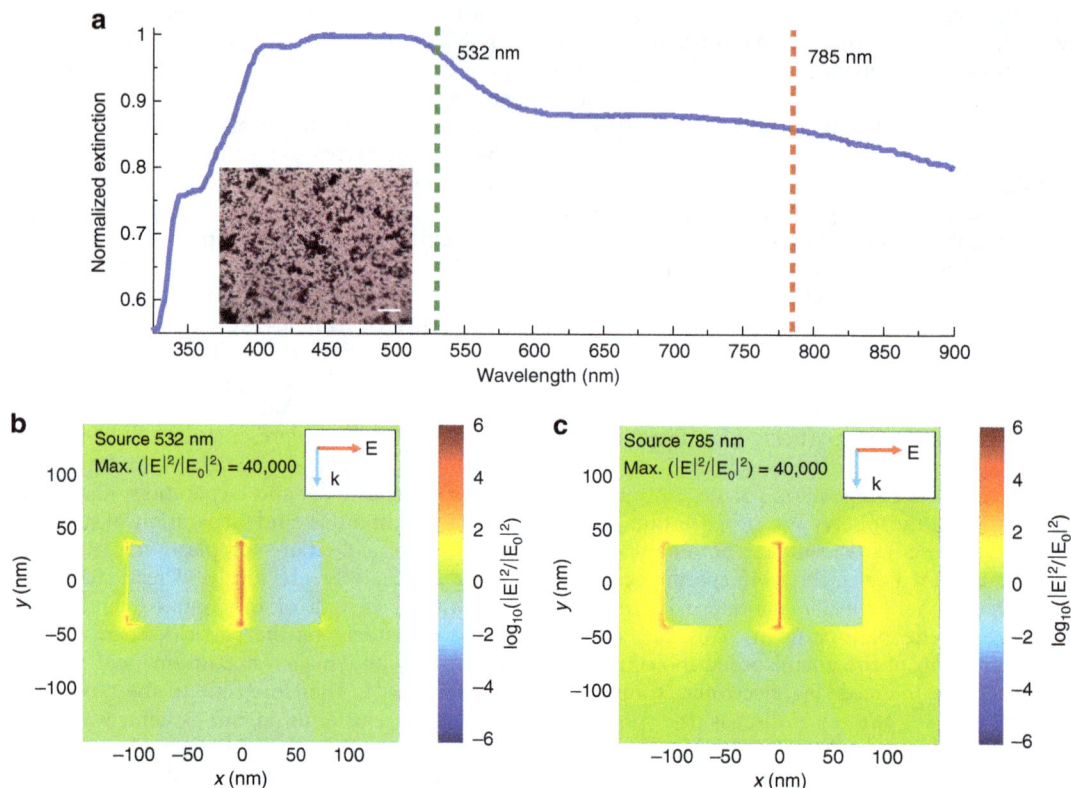

Figure 2 | Optical characterization of Ag SERS platform. (**a**) Ultraviolet-visible spectrum of silver nanocubes after deposition onto solid substrate. The two dashed lines show the wavelengths of the lasers used in this study and their overlap with the LSPR of the silver cubes (inset: optical micrograph of Ag-MB sample showing high levels of aggregation; scale bar, 20 µm). (**b,c**) Simulated spatial distribution of electric field enhancement for a silver nanocube dimer (face-to-edge orientation) with separation distance of 1 nm under (**b**) 532 nm and (**c**) 785 nm photon excitation.

SERS measurements for methylene blue on Ag. There have been a few studies of the chemisorption of methylene blue (MB) and similar molecules on Ag surfaces[23–25]. It has been suggested that MB chemisorbs through a covalent chemical bond between an Ag surface atom and the nitrogen atom in the aromatic ring of MB. We have also performed density functional theory (DFT) quantum calculations of adsorbed thiazine on Ag model slabs. We use thiazine as a surrogate for MB since similar to MB, thiazine has an N- and S-containing aromatic ring and similar bonding conformations are proposed. Thiazine is smaller than MB and therefore more easily handled computationally. Our DFT calculations, using the revised Perdew-Burke-Ernzerhof (RPBE) density functional, showed that the binding energy of thiazine to the Ag(100) surface facet is -0.29 eV (exothermic), while on the under-coordinated Ag(211) surface the binding energy is -0.37 eV. Since the RPBE functional in general underestimates the binding electronic energy and neglects van der Waals forces, we suggest that in our systems MB chemisorbs on the surface sites of Ag nanocubes, essentially coating the nanoparticles.

There are two established potential mechanisms by which LSPR can enhance Raman scattering. One mechanism involves the LSPR-induced enhancement of the local electric field intensity at the adsorbate (electromagnetic mechanism) leading to the enhancement of the Raman signal. The electromagnetic enhancement factor (Σ) of Raman signal depends on electromagnetic field enhancement at the location occupied by a probe molecule. It can be closely approximated by:

$$\Sigma = \left(|\mathbf{E}(v_i)|^2 * |\mathbf{E}(v_i + v_m)|^2 \right) \quad (1)$$

where $\mathbf{E}(v)$ is the local electromagnetic field enhancement at a photon frequency v. v_i and $(v_i + v_m)$ are the frequencies of the incident (that is, laser) photon and scattered (that is, Raman

shifted) photon, respectively[26,27]. The other mechanism involves the LSPR-induced charge excitation (or charge–exchange) mechanism[27]. Charge excitation SERS occurs as a result of the LSPR-mediated charge excitation from filled to unfilled electronic states of the nanoparticle–adsorbate complex (one or both of these electronic states are localized on the adsorbate). The propensity to undergo charge excitation is, among other things, determined by the local electronic structure of the adsorbate–metal complex[28].

Since we were interested in investigating the LSPR-induced charge excitation process and the chemical consequences of this process, we focused on identifying systems where these two mechanisms can be separated from each other, centring our studies on the charge excitation mechanism and probing its effect on chemical transformations. Below, we demonstrate that MB on Ag exhibits a LSPR-mediated charge excitation only at specific excitation wavelengths and that this process increases the rates of chemical reactions (in this case the degradation or desorption of MB on the Ag surface). These observations allow us to conclude that in this case LSPR mediates the process of charge excitation into adsorbates via the above-discussed direct mechanism.

The process of charge excitation within the metal–adsorbate complex has some key experimentally distinguishable SERS signatures. These signatures are a consequence of the fact that the charge excitation process results in an elevated population of excited vibrational modes compared with the equilibrated, thermalized system. Therefore, a key signature of the charge transfer process is an elevated anti-Stokes Raman intensity compared with that expected for a thermodynamically equilibrated system. We note that Stokes scattering signal in Raman spectroscopy is the measure of the rate of transitions of a molecule from the ground to the first excited vibrational states,

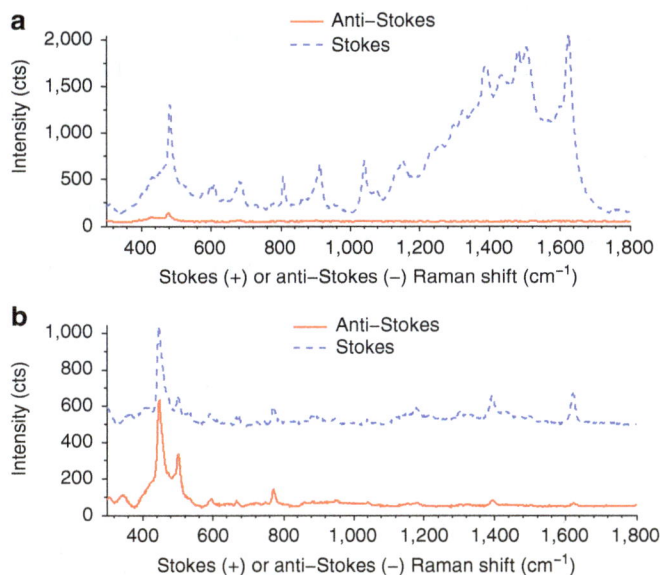

Figure 3 | SERS spectra of silver nanocube-dye structures. (**a,b**) Stokes and anti-Stokes spectra for silver nanocube–methylene blue (Ag–MB) structures gathered using a 532-nm (**a**) or 785-nm (**b**) laser. All experimental parameters other than the wavelength of incoming photons, including incident photon flux, acquisition time and sample spot size were identical. It should be noted that Stokes shifts have positive wavenumbers (cm^{-1}), while anti-Stokes shifts are negative. They are plotted on the same positive axis here to allow for direct comparison.

while the anti-Stokes scattering signal measures the reverse process. The ratio of anti-Stokes to Stokes scattering rates and thus their signal intensities for a vibrational mode can be related to the population of vibrationally excited molecules[29].

Data in Fig. 3 show Stokes and anti-Stokes SERS spectra for the Ag–MB complex at 532 nm (Fig. 3a) and 785 nm (Fig. 3b) excitation wavelengths. The spectra were collected using the same photon flux of $2.0 \times 10^{20}\,\mathrm{cm}^{-2}\mathrm{s}^{-1}$ for both lasers. Visual inspection of the data in Fig. 3 shows that the ratio of anti-Stokes to Stokes signals at 785 nm is higher than when the 532 nm laser source was used. To quantify these observations, the ratio of anti-Stokes/Stokes signal (ρ_{SERS}) for specific MB vibrational modes (v_{m}) of the Ag–MB complex under 785 and 532 nm excitations was computed and compared with the ratio for a similar energy mode in liquid toluene (ρ_{tol}). Using toluene as an external standard corrects for any wavelength-dependent biases due to the sensitivity of the charge-coupled device detector, as demonstrated previously[30,31]. Liquid toluene gives a basic non-SERS, non-resonant Raman spectrum at visible light excitation, and the anti-Stokes to Stokes intensity ratios in toluene are consistent with those predicted by Boltzmann thermal distribution[30]. The degree by which the SERS anti-Stokes signal exceeded the expectation of the Boltzmann distribution is described through the quantity (K), which was calculated using the following equation:

$$K(v_{\mathrm{m}}) = \frac{\rho_{\mathrm{SERS}}}{\rho_{\mathrm{tol}}}(v_{\mathrm{m}}) = \frac{I_{\mathrm{aS}}^{\mathrm{SERS}}(v_{\mathrm{m}})/I_{\mathrm{S}}^{\mathrm{SERS}}(v_{\mathrm{m}})}{I_{\mathrm{aS}}^{\mathrm{tol}}(v_{\mathrm{m}})/I_{\mathrm{S}}^{\mathrm{tol}}(v_{\mathrm{m}})} \qquad (2)$$

where $I_{\mathrm{aS}}^{\mathrm{SERS}}$ and $I_{\mathrm{S}}^{\mathrm{SERS}}$ are the measured anti-Stokes and Stokes intensities, respectively, for MB on Ag at a given vibrational mode energy (v_{m}). $I_{\mathrm{aS}}^{\mathrm{tol}}$ and $I_{\mathrm{S}}^{\mathrm{tol}}$ are the anti-Stokes and Stokes intensities, for the similar vibrational mode in toluene.

The data in Table 1 show the calculated K values for various modes of the Ag–MB complex at 532- and 785-nm excitation at the photon flux of $2.0 \times 10^{20}\,\mathrm{cm}^{-2}\mathrm{s}^{-1}$. The data show that the

Table 1 | K values in Ag–MB spectra.

Peak (cm^{-1})	K
532 nm	
482	0.63
807	0.6
914	1.24
1,041	1.32
785 nm	
447	3.91
771	7.67
1,394	27.75
1,621	34.45

MB, methylene blue; SERS, surface-enhanced Raman spectroscopy.
Calculated K values for selected peaks in the MB SERS spectrum excited by 532- and 785-nm lasers.

K values under 785-nm excitation range between 4 and 34 for different Ag–MB modes. On the other hand, the K values for the 532-nm excitation are close to 1. The elevated K values at 785 nm compared with 532 nm suggest that charge excitation might be taking place within the Ag–MB complex at 785 nm and that there is no charge excitation for 532-nm wavelength.

We must note however that the SERS enhancement of Stokes and anti-Stokes scattering, in addition to the vibrational populations, also depends on the field intensity at the frequency of scattered (that is, Raman shifted) photons as described in equation (1). Different field intensities at the frequencies of Stokes and anti-Stokes scattered photons can contribute to the elevated K values even without charge transfer[32,33]. To rule out this possibility, we have performed SERS studies using the same Ag nanocube SERS platform with other probe molecules (Rhodamine 6G and Acridine Orange), which under 785-nm radiation do not exhibit charge excitation (Supplementary Figure 7, Supplementary Table 3, Supplementary Note 2)[31]. These studies show that for these molecules the K values at 785 nm are significantly lower than the K values for MB, signalling that the elevated vibrational population observed for the Ag–MB complex at 785 nm is due to the charge excitation process.

Further confirmation of charge excitation taking place at 785 nm comes from observing that the ratio of anti-Stokes to Stokes signal shows a positive dependence on incident photon intensity. This is another key feature of the charge excitation process[29,34]. As discussed above, charge excitation can lead to a non-thermalized population of vibrationally excited molecules (that is, an elevated population of molecules in the first excited vibrational states, which are the initial states for $v = 1 \rightarrow 0$ transitions associated with anti-Stokes scattering). As the source intensity increases, probe molecules will have a larger fraction of modes that are excited to the first excited vibrational state and above, thus increasing the anti-Stokes to Stokes signal ratio[29,30]. This phenomenon is known as optical pumping, and in this regime the theoretically derived ratio (ρ) of anti-Stokes to Stokes signals is governed by the equation[29]:

$$\rho_{\mathrm{SERS}} = \frac{I_{\mathrm{aS}}^{\mathrm{SERS}}(v_{\mathrm{m}})}{I_{\mathrm{S}}^{\mathrm{SERS}}(v_{\mathrm{m}})} = \frac{\sigma_{\mathrm{aS}}}{\sigma_{\mathrm{S}}}\left[\frac{\tau\sigma_{\mathrm{S}}I_{\mathrm{L}}}{h\nu_{\mathrm{L}}} + e^{-\frac{h\nu_{\mathrm{m}}}{k_{\mathrm{B}}T}}\right], \qquad (3)$$

where σ_{as} and σ_{s} represent the cross-section for anti-Stokes and Stokes events respectively, τ the vibrational relaxation time constant for the molecule, I_{L} the incident light intensity, $h\nu_{\mathrm{L}}$ and $h\nu_{\mathrm{m}}$ the energy of the incident photon and vibrational mode respectively, at T the sample temperature.

The equation covers both the exponential dependence of the population of the excited vibrational states on temperature and the linear dependence on photon intensity (I_L). In the optical pumping regime (that is, when the linear term in equation (3) is of the same order or greater than the exponential term) under constant sample temperature, ρ_{SERS} and thus K (as described above, the measure of the deviations in the anti-Stokes to Stokes ratios from the ratio predicted by the equilibrium Boltzmann distribution) will show a linear dependence on the photon intensity. The data in Fig. 4 show the K values plotted against incident photon intensity for some of the most prominent bands in the Ag–MB Raman spectrum under 785-nm excitation. The intensity spans two orders of magnitude of photon flux, from 1.5×10^{19} to 2.0×10^{21} cm^{-2} s^{-1}. The data show a linearly increasing K for all modes as photon flux increases, providing compelling evidence that the samples are well within the optical pumping charge excitation regime. This is in contrast to the other measured probe molecules which showed no linear increase in K values with increasing intensity (Supplementary Fig. 8).

Discussion

The above-discussed data represent evidence that Ag–MB complexes experience charge excitation resonance under exposure to 785 nm incident photons, while the charge excitation process does not take place with 532-nm laser under identical photon fluxes. It is important to reiterate that these results cannot be explained only by the different values of the field intensities at these two wavelengths. If this was the case, any molecule could exhibit elevated anti-Stokes signals under 785-nm illumination. We have ruled out this possibility in the above-mentioned SERS studies of other probe molecules (Rhodamine 6G and Acridine Orange), which under 785-nm illumination do not exhibit charge excitation. Similarly, these studies of other probe molecules also rule out interparticle electron tunnelling experienced at very small particle separations as the only requirement for the above-mentioned charge transfer process, as if this were the case we would again expect no difference in the results when using other probe molecules.

The data presented above-shed light on some microscopic features of the charge excitation process, depicted in Fig. 1. We have postulated that the charge excitation can take place following an indirect route where excited charge carriers are first formed within the nanoparticle in a process that results in an excited electron distribution. Electrons from this distribution that gain sufficient energy can scatter through the adsorbate states forming transiently charged adsorbates. Conversely, in the direct excitation process, the presence of unoccupied adsorbate orbitals at a specific energy allows for a direct, resonant transition from occupied to unoccupied orbitals of the molecule–nanoparticle complex. We suggest that the occurrence of charge excitation only by photons of specific energy (785 nm) and not by those of higher energy (532 nm) under identical incident photon intensities shows that the mechanism of charge excitation in this system is direct. If the mechanism were indirect, one could expect to see evidence of charge excitation at any wavelength with photon energies at least as high as the difference between the metal Fermi level and the lowest unoccupied molecular orbitals of the adsorbate. Under the assumption of the indirect mechanism, as photon energies increased (for example, from 785 to 532 nm), the extent of charge excitation would be expected to rise, as the number of electrons within the metal that reached the requisite energy to scatter into the unfilled adsorbate states would increase. Because this does not occur and instead we observe charge excitation only under lower energy photon excitation, we suggest that the process of charge excitation is direct and it involves direct resonant single electron transitions from occupied to unoccupied states of the Ag–MB complex.

The evidence that points towards a direct mechanism of charge excitation also serves to rule out plasmon-induced heating of the nanoparticles as an explanation for the high-anti-Stokes data. The K values for the SERS data collected under 532-nm excitation are close to 1, suggesting that there is no significant laser induced heating of the entire system at this excitation wavelength. Equilibrated heating of the system would increase the population of excited vibrational modes of the molecule and result in elevated anti-Stokes intensity and therefore elevated K values. On the basis of the fact that for identical photon fluxes, the energy dissipated into the system would be higher for the 532 nm laser compared with the 785 laser, we argue that equilibrated heating of the entire sample does not take place for the 785-nm laser either. This suggests that the process of charge excitation observed at 785 nm results in energy being selectively deposited into the adsorbed molecule rather than into the entire system.

We can also speculate about the nature of the electronic states involved in the charge excitation process. The highest occupied molecular orbital–lowest unoccupied molecular orbital energy gap in the free MB molecule is ~ 1.86 eV (665 nm), which is above the energy of the photons from the 785 nm (1.6 eV) laser. An examination of the absorption spectrum of aqueous MB shows no interaction with light near 785 nm. This means that the electronic states involved in the charge excitation processes are associated with either strongly perturbed states localized on MB—these perturbations can be due to the chemisorption of MB in the Ag metal—or with hybridized Ag–MB electronic states.

It is also important to discuss how LSPR could affect this process of charge excitation. As discussed above, our data point us in the direction of the direct one-electron excitation by 785 nm photons among the states within the Ag–MB complex. Clearly, the existence of the electronic states that can be accessed by the 785-nm photon is a requirement for the observed charge excitation process. The SERS data suggest that there are high LSPR-mediated electron transition rates between these states. The high transition rates can only be observed under the following conditions: first, there is a large density of the electronic states corresponding to these transitions. Second, there is a high degree of overlap between the corresponding wave functions of the associated states (giving large oscillator strengths). Third, there must exist a high-intensity electric field driving the excitation process (pushing electrons from the occupied to unoccupied

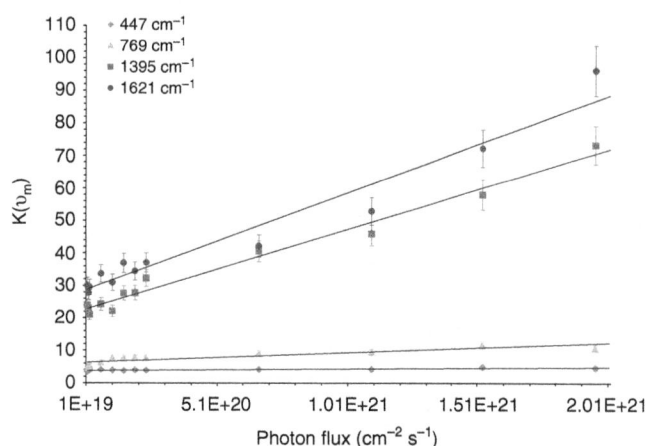

Figure 4 | Excess anti-Stokes signal as a function of incident photon flux. Excess anti-Stokes signal ratio (K) dependence on the incident photon flux for multiple prominent bands in the methylene blue spectrum gathered using the 785-nm excitation wavelength. The linear nature of the dependence is compelling evidence that optical pumping of the MB molecules is occurring. Error bars represent s.d. of three separate experimentally gathered K values for a given band and photon flux.

states) at the location of the adsorbate. Whether the first and second requirements are fulfilled depends on the local electronic structure of the Ag–MB system, the properties of which are determined by the chemisorption of MB on Ag. We hypothesize that LSPR plays a role in the third requirement, that is, LSPR is providing high-intensity fields, particularly in the small inter-particle gaps, at frequencies that can support the high rates of the electronic transitions between the states centred on the adsorbate.

This hypothesis that the role of LSPR is somewhat decoupled from the role of the local electronic structure has far reaching consequences in the potential design of selective plasmonic catalytic systems. It suggests that by engineering plasmonic nanomaterials with plasmon resonances that match the charge excitation energies of adsorbates interacting with the plasmonic nanostructure, it is possible to engineer systems that can support higher rates of electronic excitations between electronic states of a particular energy (at particular wavelengths). A natural implica-tion of this hypothesis is that plasmonic nanomaterials can be tuned in such a way that under illumination they will selectively enhance particular chemical pathways (those which are activated by the particular electronic transitions supported by plasmon-induced local fields) while suppressing other chemical pathways. One way to achieve this objective might be to combine plasmonic particles with more active metals in a single catalyst[35,36] or have bimetallic nanoparticles that combine a plasmonic metal core and a more chemically reactive metal shell. In these systems, the interaction of the chemically reactive metal with adsorbates allows for tuning of the electronic structure of adsorbates on the nanoparticle (based on the selection of the metal), while the optical properties are governed by the plasmonic metal. By matching these electronic and optical properties, it could be possible to induce preferential selective charge transfer to a particular orbital, essentially guiding photochemical transformation towards the more selective pathway.

A direct corollary of the hypothesis that nanomaterials with plasmon resonance wavelengths matching the charge excitation energies of surface adsorbates can accelerate chemical reactions of the adsorbates at these wavelengths is that MB on Ag should be more reactive under 785-nm illumination than under 532 nm. To test the corollary, we measured the rates of decomposition/desorption of MB on the surface of the Ag nanoparticles illuminated with $2.0 \times 10^{20}\,\mathrm{cm}^{-2}\,\mathrm{s}^{-1}$ photon flux of the 785- and 532-nm light sources, under inert N_2 atmosphere and at constant 100 °C reactor temperature. Data in Fig. 5 show the differences in the extent of SERS signal degradation of the most prominent Raman mode in MB (447 cm^{-1}) over a given period of laser exposure. The data show that under exposure to the 785-nm laser, MB signal degradation occurs significantly faster than under exposure to the 532-nm laser of identical intensity. The instantaneous rate (rate as time $\rightarrow 0$) of MB decomposition/desorption under 785-nm exposure was calculated to be $4.8 \times$ higher than for MB under equal intensity 532-nm exposure.

It is worth noting that the signal degradation enhancement under the 532-nm laser was not zero. The high electromagnetic fields formed on the Ag surface under 532-nm irradiation can cause instabilities in surface adsorbates and localized heating of the samples above the controlled reactor temperature can potentially occur during the prolonged laser exposure required in these studies[37]. Adsorbate instability due to high electromagnetic fields and heating cannot however explain the higher 785 nm degradation rates compared with 532 nm. The higher 785 nm degradation rates suggest that the observed charge transfer, which we have shown leads to pumping of excited vibrational modes, yields more unstable and reactive adsorbate molecules.

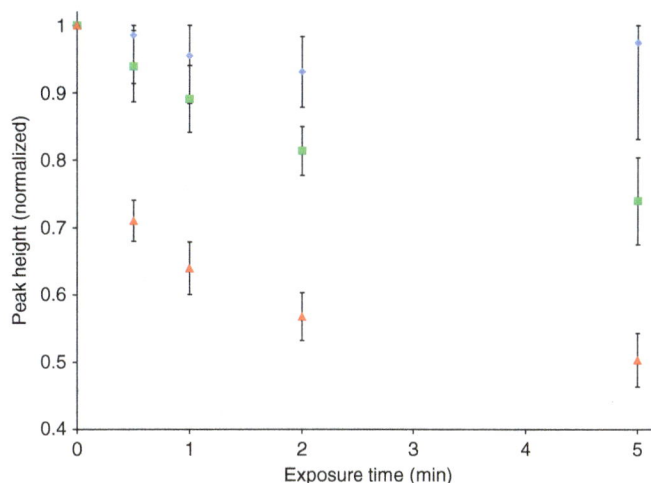

Figure 5 | Methylene blue signal degradation by prolonged laser exposure. Normalized heights of the 447 cm^{-1} Raman mode of Ag–MB gathered at 373 K and in inert N_2 atmosphere after exposure to no laser (control, blue diamonds), 532-nm laser (green squares) or 785-nm laser (red triangles) for a given periods of time. The 785-nm laser shows significant signal degradation enhancement compared with the 532 nm and control samples.

In these studies, we have coupled optical characterization methods with kinetic analysis of the rates of photocatalytic reactions to study the interaction of a probe molecule with optically excited plasmonic Ag nanoparticles. Specifically, our measurements and analysis of wavelength-dependent anti-Stokes and Stokes Raman spectra and their connection to photocatalytic decomposition/desorption of MB on Ag provide unique insight into the specifics of the mechanism of plasmon-mediated photocatalysis. We demonstrated that in the systems where the LSPR-mediated charge transfer from nanoparticles to adsorbates takes place, this charge transfer process results in elevated rates of chemical transformation. Furthermore, we showed that the mechanism of charge transfer involves the action of strong localized LSPR-induced electric field that results in an energy-specific direct electron excitation from occupied to unoccupied orbitals of the molecule–nanoparticle complex. Our findings suggest that the role of LSPR is somewhat decoupled from the role of the local electronic structure in plasmonic photocatalysis, and that by engineering plasmonic nanomaterials with resonant wavelengths matching the charge excitation energies of adsor-bates interacting with the plasmonic nanostructure, it is possible to design systems that can support very high rates of electronic excitations between electronic states of a particular energy. These observations provide a foundation for the development of plasmonic catalysts that can selectively activate targeted chemical bonds, since it is implied that plasmonic nanomaterials can be tuned in such a way that under illumination they selectively enhance particular chemical pathways.

Methods

Sample synthesis and experimental SERS studies. The plasmonic nanoparticles in these Raman studies were silver nanocubes, synthesized via the modified polyol method[38]. Synthesized cube samples were characterized by STEM imaging and found to have an average size of 71 ± 9 nm. MB, Rhodamine 6G and Acridine Orange were used as probe molecules in model systems. After synthesis, the cubes were washed 5 × by centrifugation and redispersion in a combination of acetone and ethanol. After washing, the cubes were dispersed in ethanol and small amount of NaCl was added to aid aggregation of Ag nanoparticles during the drying process. One of the dye molecules was added to the liquid solution at a concentration of 40 μM and allowed to incubate for several hours to ensure full equilibrium adsorption of the dye onto the silver surface[39]. Adding the dyes while the particles are still in liquid solution allows the molecules to access all available adsorption sites on the Ag particles. Solid samples for spectroscopic studies were prepared by drop coating the solution onto a glass or silicon chip and allowing the ethanol solvent to evaporate, leaving behind a thin layer of silver-dye aggregates. Ultraviolet–visible spectra were gathered using samples on glass slides and a Thermo Evolution 300 spectrometer. Stokes and anti-Stokes Raman spectra were gathered using a Horiba LabRAM HR system under excitation from either a 532-nm diode-pumped solid state laser or a 785-nm diode laser. The acquisition time for the spectra was 2 s to negate heating effects due to prolonged laser exposure. In solid samples, the laser was focused on nanoparticle aggregates visible through a × 50 microscope lens. In liquid samples, including toluene used for standardizing expected signal ratios, the laser focus was near the surface of the liquid.

Photocatalytic experimental studies. Photoreaction studies were conducted in a temperature-controlled Harrick Praying Mantis reaction chamber designed for Raman spectroscopy. A sample of MB on silver on a silicon chip was placed inside the chamber, with the transparent reactor window allowing Raman spectra to continue to be collected. The temperature of the sample within the reactor could then be controlled, and pure nitrogen could be flowed through the reactor to provide an inert atmosphere. Samples in MB degradation studies were heated to 373 K under N_2 atmosphere and subsequently exposed to one of two lasers (532 or 785 nm, at equal photon flux) or no light (control) for a measured period of time. Losses in Raman peak signal for prominent bands in MB were measured and interpreted as decreases in surface coverage of MB due to chemical change (decomposition or desorption).

FDTD simulations. FDTD simulations were carried out using the Meep simulation package[40]. The simulations were performed in a three-dimensional grid measuring $294 \times 294 \times 294$ nm. An additional 50 nm of perfectly matched layer was added to all sides of this grid. The space between grid points was 1 nm for all but one simulation in which the much more computationally expensive 0.5-nm mesh was used to ensure that a shrinking mesh size did not change significantly the behaviour of the field intensities within gaps between particles. The dielectric function for silver was obtained from data published by Rakic et al[41]. Nanocube dimers were arranged in a number of orientations, including face to face, face to edge, edge to edge and corner to corner; each dimer system was tested with interparticle separations of both 1 and 2 nm. In addition, Ag nanocube trimers were also simulated in edge–face–edge and edge–edge–edge orientations with 1-nm separations. The source wavelengths studied matched those of the laser in the Raman set-up, 532 and 785 nm. The energy from the source propagated perpendicular to the interparticle axis; the polarization of the source was alternately perpendicular and parallel to the interparticle axis in separate simulations. Field enhancement data were visualized on the x–y and x–z planes. Integrated field values were calculated by finding the average value of the electric field data points located on the nanoparticle surfaces only.

DFT simulations. DFT calculations were performed using real space grid-based projector augmented wavefunction DFT implemented in the GPAW software package[42]. The exchange–correlation interaction was approximated using the RPBE[43,44]. The Ag[100] and Ag[211] surfaces were modelled using 3×3 slabs and 2×3 slabs, respectively. All systems used four layer slabs, with 10 Å of vacuum space. Pyridine molecules were adsorbed vertically over a top site [100] or above a step site [211] to match the predominant facets present in the synthesized Ag nanocubes. Multiple adsorption configurations were tested for the molecules on the Ag surfaces.

References

1. Zhou, X., Liu, G., Yu, J. & Fan, W. Surface plasmon resonance-mediated photocatalysis by noble metal-based composites under visible light. *J. Mater. Chem.* **22**, 21337–21354 (2012).
2. Marimuthu, A., Zhang, J. & Linic, S. Tuning selectivity in propylene epoxidation by plasmon mediated photo-switching of Cu oxidation state. *Science* **339**, 1590–1593 (2013).
3. Linic, S., Christopher, P. & Ingram, D. B. Plasmonic-metal nanostructures for efficient conversion of solar to chemical energy. *Nat. Mater.* **10**, 911–921 (2011).
4. Mukherjee, S. *et al.* Hot electrons do the impossible: plasmon-induced dissociation of H2 on Au. *Nano Lett.* **13**, 240–247 (2013).
5. Christopher, P., Xin, H. & Linic, S. Visible-light-enhanced catalytic oxidation reactions on plasmonic silver nanostructures. *Nat. Chem.* **3**, 467–472 (2011).
6. Scholl, J. A., Koh, A. L. & Dionne, J. A. Quantum plasmon resonances of individual metallic nanoparticles. *Nature* **483**, 421–427 (2012).
7. Christopher, P., Xin, H., Marimuthu, A. & Linic, S. Singular characteristics and unique chemical bond activation mechanisms of photocatalytic reactions on plasmonic nanostructures. *Nat. Mater.* **11**, 1044–1050 (2012).
8. Mukherjee, S. *et al.* Hot-electron-induced dissociation of H2 on gold nanoparticles supported on SiO2. *J. Am. Chem. Soc.* **136**, 64–67 (2014).
9. Jeong, D. H., Jang, N. H., Suh, J. S. & Moskovits, M. Photodecomposition of Diazanaphthalenes Adsorbed on Silver Colloid Surfaces. *J. Phys. Chem. B* **104**, 3594–3600 (2000).
10. Gadzuk, J. W. Vibrational excitation in molecule–surface collisions due to temporary negative molecular ion formation. *J. Chem. Phys.* **79**, 6341 (1983).
11. Linic, S., Aslam, U., Boerigter, C. & Morabito, M. Photochemical transformations on plasmonic metal nanoparticles. *Nat. Mater.* **14**, 567–576 (2015).
12. Suh, J. S., Jang, N. H., Jeong, D. H. & Moskovits, M. Adsorbate Photochemistry on a Colloid Surface: Phthalazine on Silver. *J. Phys. Chem.* **100**, 805–813 (1996).
13. Suh, J. S., Moskovits, M. & Shakhesemampour, J. Photochemical decomposition at colloid surfaces. *J. Phys. Chem.* **97**, 1678–1683 (1993).
14. Hao, E. & Schatz, G. C. Electromagnetic fields around silver nanoparticles and dimers. *J. Chem. Phys.* **120**, 357–366 (2004).
15. Tao, A. R., Ceperley, D. P., Sinsermsuksakul, P., Neureuther, A. R. & Yang, P. Self-organized silver nanoparticles for three-dimensional plasmonic crystals. *Nano Lett.* **8**, 4033–4038 (2008).
16. Rycenga, M. *et al.* Controlling the synthesis and assembly of silver nanostructures for plasmonic applications. *Chem. Rev.* **111**, 3669–3712 (2011).
17. Sherry, L. J. *et al.* Localized surface plasmon resonance spectroscopy of single silver nanocubes. *Nano Lett.* **5**, 2034–2038 (2005).
18. Zuloaga, J., Prodan, E. & Nordlander, P. Quantum description of the plasmon resonances of a nanoparticle dimer. *Nano Lett.* **9**, 887–891 (2009).
19. Tan, S. F. *et al.* Quantum plasmon resonances controlled by molecular tunnel junctions. *Science* **343**, 1496–1499 (2014).
20. Esteban, R., Borisov, A. G., Nordlander, P. & Aizpurua, J. Bridging quantum and classical plasmonics with a quantum-corrected model. *Nat. Commun.* **3**, 825 (2012).
21. Savage, K. J. *et al.* Revealing the quantum regime in tunnelling plasmonics. *Nature* **491**, 574–577 (2012).
22. Mortensen, N. A., Raza, S., Wubs, M., Søndergaard, T. & Bozhevolnyi, S. I. A generalized non-local optical response theory for plasmonic nanostructures. *Nat. Commun.* **5**, 3809 (2014).
23. Kaczor, A., Malek, K. & Baranska, M. Pyridine on colloidal silver. Polarization of surface studied by surface-enhanced Raman scattering and density functional theory methods. *J. Phys. Chem. C* **114**, 3909–3917 (2010).
24. Xiao, G.-N. & Man, S.-Q. Surface-enhanced Raman scattering of methylene blue adsorbed on cap-shaped silver nanoparticles. *Chem. Phys. Lett.* **447**, 305–309 (2007).
25. Nicolai, S. H. a. & Rubim, J. C. Surface-enhanced resonance Raman (SERR) spectra of methylene blue adsorbed on a silver electrode. *Langmuir* **19**, 4291–4294 (2003).
26. García-Vidal, F. & Pendry, J. Collective theory for surface enhanced Raman scattering. *Phys. Rev. Lett.* **77**, 1163–1166 (1996).
27. Schatz, G. C., Young, M. A. & Van Duyne, R. P. Surface-Enhanced Raman Spectroscopy—Physics and Applications. *Topics Appl. Phys.* **103**, 19–46 (2006).
28. Kambhampati, P., Child, C. M., Foster, M. C. & Campion, A. On the chemical mechanism of surface enhanced Raman scattering: experiment and theory. *J. Chem. Phys.* **108**, 5013 (1998).
29. Maher, R. C., Galloway, C. M., Le Ru, E. C., Cohen, L. F. & Etchegoin, P. G. Vibrational pumping in surface enhanced Raman scattering (SERS). *Chem. Soc. Rev.* **37**, 965–979 (2008).
30. Kneipp, K. *et al.* Population pumping of excited vibrational states by spontaneous surface-enhanced Raman scattering. *Phys. Rev. Lett.* **76**, 2444–2447 (1996).
31. Haslett, T. L., Tay, L. & Moskovits, M. Can surface-enhanced Raman scattering serve as a channel for strong optical pumping? *J. Chem. Phys.* **113**, 1641 (2000).
32. Marimuthu, A., Christopher, P. & Linic, S. Design of plasmonic platforms for selective molecular sensing based on surface-enhanced Raman spectroscopy. *J. Phys. Chem. C* **116**, 9824–9829 (2012).
33. Itoh, T. *et al.* Second enhancement in surface-enhanced resonance Raman scattering revealed by an analysis of anti-Stokes and Stokes Raman spectra. *Phys. Rev. B* **76**, 085405 (2007).
34. Hogiu, S., Werncke, W., Pfeiffer, M. & Elsaesser, T. Mode specific vibrational kinetics after intramolecular electron transfer studied by picosecond anti-Stokes Raman spectroscopy. *Chem. Phys. Lett.* **312**, 407–414 (1999).
35. Joseph, V. *et al.* Characterizing the kinetics of nanoparticle-catalyzed reactions by surface-enhanced Raman scattering. *Angew. Chem. Int. Ed. Engl.* **51**, 7592–7596 (2012).
36. Antosiewicz, T. J., Wadell, C. & Langhammer, C. Plasmon-assisted indirect light absorption engineering in small transition metal catalyst nanoparticles. *Adv. Opt. Mater* **3**, 1591–1599 (2015).
37. Adleman, J. R., Boyd, D. A., Goodwin, D. G. & Psaltis, D. Heterogenous catalysis mediated by plasmon heating. *Nano Lett.* **9**, 4417–4423 (2009).
38. Im, S. H., Lee, Y. T., Wiley, B. & Xia, Y. Large-scale synthesis of silver nanocubes: the role of HCl in promoting cube perfection and monodispersity. *Angew. Chem. Int. Ed. Engl.* **44**, 2154–2157 (2005).
39. Ghaedi, M. *et al.* Comparison of silver and palladium nanoparticles loaded on activated carbon for efficient removal of methylene blue: kinetic and isotherm study of removal process. *Powder Technol.* **228**, 18–25 (2012).

40. Oskooi, A. F. *et al.* Meep: a flexible free-software package for electromagnetic simulations by the FDTD method. *Comput. Phys. Commun.* **181,** 687–702 (2010).

41. Rakic, A. D., Djurišic, A. B., Elazar, J. M. & Majewski, M. L. Optical properties of metallic films for vertical-cavity optoelectronic devices. *Appl. Opt.* **37,** 5271 (1998).

42. Enkovaara, J. *et al.* Electronic structure calculations with GPAW: a real-space implementation of the projector augmented-wave method. *J. Phys. Condens. Matter* **22,** 253202 (2010).

43. Hammer, B., Hansen, L. B. & Nørskov, J. K. Improved adsorption energetics within density-functional theory using revised Perdew-Burke-Ernzerhof functionals. *Phys. Rev. B* **59,** 7413–7421 (1999).

44. Wellendorff, J. *et al.* Density functionals for surface science: exchange-correlation model development with Bayesian error estimation. *Phys. Rev. B* **85,** 235149 (2012).

Acknowledgements

We gratefully acknowledge support from US Department of Energy, Office of Basic Energy Science, Division of Chemical Sciences (FG-02-05ER15686) and National Science Foundation (CBET-1437601, CHE-1362120 and CHE-1111770). S.L. acknowledges the Camille Dreyfus Teacher-Scholar Award from the Camille and Henry Dreyfus Foundation.

Author contributions

C.B. and S.L. developed the project. C.B. performed all syntheses, experimental measurements and data analysis. R.C. performed all FDTD simulations, and M.M. performed all DFT calculations. C.B. and S.L. wrote the manuscript with input from R.C. and M.M.

Additional information

15

Thermal selectivity of intermolecular versus intramolecular reactions on surfaces

Borja Cirera[1], Nelson Giménez-Agulló[2], Jonas Björk[3], Francisco Martínez-Peña[1], Alberto Martin-Jimenez[1], Jonathan Rodriguez-Fernandez[4], Ana M. Pizarro[1], Roberto Otero[1,4], José M. Gallego[5], Pablo Ballester[2,6], José R. Galan-Mascaros[2,6] & David Ecija[1]

On-surface synthesis is a promising strategy for engineering heteroatomic covalent nanoarchitectures with prospects in electronics, optoelectronics and photovoltaics. Here we report the thermal tunability of reaction pathways of a molecular precursor in order to select intramolecular versus intermolecular reactions, yielding monomeric or polymeric phthalocyanine derivatives, respectively. Deposition of tetra-aza-porphyrin species bearing ethyl termini on Au(111) held at room temperature results in a close-packed assembly. Upon annealing from room temperature to 275 °C, the molecular precursors undergo a series of covalent reactions via their ethyl termini, giving rise to phthalocyanine tapes. However, deposition of the tetra-aza-porphyrin derivatives on Au(111) held at 300 °C results in the formation and self-assembly of monomeric phthalocyanines. A systematic scanning tunnelling microscopy study of reaction intermediates, combined with density functional calculations, suggests a $[2+2]$ cycloaddition as responsible for the initial linkage between molecular precursors, whereas the monomeric reaction is rationalized as an electrocyclic ring closure.

[1] IMDEA Nanoscience, c/Faraday 9, Cantoblanco, 28049 Madrid, Spain. [2] Institute of Chemical Research of Catalonia, Barcelona Institute of Science and Technology, Avinguda Països Catalans 16, Tarragona 43007, Spain. [3] Department of Physics, Chemistry and Biology, IFM, Linköping University, 58183 Linköping, Sweden. [4] Departamento de Física de la Materia Condensada, Universidad Autónoma de Madrid, c/Francisco Tomás y Valiente 7, Cantoblanco, 28049 Madrid, Spain. [5] Instituto de Ciencia de Materiales de Madrid, c/ Sor Juana Inés de la Cruz 3, Cantoblanco, 28049 Madrid, Spain. [6] Catalan Institution for Research and Advanced Studies, Passeig Lluis Companys 23, Barcelona 08010, Spain. Correspondence and requests for materials should be addressed to P.B. (email: pballester@iciq.es) or to J.R.G.-M. (email: jrgalan@iciq.es) or to D.E. (email: david.ecija@imdea.org).

Surface-mediated synthesis of low-dimensional polymers from simple molecular precursors (monomers) under ultra-clean conditions is a rapidly emerging field with relevance for molecular electronics, optoelectronic devices, magnetism, molecular recognition and sensing, catalysis, filtration and membranes[1–5]. This promising bottom-up strategy relies on a careful selection of building blocks equipped with functional groups and surfaces, whereby one-dimensional and two-dimensional nanostructures have been engineered and a plethora of new materials is envisioned[6–21].

Regarding the molecular precursors, porphyrin and phthalocyanine derivatives are the subject of increasing attention, inspired by scientific curiosity, biological relevance and potential technological impact[22].

Several chemical reactions have been exploited to produce surface-assisted lateral covalent designs inspected with ultimate spatial resolution[23]. These include cyclocondensation of boronic acids[8], tetramerization of 1,2,4,5-tetracyanobenzene with iron[16], Schiff-base reaction of aldehydes and amines to form imine[24], acylation reactions (yielding polyamides[25], polyimides[7] or polyesters[26]), Bergmann cyclization[27], azide–alkyne cyclo-additions[28,29] and carbon–carbon couplings exemplified by alkane polymerizations[30], the Ullmann[6,17], the Glaser[18] and the Sonogashira reactions[21] or dehalogenation reactions followed by radical combination[13,14,31]. Despite their importance in bulk chemistry and their growing relevance in 3D polymer science, [2 + 2] cycloaddition reactions have not yet been exploited to produce surface-confined polymeric nanostructures on surfaces.

Importantly, the selectivity of reaction pathways under distinct thermal stimuli, although being a promising strategy to increase synthetic versatility, has remained mostly elusive on surfaces[17]. Here, we introduce surface-confined thermally tunable reaction pathways as a route to select intramolecular versus intermolecular covalent reactions yielding either monomeric phthalocyanines or low-dimensional phthalocyanine polymers, respectively. To this end, we deposit 2,3,7,8,12,13,17,18-octaethyl-5,10,15,20-tetraaza-porphyrin (OETAP) under ultra-high vacuum on a pristine Au(111) crystal held at room temperature, forming close-packed supramolecular assemblies. This precursor phase is gently annealed to 300 °C giving rise to quasi-one-dimensional polymers that can be rationalized as phthalocyanine derivatives. Further insights of the polymerization process with high-resolution low-temperature scanning tunnelling microscopy (STM) and state-of-the-art density functional (DFT) calculations reveal that the linkage between two OETAPs is likely to be initiated by a [2 + 2] cycloaddition, involving the chemical transformations of the peripheral ethyl substituents of two adjacent molecular species. Importantly, by raising the temperature of the substrate prior deposition of the molecular precursors, the intermolecular covalent coupling can be precluded. For substrate temperatures of 300 °C or higher, it is observed that the mechanisms for polymeric growth are blocked and the OETAP species are transformed into individual phthalocyanines, via an electrocyclic ring closure (ERC) reaction, and then self-assembled into islands. DFT simulations support our experimental findings indicating a lower energetic barrier for dimerization as compared with the intramolecular reaction of the monomer. We envision that our results will pave the way for the development of low-dimensional materials exploiting the propensity of molecular precursors equipped with convenient peripheries to undergo tunable covalent reactions under thermal or light stimuli.

Results

Self-assembly of OETAP precursors on Au(111).
Figure 1a shows a high-resolution STM image of an OETAP array on

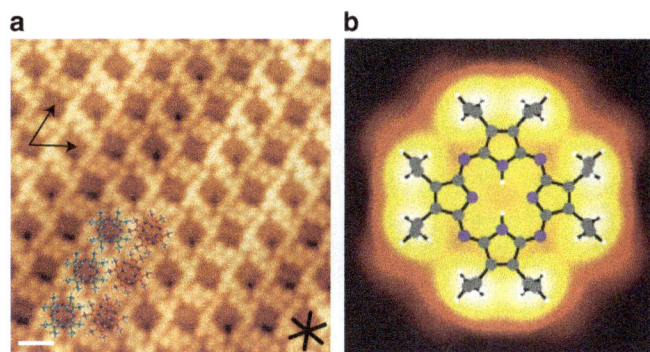

Figure 1 | Self-assembly of OETAP on Au(111) after deposition at room temperature. (a) High-resolution STM images revealing submolecular features. Black arrows depict the lattice vectors of the rhombic unit cell. Black star shows the close-packed directions of the surface. Superimposed coloured atomistic models address the two distinctly oriented molecular species (green and red, respectively). Tunnelling parameters: $V_b = 0.5$ V, $I = 0.1$ nA. Scale bar, 1 nm. **(b)** STM simulated image of an individual OETAP on Au(111) at 0.5 V, with a superimposed ball-and-stick model of the molecular species. Hydrogen, carbon and nitrogen atoms are depicted in white, grey and violet, respectively.

Au(111) self-assembled after room temperature deposition. Each molecule is visualized as a dim centre, attributed to the macrocyclic core, surrounded by eight brighter protrusions, which are assigned to the ethyl moieties, in excellent agreement with simulated STM images (cf. Fig. 1b and Supplementary Fig. 1) and consistent with surface-confined assemblies of identically substituted porphyrin compounds[32]. The dense-packed architecture displays two distinctly oriented molecular species, related by a ~16° rotation, which features a rhombic lattice ($a = 14.1 \pm 1$ Å, $b = 14.1 \pm 1$ Å, $\theta = 60°$), and is stabilized by lateral non-covalent interactions between the ethyl substituents.

Polymerization of OETAP species. Next we explore the polymerization of OETAP precursors on Au(111). After submonolayer deposition at room temperature, gentle annealing of the substrate to 75, 100 and 150 °C results in no appreciable changes. At 225 °C, the polymerization of OETAP species is initiated. High-resolution images of the assemblies after holding the substrate at 275 °C reveal the formation of molecular chains, most of them confined on the fcc regions of Au(111) (cf. Fig. 2a). A careful inspection of these chains allows us to discern submolecular features and, thus, clarify the molecular organization. The tetrapyrrolic macrocyclic cores are now visualized as dim crosses, whereas the ethyl substituents have reacted in two distinct ways: (i) by covalently linking adjacent species through the intermediacy of multiple covalent reactions (cf. below) and (ii) by undergoing an intramolecular ERC reaction. The final result is the formation of polymeric chains, in which the repeating unit (monomer) derives from OETAP. A statistical analysis shows that the average polymeric size is ~7 monomers, that the most frequent polymer comprises 6 macrocycles and that extended polymers (>20 monomers) can be formed. Importantly, the majority of the monomeric entities are joined together via two covalent structural motifs, denoted L and V (cf. Fig. 2c,d). Both structural motifs are the product of multiple covalent reactions between peripheral ethyl groups from adjacent OETAP precursors, where two molecules face each other. When the two ethyl groups that initiate the intermolecular reaction lie in opposite sides of the OETAP precursors the product is an L-type motif (cf. Fig. 2c and Supplementary Figs 1 and 2 for agreement with

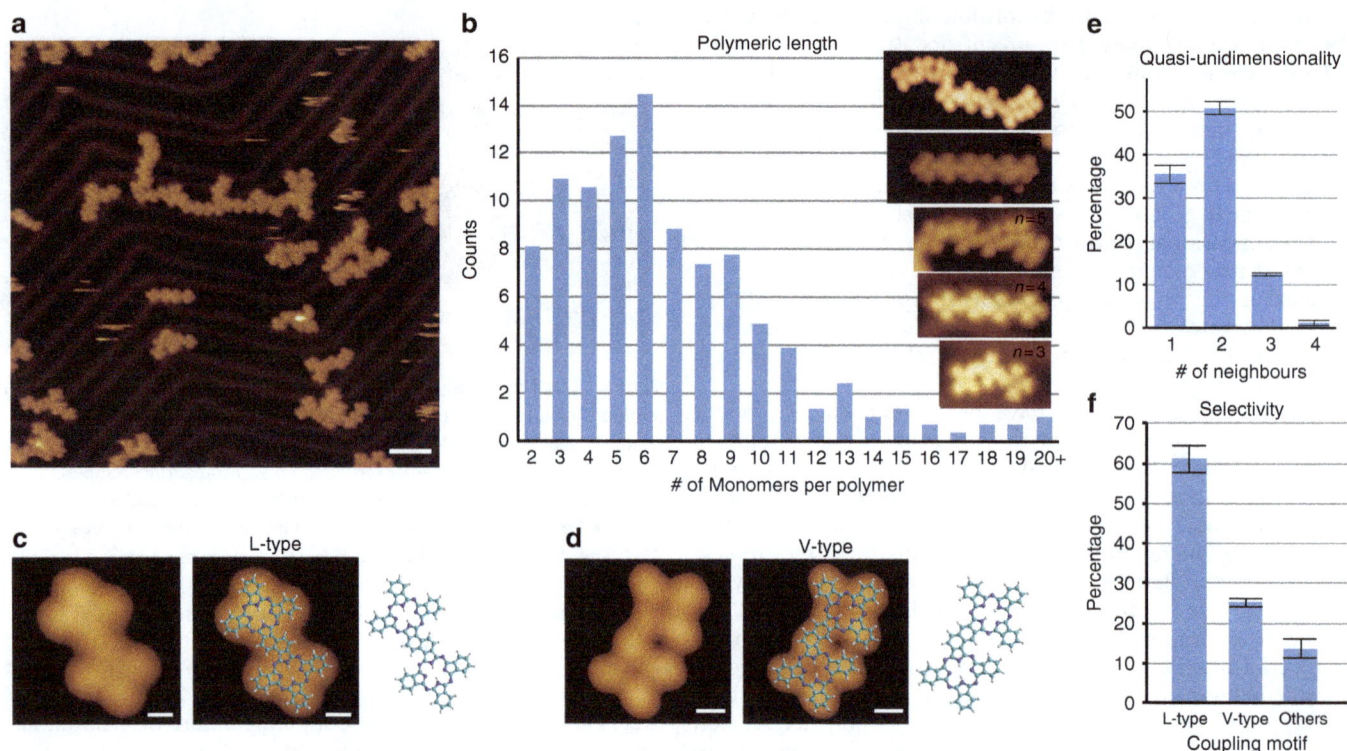

a

b
Polymeric length

$n=5$

$n=4$

$n=3$

e
Quasi-unidimensionality

c
L-type

d
V-type

f
Selectivity

Figure 2 | Surface-confined synthesis of phthalocyanine tapes. The deposition of 0.3 ML of OETAP species (precursors) on Au(111) held at room temperature and subsequent gently annealing at 275 °C affords the formation of phthalocyanine polymers. (**a**) Long-range STM image displaying the formation of quasi-unidimensional polymers mainly confined to the fcc regions. (**b**) High-resolution STM images of polymers of distinct size (3, 4, 5, 6 and 8) and histogram of the polymeric length. (**c,d**) High-resolution STM image and models of the majority of products between coupled monomers (**c**, L-type and **d**, V-type). (**e**) Histogram of the number of neighbours per monomer, highlighting the quasi-unidimensional feature of the assemblies. (**f**) Histogram of different types of coupling motifs exhibiting selectivity towards L-type-binding motif. Errors bars in **e,f** are the standard deviation of the mean. Tunnelling parameters: **a**, $V_b = -1.5$ V; **b**, $V_b = -0.5$ V for $n=3$, $V_b = -1.5$ V for $n=4$, $V_b = -1$ V for $n=5$, $V_b = -1$ V for $n=6$, $V_b = -1.2$ V for $n=8$; **c**, $V_b = -1.5$ V and **d**, $V_b = -0.7$ V. Scale bars, **a**, 5 nm; **c,d**, 0.5 nm.

DFT simulated image), whereas the V-type motif is the product when the ethyl groups lie in the same side of opposite molecules. In both cases, the resulting polymer can be interpreted as a phthalocyanine tape with locally straight (L) or slightly curved (V) appearance. Notably, the thermally induced polymerization of OETAP precursors into phthalocyanine tapes seems to present a preference for the L-type coupling motif versus the V-type motif (ca 2:1), as seen in Fig. 2f. In addition, polymers are mostly quasi-unidimensional, with over 85% of the monomers connected to only one or two neighbours. The remaining 15% of monomers are connected to three or four neighbours mainly through cross motifs based on L or V coupling schemes.

Regarding the proposed covalent binding motifs, we are certain that the observed molecular chains are polymers and not supramolecular assemblies due to the following reasons: (i) the centre-to-centre distance between macrocycles is too short for a non-covalent interaction and our DFT study of the reaction pathway (see below for details) is fully consistent with the experimental observations; and (ii) using perturbative scanning conditions, the polymeric chains can be displaced as entire units for long distances (over 10 nm) preserving their size and shape, thus revealing a strong covalent interaction between the monomers (cf. Supplementary Fig. 3 and Supplementary Movie 1).

Transformation of OETAP precursors into phthalocyanines. To test the feasibility of thermal control of the covalent reaction pathways, OETAP precursors were deposited on pristine Au(111) held at 300 °C. As depicted in Fig. 3, a new scenario is manifested. Instead of polymeric chains, we observe the formation of

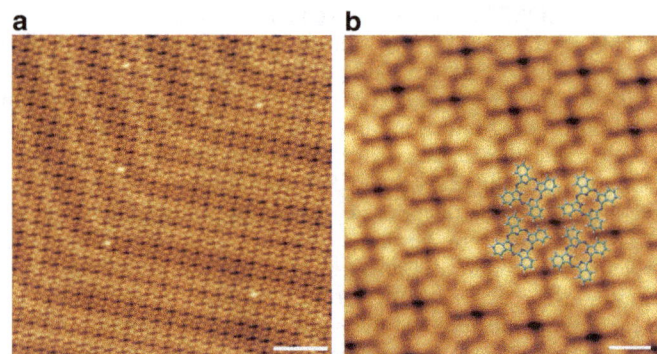

a

b

Figure 3 | Surface-assisted synthesis and self-assembly of phthalocyanines. The deposition of OETAP species on Au(111) held at 300 °C affords the on-surface chemical transformation of precursors into phthalocyanine species. (**a**) Long-range STM image, revealing the close-packed supramolecular assembly and highlighting the presence of the herringbone reconstruction of Au(111). (**b**) High-resolution STM image and superimposed modelling of the molecular constituents, which matches the reported assembly of 2H-Pc on Au(111). **a,b**, $V_b = -1$ V. Scale bars: **a**, 5 nm; **b**, 1 nm.

individual entities, which self-assemble into close-packed arrays based on a square unit cell. High-resolution STM imaging allows us to discern submolecular features of the assemblies, whereby each molecular species presents four bright protrusions and a central void forming a cross-like shape. Both the molecular appearance and the self-assembled architecture are identical to those found after deposition of 2H-Pc (free base phthalocyanine) on Au(111)[33]. Thus, based on our results and a comparison with the literature, we suggest that the deposition of OETAP precursors on Au(111) held at 300 °C gives rise to the formation of 2H-Pc species, thanks to the dehydrogenation and ring closure of their ethyl peripheries, a phenomenon observed for porphyrin derivatives[32].

Reaction intermediates during polymerization. To address the reaction mechanisms that could plausibly explain the formation of the covalent phthalocyanine polymers and phthalocyanine monomers, a detailed analysis of the STM images collected during the annealing process was carried out.

Figure 4 illustrates reaction intermediates observed after depositing OETAP species holding the substrate at room temperature, subsequently followed by gentle annealing to 225 °C. Long-range STM images show the formation of small islands mainly distributed over the fcc regions of the Au(111) surface (cf. Fig. 4a,e), thus revealing the dissolution of the former OETAPs islands and subsequent self-assembly into smaller patches. High-resolution topographs (cf. Fig. 4b–d) allow us to discern submolecular features and elucidate the molecular packing. For clarity, Fig. 4f–h display a rationalization of the distinct molecular species by superimposing a coloured

model. Here, dots and rods of the same colour represent intact ethyl moieties and ERC-reacted ethyl termini of the same molecule, respectively. In addition, bi-coloured rods indicate links between tetrapyrrole molecular units. A minority of unreacted OETAP species, characterized by the eight-dotted appearance described above, are still observed either isolated on the surface or forming part of the supramolecular islands. However, most of the molecular precursors have experienced covalent reactions, giving rise to the reaction intermediates of the polymeric chains. The majority of OETAPs have undergone one or two ring closure reactions (intramolecular electro-cyclizations) that transformed their diethyl-pyrrole units into isoindole components, which are imaged as rods. At this stage of the reaction, some of the OETAPs are also covalently linked to one another, affording distinct covalent bonding motifs, assignable to intermediates of the two main final structures (L and V). **1-L** and **2-L** represent reaction intermediates of the L-type coupling motif (cf. Fig. 4i,j), whereas **1-V** and **2-V** are the corresponding analogous reaction intermediates of the V-type-binding motif. We observe that the intermolecular bond in intermediate **1-L** comprises the reaction between two ethyl moieties in opposite sides of the OETAP precursors (cf. Fig. 4i and Supplementary Fig. 4), keeping the other two ethyl peripheries of the pyrroles intact, which are visualized as bright lobes with the same height as in individual unaltered OETAPs.

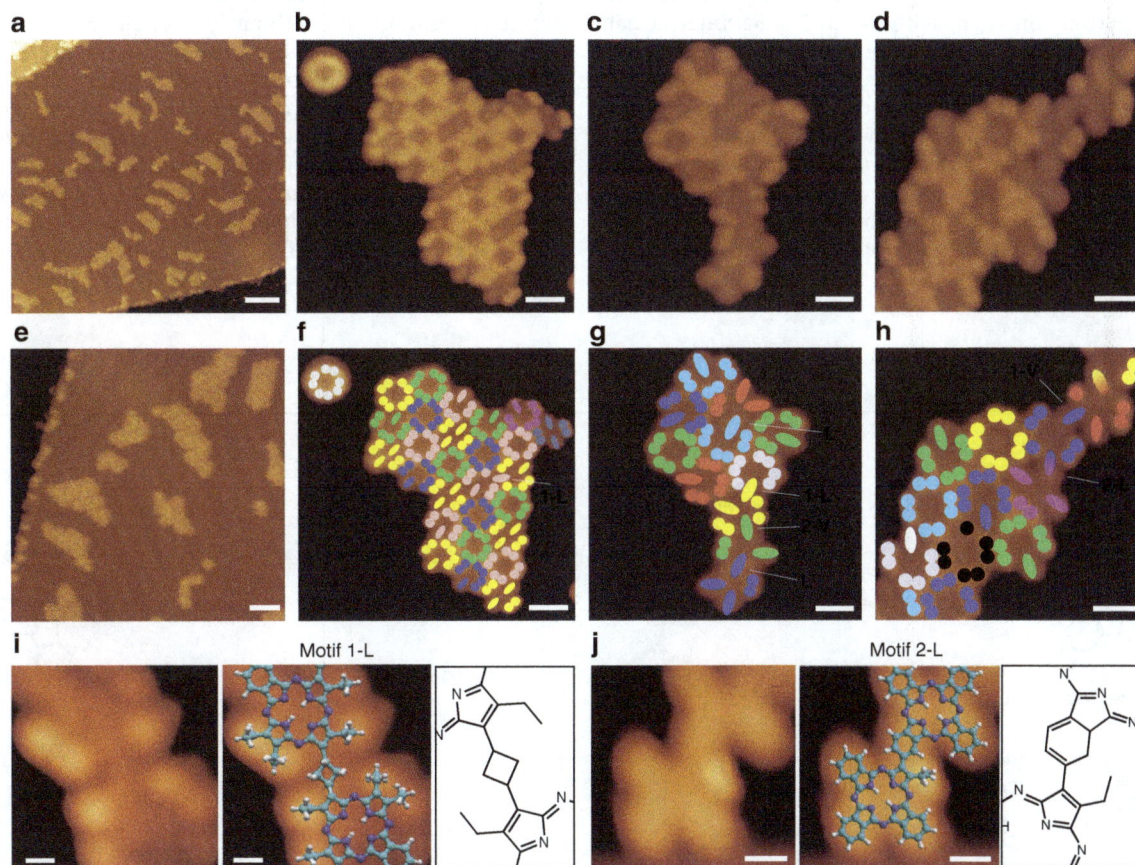

Figure 4 | Surface-assisted formation of phthalocyanine derivatives after deposition of OETAP precursors on Au(111). (**a,e**) Large-scale STM images showing the formation of patches of initially reacted molecules coexisting with unreacted species. (**b–d**) High-resolution STM images and (**f–h**) corresponding schematic colouring of the initial steps of reactions giving motifs **1-L**, **1-V**, **2-L**, **2-V** and **L**. (**i,j**) Zoom-in of motif **1-L** and **2-L**, and superimposed modelling, together with chemical drawing insets. Coloured filled dots represent ethyl moieties, bi-coloured ellipses show covalent connections between adjacent species, and mono-coloured ellipses represent full electrocyclic ring closures. Tunnelling parameters: **a**, $V_b = -0.8$ V; **b**, $V_b = -0.7$ V; **c**, $V_b = -0.7$ V; **d**, $V_b = -0.8$ V; **e**, $V_b = -1.2$ V; **i**, $V_b = 0.5$ V and **j**, $V_b = -1.5$ V. Scale bars, **a**, 10 nm; **b**, 1.5 nm; **c**, 1.0 nm; **d**, 1.0 nm; **e**, 5 nm; **f**, 1.5 nm; **g**, 1.0 nm; **h**, 1.0 nm; **i**, 0.3 nm and **j**, 0.5 nm.

In **2-L**, the reaction appears to have evolved, involving another ethyl group (cf. Fig. 4j), thus keeping only one of the four initial ethyl groups unreacted. Finally, in **L**, the four ethyl groups have reacted to produce the seed of a polymeric chain.

Discussion

Taking into account the initial precursors, the identified intermediates and the final products, we tentatively propose multi-step reaction mechanisms for both pathways, the polymerization of OETAPs into phthalocyanine quasi-unidimensional tapes (cf. Fig. 5a and Supplementary Figs 5 and 6), and the transformation of OETAP species into unsubstituted monomeric phthalocyanines (cf. Fig. 5b).

At the first stage of the polymeric reaction, two OETAPs face each other, positioning four ethyl moieties opposite each other (shown in green in Fig. 5a), which will undergo a series of complex reactions activated by the ramp annealing of the substrate. We propose that the ethyl substituents undergo an initial dehydrogenation reaction to produce an ethenyl residue per molecule, either in s-*cis* or s-*trans* conformation. Two opposite ethenyl groups adequately oriented are susceptible to react via a [2 + 2] cycloaddition affording intermediate **1-L** (or **-V**), which is repeatedly observed by STM imaging. Further work is necessary to corroborate this mechanism. The L-to-V ratio in the final coupling motifs (ca. 2:1) corresponds to a preferential ethyl-to-ethenyl dehydrogenation favouring the conformation where the ethenyl group lies s-*trans* to the bond opposite the pyrrolic nitrogen, as shown in Fig. 5a. Intermediate **1-L/V** can be considered as a dimer of two OETAP macrocycles joined together through a four-membered ring, fitting the appropriate angles and bond distances. This type of reaction, thermally forbidden by the Woodward–Hoffmann rules, is unprecedented on metallic surfaces and has only been explored on semiconductor surfaces to covalently functionalize the substrate support, whereby dangling bonds play a major role[34]. Encouragingly, DFT simulations support the tentative [2 + 2] cycloaddition (vide infra).

To explain intermediate **2-L/V**, also observed by STM imaging, a series of chemical reactions, including the loss of two methylene groups (as ethene), is tentatively suggested (cf. Supplementary Figs 5 and 6, and Supplementary Note 1). An additional dehydrogenation reaction of the remaining ethyl substituent and subsequent 6-π electrocyclization reaction followed by an aromatization process results in the formation of STM-characterized dimer **3**, observed in both L and V conformations, with a naphthalene spacer bridging the two tetrapyrrolic macrocycles.

During the polymerization process, we observe that the ethyl groups not involved in the construction of the covalent bridging spacer undergo a series of dehydrogenation and electrocyclic ring closures, giving rise to isoindole units, and thus providing the final phthalocyanine aspect to the polymers. In short, the dehydrogenation of the β-ethyl groups followed by the intramolecular electrocyclization of the resulting ethenyl resides transformed the 3,4-diethylpyrrole units into isoindole analogues (cf. Fig. 5b). We provide theoretical results (see below for details) supporting our suggestion that an analogous mechanism takes place in the transformation of OETAPs into 2H-Pcs, during the deposition process of the tetraazaporphyrins on Au(111) held at 300 °C, as previously observed for β-ethyl-substituted porphyrins[32].

To gain additional insight into reaction mechanisms, we performed DFT calculations for a model system on the Au(111) surface. We studied and compared the monomer electro-cyclization and the initial dimerization process. The complete dimerization process involves a complex multi-step sequence of events and a more in-depth analysis of all reaction steps will be accounted for elsewhere. Both the monomer electro-cyclization and the dimerization reaction of monomers require four preliminary dehydrogenation reactions that converted two ethyl substituents into ethenyl residues, with a largest energy barrier of 1.60 eV (cf. Supplementary Fig. 7 for details).

From this starting point, the monomer electro-cyclization is followed by the actual ring-closing reaction, with an effective potential energy barrier of 1.82 eV as seen in Fig. 6a–c (the ring closing reaction is associated with several barriers, which are

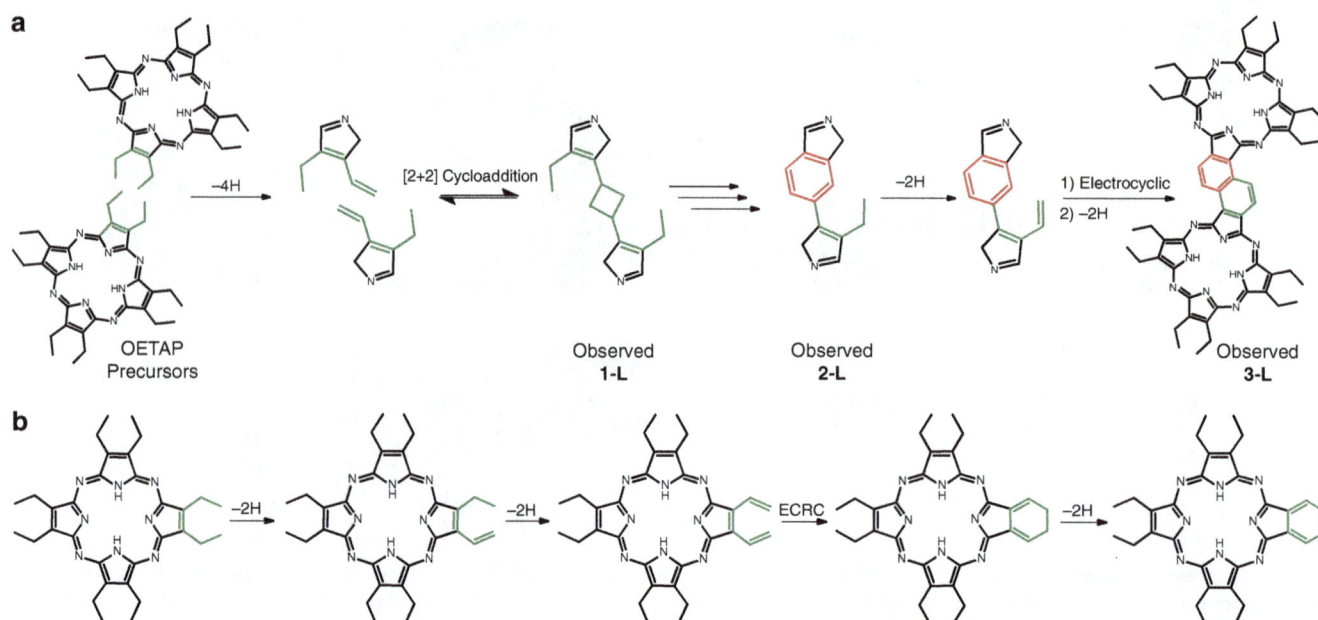

Figure 5 | Proposed reaction mechanisms describing the surface-assisted formation of phthalocyanine polymers and monomers on Au(111). (**a**) Reaction steps involving STM-imaged intermediates **1-L** and **2-L** in the phthalocyanine dimer formation (**3-L** reaction product) via L-type linking motif. For a full step-wise reaction mechanism and for the analogous reaction of the V-type motif, see Supplementary Figs 5 and 6. (**b**) Diethyl-pyrrole to isoindole moiety conversion.

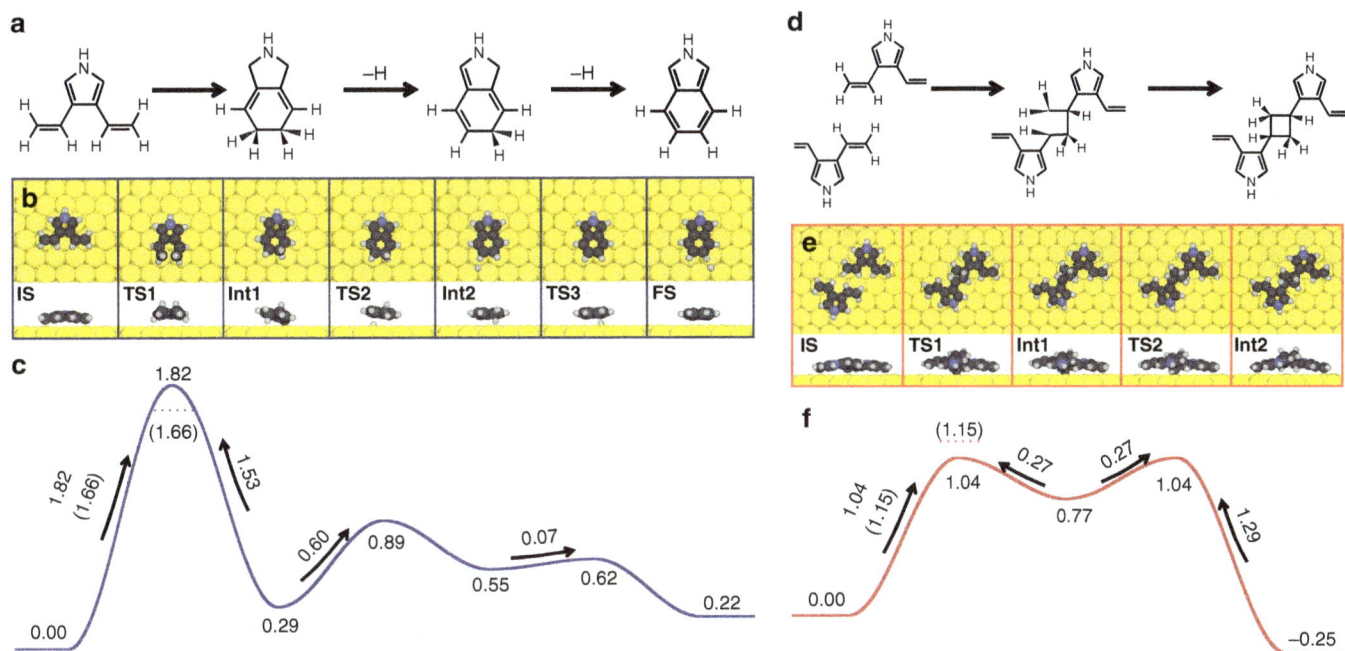

Figure 6 | Reaction pathway considerations for a model pyrrole molecule on Au(111). (**a–c**) Monomer cyclization. (**d–f**) Initial steps of the dimerization process. For both reactions, it is assumed that the ethyl legs have been transformed into ethenyl groups through dehydrogenation reactions, the details of which are shown in the Supplementary Information. (**a,d**) Chemical stick models and (**b,e**) top and side view ball models for reactants (**IS**), transition states (**TS**), reaction intermediates (**Int**) and the monomer reaction final state (**FS**). (**c,f**) Energy profiles for the monomer cyclization and initial steps of the dimerization process, respectively. Free energy barriers are shown in parentheses and indicated by the dotted lines for **TS1** of each reaction, calculated by including vibrational enthalpy and entropy at 275 °C. Energies are given in units of eV.

shown in Supplementary Fig. 8). Two dehydrogenation steps with relatively small energy barriers are required for the final aromatization process providing the isoindole unit. These results suggest the possibility to tune the reaction temperature such that the ethyl-to-ethenyl transformation is triggered, without activating the ring closuring reaction owing to its significantly higher energy barrier.

Next, we investigate the dimerization reaction, starting from two monomers that have experienced a complete dehydrogenation of the ethyl residues. The reaction is initiated by a [2 + 2] cycloaddition with an effective potential energy barrier of 1.04 eV (cf. Fig. 6d–f, **IS** to **Int2**). Notably, the [2 + 2] cycloaddition proceeds in a two-step mechanism, via an intermediate state (**Int1**) in which a carbon atom is chemically bonded to the surface. Importantly, the cycloaddition is just slightly exothermic with a reaction energy of −0.25 eV, providing a Boltzmann factor of ∼200 between **Int2** and **IS** at 275 °C, thereby substantiating the reversibility of the coupling reaction. This supports the suggested pathway in Fig. 5a, and further detailed in Supplementary Fig. 5, in which **IS** and **Int2** are in thermal equilibrium before the reaction proceeds in a Diels–Alder step. Future studies will unravel the information of the final steps of the dimerization.

In addition to potential energy barriers, we also considered the effect of zero-point and thermal contributions on the decisive steps of the monomer cyclization and dimerization by including vibrational enthalpy and entropy. Free energy barriers were evaluated at a temperature of 275 °C and are indicated for the initial monomer and dimerization steps in Fig. 4c,f, respectively. The monomer ring-closing reaction is slightly lowered by 0.16 eV, whereas the free energy barrier for the dimerization is 0.11 eV larger than the corresponding potential energy barrier. Supplementary Table 1 shows the individual contributions from vibrational enthalpy and entropy on the two barriers.

Although zero-point and thermal vibrational contributions have a small quantitative effect on barriers, both the potential energy and the free energy landscape unambiguously demonstrate that the decisive step for the dimerization has a significantly lower barrier than the monomer cyclization, with potential energy and free energy barrier differences of 0.72 and 0.51 eV, respectively. As a result, the monomer ring-closure will occur only if the dimerization is kinetically hindered, such as for extremely low coverage or in the absence of nearby non-reacted molecules. Hereby, the balance between diffusion and reaction barriers plays a crucial role, as the molecules need to meet to undergo the polymerization. These results explain why the monomer ring closure is observed for the deposition at the already heated substrate, in which situation the molecules experience the electrocyclic ring closure before meeting other species, while the intermolecular connections are formed when depositing the molecules at a room temperature substrate followed by thermal annealing allowing the precursors sufficient time to diffuse.

In summary, we have successfully introduced thermally tunable covalent reactions on surfaces to engineer quasi-unidimensional phthalocyanine tapes or self-assembled phthalocyanines by depositing molecular precursors on a well-defined Au(111) surface and regulating the substrate temperature. Our results open avenues to thermally control reaction pathways on surfaces, selecting intermolecular versus intramolecular reactions, and thus allowing to induce the growth of unprecedented polymeric heteroatomic nanoarchitectures or to produce monomeric reactions. DFT simulations corroborate the experimentally observed behaviour, demonstrating that the decisive step determining the reaction product has a lower free energy barrier for the dimerization compared with monomer intramolecular cyclization reaction. Our study discloses strategies to grow uni- and two-dimensional polymeric nano-architectures embedding heteroatomic monomers. Such systems bear prospects for molecular electronics, optoelectronics and photovoltaics.

Methods

Experiments. The experiments were performed in a custom-designed ultra-high vacuum system that hosts a low-temperature Omicron scanning tunnelling microscope, where the base pressure was below 5×10^{-10} mbar. All STM images were taken in constant-current mode with electrochemically etched tungsten tips, applying a bias (V_b) to the sample and at a temperature of ~ 80 K. The Au(111) substrate was prepared by standard cycles of Ar + sputtering (800 eV) and subsequent annealing to 723 K for 10 min. OETAP molecules were synthesized according to the procedure described by Fitzgerad and co-workers[35,36] and deposited by organic molecular-beam epitaxy from a quartz crucible held at 450 K onto a clean Au(111) at room temperature, if not stated otherwise. If necessary, in a subsequent step, the samples were annealed with a thermal gradient of $1\,°\mathrm{C}\,\mathrm{s}^{-1}$ and kept at the desired temperature for 30 min. Next, they were cooled down to room temperature with a thermal gradient of $-1\,°\mathrm{C}\,\mathrm{s}^{-1}$ and finally transferred to the STM stage held at 77 K.

Theory. Periodic DFT calculations were performed with the VASP code[37], using the projector-augmented wave method[38]. Exchange-correlation effects were described by the version of the van der Waals density functional[39] (vdWDF) introduced by Hamada[40] denoted as rev-vdWDF2. Transition states were calculated using the climbing image nudged elastic band[41] and the Dimer[42] methods. STM simulations were carried out with the Tersoff–Hamann approximation[43] using the implementation by Lorente and Persson[44]. Detailed information about DFT calculations are provided in Supplementary Methods.

References

1. Gourdon, A. On-surface covalent coupling in ultrahigh vacuum. *Angew. Chem. Int. Ed.* **47**, 6950–6953 (2008).
2. Franc, G. & Gourdon, A. Covalent networks through on-surface chemistry in ultra-high vacuum: state-of-the-art and recent developments. *Phys. Chem. Chem. Phys.* **13**, 14283–14292 (2011).
3. Zhang, X., Zeng, Q. & Wang, C. On-surface single molecule synthesis chemistry: a promising bottom-up approach towards functional surfaces. *Nanoscale* **5**, 8269–8287 (2013).
4. Liu, X.-H., Guan, C.-Z., Wang, D. & Wan, L.-J. Graphene-like single-layered covalent organic frameworks: synthesis strategies and application prospects. *Adv. Mater.* **26**, 6912–6920 (2014).
5. Zhuang, X. *et al.* Two-dimensional soft nanomaterials: a fascinating world of materials. *Adv. Mater.* **27**, 403–427 (2015).
6. Grill, L. *et al.* Nano-architectures by covalent assembly of molecular building blocks. *Nat. Nanotechnol.* **2**, 687–691 (2007).
7. Treier, M., Richardson, N. V. & Fasel, R. Fabrication of surface-supported low-dimensional polyimide networks. *J. Am. Chem. Soc.* **130**, 14054–14055 (2008).
8. Zwaneveld, N. A. A. *et al.* Organized formation of 2D extended covalent organic frameworks at surfaces. *J. Am. Chem. Soc.* **130**, 6678–6679 (2008).
9. Matena, M. *et al.* Transforming surface coordination polymers into covalent surface polymers: Linked polycondensed aromatics through oligomerization of N-heterocyclic carbene intermediates. *Angew. Chem. Int. Ed.* **47**, 2414 (2008).
10. Lafferentz, L. *et al.* Conductance of a single conjugated polymer as a continuous function of its length. *Science* **323**, 1193–1197 (2009).
11. Jensen, S., Früchtl, H. & Baddeley, C. J. Coupling of triamines with diisocyanates on Au(111) leads to the formation of polyurea networks. *J. Am. Chem. Soc.* **131**, 16706–16713 (2009).
12. Bombis, C. *et al.* Single molecular wires connecting metallic and insulating surface areas. *Angew. Chem. Int. Ed.* **48**, 9966–9970 (2009).
13. Cai, J. *et al.* Atomically precise bottom-up fabrication of graphene nanoribbons. *Nature* **466**, 470–473 (2010).
14. Bieri, M. *et al.* Two-dimensional polymer formation on surfaces: insight into the roles of precursor mobility and reactivity. *J. Am. Chem. Soc.* **132**, 16669–16676 (2010).
15. Lipton-Duffin, J. A. *et al.* Step-by-step growth of epitaxially aligned polythiophene by surface-confined reaction. *Proc. Natl Acad. Sci. USA* **107**, 11200–11204 (2010).
16. Abel, M. *et al.* Single layer of polymeric Fe-Phthalocyanine: An organometallic sheet on metal and thin insulating film. *J. Am. Chem. Soc.* **133**, 1203–1205 (2011).
17. Lafferentz, L. *et al.* Controlling on-surface polymerization by hierarchical and substrate-directed growth. *Nat. Chem.* **4**, 215–220 (2012).
18. Zhang, Y.-Q. *et al.* Homo-coupling of terminal alkynes on a noble metal surface. *Nat. Commun.* **3**, 1286 (2012).
19. Wiengarten, A. *et al.* Surface-assisted dehydrogenative homocoupling of porphine molecules. *J. Am. Chem. Soc.* **136**, 9346–9354 (2014).
20. Gao, H.-Y. *et al.* Decarboxylative polymerization of 2,6-naphthalenedicarboxylic acid at surfaces. *J. Am. Chem. Soc.* **136**, 9658–9663 (2014).
21. Sanchez-Sanchez, C. *et al.* Sonogashira cross-coupling and homocoupling on a silver surface: chlorobenzene and phenylacetylene on Ag(100). *J. Am. Chem. Soc.* **137**, 940–947 (2015).
22. Auwärter, W., Ecija, D., Klappenberger, F. & Barth, J. V. Porphyrins at interfaces. *Nat. Chem.* **7**, 105–120 (2015).
23. Lackinger, M. & Heckl, W. M. A STM perspective on covalent intermolecular coupling reactions on surfaces. *J. Phys. D Appl. Phys.* **44**, 464011 (2011).
24. Weigelt, S. *et al.* Covalent interlinking of an aldehyde and an amine on a Au(111) surface in ultrahigh vacuum. *Angew. Chem. Int. Ed.* **46**, 9227–9230 (2007).
25. Schmitz, C. H., Ikonomov, J. & Sokolowski, M. Two-dimensional ordering of poly(p-phenylene-terephthalamide) on the Ag(111) surface investigated by scanning tunneling microscopy. *J. Phys. Chem. C* **113**, 11984–11987 (2009).
26. Marele, A. C. *et al.* Formation of a surface covalent organic framework based on polyester condensation. *Chem. Commun.* **48**, 6779–6781 (2012).
27. Sun, Q. *et al.* On-surface formation of one-dimensional polyphenylene through Bergman cyclization. *J. Am. Chem. Soc.* **135**, 8448–8451 (2013).
28. Bebensee, F. *et al.* On-surface azide–alkyne cycloaddition on Cu(111): does It "Click" in ultrahigh vacuum? *J. Am. Chem. Soc.* **135**, 2136–2139 (2013).
29. Díaz Arado, O. *et al.* On-surface azide–alkyne cycloaddition on Au(111). *ACS Nano* **7**, 8509–8515 (2013).
30. Zhong, D. *et al.* Linear alkane polymerization on a gold surface. *Science* **334**, 213–216 (2011).
31. Liu, J. *et al.* Toward Cove-edged low band gap graphene nanoribbons. *J. Am. Chem. Soc.* **137**, 6097–6103 (2015).
32. Heinrich, B. W. *et al.* Change of the magnetic coupling of a metal–organic complex with the substrate by a stepwise ligand reaction. *Nano Lett.* **13**, 4840–4843 (2013).
33. Komeda, T., Isshiki, H. & Liu, J. Metal-free phthalocyanine (H₂Pc) molecule adsorbed on the Au(111) surface: formation of a wide domain along a single lattice direction. *Sci. Technol. Adv. Mater.* **11**, 054602 (2010).
34. Wong, K. T. & Bent, S. F. in *Functionalization of Semiconductor Surfaces* 51–88 (John Wiley & Sons, Inc. Hoboken, New Jersey, 2012).
35. Fitzgerald, J., Taylor, W. & Owen, H. Facile synthesis of substituted fumaronitriles and maleonitriles: precursors to soluble tetraazaporphyrins. *Synthesis* **9**, 686–688 (1991).
36. Giménez-Agulló, N. *et al.* Single-molecule-magnet behavior in the family of [Ln(OETAP)2] double-decker complexes (Ln = Lanthanide, OETAP = Octa(ethyl)tetraazaporphyrin). *Chem. Eur. J* **20**, 12817–12825 (2014).
37. Kresse, G. & Furthmüller, J. Efficient iterative schemes for ab initio total-energy calculations using a plane-wave basis set. *J. Phys. Rev. B* **54**, 11169–11186 (1996).
38. Blöchl, P. E. Projector augmented wave method. *Phys. Rev. B* **59**, 17953–17979 (1994).
39. Dion, M. *et al.* Van der Waals density functional for general geometries. *Phys. Rev. Lett.* **92**, 246401 (2004).
40. Hamada, I. van der Waals density functional made accurate. *Phys. Rev. B* **89**, 121103(R)–121107(R) (2014).
41. Henkelman, G., Uberuaga, B. P. & Jónsson, H. A climbing image nudged elastic band method for finding saddle points and minimum energy paths. *J. Chem. Phys.* **113**, 9901–9904 (2000).
42. Kästner, J. & Sherwood, P. Superlinearly conveging dimer method for transition state search. *J. Chem. Phys.* **128**, 014106 (2008).
43. Tersoff, J. & Hamann, D. R. Theory and application for the scanning tunneling microscope. *Phys. Rev. Lett.* **50**, 1998–2001 (1983).
44. Lorente, N. & Persson, M. Theoretical aspects of tunneling-current-induced bond excitation and breaking at surfaces. *Farad. Discuss* **117**, 277–290 (2000).

Acknowledgements

We thank the financial support of the EU (projects ERC Starting Grant CHEMCOMP no 279313, EC FP7-PEOPLE-2011-COFUND AMAROUT II program, EC FP7-PEOPLE-2013-CIG (no. 631396) and FEDER CTQ2014-56295-R), the Spanish Ministerio de Economía y Competitividad (MINECO) (projects RYC-2012-11133, RYC-2012-11231, CTQ2014-56295-R, CTQ2014-52974-REDC and Severo Ochoa Excellence Accreditation 2014-2018 SEV-2013-0319), the Comunidad de Madrid (project MAD2D), the Generalitat de Catalunya (2014-SGR-797), the ICIQ Foundation and the IMDEA Foundation. We acknowledge Professor R. Miranda for fruitful discussions.

Author contributions

D.E. and B.C. designed the study. N.G.-A., J.R.G.-M. and P.B. synthesized the molecular precursor. B.C., J.R.-F., A.M.-J. and D.E. performed the experiments. J.B. realized the DFT calculations. B.C., F.M.-P., A.M.P. and D.E. rationalized the tentative reaction mechanisms. B.C., J.B., J.M.G., R.O. and D.E. analysed the data. All authors co-wrote the paper. D.E. supervised the whole project.

Additional information

Competing financial interests: The authors declare no competing financial interests.

16

Dissolution and ionization of sodium superoxide in sodium-oxygen batteries

Jinsoo Kim[1,*], Hyeokjun Park[1,*], Byungju Lee[1,2], Won Mo Seong[1], Hee-Dae Lim[1], Youngjoon Bae[1], Haegyeom Kim[1,2], Won Keun Kim[3], Kyoung Han Ryu[3] & Kisuk Kang[1,2]

With the demand for high-energy-storage devices, the rechargeable metal-oxygen battery has attracted attention recently. Sodium-oxygen batteries have been regarded as the most promising candidates because of their lower-charge overpotential compared with that of lithium-oxygen system. However, conflicting observations with different discharge products have inhibited the understanding of precise reactions in the battery. Here we demonstrate that the competition between the electrochemical and chemical reactions in sodium-oxygen batteries leads to the dissolution and ionization of sodium superoxide, liberating superoxide anion and triggering the formation of sodium peroxide dihydrate ($Na_2O_2 \cdot 2H_2O$). On the formation of $Na_2O_2 \cdot 2H_2O$, the charge overpotential of sodium-oxygen cells significantly increases. This verification addresses the origin of conflicting discharge products and overpotentials observed in sodium-oxygen systems. Our proposed model provides guidelines to help direct the reactions in sodium-oxygen batteries to achieve high efficiency and rechargeability.

[1] Department of Materials Science and Engineering, Research Institute of Advanced Materials, Seoul National University, 1 Gwanak-ro, Gwanak-gu, Seoul 151-742, Republic of Korea. [2] Center for Nanoparticle Research at Institute for Basic Science, Seoul National University, 1 Gwanak-ro, Gwanak-gu, Seoul 151-742, Republic of Korea. [3] Environment and Energy Research Team, Division of Automotive Research and Development, Hyundai Motor Company, 37 Cheoldobangmulgwan-ro, Uiwang, Gyeonggi-do 437-815, Republic of Korea. * These authors contributed equally to this work. Correspondence and requests for materials should be addressed to K.K. (email: matlgen1@snu.ac.kr).

To address the increasing use of renewable energy and launch of electric vehicles, the need for rechargeable batteries with high-energy densities has been growing more rapidly than ever before[1,2]. Among the available battery chemistries, metal–oxygen systems offer the highest energy density with the largest theoretical capacities. Unlike conventional lithium-ion batteries, the direct reaction between oxygen and light metals such as lithium and sodium in metal–oxygen systems circumvents the need for a heavy transition metal redox couple in their operation, thereby making a high gravimetric energy density achievable[3–7]. The most intensively studied metal–oxygen system to date is the lithium–oxygen (Li–O$_2$) battery, which shares a similar lithium chemistry with lithium-ion batteries. However, this system suffers from poor cycle stability and efficiency, which has retarded the feasibility of its use in practical systems[8,9]. In particular, the large charge overpotential over 1 V, the main reason for the low efficiency, also accelerates the degradation of the electrode and electrolyte[9]. As an alternative, Na has been introduced to replace Li in Li–O$_2$ batteries with a few important merits[10]. Despite the reduction in the energy density resulting from the lower redox potential of Na/Na$^+$, Na resources are readily available, are less expensive than Li and can easily replace Li in the battery chemistry. It has been reported that the redox reactions in the sodium–oxygen (Na–O$_2$) battery result in an extremely low charge overpotential (~ 0.2 V) despite involving the formation of micrometre-sized sodium superoxide (NaO$_2$) cubic crystallites[11–14]. This unique phenomenon supports the idea that this system is a promising alternative not only in terms of the cost of materials but also regarding the potential practical performance advantages.

Notably, however, the reactions of Na–O$_2$ cells appear to be more diverse than those of Li–O$_2$ cells. Contrary to the initial report of NaO$_2$ discharge products, some recent lines of work could not reproduce either the formation of the discharge product NaO$_2$ or the low charge overpotential[15–17]. It was reported that sodium peroxide (Na$_2$O$_2$)[15–17] or sodium peroxide dihydrate (Na$_2$O$_2 \cdot$2H$_2$O)[18,19] was formed instead. In addition, these cells exhibited high overpotential during charge, similar to that observed in the Li–O$_2$ system. Many groups have attempted to determine the reasons for these discrepancies; however, to date, the main cause of the divergence of reactions has not been identified. Janek et al. investigated the effect of different carbon electrodes; Guo et al. and Shao-Horn et al. addressed this issue but observed no critical differences among the cases[12,20,21].

In this work, we demonstrate the interplay of the diverse reactions in Na–O$_2$ batteries involving a series of electrochemical and chemical reactions as a function of time. Under systematic control of the operating conditions, we observe that the galvanostatic charge/discharge profiles are sensitively affected by the conditions and durations of the electrochemical operations. It is also revealed that the electrochemically formed NaO$_2$ is unstable and degrades into Na$_2$O$_2 \cdot$2H$_2$O in the absence of an applied current. The spontaneous dissolution and ionization of NaO$_2$ liberates the free O$_2^-$ in the electrolyte and promotes side reactions involving the formation of Na$_2$O$_2 \cdot$2H$_2$O. On the basis of these observations, we propose reaction mechanisms of Na–O$_2$ batteries under various operating conditions. This report is the first to reveal the relationships among the different discharge products observed in Na–O$_2$ batteries, which broadens our understanding of the electrochemical and chemical reactions in Na–O$_2$ batteries. Furthermore, these discussions may offer insight and guidance to the metal–air battery community in terms of regulating the kinetics of the intertwined reactions.

Results

Electrochemical profile depending on the operating condition.

To address the previous conflicting results on the discharge products and overpotentials of Na–O$_2$ cells, we carefully assessed the effects of operating parameters on the resulting electrochemical profiles. Among the various parameters examined (Supplementary Figs 1–3 and Supplementary Notes 1–3), we observed that the charge/discharge profiles were most sensitively affected by the applied current and rest time between the discharge and charge, which was analogous to the report by Yadegari et al. as a function of discharge current or limited capacities[19]. Figure 1 presents and compares the electrochemical profiles obtained under various conditions. Although the discharge profiles are similar, with a single plateau at ~ 2.1 V, there are roughly three different charging plateaus observed at (i) ~ 2.5 V, (ii) ~ 3.0 V and (iii) 3.8 V, which agree with recent reports under certain settings[19,20]. However, the relative lengths of each plateau markedly vary under differing operating conditions. For the cases of controlled discharge currents followed by a constant current charging in Fig. 1a, it was observed that the length of the lower plateau (~ 2.5 V) in the charge profiles was reduced as the applied discharge current decreased from 0.5 to 0.02 mA. However, the lengths of the plateaus at higher voltages, that is, ~ 3.0 and 3.8 V, were substantially increased, resulting in an overall larger overpotential. Similar behaviours were observed in Fig. 1b when varying the charge currents after a constant current discharge. With the lower applied charge currents, the cell exhibited a higher charging overpotential with shortened plateau length at 2.5 V. This result contrasts with the general observation that slow charging/discharging of electrochemical cells results in a voltage close to the equilibrium potential, thereby resulting in smaller overpotentials. In addition, this result strongly indicates that the different discharge products might undergo the charging process at each case. Notably, the shapes of the electrochemical charge profiles provide important clues to determine the discharge products of Na–O$_2$ reactions[19,22]. Even though the discharge products should be identical for the cases of the same protocol of discharge, each charge profile was distinct with different charge currents. This finding implies that the initial discharge products are gradually transformed into other phases during the charge process via time-dependent reactions. To verify whether this transformation occurs via an electrochemical or chemical reaction, we also controlled the rest time between the discharge and charge processes. As observed in Fig. 1c, the lowest voltage region in the charge profiles systematically decreases on increasing the rest time from 0 to 12 h. The change in the electrochemical profile in the absence of the applied current clearly indicates that the time-dependent chemical reactions occurred during the rest period, affecting the subsequent charging. This behaviour was also confirmed in similar tests for the higher charging currents with the resting time after the discharge, which revealed the growth of the charge polarizations on increasing the rest time (Supplementary Fig. 1).

The time-dependent chemical reactions can be more clearly visualized by plotting all the voltage profiles as a function of the time. Figure 1d–j illustrates the voltage evolution of each cell after the completion of the discharge at different operating conditions. The first inflection points of the voltage profiles at charge (indicated with arrows) occur at ~ 10 h regardless of the rest or charge protocols. This result indicates that a specific time of ~ 10 h is required before observing a change of the profile, which hints at the kinetics of the chemical reactions.

Time-resolved characterization of discharge products. To confirm the time-dependent phase transformation of the discharge products via chemical reaction in Na–O$_2$ cells, we

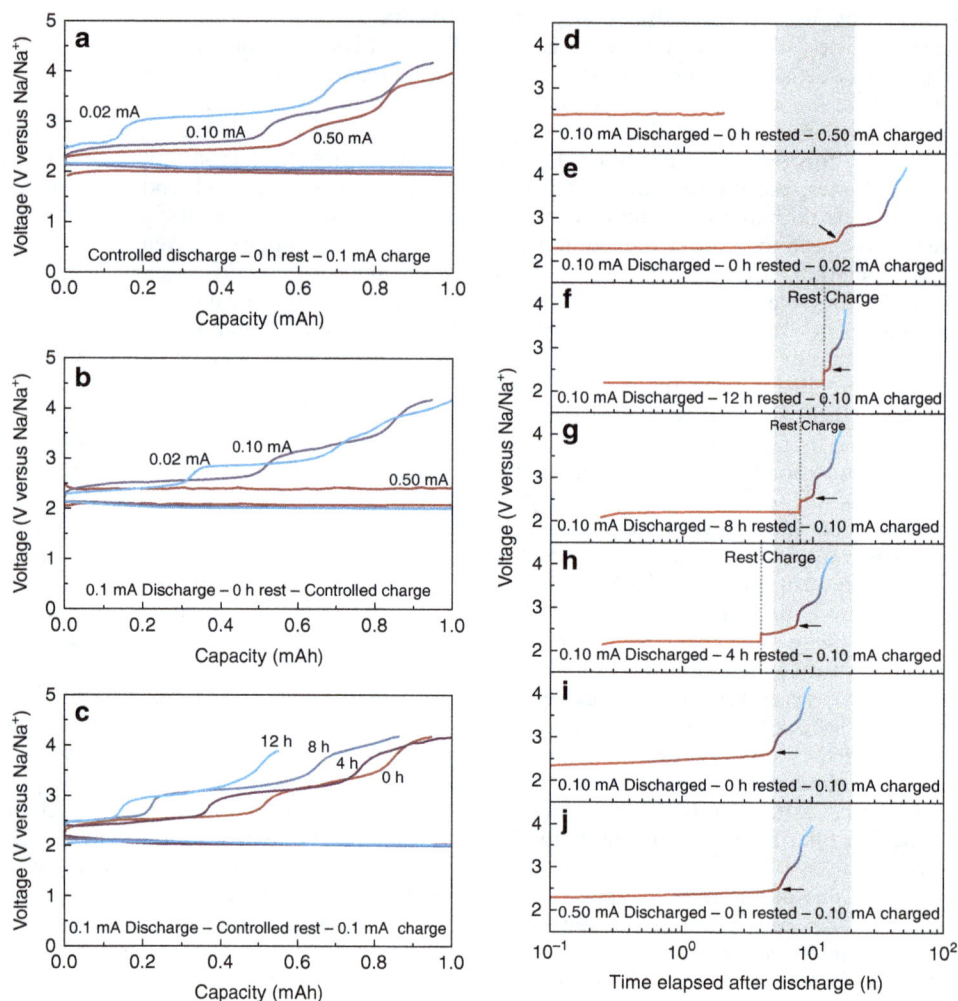

Figure 1 | Electrochemical charge/discharge profiles of Na-O$_2$ cells under various operating conditions. (**a**) Discharge currents of 0.02, 0.1 and 0.5 mA; (**b**) charge currents of 0.02, 0.1 and 0.5 mA and (**c**) rest times of 0, 4, 8 and 12 h. All the cells utilized a limited capacity of 1.0 mAh. (**d-j**) Representations of voltage profiles as a function of time corresponding to **a-c**. The shaded area indicates that range of the first points of the polarized charge potentials.

characterized the discharge products in air electrodes as a function of the rest time. The highly crystalline NaO$_2$ was observed directly after the discharge with no other phases, as demonstrated in the X-ray diffraction spectra (Fig. 2a,b)[11]. However, after being aged for several hours, the NaO$_2$ peak slowly diminished, whereas the characteristic peak of Na$_2$O$_2 \cdot$ 2H$_2$O began to appear and grew. After 12 h of resting, the initial discharge product was completely transformed into Na$_2$O$_2 \cdot$ 2H$_2$O. It should be noted that Na$_2$O$_2 \cdot$ 2H$_2$O has often been regarded as a main discharge product in previous reports of Na–O$_2$ batteries[18–20]. Recently, Ortiz-Vitoriano *et al.* reported that NaO$_2$ could convert to Na$_2$O$_2 \cdot$ 2H$_2$O on exposure to the ambient air during the characterization at room temperature[21]. However, our data show that such transformation occurs in the electrochemical cells by the intrinsic dissolving characteristics of NaO$_2$ in the electrolyte even without the exposure to the ambient atmosphere. Remarkably, the time taken for the discharge product to completely transform into Na$_2$O$_2 \cdot$ 2H$_2$O coincides with the timeline of Fig. 1d–j, which shows the inflection of the voltage rising after $\sim$10 h. When we analysed the phases of the discharge products as a function of the applied discharge currents (Supplementary Fig. 4), it was also observed that the NaO$_2$/Na$_2$O$_2 \cdot$ 2H$_2$O ratio decreased with the lower operating current, which is consistent with the time-dependent transformation of the discharge products.

Raman spectroscopy results confirmed that the initial NaO$_2$ discharge products gradually transformed into Na$_2$O$_2 \cdot$ 2H$_2$O with resting. In Fig. 2c, the two distinct peaks of NaO$_2$ and Na$_2$O$_2 \cdot$ 2H$_2$O are detected along with the characteristic bands (D/G) of the carbon electrode. The Raman signals at 1,156 and 1,136 cm^{-1} are attributed to the O–O stretch bonding in NaO$_2$ and Na$_2$O$_2 \cdot$ 2H$_2$O, respectively[21]. The systematic change in the relative ratios of NaO$_2$ and Na$_2$O$_2 \cdot$ 2H$_2$O with time is clearly illustrated in Fig. 2d, which agrees well with the results in Fig. 2b. The phase transition of NaO$_2$ to proton-containing Na$_2$O$_2 \cdot$ 2H$_2$O indicates a source of protons in the electrochemical cell. Considering the low water content in the electrolyte used for the cell (less than $\sim$5 p.p.m.), which is insufficient to form the phase (calculations provided in Supplementary Note 4)[21,23,24], the protons are likely delivered from other sources such as the electrolyte solvent. As we could expect, the rechargeability of Na–O$_2$ cell was better for the highly biased electrochemical conditions coupled with the low polarized charge profile (Supplementary Fig. 3), which is attributed to the electrochemical formation and decomposition of NaO$_2$ as shown in Supplementary Fig. 5. However, the electrochemical reversibility with the three-stepped charge profile shown from Na$_2$O$_2 \cdot$ 2H$_2$O was relatively worse compared with the former conditions. The proposed transformation mechanism will be discussed in detail later.

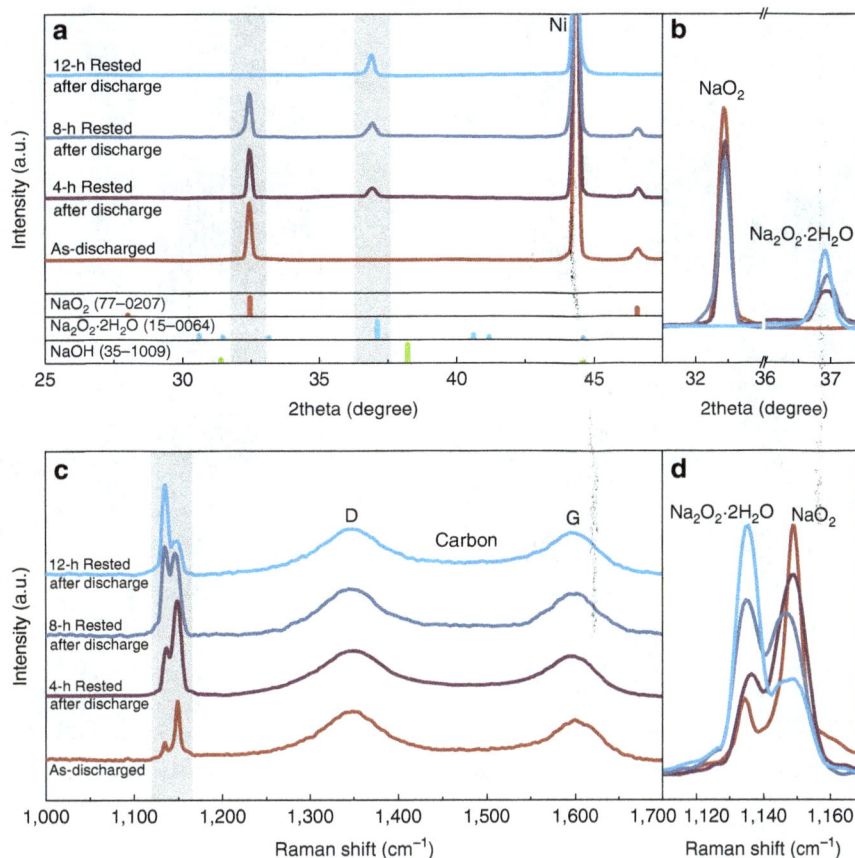

Figure 2 | Time-resolved characterization showing the phase transitions of the discharge products of the Na–O$_2$ cells. (**a,b**) X-ray diffraction spectra of the discharged cathodes of Na–O$_2$ batteries with rest times of 0, 4, 8 and 12 h. (**c,d**) Raman spectra of the discharged cathodes of Na–O$_2$ batteries with rest times of 0, 4, 8 and 12 h.

Morphological change of discharge products over time. To visualize the transition process, we examined the morphologies of the discharge products at different rest times within 12 h. In Fig. 3a, well-defined micron-sized cubic NaO$_2$ was observed immediately after the discharge, which agrees with the observation of Hartmann et al.[11]. However, the edges of the cubes became significantly dull, and the overall shapes of the cubes obtained were smudged during the rest period (Fig. 3b,c). At the end of the rest period, the cubic crystallites completely disappeared, and rod-shaped microparticles began to appear, which resemble the Na$_2$O$_2 \cdot$ 2H$_2$O in a previous report[19]. This morphological change suggests the disappearance of NaO$_2$ and the subsequent appearance of Na$_2$O$_2 \cdot$ 2H$_2$O in the cell during the rest period. Moreover, this finding implies that the transformation does not occur via a conventional solid-state or interfacial reaction between NaO$_2$ and the electrolyte to form Na$_2$O$_2 \cdot$ 2H$_2$O, which would not involve significant morphological change. Rather, it is likely to be a solution-mediated process through dissolution and nucleation[25–27].

Dissolution and ionization of NaO$_2$. We investigated the possibility of the dissolution of the solid NaO$_2$ phase in the electrolyte using electron spin resonance (ESR) spectroscopy, which is useful for detecting the magnetic responses of the unpaired electrons in radicals such as O$_2^-$ (ref. 28). Surprisingly, as observed in Fig. 4a, with the simple immersion of the pre-discharged cathodes, the ESR signal evolved within 10 min from the fresh electrolyte, indicating the presence of O$_2^-$. To avoid any effect of the remaining oxygen from the disassembled Na–O$_2$ cells, the pre-discharged cathodes were washed with fresh

electrolyte before the measurement, which led to an identical result. The calculated g-value of 2.0023 for the observed ESR signal corresponds well with the theoretical value of the unpaired electron in free O$_2^-$ (ref. 29). The solubility of NaO$_2$ in the electrolyte was roughly estimated to be $\sim$187 mM, which is in the similar order with the report by Schechter et al.[30], but has a relatively large discrepancy to the report by Hartmann et al.[31]. This discrepancy might be mainly due to the additional chemical reactions involving the precipitation of solid Na$_2$O$_2 \cdot$ 2H$_2$O. The detection of O$_2^-$ indicates that the NaO$_2$ is soluble in the ether-based electrolytes, which was also expected from the literatures with the electrochemical determinations[21,24,31]. Furthermore, this behaviour is analogous to highly soluble LiO$_2$ in the solvatable conditions of Li–O$_2$ batteries[23,32]. More importantly, the dissolution can immediately lead to the ionization of NaO$_2$, liberating O$_2^-$, the consequences of which will be discussed later.

Figure 4b shows that the peak widths of the ESR signals increased slightly with time. The broadening indicates the energy exchange of the spin with the local environments via spin–spin relaxation or spin–lattice relaxation[28]. This interaction supports the time-dependent chemical reactions associated with the dissolved O$_2^-$ with its neighbouring electrolyte solvent. The intensity of the O$_2^-$ signal is the highest $\sim$20 min after the immersion and exponentially decreases over time, indicating the instability of O$_2^-$ in the electrolyte[33]. From this behaviour, we could derive that it was a pseudo-first-order reaction that mainly relates with the concentration of O$_2^-$. On the basis of the exponential fitting of the relative intensity of ESR signals, the pseudo-first-order rate constant of H$^+$ abstraction was obtained as $\sim k' \approx 0.560$, and its corresponding half-life was

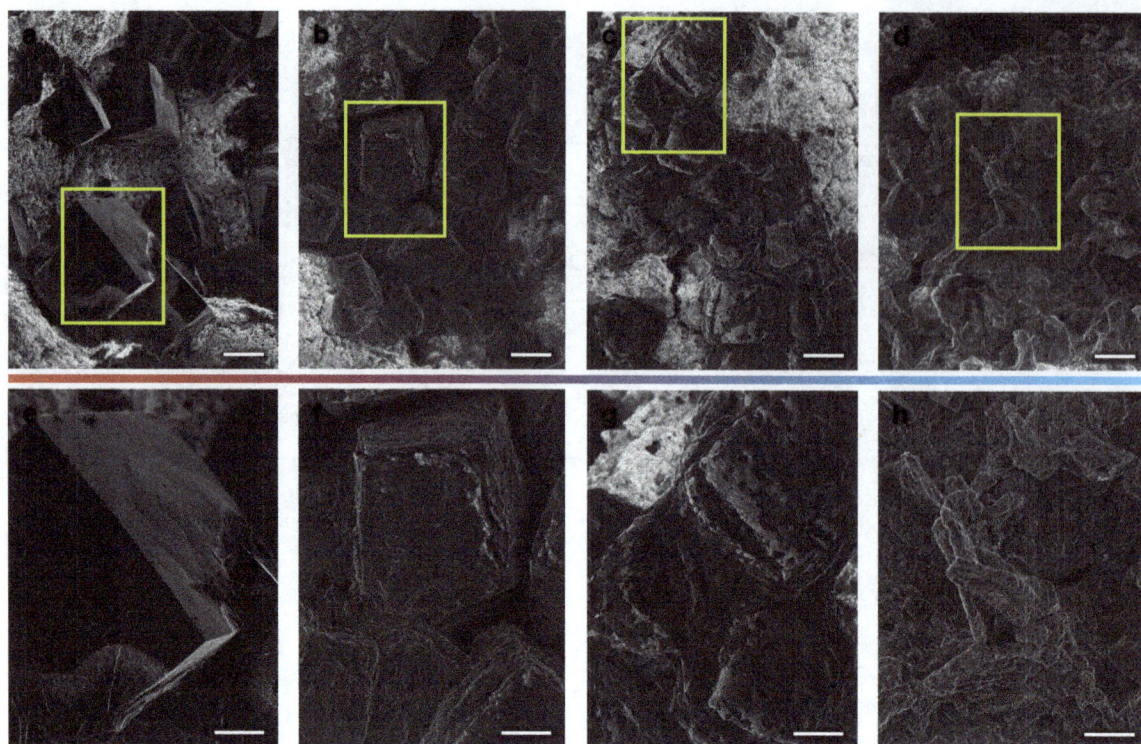

Figure 3 | Time-resolved examinations of the morphology of discharge products on the cathodes of Na-O$_2$ cells. (a-d) Morphology of the discharge products of Na-O$_2$ cells (scale bar, 10 μm). **(e-h)** Corresponding magnified scanning electron microscopy micrographs (scale bar, 5 μm); **(a,e)** as-discharged, **(b,f)** 4-h rest after discharge, **(c,g)** 8-h rest after discharge and **(d,h)** 12-h rest after discharge.

estimated as $\sim t_{1/2} = \ln(2)/k' \approx 1.24$ h (the detailed derivation is in Supplementary Note 5). Figure 4c reveals, however, that the time-dependent decay of the intensity is relatively sluggish compared with the intrinsic lifetime of normal O$_2^-$. Typically, the half-life of O$_2^-$ is $\sim$1–15 min because of its high reactivity and instability[34]. The abnormally long half-life in the electrolyte ($\sim$1.24 h) in our case is believed to occur because O$_2^-$ is continuously generated with the dissolution of NaO$_2$. The ESR signal completely vanished after $\sim$8 h, which is slightly faster than the time required for the formation of Na$_2$O$_2 \cdot$2H$_2$O in Fig. 2. Despite the evolution of O$_2^-$, the overall signal decay might be induced from the relatively dominant H$^+$ abstraction due to the reactivity of O$_2^-$. This gap in the kinetics might originate from the time required to form the Na$_2$O$_2 \cdot$2H$_2$O phase from the O$_2^-$.

To understand the dissolution and ionization behaviour of NaO$_2$, the solvation energies of various alkali-metal superoxides and peroxides were calculated for comparison using first-principle calculations with the series of dielectric constant from 7 to 30. Figure 4d reveals that generally the superoxide exhibits a lower solvation energy than the peroxide for both lithium and sodium compounds. This result is consistent with our result of NaO$_2$ dissolution and the recent experimental findings for Li-O$_2$ batteries, which indicated that LiO$_2$ is found mostly as soluble intermediates in the electrolyte in contrast to the solid phase of Li$_2$O$_2$ (refs 23,32). In addition, it is notable that the solvation energy of the sodium phases was significantly lower than that of the lithium phases, which is attributed to the weaker Lewis acidity of the Na cation compared with that of the Li cation in the polar solvent[35,36]. However, for the solvents with substantially lower dielectric constant ($\varepsilon = \sim$7), the dissolution is not favourable even in NaO$_2$. Molecular dissolution energies of NaO$_2$ in model solvents are $\sim$0.6 eV, which roughly corresponds to one molecule dissolution among 10^{10} formula units of NaO$_2$. On the other hand, it markedly diminishes to 0.17 eV (one molecule among 10^3

formula units of NaO$_2$) in $\varepsilon = 30$ (ref. 37). Note that, for the low dielectric constant solvents, dielectric constant of the solution sensitively increases with increasing salt concentration, which can result in higher solution dielectric constant than that of the pure solvent[38]. Therefore, it is expected that the dissolution of NaO$_2$ can occur when salts are present in the electrolyte, which is consistent with the observation of O$_2^-$ in the ESR analysis. It also is noted that even with the dissolving characteristics of NaO$_2$, the crystallization of NaO$_2$ is possible in the normal discharging conditions with the supersaturation of localized reactants such as Na$^+$ and O$_2^-$ (refs 21,31,39,40). In the other case where the supply of the reactants such as Na$^+$ is limited, for example, in the absence of the applied voltage, the dissolution and ionization might dominate, giving rise to the formation of Na$_2$O$_2 \cdot$2H$_2$O as a discharge product.

Proposed mechanism of Na-O$_2$ batteries. On the basis of the previous reports and our new findings, we propose a mechanism that describes the electrochemical and chemical reactions in Na-O$_2$ systems in Fig. 5. The well-established discharge process[11] can be illustrated with the reduction of an O$_2$ molecule into O$_2^-$, which reacts with Na$^+$ to form NaO$_2$ (Reaction 1), and the charge process is the reverse reaction (Reaction 2). After or during the discharge, the NaO$_2$ is prone to dissolution and ionization into the electrolyte based on the solvating energy (ΔG_{sol}) in the solvent (Reaction 3)[32]. The dissolution of NaO$_2$ generates O$_2^-$, which can degrade the surrounding molecules because of its chemical instability. Typically, the liberated O$_2^-$ is a strong reagent for the abstraction of H$^+$ from the electrolyte solvents (Reaction 4)[9], and the degree of H$^+$ abstraction[41,42] is determined by the acid-dissociation constant (pK_a) of the solvent. Some hydroperoxyl radicals (HO$_2$) might be formed during this process, resulting in the nucleophilic attack of the H$^+$-lost solvent (Reaction 5)[43]. However, the evolution of HO$_2$ can be

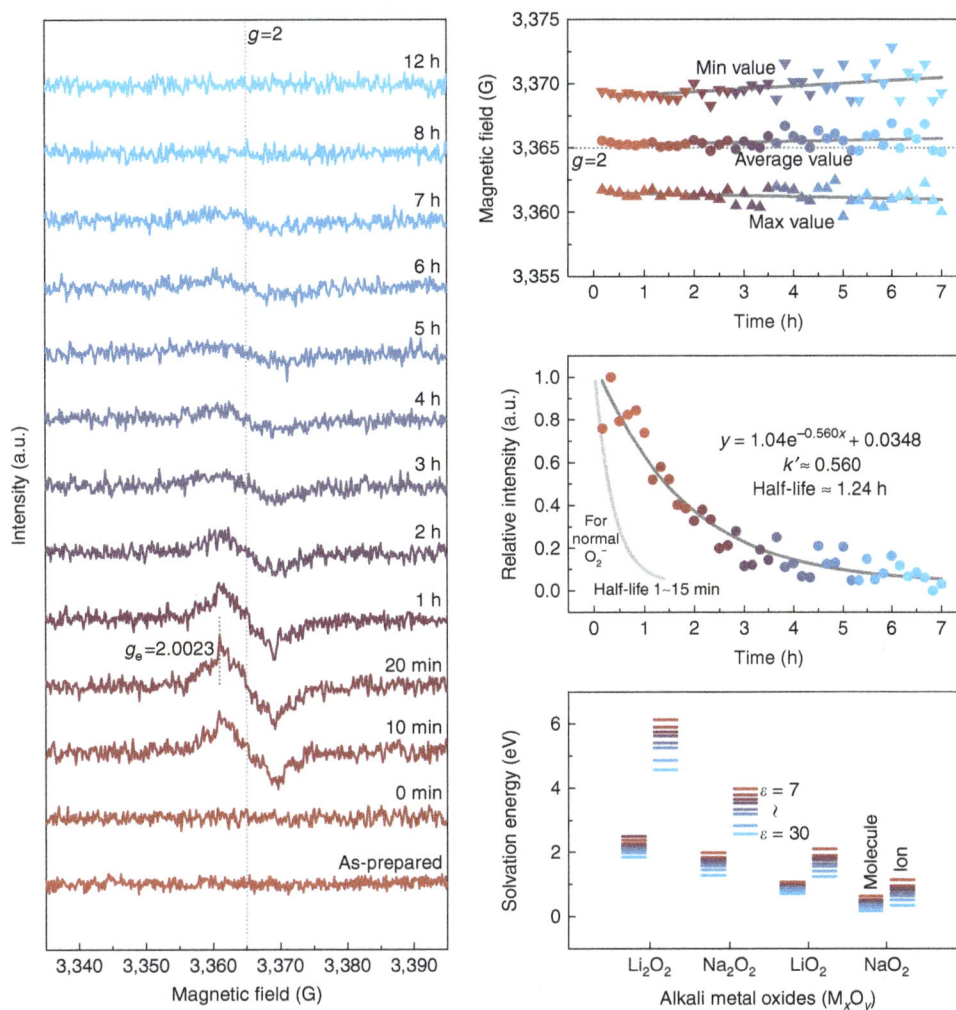

Figure 4 | ESR analysis and theoretical calculations of the dissolution and ionization of NaO$_2$ into the electrolyte. (**a**) Time-dependent ESR measurements for the fresh electrolytes (0.5 M NaCF$_3$SO$_3$ in diethylene glycol dimethyl ether) with soaking of the pre-discharged cathode without any aging. (**b**) Maximum, minimum and average values of ESR signals as a function of time. (**c**) Exponential decay of ESR signals and the common trend line of O$_2^-$. (**d**) Calculations of the solvation energy for several alkali-metal superoxides and peroxides with the various dielectric constants ($\varepsilon = 7 \sim 30$).

helpful in promoting the solution-mediated discharge/charge process as recently reported by Xia et al.[24]. Nevertheless, in a circumstance where the dissolution/ionization of NaO$_2$ is dominant, the liberation of O$_2^-$ is overwhelmingly larger than a possible HO$_2$ formation, inducing the H$^+$ abstraction from the neighbouring electrolyte solvent. Meanwhile, the solvent undergoes oxidative decompositions to produce byproducts such as carbon dioxide (CO$_2$), water (H$_2$O) and hydroxyl anions (OH$^-$; Reaction 6)[9]. It is also possible that the coupling of HO$_2$ leads to disproportionation into hydrogen peroxide (H$_2$O$_2$) and O$_2$ (Reaction 7)[44]. In the presence of both Na$^+$ and OH$^-$, which is effectively the dissolution state of sodium hydroxide (NaOH), a solid crystallite of NaOH can precipitate with a higher concentration of OH$^-$ produced. Further reaction between NaOH and H$_2$O$_2$ from Reaction 7 leads to the formation of Na$_2$O$_2 \cdot$2H$_2$O via peroxo-hydroxylation, whose reverse reaction is well known (Reaction 8)[45]. To support our proposed reaction mechanism, we chose several intermediate reactions that should be verified according to the reaction model in the Supplementary Information. Supplementary Figs 6–9 demonstrate that O$_2^-$ plays an important role after the dissolution of NaO$_2$ in converting the discharge product to Na$_2$O$_2 \cdot$2H$_2$O via degradation of the electrolyte involving OH$^-$ and H$_2$O$_2$. These identifications strongly support the proposed

mechanism of competing electrochemical and following chemical reactions in Na–O$_2$ batteries. The detailed discussions are provided in Supplementary Notes 6–9. The reaction equations are summarized below:

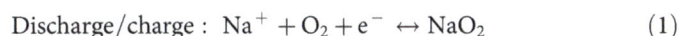

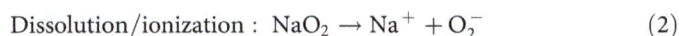

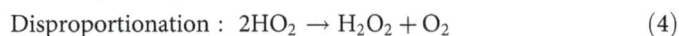

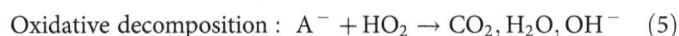

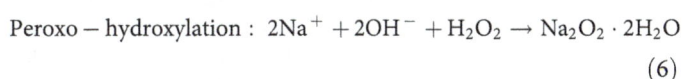

$$\text{Discharge/charge}: \ \text{Na}^+ + \text{O}_2 + \text{e}^- \leftrightarrow \text{NaO}_2 \tag{1}$$

$$\text{Dissolution/ionization}: \ \text{NaO}_2 \rightarrow \text{Na}^+ + \text{O}_2^- \tag{2}$$

$$\text{Proton abstraction}: \ \text{HA} + \text{O}_2^- \rightarrow \text{A}^- + \text{HO}_2 \tag{3}$$

$$\text{Disproportionation}: \ 2\text{HO}_2 \rightarrow \text{H}_2\text{O}_2 + \text{O}_2 \tag{4}$$

$$\text{Oxidative decomposition}: \ \text{A}^- + \text{HO}_2 \rightarrow \text{CO}_2, \text{H}_2\text{O}, \text{OH}^- \tag{5}$$

$$\text{Peroxo} - \text{hydroxylation}: \ 2\text{Na}^+ + 2\text{OH}^- + \text{H}_2\text{O}_2 \rightarrow \text{Na}_2\text{O}_2 \cdot 2\text{H}_2\text{O} \tag{6}$$

It is noteworthy that a similar behaviour has been recently reported for reactions in Li–O$_2$ batteries. The solvating environment was demonstrated to alter the stability of the intermediates, such as a lithium superoxide (LiO$_2$), thus affecting the overall reaction paths[23,32]. LiO$_2$ is a precedent phase with the direct reaction of a Li cation and superoxide anion (O$_2^-$), which readily decomposes into lithium peroxide (Li$_2$O$_2$) via either

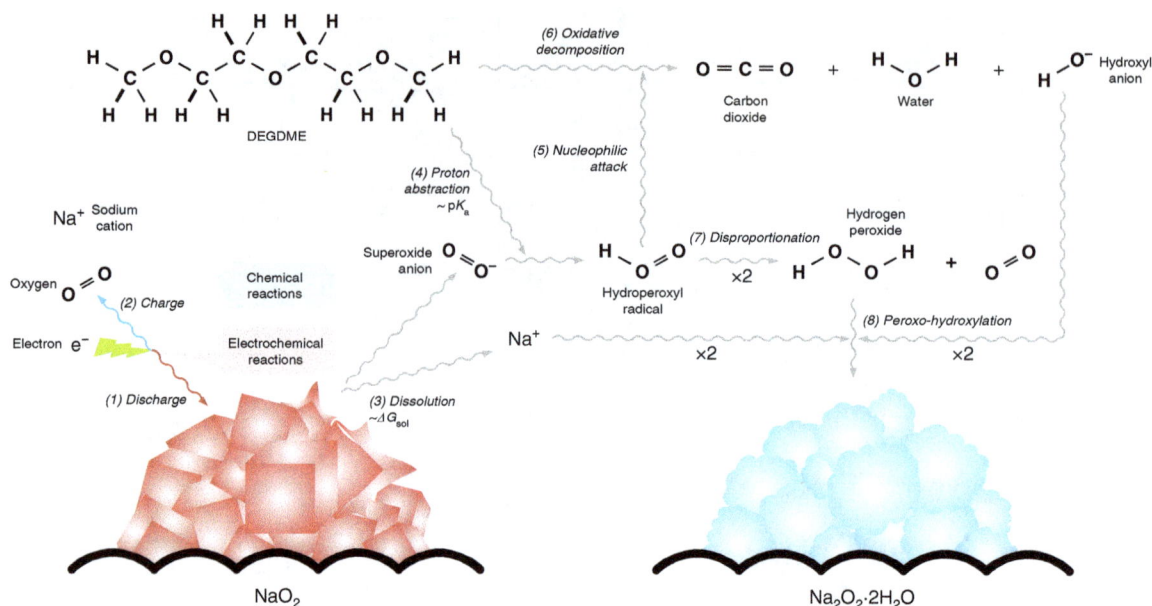

Figure 5 | Schematic of the proposed mechanism illustrating the electrochemical and chemical reactions under various operating conditions. For the electrochemical reaction, NaO_2 is formed and decomposed during discharge/charge (Reactions 1 and 2). For the chemical reaction, NaO_2 is dissolved and ionized into the electrolyte (Reaction 3), which promotes the undesired degradation of the electrolyte (Reactions 4–6). $Na_2O_2 \cdot 2H_2O$ is formed during the subsequent chemical reactions (Reactions 7 and 8).

an electrochemical surface reaction or disproportionation[23,32]. Although LiO_2 is known to be unstable[46,47], it was recently demonstrated that LiO_2 might be dissolved into the electrolyte and aid in the formation of the toroidal Li_2O_2 via a solution reaction under highly solvating conditions[23]. NaO_2 shares this dissolving nature with LiO_2 even though the thermodynamic stability of NaO_2 warrants its formation as a discharge product. The significant dissolution of NaO_2 supports the conclusion that the dominant reaction in Na–O_2 batteries relies on the solution-mediated reactions of nucleation and growth of NaO_2 (refs 21,24,31) and implies that the capacities and morphology of the reaction products would be greatly affected by the energetics of NaO_2 under various conditions (such as different electrolytes and current rates). This is also supplemented with the recently reported observations[21,31] and operating mechanism[24] in terms of the various states of electrochemical and chemical reactions.

Discussion

We successfully demonstrated the interplay of the diverse competing reactions in Na–O_2 batteries. The time-dependent chemical reactions were identified as being triggered from the dissolution and ionization of the electrochemically formed NaO_2 in the electrolyte. The liberated O_2^- reacts with the electrolyte solvent to form $Na_2O_2 \cdot 2H_2O$ following a series of intermediate steps. The $Na_2O_2 \cdot 2H_2O$ in the air electrode requires a higher energy for the decomposition, which leads to the increased charge overpotential and irreversibility of Na–O_2 cells. This report is the first to correlate the electrochemical and chemical reactions with the operating conditions in Na–O_2 batteries, and our findings concerning the relationships among different phases resolve the conflicting observations of different discharge products in previous Na–O_2 batteries. To prepare a better performing Na–O_2 battery, a strategy to prevent the transformation of NaO_2 into $Na_2O_2 \cdot 2H_2O$ while still allowing the solution-mediate discharge reaction is necessary. We hope that the findings of this study can provide a basis for researchers to navigate and direct the reactions in Na–O_2 batteries to achieve high efficiency and rechargeability.

Methods

Cell assembly and galvanostatic cycling of Na–O_2 cells. The carbon cathode was prepared by casting Ketjen Black carbon paste and polytetrafluoroethylene (60 wt% emersion in water, Sigma-Aldrich) with a mass ratio of 9:1 in a solution of isopropanol (> 99.7%, Sigma-Aldrich) and N-methyl-2-pyrrolidone (99.5%, anhydrous, Sigma-Aldrich) with a volume ratio of 1:1 on Ni mesh current collectors. The carbon-coated Ni mesh was dried at 120 °C and heated at 400 °C for 4 h in Ar to completely remove any residual H_2O impurities.

All the procedures described below were performed in an Ar-filled glove box (O_2 level < 0.5 p.p.m. and H_2O level < 0.5 p.p.m.). The Na–O_2 cells were assembled as a Swagelok-type cell with stacking of the Na metal anode, electrolyte-soaked separators and carbon cathode, which was punched with a half-inch diameter. The Na metal anode was carefully prepared by milling dry Na metal chunks (ACS Reagent, Sigma-Aldrich) after removing the contaminated surfaces. The electrolyte was prepared with diethylene glycol dimethyl ether (anhydrous, 99.5%, Sigma-Aldrich), which contains 0.5 M $NaCF_3SO_3$ (98%, Sigma-Aldrich). The solvent was dried using 3-Å molecular sieves for over 1 week, and the salt was also kept in a vacuum oven at 180 °C for the same time before use. The final water content in the electrolyte was less than 10 p.p.m. according to a Karl–Fisher titration measurement. The amount of electrolyte used for the cell was 200 µl. Two sheets of Celgard 2400 were used as separators. Electrochemical battery tests of the Na–O_2 cells were conducted using a potentio-galvanostat (WonA Tech, WBCS 3000, Korea). All the cells were relaxed under 770 torr of O_2 pressure for 10 min before the tests. After being saturated with O_2 gas, the cells were operated in the closed state with a limited capacity of 1 mAh, lower voltage cutoff of 1.6 V and upper voltage cutoff of 4.2 V. Special protocol based on a pulsed current was applied during the charge to avoid dendritic failure of the Na metal anode. The on/off time ratio of the pulse charge was 1:4 (applying current for 0.5 s and relaxing for 2 s). More details about our charge protocol are provided in Supplementary Figs 10 and 11 and following discussions in Supplementary Note 10.

Characterization of Na–O_2 cells. The discharged cathodes after the different rest times were collected from disassembled Na–O_2 cells and washed with acetonitrile (anhydrous, 99.8%, Sigma-Aldrich) in a glove box to remove any residual electrolyte. X-ray diffraction spectra of the cathodes were obtained using a Bruker D2-Phaser (Cu Kα λ = 1.5406 Å) with the aid of a specially designed air-tight holder to prevent outer atmospheric contamination. Raman spectra were obtained using a Horiba Jobin-Yvon LabRam Aramis spectrometer (the 514-nm line of an Ar-ion laser was used as the excitation source). The scattered light of the Raman signal was collected in a backscattering geometry using the × 50 microscope objective lens. Field-emission scanning electron microscopy (MERLIN Compact, ZEISS, Germany) was used for the morphological observations after Pt coating. For ESR characterization, the collected powder from the discharged cathodes after rinsing to remove the residual used electrolytes was soaked in fresh electrolyte. After immersing the powdery discharged cathodes, the ESR signal of the electrolytes was measured at room temperature using a JEOL JES-TE200 ESR spectrometer every 10 min for 12 h using a liquid quartz cell. The microwave X-band frequency was 9.42 GHz at 1-mW power.

Theoretical calculations of solvation energy. First-principles calculations were performed using the spin-polarized generalized gradient approximation. A continuum solvation model (VASPsol[48,49] code) was used to evaluate the solvation energy of the alkali-metal superoxide/peroxide (M_xO_2, M: Li, Na, $x = 1$ or 2). The following equations were used considering both the (1) molecular and (2) ionized solvated states:

$$\Delta E_{sol, mol} = E_{solvated}(M_xO_2) - E_{bulk}(M_xO_2) \tag{7}$$

$$\Delta E_{sol, ion} = x \cdot E_{solvated}(M^+) + E_{solvated}(O_2^{x-}) - E_{bulk}(M_xO_2), \tag{8}$$

where $E_{solvated}(M_xO_2)$ and $E_{bulk}(M_xO_2)$ are the total energies of the solvated and bulk M_xO_2 per formula unit, respectively. The solvated species (ions or molecules) were placed in a 13 Å × 13 Å × 13 Å cell as an isolated species. We used the plane-wave basis with an energy cutoff of 550 eV and a Monkhorst-Pack 2 × 2 × 2 k-point mesh. On the basis of previous reports[50,51] that stated that the solvation entropy term (TS) of polar molecules and ions in the standard state is less than 5% of the enthalpy term (H), we neglected the entropy effect of the solvation in these calculations.

References

1. Tarascon, J. M. & Armand, M. Issues and challenges facing rechargeable lithium batteries. *Nature* **414**, 359–367 (2001).
2. Kang, K., Meng, Y. S., Bréger, J., Grey, C. P. & Ceder, G. Electrodes with high power and high capacity for rechargeable lithium batteries. *Science* **311**, 977–980 (2006).
3. Abraham, K. M. & Jiang, Z. A polymer electrolyte-based rechargeable lithium/oxygen battery. *J. Electrochem. Soc.* **143**, 1–5 (1996).
4. Bruce, P. G., Freunberger, S. A., Hardwick, L. J. & Tarascon, J.-M. Li-O₂ and Li-S batteries with high energy storage. *Nat. Mater.* **11**, 19–29 (2012).
5. Peng, Z., Freunberger, S. A., Chen, Y. & Bruce, P. G. A reversible and higher-rate Li-O₂ battery. *Science* **337**, 563–566 (2012).
6. Ottakam Thotiyl, M. M. *et al.* A stable cathode for the aprotic Li-O₂ battery. *Nat. Mater.* **12**, 1050–1056 (2013).
7. Lim, H.-D. *et al.* Superior rechargeability and efficiency of lithium-oxygen batteries: hierarchical air electrode architecture combined with a soluble catalyst. *Angew. Chem. Int. Ed.* **126**, 4007–4012 (2014).
8. Débart, A., Paterson, A. J., Bao, J. & Bruce, P. G. α-MnO₂ nanowires: a catalyst for the O₂ electrode in rechargeable lithium batteries. *Angew. Chem. Int. Ed.* **47**, 4521–4524 (2008).
9. Freunberger, S. A. *et al.* The lithium-oxygen battery with ether-based electrolytes. *Angew. Chem. Int. Ed.* **50**, 8609–8613 (2011).
10. Yabuuchi, N., Kubota, K., Dahbi, M. & Komaba, S. Research development on sodium-ion batteries. *Chem. Rev.* **114**, 11636–11682 (2014).
11. Hartmann, P. *et al.* A rechargeable room-temperature sodium superoxide (NaO₂) battery. *Nat. Mater.* **12**, 228–232 (2013).
12. Bender, C. L., Hartmann, P., Vračar, M., Adelhelm, P. & Janek, J. On the thermodynamics, the role of the carbon cathode, and the cycle life of the sodium superoxide (NaO₂) battery. *Adv. Energy Mater.* **4**, 1301863 (2014).
13. McCloskey, B. D., Garcia, J. M. & Luntz, A. C. Chemical and electrochemical differences in nonaqueous Li-O₂ and Na-O₂ batteries. *J. Phys. Chem. Lett.* **5**, 1230–1235 (2014).
14. Hartmann, P. *et al.* Pressure dynamics in metal-oxygen (metal-air) batteries: a case study on sodium superoxide cells. *J. Phys. Chem. C* **118**, 1461–1471 (2014).
15. Liu, W., Sun, Q., Yang, Y., Xie, J.-Y. & Fu, Z.-W. An enhanced electrochemical performance of a sodium-air battery with graphene nanosheets as air electrode catalysts. *Chem. Commun.* **49**, 1951–1953 (2013).
16. Li, Y. *et al.* Superior catalytic activity of nitrogen-doped graphene cathodes for high energy capacity sodium-air batteries. *Chem. Commun.* **49**, 11731–11733 (2013).
17. Hu, Y. *et al.* Porous perovskite calcium-manganese oxide microspheres as an efficient catalyst for rechargeable sodium-oxygen batteries. *J. Mater. Chem. A* **3**, 3320–3324 (2015).
18. Kim, J., Lim, H.-D., Gwon, H. & Kang, K. Sodium-oxygen batteries with alkyl-carbonate and ether based electrolytes. *Phys. Chem. Chem. Phys.* **15**, 3623–3629 (2013).
19. Yadegari, H. *et al.* On rechargeability and reaction kinetics of sodium-air batteries. *Energy Environ. Sci.* **7**, 3747–3757 (2014).
20. Zhao, N., Li, C. & Guo, X. Long-life Na-O₂ batteries with high energy efficiency enabled by electrochemically splitting NaO₂ at a low overpotential. *Phys. Chem. Chem. Phys.* **16**, 15646–15652 (2014).
21. Ortiz-Vitoriano, N. *et al.* Rate-dependent nucleation and growth of NaO₂ in Na-O₂ batteries. *J. Phys. Chem. Lett.* **6**, 2636–2643 (2015).
22. Yadegari, H. *et al.* Three-dimensional nanostructured air electrode for sodium-oxygen batteries: a mechanism study toward the cyclability of the cell. *Chem. Mater.* **27**, 3040–3047 (2015).
23. Aetukuri, N. B. *et al.* Solvating additives drive solution-mediated electrochemistry and enhance toroid growth in non-aqueous Li-O₂ batteries. *Nat. Chem.* **7**, 50–56 (2015).
24. Xia, C., Black, R., Fernandes, R., Adams, B. & Nazar, L. F. The critical role of phase-transfer catalysis in aprotic sodium-oxygen batteries. *Nat. Chem.* **7**, 496–501 (2015).
25. Schroeder, M. A. *et al.* DMSO-Li₂O₂ interface in the rechargeable Li-O₂ battery cathode: theoretical and experimental perspectives on stability. *ACS Appl. Mater. Interfaces* **7**, 11402–11411 (2015).
26. Kumar, N., Radin, M. D., Wood, B. C., Ogitsu, T. & Siegel, D. J. Surface-mediated solvent decomposition in li-air batteries: impact of peroxide and superoxide surface terminations. *J. Phys. Chem. C* **119**, 9050–9060 (2015).
27. Kwabi, D. G. *et al.* Chemical instability of dimethyl sulfoxide in lithium-air batteries. *J. Phys. Chem. Lett.* **5**, 2850–2856 (2014).
28. Wang, Q., Yang, X.-Q. & Qu, D. In situ ESR spectro-electrochemical investigation of the superoxide anion radical during the electrochemical O₂ reduction reaction in aprotic electrolyte. *Carbon* **61**, 336–341 (2013).
29. Eastland, G. W. & Symons, M. C. Electron spin resonance studies of superoxide ions produced by radiolysis in alcoholic media. *J. Phys. Chem.* **81**, 1502–1504 (1977).
30. Schechter, D. L. & Kleinberg, J. Reactions of some metal salts with alkali superoxides in liquid ammonia. *J. Am. Chem. Soc.* **76**, 3297–3300 (1954).
31. Hartmann, P. *et al.* Discharge and charge reaction paths in sodium-oxygen batteries: does NaO₂ form by direct electrochemical growth or by precipitation from solution? *J. Phys. Chem. C* **119**, 22778–22786 (2015).
32. Johnson, L. *et al.* The role of LiO₂ solubility in O₂ reduction in aprotic solvents and its consequences for Li-O₂ batteries. *Nat. Chem.* **6**, 1091–1099 (2014).
33. Hall, P. & Selinger, B. Better estimates of exponential decay parameters. *J. Phys. Chem.* **85**, 2941–2946 (1981).
34. Maricle, D. L. & Hodgson, W. G. Reducion of oxygen to superoxide anion in aprotic solvents. *Anal. Chem.* **37**, 1562–1565 (1965).
35. Okoshi, M., Yamada, Y., Yamada, A. & Nakai, H. Theoretical analysis on de-solvation of lithium, sodium, and magnesium cations to organic electrolyte solvents. *J. Electrochem. Soc.* **160**, A2160–A2165 (2013).
36. Ziegler, M. & Madura, J. Solvation of metal cations in non-aqueous liquids. *J. Solution Chem.* **40**, 1383–1398 (2011).
37. Lee, B. *et al.* Theoretical evidence for low charging overpotentials of superoxide discharge products in metal-oxygen batteries. *Chem. Mater.* **27**, 8406–8413 (2015).
38. Petrowsky, M. A. *Ion Transport in Liquid Electrolytes* (PhD thesis, Univ. of Oklahoma, 2008).
39. Kashchiev, D. & Van Rosmalen, G. Review: nucleation in solutions revisited. *Cryst. Res. Technol.* **38**, 555–574 (2003).
40. Takiyama, H. Supersaturation operation for quality control of crystalline particles in solution crystallization. *Adv. Powder Technol.* **23**, 273–278 (2012).
41. Khetan, A., Pitsch, H. & Viswanathan, V. Solvent degradation in nonaqueous Li-O₂ batteries: oxidative stability versus H-abstraction. *J. Phys. Chem. Lett.* **5**, 2419–2424 (2014).
42. Khetan, A., Luntz, A. & Viswanathan, V. Trade-offs in capacity and rechargeability in nonaqueous Li-O₂ batteries: solution-driven growth versus nucleophilic stability. *J. Phys. Chem. Lett.* **6**, 1254–1259 (2015).
43. Sawyer, D. T. & Valentine, J. S. How super is superoxide? *Acc. Chem. Res.* **14**, 393–400 (1981).
44. Chin, D. H., Chiericato, G., Nanni, E. J. & Sawyer, D. T. Proton-induced disproportionation of superoxide ion in aprotic media. *J. Am. Chem. Soc.* **104**, 1296–1299 (1982).
45. Hill, G. S. *et al.* The X-ray structure of a sodium peroxide hydrate, Na₂O₂·8H₂O, and its reactions with carbon dioxide: relevance to the brightening of mechanical pulps. *Can. J. Chem.* **75**, 46–51 (1997).
46. Zhai, D. *et al.* Raman evidence for late stage disproportionation in a Li-O₂ battery. *J. Phys. Chem. Lett.* **5**, 2705–2710 (2014).
47. Zhai, D. *et al.* Interfacial effects on lithium superoxide disproportionation in Li-O₂ batteries. *Nano Lett.* **15**, 1041–1046 (2015).
48. Fishman, M., Zhuang, H. L., Mathew, K., Dirschka, W. & Hennig, R. G. Accuracy of exchange-correlation functionals and effect of solvation on the surface energy of copper. *Phys. Rev. B* **87**, 245402 (2013).
49. Mathew, K., Sundararaman, R., Letchworth-Weaver, K., Arias, T. A. & Hennig, R. G. Implicit solvation model for density-functional study of nanocrystal surfaces and reaction pathways. *J. Chem. Phys.* **140**, 084106 (2014).
50. Burgess, J. *Metal Ions in Solution* (Chichester: Horwood Ellis, 1978).
51. Ben-Amotz, D., Raineri, F. O. & Stell, G. Solvation thermodynamics: theory and applications. *J. Phys. Chem. B* **109**, 6866–6878 (2005).

Acknowledgements

This work was supported by the National Research Foundation of Korea (NRF) grand funded by the Korea government (MSIP; No. 2015R1A2A1A10055991) and by the HMC (Hyundai Motor Company). This work was also supported by the BK21PLUS (No. 21A20131912052) and by Project Code (IBS-R006-G1). Y.B. and K.K. also acknowledge to the support of Samsung Advanced Institute of Technology.

Author contributions

J.K. and H.P. designed and performed the experiments. B.L. calculated and predicted the solvation energies. W.M.S. examined the morphologies of the discharge products. H.-D.L. and Y.B. contributed to discussions and assisted in integration of the results. H.K. prepared the electrolyte. W.K.K. and K.H.R. supported the electrochemical measurements and characterization under dried conditions. K.K. conceived the original idea, supervised the research, contributed to scientific discussions and writing of manuscript.

Additional information

Competing financial interests: The authors declare no competing financial interests.

17

High-efficiency and air-stable P3HT-based polymer solar cells with a new non-fullerene acceptor

Sarah Holliday[1], Raja Shahid Ashraf[1], Andrew Wadsworth[1], Derya Baran[1], Syeda Amber Yousaf[2], Christian B. Nielsen[1], Ching-Hong Tan[1], Stoichko D. Dimitrov[1], Zhengrong Shang[3], Nicola Gasparini[4], Maha Alamoudi[5], Frédéric Laquai[5], Christoph J. Brabec[4], Alberto Salleo[3], James R. Durrant[1] & Iain McCulloch[1,5]

Solution-processed organic photovoltaics (OPV) offer the attractive prospect of low-cost, light-weight and environmentally benign solar energy production. The highest efficiency OPV at present use low-bandgap donor polymers, many of which suffer from problems with stability and synthetic scalability. They also rely on fullerene-based acceptors, which themselves have issues with cost, stability and limited spectral absorption. Here we present a new non-fullerene acceptor that has been specifically designed to give improved performance alongside the wide bandgap donor poly(3-hexylthiophene), a polymer with significantly better prospects for commercial OPV due to its relative scalability and stability. Thanks to the well-matched optoelectronic and morphological properties of these materials, efficiencies of 6.4% are achieved which is the highest reported for fullerene-free P3HT devices. In addition, dramatically improved air stability is demonstrated relative to other high-efficiency OPV, showing the excellent potential of this new material combination for future technological applications.

[1] Department of Chemistry and Centre for Plastic Electronics, Imperial College London, London SW7 2AZ, UK. [2] Department of Physics, Government College University, Lahore 54000, Pakistan. [3] Department of Materials Science and Engineering, Stanford University, 476 Lomita Mall, Stanford, California 94305, USA. [4] Institute of Materials for Electronics and Energy Technology (I-MEET), Friedrich-Alexander-University Erlangen-Nuremberg, 91058 Erlangen, Germany. [5] King Abdullah University of Science and Technology (KAUST), Solar and Photovoltaics Engineering Research Center (SPERC), Thuwal 23955-6900, Saudi Arabia. Correspondence and requests for materials should be addressed to S.H. (email: s.holliday@imperial.ac.uk) or to I.M. (email: i.mcculloch@imperial.ac.uk).

The efficiency of solution-processed organic photovoltaics (OPV) has been increasing rapidly, with the development of new high-performing benzodithiophene[1-4] and difluorobenzothiadiazole[5] -based donor polymers in particular that give up to 10% power conversion efficiency (PCE) combined with fullerene acceptors in single junction cells, and over 11% PCE in tandem devices[6,7]. Meanwhile, fullerene-free OPV has also been advancing, driven by the need to find alternative acceptors that overcome the high synthetic costs, limited optical absorption, poor bandgap tunability and morphological instability of fullerene-based acceptors such as phenyl-C_{61}-butyric acid methyl ester ($PC_{60}BM$) and its C_{71} analogue $PC_{70}BM$ (refs 8–10). Multiple reports of efficiencies over 6% have now been published with acceptors based on fused ring diimide[11-15] and 1,1-dicyanomethylene-3-indanone[16,17] structures. However, the majority of these record efficiencies are achieved with low-bandgap donor–acceptor polymers such as polythieno[3,4-b]-thiophene-*alt*-benzodithiophene (PTB7), which are known to present intrinsic difficulties to scale-up (thereby increasing costs) as well as suffering from issues with solubility[18], device irreproducibility and photochemical instability[19,20]. Meanwhile, the simple homo-polymer poly(3-hexylthiophene) (P3HT), one of the most extensively used and best understood polymers in OPV research for some time[21-23], is relatively stable[24,25] and readily scalable due to its straightforward synthesis[26] and compatibility with high-throughput production techniques[27]. Indeed, P3HT is currently one of the only polymers available in quantities over 10 kg (ref. 23), making it one of the few feasible candidates for commercial OPV, and its use in large-area, roll-to-roll printed solar cells has already been widely demonstrated[28]. Furthermore, the semi-crystalline nature of P3HT, compared with more amorphous polymers, is almost unique in setting an appropriate morphology lengthscale for bulk heterojunction OPV from a range of solvents and processing conditions, as well as providing it with good charge transport properties[29-31]. Despite this, P3HT has been somewhat marginalized in recent years since the introduction of higher efficiency donor–acceptor polymers. For $P3HT:PC_{60}BM$ devices, the average efficiency is only around 3% (ref. 21), with a maximum efficiency of 7.4% reported with the

more expensive fullerene indene-C_{60}-bisadduct (ICBA)[32]. We recently published a new non-fullerene acceptor called (5Z,5′Z)-5, 5′-{(9,9-dioctyl-9H-fluorene-2,7-diyl)bis[2,1,3-benzothiadiazole-7, 4-diyl(Z)methylylidene]}bis(3-ethyl-2-thioxo-1,3-thiazolidin-4-one) (FBR) that had a straightforward and scalable synthesis and gave 4.1% PCE in P3HT devices, which at the time of writing was the highest reported efficiency for a fullerene-free device with P3HT[33]. However, the short-circuit current (J_{sc}) in these devices was limited by recombination losses arising from the highly intermixed donor and acceptor phases, with FBR apparently unable to aggregate enough to form pure domains that would provide an appropriate charge percolation pathway. In addition, the large extent of spectral overlap of FBR with P3HT and lack of long-wavelength absorption reduced the ability to harvest photons across the spectrum, further limiting the generated photocurrent.

We now present a new acceptor derivative that has been designed to address both the spectral overlap and morphological issues with FBR via replacement of the fluorene core with an indacenodithiophene unit. This has the effect of planarizing the molecular structure and thus significantly red-shifting the absorption as well as increasing the tendency to crystallize on length scales commensurate with charge separation and extraction. We show how these properties can be further tuned via side-chain engineering, with linear (*n*-octyl) alkyl chains yielding a more crystalline material with a further red-shifted absorption onset relative to branched (2-ethylhexyl) chains, resulting in higher J_{sc} and PCE. Power conversion efficiencies of up to 6.4% were achieved, which is, to the best of our knowledge, the highest reported for fullerene-free P3HT solar cells. The oxidative stability of these devices is also found to be superior to the benchmark $P3HT:PC_{60}BM$ devices, as well as devices with several of the high-performance polymers tested alongside, demonstrating this to be a robust and highly promising new materials combination for OPV.

Results

Physical properties. The structure of the new IDTBR acceptors is shown in Fig. 1a. The indacenodithiophene (IDT) core was

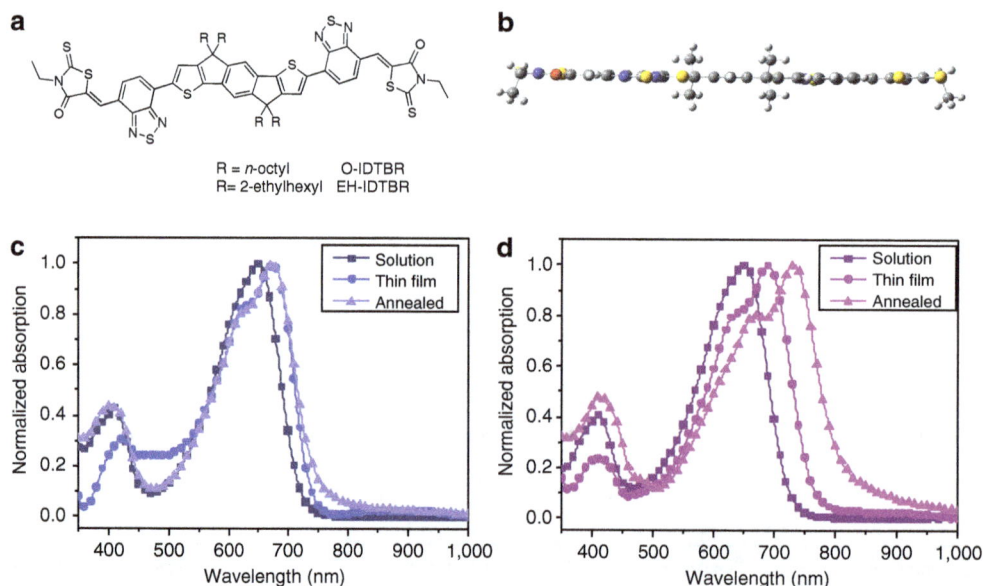

Figure 1 | Structure and UV–vis absorption of IDTBR acceptors. (a) Chemical structures of O-IDTBR and EH-IDTBR; **(b)** Optimized conformation of IDTBR as calculated by DFT (B3LYP/6-31G*)) with methyl groups replacing alkyl chains for clarity; **(c,d)** UV–vis absorption spectra of **(c)** EH-IDTBR and **(d)** O-IDTBR in chloroform solution (1.5×10^{-5} mol l^{-1}), thin film (spin-coated from 10 mg ml^{-1} chlorobenzene solution) and thin film annealed at 130 °C for 10 min. DFT, density functional theory.

synthesized according to literature procedures[34,35] and alkylated using either linear *n*-octyl (O-IDTBR) or branched 2-ethylhexyl (EH-IDTBR) side chains as shown in Fig. 2. Stille coupling of the stannylated IDT with 7-bromo-2,1,3-benzothiadiazole-4-carboxaldehyde was then followed by Knoevenagel condensation with 3-ethylrhodanine to give O-IDTBR and EH-IDTBR in 60% and 30% final yields, respectively. The acceptors are both stable up to 350 °C (Supplementary Fig. 1) and highly soluble in common organic solvents such as chloroform at room temperature, as well as non-halogenated solvents such as *o*-xylene (60 °C), enabling facile solution processing of OPV devices. In the case of FBR, a torsional angle of 33.7° was calculated between the fluorene core and the adjacent benzothiadiazole unit by density functional theory (DFT) methods. By contrast, IDTBR was calculated to be essentially planar (Fig. 1b) due to the increased quinoidal character of the phenyl-thienyl bond compared with the phenyl–phenyl bond, and the reduced steric twisting from adjacent α-C–H bonds on the coupled phenyl rings[35,36]. This enhanced planarity increases conjugation which, when combined with the more electron-rich thiophene-based core, acts to raise the highest occupied molecular orbital (HOMO). This is manifested in a significantly red-shifted UV–visible (UV–vis) absorption spectrum relative to that of FBR. Furthermore, whereas the lowest unoccupied molecular orbital (LUMO) of FBR was localized on the periphery of the molecule, the increased conjugation of IDTBR allows for slightly more delocalization of the LUMO across the central unit (Supplementary Fig. 2), which may be beneficial in terms of molecular oscillator strength and therefore molar absorption coefficient. However, the LUMO of IDTBR is still predominantly located on the periphery of the molecule, which was an important feature in the molecular design as it allows the energy of the highest occupied molecular orbital to be tuned by changing the central

unit while preserving the relatively high-lying LUMO energy and thus maintaining a high open-circuit voltage. The molar absorption coefficient of $1 \times 10^5 \, M^{-1} \, cm^{-1}$ (measured in solution) is over twice the value of FBR and demonstrates the potential of these molecules to contribute significantly more to the photocurrent relative to $PC_{60}BM$ for which the maximum extinction coefficient in the visible region (400 nm) was measured alongside to be only $3.9 \times 10^3 \, M^{-1} \, cm^{-1}$ in $CHCl_3$ (Supplementary Table 1). Furthermore, IDTBR demonstrates significantly stronger absorption in the thin film relative to typical low-bandgap polymers such as PTB7 that absorb at similar wavelengths, as shown from the extinction coefficients plotted in Supplementary Fig. 3a. The absorption coefficient of IDTBR is also higher than those values reported for P3HT[37,38]. This introduces an exciting new concept in the design of active layer materials for OPV, where the acceptor can be used as the primary low-bandgap light absorber, able to donate holes on light absorption in at least an equally efficient way as donor polymers traditionally donate electrons on light absorption.

It has been previously shown that the alkyl chain length and degree of branching can have a significant effect on the optoelectronic and aggregation properties in other IDT-BT-based systems[34] and hence the investigation of both *n*-octyl and 2-ethylhexyl chains with IDTBR. Figure 1c,d compare the UV–vis absorption spectra of the linear O-IDTBR and branched EH-IDTBR. The acceptors have very similar absorption profiles in solution with absorption maxima at 650 nm, and evidently both materials demonstrate greater absorption in the visible region relative to $PC_{60}BM$ (Supplementary Fig. 4 and Supplementary Table 1), which further improves their ability to contribute to photocurrent through absorption. In the thin film, the absorption maximum of O-IDTBR is red-shifted by 40 nm relative to that of EH-IDTBR, with a further bathochromic shift

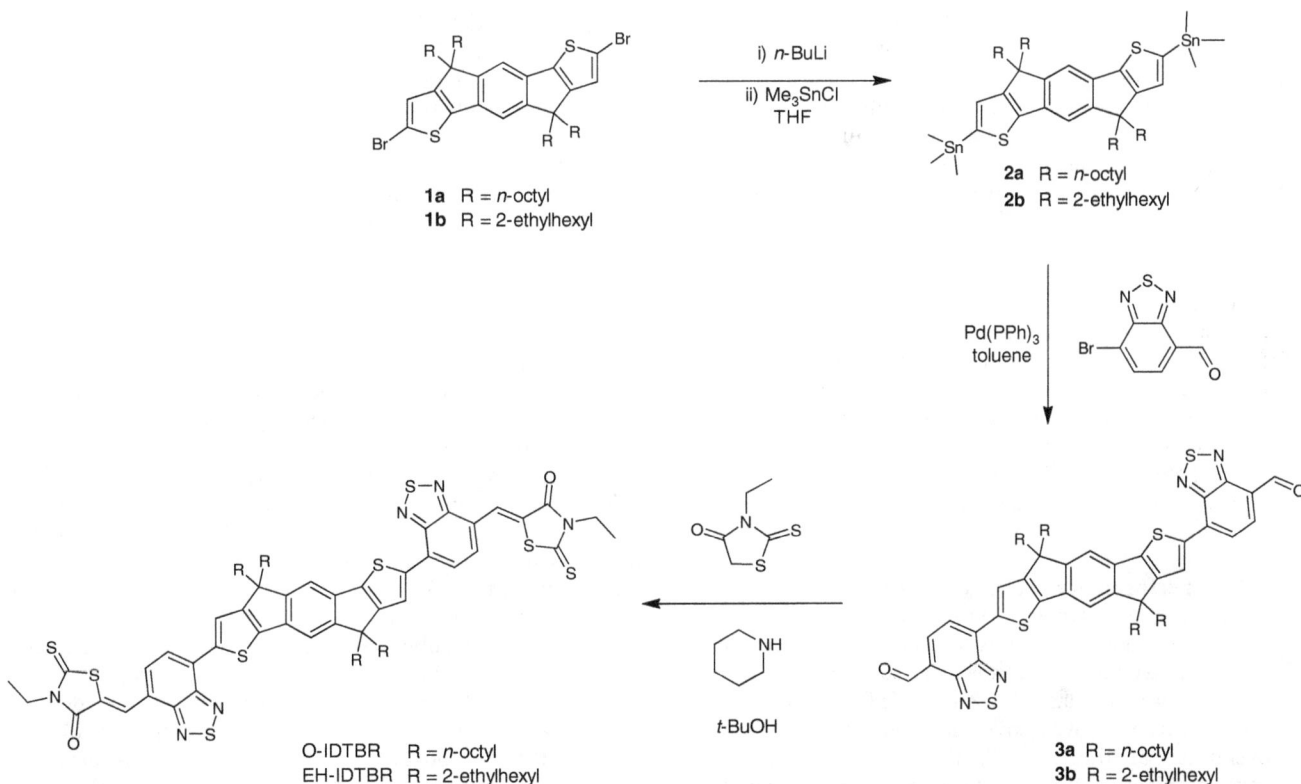

Figure 2 | Synthesis of O-IDTBR and EH-IDTBR acceptors. The brominated indacenodithiophene core is first stannylated with trimethyltin chloride, then reacted via Stille coupling with 7-bromo-2,1,3-benzothiadiazole-4-carboxaldehyde. Knoevenagel condensation with 3-ethylrhodanine yields the final product.

Table 1 | Optoelectronic properties of O-IDTBR and EH-IDTBR acceptors.

	ε (10^4 M^{-1} cm^{-1})*	λ_{max} solution (nm)*	λ_{max} film (nm)[†]	λ_{max} ann. (nm)[‡]	E_g opt. (eV)[†]	EA (eV)[§]	IP (eV)[||]
O-IDTBR	9.9 ± 0.1	650	690	731	1.63 ± 0.1	3.88 ± 0.05	5.51 ± 0.05
EH-IDTBR	10.3 ± 0.1	650	673	675	1.68 ± 0.1	3.90 ± 0.05	5.58 ± 0.05

EA, electron affinity; *IP*, ionization potential.
Measurements were carried out in:
*CHCl$_3$ solution.
[†]Thin film spin-coated from 10 mg ml^{-1} chlorobenzene solution.
[‡]Thin film annealed at 130 °C for 10 min.
[§]Cyclic voltammetry carried out on the as-cast thin film with 0.1 M TBAPF$_6$ electrolyte in acetonitrile.
[||]Estimated from the electrochemical *EA* and the optical E_g.

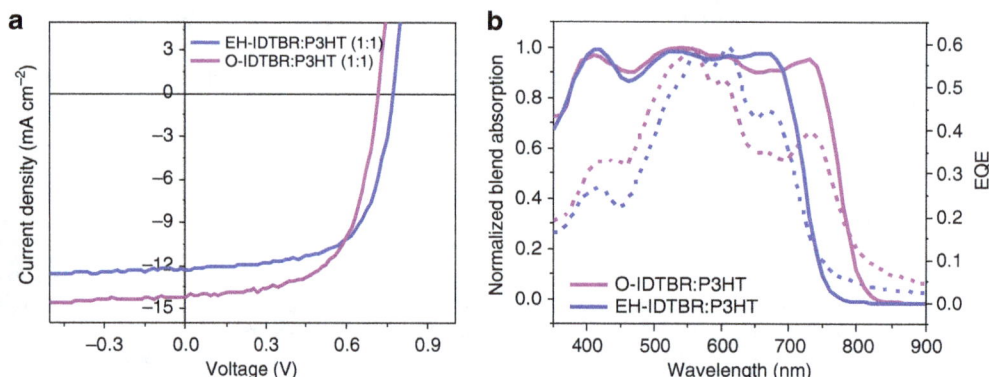

Figure 3 | J–V characteristics and EQE of IDTBR devices with P3HT. (**a**) *J–V* curves of optimized EH-IDTBR:P3HT and O-IDTBR:P3HT solar cells; (**b**) EQE spectra of optimized EH-IDTBR:P3HT and O-IDTBR:P3HT solar cells (solid lines) alongside normalized thin film absorption spectra of blends (dotted lines).

of 41 nm for O-IDTBR upon annealing (above 110 °C, see Table 1 and Supplementary Fig. 3b). The shoulder observed at shorter wavelengths, which has been previously attributed to solid-state aggregation in IDT-BT polymers[34], also becomes more pronounced with thermal annealing. By contrast, the absorption of EH-IDTBR is not affected by annealing (Table 1, Fig. 1c), indicating that the alkyl chains have a significant effect on the tendency of the material to crystallize in the thin film and this in turn strongly affects the absorption properties.

Cyclic voltammetry (CV) in the thin film shows that both EH-IDTBR and O-IDTBR have electron affinity (*EA*) values close to 3.9 eV. The *EA* of P3HT was measured for comparison to be 3.2 eV, allowing sufficient energetic offset for electron transfer between the donor and acceptor. The ionization potential (*IP*) of O-IDTBR was measured to be slightly smaller than that of EH-IDTBR, which accounts for the small difference in optical bandgap (Table 1). This may be due to the enhanced planarization effect of O-IDTBR arising from the additional intermolecular interactions of the more aggregated material. The energy offset between the *IP* of P3HT and both acceptors also appears to be suitable for efficient hole transfer.

Photovoltaic performance. Solar cells were fabricated using P3HT as the donor polymer due to the favourable energetic offsets mentioned above, as well as its widespread availability of P3HT and its potential for technological scale-up. An inverted device architecture of glass/ITO/ZnO/P3HT:IDTBR/MoO$_3$/Ag was chosen for its improved environmental stability relative to the conventional architecture[39,40], allowing for devices to be tested under ambient conditions. The active layer blends (donor-to-acceptor ratio of 1:1) were spin-coated from chlorobenzene solution under ambient conditions without the use of additives. Thermal annealing (10 min at 130 °C) of these films was used to

promote ordering of the polymer, as is typical in P3HT solar cells, as well as to induce acceptor crystallization which will be discussed later. Figure 3 and Table 2 show current density–voltage (*J–V*) data for the optimized devices with an active device area of 0.045 cm^2, which were measured under simulated AM1.5G illumination at 100 mW cm^{-2}. Both acceptors yielded high open-circuit voltage (*V*$_{oc}$) values (0.7–0.8 V) relative to reference devices with PC$_{60}$BM as the acceptor, which gave 0.58 V (Supplementary Fig. 5 and Supplementary Table 2) and this difference is accounted for by the smaller electron affinities of the IDTBR acceptors. IDTBR also generates higher short-circuit currents compared to PC$_{60}$BM with P3HT, which may be related to the increased visible wavelength absorption, and therefore greater photocurrent generation, of these new acceptors. A higher average *J*$_{sc}$ of 13.9 mA cm^{-2} is achieved from the O-IDTBR device, compared with 12.1 mA cm^{-2} for EH-IDTBR. This can be understood, at least in part, by the broader external quantum efficiency (EQE) profile of O-IDTBR, which extends beyond 800 nm due to the red-shifted absorption of the acceptor after annealing. Although the *V*$_{oc}$ and fill factor (FF) are both slightly lower for the linear chain analogue, this significantly larger *J*$_{sc}$ leads to an overall increase in average PCE from 6.0% for EH-IDTBR to 6.3% for O-IDTBR, with a maximum PCE of 6.4% for the best performing device. This is among the highest efficiencies for fullerene-free devices as well as being the highest published efficiency for non-fullerene acceptor devices with P3HT. It is also significantly higher than the reference PC$_{60}$BM:P3HT device efficiency of 3.7%, despite the reduced active layer thickness of 75 nm for the IDTBR devices compared with 150 nm for the fullerene-based device. This difference in active layer thickness can also explain the increased peak EQE in the PC$_{60}$BM:P3HT devices as shown in Supplementary Fig. 5b. To explore the compatibility of our new materials with large-area device fabrication, the dependency of *J*–

Table 2 | Photovoltaic performance of optimized EH-IDTBR:P3HT and O-IDTBR:P3HT solar cells.

	J_{sc} (mA cm^{-2})	V_{oc} (V)	FF	PCE (%)
O-IDTBR:P3HT	13.9 ± 0.2	0.72 ± 0.01	0.60 ± 0.03	6.30 ± 0.1
EH-IDTBR:P3HT	12.1 ± 0.1	0.76 ± 0.01	0.62 ± 0.02	6.00 ± 0.05

FF, fill factor; PCE, power conversion efficiency. Devices were measured under simulated AM1.5G illumination at 100 mW cm^{-2} with average values obtained from 8 to 10 devices.

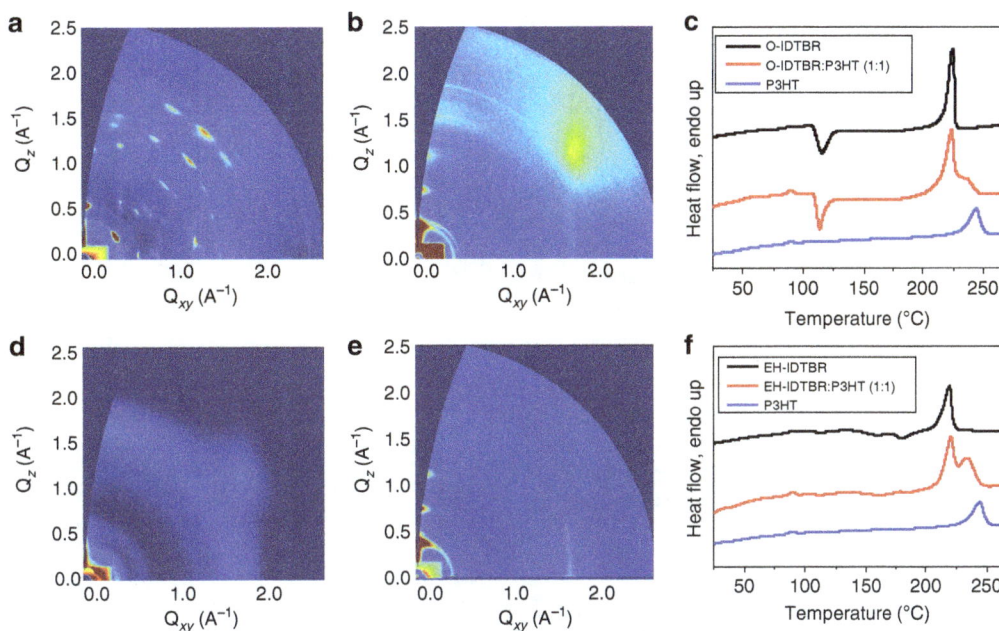

Figure 4 | Morphology of acceptors and IDTBR:P3HT blends. (a) 2D GIXRD of O-IDTBR; (b) 2D GIXRD of O-IDTBR:P3HT (1:1); (c) DSC first heating cycles of O-IDTBR, P3HT and 1:1 blend; (d) 2D GIXRD of EH-IDTBR thin film; (e); 2D GIXRD of EH-IDTBR:P3HT (1:1); and (f) DSC first heating cycles of EH-IDTBR, P3HT and 1:1 blend. Thin films for GIXRD were processed using the same conditions as described for optimized devices and DSC drop-cast samples were measured at 5 °C min^{-1}. Thermograms are offset vertically for clarity.

V properties on active area was analysed for O-IDTBR:P3HT devices, as shown in Supplementary Fig. 6 and Supplementary Table 3. For active layers of 0.15 cm^2, the PCE is maintained at 6.3% and for areas as large as 1.5 cm^2, the PCE is still relatively high as 4.5%, owing to a slight reduction in J_{sc} and FF. It should be noted that these larger area devices were prepared using procedures optimized for the 0.045 cm^2 cells, and that with further optimization of large-area devices their performance may be further improved, demonstrating these materials to be promising candidates for large-area, scalable OPV.

Crystal packing. As discussed above, one of the limiting factors of the previously published FBR acceptor was the intimately mixed morphology with P3HT due to the amorphous nature of the acceptor, leading to charge recombination losses and limiting device performance. One of the design principles of IDTBR was therefore to increase the planarity of the backbone in order to induce crystallization and the formation of pure acceptor domains. Specular X-ray diffraction (XRD) was used to compare the crystallinity of the acceptors in films that were slightly thicker than those used in device fabrication (280–290 nm) in order to provide enough resolution to observe crystalline reflections by this method. Supplementary Figs 7 and 8 show that, while FBR showed no sign of crystallinity in this case even with annealing, both O-IDTBR and EH-IDTBR give strong diffraction peaks. A clear increase in crystalline order is observed for O-IDTBR after

annealing, in accordance with the red-shifted UV–vis absorption. From differential scanning calorimetry (DSC) measurements (Fig. 4) it is apparent that, during the first heating cycle, O-IDTBR undergoes an exothermic crystallization transition with an onset temperature of 108 °C and T_c of 115 °C. No such thermally induced crystallization occurs during the heating cycle of EH-IDTBR, explaining the different optical response of the acceptors to thermal annealing. DSC measurements were also carried out on drop-cast blends of the acceptors with P3HT to determine the extent of crystallization within the blend. The blend of FBR:P3HT (Supplementary Fig. 8) shows only the melting endotherm for P3HT upon heating, which has been depressed (by 20 °C) and broadened due to the disruption in packing caused by the acceptor; however, no transition for the acceptor is observed which indicates a lack of pure acceptor domains in this blend. By contrast, the heating cycles of O-IDTBR and EH-IDTBR blends with P3HT show the endo-thermic (and exothermic, in the case of O-IDTBR) transitions from the acceptor as well as the P3HT melt transition, demonstrating that these acceptors are more able to crystallize in the blend than FBR. Furthermore, the melting temperature of P3HT is only depressed by 10 °C in the IDTBR blends, at the same heating rate, suggesting that the crystallization of P3HT is less disrupted by these acceptors.

Grazing incidence XRD (GIXRD) was used to further investigate the formation of pure donor and acceptor domains in the thin-film blends. Figure 4 shows the GIXRD patterns of

O-IDTBR and EH-IDTBR in both neat films and in 1:1 blends with P3HT, for which samples were prepared using the same conditions used for solar cells. It is evident that O-IDTBR forms a more ordered film than EH-IDTBR, with a narrow out-of-plane distribution of crystallites as given by the narrow width of the diffraction peaks. In O-IDTBR:P3HT blends, the O-IDTBR crystallites become isotropically distributed and exhibit polycrystalline rings in the diffractogram. The magnitude of the scattering wave vectors of the rings match with the diffraction peaks of neat O-IDTBR as is apparent from the peak analysis shown in Supplementary Figs 9 and 10. This suggests that the presence of P3HT may change the crystallite size and distribution of O-IDTBR but not its lattice structure.

EH-IDTBR has an out-of-plane peak centred at $Q_z = 1.69 \text{ Å}^{-1}$, and several rings in its diffraction pattern. The peak most probably results from a portion of face-on π–π stacking of EH-IDTBR aggregates. The rings indicate that besides the aggregates with face-on orientation, the film also has a considerable amorphous fraction. When EH-IDTBR is blended with P3HT, a new peak at $Q_z = 0.48 \text{ Å}^{-1}$ appears, partly overlapping with the broad P3HT (001) alkyl peak at 0.39 Å^{-1}. This peak does not correspond to any features seen in the diffraction pattern of neat EH-IDTBR, suggesting that in the presence of P3HT, EH-IDTBR crystallizes in a different orientation or a different polymorph than in neat form, although the diffraction data is not complete enough to allow us to distinguish between these two hypotheses. It should also be noted that the diffraction pattern of P3HT in the blends is the same as that of a pure P3HT film[41].

Charge-carrier mobilities. It is well known that charge transport is crucial for efficient OPV devices. Carrier mobility of both donor and acceptor materials can be affected by morphology, field or carrier densities in bulk heterojunction active layers under operating conditions[42,43]. To get a reliable charge-carrier mobility of the blend systems, photo-induced charge-carrier extraction in a linearly increasing voltage (photo-CELIV) measurements were conducted. As these photo-CELIV measurements are conducted at 1 sun illumination on actual solar cells, they can provide important information on the transport properties in working devices[44,45]. In contrast to single-carrier measurements, the CELIV technique is more sensitive to the faster carrier component in the blend. The average performing EH-IDBTR:P3HT and O-IDTBR:P3HT devices were used in this experiment, having 80–90 nm active layer thickness and 4 mm^2 active area (see Supplementary Table 4). Figure 5a shows the photo-CELIV transients of the two systems, which were recorded by applying a 2 V per 60 μs linearly increasing reverse bias pulse and a delay time (t_d) of 1 μs. From the measured photocurrent transients, the charge carrier mobility (μ) is calculated using the following equation (1):

$$\mu = \frac{2d^2}{3At_{max}^2 \left[1 + 0.36\frac{\Delta j}{j(0)}\right]} \quad if \ \Delta j \le j(0), \quad (1)$$

where d is the active layer thickness, A is the voltage rise speed $A = dU/dt$, U is the applied voltage, t_{max} is the time corresponding to the maximum of the extraction peak, and $j(0)$ is the displacement current. The photo-CELIV mobilities for the charge carriers in the O-IDTBR and EH-IDTBR blends with P3HT is found to be $5.4 \pm 0.4 \times 10^{-5}$ and $5.0 \pm 0.3 \times 10^{-5} \text{ cm}^2 \text{V}^{-1} \text{s}^{-1}$ after averaging over various delay times, respectively. The O-IDTBR:P3HT blend shows slightly higher charge-carrier density (which is the integrated area of the photo-CELIV curve at 1 μs delay time) than the branched chain analogue system. In addition to photo-CELIV, the electron mobility of EH-IDTBR:P3HT and O-IDTBR:P3HT blends was determined by space charge-limited current (SCLC) measurements on electron-only devices as well as the hole mobility of EH-IDTBR:P3HT blends on hole-only devices. Both acceptors exhibited electron mobilities ~ 3–$6 \times 10^{-6} \text{ cm}^2 \text{V}^{-1} \text{s}^{-1}$, while the hole mobility of EH-IDTBR:P3HT was found to be ~ 3–$7 \times 10^{-4} \text{ cm}^2 \text{V}^{-1} \text{s}^{-1}$ (see Supplementary Fig. 11). Both methods therefore indicate relatively low electron mobilities for these blends. It is interesting to note that in spite of this rather low mobility, IDTBR:P3HT devices display FFs of up to 64% which is within the range of the majority of high-efficiency OPV devices reported in literature[46]. This indicates that non-geminate recombination may be severely suppressed in this system and also that charge generation is not strongly field dependent. However, a more in-depth investigation into the charge recombination dynamics would be needed to determine the exact mechanism behind these high FF values, and these studies are currently on-going.

Charge extraction. Charge-carrier density (n) using charge extraction (CE)[45–47] measurements were conducted for detailed investigation of the origin of reduced V_{oc} in O-IDTBR solar cells

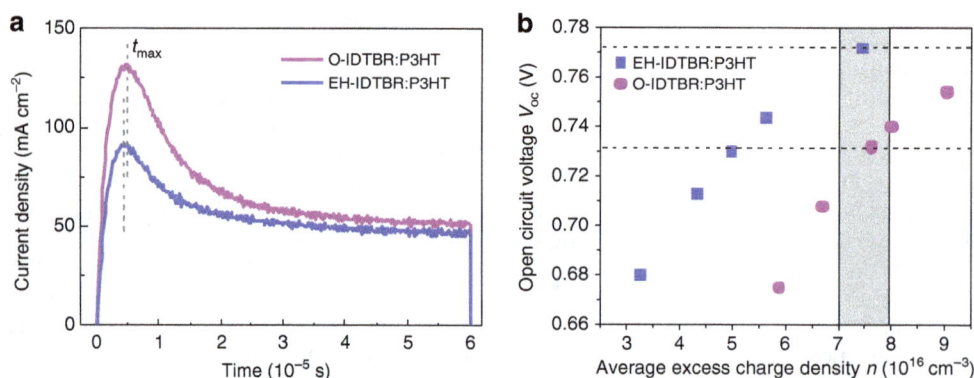

Figure 5 | Charge transport and CE of IDTBR:P3HT blends. (**a**) Photo-CELIV of the O-IDTBR:P3HT and the EH-IDTBR:P3HT solar cells at 1 μs delay times; t_{max} (the time when the extraction current reaches its maximum value) for O-IDTBR:P3HT and EH-IDTBR:P3HT is 4.7 and 4.3 μs, respectively; (**b**) average charge densities measured in O-IDTBR:P3HT and EH-IDTBR:P3HT devices operating at open circuit as a function of V_{oc} determined by CE for different bias light intensities. The grey area marks the data points corresponding ~1 sun light intensity, and dashed lines correspond to the approximate device V_{oc} values (upper line for EH-IDTBR, lower line for O-IDTBR) at 1 sun.

compared with branched EH-IDTBR cells with P3HT. All samples were operated at V_{oc}, but under different background illumination intensities and then shorted in the dark to enable CE. The measured average n as a function of V_{oc} is depicted in Fig. 5b. It is apparent that, at equivalent charge densities, O-IDTBR devices exhibit ~ 40 meV lower open-circuit voltages (see shaded region, corresponding to around 1 sun irradiation) relative to EH-IDTBR. This shift in $n(V_{oc})$ indicates a 40 meV smaller electronic bandgap for O-IDTBR devices, which is consistent with the reduced open-circuit value (0.73 V) for O-IDTBR:P3HT devices compared with EH-IDTBR:P3HT solar cells (0.77 V). This reduced V_{oc} can be explained by the more ordered microstructure of O-IDTBR:P3HT blends, as confirmed with GIXRD measurements, which results in a reduced electronic bandgap in the bulk.

Photoluminescence (PL) quenching of blends. Photoluminescence (PL) studies were carried out on the EH-IDTBR:P3HT and O-IDTBR:P3HT blends relative to neat reference films of EH-IDTBR, O-IDTBR and P3HT to compare the PL quenching efficiency (PLQE) as shown in Supplementary Fig. 12. The selected range in the PL measurement is mainly focused on the emission of the acceptor. The films were excited at 680 nm to excite selectively the IDTBR acceptors, with the PL quenching being assigned to hole transfer from IDTBR excitons to P3HT. It can be seen that the PL quenching is reasonably efficient for both systems, suggesting efficient hole transfer from acceptor excitons to the P3HT donor polymer. Qualitatively it can be seen that the PLQE is slightly larger for the linear compared with the branched chain system, which further affirms that the increased film crystallinity of O-IDTBR allows for the formation of pure acceptor domains on a lengthscale comparable to the exciton diffusion length of O-IDTBR. We note that this PL quenching contrasts with the almost quantitative acceptor PL quenching that was observed for FBR:P3HT blends, and that this is indicative of more pronounced phase segregation with both IDTBR acceptors compared with FBR[33].

Charge generation and recombination dynamics. The charge generation process was studied with femtosecond–nanosecond transient absorption spectroscopy (TAS). Transient spectra of EH-IDTBR and O-IDTBR blends, measured with the acceptors excited selectively at 680 nm, are shown in Supplementary Fig. 13. The spectra of neat EH-IDTBR and O-IDTBR films were collected using the same excitation wavelength and density. Because of the spectral overlap of exciton and polaron signals, these spectra were analysed by deconvoluting the blend spectra from the neat P3HT, neat IDTBR and polaron spectra at selected time delays. Successful deconvolution of the blend spectra using the neat data allowed the temporal evolution of the polaron signal to be extracted for both blends studied herein, as shown in Fig. 6. For both blends, polaron growth kinetics were observed on a similar timescale to acceptor exciton decay. This indicates reasonably efficient charge separation from IDTBR excitons and is also consistent with the photocurrent generation from IDTBR light absorption observed in the EQE data (Fig. 3b). The rise of the polaron signal, and decay of acceptor absorption, fitted reasonably well to single exponential functions. For EH-IDTBR:P3HT, the polaron rise kinetics, and decay kinetics of EH-IDTBR exciton absorption, primarily exhibit time constants of 10–20 ps. Only a small fraction (10–20%) of the polaron generation appears to occur within our instrument response. This contrasts with FBR:P3HT blends, where at least 50% of polaron generation was observed to be instrument response limited[33], consistent with more complete phase segregation compared with FBR. Slower polaron formation and exciton decay is observed for

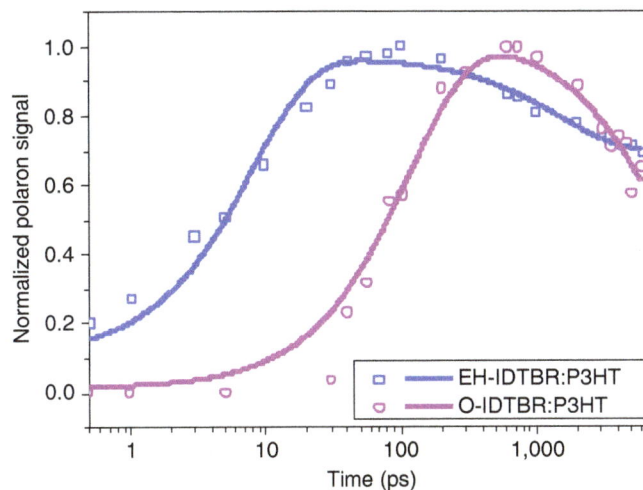

Figure 6 | Charge generation and recombination dynamics of IDTBR:P3HT blends. Rise and decay of photogenerated EH-IDTBR and O-IDTBR polaron absorption, obtained by deconvolution of the ultrafast transient absorption spectra of the EH-IDTBR:P3HT and O-IDTBR:P3HT blend films excited at 680 nm, 2 μJ cm^{-2}. Symbols correspond to deconvoluted polaron signals and the lines correspond to fitting of the data.

O-IDTBR:P3HT (60–120 ps), indicating more delayed polaron generation for this blend which is consistent with our PLQE results. We have previously reported relatively slow (hundreds of picoseconds) polaron generation from acceptor excitons in polymer:PCBM blends, and correlated these with exciton diffusion within pure PCBM domains to the donor/acceptor interface[47]. It appears likely that the slow polaron generation kinetics we observe herein are also limited by the kinetics of exciton diffusion within pure IDTBR domains, with the slower kinetics observed for O-IDTBR being consistent with increased phase separation for this blend as discussed above. Charge recombination is also apparent in Fig. 6 as a decay of the polaron signal at longer time delays. It is apparent that these kinetics are slower for O-IDTBR compared with EH-IDTBR, again most probably associated with great phase segregation in the O-IDTBR blend.

Solar cell stability. Oxidative stability is an essential consideration for the technological implantation of OPV materials[24]. For many of the record high efficiencies reported with low-bandgap polymers, all device fabrication and measurement must be carried out in inert conditions to maintain this performance. By contrast, the efficiencies reported herein for IDTBR:P3HT were obtained with device processing and measurement carried out in air (except for active layer annealing in a nitrogen glovebox). This improved stability is partially attributed to the inverted architecture used, which means that no encapsulation steps are needed for these devices. To further investigate the stability of IDTBR:P3HT devices to air, aging measurements were carried out alongside reference devices of PC$_{60}$BM:P3HT as well as three of the most widely reported high-efficiency polymers PTB7, PCE-10 (PTB7-Th) and PCE-11 (PffBT4T-2OD)[5,48,49] For a fair comparison, all devices were prepared in the same inverted architecture as for IDTBR devices. After the initial (stabilized PCE) measurement was taken, devices were stored in the dark under ambient conditions between measurements, which were taken at intervals over the course of 1,200 h. The corresponding PCE data is shown in Fig. 7, with normalized data given in Supplementary Fig. 14 along with the polymer structures. It is clear from this data that O-IDTBR:P3HT devices demonstrate the

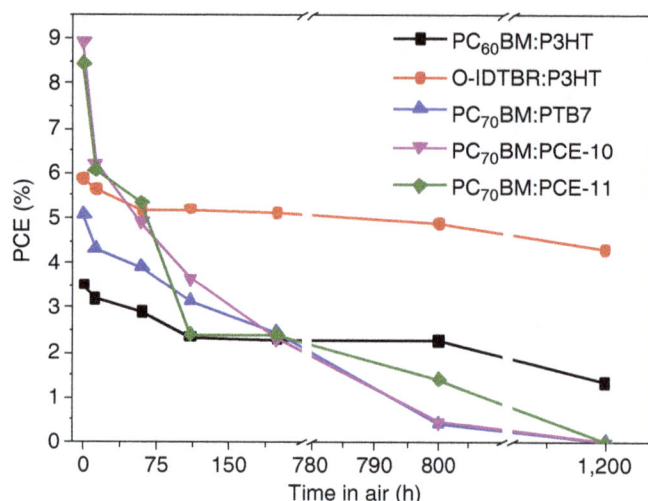

Figure 7 | Solar cell stability. Oxidative stability of O-IDTBR:P3HT device efficiencies (PCE) compared with other high-performance polymer:fullerene systems. Devices were stored in the dark under ambient conditions between measurements.

least degradation out of the materials studied, and that after an initial small drop in performance within the first 60 h, the PCE remains relatively stable and still gives 73% of the initial PCE even after 1,200 h. By contrast, the efficiency of the high-performance donor polymer devices deteriorates remarkably quickly and has fallen to zero by the end of the period of study. This further demonstrates the potential of our new acceptor design for stable, scalable solar cells with practical operating lifetimes, and also gives strong support for the choice of P3HT as donor polymer in these devices.

In addition to oxidative stability, the morphological stability of the O-IDTBR:P3HT blends was investigated. One of the main issues with fullerene-based acceptors like $PC_{60}BM$ is that large-scale aggregates and crystals emerge from the meta-stable blend morphology over time. This process can be monitored by polarized optical microscopy during accelerated aging of the films upon annealing[29,50]. To compare the thermal aging of the IDTBR blends with fullerene blends, films of O-IDTBR:P3HT and $PC_{60}BM$:P3HT were prepared on ZnO/ITO substrates and these were subjected to annealing at 140 °C for 1 h. As the micrographs in Supplementary Fig. 15 show, large (1–20 μm) aggregates appear after 1 h annealing of the fullerene blend, whereas the O-IDTBR blend remains smooth and featureless after annealing, suggesting that this new acceptor offers improved morphological stability over fullerene acceptors, at least in terms of lateral diffusion.

Discussion

In this work, we present a new small molecule electron acceptor IDTBR that is based on an indacenodithiophene core with benzothiadiazole and rhodanine flanking groups. IDTBR is designed to give high performance with the donor polymer P3HT, chosen for its commercial scale-up potential both in terms of cost, scalability and stability. In comparison with our previously published acceptor FBR, which had an essentially overlapping absorption profile with P3HT, this new acceptor has a significantly reduced optical bandgap owing to the more planar molecular backbone, delocalized electronic structure and push–pull molecular orbital hybridization, resulting in a UV–vis absorption profile that is now highly complementary to that of P3HT. This gives broader photon harvesting across the incident

solar spectrum within the active layer, which is reflected in higher short-circuit currents and power conversion efficiencies relative to FBR:P3HT devices. Furthermore, the absorption onset of this new IDTBR acceptor can be tuned by judicial choice of solubilizing alkyl chains on the IDT unit. Linear (O-IDTBR) chains promote stronger intermolecular packing, which is particularly enhanced by thermal annealing, relative to branched (EH-IDTBR) chains. One effect of this is to further red-shift the absorption of O-IDTBR relative to the branched counterpart, which results in a broader EQE profile, higher J_{sc} and an increase in PCE from 6.0 to 6.4%. CE measurements at the same light intensity reveal a reduced electronic bandgap for O-IDTBR relative to EH-IDTBR, which explains the difference in V_{oc} measured for these devices. As well as affecting the optoelectronic properties, the enhanced intermolecular interactions of the linear alkyl chain also have an effect on the blend morphology. Relative to FBR, both IDTBR acceptors exhibit increased crystallinity and, crucially, formation of pure acceptor domains as evidenced by GIXRD and DSC studies. O-IDTBR in particular shows pronounced crystal packing upon annealing, which is consistent with the reduced optical bandgap. This results in greater phase segregation for the linear analogue which is manifested in reduced PL quenching of the acceptor emission, as well as a delayed polaron generation and slower recombination dynamics in the O-IDTBR:P3HT blend. Interestingly, the charge-carrier mobilities measured for the IDTBR:P3HT blends appear quite low, considering the reasonably high FFs obtained from devices (up to 64%) and the charge recombination dynamics of these systems therefore warrant further investigation to determine whether non-geminate recombination is significantly suppressed. In addition to high efficiencies, IDTBR:P3HT devices demonstrate improved stability in ambient conditions compared with the benchmark $PC_{60}BM$:P3HT device, as well as several systems with typical low-bandgap, high-performance polymers, which were found to degrade at a dramatic rate when exposed to air. IDTBR devices also showed improved morphological stability to fullerene devices in accelerated aging studies. These results strongly supports the use of P3HT, in conjunction with non-fullerene acceptors such as IDTBR, for high-efficiency, scalable and stable OPV for future technological applications.

Methods

General characterization. ^{1}H and ^{13}C NMR spectra were collected on a Bruker AV-400 spectrometer at 298 K and are reported in p.p.m. UV–vis absorption spectra were recorded on a UV-1601 Shimadzu UV–vis spectrometer. DSC experiments were carried out with a Mettler Toledo DSC822 instrument at a heating rate of $5 °C min^{-1}$ under nitrogen. Samples were prepared by drop-casting the materials from $CHCl_3$ solution directly into the DSC pan and allowing the solvent to evaporate under Ar. Specular XRD was carried out on thin films of the acceptors spin-coated from $CHCl_3$ solutions ($30 mg ml^{-1}$, 600 r.p.m.) using a PANalytical X'Pert PRO MRD diffractometer equipped with a nickel-filtered Cu-Kα1 beam and X'Celerator detector, with a current $I = 40$ mA and accelerating voltage $U = 40$ kV. Samples for GIXRD were spin-coated on Si (100) substrates following the same spin-coating and annealing procedures as were used in fabricating solar cells.

Synthesis. The compounds 1a and 1b were prepared according to literature procedure[34,35], as was 7-bromo-2,1,3-benzothiadiazole-4-carboxaldehyde[33]. P3HT was obtained from Flexink Ltd. All other reagents and solvents were purchased from Sigma Aldrich or Acros Organics and used as received. All reactions were carried out using conventional Schlenk techniques in an inert argon atmosphere.

2a. A solution of 1a (2.11 g, 2.42 mmol) in anhydrous tetrahydrofuran (200 ml) was stirred at −78 °C for 30 min. n-BuLi (2.42 ml, 6.04 mmol, 2.5 M in hexanes) was added dropwise and the solution was stirred at −78 °C for 30 min followed by −10 °C for 30 min. After cooling again to −78 °C, trimethyltin chloride was added (7.26 ml, 7.56 mmol, 1 M in hexanes) and the solution was allowed to return to room temperature overnight. The reaction was then poured into water and extracted with hexane, washed successively with acetonitrile to remove excess trimethyltin chloride and dried over $MgSO_4$ to yield 2a as a yellow oil (2.18 g, 86%). ^{1}H NMR (400 MHz, $CDCl_3$) δ: 7.25 (s, 2H), 6.97 (s, 2H), 1.97–1.91 (m, 4H), 1.86–1.78 (m, 4H), 1.23–1.05 (m, 48H), 0.83–0.80 (t, 12H, J = 7 Hz), 0.39 (s, 18H);

^{13}C NMR (101 MHz, CDCl$_3$) δ: 157.15, 153.47, 147.71, 139.24, 135.31, 129.55, 113.42, 53.06, 39.20, 31.87, 30.07, 30.03, 29.31, 24.17, 22.68, 14.14 and $-$8.02. MS (ES-ToF): m/z calculated for C$_{54}$H$_{90}$S$_2$Sn: 1,040.45; m/z found 1,041.40 (M + H)$^+$.

3a. A solution of **2a** (1.04 g, 1.0 mmol) and 2,1,3-benzothiadiazole-4-carboxaldehyde (0.73 g, 3.0 mmol) in anhydrous toluene (40 ml) was degassed for 45 min before Pd(PPh$_3$)$_4$ (58 mg, 0.05 mmol) was added and this solution was heated at 100 °C overnight. The reaction mixture was then cooled and purified by flash column chromatography on silica mixed with potassium fluoride using CHCl$_3$ as the eluent. Further purification by column chromatography on silica using CH$_2$Cl$_2$/pentane (1:1) followed by precipitation from methanol yielded **3a** as a dark purple solid (0.93 g, 90%). ^{1}H NMR (400 MHz, CDCl$_3$) δ: 10.72 (s, 2H), 8.27 (s, 2H), 8.25 (d, J = 7.7 Hz, 2H), 8.06 (d, J = 7.5 Hz, 2H), 7.45 (s, 2H), 2.05 (dtd, J = 59.3, 12.9, 4.6 Hz, 8H), 1.05–1.2 (m, 38H), 0.99–0.81 (m, 10H), 0.77 (t, J = 6.8 Hz, 12H); ^{13}C NMR (101 MHz, CDCl$_3$) δ: 188.44, 157.04, 154.02, 152.29, 147.00, 140.67, 136.44, 134.14, 132.87, 131.62, 124.87, 124.8, 122.80, 114.12, 54.43, 39.16, 31.79, 29.98, 29.29, 29.20, 24.29, 22.58, 14.04. MS (ES-ToF): m/z calculated for C$_{62}$H$_{78}$N$_4$O$_2$S$_4$: 1,038.5; m/z found 1,041.40.

O-IDTBR. **3a** (0.40 g, 0.39 mmol) and 3-ethylrhodanine (186 mg, 1.16 mmol) were dissolved in *tert*-butyl alcohol (30 ml). Two drops of piperidine were added and the solution was left to stir at 85 °C overnight. The product was extracted with CHCl$_3$ and dried over MgSO$_4$. The crude product was purified by flash column chromatography on silica in CH$_2$Cl$_2$ and precipitated from methanol. The precipitate was collected and dried by vacuum filtration to afford **O-IDTBR** a dark blue solid (0.40 g, 78%). mp = 219–221 °C. ^{1}H NMR (400 MHz, CDCl$_3$) δ: 8.54 (s, 2H), 8.24 (s, 2H), 8.03 (d, J = 8.0 Hz, 2H), 7.74 (d, J = 7.9 Hz, 2H), 7.45 (s, 2H), 4.27 (q, J = 8.0 Hz, 4H), 2.18–1.96 (m, 8H), 1.35 (t, J = 8.1 Hz, 6H), 1.22–1.12 (m, 40H), 0.99–0.90 (m, 8H), 0.80 (m, 12H). ^{13}C NMR (101 MHz, CDCl$_3$) δ: 193.04, 167.59, 157.05, 154.63, 154.22, 151.77, 146.15, 141.02, 136.41, 131.37, 130.54, 127.29, 124.49, 124.25, 124.08, 123.82, 113.97, 54.38, 39.94, 39.19, 31.82, 30.02, 29.33, 29.24, 24.30, 22.61, 14.08 and 12.35. MS (matrix-assisted laser desorption/ionization–time of flight): m/z calculated for C$_{72}$H$_{88}$N$_6$O$_2$S$_8$: 1,324.5; m/z found 1,326.0 (M + H)$^+$.

2b. A solution of **1b** (1.09 g, 1.25 mmol) in anhydrous tetrahydrofuran (40 ml) was stirred at $-$78 °C for 30 min. *n*-BuLi (1.25 ml, 3.12 mmol, 2.5 M in hexanes) was added dropwise and the solution was stirred at $-$78 °C for 1 h. Trimethyltin chloride was then added (3.75 ml, 3.75 mmol, 1 M in hexanes) and the solution was allowed to return to room temperature overnight. The reaction was then poured into water and extracted with hexane, washed successively with acetonitrile to remove excess trimethyltin chloride and dried over MgSO$_4$ to yield **2b** as a yellow oil (1.16 g, 89%). ^{1}H NMR (400 MHz, CDCl$_3$) δ: 7.28 (s, 2H), 6.99 (s, 2H), 1.96–1.88 (m, 8H), 1.87–1.82 (m, 8H), 0.99–0.46 (m, 60H), 0.37 (s, 18H). ^{13}C NMR (101 MHz, CDCl$_3$) δ: 157.40, 153.43, 147.51, 140.73, 135.20, 130.04, 113.95, 53.52, 43.59, 34.89, 32.20, 29.75, 28.74, 28.10, 22.67, 14.16 and $-$8.16.

3b. A solution of **2b** (0.94 g, 0.90 mmol) and 2,1,3-benzothiadiazole-4-carboxaldehyde (0.53 g, 2.17 mmol) in anhydrous toluene (30 ml) was degassed for 45 min before Pd(PPh$_3$)$_4$ (52 mg, 0.05 mmol) was added and this solution was heated at 110 °C overnight. The reaction mixture was then cooled and purified by flash column chromatography on silica mixed with potassium fluoride using CHCl$_3$ as the eluent. Further purification by column chromatography on silica using CH$_2$Cl$_2$/pentane (1:1) followed by precipitation from methanol yielded **3b** as a dark purple solid (0.40 g, 43%). ^{1}H NMR (400 MHz, CDCl$_3$) δ: 10.72 (s, 2H), 8.37–8.30 (m, 2H), 8.25 (d, J = 7.6 Hz, 2H), 8.03 (d, J = 7.5 Hz, 2H), 7.49 (s, 2H), 2.15–2.05 (m, 8H), 1.05–0.85 (m, 40H), 0.74–0.50 (m, 20H). MS (ES-ToF): m/z calculated for C$_{62}$H$_{78}$N$_4$O$_2$S$_4$: 1,038.50; m/z found 1,038.50 (M$^+$).

EH-IDTBR. **3b** (0.20 g, 0.19 mmol) and 3-ethylrhodanine (93 mg, 0.58 mmol) were dissolved in *tert*-butyl alcohol (15 ml). 1 drop of piperidine was added and the solution was left to stir at 85 °C overnight. The product was extracted with CHCl$_3$ and dried over MgSO$_4$. The crude product was purified by flash column chromatography on silica with CH$_2$Cl$_2$ as the eluent followed by precipitation from methanol to yield **EH-IDTBR** as a dark blue solid (0.20 g, 80%). mp = 218–220 °C. ^{1}H NMR (400 MHz, CDCl$_3$) δ: 8.53 (s, 2H), 8.27 (m, 2H), 7.99 (m, 2H), 7.73 (d, J = 8.1 Hz, 2H), 7.47 (s, 2H), 4.25 (q, J = 8.0 Hz, 4H), 2.07 (m, 8H), 1.34 (t, J = 8.0 Hz, 6H), 0.95–0.90 (m, 36H), 0.69–0.50 (m, 24H). ^{13}C NMR (101 MHz, CDCl$_3$) δ: 193.07, 167.58, 156.76, 154.63, 153.93, 151.80, 146.14, 140.46, 136.38, 131.37, 130.64, 127.31, 125.08, 124.51, 124.30, 123.73, 114.82, 54.19, 39.94, 35.13, 34.16, 28.64, 28.25, 27.26, 22.86, 14.18, 12.33 and 10.60. MS (matrix-assisted laser desorption/ionization–time of flight): m/z calculated for C$_{72}$H$_{88}$N$_6$O$_2$S$_8$: 1,324.5; m/z found 1,325.9 (M + H)$^+$.

Cyclic voltammetry. CV measurements were performed using an Autolab PGSTAT101 potentiostat. Thin films of the acceptor were spin-coated onto ITO-coated glass substrates to be used as the working electrode, alongside a platinum mesh counter electrode and Ag/Ag$^+$ reference electrode. Measurements were carried out in anhydrous and deoxygenated acetonitrile with 0.1 M of tetrabutylammonium hexafluorophosphate (TBA PF$_6$) as the supporting electrolyte, and calibrated against ferrocene in solution using a cylindrical Pt working electrode. IP and EA values were calculated from the following equations:

$$EA = (E_{red} - E_{Fc} + 4.8) \text{ eV} \tag{2}$$

$$IP = (E_{ox} - E_{Fc} + 4.8) \text{ eV} \tag{3}$$

where E_{red} and E_{ox} are taken from the onset of reduction and oxidation, respectively, and E_{Fc} is taken as the half-wave potential of ferrocene.

OPV devices. Bulk heterojunction solar cells were fabricated with an inverted architecture (glass/ITO/ZnO/P3HT:Acceptor/MoO$_3$/Ag). Glass substrates were used with pre-patterned indium tin oxide (ITO). These were cleaned by sonication in detergent, deionized water, acetone and isopropanol, followed by oxygen plasma treatment. ZnO layers were deposited by spin-coating a zinc acetate dihydrate precursor solution (60 μl monoethanolamine in 2 ml 2-methoxyethanol) followed by annealing at 150 °C for 10–15 min, giving layers of 30 nm. The P3HT:IDTBR (1:1 ratio by mass) active layers were deposited from 24 mg ml^{-1} solutions in chlorobenzene by spin-coating at 2,000 r.p.m., followed by annealing at 130 °C for 10 min. Active layer thicknesses were 75 nm (averaged over six devices) for both acceptor blends. P3HT:PC$_{60}$BM (1:1 ratio by mass) layers were spin-coated at 1,500 r.p.m. from 40 mg ml^{-1} solutions in o-dichlorobenzene, followed by annealing at 130 °C for 20 min, resulting in active layer thicknesses of 148 nm. MoO$_3$ (10 nm) and Ag (100 nm) layers were deposited by evaporation through a shadow mask yielding active areas of 0.045 cm^2 in each device. (J–V) characteristics were measured using a Xenon lamp at AM1.5 solar illumination (Oriel Instruments) calibrated to a silicon reference cell with a Keithley 2400 source meter, correcting for spectral mismatch. Incident photon conversion efficiency was measured by a 100 W tungsten halogen lamp (Bentham IL1 with Bentham 605 stabilized current power supply) coupled to a monochromator with computer controlled stepper motor. The photon flux of light incident on the samples was calibrated using a UV-enhanced silicon photodiode. A 590-nm long-pass glass filter was inserted into the beam at illumination wavelengths longer than 580 nm to remove light from second-order diffraction. Measurement duration for a given wavelength was sufficient to ensure the current had stabilized.

The low-bandgap polymers PTB7, PCE-10 (PTB7-Th) and PCE-11 (PffBT4T-2OD) used in stability studies were obtained from Ossila, and the active layers for these devices were prepared as follows, with the same architecture a used for the IDTBR:P3HT devices.

PTB7:PC$_{70}$BM. Active layer solutions (D:A ratio 1:1.5) were prepared in CB with 3 wt% 1,8-diiodooctane (total concentration 25 mg ml^{-1}). To completely dissolve the polymer, the active layer solution was stirred on a hot plate at 80 °C for at least 3 h. Active layers were spin-coated from the warm polymer solution on preheated substrates in a nitrogen glove box at 1,500 r.p.m.

PCE-10:PC$_{70}$BM. Active layer solutions (D:A ratio 1:1.5) were prepared in CB with 3 wt% 8-diiodooctane (total concentration 35 mg ml^{-1}). To completely dissolve the polymer, the active layer solution was stirred on a hot plate at 80 °C for at least 3 h. Active layers were spin-coated from the warm polymer solution onto preheated substrates in a nitrogen glove box at 1,500 r.p.m.

PCE-11:PC$_{70}$BM. Active layer solutions (D:A ratio 1:1.4) were prepared in CB/o-DCB (1:1 volume ratio) with 3 wt% 8-diiodooctane (polymer concentration: 10 mg ml^{-1}). To completely dissolve the polymer, the active layer solution was stirred on a hot plate at 110 °C for at least 3 h. Active layers were spin-coated from the warm polymer solution onto preheated substrates in a nitrogen glove box at 1,000 r.p.m.

Photo-CELIV. In photo-CELIV measurements, the devices were illuminated with a 405 nm laser-diode. Current transients were recorded across an internal 50 Ω resistor on an oscilloscope (Agilent Technologies DSO-X 2024A). A fast electrical switch was used to isolate the cell and prevent CE or sweep out during the laser pulse and the delay time. After a variable delay time, a linear extraction ramp was applied via a function generator. The ramp, which was 20 μs long and 2 V in amplitude, was set to start with an offset matching the V_{oc} of the cell for each delay time. The geometrical capacitance is calculated as:

$$C = \varepsilon \varepsilon_0 A / d \tag{4}$$

where A is the device area (4 mm^2), $\varepsilon = 3$ and, $\varepsilon_0 = 8.85 \times 10^{-12}$ F m^{-1} are the relative and absolute dielectric permittivity, respectively, and d is the active layer thickness (90 nm). C is then calculated as 1 nF. Assuming $R_{load} = 50$ nm, the RC value is 5.9×10^{-8} s. Assuming the electrical field (E) is 1×10^5 V m^{-1}, the transient time (t) is calculated with the following formula:

$$t = t_{max} * \sqrt{3} \quad t = t_{max} * \sqrt{3} = 8 \times 10^{-6} \text{ s} \tag{5}$$

Charge extraction. In CE measurements, the devices were illuminated in air with a 405 nm laser diode for 200 μs, which was sufficient to reach a constant open-circuit voltage with steady state conditions. At the end of the illumination period, an analogue switch was triggered that switched the solar cell from open-circuit to short-circuit (50 ω) conditions within less than 50 ns. By adjusting the laser intensity, different open-circuit voltages were obtained which allowed a plot to be generated of charge-carrier density over voltage. As described by Shuttle *et al.*[51], a correction was applied for the charge on the electrodes that results from the geometric capacity of the device[52].

Space charge-limited current. SCLC measurements were performed on electron-only devices of the structure ITO/PEDOT:PSS/Al/P3HT:acceptor/Al and on hole-only devices of the structure ITO/PEDOT:PSS/P3HT:acceptor/Au using a Paios (FLUXiM AG) measurement system. The current–voltage characteristics were fitted by the Mott–Gurney law in the region where the current follows the square of the voltage to extract the carrier mobility.

PL spectroscopy and transient absorption spectroscopy (TAS). Samples for TAS and PL spectroscopy were spin-coated onto glass using the same conditions as for solar cells. Spectra were measured using a steady state spectrofluorimeter (Horiba Jobin Yvon, Spex Fluoromax 1). The spin-coated films were excited at 680 nm. Sub-picosecond TAS was carried out at 800 nm laser pulse (1 kHz, 90 fs) by using a Solstice (Newport Corporation) Ti:sapphire regenerative amplifier. A part of the laser pulse was used to generate the pump laser at 680 nm, $2 \mu J \, cm^{-2}$ with a TOPAS-Prime (light conversion) optical parametric amplifier. The other laser output was used to generate the probe light in near visible continuum (450–800 nm) by a sapphire crystal. The spectra and decays were obtained by a HELIOS transient absorption spectrometer (450–1,450 nm) and decays to 6 ns. The samples were measured in N_2 atmosphere. Deconvolution of the blend spectra was conducted by fitting the singlet EH-IDTBR exciton spectrum ($S_{exciton}$) and the P3HT:EH-IDTBR polaron spectrum ($S_{polaron}$) at 6 ns to the blend spectra for 20 different time delays using the equation:

$$\Delta_{OD} = A_1 * S_{exciton}(1,150 \, nm) + A_2 * S_{polaron}(1,000 \, nm) \qquad (6)$$

where A_1 and A_2 are linear coefficients that estimate the percentage contribution of the spectra to the experimental blend spectra. $S_{exciton}$ was derived from the transient absorption spectra of the EH-IDTBR, which peaks at 1,150 nm at selected time delays (Supplementary Fig. 13). This signal, assigned to singlet exciton absorption, disappeared at $\sim 20 \, ns$ (Fig. 6). The polaron spectrum was derived from the TA spectra of the blend at 6 ns, where no exciton contributions are expected (Supplementary Fig. 13). Note that an exciton signal from P3HT is not expected, as supported by our PL measurements.

References

1. Huang, J. *et al.* 10.4% Power conversion efficiency of ITO-free organic photovoltaics through enhanced light trapping configuration. *Adv. Energy Mater.* **5**, 1500406 (2015).
2. He, Z. *et al.* Enhanced power-conversion efficiency in polymer solar cells using an inverted device structure. *Nat. Photon.* **6**, 593–597 (2012).
3. Liao, S.-H., Jhuo, H.-J., Cheng, Y.-S. & Chen, S.-A. Fullerene derivative-doped zinc oxide nanofilm as the cathode of inverted polymer solar cells with low-bandgap polymer (PTB7-Th) for high performance. *Adv. Mater.* **25**, 4766–4771 (2013).
4. Ye, L., Zhang, S., Zhao, W., Yao, H. & Hou, J. Highly efficient 2D-conjugated benzodithiophene-based photovoltaic polymer with linear alkylthio side chain. *Chem. Mater.* **26**, 3603–3605 (2014).
5. Liu, Y. *et al.* Aggregation and morphology control enables multiple cases of high-efficiency polymer solar cells. *Nat. Commun.* **5**, 5293 (2014).
6. Zhou, H. *et al.* Polymer homo-tandem solar cells with best efficiency of 11.3%. *Adv. Mater.* **27**, 1767–1773 (2015).
7. Chen, C. C. *et al.* An efficient triple-junction polymer solar cell having a power conversion efficiency exceeding 11%. *Adv. Mater.* **26**, 5670–5677 (2014).
8. Lin, Y. & Zhan, X. Non-fullerene acceptors for organic photovoltaics: an emerging horizon. *Mater. Horizons* **1**, 470–488 (2014).
9. Eftaiha, A. F., Sun, J., Hill, I. G. & Welch, G. C. Recent advances of non-fullerene, small molecular acceptors for solution processed bulk heterojunction solar cells. *J. Mater. Chem. A* **2**, 1201 (2014).
10. Nielsen, C. B., Holliday, S., Chen, H.-Y., Cryer, S. J. & McCulloch, I. Non-fullerene electron acceptors for use in organic solar cells. *Acc. Chem. Res.* **48**, 2803–2812 (2015).
11. Zhong, Y. *et al.* Efficient organic solar cells with helical perylene diimide electron acceptors. *J. Am. Chem. Soc.* **136**, 15215–15221 (2014).
12. Zhang, X., Zhan, C. & Yao, J. Non-fullerene organic solar cells with 6.1% efficiency through fine-tuning parameters of the film-forming process. *Chem. Mater.* **27**, 166–173 (2015).
13. Zhao, J. *et al.* High-efficiency non-fullerene organic solar cells enabled by a difluorobenzothiadiazole-based donor polymer combined with a properly matched small molecule acceptor. *Energy Environ. Sci.* **8**, 520–525 (2015).
14. Li, H. *et al.* Fine-tuning the 3D structure of nonfullerene electron acceptors toward high-performance polymer solar cells. *Adv. Mater.* **27**, 3266–3272 (2015).
15. Zhong, Y. *et al.* Molecular helices as electron acceptors in high-performance bulk heterojunction solar cells. *Nat. Commun.* **6**, 8242 (2015).
16. Lin, Y. *et al.* An electron acceptor challenging fullerenes for efficient polymer solar cells. *Adv. Mater.* **27**, 1170–1174 (2015).
17. Lin, Y. *et al.* High-performance fullerene-free polymer solar cells with 6.31% efficiency. *Energy Environ. Sci.* **8**, 610–616 (2015).
18. Lou, S. J. *et al.* Effects of additives on the morphology of solution phase aggregates formed by active layer components of high-efficiency organic solar cells. *J. Am. Chem. Soc.* **133**, 20661–20663 (2011).
19. Soon, Y. W. *et al.* Correlating triplet yield, singlet oxygen generation and photochemical stability in polymer/fullerene blend films. *Chem. Commun.* **49**, 1291–1293 (2013).
20. Razzell-Hollis, J. *et al.* Photochemical stability of high efficiency PTB7:PC70BM solar cell blends. *J. Mater. Chem. A* **2**, 20189–20195 (2014).
21. Dang, M. T., Hirsch, L. & Wantz, G. P3HT:PCBM, best seller in polymer photovoltaic research. *Adv. Mater.* **23**, 3597–3602 (2011).
22. Mulligan, C. J. *et al.* A projection of commercial-scale organic photovoltaic module costs. *Sol. Energy Mater. Sol. Cells* **120**, 9–17 (2014).
23. Po, R. *et al.* From lab to fab: how must the polymer solar cell materials design change?—an industrial perspective. *Energy Environ. Sci.* **7**, 925–943 (2014).
24. Jørgensen, M. *et al.* Stability of polymer solar cells. *Adv. Mater.* **24**, 580–612 (2012).
25. Manceau, M. *et al.* Effects of long-term UVvisible light irradiation in the absence of oxygen on P3HT and P3HT: PCBM blend. *Sol. Energy Mater. Sol. Cells* **94**, 1572–1577 (2010).
26. Tremel, K. & Ludwigs, S. in *P3HT Revisited—From Molecular Scale to Solar Cell Devices.* (ed. Ludwigs, S.) Ch. 2 (Springer, 2014).
27. Bannock, J. H. *et al.* Continuous synthesis of device-grade semiconducting polymers in droplet-based microreactors. *Adv. Funct. Mater.* **23**, 2123–2129 (2013).
28. Espinosa, N., Hosel, M., Jorgensen, M. & Krebs, F. C. Large scale deployment of polymer solar cells on land, on sea and in the air. *Energy Environ. Sci.* **7**, 855–866 (2014).
29. Campoy-Quiles, M. *et al.* Morphology evolution via self-organization and lateral and vertical diffusion in polymer:fullerene solar cell blends. *Nat. Mater.* **7**, 158–164 (2008).
30. Brabec, C. J., Heeney, M., McCulloch, I. & Nelson, J. Influence of blend microstructure on bulk heterojunction organic photovoltaic performance. *Chem. Soc. Rev.* **40**, 1185–1199 (2011).
31. Holliday, S., Donaghey, J. E. & McCulloch, I. Advances in charge carrier mobilities of semiconducting polymers used in organic transistors. *Chem. Mater.* **26**, 647–663 (2014).
32. Guo, X. *et al.* High efficiency polymer solar cells based on poly(3-hexylthiophene)/indene-C70 bisadduct with solvent additive. *Energy Environ. Sci.* **5**, 7943–7949 (2012).
33. Holliday, S. *et al.* A rhodanine flanked nonfullerene acceptor for solution-processed organic photovoltaics. *J. Am. Chem. Soc.* **137**, 898–904 (2015).
34. Bronstein, H. *et al.* Indacenodithiophene-co-benzothiadiazole copolymers for high performance solar cells or transistors via alkyl chain optimization. *Macromolecules* **44**, 6649–6652 (2011).
35. Zhang, W. *et al.* Indacenodithiophene semiconducting polymers for high-performance, air-stable transistors. *J. Am. Chem. Soc.* **132**, 11437–11439 (2010).
36. McCulloch, I. *et al.* Design of semiconducting indacenodithiophene polymers for high performance transistors and solar cells. *Acc. Chem. Res.* **45**, 714–722 (2012).
37. Zhang, M., Guo, X., Ma, W., Ade, H. & Hou, J. A polythiophene derivative with superior properties for practical application in polymer solar cells. *Adv. Mater.* **26**, 5880–5885 (2014).
38. Cook, S., Furube, A. & Katoh, R. Analysis of the excited states of regioregular polythiophene P3HT. *Energy Environ. Sci.* **1**, 294–299 (2008).
39. Sun, Y., Seo, J. H., Takacs, C. J., Seifter, J. & Heeger, A. J. Inverted polymer solar cells integrated with a low-temperature-annealed sol-gel-derived ZnO film as an electron transport layer. *Adv. Mater.* **23**, 1679–1683 (2011).
40. Xu, Z. *et al.* Vertical phase separation in poly(3-hexylthiophene): fullerene derivative blends and its advantage for inverted structure solar cells. *Adv. Funct. Mater.* **19**, 1227–1234 (2009).
41. Shao, M. *et al.* The isotopic effects of deuteration on optoelectronic properties of conducting polymers. *Nat. Commun.* **5**, 3180 (2014).
42. You, J. *et al.* A polymer tandem solar cell with 10.6% power conversion efficiency. *Nat. Commun.* **4**, 1446 (2013).
43. Gasparini, N. *et al.* Photophysics of molecular-weight-induced losses in indacenodithienothiophene-based solar cells. *Adv. Funct. Mater.* **25**, 4898–4907 (2015).
44. Chen, S. *et al.* Photo-carrier recombination in polymer solar cells based on P3HT and silole-based copolymer. *Adv. Energy Mater.* **1**, 963–969 (2011).
45. Clarke, T. M., Lungenschmied, C., Peet, J., Drolet, N. & Mozer, A. J. A comparison of five experimental techniques to measure charge carrier lifetime in polymer/fullerene solar cells. *Adv. Energy Mater.* **5**, 1401345 (2015).

46. Bartesaghi, D. *et al.* Competition between recombination and extraction of free charges determines the fill factor of organic solar cells. *Nat. Commun.* **6**, 7083 (2015).

47. Dimitrov, S. D. *et al.* Towards optimisation of photocurrent from fullerene excitons in organic solar cells. *Energy Environ. Sci.* **7**, 1037–1043 (2014).

48. Lu, L. & Yu, L. Understanding low bandgap polymer PTB7 and optimizing polymer solar cells based on it. *Adv. Mater.* **26**, 4413–4430 (2014).

49. He, Z. *et al.* Single-junction polymer solar cells with high efficiency and photovoltage. *Nat. Photon.* **9**, 174–179 (2015).

50. Schroeder, B. C. *et al.* Enhancing fullerene-based solar cell lifetimes by addition of a fullerene dumbbell. *Angew. Chemie Int. Ed.* **53**, 12870–12875 (2014).

51. Shuttle, C. G. *et al.* Experimental determination of the rate law for charge carrier decay in a polythiophene: fullerene solar cell. *Appl. Phys. Lett.* **92**, 093311 (2008).

52. Heumueller, T. *et al.* Disorder-induced open-circuit voltage losses in organic solar cells during photoinduced burn-in. *Adv. Energy Mater.* **5**, 1500111 (2015).

Acknowledgements

We thank BASF for partial financial support, as well as EPSRC Projects EP/G037515/1 and EP/M023532/1, EC FP7 Project SC2 (610115), EC FP7 Project ArtESun (604397), EC FP7 Project POLYMED (612538), Project Synthetic carbon allotropes project SFB 953 and the King Abdullah University of Science and Technology (KAUST). George Richardson is gratefully acknowledged for his assistance with optical microscopy. M.A. thanks Z. Kan and Y. Firdaus for helpful discussions.

Author contributions

S.H. and A.W. synthesized the IDTBR acceptors and carried out DSC measurements. S.H. and D.B. carried out optoelectronic characterization. S.H. ran DFT calculations. R.S.A. and S.A.Y. fabricated and characterized solar cell devices. D.B. and N.G. carried out photo-CELIV and CE measurements. C.B.N. conducted specular XRD measurements. M.A. carried out SCLC experiments. C.-H.T. and S.D. carried out PL and TAS experiments. Z.S. carried out GIXRD measurements. S.H. prepared the manuscript with contributions from D.B., C.-H. T, Z. S. and F. L. All authors discussed the results and commented on the manuscript. F.L. supervised SCLC, C.J.B. supervised photo-CELIV and CE, A.S. supervised GIXRD and J.R.D. supervised PL and TAS and revised the manuscript. I.M. revised the manuscript and supervised and directed the project.

Additional information

18

Design and synthesis of the superionic conductor Na$_{10}$SnP$_2$S$_{12}$

William D. Richards[1], Tomoyuki Tsujimura[2], Lincoln J. Miara[3], Yan Wang[1], Jae Chul Kim[1,4], Shyue Ping Ong[5], Ichiro Uechi[2], Naoki Suzuki[2] & Gerbrand Ceder[1,4,6]

Sodium-ion batteries are emerging as candidates for large-scale energy storage due to their low cost and the wide variety of cathode materials available. As battery size and adoption in critical applications increases, safety concerns are resurfacing due to the inherent flammability of organic electrolytes currently in use in both lithium and sodium battery chemistries. Development of solid-state batteries with ionic electrolytes eliminates this concern, while also allowing novel device architectures and potentially improving cycle life. Here we report the computation-assisted discovery and synthesis of a high-performance solid-state electrolyte material: Na$_{10}$SnP$_2$S$_{12}$, with room temperature ionic conductivity of 0.4 mS cm^{-1} rivalling the conductivity of the best sodium sulfide solid electrolytes to date. We also computationally investigate the variants of this compound where tin is substituted by germanium or silicon and find that the latter may achieve even higher conductivity.

[1] Department of Materials Science and Engineering, Massachusetts Institute of Technology, Cambridge, Massachusetts 02139, USA. [2] Samsung R&D Institute Japan, Minoh Semba Center Building 13F, Semba Nishi 2-1-11, Minoh, Osaka 562-0036, Japan. [3] Samsung Advanced Institute of Technology—USA, 255 Main Street, Suite 702, Cambridge, Massachusetts 02142, USA. [4] Materials Sciences Division, Lawrence Berkeley National Laboratory, Berkeley, California 94720, USA. [5] Department of NanoEngineering, University of California San Diego, La Jolla, California 92093-0448, USA. [6] Department of Materials Science and Engineering, University of California Berkeley, Berkeley, California 94720, USA. Correspondence and requests for materials should be addressed to G.C. (email: gceder@berkeley.edu).

The energy density and cycle life of intercalation batteries has made them the dominant technology in all manner of applications, from consumer electronics to electric vehicles and high-performance batteries in commercial aircraft. A potential key application of sodium (Na)-ion battery technology is grid-scale energy storage due to its lower cost relative to lithium (Li)-ion. In addition to sodium being considerably more abundant than Li, Na-ion batteries have the advantage of a broader range of available cathode materials since many layered Li-transition metal oxides show improved performance in their sodium versions[1–3]. In addition, many high capacity Na-cathodes do not contain cobalt, an expensive and scarce component of many commercial Li-ion cathodes. The choice of electrolyte for Na-ion systems is not as well established as for Li-ion, but the vast majority of electrolytes in development are based on organic solvents[4]. These suffer from the same flammability concerns as their counterparts in lithium batteries, and are exacerbated by the presence of a much more reactive metal. In addition, heat dissipation properties of the battery are worsened by the size of installations required for grid storage, and thermal runaway is therefore an even greater concern than at smaller scales. The creation of low-temperature sodium-solid electrolytes would go a long way towards the development of safe solid-state Na-ion batteries, free of flammable solvents.

Despite much work in the area, no anode materials for sodium ion systems have been found that can match the conductivity, energy density and price of graphite anodes in Li-ion batteries. Though hard carbon anodes have been shown to reversibly intercalate sodium[5], capacity is very low compared with that allowed by intercalation to LiC_6 (ref. 6). Solid electrolytes may also suppress dendrite formation, enabling the use of sodium metal anodes and improving battery capacity considerably.

Na-solid electrolytes have been commercialized in high-temperature batteries such as β-alumina for sodium–sulfur (NAS) batteries[7], yet few materials with high conductivities at low temperature have been reported. Conductivities over $1 \, mS \, cm^{-1}$ have been shown in NASICON-type oxide crystals[8,9], but processing of these materials at high temperatures (typically $>1,000 \, °C$ (refs 7,10)) is required to reduce grain boundary resistance, which is incompatible with typical cathode materials and complicates battery fabrication. In both sodium and lithium systems, thiophosphate materials are promising candidates as solid electrolytes as they are soft and can be incorporated into batteries by cold pressing without requiring high-temperature sintering. In 2011, Kamaya et al. reported the synthesis of $Li_{10}GeP_2S_{12}$ (LGPS), a tetragonal structure within the $Li_{3+x}Ge_xP_{1-x}S_4$ thiophosphate system that achieves $12 \, mS \, cm^{-1}$ conductivity at room temperature[11], and in 2014 Seino et al. reported a conductivity of $27 \, mS \, cm^{-1}$ with a glass-ceramic electrolyte composed of $Li_7P_3S_{11}$ crystals precipitated from a glass[12]. There has been relatively less experimental work done on sodium systems, though recently the cubic phase of Na_3PS_4 has been reported to have conductivity as high as $0.46 \, mS \, cm^{-1}$, and has been used in an all-solid-state battery[13,14]. Similar to lithium materials, Si doping has recently been used to increase defect concentrations in Na_3PS_4, resulting in a conductivity of $0.7 \, mS \, cm^{-1}$ (ref. 15). For construction of solid-state cells, low-strain electrodes[16,17] are also important to minimize delamination of the electrolyte, especially when using harder electrolyte materials such as oxides.

Ab initio calculation of material properties can rapidly speed up the search for new solid electrolyte materials, allowing prediction of as-yet-undiscovered materials through calculations of phase diagrams[18] and evaluation of the diffusivity of carefully controlled structures, which normally can only be achieved experimentally after a long process of synthesis optimization. Even when new hypothetical materials show similarity to existing

materials, chemical intuition alone cannot reliably determine which chemical modifications will result in experimentally realizable structures, for example, the high conductivity lithium garnets[19] have no direct sodium analogues. Ab initio calculations have shown remarkable accuracy in predicting the properties of Li-solid-state conductors[11,20–24]. Here we present computational predictions of three Na-ion conductors crystallizing in the high-conductivity tetragonal structure and with conductivities equalling and exceeding the current best performing materials. We also show experimental results confirming our prediction of the $Na_{10}SnP_2S_{12}$ material, with experimentally measured conductivity and activation energy in excellent agreement with the ab initio result. In addition to rigorously probing the energetics and hence feasibility of synthesis of these materials, computational techniques are able to give an indication of their performance. They are thus able to focus experimental efforts on systems with a high probability of success. These results highlight the predictive nature of first principles calculations—in addition to explaining difficult to observe phenomena, they can be used to discover compounds with extraordinary physical properties.

Results

General considerations. Using first principles computation, we evaluate three key properties of the tetragonal phases of $Na_{10}MP_2S_{12}$ (M = Si, Ge, Sn) to determine their suitability as a solid-state electrolyte materials: (1) we determine the Na^+ conductivity and its activation energy from ab initio molecular dynamics (AIMD) simulations, (2) using high-throughput computations and structure prediction methods, we comprehensively calculate the ground-state phase diagram of each system to gauge the stability and synthesizability of each compound and (3) we extract the electrochemical anodic and cathodic stability limits from the grand canonical equilibrium at various potentials similar to the approach described in an earlier work[25]. Based on this data, we then proceeded to synthesize and test $Na_{10}SnP_2S_{12}$.

Ground-state energy calculations. Since there is typically considerable cation site disorder in these conductors, we used an electrostatic energy criterion to pre-sceen Na/Vacancy orderings on the experimentally reported structure of LGPS[26]. For each of the three symmetrically distinct M/P orderings and for full and half Na4 site occupancy, we relaxed the structures of the lowest electrostatic energy arrangements using density functional theory (DFT), taking the lowest energy of these as the 0 K enthalpy and structure. The structure of $Na_{10}MP_2S_{12}$ (NMPS) can be described as consisting of three symmetrically distinct chains of cations oriented parallel to the c-axis (Fig. 1). At unit cell coordinates $x = 0.25, y = 0.25$, tetrahedral Na sites (Na1, Na3) form a chain of partially occupied edge-sharing sites. At $x = 0, y = 0.5$ there is an edge-sharing chain of alternating Na_{oct} and $(M/P)_{tet}$ sites. At $x = 0, y = 0$, a similar chain but with a vacancy instead of M cation and more distorted Na_{oct} site is present (with repeat unit Na_{oct}-P_{tet}-Na_{oct}-Vac_{tet}). The ab initio MD results will demonstrate that the (Na1, Na3) chains carry most of the Na conductivity with occasional crossover through the Na sites in the chain at $x = 0, y = 0$. The ground state M/P ordering, which is found to be shared among all studied chemistries, is shown in Fig. 1 and the ground-state Na-ion arrangement (C222 space group) in Supplementary Fig. 2.

Ab initio molecular dynamics. The Na-ionic conductivity (σ), and activation energy (E_a) were determined from AIMD simulations between 600 and 1,300 K and extrapolated to room temperature. Ionic conductivity is calculated from AIMD through the intermediate calculation of D_σ, which has the units of a diffusivity but takes into account correlations between Na-ions (see Methods). The results are shown in Fig. 2a, and compared with similar Li compounds in Table 1. The self diffusivity (D_{self}) of the Na-ions was also calculated for comparison, with results included in Supplementary Table 1 and Supplementary Fig. 1. For both the Li and Na materials, activation energy slightly increases as M changes from Si → Ge → Sn. Somewhat surprisingly, given the size difference between Na and Li ions, Na and Li materials have similar activation energies, resulting in high room temperature conductivities particularly for the Ge and Si materials, which are predicted to have room temperature conductivities comparable to those of organic electrolytes[4]. Our result for $Na_{10}GeP_2S_{12}$ is similar to the result of ref. 27. These conductivities are more impressive, given that they are entirely due to Na^+ motion, and so the transference number is equal to 1. The degree of cooperativity of ionic motion is described by the Haven ratio H_r (ref. 28), which we calculate from the ratio of D_{self} to D_σ. This value is calculated to be ~0.56 in all of our simulations, which is slightly smaller than that observed experimentally for the lithium versions of these materials[22,23], indicating a larger degree of cooperative motion.

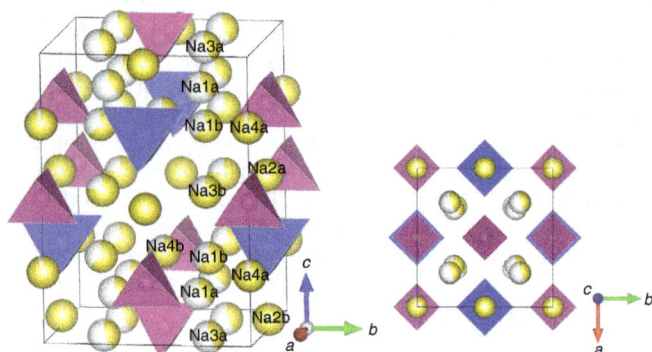

Figure 1 | Structure of $Na_{10}SnP_2S_{12}$ from DFT calculations. Sodium occupancies are calculated from 600 K AIMD simulation (see Methods). All ground-state NMPS structures share this M/P ordering, which reduces the symmetry from the $P4_2/nmc$ space group to $P\bar{4}m2$, separating each Na-site into two symmetrically distinct but similar sites marked as a and b. PS_4 tetrahedra are marked in purple, SnS_4 tetrahedra in blue and Na-sites in yellow. The ground-state Na-ordering is shown in Supplementary Fig. 2.

Phase diagrams and stability limits. To determine the feasibility of synthesizing these high conductivity tetragonal phases of $Na_{10}MP_2S_{12}$ (M = Si, Ge, Sn), we used DFT to evaluate the energies of materials and generate their respective quaternary phase diagrams. To obtain appropriate competing phases in the quaternary phase diagrams, we calculated the energy of a very large number of compounds in their relevant chemical spaces, including all known materials present in the Inorganic Crystal Structure Database (ICSD)[29] containing some or all of the four elements, all relevant materials derived from substituting sodium for lithium in all ICSD materials and the $Li_xP_yS_z$ structures compiled by Lepley et al.[30]. To further improve the coverage of these chemical spaces, we also applied the data-mined substitution methodology of Hautier et al.[18] to predict possible structures from a broader range of chemistries in the ICSD. The 0 K phase diagram for Na-Sn-P-S is shown in Fig. 3. Na-Ge-P-S and Na-Si-P-S phase diagrams are available in Supplementary Fig. 3.

No quaternary ground states are found in any of the three systems. Decomposition energy (E_{decomp}) to the equilibrium ground-state structures is calculated using the convex hull method implemented in pymatgen[31] and is shown in Table 2, and compared with their lithium counterparts. For example, the stability of the $Na_{10}SnP_2S_{12}$ phase is given by the calculated enthalpy of the decomposition reaction $Na_{10}SnP_2S_{12} \rightarrow 2 Na_3PS_4 + Na_4SnS_4$. Even though all the considered electrolyte structures show a small driving force at 0 K to decompose to $(Li/Na)_4MS_4$ (M = Si, Ge, Sn) and $(Li/Na)_3PS_4$, this is similar to the Li-analogues that have similar decomposition energies, and have all been synthesized[11,21,23,24]. We expect high configurational entropy on the cation sites to result in their stabilization at moderate temperatures. An approximation of this entropy, neglecting the ion–ion interactions, can be obtained using the formula $S = -k_B \sum_i p_i \ln p_i$, where k_B is the Boltzmann constant, p_i is the probability of each state (occupied or unoccupied) and the sum is over all states for each site. Using a value of 50% occupancy of the Na-atoms in the edge-sharing c-axis chains and 50% M/P occupancy (28 sites with 50% occupancy per 50 atom unit cell) yields a value of 0.0334 meV K^{-1} per atom, which at 300 K already would stabilize the Sn and Ge compositions. This is an upper bound on the configurational entropy but vibrational entropy, particularly the soft phonon modes of the diffusing ions, is also expected to contribute to the structure's stabilization. Table 2 also shows the calculated anodic and cathodic stability limits evaluated from the chemical potentials of Na at which the compound decomposes, following the methods of ref. 25. Since

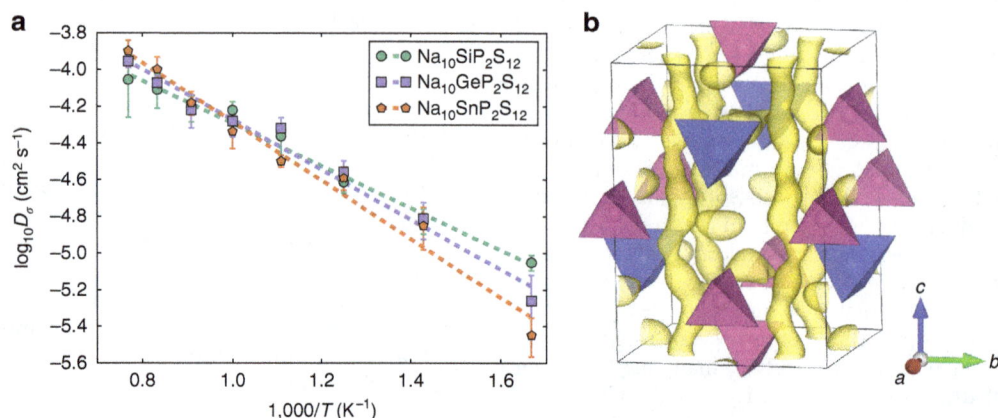

Figure 2 | DFT computed diffusivity of $Na_{10}SnP_2S_{12}$. (**a**) Na-diffusivity in $Na_{10}SiP_2S_{12}$, $Na_{10}GeP_2S_{12}$ and $Na_{10}SnP_2S_{12}$ from AIMD simulation. Dashed lines are Arrhenius fits to the data, and error bars are standard error of the mean. (**b**) Na-ion probability density isosurface (yellow) of $Na_{10}SnP_2S_{12}$ from 600 K AIMD simulation. SnS_4 tetrahedra are marked in blue, PS_4 tetrahedra in purple.

Table 1 | Ionic conductivity of cation-substituted compounds $X_{10}MP_2S_{12}$.

Compound	DFT simulation		Experimental	
	E_a (eV)	Conductivity, 298 K (mS cm^{-1})	E_a (eV)	Conductivity, 298 K (mS cm^{-1})
$Na_{10}SiP_2S_{12}$	0.229	10.28	NA	NA
$Na_{10}GeP_2S_{12}$	0.270	3.50	NA	NA
$Na_{10}SnP_2S_{12}$	0.317	0.94	0.356	0.4 (this work)
$Li_{10}SiP_2S_{12}$	0.20	23 (ref. 20)	0.196	2.3 (ref. 21)
$Li_{10}GeP_2S_{12}$	0.21	13 (ref. 20)	0.22–0.25	9–12 (refs 11,22)
$Li_{10}SnP_2S_{12}$	0.24	6 (ref. 20)	0.24–0.27	4–7 (refs 23,24)

NA, not applicable. DFT simulation and experimental results on the sodium structures are from this work. Experimental and calculated values for the Li compounds are taken from the literature.

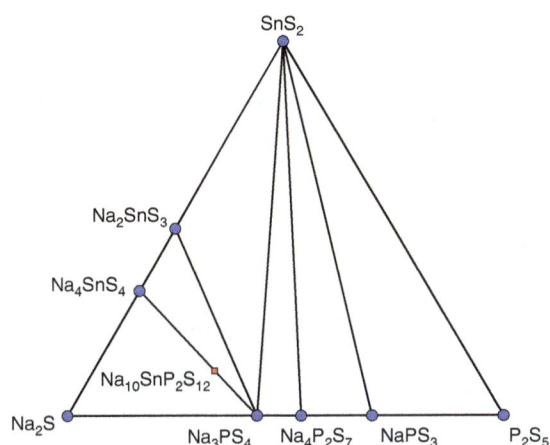

Figure 3 | Na-Sn-P-S phase diagram. Pseudo-ternary 0 K Na-Sn-P-S phase diagram constructed from DFT energy calculations, with location of $Na_{10}SnP_2S_{12}$. Stable phases marked with blue dot.

these materials by our calculations are metastable at 0 K, we instead consider the potentials at which the ground-state materials equilibrium becomes unstable, for example, for $Na_{10}SnP_2S_{12}$, when either Na_4SnS_4 or Na_3PS_4 becomes unstable.

When the chemical potential (voltage) of the alkali is below (above) the stable region (as can be experienced at the cathode interface during charging), the ion and its associated electron is pulled from the electrolyte, which decomposes into a mixture of sulfides and elemental sulfur (for example, Na_4SnS_4 decomposes to S and Na_2SnS_3 above 1.82 V versus Na metal, and S and SnS_2 above 2 V). In contact with the anode (cathodic limit), the Li/Na metal may reduce the metal or phosphorus in the electrolyte, potentially leading to electron conductivity through the electrolyte if this reaction continues without passivation. The cathodic limit for Na and Li compounds is set by the partial reduction of phosphorus to form Na_2PS_3, and the calculated cathodic stability is thus unaffected by the choice of metal (M) cation. The potentials at which the metal cation is fully reduced by the alkali are also listed in Table 2, and indicate the potential at which the decomposition reaction is no longer passivating. The shift in the stability window between the Na and Li materials is due to the differing reduction potentials of the alkali metal. Previous DFT studies have shown that this reduction reaction can be passivated in some systems by the formation of a thin layer of Li_2S (ref. 30), though in practice insulating barrier coatings are typically employed at the anode/cathode interfaces[11,32,33]. The anodic voltage stability limit is set primarily by the reaction energy of the alkali metal with elemental sulfur, though in compounds with highly negative enthalpies of mixing from the binary sulfides the stability range is extended slightly. This effect is small in the considered electrolyte materials,

with the anodic stability only changing on the order of 0.1 V between materials with different (M) cations.

Synthesis and experimental verification. In validation of our computational predictions, we report successful synthesis of $Na_{10}SnP_2S_{12}$, which was chosen due to its low materials cost and E_{decomp} of 7.1 meV per atom, which is lower than comparable materials that have been synthesized. $Na_{10}SnP_2S_{12}$ was prepared from the binary sulfide phases (see Methods), under a range of cooling rates. The lattice volume and conductivity of the synthesized phase increase as the cooling rate is lowered (Supplementary Fig. 4), with the highest conductivity achieved by cooling from 700 °C over 99 h. To compare the experimental XRD pattern with that predicted from DFT calculation, we used the Na and Sn/P site disordered structure with positions and fractional occupancies of each site generated from k-means clustering of Na-position data from the 600 K AIMD simulation as a starting point for powder XRD simulation of the structure. Comparison of the simulated and experimental XRD patterns is shown in Fig. 4a. The obtained material is predominantly the expected tetragonal $Na_{10}SnP_2S_{12}$, with small amounts of P_2S_5, Na_3PS_4, primarily in the tetragonal α-phase as indicated by the peak splitting at 31 and 36 degrees[13], and Na_2S, which formed during the slow cooling. At faster cooling rates, these impurity phases do not form but the resulting material has lower conductivity due to the lower lattice volume. The change in the lattice volume and conductivity is likely a result of the structure in the slow cooled sample having a higher ratio of Sn to P, since the observed impurities contain no Sn. Similar dependency of conductivity and lattice volume on this ratio are seen in the lithium systems[34]. The low conductivity of the impurity phases the slow-cooled sample are expected to reduce the measured conductivity by reduction in the effective cross-sectional area. The strong relation between lattice volume and conductivity also support the conductivity measured in the slow-cooled sample being that of $Na_{10}SnP_2S_{12}$.

The intensities of the 011 and (110 and 002) reflections, producing XRD peaks at 12 and 16 degrees, vary as a function with cooling rate, but are not strongly correlated with conductivity. Supplementary Fig. 5 shows the XRD spectrum of a quenched sample in which these low-angle peaks are more clearly visible. The variation in these peak intensities may be caused either by slight disorder between the P_{tet}, M_{tet} and Vac_{tet} sites, or by changes in average size of the $(Sn/P)S_4$ tetrahedra from slight compositional variation.

Considering that AIMD simulations were performed at elevated temperatures and extrapolated to experimental conditions, the conductivity predicted from these simulations is in remarkable agreement to our experimental electrochemical impedance spectroscopy results (Fig. 4b). We predicted a room temperature conductivity of 0.94 mS cm^{-1} with activation energy

Table 2 | Phase equilibria decomposition enthalpies and stability ranges for $X_{10}MP_2S_{12}$.

Cation(X)	Cation(M)	Decomposition products	E_{decomp} (meV per atom)	Metal reduction (V versus metal anode)	Cathodic stability (V versus metal anode)	Anodic stability (V versus metal anode)
	Si	$Na_4SiS_4 + 2\ Na_3PS_4$	13.6	0.80	1.25	1.77
Na	Ge	$Na_4GeS_4 + 2\ Na_3PS_4$	7.2	1.10	1.25	1.70
	Sn	$Na_4SnS_4 + 2\ Na_3PS_4$	7.1	1.09	1.25	1.82
	Si	$Li_4SiS_4 + 2\ Li_3PS_4$	14.9	1.36	1.78	2.14
Li	Ge	$Li_4GeS_4 + 2\ Li_3PS_4$	14.7	1.64	1.78	2.06
	Sn	$Li_4SnS_4 + 2\ Li_3PS_4$	13.4	1.57	1.78	2.02

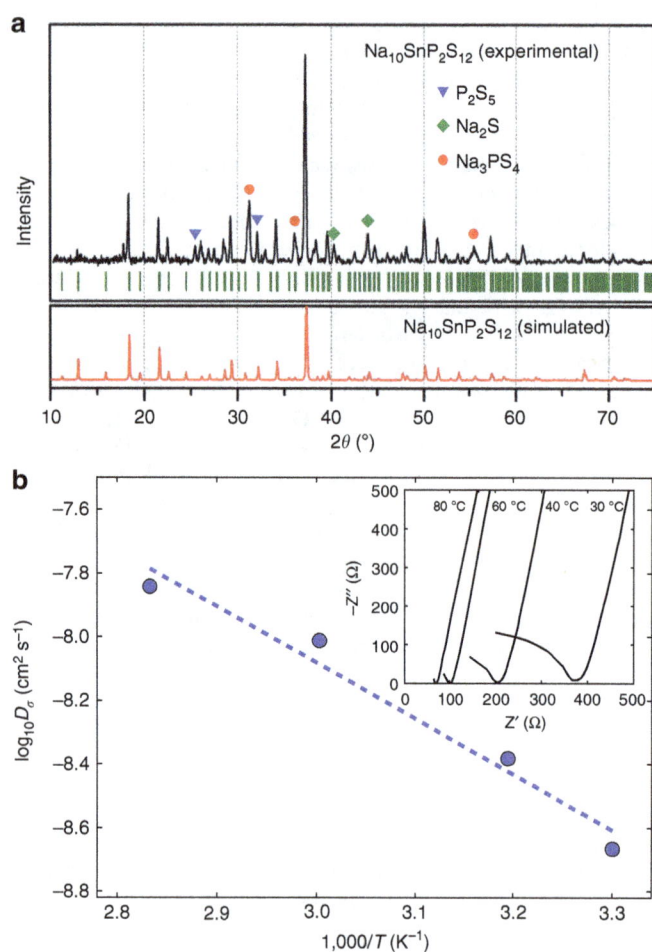

Figure 4 | Experimental crystal structure and diffusivity of $Na_{10}SnP_2S_{12}$. (a) Experimental and simulated XRD patterns of $Na_{10}SnP_2S_{12}$, showing small amounts of recrystallized P_2S_5, Na_3PS_4 and Na_2S. (b) Diffusivity calculated from experimentally measured ionic conductivity versus temperature. Dashed line is an Arrhenius fit to the data. (inset) Electrochemical impedance spectroscopy measurements.

best performing sulfide electrolyte to date—cubic Na_3PS_4, which achieves conductivities of between 0.2 and $0.7\ \mathrm{mS\ cm^{-1}}$ depending on doping and processing conditions[13–15]. These thiophosphate electrolytes benefit from improved processability relative to the oxide β-alumina and NASICON-based compounds, which can have higher conductivities but require high-temperature sintering, making them difficult to incorporate into room temperature batteries. To evaluate the potential for even better conductors in this family of compounds, we investigate in more detail the conductivity mechanism in these compounds and the effect of the main group metal (Si, Ge, Sn) on it.

From our DFT calculations, we see that the activation energy for Na diffusion in the NMPS materials shown in Fig. 2a increases as the ionic radius of the (M)etal in the compound increases, with $E_a^{Si} < E_a^{Ge} < E_a^{Sn}$. This trend is also seen in activation energies for the Li conductors, both in experimental and DFT studies (Table 1). This is surprising since often the activation energy barrier between adjacent sites in a structure decreases as the size of the anion framework increases. In the NMPS conductors, however, the lattice parameter differences are small (<1%, Supplementary Table 2), and the activation energy actually increases as the cell volume increases. The valence of the other cations near the transition state has been pointed to as an important factor as it can increase the activation energy by strong repulsion of the alkali in the activated state[35,36], but this is unlikely to play a role here as Si, Ge and Sn all have valence 4 +. Hence, because of their similar volume and cation valence, these three compounds form a good data set to evaluate potentially more subtle chemical influences on the conductivity. To understand the somewhat counterintuitive result, we examine the diffusion paths and site occupancies in each compound as a measure of the free-energy landscape of the structures.

From the AIMD Na-ion trajectories, we calculate the Na-ion probability density, defined as the time-averaged Na-ion occupancy, allowing visualization of the Na-ion diffusion mechanism. The probability density from AIMD simulation of $Na_{10}SnP_2S_{12}$ at 600 K in Fig. 2b is representative of all of our AIMD simulations, and shows that the majority of the Na diffusion occurs within the c-axis chain of partially occupied Na sites at $x = 0.25$ and $y = 0.25$, with some crossover between these channels. These results are in good qualitative agreement with the highly anisotropic Li sites seen in previous spectroscopic studies on $LGPS^{11,26}$.

The Na-site occupancies of the three materials as a function of simulation temperature are shown in Fig. 5. $P4_2/nmc$ spacegroup operations are applied to the Na-positions before analysis to undo the splitting of Na sites caused by the M/P ordering and shown in

of 317 meV, while experimentally $Na_{10}SnP_2S_{12}$ shows a conductivity of $0.4\ \mathrm{mS\ cm^{-1}}$ with an activation energy of 356 meV.

Discussion

$Na_{10}SnP_2S_{12}$ is a remarkably good ionic conductor; its room temperature conductivity of $0.4\ \mathrm{mS\ cm^{-1}}$ is comparable to the

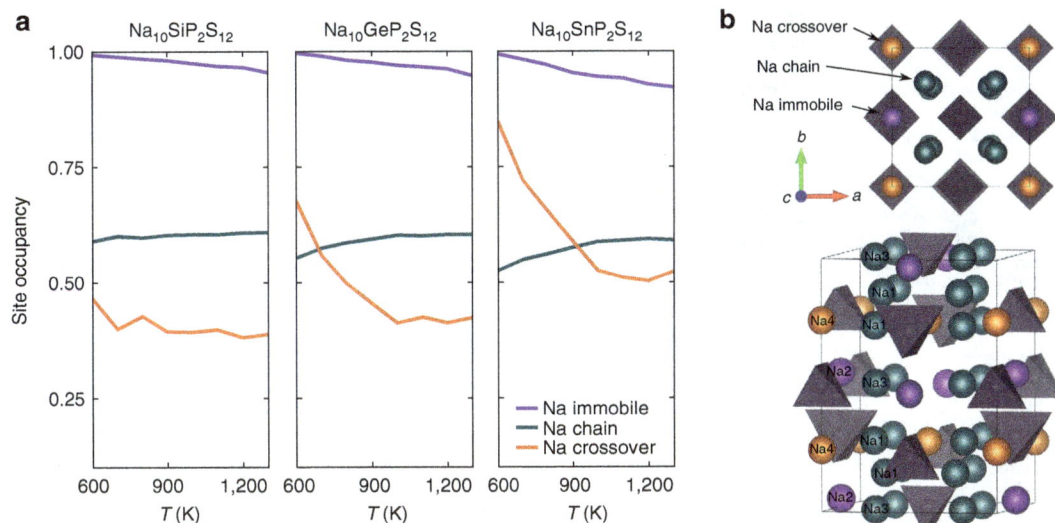

Figure 5 | Na-site occupancy analysis of NMPS structures. (a) Occupancy of Na sites in $Na_{10}SiP_2S_{12}$, $Na_{10}GeP_2S_{12}$ and $Na_{10}SnP_2S_{12}$ from AIMD simulation between 600 and 1,300 K, after imposing $P4_2/nmc$ spacegroup operations. The site occupancies in the Na-chain (Na1 and Na3) have been combined for clarity. **(b)** Illustration of the Na-chain, Na-crossover and Na-immobile sites. SnS_4 and PS_4 tetrahedra (grey), all spheres are Na-sites.

Fig. 1. Trends in occupancy are similar for Na-sites that are part of the same c-axis cation chain, again confirming a flat energy landscape and high mobility along it. These Na-ion diffusion pathways are connected to each other through the Na4 (Na-crossover) sites, which are part of the Na_{oct}-P_{tet}-Na_{oct}-Vac_{tet} chain along the c-axis at $x=0$, $y=0$. The Na-sites in the fully occupied Na_{oct}-P_{tet}-Na_{oct}-M_{tet} c-axis chain at $x=0$, $y=0.5$ have low energy and high occupancy, and are labelled as Na-immobile sites in Fig. 5 as they are not expected to contribute strongly to diffusion at low temperatures.

At high temperatures, the occupancies of each Na-site are almost identical across the three chemistries, indicating that they are dominated by entropic effects and not by the specific enthalpic differences between the compounds. At low temperatures, relative occupancies are more dependent on differences in site enthalpy. Considering first the Sn material, the occupancy of the Na-crossover sites dramatically increases as temperature is reduced, indicating that the enthalpy of the Na-crossover sites is significantly lower than the Na-chain sites. In contrast to $Na_{10}SnP_2S_{12}$, occupancy of the Na-crossover sites in the Si material is relatively unaffected by temperature, indicating minimal site enthalpy difference between the Na-chain and Na-crossover sites. The behaviour of occupancies in $Na_{10}GeP_2S_{12}$ is between these two extrema.

The diffusivity of Na-ions is determined primarily by the smoothness of their free-energy landscape. In materials where atoms can be trapped in very low-energy minima, activation energy for moving between these sites is increased, and thus diffusivity is reduced. The trends in Na-crossover site energy correlate well with the activation energies observed in simulation and explain why $Na_{10}SiP_2S_{12}$ has the highest predicted conductivity. At low T, the energy of the Na in the chain and crossover sites are almost equal, allowing Na to migrate in three dimensions with a very low barrier.

The good correspondence between the simulated and experimental results highlight the value of DFT as a predictive tool for the identification of new electrolyte materials. In this work, we focused synthesis efforts on the Sn material due to its affordability relative to the Ge version as well as its low E_{decomp} of 7.1 meV per atom, which is lower than comparable materials that have been synthesized. In LGPS and related lithium electrolytes,

contact with the highly reducing lithium metal or graphite anode can cause electrolyte decomposition by reduction of the transition metal. For these sodium electrolytes this may be less of a concern due to the lower reduction potential of sodium. These newly predicted materials may also prove to be more stable in battery applications than cubic Na_3PS_4 material, since decomposition of $Na_{10}MP_2S_{12}$ requires diffusion of high-valent cations to form Na_3PS_4 and Na_4MS_4, in contrast to cubic Na_3PS_4, which can convert to a low conductivity tetragonal phase[37] at the same composition. The conductivity of the new $Na_{10}SnP_2S_{12}$ electrolyte rivals that of the best known sulfide sodium-conductors, and the predicted Ge and Si materials, if confirmed, have the potential to surpass the conductivities of all known Na-electrolytes in a system much more compatible with all solid-state battery fabrication than NASICON-based and other oxide electrolytes.

In this work, we used first principles calculation to predict the existence of several new high-performance sodium electrolyte materials, with excellent agreement to subsequent experimental results. This marks the first use of computational prediction to design novel sodium electrolytes. The resulting $Na_{10}SnP_2S_{12}$ electrolyte, with a conductivity of 0.4 mS cm^{-1} at room temperature and activation energy of 0.35 eV, rivals the best known sulfide sodium-electrolytes and our predicted materials have the potential to surpass this conductivity. Our study highlights the benefits that can be gained by using *ab initio* approaches to guide material discovery.

Methods

Density functional theory calculations. All *ab initio* structure calculations were performed with calculations implemented in VASP[38], using the projector augmented-wave method[39]. Calculations used the Perdew–Burke–Ernzerhof generalized-gradient approximation[40]. For energy calculations of NMPS structures, a Monkhorst–Pack k-point grid of $4 \times 4 \times 4$ was used, for other competing phases, k-points were chosen such that $n_{kpoints} \times n_{atoms} > 1,000$. The VASP pseudopotential set of Li (PAW_PBE Li 17Jan2003), Na (PAW_PBE Na 08Apr2002), Ge (PAW_PBE Ge 05Jan2001), Si (PAW_PBE Si 05Jan2001), Sn (PAW_PBE Sn_d 06Sep2000), P (PAW_PBE P 17Jan2003) and S (PAW_PBE S 17Jan2003) was used.

Phase diagrams and stability limits. Phase diagrams are constructed using pymatgen[31] by computing the lower convex hull of DFT computed energy per atom in composition space, similar to the work in ref. 25. Materials on this convex hull cannot decompose to a lower energy combination of phases and are therefore stable. Grand potential phase diagrams are constructed by computing the lower convex hull of the DFT computed grand potential ($\Phi[c,\mu_{Li}] = E[c] - n_{Li}[c]\mu_{Li}$, where $E[c]$, $n_{Li}[c]$ and μ_{Li} are the DFT computed energy, lithium content and lithium chemical potential of composition c) in composition space. Anodic and cathodic stability limits are given by the maximum and minimum lithium chemical potentials for which a material (or its 0 K decomposition products in the case of metastable structures) is found to be stable.

Conductivity simulations. We performed AIMD simulations under the Born–Oppenheimer approximation to determine Na diffusivity in the NMPS system using VASP[38]. Atom trajectories are calculated using Newtonian dynamics with Verlet integration in an NVT ensemble. A Nose–Hoover thermostat with a period of 40 timesteps (80 fs) was used for all simulations. Na-atom displacements are calculated with respect to the centre of mass of the framework (non-Na) atoms. Self-diffusivities from these simulations were calculated by fitting the Einstein relation of mean-squared displacements to time ($\langle\|\Delta x_i\|^2\rangle = 2dD_{self}t$), where d is the dimensionality, using tools implemented in the pymatgen software package[31]. Ionic conductivities taking into account correlations between Na-ions were calculated from the mean square displacement of the net Na-ion motion $\left\langle\|\sum_{i=1}^{n}\Delta x_i\|^2\right\rangle = 2dD_\sigma nt$. Inserting D_σ into the Nernst–Einstein equation is equivalent to using the Green–Kubo expression for ionic conductivity when Na-ions are the only mobile charge carriers[41,42]. The AIMD simulations were performed on a single unit cell of NMPS, with 50 ions (2 formula units). The volume and shape of the cells were obtained from the fully relaxed cells used for the energy calculations by enforcing tetragonal symmetry (equality of the a and b lattice parameters). The time step of the simulation was 2 fs. To reduce the computational cost of the calculation, forces were calculated using a single k-point. Temperatures were initialized at 300 K and scaled to the appropriate temperature over 1,000 time steps (2 ps), starting with the ground-state structure. Simulations between 600 and 900 K were 350,000 time steps (700 ps), and simulations above 900 K were 250,000 time steps (500 ps).

Calculation of the activation energy (E_a) and extrapolation of results to room temperature was performed with an Arrhenius fit to the diffusivity data. The Haven ratio, H_r, an indication of the cooperativity of ionic motion, is calculated from the ratio of D_{self} to the D_σ in each simulation.

Ionic probability density. Na-ion probability densities were calculated from the AIMD simulations. After enforcing $P4_2/nmc$ symmetry, Na-ion positions relative to the centre of mass of the framework (P, M, S) atoms were smoothed using a Gaussian kernel with s.d. of 0.2 Å, and the resulting density visualized using Vesta[43].

Fractional occupancies. Fractional occupancies were calculated using a k-means clustering algorithm[44], initialized with atomic positions from the structure of LGPS[26]. At each clustering step, the shortest distance (taking into account periodic boundary conditions) to each mean was calculated, and a linear assignment algorithm[45] as implemented in pymatgen[31] was used at each simulation time step to assign each Na-ion position to the nearest mean, ensuring that at most a single Na atom from each time step is assigned to any given mean. The resulting cluster sizes and centroids were used to define the occupancy and location of Na sites.

Synthesis. $Na_{10}SnP_2S_{12}$ was synthesized by mixing stoichiometric amounts of Na_2S (Kojundo Chemical Laboratory Co. Ltd, 99%), P_2S_5 (Sigma-Aldrich Co., 99%) and SnS_2 (Kojundo Chemical Laboratory Co. Ltd, 99.9%) with a planetary ballmill (380 r.p.m. for 17 h). The pelletized mixture was wrapped in gold foil and heated at 700 °C for 12 h in an evacuated quartz tube and slow-cooled down to room temperature for 99 h (approximately -0.1° min^{-1}).

X-ray diffraction. The X-ray diffraction pattern is obtained with Cu–K$_\alpha$ radiation (40 kV, 40 mA) from 10–90° 2θ with 0.03° step intervals.

Conductivity measurement. Na-ion conductivity was measured with electrochemical impedance spectroscopy using an AUTOLAB PGSTAT30 (Metrohm Autolab, Utrecht) at 30, 40, 60 and 80 °C with a frequency ranging from 1 MHz to 100 mHz and an amplitude of 10 mV under normal pressure. An indium foil-blocking electrode was pressed onto both sides of the $Na_{10}SnP_2S_{12}$ pellet (11.5 mm diameter and 0.75 mm thickness). The conductivity values were obtained from the Cole–Cole plot of the data.

References

1. Ellis, B. L. & Nazar, L. F. Sodium and sodium-ion energy storage batteries. *Curr. Opin. Solid State Mater. Sci.* **16**, 168–177 (2012).
2. Kim, S.-W., Seo, D.-H., Ma, X., Ceder, G. & Kang, K. Electrode materials for rechargeable sodium-ion batteries: potential alternatives to current lithium-ion batteries. *Adv. Energy Mater.* **2**, 710–721 (2012).
3. Pan, H., Hu, Y.-S. & Chen, L. Room-temperature stationary sodium-ion batteries for large-scale electric energy storage. *Energy Environ. Sci.* **6**, 2338 (2013).
4. Ponrouch, A., Marchante, E., Courty, M., Tarascon, J.-M. & Palacn, M. R. In search of an optimized electrolyte for Na-ion batteries. *Energy Environ. Sci.* **5**, 8572 (2012).
5. Komaba, S. *et al.* Electrochemical Na insertion and solid electrolyte interphase for hard-carbon electrodes and application to Na-ion batteries. *Adv. Funct. Mater.* **21**, 3859–3867 (2011).
6. Palomares, V., Casas-Cabanas, M., Castillo-Martnez, E., Han, M. H. & Rojo, T. Update on Na-based battery materials. A growing research path. *Energy Environ. Sci.* **6**, 2312–2337 (2013).
7. Hueso, K. B., Armand, M. & Rojo, T. High temperature sodium batteries: status, challenges and future trends. *Energy Environ. Sci.* **6**, 734–749 (2013).
8. Khireddine, H., Fabry, P., Caineiro, A. & Bochu, B. Optimization of NASICON composition for Na$^+$ recognition. *Sensors Actuators B Chem.* **40**, 223–230 (1997).
9. Fergus, J. W. Ion transport in sodium ion conducting solid electrolytes. *Solid State Ionics* **227**, 102–112 (2012).
10. Zhang, P., Matsui, M., Takeda, Y., Yamamoto, O. & Imanishi, N. Water-stable lithium ion conducting solid electrolyte of iron and aluminum doped NASICON-type $LiTi_2(PO_4)_3$. *Solid State Ionics* **263**, 27–32 (2014).
11. Kamaya, N. *et al.* A lithium superionic conductor. *Nat. Mater.* **10**, 682–686 (2011).
12. Seino, Y., Ota, T., Takada, K., Hayashi, A. & Tatsumisago., M. A sulphide lithium super ion conductor is superior to liquid ion conductors for use in rechargeable batteries. *Energy Environ. Sci.* **7**, 627–631 (2014).
13. Hayashi, A., Noi, K., Sakuda, A. & Tatsumisago, M. Superionic glass-ceramic electrolytes for room-temperature rechargeable sodium batteries. *Nat. Commun.* **3**, 856 (2012).
14. Hayashi, A., Noi, K., Tanibata, N., Nagao, M. & Tatsumisago, M. High sodium ion conductivity of glass-ceramic electrolytes with cubic Na_3PS_4. *J. Power Sources* **258**, 420–423 (2014).
15. Tanibata, N. *et al.* X-ray crystal structure analysis of sodium-ion conductivity in 94 $Na_3PS_4 \cdot 6$ Na_4SiS_4 glass-ceramic electrolytes. *ChemElectroChem.* **1**, 1130–1132 (2014).
16. Wang, Y., Xiao, R., Hu, Y.-S., Avdeev, M. & Chen., L. P2-$Na_{0.6}[Cr_{0.6}Ti_{0.4}]O_2$ cation-disordered electrode for high-rate symmetric rechargeable sodium-ion batteries. *Nat. Commun.* **6**, 6954 (2015).
17. Wang, Y. *et al.* A zero-strain layered metal oxide as the negative electrode for long-life sodium-ion batteries. *Nat. Commun.* **4**, 2635 (2013).
18. Hautier, G., Fischer, C. C., Jain, A., Mueller, T. & Ceder, G. Finding nature's missing ternary oxide compounds using machine learning and density functional theory. *Chem. Mater.* **22**, 3762–3767 (2010).
19. Thangadurai, V., Narayanan, S. & Pinzaru, D. Garnet-type solid-state fast Li ion conductors for Li batteries: critical review. *Chem. Soc. Rev.* **43**, 4714–4727 (2014).
20. Ong, S. P. *et al.* Phase stability, electrochemical stability and ionic conductivity of the $Li_{10\pm1}MP_2X_{12}$ (M = Ge, Si, Sn, Al or P, and X = O, S or Se) family of superionic conductors. *Energy Environ. Sci.* **6**, 148–156 (2013).
21. Whiteley, J. M., Woo, J. H., Hu, E., Nam, K.-W. & Lee, S.-H. Empowering the lithium metal battery through a silicon-based superionic conductor. *J. Electrochem. Soc.* **161**, A1812–A1817 (2014).
22. Kuhn, A., Duppel, V. & Lotsch., B. V. Tetragonal $Li_{10}GeP_2S_{12}$ and Li_7GePS_8 - exploring the Li ion dynamics in LGPS Li electrolytes. *Energy Environ. Sci.* **6**, 3548–3552 (2013).
23. Kuhn, A. *et al.* A new ultrafast superionic Li-conductor: ion dynamics in $Li_{11}Si_2PS_{12}$ and comparison with other tetragonal LGPS-type electrolytes. *Phys. Chem. Chem. Phys.* **16**, 14669–14674 (2014).
24. Bron, P. *et al.* $Li_{10}SnP_2S_{12}$: an affordable lithium superionic conductor. *J. Am. Chem. Soc.* **135**, 15694–15697 (2013).
25. Ong, S. P., Wang, L., Kang, B. & Ceder., G. Li-Fe-P-O$_2$ phase diagram from first principles calculations. *Chem. Mater.* **20**, 1798–1807 (2008).
26. Kuhn, A., Koehler, J. & Lotsch., B. V. Single-crystal X-ray structure analysis of the superionic conductor $Li_{10}GeP_2S_{12}$. *Phys. Chem. Chem. Phys.* **15**, 11620–11622 (2013).
27. Kandagal, V. S., Bharadwaj, M. D. & Waghmare, U. V. Theoretical prediction of a highly conducting solid electrolyte for sodium batteries: $Na_{10}GeP_2S_{12}$. *J. Mater. Chem. A* **3**, 12992–12999 (2015).
28. Murch, G. The Haven ratio in fast ionic conductors. *Solid State Ionics* **7**, 177–198 (1982).
29. Belsky, A., Hellenbrandt, M., Karen, V. L. & Luksch, P. New developments in the Inorganic Crystal Structure Database (ICSD): accessibility in support of materials research and design. *Acta Crystallogr. Sect. B Struct. Sci.* **58**, 364–369 (2002).
30. Lepley, N. & Holzwarth, N. Computer modeling of crystalline electrolytes - lithium thiophosphates and phosphates. *ECS Trans.* **35**, 39–51 (2011).

31. Ong, S. P. *et al.* Python materials genomics (pymatgen): a robust, open-source python library for materials analysis. *Comput. Mater. Sci.* **68**, 314–319 (2013).

32. Sahu, G. *et al.* Air-stable, high-conduction solid electrolytes of arsenic-substituted Li_4SnS_4. *Energy Environ. Sci.* **7**, 1053–1058 (2014).

33. Takada, K. *et al.* Interfacial modification for high-power solid-state lithium batteries. *Solid State Ionics* **179**, 1333–1337 (2008).

34. Hori, S. *et al.* Synthesis, structure, and ionic conductivity of solid solution, $Li_{10+\delta}M_{1+\delta}P_{2-\delta}S_{12}(M = Si, Sn)$. *Faraday Discuss* **176**, 83–94 (2014).

35. Kang, K. & Ceder, G. Factors that affect Li mobility in layered lithium transition metal oxides. *Phys. Rev. B* **74**, 094105 (2006).

36. Van der Ven, A. & Ceder, G. Lithium diffusion mechanisms in layered intercalation compounds. *J. Power Sources* **97-98**, 529–531 (2001).

37. Jansen, M. & Henseler, U. Synthesis, structure determination, and ionic conductivity of sodium tetrathiophosphate. *J. Solid State Chem.* **99**, 110–119 (1992).

38. Kresse, G. & Furthmüller, J. Efficient iterative schemes for ab initio total-energy calculations using a plane-wave basis set. *Phys. Rev. B* **54**, 11169–11186 (1996).

39. Blöchl, P. E. Projector augmented-wave method. *Phys. Rev. B* **50**, 17953–17979 (1994).

40. Perdew, J. P., Burke, K. & Ernzerhof, M. Generalized gradient approximation made simple. *Phys. Rev. Lett.* **77**, 3865–3868 (1996).

41. Helfand, E. Transport coefficients from dissipation in a canonical ensemble. *Phys. Rev.* **119**, 1–9 (1960).

42. Uebing, C. & Gomer, R. Determination of surface-diffusion coefficients by Monte-Carlo methods - comparison of fluctuation and Kubo-Green methods. *J. Chem. Phys.* **100**, 7759–7766 (1994).

43. Momma, K. & Izumi., F. VESTA: a three-dimensional visualization system for electronic and structural analysis. *J. Appl. Crystallogr.* **41**, 653–658 (2008).

44. Lloyd, S. Least squares quantization in PCM. *Inf. Theory, IEEE Trans.* **28**, 129–137 (1982).

45. Jonker, R. & Volgenant., A. A shortest augmenting path algorithm for dense and sparse linear assignment problems. *Computing* **340**, 325–340 (1987).

Acknowledgements

We thank Dr Rahul Malik for comments on an early version of the manuscript. This work was supported by Samsung Advanced Institute of Technology and computational resources were provided by the Extreme Science and Engineering Discovery Environment (XSEDE), which is supported by National Science Foundation grant number ACI-1053575.

Author contributions

W.D.R., S.P.O. and G.C. designed the computational study. Synthesis and experimental analysis was performed by T.T., I.U. and N.S. W.D.R. performed the computational experiments and prepared the manuscript. W.D.R., L.J.M., J.K. and Y.W. analysed and interpreted the data. All authors discussed results and commented on the manuscript.

Additional information

19

Water-mediated cation intercalation of open-framework indium hexacyanoferrate with high voltage and fast kinetics

Liang Chen[1,*], Hezhu Shao[1,*], Xufeng Zhou[1], Guoqiang Liu[1], Jun Jiang[1] & Zhaoping Liu[1]

Rechargeable aqueous metal-ion batteries made from non-flammable and low-cost materials offer promising opportunities in large-scale utility grid applications, yet low voltage and energy output, as well as limited cycle life remain critical drawbacks in their electrochemical operation. Here we develop a series of high-voltage aqueous metal-ion batteries based on 'M$^+$/N$^+$-dual shuttles' to overcome these drawbacks. They utilize open-framework indium hexacyanoferrates as cathode materials, and TiP$_2$O$_7$ and NaTi$_2$(PO$_4$)$_3$ as anode materials, respectively. All of them possess strong rate capability as ultra-capacitors. Through multiple characterization techniques combined with *ab initio* calculations, water-mediated cation intercalation of indium hexacyanoferrate is unveiled. Water is supposed to be co-inserted with Li$^+$ or Na$^+$, which evidently raises the intercalation voltage and reduces diffusion kinetics. As for K$^+$, water is not involved in the intercalation because of the channel space limitation.

[1] Ningbo Institute of Materials Technology and Engineering, Chinese Academy of Sciences, Ningbo 315201, China. * These authors contributed equally to this work. Correspondence and requests for materials should be addressed to Z.L. (email: liuzp@nimte.ac.cn) or to L.C. (email: cl@nimte.ac.cn).

With the increased demand for clean energies such as solar and wind power that are integrated into the utility grid, rechargeable aqueous metal-ion batteries (RAMB) have drawn a great deal of attention because of their better safety, higher-rate capability, lower cost and more eco-friendliness relative to organic counterparts[1–3]. To date, a variety of RAMB on the basis of metal-ion intercalation chemistry have been explored[4–16]. Particularly, RAMB that use alkali cations (Li^+, Na^+ and K^+) as electrochemical shuttles have garnered great interest[8–14]. As sodium and potassium are more abundant than lithium, RAMB with Na^+ and K^+ shuttles are considered as more competitive power sources for large-scale energy storage. Nevertheless, due to the larger radii of Na^+ (102 pm) and K^+ (138 pm) in contrast to Li^+ (76 pm), a few intercalation compounds, including tunnel-structured oxides, polyanionic phosphates, hexacyanometallates and layer oxides that are suitable for reversible insertion/extraction of Na^+ or K^+ in aqueous media have been identified so far[8–14,17–22]. Among them, only $NaTi_2(PO_4)$ and $NaMn_2(CN)_6$ can be used as anode materials[11,13]. Thus, the known RAMB based on sodium-ion and/or potassium-ion technology are quite limited, and most of them suffer the problem of low-energy density due to the low-voltage output (<1.2 V). Seeking high-voltage RAMB as viable alternatives to commercial aqueous batteries for large-scale energy storage still remains a big challenge.

Recently, we have proposed the fundamental concept of 'M^+/N^+-dual shuttles' to construct RAMB, such as $Na_{0.44}MnO_2/TiP_2O_7$, $LiMn_2O_4/Na_{0.22}MnO_2$, Ni_1Zn_1HCF/TiP_2O_7 and $Ni_1Zn_1HCF/NaTi_2(PO_4)_3$, which are the so-called aqueous mixed-ion batteries (AMIB)[23,24]. Unlike the traditional 'rocking-chair' lithium-ion battery, such batteries operate based on the migration of dual shuttles between electrode and electrolytes. Their one important characteristic is that the total concentration of M^+ and N^+ is fixed during charging/discharging, but the M^+/N^+ ratio is changed. Unfortunately, all of them are still limited to low-voltage output (<1.5 V), as well as rapid capacity fading, which greatly hinder their practical application. Interestingly, it is also found that the metal hexacyanoferrate (MeHCF) with open-framework structure allowing for co-insertion/extraction of alkali cations can act as a promising cathode candidate for AMIB. MeHCF with the general formula $A_xM[Fe(CN)_6]_y$ always belongs to the face-centred cubic structure, in which MC_6 and FeN_6 octahedra are bridged through CN ligands. Its unit cell consists of eight CN-bridged transition metal cubes, and provides lots of interstitial sites that can host a wide range of cations, including Rb^+ and Cs^+, or even water molecules. Such an open-framework with large interstitial sites enables rapid transport of cations with different ionic radii throughout the lattice. But the fundamental knowledge about how alkali cations are intercalated into MeHCF, especially in aqueous electrolytes, which is of interest scientifically, is still lacking at present. In addition, we and others have found that the electrode potential of MeHCF can be easily tuned by altering the transition metal M, which offers a new routine for viable cathode materials with high voltage and long lifetime to make functional AMIB in the near future[24–30].

In this article, we introduce open-framework indium HCF (InHCF) as a cathode material for AMIB. Its unique electrochemical behaviour with various single and dual alkali cations A^+ are uncovered. The roles of orientation, graphene and guest species on its rate capability are also extensively studied. By combining it with carbon-coated TiP_2O_7 and $NaTi_2(PO_4)_3$ as anode materials, a series of AMIB with high-voltage output ≥ 1.2 V are demonstrated. All of them possess high power density as an ultra-capacitor, but with higher-energy density. Finally, water-mediated cation intercalation in InHCF is revealed, using multiple characterization techniques coupled with density functional theory (DFT) calculations.

Results

Structure of indium hexacyanoferrate.
Indium HCF with and without graphene modification (InHCF and InHCF/Gr) were synthesized by a simple precipitation method described in the Methods section below. The fresh InHCF is yellow, and turns into green after vacuum drying. When modified with graphene, it changes into a black solid. X-ray diffraction analyses of above three compounds show that they all exhibit the same face-centred cubic structure as other MeHCFs[9,24,25,31]. Through the Rietveld refinement (Fig. 1a), InHCF is found to adapt the Fm-$3m$ space group with a lattice parameter of 10.51 Å, larger than those of NiHCF (10.09 Å), CuHCF (10.1 Å) and FeHCF (10.18 Å)[9,25,31]. Figure 1b displays its structure with a formula of $InFe(CN)_6$, a three-dimensional (3D) framework made up of FeC_6 octahedra and InN_6 octahedra linked together by CN ligands. The framework contains open [100] channels with a size of ca. 5.3 Å as shown in Fig. 1c, which allows for a rapid diffusion of a wide variety of guest cations. The $\angle Fe-C-N$ and $\angle In-N-C$ bond angles are both 180°, while the bond distances of $Fe-C$, $In-N$ and $C-N$ are 1.97, 2.13 and 1.15 Å, respectively. As well, carbon-coordinated Fe (III) in a strong crystal field favours the low-spin configuration (t_{2g}^5, $S=1/2$), while nitrogen-coordinated In (III) has zero unpaired electrons at no spin state. With the insertion of alkali cations A^+, Fe (III) is supposed to be reduced to Fe (II) with zero unpaired electrons (t_{2g}^6, $S=0$), while In (III) maintains the same valence. In our DFT calculations, the lattice parameter of InHCF is optimized to be 10.61 Å in excellent agreement with the experimental value (10.51 Å). Other calculated structural parameters including bond distances, bond angles and fractional coordinates summarized in Supplementary Tables 1 and 2 are in accord with the experimental values, suggesting that GGA + U calculations offer adequate accuracy in the description of InHCF.

The morphologies of InHCF and InHCF/Gr are further characterized by scanning electron microscopy (SEM). InHCF looks like 'toy bricks' assembled by well-crystallized nanocubes with sizes between 50 and 200 nm. Each nanocube with six [100] orientations shows a single-crystalline state (Fig. 1d). Elemental analyses show that In, Fe and N elements with atomic ratios of 15:21:64 are found. From their transmission electron microscopy (TEM) images, cube-like nanoparticles with sizes ranging from 50 to 200 nm are also observed (Fig. 1e), which are consistent with SEM characterizations. Periodic diffraction spots taken from a single particle correspond to a single crystal with [100] orientations (Supplementary Fig. 1). However, irregular nanoparticles randomly deposited on graphene are observed for InHCF/Gr (Fig. 1e; Supplementary Fig. 1). The average particle size is ca. 50 nm, smaller than that of InHCF. It is anticipated that two-dimensional graphene can adsorb InHCF nanocrystals during the nucleation stage of precipitation reaction, which blocks their further growth up. Elemental analysis from TEM also shows that In, Fe and N elements coexist in InHCF/Gr.

Electrochemical behaviour of InHCF.
The electrochemical behaviour of electrodeposited MeHCF thin films with a variety of anions have been investigated by Neff's and other groups since 1980s (refs 32–34). However, such thin films have mass loading as low as several $\mu g\,cm^{-2}$, which is too far from their practical battery application. Herein, our bulk InHCF electrodes have mass loadings of ca. 10 mg cm^{-2}, 100 times over previous electrodeposited films. As shown in Fig. 2a–c, InHCF/Gr exhibits one pair of well-defined and symmetric redox peaks with either

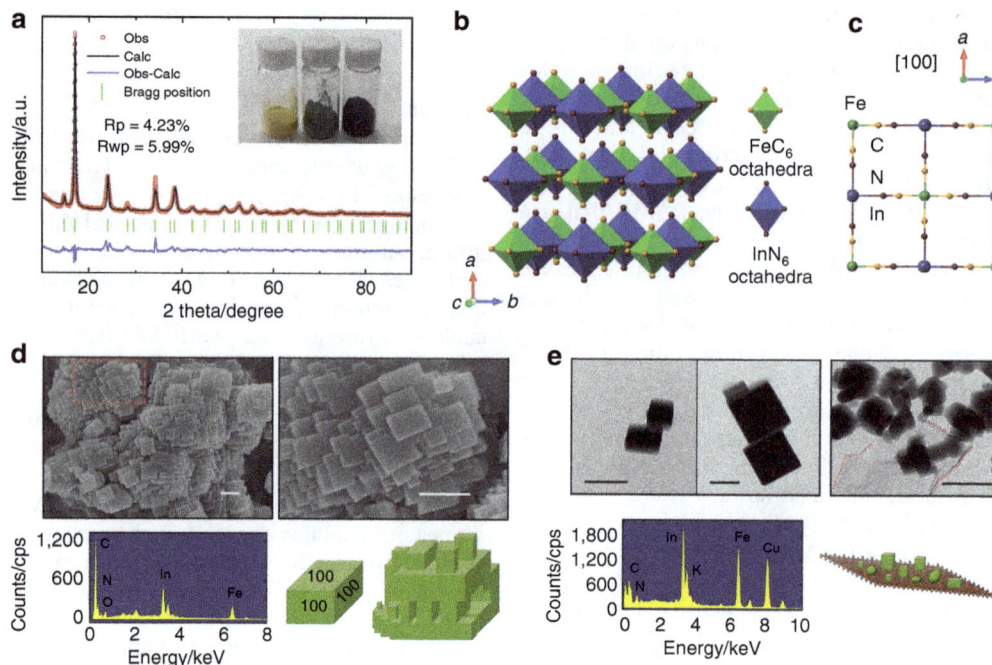

Figure 1 | Cubic InHCF. (**a**) Rietveld refinement of X-ray power diffraction pattern for InHCF with cubic structure. Insets are the photos of fresh InHCF, InHCF after vacuum drying and InHCF/Gr. (**b**) A unit cell of cubic InHCF with a formula of $InFe(CN)_6$. (**c**) Ball stick pattern of $InFe(CN)_6$ viewed down the [100] zone axis. (**d**) SEM images of InHCF nanocube with different magnification (top of panel), energy-dispersive X-ray spectroscopy (EDS) of InHCF nanocubes and a schematic of 'toy bricks' assembled by InHCF nanocube with six facets of [100] (bottom of panel). Scale bar, 200 nm. (**e**) TEM images of InHCF nanocube and InHCF/Gr (top of panel), EDS of InHCF/Gr (bottom of panel) and a schematic of InHCF with irregular shapes deposited on graphene. Scale bar, 100 nm.

alkali cations A^+ (Li^+, Na^+ or K^+). Such characteristic peaks are associated with the reversible conversion between Fe(II) and Fe(III) accompanied by the storage/release of A^+, which are demonstrated in the Discussion section below. The formal potential E_f follows the order: Li^+ (0.79 V) $< K^+$ (0.97 V) $< Na^+$ (1.03 V). This trend is different from CuHCF, NiHCF and Ni_xZn_yHCF, where the intercalation of A^+ with larger ionic radius happens at higher voltage ($Li^+ < Na^+ < K^+$)[12,24]. ΔE_p (the difference between anode and cathode peak potentials, $E_{pa} - E_{pc}$) is an important factor for energy loss in a battery, and small ΔE_p means low polarization overpotential and fast reaction kinetics. At a scan rate of $2 \, mV \, s^{-1}$, ΔE_p of InHCF/Gr with A^+ are within 200 mV, indicating a fast reaction between Fe(II) and Fe(III). And ΔE_p follows another order as E_f: 130 mV (K^+) < 180 mV (Li^+) < 200 mV (Na^+). Furthermore, a specific capacity of $52 \pm 1 \, mAh \, g^{-1}$ (1C rate) is measured for InHCF/Gr with either A^+, showing that it has the same storage ability with alkali cations, regardless of their different ionic radii (Fig. 2d–f; Supplementary Fig. 2 and Supplementary Note 1). Its rate capability with A^+ follows the opposite order as ΔE_p: $K^+ > Li^+ > Na^+$. The high-rate capacity retention of InHCF/Gr is greatest when it is cycled with K^+ for its 93% capacity retention at 10C and 79% at 40C (Supplementary Fig. 3). Intercalation of Li^+ and Na^+ into InHCF/Gr occurs at moderate high rates, as it retains 64 and 44% of its discharge capacity at 40C during their respective reactions. In addition, the rate behaviours of InHCF without graphene modification are also tested for comparison. It is found to be less impressive, with 59% (Li^+), 37% (Na^+) and 73% (K^+) of the discharge capacities retained at 40C in those cases (Supplementary Fig. 3).

The electrochemical behaviours of InHCF/Gr in mixed-ion electrolytes are also revealed. In the case of Li_2SO_4/Na_2SO_4 electrolytes, one pair of redox peaks still appears. But their potentials are close to the ones in Na_2SO_4. Meanwhile, the charge/discharge voltage plateaus in mixed-electrolytes are also

close to the ones in Na_2SO_4, which are about 200 mV higher than the ones in Li_2SO_4. The discharging capacities in mixed-electrolytes are $\sim 52 \, mAh \, g^{-1}$ (1C rate), identical to the one in Li_2SO_4 or Na_2SO_4. The E_{ocp} and $E_{1/2}$ increase with an increase of Na^+ concentration in electrolytes (Supplementary Fig. 4), implying that the existence of Na^+ gives rise to high working voltage, which is desirable for high-energy density. Moreover, the capacity retentions of InHCF/Gr at 40C in 0.25 M $Li_2SO_4 + 0.25$ M Na_2SO_4 and 0.1 M $Li_2SO_4 + 0.4$ M Na_2SO_4 are 49 and 51%, between those in 0.5 M Li_2SO_4 and 0.5 M Na_2SO_4 (Supplementary Figs 5–6). Obviously, the introducing of Li^+ into Na_2SO_4 electrolytes is beneficial to the rate capability of InHCF/Gr.

With regard to Li_2SO_4/K_2SO_4 electrolytes, similar behaviours are observed. Figure 2b displays that the E_{pa} and E_{pc} in Li_2SO_4/K_2SO_4 electrolytes are higher than those in Li_2SO_4, while the ΔE_p becomes smaller. Moreover, the charge/discharge voltage plateaus in Li_2SO_4/K_2SO_4 are ~ 150 mV higher than those in Li_2SO_4. Both $E_{1/2}$ and E_{ocp} are electrolytes dependent, which go up with an increase of K^+ concentration in electrolytes (Supplementary Fig. 4). In addition, the capacity retention of InHCF/Gr at 40C in Li_2SO_4/K_2SO_4 electrolytes is up to 85%, higher than those in Li_2SO_4 and K_2SO_4 (Supplementary Fig. 5–6). It is illustrated that the addition of K^+ into Li_2SO_4 electrolytes can enhance both the working voltage and rate capability of InHCF/Gr.

In Na_2SO_4/K_2SO_4 electrolytes, these behaviours are not as notable as those in Li_2SO_4/Na_2SO_4 and Li_2SO_4/K_2SO_4 electrolytes. When InHCF/Gr is cycled in 0.4 M $Na_2SO_4/0.1$ M K_2SO_4 (1C rate), the discharge voltage plateau is almost identical to that in 0.5 M Na_2SO_4, which are 50 mV higher than that in 0.5 M K_2SO_4. At the same time, the polarization (voltage difference between charging and discharging profiles) is only 16 mV at 1C, smaller than that in Na_2SO_4 (32 mV). When it is cycled at 40C, the capacity retention remains 64% much higher

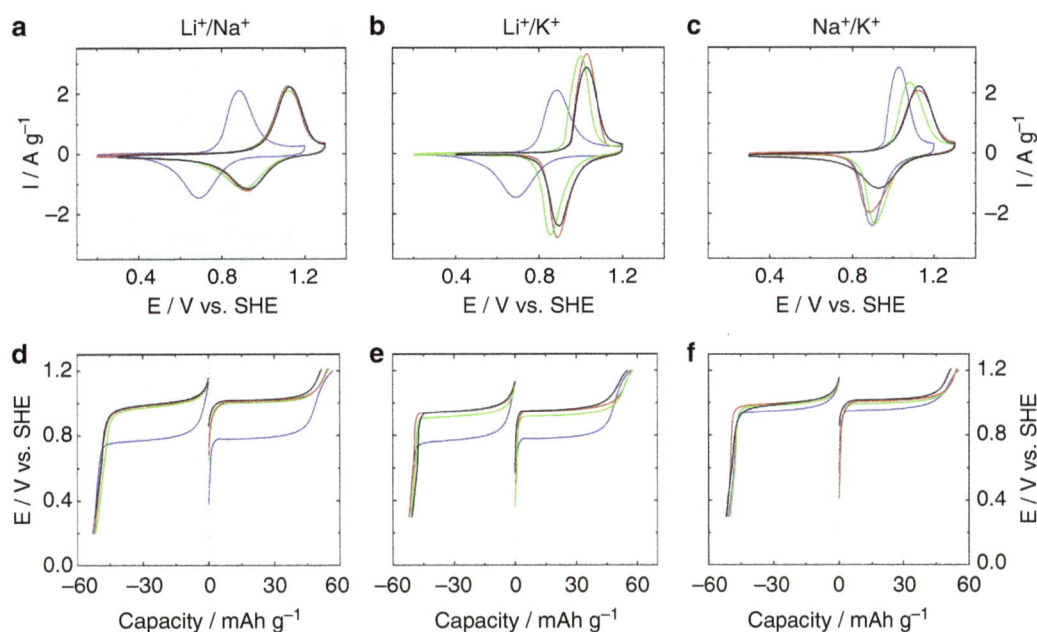

Figure 2 | Electrochemical properties of InHCF/Gr. (a–f) Cyclic voltammograms (a–c) at a scan rate of $2\,mV\,s^{-1}$ and galvanostatic profiles (d–f) measured at a rate of 1C for InHCF/Gr in various electrolytes ($1C = 60\,mA\,g^{-1}$). All electrode potentials are versus standard hydrogen electrode (SHE). Blue line is corresponding to $0.5\,M\,Li_2SO_4$ in a,b,d and e) or $0.5\,M\,K_2SO_4$ in c and f Green line is corresponding to $0.25\,M\,Li_2SO_4/0.25\,M\,Na_2SO_4$ in a and d or $0.25\,M\,Li_2SO_4/0.25\,M\,K_2SO_4$ in b and e or $0.25\,M\,Na_2SO_4/0.25\,M\,K_2SO_4$ in c and f. Red line is corresponding to $0.1\,M\,Li_2SO_4/0.4\,M\,Na_2SO_4$ in a and d or $0.1\,M\,Li_2SO_4/0.4\,M\,K_2SO_4$ in b and e or $0.4\,M\,Na_2SO_4/0.1\,M\,K_2SO_4$ in c and f. Black line is corresponding to $0.5\,M\,Na_2SO_4$ in a,c,d and f or $0.5\,M\,K_2SO_4$ in b and e.

than 44% in Na_2SO_4 (Supplementary Figs 5–6). These results demonstrate that the existence of K^+ in Na_2SO_4 electrolytes can alleviate the polarization without the loss of the working voltage.

Aqueous mixed-ion batteries. On the basis of the unique intercalation chemistry of InHCF/Gr, we assembled three practical prototypes of batteries operating in mixed-ion electrolytes (B-I, B-II and B-III). Figure 3a illustrates the principle of such batteries: the release and storage of M^+ with small ionic radius occur at anode, and the co-insertion/extraction reactions of M^+ and N^+ (large ionic radius) take place at cathode. During charging/discharging, the total concentration of M^+ and N^+ is fixed to ensure the charge neutrality of the electrolytes, but the M^+/N^+ ratio is changed. Our previous studies show that the cubic TiP_2O_7 and rhombohedral $NaTi_2(PO_4)_3$ can be used as the anodes for AMIB, as a result of their specific ion-selectivity properties and reasonable working voltages[23,24]. Herein, carbon-coating TiP_2O_7 and $NaTi_2(PO_4)_3$ synthesized by solid-state reactions are selected as the anodes for B-I (InHCF/$Li^+ + Na^+/TiP_2O_7$), B-II (InHCF/$Li^+ + K^+/TiP_2O_7$) and B-III (InHCF/$Na^+ + K^+/NaTi_2(PO_4)_3$). Through X-ray diffraction and Rietveld refinements, a pure cubic phase for TiP_2O_7 and a pure rhombohedral phase for $NaTi_2(PO_4)_3$ are obtained (Supplementary Figs 7–8). High-resolution TEM (HR-TEM) analyses show that uniform carbon layers with a thickness of ca. 10 nm are coated on them, ensuring their good electrochemical performances (Supplementary Figs 7–8). The galvanostatic profiles of B-I, B-II and B-III along with the voltage profiles of their individual anode and cathode electrodes versus standard hydrogen electrodes (SHE) are displayed in Fig. 3c–e. An average working voltage (E_{av}) is found to be 1.45 V for B-I, 1.4 V for B-II and 1.6 V for B-III, which is beyond a theoretical water electrolysis voltage of 1.23 V. The galvanostatic profiles of InHCF/Li^+/TiP_2O_7 (B-IV) and InHCF/$Na^+/NaTi_2(PO_4)_3$ (B-V) batteries are added for comparison (Fig. 3f). Their E_{av}s are only 1.2 and 1.55 V, lower than those of their corresponding AMIB.

In Supplementary Table 3, it shows that B-I, B-II and B-III compare favourably with our previous AMIB and many other RAMB, in terms of E_{av} and energy density. In particular, E_{av} of B-III, as well as B-V is up to 1.6 V, which has been the highest record among RAMB based on sodium-ion technology until now.

In addition to high E_{av}, B-III also possesses strong high-rate capability. At current rates as high as 40C, the discharge capacity remains >73% of that at 1.5C (Fig. 4a). As well, its cycle life is good. During galvanostatic cycling between 0.4 and 1.83 V at 1.5C, ~90% of the initial discharge capacity are retained after 200 cycles (Fig. 4b). The coulombic efficiency is ~100% during cycling except the first cycle, suggesting that many side reactions (for example, water electrolysis reactions) are eliminated during cycling despite the high-voltage charge limit (1.83 V) for B-III (Supplementary Fig. 9; Supplementary Table 4; Supplementary Notes 2 and 3). A summary of the discharged energy and power derived from B-III along with B-I and B-II are displayed in Fig. 4c. A specific energy of 49, 48 and $56\,Wh\,kg^{-1}$ based on the total mass of active electrode materials are obtained for B-I, B-II and B-III, respectively. Typically, the electrode materials are ~ 60% of the total weight of the practical cells with large size[4,7]. Thus, a practical specific energy of B-III close to $34\,Wh\,kg^{-1}$ can be expected, which can compete with current Pb-acid battery with ca. $35\,Wh\,kg^{-1}$ for energy storage. They also demonstrate high power density comparable to that of an ultra-capacitor ($>1,000\,W\,kg^{-1}$), but with much higher-energy density ($>20\,Wh\,kg^{-1}$). For instance, B-II can deliver a specific power of $2,700\,W\,kg^{-1}$ at a specific energy of $36\,Wh\,kg^{-1}$. Their capacity retentions at various rates follow the same order as InHCF/Gr: B-II (Li^+/K^+) > B-III (Na^+/K^+) > B-I (Li^+/Na^+), implying that B-II is the most promising system to meet high power energy storage among them. The Ragone plots of B-IV and B-V are also displayed in Fig. 4c. As seen, their performances are much less impressive, when compared with their corresponding AMIB.

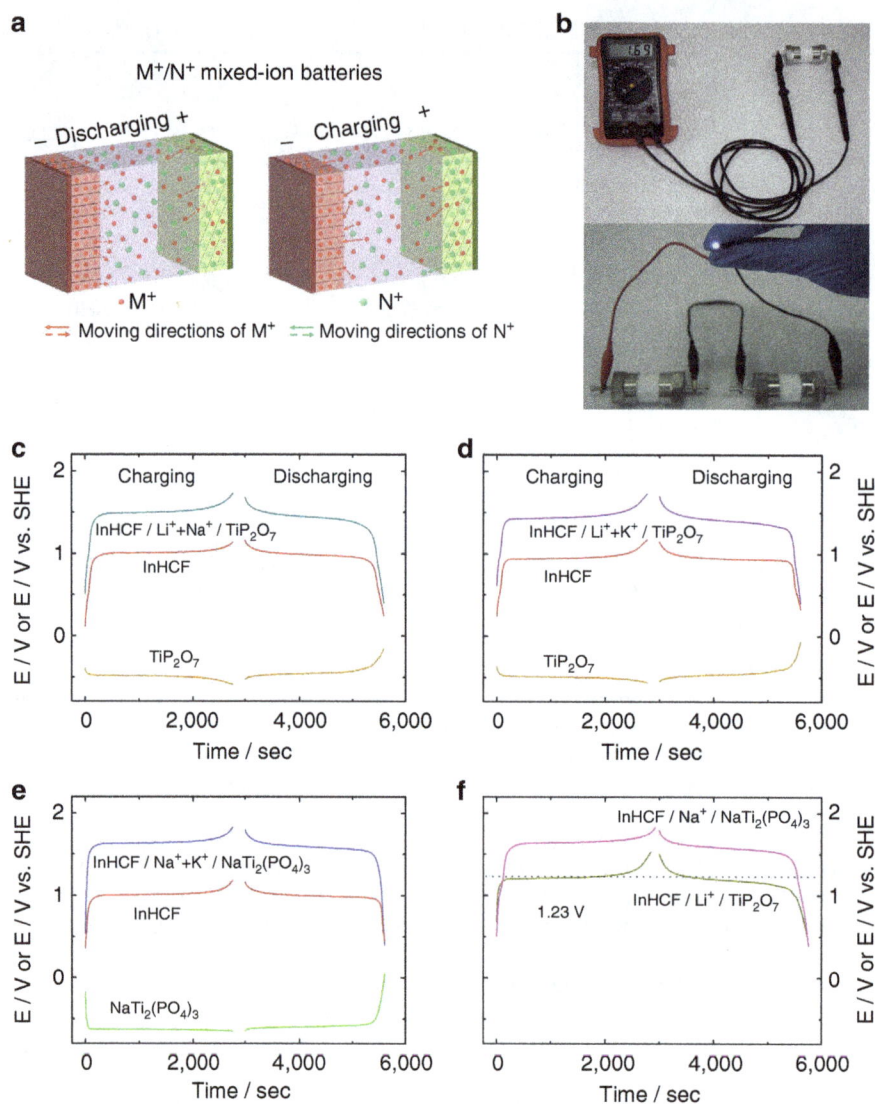

Figure 3 | Aqueous mixed-ion batteries. (a) A schematic of M^+/N^+ aqueous mixed-ion battery. (b) Photos of $InHCF/Na^+ + K^+/NaTi_2(PO_4)_3$ (B-III) battery in a Swagelok-type cell. One is connected with a voltage metre, and two in series are lighting up a 3-V blue LED. (c-e) Galvanostatic profiles of $InHCF/Li^+ + Na^+/TiP_2O_7$ (B-I), $InHCF/Li^+ + K^+/TiP_2O_7$ (B-II) and $InHCF/Na^+ + K^+/NaTi_2(PO_4)_3$ (B-III) batteries along with the voltage profiles of their individual anode and cathode electrodes versus SHE at a rate of 1.5C. (f) Galvanostatic profiles of $InHCF/Li^+/TiP_2O_7$ (B-IV) and $InHCF/Na^+/NaTi_2(PO_4)_3$ (B-V) batteries measured at a rate of 1.5C. $1C = 60\,mA\,g^{-1}$.

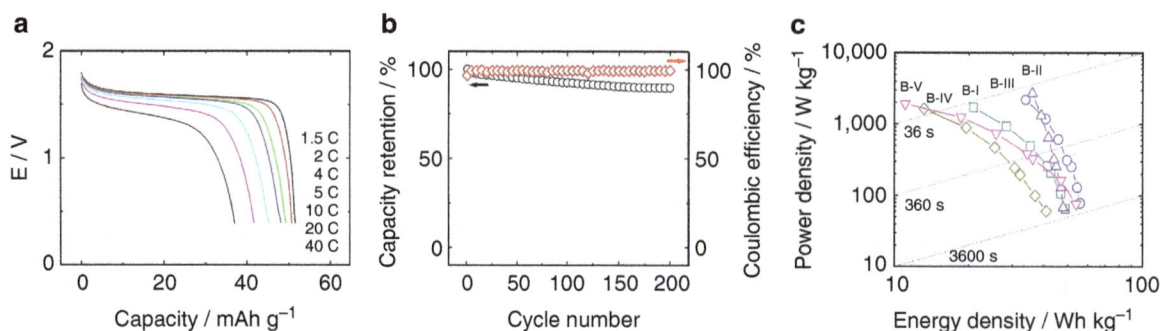

Figure 4 | Rate capability and cycle life of mixed-ion batteries. (a) Rate capability of B-III battery. (b) Long-term cycling performance for B-III battery at a rate of 1.5C. (c) A Ragone plot of B-I, B-II, B-III, B-IV and B-V.

Roles of orientation, graphene and mixed-ion electrolytes. Both the dimensionality and activation energy of cation migration within the crystal structures of electrode materials are critical for their rate capabilities. The high-rate capability of InHCF arises from its special structure. InHCF is composed of nanocubes with

six [100] facets, which can behave as 3D diffusion pathways for A^+ migration. Meantime, A^+ immigration through [100] facets is supposed to have low activation energy, due to the large open space of [100] channel. Compared with InHCF, the higher-rate capability for InHCF/Gr is ascribed to another two factors. One is that graphene in InHCF/Gr serves as a 3D electronic network to diminish the resistance between InHCF nanoparticles, which facilitates the electron transfer (Supplementary Note 4). The other is size effect of InHCF/Gr whose average size is smaller than that of InHCF, giving shorter diffusion path for A^+ migration (Supplementary Fig. 10).

Owning to the unique electrochemical properties of InHCF/Gr cathode in mixed-ion electrolytes as previously described, the utilization of mixed-ion electrolytes have two merits: raising the reaction potential (E) and speeding up reaction kinetics. For a battery cathode, raising E can lead to high-voltage output that is desirable for high-energy density. When an intercalation reaction occurs at H with ion-selectivity towards A^+ as shown in equation (1), E can be calculated using the Nernst equation (2):

$$H_{(s)} + nA^+_{aq} + ne^- \leftrightarrow A_nH_{(s)} \qquad (1)$$

$$E = E^\ominus + \frac{2.303RT}{nF}\lg(a^n_{A^+} a_H / a_{A_nH}) \qquad (2)$$

where a_{A^+}, a_H and a_{A_nH} refer to the activities of A^+, H and A^+-intercalated H (A_nH). From equation (2), E is found to be strongly dependent on A^+ activity. For a battery anode, reducing E can expand the working voltage that is preferable for high-energy density. Using mixed-ion electrolytes that can decrease A^+ activity (a_{A^+}) is a good tool for reducing E, which has been demonstrated on TiP_2O_7 and $NaTi_2(PO_4)_3$ materials in our previous works. Therefore, utilizing mixed-ion electrolytes instead of one single electrolyte for InHCF/TiP_2O_7 and InHCF/$NaTi_2(PO_4)_3$ batteries have two merits: one is to raise the voltage output and the other is to enhance the rate capability, which have been demonstrated by the electrochemical data (Supplementary Note 5).

Water-mediated cation intercalation in InHCF. Due to the highest intercalation voltage of Na^+ among A^+, the Na^+-intercalation reaction is selected as a probe to understand the intercalation mechanisms of alkali cations in InHCF. First, we used ex situ X-ray photoelectron spectroscopy (XPS) to analyse the variations of oxidation states of Fe, In and Na elements in InHCF/Gr upon electrochemical cycling. Fe 2p core level of the material at a − g states are displayed in Fig. 5a. The peaks at 709.2 and 722.1 eV for InHCF/Gr can be attributed to the $2p_{3/2}$ and $2p_{1/2}$ of Fe (II). During charging (from a to d), the intensities of these peaks decrease, and two new peaks at 710.8 and 724.4 eV are found for InHCF/Gr with c and d states. The appearances of the new peaks are related to Fe(III). During discharging, these new peaks for InHCF/Gr with e, f and g states are diminished, and the Fe(II) 2p peak intensities increase as InHCF/Gr is discharged from d to g. So it is concluded that the reversible redox reactions between Fe(II) and Fe(III) take place during the entire charging/discharging process. Ex situ XPS of In $3p_{1/2}$ and 3d were also collected (Fig. 5a,b). The peaks centred at 705.0, 445.7 and 453.3 eV, respectively, can be ascribed to the $3p_{1/2}$, $3d_{5/2}$ and $3d_{3/2}$ of In(III). Unlike Fe 2p, all spectra remain almost the same, indicating that In maintain the same valence upon cycling. Similar results were obtained for N 1s and F 1s, implying that N and F elements like In element do not undergo redox reactions upon cycling (Supplementary Fig. 11)[35]. Meanwhile, the Na 1s peak density as shown in Fig. 5c decreases from a to d, and increases from d to g. It is inferred that Na^+ extraction/insertion reactions happen during their respective charging/discharging process. The atomic ratios of different elements upon cycling determined by XPS are also summarized in Fig. 5e and Supplementary Table 5. As shown, the Na/Fe ratio decreases from 0.77 to 0.26 during charging, and then goes back to original 0.77 after discharging. However, during the electrochemical cycle,

the In/Fe and N/Fe atomic ratios remain within 1.23 ± 0.08 and 6.37 ± 0.38, respectively. Compared with In, Fe and N, the obvious composition changes for Na in InHCF/Gr are linked to the Na^+-insertion/extraction reaction during the cycle. All the above XPS results illustrate that Fe is the redox centre of InHCF that involves the reversible conversion between Fe(II) and Fe(III) accompanying the extraction/insertion of Na^+.

X-ray diffraction was also employed to monitor the structural characteristics of InHCF/Gr with cycling. As shown in Fig. 5d, InHCF/Gr maintains its cubic structure upon cycling, suggesting a quasi-solid-solution reaction. Interestingly, the X-ray diffraction reflections (for example, (004), (133) and (024)) are found to shift to smaller angles during charging (Na^+-extraction) and move to higher angles during discharging (Na^+-insertion). For InHCF/Gr, the lattice contraction induced by Na^+-insertion is opposite to the lattice expansion of other intercalation compounds, such as graphite, $LiFePO_4$, TiP_2O_7 and $Na_{0.44}[Mn_{1-x}Ti_x]O_2$ caused by cation intercalation. But InHCF/Gr undergoes a very small lattice contraction (from 10.51 to 10.49 Å, only $\sim 0.2\%$) after Na^+-insertion, which is close to zero-strain characteristic. Such a small lattice contraction is mainly linked to the slightly decrease in the Fe − C bond distance during reduction, which is confirmed by our DFT calculations (from 1.91 to 1.89 Å). And it is also associated with the measured small polarization and good cycle life for InHCF/Gr.

In the cubic structure of $InFe(CN)_6$ as shown in Fig. 6a, A^+ can occupy five possible interstitial sites, which are denoted with Wyckoff notations as 8c, 24d, 32f(n), 32f(c) and 48g. Among them, 24d site provides the smallest free space, while 8c site gives the largest free space. To avoid significant expansion of the lattice framework, the size of inserted ions to occupy the 24d site and 8c site should be limited to 140 and 260 pm, respectively. To assess the relative stability of the interstitial sites that occupied by different A^+, the binding energy (E_b) are calculated using DFT (Supplementary Methods)[31,36–38]. In terms of Li^+ and Na^+, 24d and/or 48g sites give the lowest E_b, while 8c site gives the highest E_b in Table 1. This trend changes for large cation K^+. It prefers to occupy 8c site or 48g site in comparison with 24d site. The reason is that small cations (Li^+ and Na^+) are more stable at 24d and/or 48g for shorter cation–anion-bonding distances, while large cation K^+ is less favourable to occupy 24d site because of the space limitation. Another observation is that the 32f(n) site is always more energetically stable than the 32f(c) site, and the nearest A − N distance is always smaller than the A − C ones. It can be interpreted as the electronegative nitrogen is more attractive to A^+ than carbon. The calculated volumes of the $AInFe(CN)_6$ unit with A^+ occupying different sites (primitive cell) are also summarized in Supplementary Fig. 12. It can be seen that almost all A^+-intercalation tend to shrink the volume except K^+-intercalation into 24d site. This can be probably because K^+ in 24d site has strong steric repulsion with its surrounding atoms, resulting in volume expansion. When considering the most stable interstitial site, the volume contractions after the intercalation of Li^+, Na^+ and K^+ are within 3%, 0.8% and 1%, respectively, which are close to zero-strain characteristic. Such a phenomenon is consistent with our above X-ray diffraction analysis.

It is previously mentioned that a variety of intercalants, including alkali cations and H_2O can be intercalated into InHCF. However, as a result of the large size of H_2O (3 Å), the stable interstitial sites for H_2O are quite limited. Among them, 8c site is ideal for H_2O occupation, owning to its largest free space. Our DFT calculation shows that the E_b between H_2O and InHCF is − 0.67 eV, suggesting their much weaker interaction than E_b between A^+ and InHCF (Supplementary Fig. 13). As small A^+ occupying the 24d site in the lattice, the empty 8c site is supposed to accommodate H_2O, in which O atom has coulomb attraction

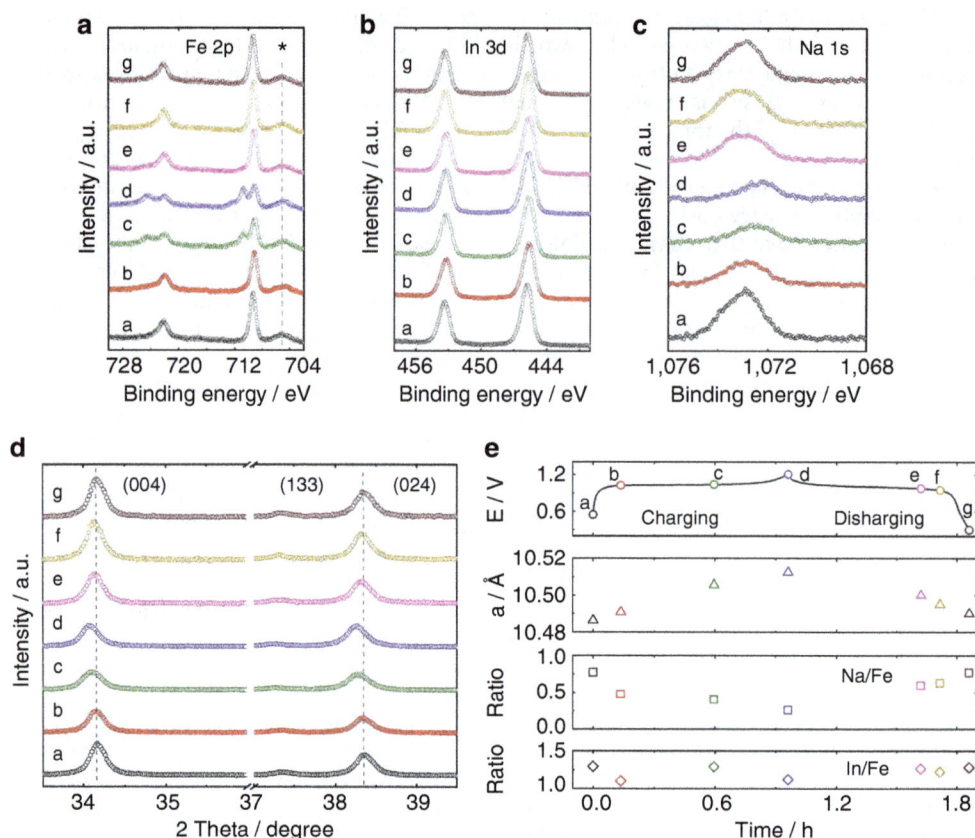

Figure 5 | Sodium-ion intercalation mechanism in InHCF. (**a–c**) *Ex situ* XPS spectra recorded from Fe 2p, In 3d and Na 1s core level of InHCF/Gr upon electrochemical cycling (a–g states as shown in **e** are selected). A binding energy of 688.2 eV for the F 1s (from polyvinylidene fluoride (PVDF)) was used as reference. The peaks labelled by * symbol (705.0 ± 0.1 eV) are related to In 3$p_{1/2}$. (**d**) *Ex situ* X-ray diffraction patterns of InHCF/Gr with cycling. (**e**) Changes of lattice parameter (*a*), surface atomic ratio (Na/Fe and In/Fe) in InHCF/Gr during electrochemical cycling. a, 0.55 V, b, 1.02 V, c, 1.035 V, d, 1.2 V, e, 0.975 V, f, 0.945 V and g, 0.3 V seven states are selected for comparison.

with positively charged A^+. To confirm the assumption, a water molecule and A^+ are initially put into 8c site and 24d site, respectively. After geometry optimization, H_2O almost stay at original 8c site and A^+ move to another site displaced from 24d site. When the sites of H_2O and A^+ are exchanged by each other, it is also found that after geometry optimization, H_2O moves from 24d site to a site near 8c site and A^+ moves towards 24d site. The distance between Na and O (or Li and O) is found to be 2.24 Å (or 1.88 Å), indicative of the interaction between Na^+ (or Li^+) and H_2O (Fig. 6b). $E_b([Na-OH_2]^+)$ and $E_b([Li-OH_2]^+)$ are calculated to be -3.94 and -4.0 eV, respectively, which are 0.61 and 0.87 eV lower than their corresponding E_b at 24d site. The more remarkable interaction between Li^+ and H_2O is due to the higher ionic potential (charge/radius) of Li^+ relative to Na^+. All above results suggest that in terms of Li^+ and Na^+, H_2O can be co-inserted into InHCF with them. As for K^+, the size of $[K-OH_2]^+$ is up to 288 pm that exceeds the size limit of 8c site, which infers that K^+ is solely inserted into InHCF without water (Supplementary Fig. 14; Supplementary Note 6). The rate capability of intercalation compounds that critically depends on cation diffusion is related to cation size. It is well known that large cation always suffers from slow migration kinetics, leading to low-rate capability. The sizes of cations follow the order: $r(K^+)$ $<r(Li-OH_2^+)<r(Na-OH_2^+)$, providing a good explanation for why the rate capability of InHCF in aqueous electrolytes follow the order: $K^+ > Li^+ > Na^+$. In mixed-ion electrolytes, co-insertion/extraction of two alkali cations takes place at cathode side. So it is not surprising that the rate capabilities of InHCF/Gr with Li^+/Na^+, Li^+/K^+ and Na^+/K^+ are better than those

with Na^+, Li^+ and Na^+, respectively. Through the equation $V = -E_b/ne$ ($n = 1$ in this case), we can calculate the intercalation potential $V(Li-OH_2^+) = 4.0$ V versus Li/Li^+, $V(Na-OH_2^+) = 3.94$ V versus Na/Na^+ and $V(K^+) = 3.86$ V versus K/K^+. When normalized to SHE, $V(Li-OH_2^+) = 0.96$ V, $V(Na-OH_2^+) = 1.23$ V and $V(K^+) = 0.97$ V are obtained (Fig. 6c). These values agree well with our experimental E_fs (0.79, 1.03 and 0.97 V versus SHE). Clearly, the calculated and experimental intercalation potential (*V*) follow the same order: $V(Li-OH_2^+)$ $< V(K^+) < V(Na-OH_2^+)$. This trend is quite a striking result. In typical intercalation compounds for organic batteries, the voltage to intercalate Na^+ is ~ 0.1–0.5 V lower than that for the intercalation of Li^+, due to the weaker bonding between Na^+ and host, while the voltage to intercalate K^+ can be comparable to that of Li^+ (ref. 37).

The intercalation mechanism is further verified by analysing the electronic structure of InHCF before and after cation insertion. As illustrated in Fig. 6d, Fe^{3+} in InHCF has one unpaired electron at low-spin state, while Fe^{2+} in AInHCF has zero unpaired electrons. Therefore, the magnetic momentum of Fe atom correlates to its oxidation state and spin state. In InHCF, the calculated magnetic momentum is 1.0 μB for low-spin Fe^{3+}. When InHCF changes into AInHCF, the magnetic momentum of Fe atom reduce to 0 μB, corresponding well to the reduction from Fe^{3+} (t_{2g}^5) to Fe^{2+} (t_{2g}^6). Here, the density of states projected on atomic orbitals (density of states projected on atomic orbitals (PDOS)) for InHCF, NaInHCF and NaInHCF-H_2O are shown as examples. From InHCF to NaInHCF, the major change of the PDOS is observed for Fe atom, while the main characteristic of In

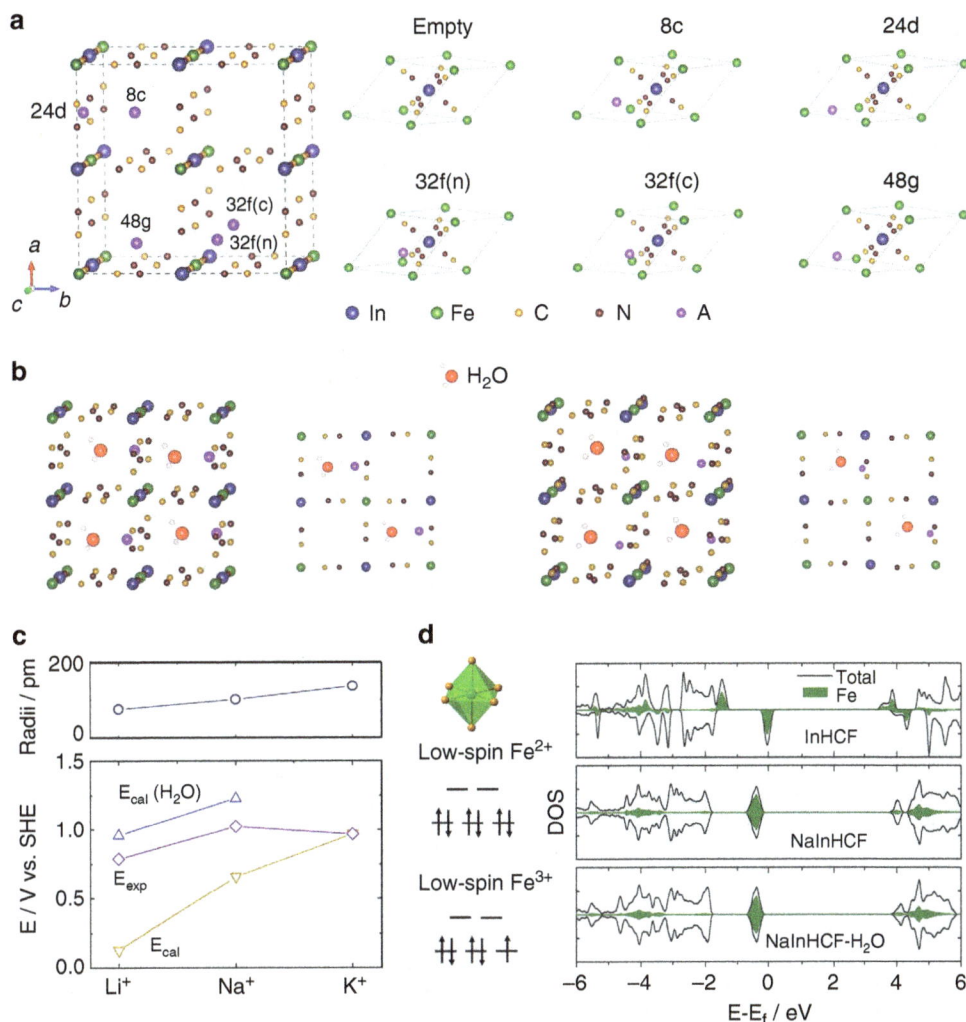

Figure 6 | DFT calculations. (a) Crystal structure of InHCF with five possible interstitial sites and primitive cells of InHCF with A$^+$ occupying the empty, 8c (body centre), 24d (face centre), 32f(n) (displaced from 8c sites toward N-coordinated corner), 32f(c) (displaced from 8c sites toward C-coordinated corner) and 48g (displaced between 8c and 24d) sites. **(b)** Crystal structures of InHCF with intercalation of [Na-OH$_2$]$^+$ (left of panel) and [Li-OH$_2$]$^+$ (right of panel). **(c)** Comparison of ionic radii and intercalation voltage for Li$^+$, Na$^+$ and K$^+$. **(d)** Schematics of the electronic states of Fe at different oxidation states and spin states, and electronic density of states projected on Fe ions in InHCF, NaInHCF and NaInHCF-H$_2$O.

Table 1 | Radii (unit: pm) of intercalants and the calculated binding energies (E_b, unit: eV) when intercalant occupying different interstitial sites.

Intercalant	Radius*	E_b				
		8c	24d	32f(n)	32f(c)	48g
Li$^+$	76	− 2.01	**− 3.13**†	− 3.05	− 2.81	**− 3.13**
Na$^+$	102	− 2.73	**− 3.33**	− 3.18	− 3.17	**− 3.33**
K$^+$	138	− 3.8	− 3.41	**− 3.86**	− 3.85	− 3.85
[Li-OH$_2$]$^+$	226			− 4.00‡		
[Na-OH$_2$]$^+$	252			− 3.94‡		
H$_2$O	150			− 0.67‡		

*The ionic radius of A$^+$ is adapted from Shannon's report[17]. The radius of [A-OH$_2$]$^+$ is estimated by adding the radii of A$^+$ and H$_2$O. r (H$_2$O) is supposed to be 150 pm.
†The bold font is used to emphasize the most stable site for each cation.
‡See details about how to calculate E_b for [A-OH$_2$]$^+$ and H$_2$O in Supplementary Methods.

with InHCF. In short, the experimental and theoretical studies on the intercalation mechanism suggest that in terms of the intercalation compounds for RAMB, the role of water that has been ignored before should be put into consideration.

Discussion

In summary, a series of high-voltage AMIB (B-I, B-II and B-III) based on unique electrochemical properties of InHCF, TiP$_2$O$_7$ and NaTi$_2$(PO$_4$)$_3$ in mixed-ion electrolytes are validated. They have high power density comparable to that of an ultra-capacitor, but with much higher-energy density. Among them, B-III exhibits a specific energy of 56 Wh kg^{-1} with E_{av} of 1.6 V, and B-II deliver a specific power of 2700 W kg^{-1} as an ultra-capacitor. The cation intercalation chemistry in InHCF is also unveiled. Iron not indium atom in InHCF is confirmed as the centre involved in the redox reaction. Such a reaction is accompanied by the intercalation of alkali cations that induces negligible structural deformations (zero-strain characteristic). DFT calculations also show that the intercalation is strongly affected by the ionic radii of inserted species. In terms of Li$^+$ or Na$^+$ with small size, they are supposed to be co-inserted into InHCF with H$_2$O, while H$_2$O occupying the site near 8c site in InHCF. K$^+$ with large size is

atom is almost unaltered (Supplementary Fig. 15). This result coincides with our XPS examination. The PDOS of Fe and In atoms in NaInHCF-H$_2$O are almost identical to those in NaInHCF, implying that water has little effect on the chemical environments of Fe and In atoms, resulting in a weak interaction

thought to be solely inserted into the 8c site without water, as a result of the space limitation in the [100] channel of InHCF. The calculated voltages for the intercalation of A^+ are in good agreement with our experimental values. Water-mediated cation intercalation also explains why the rate capability of InHCF with A^+ in aqueous electrolytes follows the order: $K^+ > Li^+ > Na^+$. Overall, our studies not only enrich the family of RAMB, but also offer a supplement to the existing intercalation chemistry, broadening our horizons for battery research.

Methods

Material syntheses. Graphene was obtained through thermal exfoliation of graphene oxide (Hummer's method) at 800 °C. InHCF/Gr was synthesized by *in situ* deposition method from our previous literature[16,24]. Typically, 100 ml of 0.085 M $InCl_3$ (pH 3.0, adjusted by aq. HCl) and 100 ml of 0.043 M $K_3Fe(CN)_6$ were simultaneously added into 75 ml of graphene water suspension (1 mg ml^{-1}) under vigorous stirring. The dropping rates of $InCl_3$ and $K_3Fe(CN)_6$ solutions were precisely controlled by peristaltic pump (0.6 ml min^{-1}). After the reaction was completed, the black InHCF/Gr slurry was formed. Finally, the precipitates were washed with de-ionized water several times and then dried at 80 °C overnight. Orange precipitates were obtained for InHCF by using de-ionized water instead of graphene suspension. TiP_2O_7 and $NaTi_2(PO_4)_3$ were prepared by solid-state reaction method. The high purity TiO_2 (up to 99.9%) and $NH_4H_2PO_4$ in a 1:2 ratio were added into ethanol to form paste. This paste was ball-milled in a planetary ball mill at 400 r.p.m. for 6 h. After ball milling, the mixture was dried at 80 °C to evaporate ethanol. Then it was successively heated for 5 h at 300 °C and for 12 h at 900 °C under air with intermediate grindings. The obtained white TiP_2O_7 solids, referred as as-prepared TiP_2O_7, were homogenized in a mortar for further use. For carbon-coated TiP_2O_7, 2.5 g of as-prepared TiP_2O_7, 2 g of glucose and appropriate amount of ethanol were mixed and then ball-milled for 4 h. The solids were obtained by evaporating the ethanol at 80 °C. After calcined at 800 °C for 3 h in a mixture gas (C_2H_4/Ar, 5/95), the solids became black, implying that TiP_2O_7 were coated by carbon. Carbon-coated $NaTi_2(PO_4)_3$ are prepared by the similar method for carbon-coated TiP_2O_7 as stated above, except using the different precursor for $NaTi_2(PO_4)_3$ (Na_2CO_3:TiO_2:$NH_4H_2PO_4$ = 1:4:6).

Characterization. Powder X-ray diffraction patterns were collected using an AXS D8 Advance diffractometer (Cu Kα radiation; receiving slit, 0.2 mm; scintillation counter, 40 mA; 40 kV) from Bruker Inc. The morphology and structure of samples were analysed by a Hitachi S-4800 field emission SEM and an FEI Tecnai G2 F20 TEM at an accelerating voltage of 200 kV. Thermal gravimetric analysis was performed on a Pyris Diamond thermogravimetric/differential thermal analyser (Perkin-Elmer) to analyse the cabon content in carbon-coated TiP_2O_7 and $NaTi_2(PO_4)_3$. X-ray photoelectron spectra (XPS) were collected by a Shimadzu/Kratos AXIS Ultra XPS spectrometer. All binding energies were referenced to the F 1*s* peak (from polyvinylidene fluoride) of 688.2 eV.

Electrochemical measurements. Electrochemical measurements were carried on Solartron 1470E multi-channel potentiostats using either a two-electrode or a three-electrode cell set-up. For three-electrode set-up, an Ag/AgCl electrode (0.2 V versus SHE) and Pt gauze were employed as reference and counter electrodes, respectively. Electrode were prepared by casting slurries of active materials (75 wt%), Super P (15 wt%) and polyvinylidene fluoride (10 wt%) in *N*-methyl-2-pyrrolidinone on steel iron grid, and air drying at 80 °C for 12 h. Discs with diameter of 1.3 cm were cut for electrochemical tests. Mass loadings for the electrodes were determined by comparing the mass of the electrode with that of the original blank one.

DFT calculations. The calculations were based on DFT method in the generalized gradient approximation with the Perdew, Burke and Ernzerhof functional[39], as implemented in the Vienna *ab initio* Simulation Package, which employed a plane-wave basis[40,41]. The plane-wave energy cutoff was set to be 500.0 eV, and the electronic energy convergence was 10^{-5} eV. During relaxations, the force convergence for ions was 10^{-2} eV/Å. A Γ—centred $6 \times 6 \times 6$ Monkhorst–Pack k-point mesh for primitive cell was used to sample the irreducible Brillouin zone. In $InFe(CN)_6$, Fe ions locate in the octahedral environment and bond to carbon, and such carbon-coordinated Fe in a strong crystal field favours the low-spin configuration. To correctly characterize the localization of Fe *d*-electrons, GGA + U method with a Hubbard-type potential to describe the d-part of the Hamiltonian was applied. Previous studies using the U = 3.0 eV for the low-spin Fe had demonstrated that GGA + U calculations could provide good predictions about the structural and electronic properties of HCF compounds[31,42,43]. Then, we set the U value of 3.0 eV for Fe atom during the calculations of structural relaxation and electronic structures.

References

1. Dunn, B., Kamath, H. & Tarascon, J. M. Electrical energy storage for the grid: a battery of choices. *Science* **334**, 928–935 (2011).
2. Kim, H. *et al.* Aqueous rechargeable Li and Na ion batteries. *Chem. Rev.* **114**, 11788–11827 (2014).
3. Suo, L. M. *et al.* 'Water-in-salt' electrolyte enables high-voltage aqueous lithium-ion chemistries. *Science* **350**, 938–943 (2015).
4. Li, W., Dahn, J. R. & Wainwright, D. S. Rechargeable lithium batteries with aqueous electrolytes. *Science* **264**, 1115–1118 (1994).
5. Wang, G. J. *et al.* An aqueous rechargeable lithium battery with good cycling performance. *Angew. Chem. Int. Ed.* **46**, 295–297 (2007).
6. Wang, H. B., Huang, K. L., Zeng, Y. Q., Yang, S. & Chen, L. Q. Electrochemical properties of TiP_2O_7 and $LiTi_2(PO_4)_3$ as anode material for lithium ion battery with aqueous solution electrolyte. *Electrochim. Acta* **52**, 3280–3285 (2007).
7. Luo, J. Y., Cui, W. J., He, P. & Xia, Y. Y. Raising the cycling stability of aqueous lithium-ion batteries by eliminating oxygen in the electrolyte. *Nat. Chem.* **2**, 760–765 (2010).
8. Li, Z., Young, D., Xiang, K., Caeter, W. C. & Chiang, Y. M. Towards high power high energy aqueous sodium-ion batteries: the $NaTi_2(PO_4)_3$/$Na_{0.44}MnO_2$ system. *Adv. Energy Mater.* **3**, 290–294 (2013).
9. Wessells, C. D., Huggins, R. A. & Cui, Y. Copper hexacyanoferrate battery electrodes with long cycle life and high power. *Nat. Commun.* **2**, 550 (2011).
10. Whitacre, J. F., Tevar, A. & Sharma, S. $Na_4Mn_9O_{18}$ as a positive electrode material for an aqueous electrolyte sodium-ion energy storage device. *Electrochem. Commun.* **12**, 463–466 (2010).
11. Park, S. I., Gocheva, I., Okada, S. & Yamaki, J. I. Electrochemical properties of $NaTi_2(PO_4)_3$ anode for rechargeable aqueous sodium-ion batteries. *J. Electrochem. Soc.* **158**, A1067–A1070 (2011).
12. Wessells, C. D., Peddada, S. V., McDowell, M. T., Huggins, R. A. & Cui, Y. The effect of insertion species on nanostructured open framework hexacyanoferrate battery electrodes. *J. Electrochem. Soc.* **159**, A98–A103 (2012).
13. Pasta, M. *et al.* Full open-framework batteries for stationary energy storage. *Nat. Commun.* **5**, 3007 (2014).
14. Wu, X. Y. *et al.* Energetic aqueous rechargeable sodium-ion battery based on $Na_2CuFe(CN)_6$-$NaTi_2(PO_4)_3$ intercalation chemistry. *ChemSusChem* **7**, 407–411 (2014).
15. Xu, C. J., Li, B. H., Du, H. D. & Kang, F. Y. Energetic zinc ion chemistry: the rechargeable zinc ion battery. *Angew. Chem. Int. Ed.* **51**, 933–935 (2012).
16. Zhang, L. Y., Chen, L., Zhou, X. F. & Liu, Z. P. Towards high-voltage aqueous metal-ion batteries beyond 1.5 V: the zinc/zinc hexacyanoferrate system. *Adv. Energy Mater.* **5**, 1400930 (2015).
17. Shannon, R. D. Revised effective ionic radii and systematic studies of interatomic distances in halides and chalcogenides. *Acta. Crystallogr.* **A32**, 751–767 (1976).
18. Wu, X. Y. *et al.* Vacancy-free Prussian blue nanocrystals with high capacity and superior cyclability for aqueous sodium-ion batteries. *ChemNanoMat* **1**, 188–193 (2015).
19. Wang, Y. S. *et al.* Ti-substituted tunnel-type $Na_{0.44}MnO_2$ oxide as a negative electrode for aqueous sodium-ion batteries. *Nat. Commun.* **6**, 6401 (2015).
20. Wang, Y. S. *et al.* A novel high capacity positive electrode material with tunnel-type structure for aqueous sodium-ion batteries. *Adv. Energy Mater.* **5**, 1501005 (2015).
21. Liu, Y. *et al.* High-performance aqueous sodium-ion batteries with $K_{0.27}MnO_2$ cathode and their sodium storage mechanism. *Nano Energy* **5**, 97–104 (2014).
22. Deng, C., Zhang, S., Dong, Z. & Shang, Y. 1D nanostructured sodium vanadium oxide as a novel anode material for aqueous sodium ion batteries. *Nano Energy* **4**, 49–55 (2014).
23. Chen, L. *et al.* New-concept batteries based on aqueous Li^+/Na^+ mixed-ion electrolytes. *Sci. Rep.* **3**, 1946 (2013).
24. Chen, L., Zhang, L. Y., Zhou, X. F. & Liu, Z. P. Aqueous batteries based on mixed monovalence metal Ions: a new battery family. *ChemSusChem* **7**, 2295–2302 (2014).
25. Wessells, C. D., Peddada, S. V., Huggins, R. A. & Cui, Y. Nickel hexacyanoferrate nanoparticle electrodes for aqueous sodium and potassium ion batteries. *Nano Lett.* **11**, 5421–5425 (2011).
26. Wessells, C. D. *et al.* Tunable reaction potentials in open framework nanoparticle battery electrodes for grid-scale energy storage. *ACS Nano* **6**, 1688–1694 (2012).
27. Wu, X. Y. *et al.* Low-defect Prussian blue nanocubes as high capacity and long life cathodes for aqueous Na-ion batteries. *Nano Energy* **13**, 117–123 (2015).
28. Wang, L. *et al.* A superior low-cost cathode for a Na-ion battery. *Angew. Chem. Int. Ed.* **52**, 1964–1967 (2013).

29. Song, J. *et al.* Removal of interstitial H_2O in hexacyanometallates for a superior cathode of a sodium-ion battery. *J. Am. Chem. Soc.* **137**, 2658–2664 (2015).

30. Lee, H. W. *et al.* Manganese hexacyanomanganate open framework as a high-capacity positive electrode material for sodium-ion batteries. *Nat. Commun.* **5**, 5280 (2014).

31. Ling, C., Chen, J. J. & Mizuno, F. First-principles study of alkali and alkaline earth ion intercalation in iron hexacyanoferrate: the important role of ionic radius. *J. Phys. Chem. C* **117**, 21158–21165 (2013).

32. Ellis, D., Eckhoff, M. & Neff, V. D. Electrochromism in the mixed-valence hexacyanides. 1. Voltammetric and spectral studies of the oxidation and reduction of thin films of Prussian blue. *J. Phys. Chem.* **85**, 1225–1231 (1981).

33. Dong, S. J. & Jie, Z. Electrochemistry of indium hexacyanoferrate film modified electrodes. *Electrochimi. Acta* **34**, 963–968 (1989).

34. de Tacconi, N. R. & Rajeshwar, K. Metal hexacyanoferrates: electrosynthesis, *in situ* characterization, and applications. *Chem. Mater.* **15**, 3046–3062 (2003).

35. Younesi, R., Hahlin, M., Björefors, F., Johansson, P. & Edström, K. Li–O_2 battery degradation by lithium peroxide (Li_2O_2): a model study. *Chem. Mater.* **25**, 77–84 (2013).

36. Aydinol, M. K., Kohan, A. F., Ceder, G., Cho, K. & Joannopoulos, J. *Ab initio* study of lithium intercalation in metal oxides and metal dichalcogenides. *Phys. Rev. B* **56**, 1354–1365 (1997).

37. Ong, S. P. *et al.* Voltage, stability and diffusion barrier differences between sodium-ion and lithium-ion intercalation materials. *Energy Environ. Sci.* **4**, 3680–3688 (2011).

38. Islam, M. S. & Fisher, C. A. J. Lithium and sodium battery cathode materials: computational insights into voltage, diffusion and nanostructural properties. *Chem. Soc. Rev.* **43**, 185–204 (2014).

39. Perdew, J. P., Burke, K. & Ernzerhof, M. Generalized gradient approximation made simple. *Phys. Rev. Lett.* **77**, 3865–3868 (1996).

40. Kresse, G. & Hafner, J. *Ab initio* molecular dynamics for liquid metals. *Phys. Rev. B* **47**, 558–561 (1993).

41. Kresse, G. & Furthmüller, J. Efficient iterative schemes for *ab initio* total-energy calculations using a plane-wave basis set. *Phys. Rev. B* **54**, 11169–11186 (1996).

42. Wojdel, J. C., Moreira, I. d. P. R., Bromley, S. T. & Illas, F. Prediction of half-metallic conductivity in Prussian blue derivatives. *J. Mater. Chem.* **19**, 2032–2036 (2008).

43. Wojdel, J. C., Moreira, I. d. P. R., Bromley, S. T. & Illas, F. On the prediction of the crystal and electronic structure of mixed-valence materials by periodic density functional calculations: the case of Prussian Blue. *J. Chem. Phys.* **128**, 044713 (2008).

Acknowledgements

We acknowledge financial supports from National Natural Science Foundation of China (grant nos 51404233 and 11404348), Key Research Program of Chinese Academy of Sciences (grant no. KGZD-EW-T08-2), Natural Science Foundation of Zhejiang Province and Ningbo city (grant nos LY15B030004 and 2014A610044), Science Technology Department of Zhejiang Province (grant no. 2014C31009).

Author contributions

L.C. and Z.P.L. designed the experiments. L.C. carried out the experiments. H.Z.S. and G.Q.L. performed the DFT calculations. All authors contributed to the data analysis and co-wrote the paper. L.C. and Z.P.L. proposed and supervised the project.

Additional information

20

Non-equilibrium behaviour in coacervate-based protocells under electric-field-induced excitation

Yudan Yin[1], Lin Niu[1], Xiaocui Zhu[2], Meiping Zhao[2], Zexin Zhang[3], Stephen Mann[4] & Dehai Liang[1]

Although numerous strategies are now available to generate rudimentary forms of synthetic cell-like entities, minimal progress has been made in the sustained excitation of artificial protocells under non-equilibrium conditions. Here we demonstrate that the electric field energization of coacervate microdroplets comprising polylysine and short single strands of DNA generates membrane-free protocells with complex, dynamical behaviours. By confining the droplets within a microfluidic channel and applying a range of electric field strengths, we produce protocells that exhibit repetitive cycles of vacuolarization, dynamical fluctuations in size and shape, chaotic growth and fusion, spontaneous ejection and sequestration of matter, directional capture of solute molecules, and pulsed enhancement of enzyme cascade reactions. Our results highlight new opportunities for the study of non-equilibrium phenomena in synthetic protocells, provide a strategy for inducing complex behaviour in electrostatically assembled soft matter microsystems and illustrate how dynamical properties can be activated and sustained in microcompartmentalized media.

[1] Beijing National Laboratory for Molecular Sciences, MOE Key Laboratory of Polymer Chemistry and Physics, College of Chemistry and Molecular Engineering, Peking University, Beijing 100871, China. [2] Beijing National Laboratory for Molecular Sciences, MOE Key Laboratory of Bioorganic Chemistry and Molecular Engineering, College of Chemistry and Molecular Engineering, Peking University, Beijing 100871, China. [3] Center for Soft Condensed Matter Physics and Interdisciplinary Research, Soochow University, Suzhou 215006, China. [4] Centre for Protolife Research, School of Chemistry, University of Bristol, Bristol BS8 1TS, UK. Correspondence and requests for materials should be addressed to S.M. (email: s.mann@bristol.ac.uk) or to D.L. (email: dliang@pku.edu.cn).

The design and construction of rudimentary forms of synthetic cell-like entities (protocells) exhibiting controlled reactivity, internal structuration and adaptive functionality are providing new approaches to protocell modelling[1–3], the fabrication of integrated materials micro-ensembles[4–6], and mechanisms related to prebiotic organization and the origin of life on the early Earth[7]. Functional protocells have been prepared by membrane self-assembly of amphiphilic organic molecules (polymer[8,9] or lipid[10,11] vesicles), surface-functionalized inorganic nanoparticles (colloidosomes)[12,13] and protein–polymer nano-conjugates (proteinosomes)[14]; layer-by-layer membrane assembly of counter-charged polyelectrolytes[15,16]; and microphase separation of membrane-free liquid micro-droplets by complex coacervation[17]. Hybrid protocells based on membrane-coated coacervate droplets with homogenous[18] or subdivided interiors[19,20], or lipid vesicles containing discrete polymer-enriched internalized domains[21,22], have also been recently reported. Significantly, these strategies are almost exclusively based on the static assembly and confinement of molecular and nanoscale components that operate under close to equilibrium conditions. As a consequence, even the most highly integrated protocellular systems are functionally compromised compared with basic life processes, which occur under a continuous flux of energy and matter exchange.

Thus, a major challenge in synthetic protocell research involves the development of methodologies that enable the sustained activation of chemical micro-ensembles such that they persist and function under non-equilibrium conditions. While modularized systems can be locally excited, for example, by light-driven pumping of protons across a targeted vesicle membrane to induce gene expression[23], energization at a non-local level requires the maintenance and coupling of chemical fuel gradients to internalized protometabolic processes, or alternatively, the indiscriminate excitation of the protocell medium by an externally applied field. In this regard, protocell models based on coacervate liquid microdroplets are distinctive for their molecularly crowded, reduced dielectric constant aqueous interiors that consist of a highly enriched matrix of electrostatically interacting counter-charged polyelectrolytes[17,24]. These protocell models are established under equilibrium conditions, and therefore exhibit no complex dynamical properties. However, as charged macromolecular complexes are sensitive to electrohydrodynamic forces[25,26], coacervates can be globally excited by exposure to an applied electric field[27–29], suggesting that it should be possible to use electric fields to sustain and control the activation of coacervate-based protocells under non-equilibrium conditions. Previous studies on the induced mobility of coacervate microdroplets prepared from high–molecular-weight (2,000 bp) double-stranded DNA have been reported[29], although no complex internalized behaviour was observed.

Here we demonstrate that the electric-field-induced energization of spatially confined coacervate microdroplets containing a mixture of oppositely charge polypeptides and short single strands of DNA provides a step towards the design and construction of synthetic protocells capable of exhibiting a range of complex behaviours under non-equilibrium conditions. By confining the droplets within a microfluidic channel and applying a range of electric field strengths, we produce linear arrays of membrane-free protocells that undergo cycles of transient subcompartmentalization, dynamical fluctuations in size and shape, chaotic processes of growth and fusion, and unidirectional movement. We show that excitation of the protocells leads to a continuous exchange of matter with the environment via repetitive cycles of vacuole nucleation, growth and expulsion, spontaneous ejection and sequestration of microdomains of the coacervate matrix, directional capture of solute molecules, and pulsed enhancement of enzyme cascade reactions. Switching off the electric field immediately arrests the dynamical behaviour, and recapitulates the equilibrium structure and form of the protocells. Taken together, our results highlight new opportunities for the study of non-equilibrium phenomena in synthetic protocell research, provide a novel strategy for inducing complex behaviour in electrostatically assembled soft matter microsystems and illustrate how dynamical properties can be activated and sustained in microcompartmentalized media.

Results

Electric field excitation of coacervate-based protocells. We prepare coacervate microdroplets within a microfluidic channel by flow-induced mixing of cationic poly-L-lysine (PLL, $M_W = 30{,}000$–$70{,}000$) and a negatively charged single-stranded oligodeoxynucleotide (ss-oligo, $M_W = 6{,}387$, 21 nt) in 0.01 M phosphate buffer, and then apply an electric field of variable strength along the microfluidic channel (Fig. 1a). The PLL and

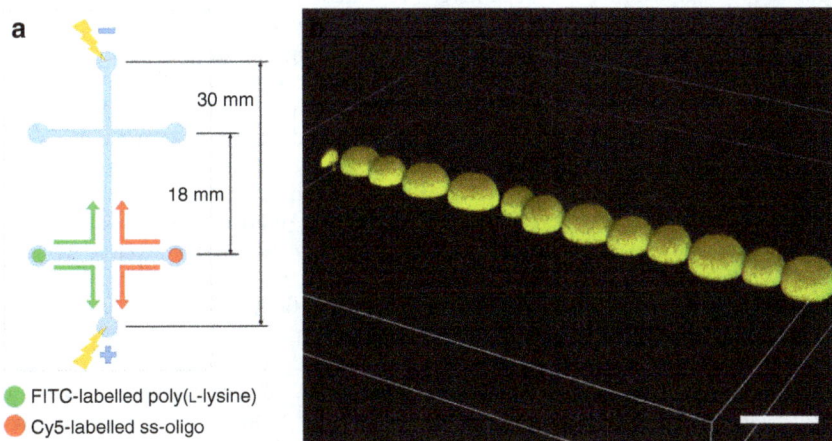

Figure 1 | Experimental set-up. (a) Schematic showing microfluidic chip layout and flow of solutions. The microchannels were $80 \times 25\,\mu m$ in width and depth, respectively, and etched on a glass chip. FITC–PLL and Cy5–ss-oligo were placed in the opposite sample reservoirs and allowed to mix under flow at room temperature to produce an array of fluorescent PLL/ss-oligo coacervate microdroplets along the central channel. The electric field was then applied along the microfluidic channel. **(b)** 3D confocal fluorescence microscopy reconstruction showing linear array of discrete PLL/ss-oligo droplets prepared at a charge ratio of 1:1 in a microfluidic channel in the absence of an electric field. Adherence of the coacervate droplets to the channel wall due to gravity gives rise to the observed hemispherical morphology. Scale bar, $20\,\mu m$.

ss-oligo are fluorescently labelled with fluorescein isothiocyanate (FITC) and 1,1-*bis*(3-hydroxypropyl)-3,3,3,3-tetramethylindodicarbocyanine (Cy5), respectively, to assist imaging of the droplets within the microfluidic channel. In the absence of an electric field, room temperature mixing of the PLL and ss-oligo at a charge ratio of $\approx 1{:}1$ produces a linear array of discrete coacervate microdroplets that are typically $20\,\mu m$ in diameter and stable with respect to coalescence. Three-dimensional (3D) confocal fluorescence microscopy images reveal uniform fluorescence intensity throughout the droplets, indicating that the complex coacervates are structurally homogeneous in the absence of an electric field (Fig. 1b).

Switching on the electric field (E) above a critical value has a remarkable influence on the structure and dynamics of the preformed PLL/ss-oligo coacervate microdroplets. While the microdroplets remain essentially unchanged at $E < 20\,\mathrm{V\,cm^{-1}}$ field, increasing the electric field to 30 or $40\,\mathrm{V\,cm^{-1}}$ results in the cyclical appearance and disappearance of non-fluorescent vacuoles within the interior of the coacervate phase (Fig. 2a–c; Supplementary Movies 1–3). Under these conditions, subcompartmentalization is repeatedly initiated at a small number (typically $n = 1$–3) of randomly located sites in droplets that remain effectively unchanged in overall size and shape, and immobile within the microfluidic channel. In contrast, major fluctuations in droplet morphology and dynamics occur at $E = 50$ and $70\,\mathrm{V\,cm^{-1}}$ due to an increased intensity in the growth and expulsion of the vacuoles, which are associated, respectively, with cycles of expansion and contraction of the coacervate droplets. As a consequence, the energized droplets often undergo fusion to produce larger membrane-free protocells with co-existing multiple subcompartments (Fig. 2d). The structural and morphological fluctuations are sufficiently intense that fluorescent microparticles of the coacervate medium are ejected from and then consumed by the larger droplets during contraction and swelling, respectively (Fig. 2e). Phenomenologically, the energized protocells exhibit 'life-like' chaotic behaviour involving unidirectional movement (electric field induced), morphogenesis (rapid fluctuations in size and shape), transient subcompartmentalization, fusion and growth, and continuous exchange of matter with the environment (vacuole growth/expulsion, matrix ejection/sequestration) (Supplementary Movies 4 and 5). Furthermore, at $100\,\mathrm{V\,cm^{-1}}$, the electrohydrodynamic force is strong enough to stretch the subdivided droplets in a direction normal to the electric field (Fig. 2f; Supplementary Movie 6). Significantly, the repetitive formation of the vacuoles is stopped immediately the electric field is switched off.

Vacuole formation and protocell dynamics. We use fluorescence microscopy to investigate the nature of vacuole formation within the PLL/ss-oligo protocells. 3D confocal fluorescence microscopy images confirm the presence of internal cavities in the subdivided droplets (Fig. 3a). The vacuoles nucleate randomly inside the protocells, rapidly grow in size, fuse if in the presence of more than one subcompartment and then disappear on contact with the surface of the coacervate microdroplet (Fig. 3b). Release of the vacuole contents into the bulk aqueous phase is followed by a further cycle of nucleation, growth and ejection of the subcompartments, which continue without loss of droplet integrity until the electric field is switched off. The lifetime associated with vacuole growth and expulsion is determined within individual droplets from the video recordings. As the fluorescence intensity associated with the vacuoles is close to the background level, fluorescence line scans across subdivided droplets give a series of peaks corresponding to the edges of the coacervate-rich areas (Supplementary Fig. 1). The size of the vacuoles is determined by

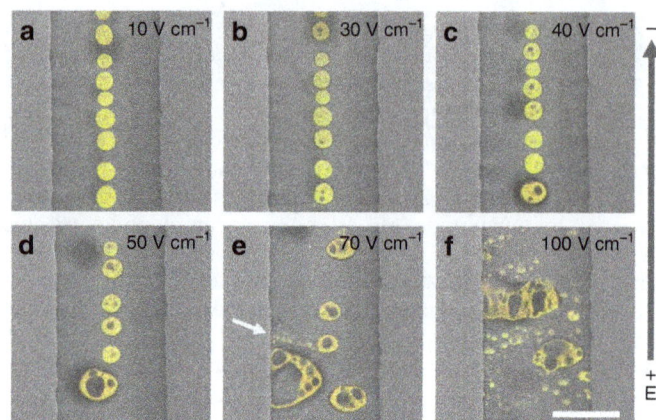

Figure 2 | Electric field excitation of coacervate-based protocells. Video images showing PLL/ss-oligo coacervate droplets prepared in a microfluidic channel and then exposed to an applied electric field (E) of varying strength. (**a**) At $E = 10\,\mathrm{V\,cm^{-1}}$, showing a stable array of fluorescent coacervate droplets with homogeneous size, structure and morphology. (**b,c**) At $E = 30$ or $40\,\mathrm{V\,cm^{-1}}$, showing immobile arrays of coacervate microdroplets containing small numbers of subcompartments (localized non-fluorescent regions). The vacuoles nucleate, grow and disappear in a cyclical process of matter exchange with the environment without significantly influencing droplet size and shape (see Supplementary Movies 1–3 for more details). (**d,e**) At $E = 50$ or $70\,\mathrm{V\,cm^{-1}}$, showing a destabilized array of motile protocells with co-existing multiple vacuoles and increased polydispersity due to droplet fusion. Corresponding videos (Supplementary Movies 4 and 5) show rapidly fluctuating changes in the shape and size of individual droplets as they move towards the negative electrode. Arrow in **e** highlights the presence of fluorescent ejectiles that were released in the direction of the electric field from the adjacent multi-compartmentalized droplet; the ejectiles were subsequently sequestered by the same droplet at a later time interval. (**f**) At $100\,\mathrm{V\,cm^{-1}}$, showing transversely stretched multi-vacuole-containing protocells and extensive amounts of ejected coacervate matrix due to the high electrohydrodynamic forces. Scale bar, $50\,\mu m$.

measuring the distance between the fluorescence intensity peaks at various time intervals. As shown in Fig. 3c, the vacuole lifetime is of the order of around 5 s in an electric field of $80\,\mathrm{V\,cm^{-1}}$.

Changes in the composition of the coacervate matrix during repeated cycles of swelling and contraction accompanying vacuole growth and expulsion at electric fields of 10–$70\,\mathrm{V\,cm^{-1}}$ are investigated by fluorescence microscopy measurements on individual droplets. Time-dependent measurements of the relative fluorescence intensity, $I_r = I_{\mathrm{PLL}}/I_{\mathrm{oligo}}$, are used as an approximate index of how the PLL/ss-oligo ratio in the droplets changes over a period of 50 s (Fig. 3d). No significant change in the normalized ratio is observed at $10\,\mathrm{V\,cm^{-1}}$, indicating that the coacervate matrix is compositional stable in the vacuole-free protocells. In contrast, onset of subcompartmentalization in droplets at $E > 30\,\mathrm{V\,cm^{-1}}$ results in a time-dependent reduction in I_r that fluctuates probably due to matter exchange with the external environment during the swelling/contraction cycles, and becomes more pronounced as the electric field is increased. For example, a 10% drop in the value of I_r is observed after 50 s at $70\,\mathrm{V\,cm^{-1}}$, indicating that a considerable amount of PLL is released as the protocells undergo rapid fluctuations in size and shape. To determine when and how the PLL is released, we measure the I_r values associated with the fluorescent microparticles produced *in situ* by ejection from a neighbouring parent droplet at $70\,\mathrm{V\,cm^{-1}}$ (Fig. 3e). The average I_r value for the ejected particles (region 2 in Fig. 3e and Supplementary Fig. 2) is $\sim 40\%$ higher than that of the parent droplet (region 1 in Fig. 3e),

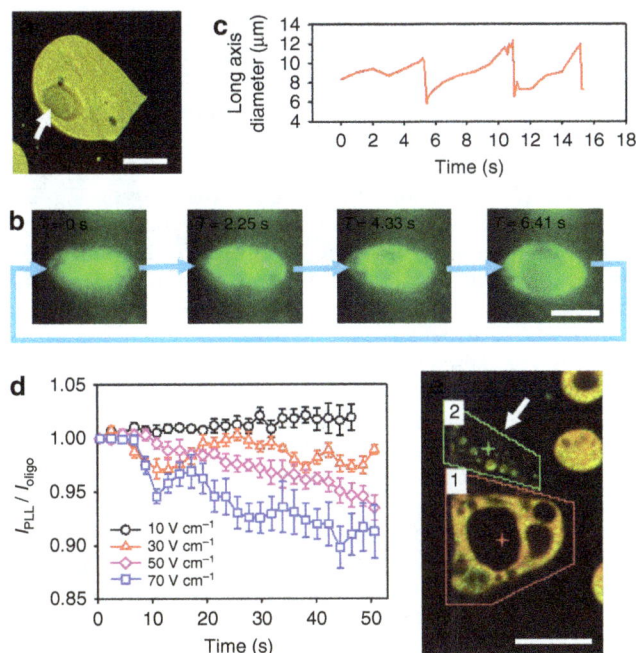

Figure 3 | Vacuole formation and protocell dynamics in applied electric fields. (**a**) 3D confocal fluorescence microscopy reconstruction of a PLL/ss-oligo coacervate microdroplet exposed to an electric field of 70 V cm^{-1} showing the presence of a single 18-μm-sized vacuole (solid arrow). Scale bar, 20 μm. (**b**) Fluorescence microscopy images showing a time sequence of events leading to vacuole nucleation, fusion and growth within a single PLL/ss-oligo microdroplet placed in an electric field of 80 V cm^{-1}. Contact of the vacuole with the droplet surface releases the subcompartment into the external medium and initiates a new cycle of matter transfer. Scale bar, 10 μm. (**c**) Plot of droplet size against time for three cycles of vacuole growth within and expulsion from an individual protocell at $E = 80$ V. (**d**) Plots of relative fluorescence intensity ($I_{PLL}/I_{ss\text{-}oligo}$) against time for individual PLL/ss-oligo microdroplets exposed to electric fields ranging from 10 to 70 V cm^{-1}. Values for I_{PLL} and I_{oligo} were normalized to their initial values recorded before the electric field was switched on ($t = 0$). Error bar: s.e. mean. (**e**) Confocal fluorescence microscopy image showing large multi-compartmentalized PLL/ss-oligo protocell (region 1) and ejected microparticles (region 2) at 70 V cm^{-1}. Fluorescence channels for FITC–PLL (green) and Cy5-tagged ss-oligo (red) are combined in the images. Analyses of the relative fluorescence intensities indicate that the concentration of PLL is highest in the expelled particles. Scale bar, 20 μm.

indicating that PLL is preferentially released from the coacervate matrix during repeated cycles of vacuolization. As the PLL chain length is considerably greater than that for the ss-oligo, uncomplexed loops of PLL within the coacervate matrix will be subjected to higher dragging forces, and therefore preferentially expelled during the excessive fluctuations in droplet morphology observed at relatively high electric fields.

Molecular uptake during vacuolarization. Other studies indicate that not only are components of the droplet selectively expelled but also solute molecules in the external medium can be taken up directionally and accumulated inside the droplets in the presence of an electric field. Using a water-soluble fluorescent dye (calcein) as a model molecule, we monitor the migration behaviour and spatial distribution of the solute during the vacuolarization process. Negligible uptake of calcein into the coacervate matrix of the microdroplets is observed by confocal fluorescence microscopy in the absence of an electric field (Fig. 4a,b), consistent with the

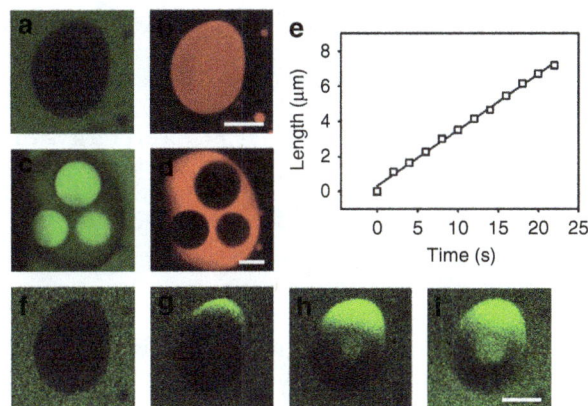

Figure 4 | Molecular uptake during vacuolarization. (**a-d**) Confocal fluorescence microscopy images showing a single PLL/Cy5-labelled ss-oligo coacervate microdroplet prepared in a microfluidic channel with addition of the water-soluble, green fluorescent dye, calcein (20 μM) before (**a,b**) and after (**c,d**) activation in an applied electric field at 30 V cm^{-1}. No uptake of calcein is observed in the microdroplet before excitation (**a**), which shows only homogeneous red fluorescence associated with the Cy5-ssDNA component (**b**). In contrast, calcein is observed in the vacuoles but not in the coacervate matrix when the droplet is exposed to an electric field (**c,d**), indicating that the vacuoles are formed by water ingress from the external environment. (**e-i**) Time-dependent uptake of calcein in the microdroplets under energization after 1 (**f**), 15 (**g**), 30 (**h**) and 45 s (**i**). Positions of the calcein wave front (**e**) were used to calculate dye mobility in the droplets. Scale bar, 10 μm.

relative hydrophobic nature of the PLL/ss-oligo phase. In contrast, calcein is observed in the vacuoles but not in the coacervate matrix when the droplet is exposed to an electric field (Fig. 4c,d). As the vacuolization process is cyclical, transportation of the external aqueous phase into and out of the droplet occurs as long as the electric field is switched on. Moreover, once the electric field is applied, calcein molecules migrate into the coacervate phase from one end of the droplet along the direction of electric field (Fig. 4e–i; Supplementary Movie 7). A linear time-dependent relationship is observed for the advancing wave front of dye molecules, indicating an almost constant velocity for calcein uptake (Fig. 4e). The calculated mobility within the droplets (-3.2×10^{-10} m^2 s^{-1} V^{-1}) is significantly lower than the value observed in free solution (-6.0×10^{-8} m^2 s^{-1} V^{-1})[30]. As a consequence, the concentration of calcein in the droplet as indicated by the fluorescence intensity is considerably higher than that in the surrounding phase.

Transport of the external aqueous phase during vacuolization is driven by osmotic pressure generated by an elevated local ion concentration in the droplets exposed to an electric field. Droplets prepared at PLL and ss-oligo at higher concentrations (PLL: 4.0 mg ml^{-1}, ss-oligo: 6.0 mg ml^{-1}) contain domains rich in ions due to incomplete complexation[31]. Indeed, under these conditions, vacuolization is observed in the absence of an electric field when the continuous phase is replaced by a buffer solution (Supplementary Fig. 3). The vacuoles are mainly located in the PLL-enriched domain of the forming droplets due to the longer chain length of the cationic polymer, which gives rise to an increased number of unneutralized loops[31], and higher ionic concentration. Significantly, enhanced dissociation of the polyelectrolyte complex occurs in the presence of an electric field, which in turn increases the local concentration of free ions within the droplets, enabling vacuolization in the droplet at lower polymer concentrations. As the imbibed water does not mix with the viscoelastic coacervate phase, the vacuole grows under a compressive force, but because there is no restraining membrane at the droplet surface is stochastically released back into the continuous phase, contracting the droplet and initiating the next cycle of matter transfer. The presence of elevated salt concentrations in the continuous phase or substitution of the ss-oligo for a more strongly bonded double-stranded oligo-DNA counterpart suppresses vacuole formation (data not shown) by offsetting the osmotic pressure or curtailing disassociation of the complex, respectively.

Electric-field-mediated enhancement of enzyme reactions. Inspired by the directional movement, accumulation and throughput of molecules observed for microdroplets under electric field energization, we explore the possibility of undertaking enzyme-mediated chemical transformations[19] in the activated medium. Glucose oxidase (GOx) and horseradish peroxidase (HRP) are encapsulated in the coacervate droplets during preparation, and a mixture of o-phenylenediamine (oPD, substrate for HRP) and β-D-glucose (substrate for GOx) is injected into the microfluidic channel to replace the original buffer. Diffusion of oPD and β-D-glucose into the coacervate droplets initiates the enzyme-mediated tandem reaction to produce a fluorescent final product, 2,3-diaminophenazine (2,3-DAP) that is monitored in individual droplets over time (Supplementary Table 1). A steady increase in the 2,3-DAP fluorescence intensity followed by a slow decrease after 35 min is observed for enzyme-containing droplets in the absence of an electric field (Fig. 5a–d; Fig. 5e (blue curve)). We attribute the fall off in intensity to the formation of a quasi-steady state between

2,3-DAP production within the droplet and release of the product to the external environment. This behaviour is not observed for cascade reactions undertaken in bulk solution; in this case, a continuous increase in intensity is observed over a time period of 60 min. Significantly, a sharp increase in 2,3-DAP fluorescence intensity in the droplets is observed when an electric field of $50 \, V \, cm^{-1}$ is applied for 1 min (Fig. 5e (red curve); Fig. 5f–i). The electric field also redistributes the positively charged 2,3-DAP molecules towards one region of the droplet (Fig. 5g; Supplementary Fig. 4), and the fluorescence intensity decreases to a steady state value within a few minutes after the electric field $(50 \, V \, cm^{-1})$ is switched off (Fig. 5e), such that the enzyme reaction can be pulsed by intermittent switching on and off of the electric field (Fig. 5e). Control experiments undertaken in the absence of the enzymes show negligible changes in fluorescence, indicating that the observed increase in fluorescence intensity originates specifically from the formation of 2,3-DAP (Supplementary Fig. 5).

We attribute the electric-field-mediated enhancement of the enzyme cascade reaction rate to the triggered rapid uptake and accumulation of oPD and β-D-glucose within the droplets due to electric-field-induced vacuolization. Interestingly, at a higher voltage of $80 \, V \, cm^{-1}$, where dynamic fluctuations in the size and shape of the droplet occur, the overall instantaneous intensity of 2,3-DAP is considerably higher than at $50 \, V \, cm^{-1}$ after the droplets are energized for 1 min. Moreover, the 2,3-DAP product is mainly distributed outside the substrate-containing aqueous vacuole and within the enzyme-containing coacervate matrix (Fig. 5f–i). Overall, these experiments demonstrate that electric field energization of the microdroplets can be used to trigger immediate rate enhancements in spatially confined enzyme reactions, and that vacuolization offers a transient means to partially separate reactants and product molecules associated with the biocatalytic cascade.

Discussion

Our results highlight a novel experimental protocol for the controlled electric field excitation of coacervate-based microdroplets that offers new opportunities for the study of non-equilibrium phenomena in synthetic protocells. The ability to study protocells under sustained energization is in marked contrast to current models that are limited by their design and function under close to equilibrium conditions. Our studies therefore offer a step towards a better representation of minimal life processes operating under a continuous flux of energy and exchange of matter with the environment. In particular, the possibility of controlling the onset, duration and termination of complex dynamical behaviour such as morphogenesis, transient subcompartmentalization (vacuolization), growth and fusion, spontaneous ejection and sequestration of matter, directional capture of solute molecules, and pulsed enhancement of enzyme cascades should provide an interesting context in which to study chemical reactivity and signalling in protocell communities. For example, coacervate microdroplets exhibit key biomimetic characteristics such as high levels of chemical enrichment[17,24], uptake and retention of enzymes and DNA without droplet transfer or exchange[32,33], increased rates of enzymatic transformations[17,24,34] and surface properties compatible with facilitating membrane assembly[18–20]. The exploitation of these properties in energized protocells should provide a route towards increased complexity and functionality that could have implications for new developments in activated microstorage and delivery, and non-equilibrium microreactor technologies.

The demonstration of diverse non-equilibrium phenomena in coacervate microdroplets exposed to an electric field offers a step

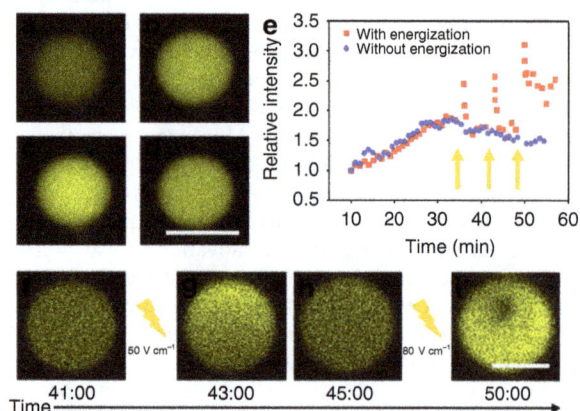

Figure 5 | Electric-field-mediated enhancement of an enzyme cascade reaction. (**a–d**) Confocal fluorescence microscopy images of a single PLL/ss-oligo coacervate microdroplet containing GOx/HRP enzymes and showing 2,3-DAP fluorescence in the absence of an electric field 10 (**a**), 20 (**b**), 30 (**c**) and 50 min (**d**) after addition of oPD and β-D-glucose substrates; scale bar, 10 μm. (**e**) Relative intensities of 2,3-DAP produced in droplets subjected to intermittent energization (red squares) or without energization (blue dots). An electric field of $50 \, V \, cm^{-1}$ was applied at 35 and 42 min for a period of 1 min, and at 49 min using a field of $80 \, V \, cm^{-1}$ for 1 min (arrows). (**f–i**) Confocal fluorescence microscopy images of 2,3-DAP fluorescence intensity just before energization (**f**), immediately after energization in an electric field of $50 \, V \, cm^{-1}$ showing increase in fluorescence intensity and redistribution of 2,3-DAP (upper region of the droplet) (**g**), 2 min after the $50 \, V \, cm^{-1}$ electric field is switched off showing decrease in fluorescence intensity (**h**), and immediately after energization in an electric field of $80 \, V \, cm^{-1}$ showing high-intensity 2,3-DAP fluorescence in the coacervate matrix but not in the vacuole (**i**). Scale bar, 5 μm.

towards novel examples of reactive soft matter. Electrohydrodynamic forces within the molecularly crowded interior of the protocells facilitate disassociation the PLL/ss-oligo complex to produce a charge-activated matrix and osmotic gradient at the droplet surface that drive a cyclical exchange of matter with the environment. Our results suggest that loosely associated polyelectrolytes are particularly sensitive to the electric field, and that the composition and structural heterogeneity of the coacervate matrix are important criteria in determining their non-equilibrium behaviour. As coacervate microdroplets can be prepared across a range of polyelectrolyte charge ratios[24], and kinetically trapped structural states[29,35], changes in experimental methodology (non-stoichiometric mixtures, ageing times) need to be considered in detail when investigating their dynamical properties.

Finally, we speculate that our results could be of interest to areas of origin of life research, particularly in stimulating new ideas concerning the activation and subdivision of prebiotic assemblies on the early Earth. The spontaneous microphase separation and chemical enrichment in water of counter-charged biologically relevant molecules such as ATP, DNA, RNA and polypeptides[17,18,33] to produce coacervate droplets is consistent with a possible mechanism of membrane-free prebiotic compartmentalization[17,36]. Condensates of these charged components are prevalent in extant organisms[37–39], and electric fields exist both extracellularly and intracellularly[40,41], and play a role in dynamical processes such as tissue morphogenesis and regeneration[42]. However, although electric discharges (lightning) have been considered a plausible mechanism for the synthesis of certain prebiotic molecules[43,44], the possibility that electric fields generated in the ground from earthquake activity[45], or streaming potentials in porous rocks[46] could facilitate the energization of early cells, or influence their dynamical behaviour and trafficking of matter remains unknown.

Methods

Experimental set-up. A ss-oligo ($M_W = 6,387$) with a 21 nt random sequence and a Cy5 fluorescent tag linked at the 5′-terminus was purchased from Invitrogen Inc. FITC-labelled PLL ($M_W = 30,000$–$70,000$) was purchased from Sigma-Aldrich (St Louis, MI, USA) and used as received. Microchannels with a double-cross layout were etched on a glass chips by soft lithography. The ss-oligo (1.5 mg ml^{-1}) and PLL (1.0 mg ml^{-1}) were dissolved in phosphate buffer (0.01 M, pH 6.86) and loaded into the opposite wells of a microfluidic channel device (channel dimensions; 80-μm width × 25-μm depth) that was previously washed in sequence with NaCl (1 M), NaOH (1 M), HCl (1 M), deionized water and 1% (w/w) phosphate-buffered polyvinylpyrrolidone ($M_W = 30,000$; 0.01 M KH$_2$PO$_4$–Na$_2$HPO$_4$, pH 6.86, 2 h) to minimize electro-osmosis flow and to prevent adsorption of analytes. The flow of the solution was driven by gravity. At the same height, the flow rates of ss-oligo and PLL were similar. They were thus mixed in the central channel of the chip with equal volume. Discrete coacervate microdroplets with 1:1 charge ratio were formed by mixing ss-oligo and PLL at 1.5 and 1.0 mg ml^{-1}, respectively. (Varying the charge ratio or aging time after mixing of the polyelectrolytes was used to change the mean droplet diameter if required.) After 3 min, residual aqueous oligoDNA and PLL were replaced in the channel by phosphate buffer. An electric field, typically between 10 and 200 V cm^{-1}, was then applied along the microfluidic channel by using a ZDMCI6-1 Microfluidic Chip Detection System (Hang-zhou Syltech Technology Co., Ltd). The changes in droplet morphology, organization and dynamics were recorded by using an Inverted Fluorescence Microscope (IX71, Olympus, USA) with a U-MWIBA module (excitation filter BP460–495, emission filter BA510–550) equipped with an EMCCD (Evolve512, Photometric, USA), or a Laser Scanning Confocal Microscope (LSCM, A1R-si, Nikon, Japan) for video imaging or 3D image reconstruction, respectively. The excitation and emission wavelengths were 638 and 700 nm (ss-oligo), and 488 and 515 nm (PLL), respectively.

Similar experiments were undertaken using coacervate droplets prepared from a mixture of PLL and a double-stranded oligodeoxynucleotide. The latter was prepared by mixing two ss-oligos with complementary sequences (5′-CTTACGCTGAGTACTTCGATT-3′ and 5′-AATCGAAGTACTCAGCGT AAG-3′; Invitrogen Inc) at equal molar ratio, followed by heating at 95 °C for 5 min and slow cooling to room temperature.

Monitoring of dynamical behaviour. Cyclical changes in the size of the internal compartments produced within individual coacervate microdroplets exposed to an

electric field of 80 V cm^{-1} were determined by time-dependent video measurements of the distance between the two inner edges of selected vacuoles. As the fluorescence intensity associated with the vacuoles was close to the background level, fluorescence line scans across subdivided droplets gave a series of peaks corresponding to the edges of the coacervate-rich areas. The size of the vacuoles was determined by measuring the distance between the fluorescence intensity peaks at various time intervals.

Experiments to confirm the aqueous content of the vacuoles formed by the excitation of electric field were carried out with PLL/ss-oligo (Cy5 labelled) coacervate microdroplets produced as described above but in the presence of 20 μM calcein. Uptake experiments were undertaken using a similar procedure.

Time-dependent changes in the macromolecular composition of individual coacervate microdroplets exposed to electric fields ranging from 10 to 70 V cm^{-1} were monitored by determining the relative fluorescence intensity, $I_r = I_{PLL}/I_{oligo}$ as an index of the PLL/ss-oligo ratio, where I_{PLL} and I_{oligo} were the fluorescence intensities of FITC-labelled PLL and Cy5-labelled DNA, respectively. Both intensities were normalized to their initial values recorded before the electric field was switched on ($t = 0$). Changes in fluorescence for more than five droplets were determined, and the average values used for comparison.

Enzyme cascade experiments. Experiments to study the influence of electric field energization of the coacervate microdroplets on spatially confined enzyme-mediated reactions were undertaken as follows. Coacervate droplets co-mixed with GOx (GOx from *Aspergillus niger*) and HRP (FITC-labelled GOx or rhodamine isothiocyanate (RITC)-labelled HRP were used in some experiments) were prepared from PLL (1.0 mg ml^{-1}) and a mixture of ss-oligo (1.5 mg ml^{-1}), GOx (20 μg ml^{-1}) and HRP (10 μg ml^{-1}). A double-component mixture of *o*PD (1 mM) and β-D-glucose (2 mM) (substrates for HRP and GOx) in 20 mM MES buffer (pH 6) was injected into the channel replacing the buffer and initiating the enzyme-mediated tandem reaction in the coacervate droplets. The intensity of the fluorescent product 2,3-DAP was measured using laser scanning confocal microscope (405 nm excitation and 530 (±30) nm emission). Unlabelled GOx or HRP was used in some experiments to avoid fluorescence resonance energy transfer.

References

1. Miller, D. M. & Gulbis, J. M. Engineering protocells: prospects for self-assembly and nanoscale production-lines. *Life* **5**, 1019–1053 (2015).
2. Li, M., Huang, X., Tang, T. Y. D. & Mann, S. Synthetic cellularity based on non-lipid micro-compartments and protocell models. *Curr. Opin. Chem. Biol.* **22**, 1–11 (2014).
3. Caschera, F. & Noireaux, V. Integration of biological parts toward the synthesis of a minimal cell. *Curr. Opin. Chem. Biol.* **22**, 85–91 (2014).
4. Stäedler, B. *et al.* Polymer hydrogel capsules: en route toward synthetic cellular systems. *Nanoscale* **1**, 68–73 (2009).
5. Renggli, K. *et al.* Selective and responsive nanoreactors. *Adv. Funct. Mater.* **21**, 1241–1259 (2011).
6. Huang, X., Patil, A. J., Li, M. & Mann, S. Design and construction of higher-order structure and function in proteinosome-based protocells. *J. Am. Chem. Soc.* **136**, 9225–9234 (2014).
7. Monnard, P.-A. & Walde, P. Current ideas about prebiological compartmentalization. *Life* **5**, 1239–1263 (2015).
8. Peters, R. J. R. W. *et al.* Cascade reactions in multicompartmentalized polymersomes. *Angew. Chem. Int. Ed. Engl.* **53**, 146–150 (2014).
9. Peters, R. J. R. W., Louzao, I. & van Hest, J. C. M. From polymeric nanoreactors to artificial organelles. *Chem. Sci.* **3**, 335–342 (2012).
10. Martini, L. & Mansy, S. S. Cell-like systems with riboswitch controlled gene expression. *Chem. Commun.* **47**, 10734–10736 (2011).
11. Nourian, Z. & Danelon, C. Linking genotype and phenotype in protein synthesizing liposomes with external supply of resources. *ACS Synth. Biol.* **2**, 186–193 (2013).
12. Li, M., Harbron, R. L., Weaver, J. V. M., Binks, B. P. & Mann, S. Electrostatically gated membrane permeability in inorganic protocells. *Nat. Chem.* **5**, 529–536 (2013).
13. Li, M., Huang, X. & Mann, S. Spontaneous growth and division in self-reproducing inorganic colloidosomes. *Small* **10**, 3291–3298 (2014).
14. Huang, X. *et al.* Interfacial assembly of protein-polymer nano-conjugates into stimulus-responsive biomimetic protocells. *Nat. Commun.* **4**, 2239 (2013).
15. Chandrawati, R. *et al.* Engineering advanced capsosomes: maximizing the number of subcompartments, cargo retention, and temperature-triggered reaction. *ACS Nano* **4**, 1351–1361 (2010).
16. Chandrawati, R. & Caruso, F. Biomimetic liposome- and polymersome-based multicompartmentalized assemblies. *Langmuir* **28**, 13798–13807 (2012).
17. Koga, S., Williams, D. S., Perriman, A. W. & Mann, S. Peptide-nucleotide microdroplets as a step towards a membrane-free protocell model. *Nat. Chem.* **3**, 720–724 (2011).

18. Tang, T. Y. D. *et al.* Fatty acid membrane assembly on coacervate microdroplets as a step towards a hybrid protocell model. *Nat. Chem.* **6**, 527–533 (2014).

19. Williams, D. S., Patil, A. J. & Mann, S. Spontaneous structuration in coacervate-based protocells by polyoxometalate-mediated membrane assembly. *Small* **10**, 1830–1840 (2014).

20. Fothergill, J., Li, M., Davis, S. A., Cunningham, J. A. & Mann, S. Nanoparticle-based membrane assembly and silicification in coacervate microdroplets as a route to complex colloidosomes. *Langmuir* **30**, 14591–14596 (2014).

21. Dominak, L. M., Omiatek, D. M., Gundermann, E. L., Heien, M. L. & Keating, C. D. Polymeric crowding agents improve passive biomacromolecule encapsulation in lipid vesicles. *Langmuir* **26**, 13195–13200 (2010).

22. Keating, C. D. Aqueous phase separation as a possible route to compartmentalization of biological molecules. *Acc. Chem. Res.* **45**, 2114–2124 (2012).

23. Miller, D. *et al.* Protocell design through modular compartmentalization. *J. Roy. Soc. Interface* **10**, 20130496 (2013).

24. Williams, D. S. *et al.* Polymer/nucleotide droplets as bio-inspired functional micro-compartments. *Soft Matter* **8**, 6004–6014 (2012).

25. Salipante, P. F. & Vlahovska, P. M. Electrohydrodynamics of drops in strong uniform dc electric fields. *Phys. Fluids* **22**, 112110 (2010).

26. Manning, G. S. Molecular theory of polyelectrolyte solutions with applications to electrostatic properties of polynucleotides. *Q. Rev. Biophys.* **11**, 179–246 (1978).

27. Bungenberg De Jong, H. G. & Bank, O. Zur morphologie von komplexgelkörpern. *Protoplasma* **33**, 321–340 (1939).

28. Smith, A. E. & Chance, M. A. C. Coacervate behaviour in an alternating electric field. *Nature* **209**, 74–75 (1966).

29. Niu, L. *et al.* Dynamic assembly of DNA and polylysine mediated by electric energy. *Chem. Commun.* **51**, 1506–1509 (2015).

30. Prausnitz, M. R. *et al.* Transdermal transport efficiency during skin electroporation and iontophoresis. *J. Controlled Release* **38**, 205–217 (1996).

31. Zheng, C. *et al.* Long-term kinetics of DNA interacting with polycations. *Polymer* **55**, 2464–2471 (2014).

32. Tang, T. Y. D., Antognozzi, M., Vicary, J. A., Perriman, A. W. & Mann, S. Small-molecule uptake in membrane-free peptide/nucleotide protocells. *Soft Matter* **9**, 7647–7656 (2013).

33. van Swaay, D., Tang, T. Y. D., Mann, S. & de Mello, A. Microfluidic formation of membrane-free aqueous coacervate droplets in water. *Angew. Chem. Int. Ed. Engl.* **54**, 8398–8401 (2015).

34. Crosby, J. *et al.* Stabilization and enhanced reactivity of actinorhodin polyketide synthase minimal complex in polymer-nucleotide coacervate droplets. *Chem. Commun.* **48**, 11832–11834 (2012).

35. Zheng, C. *et al.* Structure and stability of the complex formed by oligonucleotides. *Phys. Chem. Chem. Phys.* **14**, 7352–7359 (2012).

36. Oparin, A. I. *The Origin of Life* 2nd edn (Dover Publications, Inc., 1953).

37. Brangwynne, C. P., Mitchison, T. J. & Hyman, A. A. Active liquid-like behavior of nucleoli determines their size and shape in Xenopus laevis oocytes. *Proc. Natl Acad. Sci. USA* **108**, 4334–4339 (2011).

38. Hyman, A. A. & Simons, K. Beyond oil and water-phase transitions in cells. *Science* **337**, 1047–1049 (2012).

39. Korolev, N., Allahverdi, A., Lyubartsev, A. P. & Nordenskioeld, L. The polyelectrolyte properties of chromatin. *Soft Matter* **8**, 9322–9333 (2012).

40. Jaffe, L. F. Electrical currents through developing fucus egg. *Proc. Natl Acad. Sci. USA* **56**, 1102–1109 (1966).

41. Tyner, K. M., Kopelman, R. & Philbert, M. A. 'Nanosized voltmeter' enables cellular-wide electric field mapping. *Biophys. J.* **93**, 1163–1174 (2007).

42. Ojingwa, J. C. & Isseroff, R. R. Electrical stimulation of wound healing. *J. Invest. Dermatol.* **121**, 1–12 (2003).

43. Demaneche, S. *et al.* Laboratory-scale evidence for lightning-mediated gene transfer in soil. *Appl. Environ. Microbiol.* **67**, 3440–3444 (2001).

44. Ruiz-Mirazo, K., Briones, C. & de la Escosura, A. Prebiotic systems chemistry: new perspectives for the origins of life. *Chem. Rev.* **114**, 285–366 (2014).

45. Freund, F. T. *et al.* Stress-induced changes in the electrical conductivity of igneous rocks and the generation of ground currents. *Terr. Atoms. Oceanic Sci.* **15**, 437–467 (2004).

46. Revil, A., Schwaeger, H., Cathles, L. M. & Manhardt, P. D. Streaming potential in porous media 2. Theory and application to geothermal systems. *J. Geophys. Res.* **104**, 20033–20048 (1999).

Acknowledgements

D.L. thanks the financial support from the National Natural Science Foundation of China (21174007) and the National Basic Research Program of China (973 Program, 2012CB821500). S.M. thanks the Engineering and Physical Sciences Research Council (UK) and European Research Council (advanced grant) for financial support.

Author contributions

Y.Y., L.N., D.L. and S.M. conceived the experiments; Y.Y., L.N. and X.Z. performed the experiments; Z.Z. and M.Z. undertook the data analysis; Y.Y., D.L. and S.M. wrote the manuscript.

Additional information

21

Photoacoustics of single laser-trapped nanodroplets for the direct observation of nanofocusing in aerosol photokinetics

Johannes W. Cremer[1], Klemens M. Thaler[2], Christoph Haisch[2] & Ruth Signorell[1]

Photochemistry taking place in atmospheric aerosol droplets has a significant impact on the Earth's climate. Nanofocusing of electromagnetic radiation inside aerosols plays a crucial role in their absorption behaviour, since the radiation flux inside the droplet strongly affects the activation rate of photochemically active species. However, size-dependent nanofocusing effects in the photokinetics of small aerosols have escaped direct observation due to the inability to measure absorption signatures from single droplets. Here we show that photoacoustic measurements on optically trapped single nanodroplets provide a direct, broadly applicable method to measure absorption with attolitre sensitivity. We demonstrate for a model aerosol that the photolysis is accelerated by an order of magnitude in the sub-micron to micron size range, compared with larger droplets. The versatility of our technique promises broad applicability to absorption studies of aerosol particles, such as atmospheric aerosols where quantitative photokinetic data are critical for climate predictions.

[1] Department of Chemistry and Applied Biosciences, Laboratory of Physical Chemistry, ETH Zurich, Vladimir-Prelog-Weg 2, CH-8093 Zurich, Switzerland. [2] Laboratory for Applied Laser Spectroscopy, Chair of Analytical Chemistry, Technical University of Munich, Marchioninistrasse 17, D-81377 Munich, Germany. Correspondence and requests for materials should be addressed to R.S. (email: rsignorell@ethz.ch).

Understanding fundamental processes that govern the reaction dynamics of gas phase, aerosol and cloud processes is crucial for the advancement of global atmospheric chemistry modelling[1-14]. Much of the chemistry occurring in the Earth's atmosphere is driven by sunlight. Photochemical reactions, in which aerosol particles or droplets act as the active reaction medium, can be highly complex because they are influenced by optical phenomena, transport properties and surface effects[2]. Optical phenomena play a fundamental role in light-initiated particle processes since the radiation flux within the particles determines the activation rate of the photochemically active species.

Focusing of electromagnetic radiation inside small particles leads to an enhancement of the overall light intensity, compared with the intensity of the incident radiation and to structuring and localization of the internal optical fields[15-23]. These phenomena depend strongly on the particle size, the particle composition and the wavelength of electromagnetic radiation. The fundamental influence of the enhanced electromagnetic energy density on the rate of photochemical reactions in micro- and nanodroplets has been recognized and calculations have provided limited evidence for enhanced photochemical rates[24-27]. Experimental results remain inconclusive concerning the influence of light enhancement on the kinetics, mainly because direct observation of the actual photoactive step was not possible[23,28-34]. The observation of size-dependent effects in ensembles of aerosol or emulsion droplets is often hindered because the droplet size distribution cannot be varied and determined with the necessary accuracy. However, even single-droplet techniques have so far not provided size-dependent photolysis rates because the direct measurement of the population decay of the photoactive substance was not possible.

Elastic light scattering is sensitive enough to allow measurements on single sub-micron-sized droplets, but the information content is not specific enough to extract size-dependent rates. Raman spectroscopy, by contrast, could provide specific information but it comes with the disadvantage of low sensitivity (long averaging times), which would make its application to study processes in single submicrometre droplets where nanofocusing becomes important very challenging. Single-droplet fluorescence studies require a fluorescing compound, which strongly restricts its applicability. Furthermore, the fluorescence depletion is not always a reliable measure of the population decay of the photoactive species because of varying quenching efficiencies. The recently presented cavity ring-down studies on single droplets provide information on the extinction but not directly on the droplet absorption[35,36]. Even in combination with light scattering measurements, the determination of rates in nanodroplets is likely prohibited by the uncertainty of the derived absorption.

This study reports the direct observation of light nanofocusing on the photokinetics in nanometre- to micron-sized droplets in the ultraviolet/visible (UV/VIS) range. To this end, we introduce single-droplet photoacoustic (PA) absorption spectroscopy, allowing the direct detection of the population decay of the photoactive substance. PA spectroscopy has been successfully used for the investigation of ensembles of aerosol particles[37-40], but its applicability to single aerosol particle studies has been controversial and has not previously been realised experimentally. Here we demonstrate the feasibility of single-droplet PA spectroscopy in combination with laser trapping, and provide direct experimental evidence for the size-dependence of the photolysis rate in model aerosol droplets due to nanofocusing effects. The results are compared with simulations using classical cavity electrodynamics.

Results

Principle of single-droplet PAs.

The two experimental set-ups, using a microphone and a quartz tuning fork, respectively, for resonant single-droplet PA measurements, are sketched in Fig. 1a,b (see Methods). For the droplet absorption experiments, we use a $\lambda = 445$ nm excitation laser (Nichia laser diode NDB7112E) of variable power (0.3–40 mW) modulated at the resonance frequency of the PA-resonator and the tuning fork, respectively. The resonance frequency and the Q-factor of the PA-resonator and the tuning fork are 3.97 kHz and ~ 8.9, and 32.7 kHz and $\sim 8,000$, respectively (Supplementary Fig. 1). The power of the excitation laser is recorded by a power meter after passing the PA cell and the tuning fork, respectively (Supplementary Fig. 2). The amplified PA signals are averaged over either 30 ms or 200 ms. For single-particle trapping, we use a counter-propagating optical tweezer built from a continuous laser beam of $\lambda = 660$ nm of ~ 200 mW (Laser Quantum, Opus 660) (Supplementary Fig. 2). Such multiple beam optical traps allow trapping of sub-micron droplets, and combine the advantage of a comparatively simple set-up with high trapping stability and tight particle confinement (< 100 nm)[41-43]. Droplets are trapped by gradient forces pointing towards the trap centre for all translational degrees of freedom. Single droplets are captured in the trap centre from a plume of aerosol generated by a nebulizer (see Methods). The droplet size is determined from laser light elastically scattered by the droplet[41,44] (see Methods, Supplementary Fig. 3). Figure 1e shows an example for light scattering measurements for droplet sizing. In the microphone set-up (Fig. 1a), the trap centre is located in the middle of the PA-resonator above the microphone[45]. The trapping and excitation lasers enter and exit the cell through wide-band, anti-reflective windows coated for the respective wavelengths. The CMOS camera for particle imaging and light scattering measurements is placed perpendicular to the excitation and trapping laser. The aerosol inlet and outlet are on the side of the PA cell outside the resonator.

In the tuning fork set-up (Fig. 1b), the droplet is trapped between the tines of the fork with collinearly aligned excitation laser and trapping laser beams. The CMOS camera for particle imaging and light scattering measurements is placed opposite the tines of the fork. Figure 1c,d shows images of a single droplet trapped in between the tines and of a droplet ensemble flowing through the tines, respectively. The principal attractiveness of the tuning fork derives from its high detection sensitivity (very high Q-factors) and low sensitivity to environmental acoustic noise[46]. In our set-up, we mainly profit from the ease of combining it with laser trapping and light scattering measurements, as well as from the fact that it is chemically inert.

PA response of a single droplet.

Figure 2 provides typical noise levels, background signals and a proof for single-droplet detection. Figure 2a illustrates the noise level and the background signal for the empty trap with the trapping laser on. With the excitation laser off (-5 s $< t < 0$ s), the background signal (average) and noise level (1 s.d.) are $\sim 2.2 \pm 1.2\,\mu V$. Once the excitation laser is turned on at time $t = 0$ s, a background signal of $BS = 5.3\,\mu V$ with a noise level of $NL = 1.7\,\mu V$ is recorded. The background signal is caused by excitation laser light scattered from the cell walls and hitting the microphone. Blocking the trapping laser (that is, disabling the trap) leaves the noise, as well as the background unchanged.

Figure 2b shows the same as Fig. 2a but with a single VIS441/tetraethylene glycol (TEG) solution droplet in the trap. The PA signal reaches a maximum (S_{max}) just after the excitation laser is turned on and then decreases exponentially as the VIS441 absorber undergoes photolysis. In Fig. 2b, the trap is disabled at

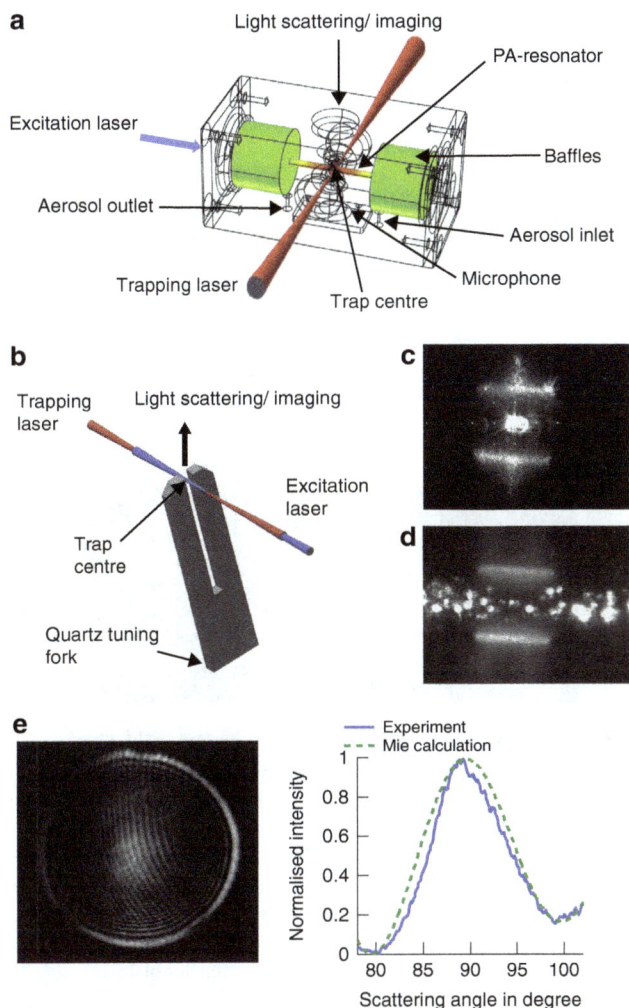

Figure 1 | Sketch of the experimental single-droplet PA set-ups.
(**a**) Microphone set-up with PA-resonator, excitation laser, trapping laser and light scattering measurements. The colours in the PA cell indicate that the acoustic mode has its maximum amplitude (red) in the vicinity of the microphone and a value close to zero (green) in the region of the acoustical baffles. (**b**) Tuning fork set-up. (**c**) Snapshot of a single droplet trapped between the tines of the tuning fork (view from top). Note that the droplet ($\sim 1\,\mu m$) is much smaller than the detection volume between the tines ($\sim 0.3 \times 0.34 \times 2\,mm$). (**d**) Snapshot of an ensemble of droplets flowing in between the tines from left to right. (**e**) Light scattering image as recorded by the CMOS camera (left) with experimental and calculated phase function (right) for a droplet with a radius of $a = 530\,nm$.

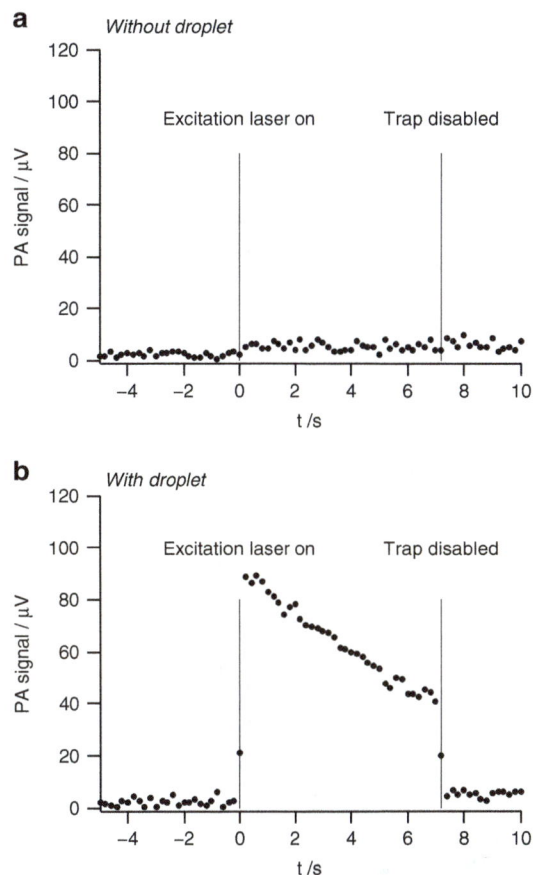

Figure 2 | Typical noise levels and background signals for the microphone set-up. (**a**) For the empty trap, identical noise levels and background signals are recorded for a pure TEG solvent droplet in the trap (data not shown). (**b**) With a VIS441/TEG solution droplet in the trap. At $t = 0\,s$, the excitation laser (445 nm) is switched on and at $t = 7\,s$ the trapping laser is switched off.

$t = 7\,s$ which leads to the immediate loss of the droplet and to a decrease of the PA signal to the background signal BS. This proves that the signal between $0\,s < t < 7\,s$ indeed comes from a single droplet. The signal to noise ratio $\frac{S}{NL} = \frac{S_{max} - BS}{NL}$ depends on the power P of the excitation laser, the concentration of the solution and the droplet size. Figure 3 shows exemplary experimental data for solution droplets of different size excited with different laser powers P. With the tuning fork set-up, we find an improvement in the $\frac{S}{NL}$ ratio of a factor of ~ 3 compared with the microphone set-up. Note that the PA signal is caused by absorptive heating of the droplet and subsequent heat transfer to the surroundings. Evaporation of the solvent does not occur.

Detection limit. The minimum absorbance detectable with the single-particle PA set-ups at a given power of the excitation laser

can be estimated from the PA signal and the single-droplet absorbance assuming that the noise level NL is the detection limit (see Methods, Calculation of droplet absorbance). As an example, we use the PA signal at $t = 0$ (S_{max}) of the 530 nm droplet shown in Fig. 3a recorded at a laser power of 2.8 mW with an averaging time of 200 ms. The refractive index of this droplet, $n + ik = 1.463 + i \cdot 0.0062$ (Methods and Supplementary Figs 4 and 5), yields an absorption cross-section of $C_{abs} = 0.22\,\mu m^2$. $(S_{max} - BS) = 33.3\,\mu V$ corresponds to an equivalent absorbance of $A = 1.8 \times 10^{-5}$. From the measured NL of the 530 nm droplet of $NL = 1.7\,\mu V$, we derive a minimum absorbance $A_{min} = 9 \times 10^{-7}$ detectable with the microphone set-up. The improvement in $\frac{S}{NL}$ by a factor of ~ 3 for the tuning fork set-up reduces the detection limit to $A_{min} \sim 3 \times 10^{-7}$, a minimum detectable absorption coefficient of $\alpha_{min} = 0.0074 \times 10^{-6}\,m^{-1}$ or a minimum detectable absorption cross-section $C_{abs,min} = 0.0037\,\mu m^2$ (laser power of 2.8 mW and averaging time of 200 ms) (see Methods equations (2)–(4)). The equivalent particle radius of 146 nm corresponds to a probe volume of 13 al. This far exceeds the performance of typical spectrometers ($A_{min} \sim 10^{-3}$–10^{-4}), and is at least comparable to the most sensitive laser spectroscopic absorption measurements for macroscopic probe volumes. Note that in our set-up, this sensitivity is achieved for small (attolitre) probe volumes and very short ($\ll 1\,s$) measurement times. Both can be further improved by increasing the laser power.

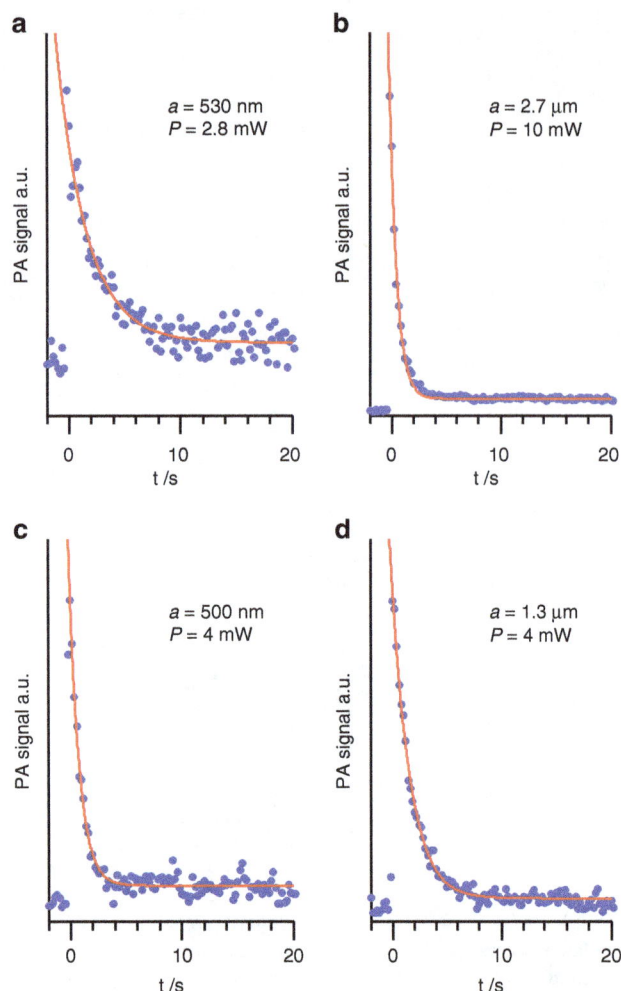

Figure 3 | Exemplary PA signals of VIS441/TEG solution droplets as a function of time. The decay of the signal is caused by photolysis of the solute. The experimental data (blue dots) are recorded for different droplet radii a and different power P of the excitation laser. (**a,b**) Recorded with the microphone set-up. (**c,d**) Recorded with the tuning fork setup. The red lines are fits to the experimental data providing experimental first half-lives $t_{1/2}$ (see Fig. 4).

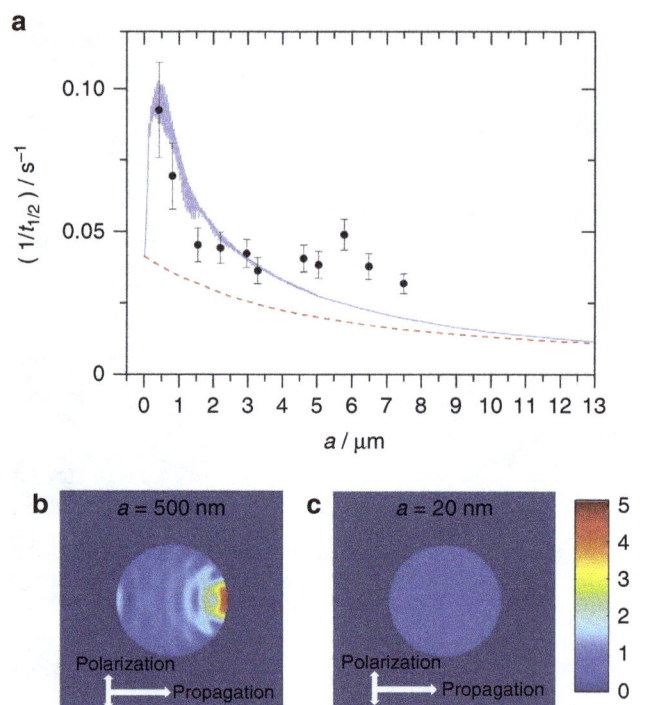

Figure 4 | Size-dependent photokinetics. (**a**) Inverse first half-lives $\frac{1}{t_{1/2}}$ as a function of the droplet radius for a laser power of 1 mW. Black circles: statistically evaluated experimental data. Error bars show 1 s.d. Full blue line: model prediction including nanofocusing and scattering effects. The calculations are for a quantum yield of 7×10^{-6}. Dashed red line: model prediction for a hypothetical bulk limit, that is, excluding contributions from nanofocusing and scattering. Distribution of the light intensity inside droplets at $t = 0$ s for (**b**) a 0.5 μm droplet and (**c**) a 20 nm droplet. The colour scheme is relative to an incident light intensity of 1.

Size-dependent photokinetics. The photokinetics in small droplets do not follow simple pseudo first order kinetics because the light intensity distribution inside the droplets is time dependent; that is, because of the concentration dependence of the nanofocusing. Therefore, we use the same initial concentration for all experiments so that the first half-life can be used as a measure for the size-dependence of the photokinetics. With our PA set-up, we directly measure the decay in absorption resulting from the population decay of the photoactive substance. Diffusion is so fast in the droplets ($\sim 10\,\mu m^2\,s^{-1}$) that concentration gradients cannot build up and homogeneous concentrations for the solute can be assumed at all times. A model for the droplet photokinetics under these conditions is provided in Methods (Calculation of droplet photokinetics).

The model prediction (full blue line in Fig. 4a) shows a strong droplet size-dependence with a maximum in the inverse first half-life $\frac{1}{t_{1/2}}$ at a droplet radius of $\sim 0.5\,\mu m$. Pronounced increases in $\frac{1}{t_{1/2}}$ and fluctuations due to resonances are observed over the droplet size range from ~ 50 nm to $\sim 1.2\,\mu m$. In this size range, the increase of the rate is caused by the enhancement of the internal electromagnetic field intensity through focusing of the light inside the droplet. Figure 4b shows the distribution of the light intensity inside a $\sim 0.5\,\mu m$ droplet at $t = 0$ s. The enhancement of the overall intensity and the local variation of the intensity are pronounced. The inverse half-life at a droplet radius of $\sim 0.5\,\mu m$ is increased by a factor of ~ 2.5 compared with the infinitely small droplet limit. The kinetics in these small droplets is no longer influenced by nanofocusing inside the droplet or light scattering by the droplet as visualized by the internal field intensity in Fig. 4c. The inverse half-life of larger droplets ($> 6\,\mu m$) exhibits only a weak size-dependence but decreases continuously (towards zero for infinitely large particles). The rate for these large droplets is determined by the balance between the decay of the absorber and the rise of the decay rate with time. As the photolysis proceeds, the penetration depth of the light and hence the internal field intensity increases. As in the case of very small droplets, nanofocusing is not important for very large droplets. Large droplets essentially represent the behaviour of thin bulk films with the same effective thickness as the droplet. The $\frac{1}{t_{1/2}}$ increases by a factor of ~ 10 for a 0.5 μm compared with a 13 μm droplet, which implies a substantial increase in the rate of sub-micron-size droplets relative to bulk. The dashed red line in Fig. 4a simulates the behaviour of a hypothetical droplet excluding the influence of nanofocusing and light scattering but still accounting for the droplet-size-dependent absorption (see Methods, Calculation of droplet photokinetics). This curve represents bulk behaviour. The comparison with the full blue line clearly shows the pronounced influence of light focusing on the rate in the sub-micron to micron size range.

The statistically evaluated experimental first half-lives (black circles in Fig. 4a) are determined from time-dependent PA measurements (see Methods, Statistical analysis). The experimental results clearly follow the size-trend predicted by the model (full blue line). The pronounced maximum of $\frac{1}{t_{1/2}}$ at a droplet size of $\sim 0.5\,\mu m$ is clearly visible even though the data scatter notably below $\sim 1\,\mu m$ (Supplementary Fig. 6) mainly because of the uncertainty in the droplet size determination (Methods, Droplet Sizing). Our experimental data show somewhat higher values of the inverse half-life for larger droplets than the model prediction. Deviations from the model assumptions including modified PA response in large droplets[47] could potentially account for this. We have recently introduced a broad-band scattering method for accurate sizing of submicrometre particles, which will allow us in future to significantly reduce the size uncertainty in the submicrometre range (unpublished data). However, already at the current level of accuracy, the data in Fig. 4a provide the first direct observation of the strong influence of nanofocusing of light on the photokinetics in droplets.

Discussion

The experimental results in Fig. 4 confirm a strong size-dependence of the rate of photochemical reactions in droplets. This optical phenomenon shows the most pronounced effect in the submicrometre to micrometre droplet size range for electromagnetic radiation in the UV to VIS range, that is, for the relevant frequency range in atmospheric processes. Classical cavity electrodynamics provides a semi-quantitative description of the kinetics for our ideal model system. The photokinetics of our model system is representative of typical atmospheric aerosols; that is, of typical optical properties of these particles. For example, similar quantitative results are predicted for aqueous droplets (Supplementary Fig. 7). The acceleration of the kinetics we find in the visible range is predicted to be even more pronounced in the UV range of the solar spectrum (Supplementary Fig. 7). Many aerosol particles are non-spherical. However, for particles with different shapes but the same volumes one finds quantitatively similar nanofocusing effects as for droplets. Nanofocusing also affects surface reactions since the strong intensity enhancement in forward direction shown for the internal field in Fig. 4b extends to the external field near the surface (not shown). The ability to measure and thus quantify the kinetics of the light-induced step in photochemical reactions in aerosol particles is of fundamental importance for atmospheric chemistry, where chemical processes are largely driven by sunlight. The diverse and complex processes (for example, transport and surface phenomena) in atmospheric aerosol particles require direct measurement methods as the one introduced here because simple models are of limited applicability.

The introduction of single-droplet PA spectroscopy in the present study finally makes the direct observation of the photoactive step possible. Single-droplet PA was previously deemed not feasible because of sensitivity and background issues. Here we demonstrate the viability of this new method and its very high sensitivity ($C_{abs,min} = 0.0037\,\mu m^2$) enabling studies even of single nanodroplets (10 al). PA spectroscopy provides a general absorption method that can be used in any frequency range. The combination with laser trapping lets us follow the evolution of individual droplets under controlled conditions over extended periods of time (up to several days). This versatility enables fundamental studies on many different droplet systems relevant to atmospheric and technical processes. The investigation of droplet photokinetics is just one example where this new broadly applicable single-droplet method can make an important contribution.

Methods

PA measurements with microphone. The PA cell is made of brass and consists of a longitudinal PA-resonator (length 40 mm, diameter 4 mm), which is connected to two buffer volumes with acoustical baffles for sound insulation (Fig. 1a)[45]. A sensitive microphone (EK 23029, Knowles) is used with a custom-made preamplifier. The output signal is recorded by a lock-in amplifier (Stanford, SR 830).

PA measurements with tuning fork. The distance between the two tines of the tuning fork (Q 32.768 kHz TC 38, AURIS) is 300 μm. Each tine has a width of 600 μm, a thickness of 340 μm and a length of 3.8 mm. The quartz tuning fork acts as the resonant acoustic transducer, which generates an electric signal on resonant excitation by an acoustic wave due to the piezoelectric effect[46]. The signal recording is identical to the microphone set-up except for the more precise reference frequency adapted to the higher Q-factor.

Aerosol generation and materials. To study photokinetics in single droplets, solutions of the photoactive dye VIS441 (Cyanine dye with formula $NaC_{17}H_{25}N_3O_5S_3$ and molar mass 470, QCR solutions) in TEG solvent (ACROS organics, 99.5%) are nebulized with a medical nebulizer (Pari, PARI Boy SX). A concentration of $4.55\,gl^{-1}$ VIS441 in TEG is used. For measurements on pure solvent droplets, pure TEG is nebulized. The Supplementary Fig. 4 shows an UV/VIS spectrum of a bulk solution of VIS441 in TEG and of pure TEG solvent, respectively.

Droplet sizing. The particle size is determined from excitation laser light scattered elastically by the droplet. The scattered light intensity is collected for scattering angles between 76.5° and 103.5° and focused onto a CMOS camera (Thorlabs, DCC1645C, 1,280 × 1,024 pixels) using a camera objective (Super Carenar, focal length = 50 mm, f-number = 1.7). The particle size is retrieved by fitting calculated phase functions to experimental ones using Mie theory[41,44]. The sizing of sub-micron-sized droplets is difficult because only few fringes are left in the scattering pattern (for example, Fig. 1e). Larger particles exhibit brighter scattering images and many more fringes (Supplementary Fig. 3), which makes sizing easier. We estimate uncertainties in the droplet radius of about half the wavelength.

Calculation of droplet absorbance. The PA signal S is assumed to be proportional to the power P_{abs} absorbed by the droplet, which is located at the centre of a Gaussian excitation laser beam (beam waist radius of 87 μm and cross-section $q_L = 11,889\,\mu m^2$) with incident power P

$$S \propto P_{abs} = I_0 \cdot C_{abs} = P \cdot \frac{C_{abs}}{q_L} \qquad (1)$$

I_0 is the intensity incident on the droplet and C_{abs} is the droplet's absorption cross-section. The equivalent absorbance A due to absorption is given by

$$A = \ln\left(\frac{P}{P - P_{abs}}\right) = \ln\left(\frac{q_L}{q_L - C_{abs}}\right) \approx \frac{C_{abs}}{q_L} \qquad (2)$$

For a single droplet in the PA cell, the equivalent absorption coefficient is given by,

$$\alpha = \frac{C_{abs}}{V_{res}} \qquad (3)$$

where $V_{res} = 0.5\,cm^3$ is the volume of the PA-resonator. The absorption cross-section of the droplet is calculated from the Mie theory[44] with a refractive index of the surroundings equal to 1:

$$C_{abs} = \frac{2\pi}{\varepsilon \cdot \mu \cdot \omega^2}\left[\sum_j (2j+1)\left(\text{Re}\{a_j(x,m) + b_j(x,m)\} - |a_j(x,m)|^2 - |b_j(x,m)|^2\right)\right] \qquad (4)$$

Here a_n and b_n are the scattering coefficients, $x = \frac{2\pi a}{\lambda}$ is the size parameter, a is the droplet radius, λ its wavelength of light in vacuum, ω is the angular frequency of the light, ε and μ are the permittivity and the permeability, respectively, of the droplet, and $m = n + ik$ is the droplet's complex index of refraction at the wavelengths of the excitation laser ($\lambda = 445\,nm$). The latter is determined from UV/VIS absorption and refractometric measurements of VIS441/TEG bulk solution and a pure TEG solution using Kramers–Kronig inversion. The refractive index of the VIS441/TEG solution for a dye concentration $4.55\,gl^{-1}$ and the pure TEG solvent are $n + ik = 1.463 + i \cdot 0.0062$ and $n_s + ik_s = 1.460 + i \cdot 0.0000$, respectively. The refractive index of the VIS441/TEG solution (dye concentration $4.55\,gl^{-1}$) in the UV/VIS range is provided in the Supplementary Fig. 5. For other dye concentrations (see photokinetics), it is assumed to depend linearly on the dye concentration.

Calculation of droplet photokinetics. The droplet photokinetics is described by the following rate equation

$$\frac{dN}{dt} = -p \cdot \frac{I(r)}{h\nu} \cdot \sigma(r) \cdot N(r) \qquad (5)$$

with the number density of reactant molecules N, Planck's constant h, excitation laser frequency ν, molecular absorption cross-section σ and the quantum yield *p*. Here r denotes the location within the droplet and I is the local field intensity. Both I and σ depend on the complex index of refraction, which in turn depends on the number density N, so that the rate law is no longer pseudo first order. The power absorbed by the droplet is given by the rate of absorption integrated over the droplet's volume V

$$P_{abs}(t) = I_0 \cdot C_{abs}(N) = -hv \cdot \int p^{-1} \frac{dN}{dt} dV \qquad (6)$$

Assuming fast diffusion, that is, $N \neq N(r)$, we obtain:

$$\frac{dN}{dt} = -f \cdot V^{-1} C_{abs}(N) \qquad (7)$$

where $f = p \cdot I_0 / hv$ is the product of incident photon flux and reaction probability. Equation (7) is integrated using a 4th order Runge–Kutta method with the time-dependent PA signal given by equation (1). The corresponding inverse first half-lives of the PA signal as a function of droplet radius are shown as a full blue line in Fig. 4a.

To illustrate the effect of nanofocusing, we compare the above model (full blue line in Fig. 4a) with a model that neglects the influence of nanofocusing (dashed red line in Fig. 4a). This model is obtained from equations (5) and (6)

$$C_{abs} = N(t) \int \frac{I(r)}{I_0} \sigma(r) dV \qquad (8)$$

by inserting the small particle limit[44] for σ

$$\sigma V^{-1} = \frac{\pi}{\lambda} \text{Im}\left\{ \frac{-18}{m^2 + 2} \right\} \qquad (9)$$

and a Beer–Lambert expression for the intensity distribution within the particle

$$I(r) = I_0 \cdot \exp\left\{ -\frac{4\pi k}{\lambda} \ell(r) \right\} \qquad (10)$$

where $\ell(r) = r \cos \theta + \sqrt{a^2 - (r \sin \theta)^2}$ is the absorption path length at distance r from the centre of the particle and at polar angle θ relative to the incident beam direction.

Statistical analysis of experimental photolysis data. To account for the uncertainties both in the particle radii and in the decay half-lives of the experimental data set (Supplementary Fig. 6), we perform a two-step maximum likelihood analysis. First, the distribution of particle radii $D(a) = \sum g_i(a)$ is analysed assuming a normally distributed error for the size determination,

$$g_i(a) = \frac{\exp\left\{ -(a - a_i)^2 / 2\sigma_a^2 \right\}}{\sigma_a \sqrt{2\pi}} \qquad (11)$$

with a constant s.d. of $\sigma_a = 220$ nm. The local extrema in D at a_k divide the size range into sections with a lower and an upper half for each cluster of data, which are combined into a single section for isolated data points. For each section, we finally obtain the most probable values for particle radius and the inverse half-life as weighted averages over the particles with weights given by,

$$w_{i,k} = \sigma_{t,i}^{-2} \int_{a_{k-1}}^{a_k} g_i(a) da \Big/ \sum_i \sigma_{t,i}^{-2} \int_{a_{k-1}}^{a_k} g_i(a) da \qquad (12)$$

This implies normally distributed errors for the experimental inverse half-lives with s.d. $\sigma_{t,i}$ ranging from about 10% for the most accurate measurements to about 50% for measurements with $\frac{S}{NL} < 10$ (typically small particles). The error bars in Fig. 4a were obtained by s.e. propagation.

References

1. Nie, W. *et al.* Polluted dust promotes new particle formation and growth. *Sci. Rep.* **4**, 1–6 (2014).
2. George, C., Ammann, M., D'Anna, B., Donaldson, D. J. & Nizkorodov, S. A. Heterogeneous photochemistry in the atmosphere. *Chem. Rev.* **115**, 4218–4258 (2015).
3. Liu, F., Beames, J. M., Petit, A. S., McCoy, A. B. & Lester, M. I. Infrared-driven unimolecular reaction of CH_3CHOO Criegee intermediates to OH radical products. *Science* **345**, 1596–1598 (2014).
4. Reed Harris, A. E. *et al.* Photochemical kinetics of pyruvic acid in aqueous solution. *J. Phys. Chem. A* **118**, 8505–8516 (2014).
5. Finlayson-Pitts, B. J. Chlorine chronicles. *Nat. Chem.* **5**, 724 (2013).
6. Monge, M. E. *et al.* Alternative pathway for atmospheric particles growth. *Proc. Natl Acad. Sci. USA* **109**, 6840–6844 (2012).
7. Jimenez, J. L. *et al.* Evolution of organic aerosols in the atmosphere. *Science* **326**, 1525–1529 (2009).
8. Lack, D. A. *et al.* Relative humidity dependence of light absorption by mineral dust after long-range atmospheric transport from the Sahara. *Geophys. Res. Lett.* **36**, L24805 (2009).
9. Laskin, A. *et al.* Reactions at interfaces as a source of sulfate formation in sea-salt particles. *Science* **301**, 340–344 (2003).
10. Vaida, V., Kjaergaard, H. G., Hintze, P. E. & Donaldson, D. J. Photolysis of sulfuric acid vapor by visible solar radiation. *Science* **299**, 1566–1568 (2003).
11. Jacobson, M. Z. Strong radiative heating due to the mixing state of black carbon in atmospheric aerosols. *Nature* **409**, 695–697 (2001).
12. Lelieveld, J. & Crutzen, P. J. Influences of cloud photochemical processes on tropospheric ozone. *Nature* **343**, 227–233 (1990).
13. Crutzen, P. J. & Arnold, F. Nitric acid cloud formation in the cold Antarctic stratosphere: a major cause for the springtime 'ozone hole'. *Nature* **324**, 651–655 (1986).
14. Tyndall, J. On the blue colour of the sky, and the polarization of light. *Proc. R. Soc.* **37**, 384–394 (1869).
15. Benincasa, D. S., Barber, P. W., Zhang, J.-Z., Hsieh, W.-F. & Chang, R. K. Spatial distribution of the internal and near-field intensities of large cylindrical and spherical scatterers. *Appl. Opt.* **26**, 1348–1356 (1987).
16. Symes, R., Sayer, R. M. & Reid, J. P. Cavity enhanced droplet spectroscopy: principles, perspectives and prospects. *Phys. Chem. Chem. Phys.* **6**, 474–487 (2004).
17. Cappa, C. D., Wilson, K. R., Messer, B. M., Saykally, R. J. & Cohen, R. C. Optical cavity resonances in water micro-droplets: implications for shortwave cloud forcing. *Geophys. Res. Lett.* **31**, L10205 (2004).
18. Brem, B. T., Gonzalez, F. C. M., Meyers, S. R., Bond, T. C. & Rood, M. J. Laboratory-measured optical properties of inorganic and organic aerosols at relative humidities up to 95%. *Aerosol. Sci. Technol.* **46**, 178–190 (2012).
19. Preston, T. C. & Signorell, R. Vibron and phonon hybridization in dielectric nanostructures. *Proc. Natl Acad. Sci. USA* **108**, 5532–5536 (2011).
20. Preston, T. C. & Signorell, R. From plasmon spectra of metallic to vibron spectra of dielectric nanoparticles. *Acc. Chem. Res.* **45**, 1501–1510 (2012).
21. Hickstein, D. D. *et al.* Mapping nanoscale absorption of femtosecond laser pulses using plasma explosion imaging. *Am. Chem. Soc. Nano* **8**, 8810–8818 (2014).
22. Goldmann, M., Miguel-Sánchez, J., West, A. H. C., Yoder, B. L. & Signorell, R. Electron mean free path from angle-dependent photoelectron spectroscopy of aerosol particles. *J. Chem. Phys.* **142**, 224304 (2015).
23. Ruth Signorell & Jonathan, P. Reid (eds) *Fundamentals and Applications in Aerosol Spectroscopy* (CRC Press, 2011).
24. Mayer, B. & Madronich, S. Actinic flux and photolysis in water droplets: Mie calculations and geometrical optics limit. *Atmos. Chem. Phys.* **4**, 2241–2250 (2004).
25. Ray, A. K. & Bhanti, D. D. Effect of optical resonances on photochemical reactions in microdroplets. *Appl. Opt.* **36**, 2663–2674 (1997).
26. Ruggaber, A. *et al.* Modelling of radiation quantities and photolysis frequencies in the aqueous phase in the troposphere. *Atmos. Environ.* **31**, 3137–3150 (1997).
27. Bott, A. & Zdunkowski, W. Electromagnetic energy within dielectric spheres. *J. Opt. Soc. Am. A* **4**, 1361–1365 (1987).
28. Penconi, M. *et al.* The use of chemical actinometry for the evaluation of the light absorption efficiency in scattering photopolymerizable miniemulsions. *Photochem. Photobiol. Sci.* **14**, 308–319 (2015).
29. Nissenson, P., Knox, C. J. H., Finlayson-Pitts, B. J., Phillips, L. F. & Dabdub, D. Enhanced photolysis in aerosols: evidence for important surface effects. *Phys. Chem. Chem. Phys.* **8**, 4700–4710 (2006).
30. Mills, C. T., Rowland, G. A., Westergren, J. & Phillips, L. F. Quantum yields of CO_2 and SO_2 formation from 193 nm photo-oxidation of CO in a sulfuric acid aerosol. *J. Photochem. Photobiol. A* **93**, 83–87 (1996).
31. Kitagawa, F. & Kitamura, N. A laser trapping-spectroscopy study on the photocyanation of perylene across a single micrometre-sized oil droplets/water interface: droplet-size effects on photoreaction quantum yield. *Phys. Chem. Chem. Phys.* **4**, 4495–4503 (2002).
32. Barnes, M. D., Whitten, W. B. & Ramsey, J. M. Enhanced fluorescence yields through cavity quantum-electrodynamic effects in microdroplets. *J. Opt. Soc. Am. B* **11**, 1297–1304 (1994).
33. Taflin, D. C. & Davis, E. J. A study of aerosol chemical reactions by optical resonance spectroscopy. *J. Aerosol. Sci.* **21**, 73–86 (1990).
34. Ward, T. L., Zhang, S. H., Allen, T. & Davis, E. J. Photochemical polymerization of acrylamide aerosol particles. *J. Colloid Interface Sci.* **118**, 343–355 (1987).
35. Sanford, T. J., Murphy, D. M., Thomson, D. S. & Fox, R. W. Albedo measurements and optical sizing of single aerosol particles. *Aerosol. Sci. Technol.* **42**, 958–969 (2008).
36. Cotterell, M. I., Mason, B. J., Preston, T. C., Orr-Ewing, A. J. & Reid, J. P. Optical extinction efficiency measurements on fine and accumulation mode aerosol using single particle cavity ring-down spectroscopy. *Phys. Chem. Chem. Phys.* **17**, 15843–15856 (2015).
37. Haisch, C. Photoacoustic spectroscopy for analytical measurements. *Meas. Sci. Technol.* **23**, 012001 (2012).

38. Haisch, C., Menzenbach, P., Bladt, H. & Niessner, R. A wide spectral range photoacoustic aerosol absorption spectrometer. *Anal. Chem.* **84,** 8941–8945 (2012).

39. Lack, D. A. *et al.* Aircraft instrument for comprehensive characterization of aerosol optical properties, part 2: black and brown carbon absorption and absorption enhancement measured with photo acoustic spectroscopy. *Aerosol. Sci. Technol.* **46,** 555–568 (2012).

40. Gyawali, M. *et al.* Photoacoustic optical properties at UV, VIS, and near IR wavelengths for laboratory generated and winter time ambient urban aerosols. *Atmos. Chem. Phys.* **12,** 2587–2601 (2012).

41. David, G. *et al.* Stability of aerosol droplets in Bessel beam optical traps under constant and pulsed external forces. *J. Chem. Phys.* **142,** 154506 (2015).

42. Thanopulos, I., Luckhaus, D., Preston, T. C. & Signorell, R. Dynamics of submicron aerosol droplets in a robust optical trap formed by multiple Bessel beams. *J. Appl. Phys.* **115,** 154304 (2014).

43. Li, T., Kheifets, S., Medellin, D. & Raizen, M. G. Measurement of the instantaneous velocity of a brownian particle. *Science* **328,** 1673–1675 (2010).

44. Bohren, C. F. & Huffman, D. R. *Absorption and Scattering of Light by Small Particles* (John Wiley & Sons, 1998).

45. Beck, H. A., Niessner, R. & Haisch, C. Development and characterization of a mobile photoacoustic sensor for on-line soot emission monitoring in diesel exhaust gas. *Anal. Bioanal. Chem.* **375,** 1136–1143 (2003).

46. Kosterev, A. A., Tittel, F. K., Serebryakov, D. V., Malinovsky, A. L. & Morozov, I. V. Applications of quartz tuning forks in spectroscopic gas sensing. *Rev. Sci. Instrum.* **76,** 043105 (2005).

47. Raspet, R., Slaton, W. V., Arnott, W. P. & Moosmüller, H. Evaporation-condensation effects on resonant photoacoustics of volatile aerosols. *J. Atmos. Oceanic Technol.* **20,** 685–695 (2003).

Acknowledgements

This work was supported by the Swiss National Science Foundation (SNSF grant nr. 200020_159205), ETH Zurich and the TUM International Graduate School of Science and Engineering (IGSSE). We are very grateful to Dr David Luckhaus for support in the data analysis and to Prof. Markus Sigrist for helpful discussions concerning the PA set-up. We would like to thank Guido Grassi and Daniel Zindel from the analytical service at ETH and the electronic and mechanical workshops at ETH and TUM for their technical support.

Author contributions

J.W.C. implemented the experimental set-up and performed the measurements. K.M.T. and C.H. designed the PA cell and contributed to the experimental set-up and initial test measurements. J.W.C. and R.S. analysed the data. R.S. conceived the project, performed the calculations and wrote the manuscript.

Additional information

22

Spontaneous incorporation of gold in palladium-based ternary nanoparticles makes durable electrocatalysts for oxygen reduction reaction

Deli Wang[1], Sufen Liu[1], Jie Wang[1], Ruoqian Lin[2], Masahiro Kawasaki[3], Eric Rus[4], Katharine E. Silberstein[4], Michael A. Lowe[4], Feng Lin[5], Dennis Nordlund[6], Hongfang Liu[1], David A. Muller[7,8], Huolin L. Xin[2] & Héctor D. Abruña[4]

Replacing platinum by a less precious metal such as palladium, is highly desirable for lowering the cost of fuel-cell electrocatalysts. However, the instability of palladium in the harsh environment of fuel-cell cathodes renders its commercial future bleak. Here we show that by incorporating trace amounts of gold in palladium-based ternary (Pd_6CoCu) nanocatalysts, the durability of the catalysts improves markedly. Using aberration-corrected analytical transmission electron microscopy in conjunction with synchrotron X-ray absorption spectroscopy, we show that gold not only galvanically replaces cobalt and copper on the surface, but also penetrates through the Pd–Co–Cu lattice and distributes uniformly within the particles. The uniform incorporation of Au provides a stability boost to the entire host particle, from the surface to the interior. The spontaneous replacement method we have developed is scalable and commercially viable. This work may provide new insight for the large-scale production of non-platinum electrocatalysts for fuel-cell applications.

[1] Key laboratory of Material Chemistry for Energy Conversion and Storage (Huazhong University of Science and Technology), Ministry of Education, Hubei Key Laboratory of Material Chemistry and Service Failure, School of Chemistry and Chemical Engineering, Huazhong University of Science and Technology, Wuhan 430074, China. [2] Center for Functional Nanomaterials, Brookhaven National Laboratory, Upton, New York 11973, USA. [3] JEOL USA, Inc., Peabody, Massachusetts 01960, USA. [4] Department of Chemistry and Chemical Biology, Cornell University, Ithaca, New York 14853, USA. [5] Energy Storage and Distributed Resources Division, Lawrence Berkeley National Laboratory, Berkeley, California 94720, USA. [6] Stanford Synchrotron Radiation Lightsource, SLAC National Accelerator Laboratory, Menlo Park, California 94025, USA. [7] School of Applied and Engineering Physics, Cornell University, Ithaca, New York 14853, USA. [8] Kavli Institute at Cornell for Nanoscale Science, Cornell University, Ithaca, New York 14853, USA. Correspondence and requests for materials should be addressed to D.W. (email: wangdl81125@hust.edu.cn) or to H.L.X. (email: hxin@bnl.gov) or to H.D.A. (email: hda1@cornell.edu).

Pt-based nanoparticles have been extensively studied as electrocatalysts for the oxygen reduction reaction (ORR) in both acid and alkaline media[1-7]. However, the high cost and the limited reserve of Pt hinder the widespread deployment of fuel-cell technologies. Numerous strategies have been employed including developing new structures and morphologies for Pt-based nanocatalysts[3,8-13], Pt monolayer decorated nanoparticles[14,15], other platinum group nanomaterials[16-19], as well as non-precious-metal nanoparticles[16,20-24]. Among these alternatives, Pd is very attractive because it is much more abundant and less expensive than Pt. However, the ORR activity of pure Pd in acid is much lower than that of Pt. Previous studies have focused on alloying Pd with 3d transition metals, such as Fe[25], Co[14], Ni[26], Cu[27] and others, resulting in a significant improvement in catalytic activity. Moreover, the activity of Pd-based binary nanoparticles can be further enhanced by adding a third element[28]. In spite of that, Pd-based catalysts undergo fast degradation as ORR cathodes under regular cycling conditions. Previous studies have shown that these catalysts can be stablized by galvanically replacing the surface atoms with Pt[14,29], which simultaneously increases Pt usage.

In this study, we show that Au atoms can penetrate the bulk lattice of Pd_6CoCu nanoparticles and spontaneously replace Co and Cu atoms in the interior of the nanoparticles through galvanic replacement reactions. This effect allows the uniform incorporation of trace amounts of Au in Pd_6CoCu nanoparticles, which enhances the long-term stability of the electrocatalyst for the ORR. We also show that there a is much smaller displacement (ca. 25 mV) in the half-wave potential for $Au–Pd_6CoCu$ nanoparticles over 10,000 voltage cycles in O_2-saturated 0.1 M $HClO_4$ solution, compared with a 54-mV negative potential shift for Pd_6CoCu nanoparticles without Au incorporation after only 1,000 cycles in O_2-saturated 0.1 M $HClO_4$ solution. More importantly, PEMFC single-cell performance using a $Au–Pd_6CoCu/C$ cathode electrocatalyst exhibits a maximum power-density loss of about 21% after a 100-h stress test at 80 °C, which is dramatically better than a Au-free catalyst.

Results

Synthesis and charaterization of Pd_6CoCu/C nanoparticles.

Pd_6CoCu nanoparticles supported on Vulcan XC-72 carbon (Pd_6CoCu/C) were synthesized using an impregnation reduction method followed by high temperature heat-treatment as reported previously[8,14,30]. The crystal structures of the particles were characterized by X-ray diffraction (Supplementary Fig. 1). Pd/C, Pd_3Cu/C and Pt_3Co/C nanoparticles are also shown for comparison. The X-ray diffraction patterns suggest a face-centered cubic structure. The peak positions of the bi-/tri- metallic nanoparticles shift to higher angles, as anticipated, compared to pure Pd/C. The angular shifts indicate alloy formation between Pd and Co/Cu, and reflect a lattice contraction, which is caused by the incorporation of smaller atoms, Co/Cu, into the Pd face-centered cubic lattice. The surface composition of the particles was determined by X-ray photoelectron spectroscopy (XPS) (Supplementary Fig. 2). For the Pd_6CoCu/C sample, there is almost no Co2p signal, while two distinct Cu2p peaks are evident. This suggests that Cu is likely distributed closer to the surface of the particle than Co as illustrated in Fig. 1.

To directly measure the structure and chemical composition of the nanoparticles at the atomic scale, we used aberration-corrected scanning transmission electron microscopy (STEM) in conjunction with electron energy loss spectroscopy (EELS) to characterize the samples (Fig. 2). The Pd_6CoCu/C nanoparticles appear to be uniformly distributed on the carbon support with an

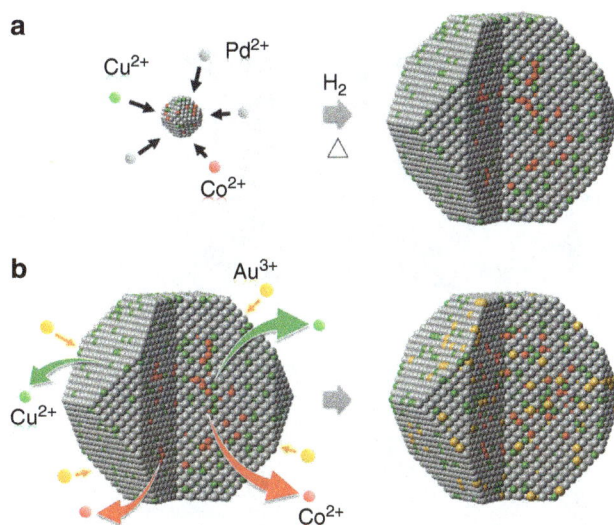

Figure 1 | Diagrammatic illustration of the synthetic strategy of nanoparticles. Schematic illustration of the formation of Pd_6CoCu/C (**a**) and $Au–Pd_6CoCu/C$ (**b**).

average particle size of 7.5 nm, consistent with the particle size calculated from X-ray diffraction (Supplementary Fig. 3 and Supplementary Table 1). Figure 2a,b show the annular dark-field (ADF)-STEM images, which reflect the atomic mass contrast of the materials, and EELS elemental mapping of the as-prepared Pd_3Cu/C and Pd_6CoCu/C nanoparticles, respectively. For Pd_3Cu/C nanoparticles, the EELS spectroscopic images in Fig. 2a indicate a uniform elemental distribution of Pd and Cu from the particle's interior to the surface. There is no surface segregation between Pd and Cu. This is in contrast to our previously reported Pd_3Co/C nanoparticles that had a ~1-nm-thick Pd-rich shell on the particles surface[14,30].

Figure 2b shows the ADF-STEM images and EELS elemental maps of Pd_6CoCu/C nanoparticles. There is no phase segregation in the composite maps of Pd and Cu, which is the same as in Pd_3Cu/C nanoparticles. However, a Pd-rich shell was observed for the composite maps of Pd and Co. To quantify the surface and subsurface chemical composition with atomic-scale resolution, we investigated a single facet of a Pd_6CoCu particle (Fig. 2c). The chemical maps in Fig. 2c clearly show that there is a ~1-nm-thick Pd–Cu-rich layer present on the surface of the Pd_6CoCu/C nanoparticle. This is consistent with the low Co 2p signal in XPS, as the Pd–Cu-rich shell blocks the photoelectrons generated from Co 2p core levels.

The difference in the segregation behaviour between Pd–Cu and Pd–Co is likely a result of two contributions. First, as has been calculated by Nørskov et al.[31] and Wang et al.[32], the surface segregation energy between Pd and Co is much higher than between Pd and Cu. This suggests that there would be a much stronger driving force for surface segregation in Pd–Co than in Pd–Cu nanoparticles. Second, the differences in the gas-adsorption energies of two metals can induce surface segregation[33]. The particles were heat-treated in a flowing stream of hydrogen. The fact that hydrogen binds strongly to Pd surfaces and can even be stored in Pd lattices, suggests that Pd can be enriched at the surface, thus pushing other elements away from the surface.

Synthesis and charaterization of $Au–Pd_6CoCu/C$ nanoparticles.

The Au cluster-modified Pt/C nanoparticles, prepared

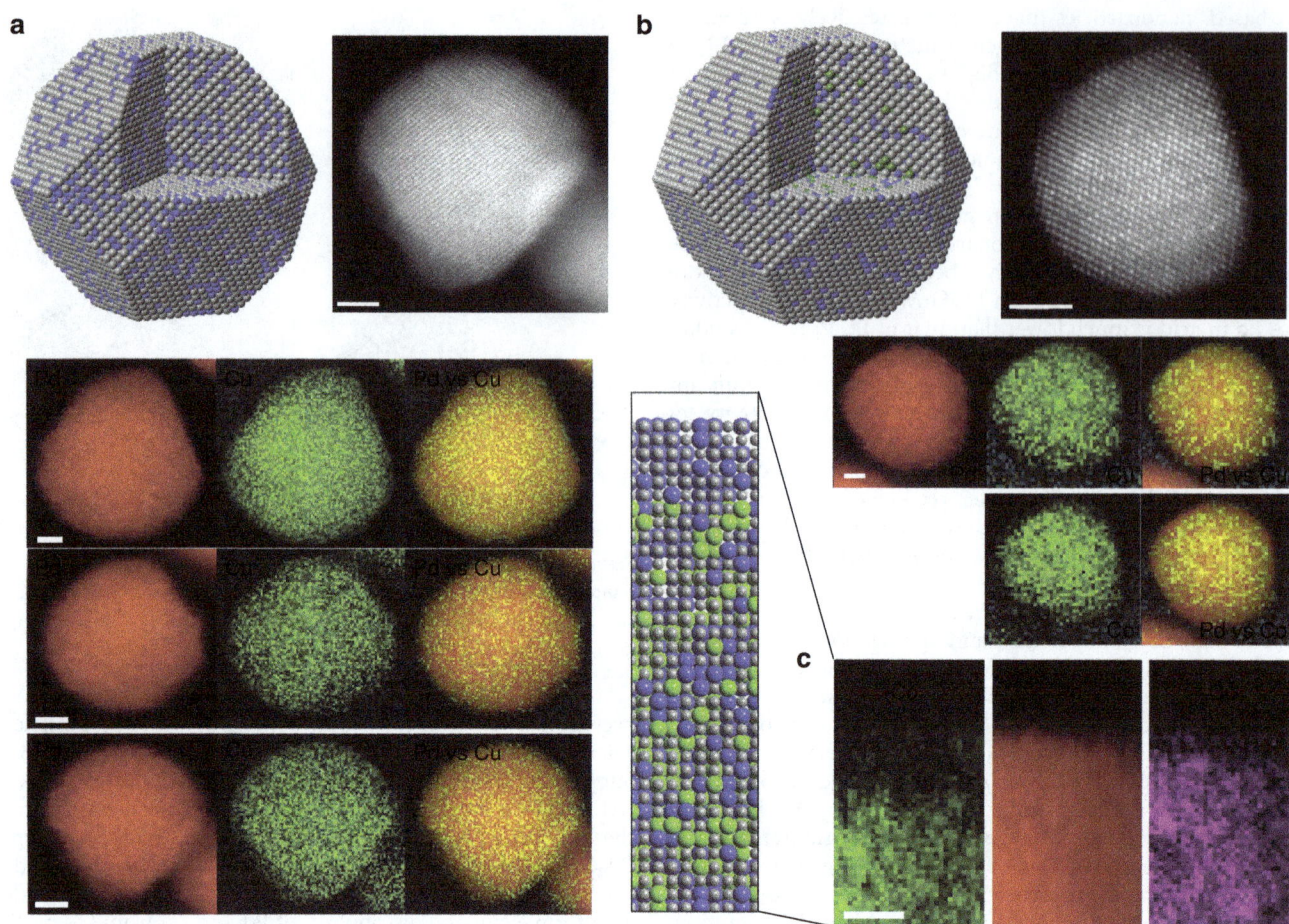

Figure 2 | ADF-STEM images and elemental mapping. Modelled structure, ADF-STEM image, and 2D EELS maps of each element and composites of Pd3Cu/C (**a**) and Pd6CoCu/C (**b**). Scale bar, 2 nm. (**c**) Elemental maps of a Pd_6CoCu particle's surface and subsurface volume extracted from an atomic-scale EELS map. Scale bar, 1 nm.

by the Cu under-potential deposition method, have been reported to significantly enhance the stability of Pt/C for the ORR[34]. Recently, Stamenkovic *et al.* reported that Au-stabilized Pt–Au–Ni nanoparticles with a multilayer profile of Ni@Au@PtNi could endure extended durability testing[35]. Adzic *et al.* showed that structurally ordered intermetallic AuPdCo nanoparticles exhibited increased activity and much better durability for the ORR[36].

Inspired by previous studies, we used a galvanic replacement method to deposit Au on Pd_6CoCu/C nanoparticles since the equilibrium electrode potential of the $AuCl_4^-$/Au couple (0.93 V versus standard hydrogen electrode (SHE)) is more positive than those of the Cu^{2+}/Cu (0.34 V versus SHE) and $PdCl_4^{2-}$/Pd (0.591 V versus SHE) couples. The stability of Pd_6CoCu/C nanoparticles for ORR could be enhanced after Au replacement. The deposition procedure is similar to the one that we have previously reported[14,30]. XPS spectra clearly show Au_{4f} peaks (inset in Supplementary Fig. 4), indicating that Au has successfully replaced atoms on/in the Pd_6CoCu/C nanoparticle. Moreover, there is no evidence of a phase transformation after Au has been deposited as shown by X-ray diffraction measurements (Supplementary Fig. 5).

Figure 3a shows energy-dispersive X-ray spectroscopic analysis of Au–Pd_6CoCu/C nanoparticles in an aberration-corrected STEM. A line-profile was acquired on a single particle to reveal the elemental distribution of Pd, Co, Cu and Au using a 200 keV aberration-corrected STEM (Fig. 3b). Interestingly and surprisingly, Au showed the same distribution profile as Pd, suggesting

that Au penetrates through the entire particle as illustrated in Fig. 1. To demonstrate the elemental distribution of Au without ambiguity, we performed a two-dimensional (2D) mapping of the particle. The Au map shows the same spatial distribution as Pd, again strongly supporting our assertion that Au atoms are uniformly incorporated into the nanoparticle (Fig. 3c).

The incorporation of Au was further examined by X-ray absorption spectroscopy (XAS). The Au-L3 X-ray absorption near-edge structure was used to discriminate between highly oxidized and reduced forms of the metal. Figure 4a compares a Au film and Au–Pd_6CoCu/C nanoparticles. It can be clearly seen that both samples exhibit the characteristic three-peak pattern following the edge jump[37], indicating that the gold in the Au–PdCuCo sample is reduced (metallic). The 'whiteline' intensity of Au, which is closely related to the d-band population, was slightly lower in Au–Pd_6CoCu/C than that in the Au film, and the edge position was shifted to slightly higher energy. Most importantly, the small but significant decrease in intensity of the whiteline demonstrates a partial charge transfer from Pd to Au as observed and described in detail for Au–Pd alloys in previous studies[38–40].

The Fourier transform of the extended X-ray absorption fine structure of the Au film and Au–Pd_6CoCu/C nanoparticles, measured at the Au-L3 edge, are shown in Fig. 4b. The two peaks in the spectra indicate that gold has two different near neighbours. We ascribe the decrease in magnitude of the first-shell peak in Au–Pd_6CoCu/C relative to the gold foil to a lower coordination number of Au in the nanoparticle, which is

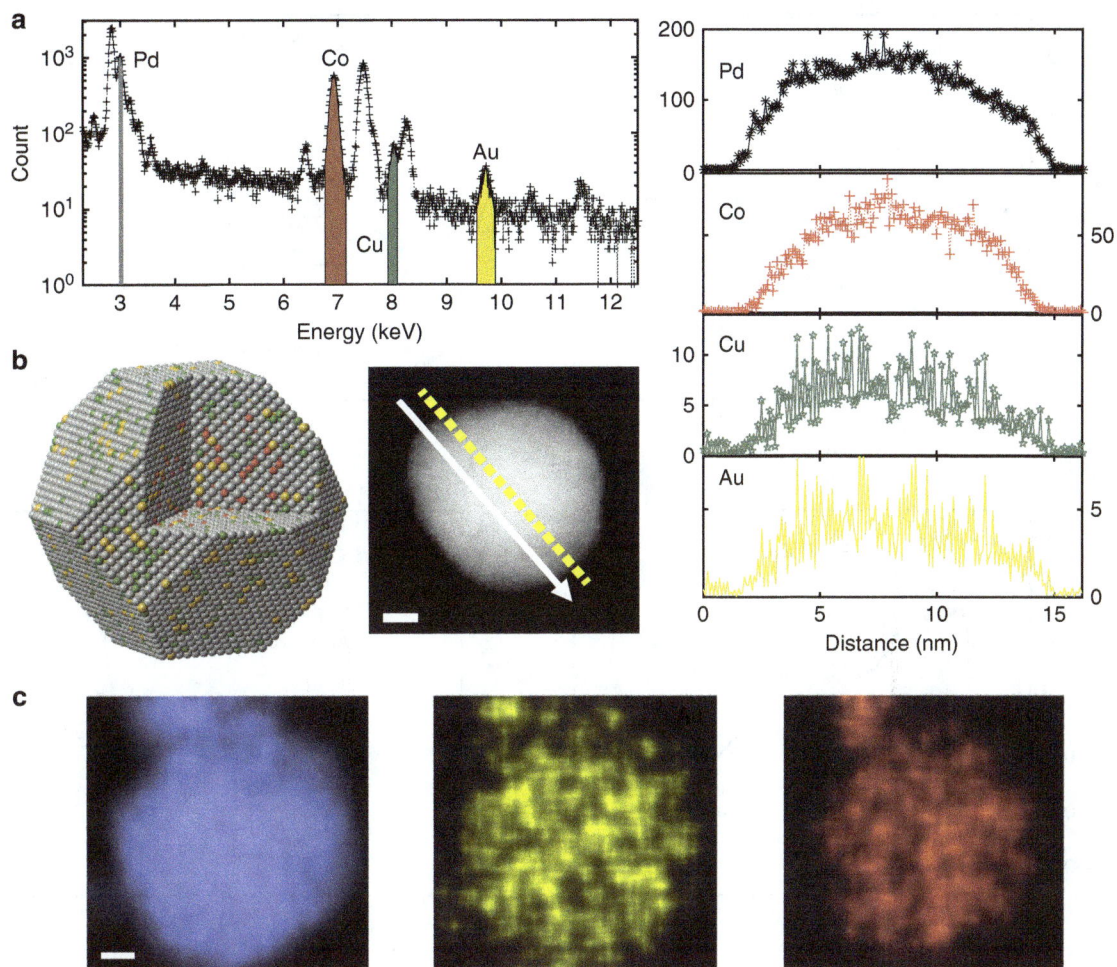

Figure 3 | Composition analysis of a Au–Pd₆CoCu/C nanoparticle. (**a**) Energy-dispersive X-ray spectroscopic (EDX) analysis of Au–Pd$_6$CoCu/C nanoparticles. (**b**) Elemental distribution of Pd, Co, Cu and Au in a single Au–Pd$_6$CoCu/C nanoparticle extracted from an aberration-corrected STEM–EDX line profile. Scale bar, 2 nm. (**c**), Aberration-corrected STEM–EDX 2D elemental maps of Pd, Co and Au. Scale bar, 5 nm.

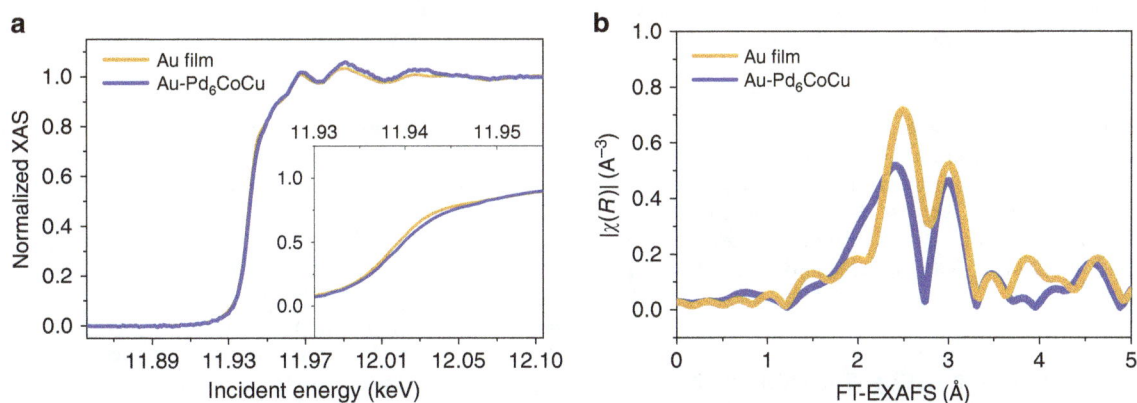

Figure 4 | X-ray absorption analysis. (**a**) XANES and (**b**) FT-EXAFS spectra of Au-L3 edge in a Au film and a Au-Pd$_6$CoCu/C nanoparticle. EXAFS, extended X-ray absorption fine structure; XANES, X-ray absorption near-edge structure.

consistent with a previous XAS study of Pd–Au alloys by Sham[38] and a consequence, at least in part, of the phase shifts of high-Z absorbers and scatterers. In a word, the XAS study demonstrates that at the bulk ensemble level, Au has been uniformly incorporated in the Pd–Co–Cu lattice, rather than forming a Au shell or Au clusters.

The observation that Au can spontaneously penetrate the Pd–Co–Cu lattice to reach the interior of the particle might be related to the fact that the as-prepared Pd$_6$CoCu/C nanoparticles were Pd and Cu rich on the surface (Fig. 2c). The equilibrium electrode potential of the Cu^{2+}/Cu couple is lower than that of the PdCl$_4^{2-}$/Pd couple. Therefore, it is likely that Au displaces Cu first, leaving the Pd atoms untouched. In addition, Au and Pd are intermixable at all proportions. Therefore, thermodynamically, it is favourable for Au to uniformly mix with Pd. This is in contrast to the previously investigated Pt–Pd$_3$Co system[14,30], where Pt can

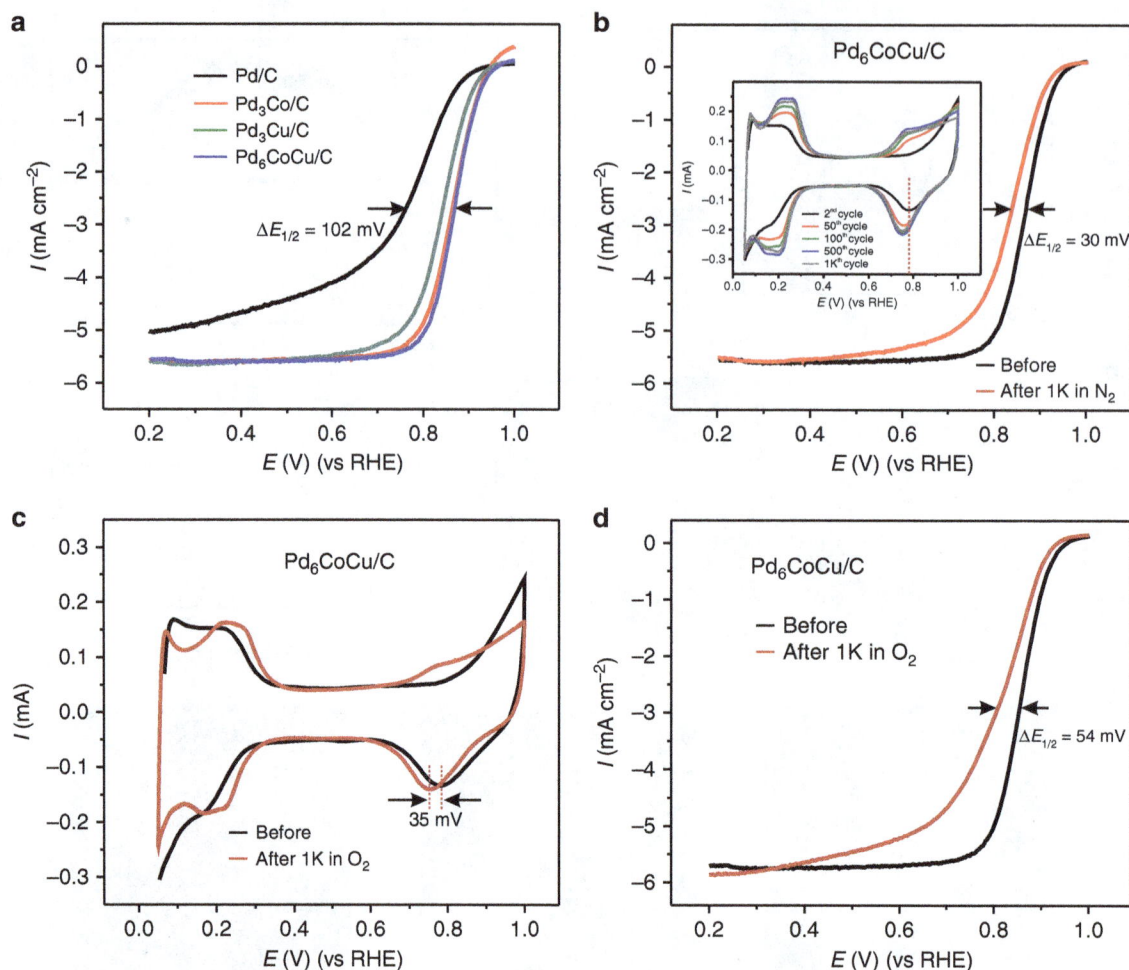

Figure 5 | Electrocatalytic activity for ORR. (a) Comparison of ORR polarization curves for Pd/C, Pd$_3$Co/C, Pd$_3$Cu/C and Pd$_6$CoCu/C in O$_2$-saturated 0.1 M HClO$_4$ at room temperature; rotation rate, 1,600 r.p.m.; and sweep rate, 5 mV s^{-1}. **(b)** ORR polarization curves for Pd$_6$CoCu/C before and after 1,000 potential cycles in N$_2$-saturated 0.1 M HClO$_4$, rotation rate, 1,600 r.p.m. and sweep rate, 5 mV s^{-1}. The inset in **b** shows the changes in the CV profiles after different numbers of cycles at a sweep rate of 50 mV s^{-1}. **(c)** CV changes in the voltammetric profile for Pd$_6$CoCu/C before and after 1,000 potential cycles in O$_2$-saturated 0.1 M HClO$_4$ between 0.6 and 1.0 V at a sweep rate of 50 mV s^{-1}. **(d)** ORR polarization curves for Pd$_6$CoCu/C before and after 1,000 potential cycles in O$_2$-saturated 0.1 M HClO$_4$, rotation rate, 1,600 r.p.m.; sweep rate, 5 mV s^{-1}.

only displace atoms on the surface of the nanoparticles and form a Pt-rich shell. The phase diagram shows that, in this case, Pt and Pd are not miscible in all proportions. With regards to the Pt-displacement experiment, we speculate that the intersolubility of Au and Pd might be one of the driving forces for Au to penetrate the Pd–Co–Cu bulk lattice and thus the entire nanoparticle.

Electrochemical and fuel-cell testing. The ORR activities of the electrocatalysts were evaluated in an O$_2$-saturated 0.1 M HClO$_4$ solution by loading the materials (with the same Pd mass loading) onto a glassy carbon rotating disk electrode with experiments performed at a rotation rate of 1,600 r.p.m. and a sweep rate of 5 mV s^{-1} at room temperature. All electrodes were pretreated by cycling the potential between +0.05 and +1.00 V at a sweep rate of 50 mV s^{-1}, for 50 cycles, before the ORR activity test. Polarization curves for the ORR of the different catalysts are shown in Fig. 5a. The ORR kinetics was markedly accelerated on the tri-metallic Pd$_6$CoCu/C alloy nanoparticles, and the half-wave potential ($E_{1/2}$) was significantly positively shifted, relative to Pd/C and Pd$_3$Cu/C nanoparticles, indicating a significant increase in the ORR activity. It can also be seen in Fig. 5a that the

Pd$_6$CoCu/C nanoparticles exhibit a slightly higher ORR activity than the Pd$_3$Co/C nanoparticles. This may be due to the smaller particle size and slightly different lattice parameter (Supplementary Table 1).

The durability of the Pd$_6$CoCu/C nanoparticles was evaluated by potential cycling between +0.05 and +1.0 V at a scan rate of 50 mV s^{-1} for 1,000 cycles in N$_2$-saturated 0.1 M HClO$_4$ solution followed by an assessment of the ORR activity. As shown in Fig. 5b, after 1,000 potential cycles, the Pd$_6$CoCu/C catalyst showed a degradation of around 30 mV in its half-wave potential for the ORR. The CVs after different numbers of cycles are presented in the inset to Fig. 5b. Durability tests were also conducted in O$_2$-saturated 0.1 M HClO$_4$ solution between +0.6 and +1.0 V at a scan rate of 50 mV s^{-1} for 1,000 potential cycles. It can be seen from changes in the CVs (Fig. 5c) that the hydrogen adsorption/desorption peaks shifted to positive potentials, while the peak potential for the reduction of the Pd oxides shifted negatively by ~35 mV. The ORR durability, presented in Fig. 5d, shows a significant negative shift in the half-wave potential of about 54 mV after 1,000 potential cycles, indicating that the decay in the catalytic activity of Pd$_6$CoCu/C was accelerated when cycling in the presence of oxygen.

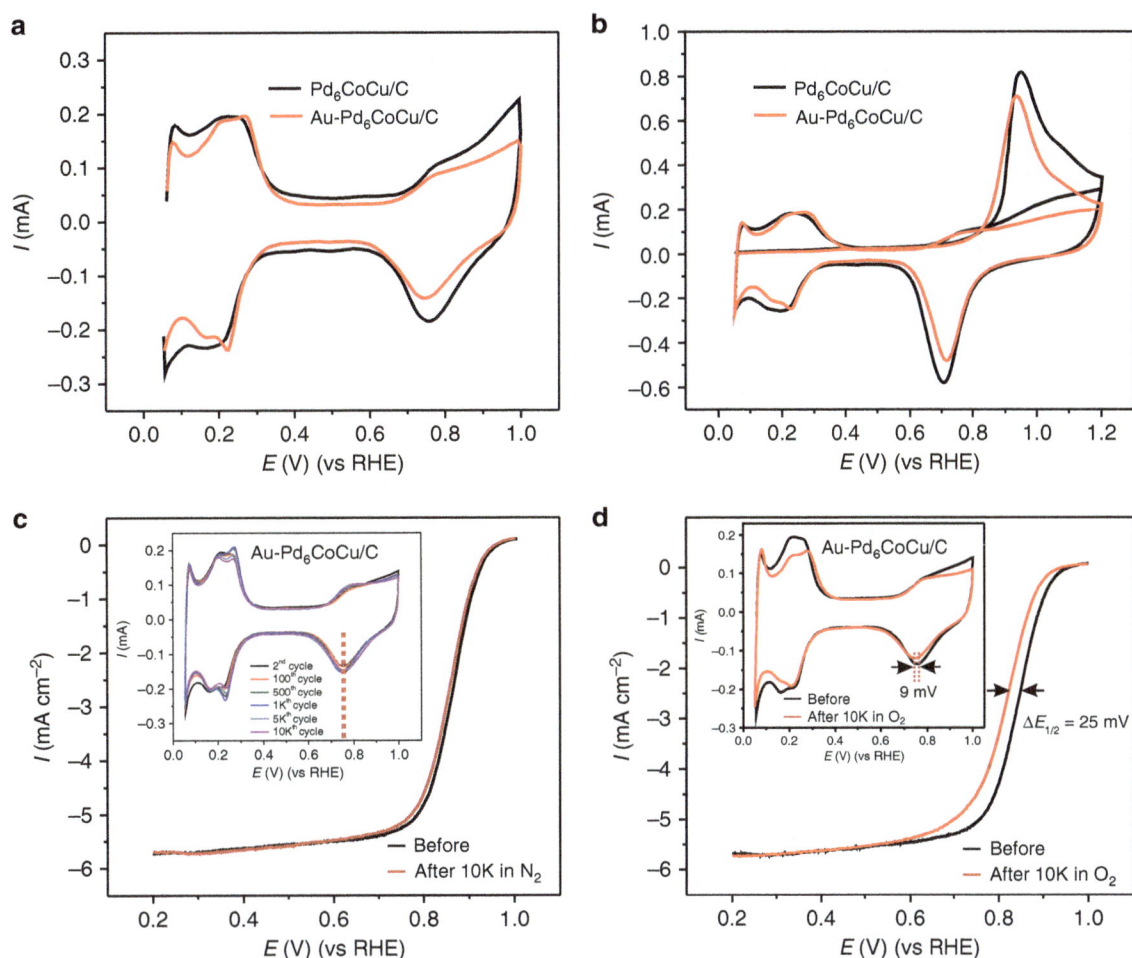

Figure 6 | Electrochemical measurements. (**a**) CVs and (**b**) CO stripping of Pd_6CoCu/C nanoparticles with and without Au decoration in 0.1 M $HClO_4$ purged with N_2, sweep rate, 50 mV s^{-1}. (**c**) ORR polarization curves of Au-Pd_6CoCu/C nanoparticles before and after cycling in N_2-saturated 0.1 M $HClO_4$; rotation rate, 1,600 r.p.m.; and sweep rate, 5 mV s^{-1}. The inset in **c** shows CVs of the Au-Pd_6CoCu/C nanoparticles after different numbers of potential cycles between 0.05 and 1.0 V in N_2-saturated 0.1 M $HClO_4$ and sweep rate, 50 mV s^{-1}. (**d**) ORR polarization curves of Au-Pd_6CoCu/C nanoparticles before and after cycling in O_2-saturated 0.1 M $HClO_4$ between 0.6 and 1.0 V; rotation rate, 1,600 r.p.m. and sweep rate, 5 mV s^{-1}. The inset in **d** shows changes in the voltammetric profile for Au-Pd_6CoCu/C before and after 10,000 potential cycles between 0.6 and 1.0 V in O_2-saturated 0.1 M $HClO_4$, sweep rate, 50 mV s^{-1}.

The decay in electrocatalytic activity for the ORR on $Pd_6CoCu/$ C electrode, after continuous cycling, most likely arises from surface changes during cycling. Since the initial surface of the Pd_6CoCu/C nanoparticles is a Pd–Cu-rich alloy (Fig. 2b,c; Pd and Cu are evenly distributed on the surface), the Cu dissolves during potential cycling, causing the lattice strained Pd surface to expand with Cu leaching out of the nanoparticle surface. The surface composition changes can be seen clearly from the CVs of the Pd_6CoCu/C nanoparticle after different cycles (inset of Fig. 5b,c). The onset of Pd oxide formation, and the Pd oxides reduction peaks shift towards negative potentials with cycling. The readily formed Pd oxides inhibit the OH_{ad} species on the Pd_6CoCu/C electrode surface and block the active sites for O_2 adsorption, causing the decay in the ORR activity.

. After replacement with Au the CV of the Pd_6CoCu/C nanoparticle changed significantly. As shown in Fig. 6a, the total surface area decreased (see also the CO stripping in Fig. 6b) and the double layer current also decreased. In the hydrogen region, the Au-decorated Pd_6CoCu nanoparticles exhibited, two well-defined hydrogen adsorption/desorption peaks compared with one broad hydrogen peak for Pd_6CoCu/C nanoparticles, from $+0.1$ to $+0.3$ V. In addition, the Pd oxides reduction peak shifted slightly negative after Au decoration. Because the Au atom

is larger than Pd, the lattice parameter of Pd on the top layer of the nanoparticle will expand after Au incorporation. The expanded lattice will modify the oxygen adsorption energy of the nanocatalyst, causing the negative shift of the oxides reduction peak. The same phenomenon can also be seen in the CO-stripping voltammetry (Fig. 6b), which shows that the onset and peak potential for CO oxidation, shifted slightly negative after Au-decoration of the Pd_6CoCu/C nanoparticles.

The stabilizing effect of Au on the Pd_6CoCu/C nanoparticles was evaluated by cycling the electrode in both N_2- and O_2-saturated 0.1 M $HClO_4$ solutions, respectively. As shown in the inset of Fig. 6c,d, the CVs of the Au–Pd_6CoCu/C nanoparticles in an N_2 atmosphere changed slightly with increasing number of potential cycles. Moreover, unlike the Pd_6CoCu/C catalysts, the peak position of the Pd oxides reduction remained unchanged in the N_2 atmosphere, although it shifted negatively by about 9 mV in the O_2 atmosphere after 10,000 potential cycles, indicating that the Pd_6CoCu/C catalyst was stabilized by the Au decoration. Besides, the morphology of the Au–Au–Pd_6CoCu/C nanoparticles remained almost spherical after potential cycles, although the average size was slightly increased (Supplementary Fig. 6). The catalytic activity for the ORR of the Au–Pd_6CoCu/C catalyst exhibited a slight

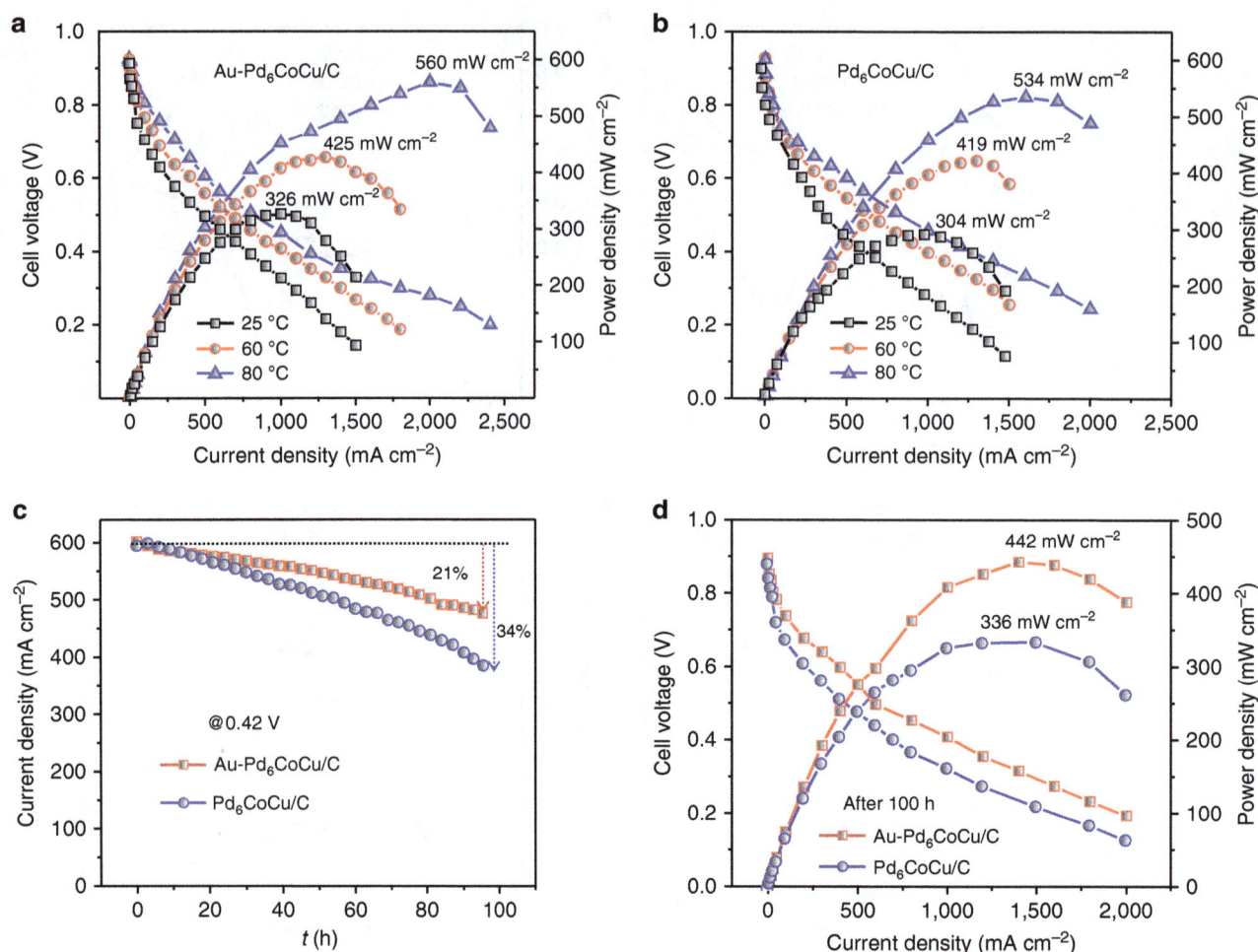

Figure 7 | Single-cell performance. (**a**) Polarization and power-density curves for PEMFCs, using a Au–Pd_6CoCu/C cathode catalyst at 25, 60 and 80 °C. (**b**) Polarization and power-density curves for PEMFCs using a Pd_6CoCu/C cathode catalyst at 25, 60 and 80 °C. (**c**) Stability evaluation of Au–Pd_6CoCu/C and Pd_6CoCu/C cathodes on polarizing the cell at 0.42 V for 100 h in a single cell at 80 °C. (**d**) Comparison of the polarization and power densities for PEMFCs using a Au–Pd_6CoCu/C and Pd_6CoCu/C cathode catalyst after life test at 80 °C. Pt/C (Alfa Aesar, 40wt%) was used as the anode catalyst, and the metal loading was 0.5 mg cm^{-2} in both the anode and cathode. The current density values given are with respect to geometric electrode area.

degradation in the half-wave potential after 10,000 cycles in N_2-saturated 0.1 M $HClO_4$ solution between +0.05 and +1.0 V (Fig. 6c). In contrast, the Pd_6CoCu/C catalyst exhibited a shift of ~30 mV (Fig. 5b). After testing the electrode for 10,000 potential cycles in O_2-saturated 0.1 M $HClO_4$ solution between +0.6 and +1.0 V at a scan rate of 50 mV s^{-1} (Fig. 6d), the electrochemical surface area decreased by only 13.8% when compared with the initial value. The half-wave potential shifted negatively by about 25 mV, a value that is much lower than that of the Pd_6CoCu/C electrode, after only 1,000 cycles (54 mV). The enhancement in the stability likely derives from the partial charge donation of Pd to Au (Fig. 4a) and slight expansion of the Pd lattice (Fig. 6a; Supplementary Table 1), thus enhancing the durability of Pd during cycling.

The stability of the Au–Pd_6CoCu/C cathode was further evaluated and compared with Pd_6CoCu/C cathode in a single-cell PEMFC with commercial Pt/C as the anode (Fig. 7). Figure 7a shows the fuel-cell performance using Au–Pd_6CoCu/C as the cathode electrocatalyst at different temperatures. It was observed that the fuel-cell performance was enhanced with increasing operating temperature. The maximum power density at 25 °C was 326 mW cm^{-2}. It increased to 425 and 560 mW cm^{-2} at 60 °C and 80 °C, which is almost 30% and 72% higher than the starting value, respectively. This behaviour suggests that the oxygen

reduction kinetics at the Au–Pd_6CoCu/C nanoparticles were enhanced by increasing the temperature. The fuel-cell performance using Pd_6CoCu/C as the cathode electrocatalyst at different temperatures exhibited similar behaviour as Au–Pd_6CoCu/C (Fig. 7b). The long-term durability of the catalysts was assessed by recording the current density with time at a constant cell voltage of 0.42 V at 80 °C for 100 h (Fig. 7c) and the polarization and power-density curves before and after polarizing the fuel cell (Fig. 7d). The Au–Pd_6CoCu/C cathode exhibited higher durability. The losses in the current density and in maximum power density were both about 21%. Compared with the losses of 34 and 37% in the current density and the maximum power density observed from the Pd_6CoCu/C cathode, it further demonstrates that Au decoration enhanced the durability of Pd_6CoCu/C cathode electrocatalysts.

Discussion

The use of gold to stabilize fuel-cell nanocatalyts for the ORR has become an increasingly attractive strategy. However, developing a synthetic procedure that can lower Pt and Au usage, while preserving scalability for industrial production, has remained challenging and elusive. The method that we have developed, and which we present here, is extremely effective and scalable and addresses the above-mentioned challenges. Compared with methods pioneered by the Adzic group, which are only applicable to laboratory scale experiments, our strategy eliminates the need to carry out lengthy synthesis steps[36] and Cu under-potential deposition[34]. In addition, we have demonstrated that only trace amounts of gold (Au:Pd = 1:100) are needed to stablize the Pd-based alloy particles. This is almost a two-orders of magnitude lower loading of gold, compared with results reported in previous Au–Pd studies[29,36]. Moreover, according to the previous study[34], we expected that the Pd_6CoCu particles would have been decorated with a Au overlayer or Au clusters. Contrary to that expectation, in the present work we found that gold atoms actually penetrate the Pd–Co–Cu lattice and uniformly distribute within the particles without disrupting the host particle structure. Therefore, our strategy offers not only technological advances, but also undiscovered new structures and synthesis routes.

In summary, a ternary Pd_6CoCu nanoparticle catalyst with ultralow amounts of Au decoration (Au–Pd_6CoCu/C) has been successfully prepared by a simple spontaneous replacement method. Contrary to all previous examples, the Au is homogeneously distributed throughout the nanoparticles (over 10,000 cycles). The Au–Pd_6CoCu/C catalysts exhibited excellent stability for the ORR under-potential cycling both in N_2- and O_2-saturated 0.1 M $HClO_4$ solution. The single fuel-cell test indicates that the durability of Au–Pd_6CoCu/C significantly enhances the 100-h life test compared with Pd_6CoCu/C. The high durability of the electrocatalyst for the ORR can be ascribed to the homogeneous distribution of gold and the charge transfer of Pd to Au, causing a reduction of the d-electron occupation of Pd with a slightly expanded lattice, which enhances the corrosion resistance of Pd. This study provides a new strategy for optimizing the stability of fuel-cell catalysts.

Methods

Material synthesis. Carbon supported Pd_3Co, Pd_3Cu and Pd_6CoCu nanoparticles with 20wt% of Pd were prepared using an impregnation method. In a typical synthesis for Pd_6CoCu/C, 67 mg of $PdCl_2$, 14.9 mg of $CoCl_2 \cdot 6H_2O$ and 10.7 mg of $CuCl_2 \cdot 2H_2O$ were first dissolved in deionized water, and 152 mg of preheated Vulcan XC-72 carbon support were dispersed in the solution. After ultrasonic blending for 30 min, the suspension was heated under magnetic stirring to allow the solvent to evaporate and to form a smooth, thick slurry. The slurry was dried in an oven at 60 °C overnight and grounded in an agate mortar, the resulting dark and free-flowing powder was then reduced in a tube furnace at 150 °C under flowing H_2/N_2 for 2 h. Pd_3Cu/C and Pd_3Co/C were prepared in the same procedure for comparison. The as-prepared Pd_3Co/C, Pd_3Cu/C and Pd_6CoCu/C nanoparticles were deposited at 500 °C under a flowing H_2 atmosphere for 10 h to form a Pd-rich shell.

The Au-decorated Pd_6CoCu/C nanoparticles were prepared by a spontaneous displacement reaction. Fifty milligrams of the as-prepared carbon-supported Pd_6CoCu/C nanoparticles were suspended in 10 ml of 0.1 mM $NaAuCl_4$ solution. One monolayer of Au (calculated from stoichiometric ratios) was deposited on the Pd_6CoCu/C nanoparticles surface (the atomic ratio of Pd to Au is ~100:1). After ultrasonic blending for 30 min, the suspension was heated to 60 °C under magnetic stirring and left to react for 5 h. The sample was then centrifuged and washed using deionized water until the pH value was close to 7, and finally dried in a vacuum oven overnight.

Characterization. The catalysts were characterized by using an X'Pert PRO diffractometer, and diffraction patterns were collected at a scanning rate of 4° per min. The STEM–EELS maps were acquired on a 5th order aberration-corrected STEM (Nion UltraSTEM) operated at 100 kV, with a convergence angle $\alpha_{max} = $ ~30 mrad. The STEM–EDX line profile was acquired on an aberration-corrected JEOL ARM200CF operated at 200 keV with a Large Angle SDD-EDX detector, DrySD100GV. The effective detection area is 100 mm^2. A nickel TEM grid was used to avoid background signals from Cu. The 2D STEM–EDX map was acquired on an aberration-corrected dedicated STEM (Hitachi HD2700C) operated at 200 keV with a Bruker SSD EDX detector. XPS data were obtained using an AXIS-ULTRA DLD-600 W Instrument.

Electrochemical testing. Electrochemical experiments were carried out in 0.1 M $HClO_4$ at room temperature using an Autolab electrochemistry station. Working electrodes were prepared by applying a thin catalyst film onto a glassy carbon electrode (GC, 5 mm in diameter). The catalyst ink was prepared by mixing 5 mg of

the catalyst with 1 ml of Nafion (0.05wt% Nafion dissolved in ethanol) solution. The mixture was sonicated and about 5.0 µl were applied onto a glassy carbon disk. After solvent evaporation, a thin layer of the Nafion-catalyst-Vulcan ink remained on the GC surface to serve as the working electrode. The Pd loading on the rotating disk electrode was calculated as 25.5 $µg_{Pd}$ per cm^2. A Pt wire was used as the counter electrode and a reversible hydrogen electrode, in the same electrolyte as the electrochemical cell, was used as the reference electrode. All potentials are referred to the reversible hydrogen electrode. The ORR polarization curves were obtained by sweeping the potential from + 0.20 to + 1.0 V at a scan rate of 5 mV s^{-1} and at a rotation rate of 1,600 r.p.m.

MEA preparation and electrochemical investigation in single cells. Membrane electrode assemblies (MEAs), with an active electrode area of 1 cm^2, were prepared using the gas-diffusion electrode method. Pt/C (Alfa Aesar, 40 wt%) was used as the anode and the as-prepared Pd_6CoCu/C and Au–Pd_6CoCu/C nanoparticles were used as the cathode catalysts. The catalyst loading was 0.2 mg_{Pt} per cm^2 on the anode and 0.25 mg_{Pd} per cm^2 on the cathode. The anode and cathode gas-diffusion electrodes were placed on the two sides of a Nafion 211 membrane (DuPont), and then were hot-pressed with a pressure of 0.5 MPa for 1 min at 140 °C to form an MEA. The MEA was inserted into a single-cell module for testing. The single cell was activated at a cell temperature of 80 °C. Fully humidified H_2 and O_2 were supplied to the anode and the cathode with a flow rate of 300 and 500 s.c.c.m., respectively. During the activation process, the current density was held constant for 30 min every time it reached 0.5, 1, 1.5 and 2 A cm^{-2}. Polarization curves were recorded in a galvanostatic mode (without ohmic; iR-correction) with a hold time of 3 min per point. To study the temperature effects on the cell performance, polarization curves were recorded at 25, 40, 60 and 80 °C. Durability tests were performed at a constant potential of 0.42 V for 100 h of polarization after conditioning as described above without interruption. The humidity and temperature were kept the same as during the polarization tests.

References

1. Strasser, P. *et al.* Lattice-strain control of the activity in dealloyed core-shell fuel cell catalysts. *Nat. Chem.* **2,** 454–460 (2010).
2. Greeley, J. *et al.* Alloys of platinum and early transition metals as oxygen reduction electrocatalysts. *Nat. Chem.* **1,** 552–556 (2009).
3. Cui, C. H. *et al.* Compositional segregation in shaped Pt alloy nanoparticles and their structural behaviour during electrocatalysis. *Nat. Mater.* **12,** 765–771 (2013).
4. Stamenkovic, V. R. *et al.* Trends in electrocatalysis on extended and nanoscale Pt-bimetallic alloy surfaces. *Nat. Mater.* **6,** 241–247 (2007).
5. Stamenkovic, V. R. *et al.* Improved oxygen reduction activity on Pt3Ni(111) via increased surface site availability. *Science* **315,** 493–497 (2007).
6. Golikand, A. N. *et al.* Electrocatalytic oxygen reduction on single-walled carbon nanotubes supported Pt alloys nanoparticles in acidic and alkaline conditions. *J. Appl. Electrochem.* **39,** 1369–1377 (2009).
7. Jiang, L. *et al.* Oxygen reduction reaction on carbon supported Pt and Pd in alkaline solutions. *J. Electrochem. Soc.* **156,** B370–B376 (2009).
8. Wang, D. L. *et al.* Structurally ordered intermetallic platinum-cobalt core-shell nanoparticles with enhanced activity and stability as oxygen reduction electrocatalysts. *Nat. Mater.* **12,** 81–87 (2013).
9. Zhou, W., Wu, J. B. & Yang, H. Highly uniform platinum icosahedra made by hot injection-assisted GRAILS method. *Nano Lett.* **13,** 2870–2874 (2013).
10. Yu, T. *et al.* Platinum concave nanocubes with high-index facets and their enhanced activity for oxygen reduction reaction. *Angew. Chem. Int. Ed. Engl.* **50,** 2773–2777 (2011).
11. Wang, D. L. *et al.* Tuning oxygen reduction reaction activity via controllable dealloying: a model study of ordered Cu3Pt/C intermetallic nanocatalysts. *Nano Lett.* **12,** 5230–5238 (2012).
12. Oezaslan, M., Heggen, M. & Strasser, P. Size-dependent morphology of dealloyed bimetallic catalysts: linking the nano to the macro scale. *J. Am. Chem. Soc.* **134,** 514–524 (2012).
13. Chen, C. *et al.* Highly crystalline multimetallic nanoframes with three-dimensional electrocatalytic surfaces. *Science* **343,** 1339–1343 (2014).
14. Wang, D. *et al.* Pt-decorated PdCo@Pd/C Core-shell nanoparticles with enhanced stability and electrocatalytic activity for the oxygen reduction reaction. *J. Am. Chem. Soc.* **132,** 17664–17666 (2010).
15. Zhou, W. P. *et al.* Improving electrocatalysts for O_2 reduction by fine-tuning the Pt-support interaction: Pt monolayer on the surfaces of a Pd3Fe(111) single-crystal alloy. *J. Am. Chem. Soc.* **131,** 12755–12762 (2009).
16. Zhang, Y. W. *et al.* Manageable N-doped graphene for high performance oxygen reduction reaction. *Sci. Rep.* **3,** 2771 (2013).
17. Yin, H. J. *et al.* Facile synthesis of surfactant-free Au cluster/graphene hybrids for high-performance oxygen reduction reaction. *Acs Nano* **6,** 8288–8297 (2012).

18. Shao, M. H. *et al.* Palladium monolayer and palladium alloy electrocatalysts for oxygen reduction. *Langmuir* **22**, 10409–10415 (2006).

19. Zhang, R. Z. & Chen, W. Non-precious Ir-V bimetallic nanoclusters assembled on reduced graphene nanosheets as catalysts for the oxygen reduction reaction. *J. Mater. Chem. A* **1**, 11457–11464 (2013).

20. Li, Y. G. *et al.* An oxygen reduction electrocatalyst based on carbon nanotube-graphene complexes. *Nat. Nanotechnol.* **7**, 394–400 (2012).

21. Liang, Y. Y. *et al.* Co_3O_4 nanocrystals on graphene as a synergistic catalyst for oxygen reduction reaction. *Nat. Mater.* **10**, 780–786 (2011).

22. Bashyam, R. & Zelenay, P. A class of non-precious metal composite catalysts for fuel cells. *Nature* **443**, 63–66 (2006).

23. Cheng, F. Y. *et al.* Rapid room-temperature synthesis of nanocrystalline spinels as oxygen reduction and evolution electrocatalysts. *Nat. Chem.* **3**, 79–84 (2011).

24. Gong, K. P. *et al.* Nitrogen-doped carbon nanotube arrays with high electrocatalytic activity for oxygen reduction. *Science* **323**, 760–764 (2009).

25. Shao, M. H., Sasaki, K. & Adzic, R. R. Pd-Fe nanoparticles as electrocatalysts for oxygen reduction. *J. Am. Chem. Soc.* **128**, 3526–3527 (2006).

26. Li, B. & Prakash, J. Oxygen reduction reaction on carbon supported Palladium-Nickel alloys in alkaline media. *Electrochem. Commun.* **11**, 1162–1165 (2009).

27. Shao, M. *et al.* Pt mono layer on porous Pd-Cu alloys as oxygen reduction electrocatalysts. *J. Am. Chem. Soc.* **132**, 9253–9255 (2010).

28. Fernandez, J. L. *et al.* Pd-Ti and Pd-Co-Au electrocatalysts as a replacement for platinum for oxygen reduction in proton exchange membrane fuel cells. *J. Am. Chem. Soc.* **127**, 13100–13101 (2005).

29. Sasaki, K. *et al.* Highly stable Pt monolayer on PdAu nanoparticle electrocatalysts for the oxygen reduction reaction. *Nat. Commun.* **3**, 1115–1124 (2012).

30. Wang, D. L. *et al.* Facile synthesis of carbon-supported Pd-Co core-shell nanoparticles as oxygen reduction electrocatalysts and their enhanced activity and stability with monolayer Pt decoration. *Chem. Mater.* **24**, 2274–2281 (2012).

31. Ruban, A. V., Skriver, H. L. & Norskov, J. K. Surface segregation energies in transition-metal alloys. *Phys. Rev. B* **59**, 15990–16000 (1999).

32. Wang, L.-L. & Johnson, D. D. Predicted trends of core — shell preferences for 132 late transition-metal binary-alloy nanoparticles. *J. Am. Chem. Soc.* **131**, 14023–14029 (2009).

33. Yin, Y. D. *et al.* Formation of hollow nanocrystals through the nanoscale Kirkendall Effect. *Science* **304**, 711–714 (2004).

34. Zhang, J. *et al.* Stabilization of platinum oxygen-reduction electrocatalysts using gold clusters. *Science* **315**, 220–222 (2007).

35. Kang, Y. *et al.* Multimetallic core/interlayer/shell nanostructures as advanced electrocatalysts. *Nano Lett.* **14**, 6361–6367 (2014).

36. Kuttiyiel, K. A. *et al.* Gold-promoted structurally ordered intermetallic palladium cobalt nanoparticles for the oxygen reduction reaction. *Nat. Commun.* **5**, 5185–5194 (2014).

37. Zhang, P. & Sham, T. K. Tuning the electronic behavior of Au nanoparticles with capping molecules. *Appl. Phys. Lett.* **81**, 736–738 (2002).

38. Zhang, P. *et al.* Organosulfur-functionalized Au, Pd, and Au-Pd nanoparticles on 1D silicon nanowire substrates: Preparation and XAFS studies. *Langmuir* **21**, 8502–8508 (2005).

39. Kuhn, M. & Sham, T. K. Charge redistribution and electronic behavior in a series of Au-Cu alloys. *Phys. Rev. B* **49**, 1647–1661 (1994).

40. Lee, Y. S. *et al.* Charge redistribution and electronic behavior in Pd-Au alloys. *J. Korean Phys. Soc.* **37**, 451–455 (2000).

Acknowledgements

This work was supported by the National Science Foundation of China (21306060, 21573083), the Program for New Century Excellent Talents in Universities of China (NCET-13-0237) and the Doctoral Fund of Ministry of Education of China (20130142120039). This research used resources of the Center for Functional Nanomaterials, Brookhaven National Laboratory, which is supported by the U.S. Department of Energy, Office of Basic Energy Sciences, under Contract No. DE-SC0012704. Experimental data recording for Fig. 2 was supported by the Energy Materials Center at Cornell (emc^2), an Energy Frontier Research Center funded by the U.S. Department of Energy, Office of Basic Energy Sciences, under Award No. DE-SC0001086. This work also made use of the facility of the Cornell Center for Materials Research (CCMR) with support from the National Science Foundation Materials Research Science and Engineering Centers (MRSEC) program (Contract No. DMR 1120296). This work also made use of the Cornell High Energy Synchrotron Source (CHESS) which is supported by the National Science Foundation and the National Institutes of Health/National Institute of General Medical Sciences under NSF award DMR-0936384. Some preliminary study was performed at Stanford Synchrotron Radiation Lightsource, a Directorate of SLAC National Accelerator Laboratory supported by the U.S. Department of Energy, Office of Science, Office of Basic Energy Sciences under Contract No. DE-AC02-76SF00515. The authors are grateful to Dr Jianguo Liu from Nanjing University for helping doing the fuel-cell performance and useful discussion and analysis of the data. D. Wang would like to thank Analytical and Testing Center of Huazhong University of Science & Technology for allowing us to use its facilities.

Author contributions

D.W., H.L.X. and H.D.A conceived the idea. All authors contributed to the design of the experiments. D.W., S.L and J. W. performed the synthesis and electrochemical tests. R.L., M.K. and H.L.X. performed TEM measurements and analysis. M.A.L., F.L., D.N. performed the synchrotron X-ray measurements and analysis. All authors discussed the results and implications at all stages. D.W., H.L.X. and H.D.A. wrote the paper with assistance from all authors.

Additional information

23

Rational design of efficient electrode–electrolyte interfaces for solid-state energy storage using ion soft landing

Venkateshkumar Prabhakaran[1], B. Layla Mehdi[1], Jeffrey J. Ditto[2], Mark H. Engelhard[3], Bingbing Wang[3,†], K. Don D. Gunaratne[1], David C. Johnson[2], Nigel D. Browning[1], Grant E. Johnson[1] & Julia Laskin[1]

The rational design of improved electrode–electrolyte interfaces (EEI) for energy storage is critically dependent on a molecular-level understanding of ionic interactions and nanoscale phenomena. The presence of non-redox active species at EEI has been shown to strongly influence Faradaic efficiency and long-term operational stability during energy storage processes. Herein, we achieve substantially higher performance and long-term stability of EEI prepared with highly dispersed discrete redox-active cluster anions (50 ng of pure ∼0.75 nm size molybdenum polyoxometalate (POM) anions on 25 μg (∼0.2 wt%) carbon nanotube (CNT) electrodes) by complete elimination of strongly coordinating non-redox species through ion soft landing (SL). Electron microscopy provides atomically resolved images of a uniform distribution of individual POM species soft landed directly on complex technologically relevant CNT electrodes. In this context, SL is established as a versatile approach for the controlled design of novel surfaces for both fundamental and applied research in energy storage.

[1] Physical Sciences Division, Pacific Northwest National Laboratory, PO Box 999, MSIN K8-88, Richland, Washington 99352, USA. [2] Department of Chemistry, University of Oregon, Eugene, Oregon 97403, USA. [3] Environmental Molecular Sciences Laboratory, Pacific Northwest National Laboratory, Richland, Washington 99352, USA. † Present address: State Key Laboratory of Marine and Environmental Science and College of Ocean and Earth Sciences, Xiamen University, Xiamen 361102, China. Correspondence and requests for materials should be addressed to J.L. (email: Julia.Laskin@pnnl.gov).

The development of improved materials for efficient energy storage is at the forefront of both fundamental and applied research in energy technology. Substantial progress both in understanding underlying phenomena and device fabrication is required to increase the charge-discharge capacity, rate and life cycle of energy storage devices while decreasing the cost associated with their manufacture[1,2]. Hence, the race to develop high performance, ecofriendly, low-cost energy storage devices is progressing along various parallel tracks. One familiar track is the development and optimization of the hybrid battery which is usually comprised of lithium ion batteries (Li-ion) coupled with supercapacitors[1,3–5]. The main challenge with Li-ion batteries is to provide improved power density and fast charge rates at a reduced price per kWh^{-1}, whereas supercapacitors need improvement in their energy densities so that the latter can respond to the coupled battery during combined operation[5–9]. In the past decade, both the power and energy density of supercapacitors have been significantly improved by combining non-Faradaic electrochemical double-layer capacitance and Faradaic pseudocapacitance[10,11] through a combination of high surface area carbon and metal oxides integrated at a molecular level[12,13]. Redox-supercapacitor electrodes have been fabricated by the direct transfer of redox pseudocapacitive materials from the solution phase onto electrodes using direct painting[14], ambient air spray[15], chemical vapour deposition[16], atomic layer deposition[17] and electrodeposition[18]. Electrodes prepared via such methods typically contain both electrochemically active components and inactive counter ions at the electrode–electrolyte interface (EEI). Polyoxometalates (POMs) are attractive redox species because of their stability, multi-electron redox activity[19,20] and strong interactions with electrodes[21–23]. Here, we demonstrate that fabrication of an energy storage device using uniformly distributed discrete POM anions without aggregate formation is an important breakthrough towards achieving superior performance not possible with conventional preparation techniques.

We employ soft landing (SL) of mass-selected ions for the preparation of well-defined electrode surfaces with an unprecedented level of control. SL enables the uniform deposition of ions with specific composition, charge state and kinetic energy[24–29] which is critical to gaining fundamental understanding of interfacial phenomena relevant to catalysis and materials science[27,30–33]. Previously, SL and reactive landing have been used to functionalize carbon nanotubes (CNTs), graphene, Si and Au surfaces to study electrochemical properties of well-defined deposited ions[34–38]. Notably, SL allows us to fabricate electrode surfaces for energy storage devices that are difficult to prepare using conventional solution- and vacuum-based deposition techniques. This provides a unique opportunity to design efficient and stable EEI by (i) studying the inherent activity and efficacy of precisely defined electro-active species; (ii) understanding the effect of electrochemically inactive counter-ions at the EEI on overall device performance; and (iii) identifying pathways to enhance the overall efficacy and selectivity of reactions of interest.

We demonstrate that SL of only small amounts of redox-active species results in dramatically enhanced total specific capacitance and superior long-term stability of carbon-based electrodes. We also show that aggregation is largely responsible for the reduced performance of POM-based supercapacitors fabricated from solution.

Results

Role of ion soft landing in electrode fabrication. Using a specially designed high-flux ion SL instrument (see Supplementary Fig. 1)[39], we fabricated macroscopic energy storage devices and investigated the role of anion–cation interactions and aggregate formation on the performance of POM-based supercapacitors. The SL instrument is equipped with a dual ion funnel interface, which increases the deposition rate of ions (refer to Methods section for a detailed description of our SL instrument). Specifically, redox-supercapacitor devices were fabricated by immobilizing mass-selected $PMo_{12}O_{40}{}^{3-}$ onto CNT-coated carbon electrodes (CNT electrodes) and compared with a non-Faradaic CNT-based device. SL can be performed either in vacuum or at ambient conditions on a benchtop[29]. In this study, vacuum-based SL was used to achieve selective deposition of $PMo_{12}O_{40}{}^{3-}$ anions while excluding other charge states of POM (2- and 1-) produced by electrospray ionization (ESI) (see Supplementary Fig. 2). Predetermined amounts of POM were deposited onto CNT electrodes either using SL or ambient electrospray deposition (ESD) of $Na_3[PMo_{12}O_{40}]$, $(NH_4)_3[PMo_{12}O_{40}]$ or $H_3[PMo_{12}O_{40}]$ solutions (Fig. 1). In ESD, micron size charged droplets of solvent containing POM anions and counter-cations, are deposited onto electrodes with a set deposition rate controlled by the flow of POM solution. Therefore, the ions deposited using ESD are neither mass-selected nor charge-selected. Ambient ESD along with its analogue, ambient air spray, that does not use high voltage are commonly used fabrication methods for thin film deposition on surfaces and electrodes[40–45]. It has been demonstrated that deposition of small microdroplets by ESD produces more uniform films in comparison with bulk solution drop casting making it a preferred deposition method. Furthermore, the ability to control the amount of deposited material by ESD enables a quantitative comparison with SL.

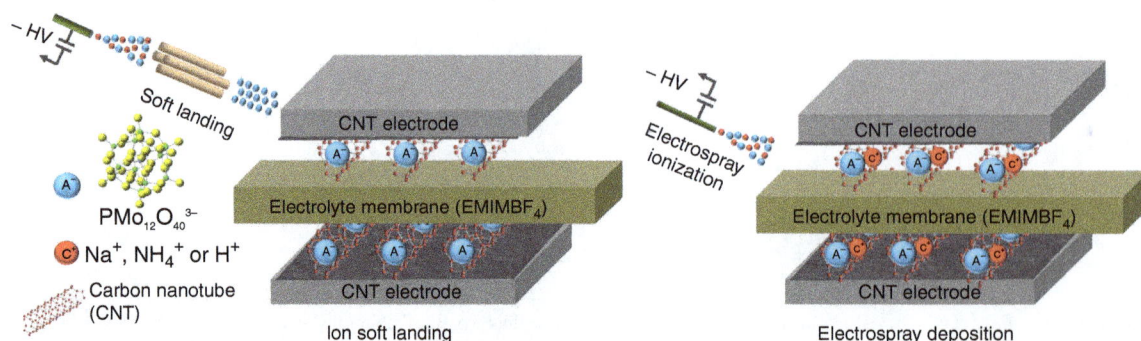

Figure 1 | Electrode–electrolyte interfaces of fabricated supercapacitor devices. Schematic representation of EEI of redox-supercapacitors fabricated using SL and ambient ESD. Note the absence of countercations in the device prepared by SL.

In addition, we fabricated supercapacitors by drop casting $Na_3[PMo_{12}O_{40}]$ solution (DRP-NaPOM-CNT) at different POM loadings and compared the total specific capacitance of these electrodes with ones prepared by ESD. Consistent with the literature[46–48], the results demonstrated the superior performance of ESD (see Supplementary Fig. 3) making it a suitable benchmarking technique for comparison with SL.

A total of five different electrode configurations were employed in this study: (i) pristine CNT electrode (pCNT); (ii) SL electrode containing a known amount of charge- and mass-selected $PMo_{12}O_{40}^{3-}$ (SL-CNT); (iii) ESD electrode with $Na_3[PMo_{12}O_{40}]$ (ESD-NaPOM-CNT); (iv) ESD electrode with $(NH_4)_3[PMo_{12}O_{40}]$ (ESD-NH$_4$POM-CNT); and (v) ESD electrode with $H_3[PMo_{12}O_{40}]$ (ESD-HPOM-CNT).

This study demonstrates that SL-CNT containing only a small amount of $PMo_{12}O_{40}^{3-}$ ($\sim$50 ng POM on $\sim$25 µg of the CNT material) have remarkably higher total specific capacitance and superior long-term stability compared with ESD electrodes. High-angle annular dark-field scanning transmission electron microscopy (HAADF-STEM) demonstrates that SL electrodes exhibit an extremely uniform and narrow distribution of discrete 0.75 nm diameter $[PMo_{12}O_{40}]$ clusters without any agglomeration. In contrast, the formation of aggregates and agglomerated POM in the presence of counter cations and solvent is observed on ESD electrodes. On the basis of these observations, we propose that aggregation is largely responsible for the reduced performance of POM-based supercapacitors fabricated from solution. Our results demonstrate that higher performance and longer stability can be achieved in redox-supercapacitors by uniform deposition of discrete redox-active species without strongly coordinating yet inactive counter ions or solvent at the EEI.

Electrochemical performance of fabricated devices. The supercapacitors were fabricated as described in the methods section using a 1-ethyl-3-methylimidazolium tetra fluoroborate (EMIMBF$_4$) ionic liquid membrane as a separator. The effect of the POM deposition technique and loading on total specific capacitance and stability of the fabricated devices was assessed. Our previous study showed that soft-landed $PMo_{12}O_{40}^{3-}$ remain intact and redox active on surfaces[49]. The redox activity of $PMo_{12}O_{40}^{3-}$ in the EMIMBF$_4$ ionic liquid was confirmed by examining the cyclic voltammograms (CVs) of different POM salts on glassy carbon (GC) electrodes (see Supplementary Fig. 4). A rectangular CV over different potential ranges was observed for supercapacitors fabricated using pCNT (see Supplementary Fig. 5) confirming the stability of the electrolyte membrane and the fabricated supercapacitor over a technologically relevant potential range[50–52].

The CVs of supercapacitors fabricated with pCNT, SL-CNT, ESD-NaPOM-CNT and ESD-NH$_4$POM-CNT (Fig. 2a) are rectangular in shape indicating that they exhibit an ideal capacitive-like behaviour. The respective galvanostatic charge-discharge (GCD) curves (Fig. 2b) are almost symmetrical triangles without any significant voltage drop related to internal resistance during the changing of polarity, suggesting fast transmission of ions at the EEI. However, ESD-HPOM-CNT showed neither rectangular CV (Fig. 2a) nor triangular GCD curves (Fig. 2b) which may be attributed to the presence of additional Faradaic reactions. The increase in current response closer to 1 V in the CV of ESD-HPOM-CNT indicates the presence of a side reaction such as oxygen evolution[53].

The GCD measurements were performed in triplicates and a total specific capacitance of 112 ± 12, 153 ± 8, 120 ± 16, 76 ± 16 and 128 ± 13 F g^{-1} and specific energy densities of 15.0 ± 1.7,

21.3 ± 1.1, 16.7 ± 2.2, 10.6 ± 2.2 and 17.75 ± 1.5 Wh kg^{-1} were obtained for the pCNT, SL-CNT, ESD-NaPOM-CNT, ESD-NH$_4$POM-CNT and ESD-HPOM-CNT, respectively, using equations 1 and 2. Similar values were also calculated from CV data using equation 3 indicating that the supercapacitors perform well both in constant potential (CV) and constant current mode (GCD). Of note, the total specific capacitance and energy density of pCNT are in agreement with literature values[54], which confirms the reliability of the baseline pCNT and the device fabrication method adopted in this study. The lack of dependence of the specific capacitance on the scan rates (see Supplementary Figs 6–9 and 11) indicates that, except for ESD-HPOM-CNT, the mobility of the ions and surface pseudocapacitance of unmodified and modified electrodes remain constant and stable over a wide range of charge-discharge rates of interest to energy storage applications. The ESD-HPOM-CNT showed an unstable capacitance over different scan rates (see Supplementary Figs 10 and 11), which may be attributed to the presence of additional Faradaic capacitance as discussed earlier.

The total specific capacitance and energy density of the SL-CNT (Fig. 2c) is $\sim$36, 27, 101 and 20% higher than the values obtained for pCNT, ESD-NaPOM-CNT, ESD-NH$_4$POM-CNT and ESD-HPOM-CNT, respectively. Therefore, the efficient participation of SL-$PMo_{12}O_{40}^{3-}$ during redox reaction in the absence of counter cations contributes to the enhanced capacitance of SL-CNT. The 30% decrease and the 7 and 14% increase in the total specific capacitance of ESD-NH$_4$POM-CNT, ESD-NaPOM-CNT and ESD-HPOM-CNT, respectively, in comparison with pCNT clearly indicate the important role of counter cations that are absent in SL-CNT. For example, NH_4^+ is known to strongly adsorb on carbon[55,56], which may decrease the electrochemically active surface area and reduce the non-Faradaic capacitance of the CNT surface along with the Faradaic capacitance of POM. Similar specific power densities of $\sim$4 kW kg^{-1} were obtained for both pristine and modified CNT electrodes using equation 4. However, the SL-CNT with a similar specific power density is characterized by a higher specific energy density demonstrating the improved performance of this electrode.

Effect of POM loading on specific capacitance. GCD measurements were also carried out with the SL-CNT, ESD-NaPOM-CNT and ESD-HPOM-CNT containing different loadings of POM (Fig. 2d). The ESD-NH$_4$POM-CNT was excluded in this further study as it showed lower capacitance than the pCNT. The SL-CNT achieved its maximum total specific capacitance of $\sim$160 F g^{-1} with 1.75×10^{13} POM ions. Meanwhile, the ESD-NaPOM-CNT required almost twice the amount of POM (3×10^{13} ions) and the ESD-HPOM-CNT required 2×10^{13} ions compared with the SL-CNT to achieve a similar maximum total specific capacitance. This demonstrates the higher efficacy of SL-CNT. For both SL and ESD electrodes, further addition of POM above the required maximum amount resulted in a decrease in the total specific capacitance.

It is remarkable that the maximum total capacitance was obtained only with 1.75×10^{13} SL-$PMo_{12}O_{40}^{3-}$ ($\sim$50 ng POM on 25 µg of the active CNT material) demonstrating the unexpectedly high contribution of nanogram quantities of pure $PMo_{12}O_{40}^{3-}$ to the total capacitance. The Faradaic component of the total specific capacitance (Fig. 2d) was calculated using equation 5. It is observed that the specific Faradaic capacitance of SL-CNT is greater than that of ESD-NaPOM-CNT and ESD-HPOM-CNT at all POM loadings. The observed decrease in capacitance at higher coverage may be attributed to formation

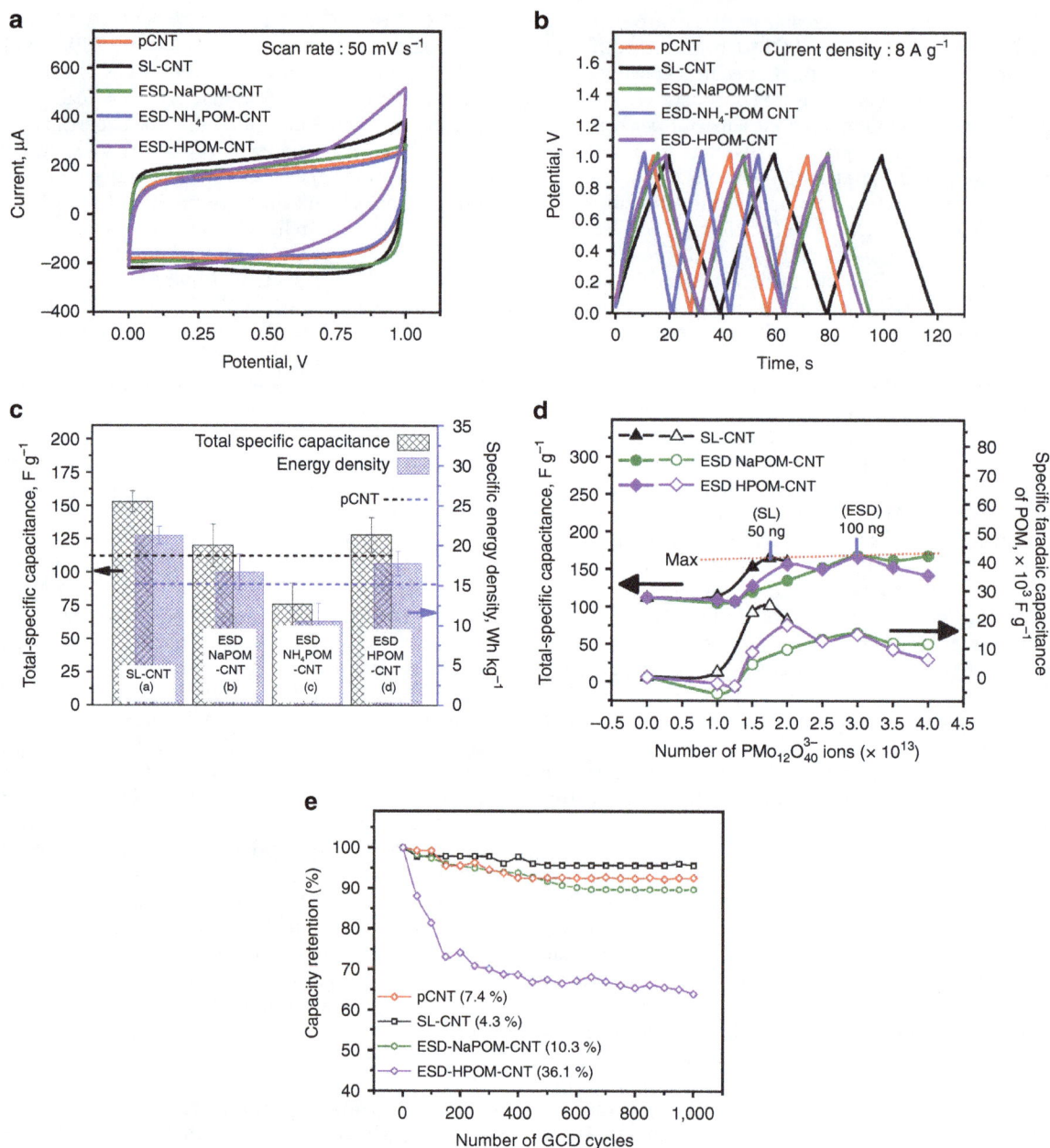

Figure 2 | Superior performance of SL supercapacitors. Characterization of supercapacitors fabricated with pCNT, SL-CNT, ESD-NaPOM-CNT, ESD-NH$_4$POM-CNT and ESD-HPOM-CNT and an EMIMBF$_4$ based membrane as a separator: (**a**) representative CV and (**b**) galvanostatic charge-discharge (GCD) curves. (**c**) Comparison of the total specific capacitance and specific energy density. Approximately 1.5×10^{13} PMo$_{12}$O$_{40}^{3-}$ were deposited in each case. (**d**) Effect of PMo$_{12}$O$_{40}^{3-}$ loading on the total specific capacitance (black) and specific Faradaic capacitance (green). (**e**) Capacity retention as a function of the number of GCD cycles. Current density: 8 A g^{-1}.

of additional POM layers in which underlying layers do not participate in redox activity.

The capacity retention (stability) of the pCNT, SL-CNT, ESD-NaPOM-CNT and ESD-HPOM-CNT was evaluated by performing 1,000 GCD cycles. After cycling (Fig. 2e), the capacitance of ESD-NaPOM-CNT decreased by 10.3 %, whereas only a 4.3% decrease was observed with SL-CNT indicating that SL-CNT have almost twice the lifetime of the ESD-NaPOM-CNT. The capacitance of pCNT was decreased by 7.4 % after cycling, which further confirmed the higher stability of SL-CNT. Interestingly, the ESD-HPOM-CNT showed ~36% decrease in capacitance after 1,000 GCD cycles. The lower stability of ESD-HPOM-CNT was attributed to degradation of EEI due to

the presence of side reactions as evidenced in the CVs (See Fig. 2a and Supplementary Fig. 10) or chemical degradation of EMIMBF$_4$ in presence of free H$^+$ ions[57]. However, the exact mechanism responsible for the lower stability of ESD-HPOM-CNT is not clear. Further characterization of the instability of ESD-HPOM-CNT is beyond the scope of this work. The poor capacity retention of ESD-HPOM-CNT made them unsuitable to compare with SL-CNT.

Physical and chemical properties of fabricated electrodes. The exceptional performance and superior long-term stability of SL-CNT may reflect the absence of interactions between

$PMo_{12}O_{40}^{3-}$ ions and their strongly coordinating counterions, which facilitates the uniform distribution of discrete $PMo_{12}O_{40}^{3-}$ on SL-CNT without agglomeration, or differences in the oxidation state of POM. To evaluate these possibilities, the distribution and chemical state of POM on CNT were analysed using HAADF-STEM, scanning electron microscopy (SEM) and X-ray photoelectron spectroscopy (XPS).

Considering the fact that the ESD-NaPOM-CNT showed a similar ideal GCD behaviour and redox activity of POM anions with respect to SL-CNT, it was selected as a reference to compare with SL-CNT. Therefore, HAADF-STEM was used to examine the morphology of pCNT, SL-CNT and ESD-NaPOM-CNT at the nanometre scale (See Fig. 3a–f and Supplementary Figs 12 and 13). It is remarkable that a uniform distribution of soft-landed POM was observed directly on the complex commercially relevant CNT electrodes. A uniform and narrow size distribution of POM was observed on SL-CNT (Fig. 3b,d). The diameter of individual clusters was measured to be $\sim 0.74 \pm 0.04$ nm (see Fig. 3f and Supplementary Figs 14 and 15) which matches the theoretical size calculated for $PMo_{12}O_{40}^{3-}$ (ref. 20). STEM images indicate that $PMo_{12}O_{40}^{3-}$ on SL-CNT are predominantly preserved as discrete ions without any significant agglomeration and the estimated uniform cluster coverage is 2.1×10^5 POM clusters per μm^2. Fig. 3f also shows the presence of a minor fraction of individual Mo atoms that are challenging to group as part of POM clusters (see Supplementary Discussion). In contrast, STEM images of ESD electrodes (see Fig. 3c,e,f and Supplementary Fig. 13) show features in the size range of 3–

10 nm with an average size of 3.66 ± 2.11 nm, indicating agglomeration of POM during ESD. Such aggregate formation on ESD electrodes may be attributed to the agglomeration of POM anions in the presence of counter cations (Na^+, NH_4^+ or H^+) as the solvent evaporates from the surface. The presence of nanoscopic aggregates consisting of low-conductivity cation-POM assemblies at the EEI may decrease the overall interfacial surface area and impart additional contact resistance, which would decrease the overall capacitance of the supercapacitor. Thus, STEM images clearly show the agglomeration of POM in the ESD electrodes and a uniform distribution of individual discrete $[PMo_{12}O_{40}]$ clusters in the SL electrodes, which explains both the lower performance of ESD-NaPOM-CNT and higher performance of SL-CNT.

In addition to STEM, we also used SEM to characterize the agglomeration of POM in ESD electrodes at the microscopic scale. SEM images of pCNT, SL-CNT, ESD-NaPOM-CNT, ESD-NH4POM-CNT and ESD-HPOM-CNT are presented in Fig. 4. The presence of CNTs on top of the bulk carbon fibres of the base carbon electrode is clearly visible in Fig. 4a. CNT deposition helps increase the overall electrode surface area and also lowers the contact resistance at the EEI. The diameter of the pristine CNTs is found to be ~ 15 nm. The SL-CNT shows similar morphology to the pristine CNT electrode and does not exhibit any characteristic presence of POM aggregates (see Fig. 4b). Given the extremely small size of individual POM clusters (diameter ~ 0.75 nm), it is not possible to image them using conventional SEM. In contrast, ESD-NaPOM-CNT, ESD-NH4POM-CNT and

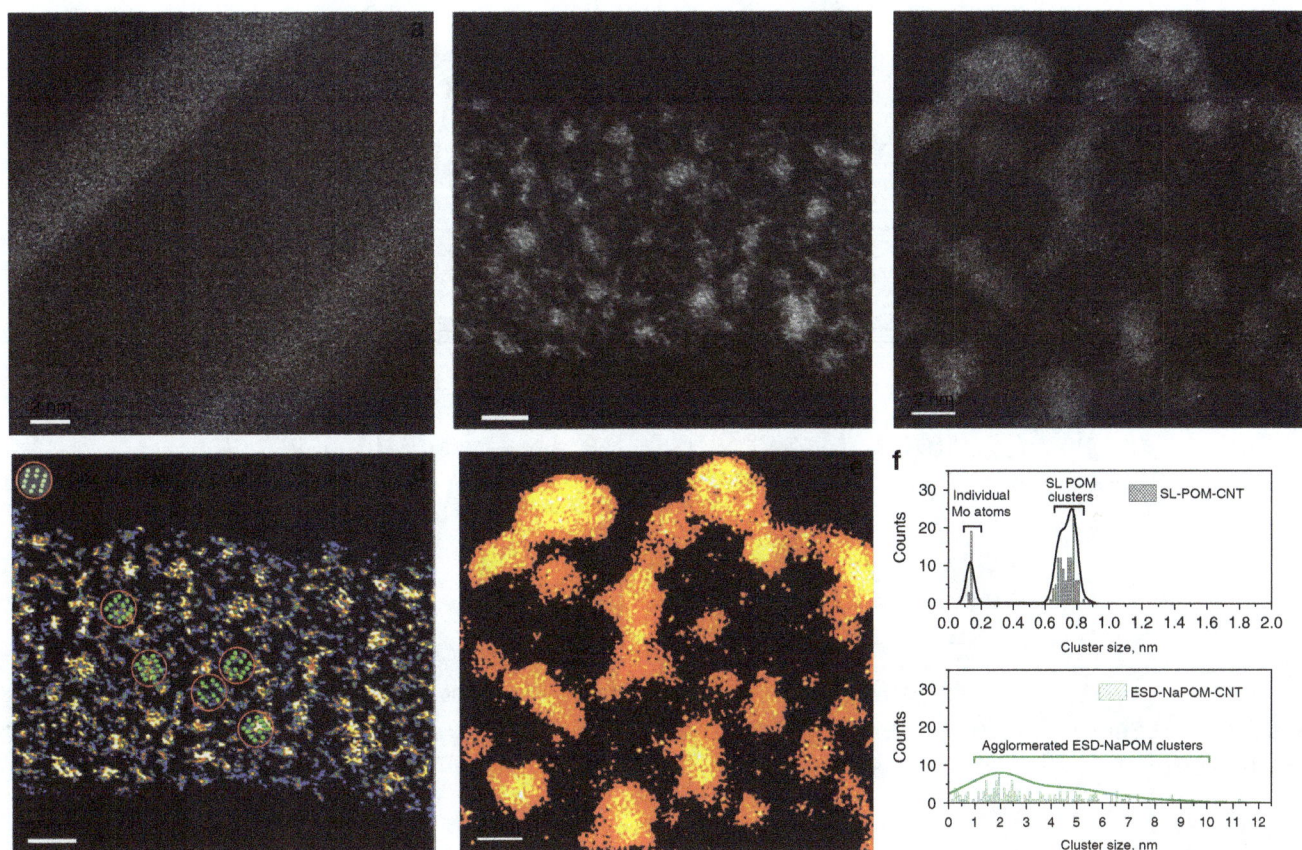

Figure 3 | Atomically resolved STEM images of monodisperse SL POM. (a–c) HAADF-STEM images of pCNT, SL-CNT and ESD-NaPOM-CNT, respectively. **(d,e)** The corresponding HAADF STEM images of SL-CNT and ESD-NaPOM-CNT, respectively, **(f)** Histogram of cluster size distribution in SL-CNT and ESD-NaPOM-CNT, respectively. Approximately 1.5×10^{13} $PMo_{12}O_{40}$ ions were loaded on the modified CNT electrodes. Selected examples of intact SL POM clusters were mapped in **d**—red circles. Note: The raw STEM images were processed using the Gatan Microscopy Suite DigitalMicrograph software to generate colour-enhanced Z-contrast images.

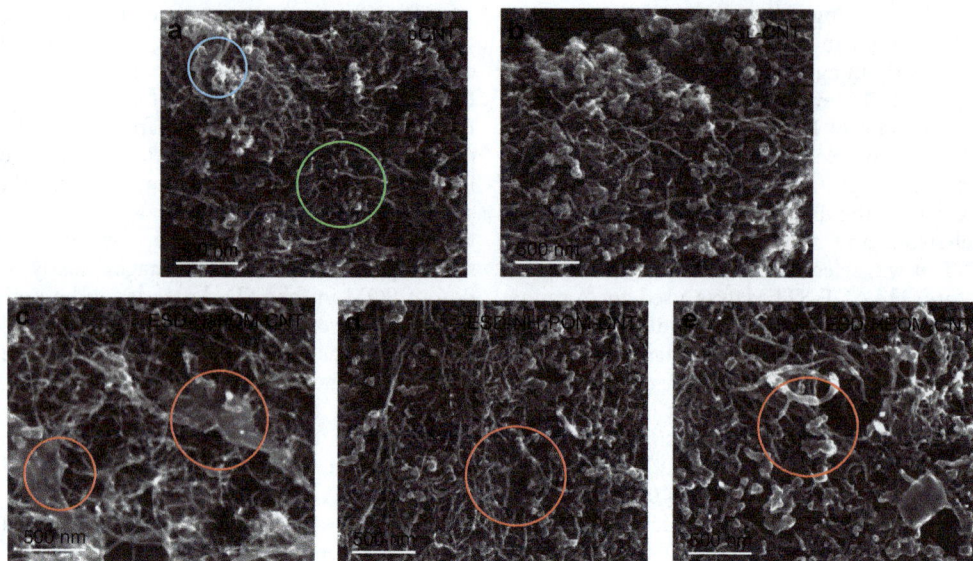

Figure 4 | POM aggregate formation at the microscopic scale. Scanning electron micrographs of (**a**) pCNT (**b**) SL-CNT (**c**) ESD-NaPOM-CNT (**d**) ESD-NH$_4$POM-CNT and (**e**) ESD-HPOM-CNT. Approximately 1.5×10^{13} PMo$_{12}$O$_{40}$ ions were loaded in the modified CNT electrodes. Green and blue circles in **a** represent the regions of CNT and carbon fibres, respectively. Red circles in **c–e** highlight microscopic aggregates observed on three ESD electrodes.

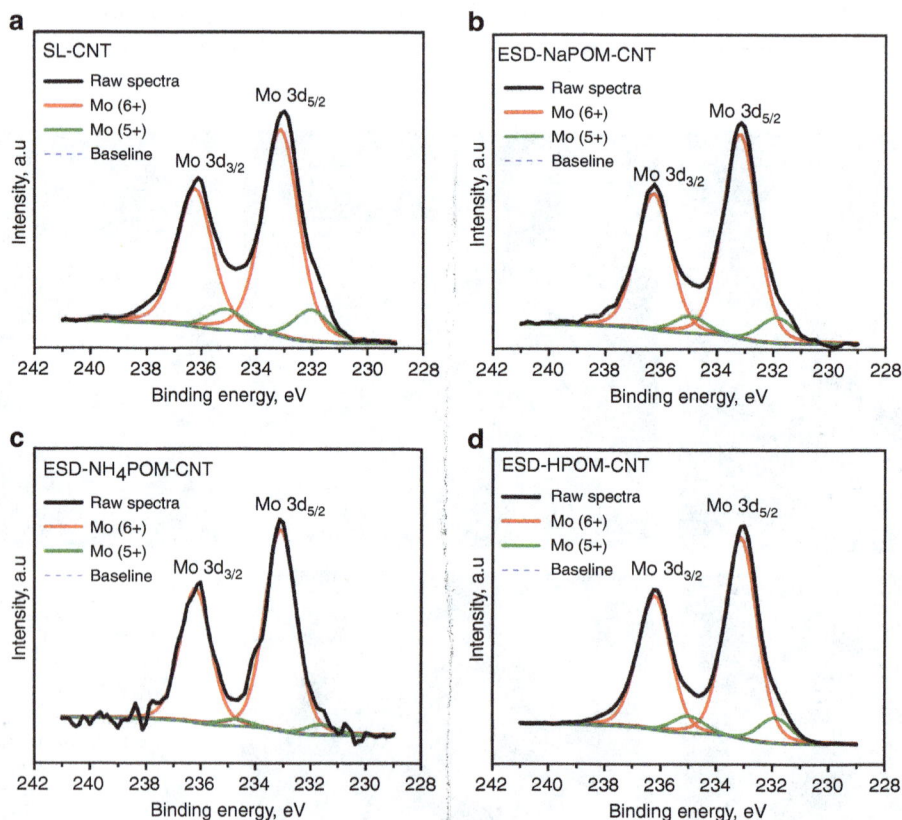

Figure 5 | Chemical state of SL and ESD electrodes. Mo 3d spectra of (**a**) SL-CNT (**b**) ESD-NaPOM-CNT (**c**) ESD-NH$_4$POM-CNT and (**d**) ESD-HPOM-CNT. Approximately 1.5×10^{13} POM ions were loaded in each case.

ESD-HPOM-CNT show the presence of distinct aggregates on the top layer. The diameter of the CNTs present in the ESD-NH$_4$POM-CNT and ESD-HPOM-CNT electrode are increased to ∼50 nm from ∼15 nm, indicating that the aggregates are preferentially formed around the outside walls of the CNTs. Formation of aggregates of POM clusters in the ESD electrodes observed with

SEM analysis is consistent with the agglomeration of POM seen in the STEM analysis. Again, such aggregate formation on the electrode surface post ESD may be attributed to the agglomeration of POM in the presence of its counter cations (Na$^+$, NH$_4^+$ and H$^+$) as the solvent evaporates. Aggregates were not observed on the SL electrode at similar coverage. Assuming that the aggregates

Table 1 | XPS analysis results of Mo 3d spectra obtained on the CNT-carbon electrodes modified with POM anions.

Binding energy (eV)	Oxidation state		Area %			
			SL-CNT	ESD-NaPOM-CNT	ESD-NH$_4$POM-CNT	ESD-HPOM-CNT
232.0	Mo 3d$_{5/2}$	5+	8.1	6.7	3.0	6.8
233.1		6+	52.0	53.4	57.0	53.4
235.1	Mo 3d$_{3/2}$	5+	5.4	4.4	2.0	4.5
236.2		6+	34.6	35.6	38.0	35.5
	FWHM (eV)		1.6 eV	1.4 eV	1.3 eV	1.4 eV

CNT, carbon nanotube; ESD, electrospray deposition; FWHM, full-width at half-maximum; POM, polyoxometalate; SL, soft landing; XPS, X-ray photoelectron spectroscopy.

are composed of POM salt with lower electrical conductivity[58,59], the presence of the microscopic aggregates at the EEI may significantly decrease the overall interfacial surface area and impart additional contact resistance, which would decrease the overall capacitance of the device.

We also used XPS to determine the oxidation state of Mo in SL and ESD electrodes. Figure 5a–c shows Mo 3d XPS spectra of SL- CNT, ESD-NaPOM-CNT, ESD-NH$_4$POM-CNT and ESD-HPOM-CNT. All Mo 3d XPS spectra exhibit the characteristic 3d$_{5/2}$ and 3d$_{3/2}$ doublet caused by spin–orbit coupling of the Mo 3d orbitals. Deconvolution of all Mo 3d XPS spectra with a fixed intensity area ratio of 2:3 (corresponding to d orbital[60,61]) reveals two peaks for each Mo 3d$_{5/2}$ and Mo 3d$_{3/2}$ spin–orbit coupling. The peaks located at 232.0 eV/235.1 eV and 233.1 eV/236.2 eV in Mo (3d$_{5/2}$/3d$_{3/2}$) spectra correspond to Mo^{5+} and Mo^{6+}, respectively, which is in agreement with literature values[62-64]. In addition, the characteristic binding energy separation (Δ Mo 3d) between the Mo 3d$_{5/2}$ and 3d$_{3/2}$ doublet of Mo^{5+} and Mo^{6+} is ~3.1 eV as observed previously in the literature for Mo^{5+} and Mo^{6+} ions[65]. These XPS observations show that similar oxidation states of POM clusters on CNT are present following SL and ESD. The calculated area % of different Mo chemical states post XPS analysis of all Mo 3d spectra are reported in Table 1. The area % of Mo^{5+} and Mo^{6+} in both Mo 3d$_{5/2}$ and 3d$_{3/2}$ of the SL-CNT and ESD-NaPOM-CNT are similar indicating that the charge state of Mo is not affected in both cases. However, a slightly higher abundance of Mo^{6+} seen in the ESD-NH$_4$POM-CNT may be due to charge (e$^-$) transfer from POM to NH$_4^+$ (NH$_4^+$ is a Lewis acid). In addition, the C 1s spectra of different CNT electrodes were also examined, but no significant changes were observed (see Supplementary Fig. 16).

The preceding binding energy and area calculations do not reveal any substantial changes in the oxidation state of Mo in SL and ESD electrodes; however, the widths of the peaks observed on ESD and SL electrodes are noticeably different. Specifically, the full-width at half-maximum (FWHM) of peaks assigned to the 5+ and 6+ oxidation states of Mo in SL-CNT, ESD-NaPOM-CNT, ESD-NH$_4$POM-CNT and ESD-HPOM-CNT are 1.6, 1.4, 1.3 and 1.4 eV, respectively. The Mo 3d$_{5/2}$ and 3d$_{3/2}$ lines originating from SL POM ions are broader than those observed on both ESD electrodes. It is reasonable to attribute the increased broadening in SL electrodes to the presence of electronic interactions between [PMo$_{12}$O$_{40}$] clusters and the CNT support for the isolated SL clusters and lower crystallinity of the material prepared by SL compared with ESD[66-68]. In other words, the XPS analysis (see Fig. 5 and Supplementary Fig. 15) of Mo 3d and C 1s peaks reveals no significant change in the oxidation state of Mo and C between SL and ESD electrodes but indicates a higher crystallinity of POM on ESD electrodes which may be caused by agglomeration. This provides further evidence, in addition to the STEM and SEM images, for the surface immobilization of discrete POM anions without counter ions or solvent using SL.

Discussion

Collectively, evidence obtained using HAADF-STEM, SEM and XPS indicates that the uniform distribution of POM, lack of agglomeration and absence of counter ions on the electrode are responsible for the exceptionally high performance of SL-CNT. Such uniform distribution of active material on the support enables efficient access to redox active species during GCD operation and improves the overall efficacy and stability of the redox-supercapacitors. In contrast, ESD results in formation of a less active, agglomerated phase that shows reduced redox activity in comparison with isolated clusters prepared by SL. Once the agglomerated phase has formed, repartitioning of individual clusters into the ionic phase in a solid state electrolyte is likely inhibited. Our results demonstrate that higher performance and longer stability can be achieved in redox-supercapacitors by uniform deposition of discrete redox active species at the EEI.

In summary, ion SL enabled the first fabrication of high-density CNT electrodes containing monodisperse POM anions and direct atomically resolved imaging of the uniform distribution of individual redox-active species on complex commercially relevant electrodes using STEM. We present evidence that agglomeration of active species due to interaction with strongly coordinating counterions is one of the key factors affecting the performance and stability of solid-state electrochemical energy storage devices such as batteries and supercapacitors. This work clearly demonstrates, for the first time, that elimination of these interactions through mass-selection and uniform deposition of small amounts of intact active species on CNT electrodes substantially improves device performance. In addition to energy storage, SL deposition can be extended to design efficient energy conversion devices where improved charge transfer across active layers[69] is desired. In this context, SL is established as a breakthrough deposition technology that enables the rational design of efficient EEI for energy storage applications through fundamental understanding of the key limiting factors.

Methods

Electrode preparation. The CNT-coated carbon fibre paper was obtained from SGL carbon GmbH (Meitingen, Germany) (SIGRACET) and cut into 1 cm^2 pieces that were directly used in this study. Hereafter, the 1 cm^2 pieces of carbon fibre paper will be referred to as CNT electrode. The thickness of CNT electrode is ~300 µm and the surface loading of CNT on the electrode is ~25 µg cm^{-2}. It is assumed that only the CNT present on the top layer contribute to the non-Faradaic double layer capacitance of the surface supercapacitor fabricated in this work and the minor contribution from the layer behind the electrode interface is disregarded. Therefore, the combined weight of the CNT loading (~25 µg cm^{-2}) and immobilized [PMo$_{12}$O$_{40}$] were used as a total weight of the active material to calculate the total specific capacitance of the supercapacitor.

Soft landing of mass-selected PMo$_{12}$O$_{40}^{3-}$ anions. The PMo$_{12}$O$_{40}^{3-}$ anions were soft-landed onto CNT electrodes using a custom-designed ion SL instrument described in detail in previous studies[39,49]. See Supplementary Fig. 1. The ion SL instrument is equipped with an ESI source, a dual ion funnel, an RF-only collision quadrupole, a quadrupole mass filter (Extrel CMS, Pittsburgh, PA), and three einzel lenses that focus the ion beam onto a deposition target. A 150 µM

$Na_3[PMo_{12}O_{40}] \times H_2O$ solution in methanol is introduced at a flow rate of $65\,\mu l\,h^{-1}$ to the ESI emitter. The charged droplets produced using ESI are transferred into the vacuum system using a heated stainless steel inlet and two electrodynamic ion funnels. The desolvated ions from the funnel are then collimated by colliding with the background gas in an RF-only quadrupole (collisional quadrupole). Subsequently, the ions are mass-filtered to allow through only $PMo_{12}O_{40}^{3-}$ anions ($m/z = 607$; $m =$ mass and $z =$ ionic charge) using a resolving quadrupole. The collisional quadrupole and resolving quadrupole are maintained at pressures of 2×10^{-2} and at 8×10^{-5} Torr, respectively. The mass-selected ions are then refocused with three einzel lenses in series before SL on the CNT electrode, which is mounted inside the vacuum deposition system. The collision energy of the soft-landed ions, determined by the difference between the CQ DC potential and the surface, was in a range of $30 - 35$ eV per charge translating to an ion kinetic energy of ~ 90 eV for $PMo_{12}O_{40}^{3-}$. An ion current of ~ 2 nA on the surface was measured with a picoammeter (model 9103, RBD Instruments, Bend, OR) and remained steady throughout the deposition. The total number of ions deposited was calculated using the measured ion current over time. The ion beam produced in the ion SL instrument is circular in shape and ~ 3 mm in diameter. Therefore, the total deposition area is $\sim 7\,mm^2$ on $100\,mm^2$ CNT electrode.

Ambient electrospray of $PMo_{12}O_{40}$. Electrodes were also prepared by direct ambient ESD of $150\,\mu M$ solutions of three different POMs ($Na_3[PMo_{12}O_{40}]$, $(NH_4)_3[PMo_{12}O_{40}]$ and $H_3[PMo_{12}O_{40}]$) with a fixed flow rate of $20\,\mu l\,h^{-1}$ for a specific time period to deliver a predetermined amount of POM onto the CNT electrodes. POM solution was delivered to the ESD emitter through a fused silica capillary (inner diameter (ID): $100\,\mu m$, outer diameter (OD): $360\,\mu m$, Polymicro Technologies, Phoenix, AZ) using a syringe pump (KD Scientific, Holliston, MA). The emitter was produced by pulling a $500\,\mu m$ OD four bore capillary (VitroCom, NJ, USA) to a final OD of $130\,\mu m$ using a micropipette puller (P-2000, Sutter Instrument Company, Novato, CA). The ESD emitter was connected to the fused silica capillary using a microtight union (Upchurch Scientific). A voltage of ~ -2.5 kV was applied to the emitter to generate charged droplets. The electrode was positioned 3 mm away from the ESD emitter. It should be noted that ESD electrodes contain both anions ($[PMo_{12}O_{40}]^{n-}$, $n = 2,3$), counter cations (Na^+, NH_4^+, H^+) and solvent molecules while the SL electrodes contain only mass-selected $[PMo_{12}O_{40}]^{3-}$ anions. See Fig. 1. The diameter of the deposition area is ~ 3 mm. We found that the performance of the supercapacitor does not change with respect to the deposition spot size. However, we maintained the same deposition area for both SL and ESD during electrode preparation. The total number of ions deposited over time during ESD was calculated based on the flow rate and concentration of the POM solution.

Preparation of electrolyte membrane.

An ionic liquid-based electrolyte membrane was prepared using $EMIMBF_4$ and copolymer poly(vinylidene fluoride-co-hexafluoropropylene) (PVDF-HFP). A typical preparation method is as follows: 2 g of PVDF-HFP was dissolved in 13 ml of DMF and stirred overnight at room temperature to make a homogenous solution. Two millilitre of $EMIMBF_4$ ionic liquid was then added to the PVDF-HFP/DMF solution and stirred continuously for 4 h. Finally, 15 ml of $EMIMBF_4$/PVDF-HFP solution was cast onto a 9 cm diameter petri dish and dried it in the oven at 70 °C for 12 h. The resulting membrane was peeled off and used as-prepared. See Supplementary Methods for detailed specifications of all chemicals used in this study.

Fabrication and testing of supercapacitor devices.

The supercapacitor devices were fabricated by placing an $EMIMBF_4$/PVDF-HFP-based electrolyte membrane between two similar 1 cm × 1 cm as-prepared electrodes. The electrode assembly was gently pressed at 100 psi using a mechanical press (International Crystal Laboratories, Garfield, NJ) for 5 min and assembled in a two electrode testing cell specially designed for performing electrochemical measurements such as CV and GCD.

The GCD characteristics and stability of the supercapacitor devices were evaluated using a VersaSTAT 3 potentiostat/galvanostat (Princeton Applied Research, Oak Ridge, TN).The GCD experiments were carried out in the voltage range of 0–1 V at a current density of 8 A g^{-1}—typical operating conditions used to study the performance of supercapacitors. Subsequently, the stability test was carried out by running the GCD experiment for 1,000 charge-discharge cycles.

Before performing GCD experiments, the intrinsic redox activity of $PMo_{12}O_{40}$ anions in $EMIMBF_4$ was evaluated using CV to make sure that $PMo_{12}O_{40}$ anions are redox active in $EMIMBF_4$ electrolyte. A CV was obtained on a GC working electrode containing 1×10^{14} $Na_3[PMo_{12}O_{40}]$ clusters deposited using ESD. Pristine $EMIMBF_4$ was used as an electrolyte; platinum and silver wires were used as a counter and pseudo-reference electrodes, respectively.

The total specific capacitance (that is, the sum of capacitance contributions from non-faradaic reactions on carbon and faradaic reactions using the redox active material) and energy density of the supercapacitor were calculated using the

following equations:

$$C_{sp(GCD),t} = \frac{i}{-\left(\frac{\Delta V}{\Delta t}\right).m} \tag{1}$$

$$E = \frac{C_{sp(GCD),t}.\Delta V^2}{2.\left(10^{-3}\frac{kg}{g}\right).(3600\,s/h)} \tag{2}$$

$$C_{sp(CV),t} = \frac{\int I dV}{v.m.V} \tag{3}$$

$$P = \frac{E}{\Delta t} \tag{4}$$

where, i—current density used in the GCD experiment (8 A g^{-1}); $\Delta V/\Delta t$—slope of the discharge curve (V s^{-1}); $C_{sp(GCD),t}$ and $C_{sp(CV),t}$- total specific capacitance of the supercapacitor calculated from GCD and CV experiments respectively; ΔV—voltage difference between the voltage at the beginning of discharging and the voltage at the end of discharge in GCD measurement; I and V—response current (A) and potential window (V) in CV measurement, respectively; v—scan rate in the CV measurement; m—total mass of the active material on the electrode surface; E—energy density of the supercapacitor (Wh kg^{-1}), P—Power density of the supercapacitor (kW kg^{-1}), Δt—discharge time (s).

To calculate the specific Faradaic capacitance of POM-based supercapacitors, it is assumed that only CNT present on the top layer contributes to the non-Faradaic double layer capacitance and the minor contribution from the layer behind the electrode interface is disregarded. Therefore, the combined weight of the CNT loading ($\sim 25\,\mu g\,cm^{-2}$) and the weight of immobilized $[PMo_{12}O_{40}]$ were used as a total weight of the active material to calculate the total specific capacitance of the supercapacitor using equation 5.

$$C_{sp,[PMo12]} = \frac{\left[(C_{sp,t}*W_{t,CNT+PMo12}) - (C_{sp,CNT}*W_{CNT})\right]}{W_{PMo12}} \tag{5}$$

$C_{sp,t}$—total specific capacitance (F/g) of modified CNT electrodes calculated using equation 1. $W_{t,CNT+PMo12}$—combined weight (g) of Faradaic and non-Faradaic active material (CNT loading $\sim 25\,\mu g$ and loading of $[PMo_{12}O_{40}]$ clusters in each case); $C_{sp,CNT}$—total specific capacitance (F/g) of pristine CNT (only CNT as an active material); W_{CNT}—weight (g) of CNT loading in a pristine CNT ($\sim 25\,\mu g$); $W_{PMo12}$$W_{PMo12}$—mass of $[PMo_{12}O_{40}]$ clusters at each loadings; $C_{sp,[PMo12]}$—specific Faradaic capacitance contributed from $[PMo_{12}O_{40}]$ clusters alone in the modified electrodes.

The oxidation state and surface characteristics of $[PMo_{12}O_{40}]$ clusters on the different CNT electrodes were examined using XPS, SEM and STEM so that structural characteristics could be correlated with the electrochemical activity.

Electrode characterization. *Scanning transmission electron microscopy.*

The size and distribution of $[PMo_{12}O_{40}]$ clusters on CNT electrodes were determined using STEM. In a typical sample preparation, a lift-out procedure was employed using a focused ion beam workstation, in which a portion of the CNT electrode was removed and transferred onto a copper TEM grid (Ted Pella, Inc., Redding, CA)[70]. STEM micrographs were then obtained using a FEI TITAN 80–300 eV TEM/STEM operated at 300 kV. The microscope is fitted with a spherical aberration corrector for the probe forming lens, enabling sub-Angstrom resolution in the STEM imaging mode. STEM analysis provides insight into the size and distribution of SL and ESD $[PMo_{12}O_{40}]$ clusters on CNT electrodes through Z-contrast imaging in the HAADF mode. Approximately 1.5×10^{13} $[PMo_{12}O_{40}]$ clusters were deposited using SL and ESD on CNT electrode and used for STEM analysis.

The STEM images were analysed as follows: the histograms of the cluster size distribution for SL and ESD samples were prepared with the Gatan Digital Micrograph software, which allows one to establish contrast threshold difference for Mo atoms. To analyse the STEM images of the SL sample (Fig. 3b), intact POM clusters and individual Mo atoms were identified and counted towards the generation of a histogram. A similar method was adopted for analysing images of ESD samples in Fig. 3c. These images were analysed by manually defining the features corresponding to both agglomerated POM clusters and individual Mo atoms and calculating the areas of these features. The average size of individual features shown in Fig. 3f was obtained assuming they have circular shapes.

Scanning electron microscopy. The surface morphology of different CNT electrodes was examined using a scanning electron microscope (Quanta 3D model, FEI, Inc.) operated at 10 kV accelerating voltage. Approximately 1.5×10^{13} $[PMo_{12}O_{40}]$ clusters were deposited using SL and ESD on CNT electrodes for SEM analysis.

X-ray photoelectron spectroscopy. The oxidation states of Mo on different CNT electrodes were evaluated using Mo 3d XPS spectra. XPS measurements were performed with a Physical Electronics Quantera Scanning X-ray Microprobe. This system uses a focused monochromatic Al Kα X-ray (1,486.7 eV) source for excitation and a spherical section analyser. The instrument has a 32 element multichannel detection system. A 100 W X-ray beam focused to 100 μm diameter was rastered over a 1.2 mm × 0.1 mm rectangular region on the sample. The X-ray beam is incident normal to the sample and the photoelectron detector is at 45° off-normal. High energy resolution spectra were collected using a pass-energy of

69.0 eV with a step size of 0.125 eV. For the Ag $3d_{5/2}$ line, these conditions produced a FWHM of 0.91 eV. Approximately 1.5×10^{13} [PMo$_{12}$O$_{40}$] clusters were deposited using SL and ESD on CNT electrode and used for XPS analysis. The XPS spectra were fitted using Multipak Spectrum software and the following peak fitting parameters were used. The intensity ratio of Mo $3d_{5/2}$ and Mo $3d_{3/2}$ peaks arising from spin–orbit coupling was fixed at 2:3 corresponding to d orbital and the spit–orbit splitting was fixed at 3.1 eV. The Shirley background subtraction method with 80% Gaussian, 20% Lorentz line shapes was employed to fit both Mo $3d$ and C $1s$ XPS spectra.

References

1. Goodenough, J. B. Electrochemical energy storage in a sustainable modern society. *Energy Environ. Sci.* **7**, 14–18 (2014).
2. Lavine, M., Szuromi, P. & Coontz, R. Electricity now and when. *Science* **334**, 921–921 (2011).
3. Suárez-Guevara, J., Ruiz, V. & Gomez-Romero, P. Hybrid energy storage: high voltage aqueous supercapacitors based on activated carbon–phosphotungstate hybrid materials. *J. Mater. Chem. A* **2**, 1014–1014 (2014).
4. Vlad, a. *et al.* Hybrid supercapacitor-battery materials for fast electrochemical charge storage. *Sci. Rep.* **4**, 4315–4315 (2014).
5. Choi, N. S. *et al.* Challenges facing lithium batteries and electrical double-layer capacitors. *Angew. Chem. Int. Ed. Engl.* **51**, 9994–10024 (2012).
6. Simon, P. & Gogotsi, Y. Materials for electrochemical capacitors. *Nat. Mater.* **7**, 845–854 (2008).
7. Armand, M. & Tarascon, J. M. Building better batteries. *Nature* **451**, 652–657 (2008).
8. Tarascon, J. M. & Armand, M. Issues and challenges facing rechargeable lithium batteries. *Nature* **414**, 359–367 (2001).
9. Goodenough, J. B. & Park, K. S. The Li-ion rechargeable battery: a perspective. *J. Am. Chem. Soc.* **135**, 1167–1176 (2013).
10. Wang, G., Zhang, L. & Zhang, J. A review of electrode materials for electrochemical supercapacitors. *Chem. Soc. Rev.* **41**, 797–828 (2012).
11. Simon, P., Gogotsi, Y. & Dunn, B. Where do batteries end and supercapacitors begin? *Science* **343**, 1210–1211 (2014).
12. Sathiya, M., Prakash, A. S., Ramesha, K., Tarascon, J. M. & Shukla, A. K. V2O5-anchored carbon nanotubes for enhanced electrochemical energy storage. *J. Am. Chem. Soc.* **133**, 16291–16299 (2011).
13. Gomez-Romero, P. Hybrid organic–inorganic materials—in search of synergic activity. *Adv. Mater.* **13**, 163–174 (2001).
14. Liu, Q., Nayfeh, M. H. & Yau, S. T. Brushed-on flexible supercapacitor sheets using a nanocomposite of polyaniline and carbon nanotubes. *J. Power Sources* **195**, 7480–7483 (2010).
15. Huang, C. & Grant, P. S. One-step spray processing of high power all-solid-state supercapacitors. *Sci. Rep.* **3**, 2393–2393 (2013).
16. Wang, X. *et al.* Chemical vapor deposition growth of crystalline monolayer MoSe2. *ACS Nano* **8**, 5125–5131 (2014).
17. Boukhalfa, S., Evanoff, K. & Yushin, G. Atomic layer deposition of vanadium oxide on carbon nanotubes for high-power supercapacitor electrodes. *Energy Environ. Sci.* **5**, 6872–6872 (2012).
18. Yang, J., Lian, L., Ruan, H., Xie, F. & Wei, M. Nanostructured porous MnO2 on Ni foam substrate with a high mass loading via a CV electrodeposition route for supercapacitor application. *Electrochim. Acta* **136**, 189–194 (2014).
19. Mattes, R. Heteropoly and isopoly oxometalates. Von M. T. Pope. Springer-Verlag, Berlin 1983. XIII, 180 S., geb. DM 124.00. *Angew. Chem. Int. Ed. Engl.* **96**, 730–730 (1984).
20. Wang, H. *et al.* In operando X-ray absorption fine structure studies of polyoxometalate molecular cluster batteries: polyoxometalates as electron sponges. *J. Am. Chem. Soc.* **134**, 4918–4924 (2012).
21. Kulesza, P. J. *et al.* Fabrication of network films of conducting polymer-linked polyoxometallate-stabilized carbon nanostructures. *Electrochim. Acta* **51**, 2373–2379 (2006).
22. Rong, C. & Anson, F. C. Spontaneous adsorption of heteropolytungstates and heteropolymolybdates on the surfaces of solid electrodes and the electrocatalytic activity of the adsorbed anions. *Inorg. Chim. Acta* **242**, 11–16 (1996).
23. Rausch, B., Symes, M. D., Chisholm, G. & Cronin, L. Decoupled catalytic hydrogen evolution from a molecular metal oxide redox mediator in water splitting. *Science* **345**, 1326–1330 (2014).
24. Franchetti, V., Solka, B. H., Baitinger, W. E., Amy, J. W. & Cooks, R. G. Soft landing of ions as a means of surface modification. *Int. J. Mass Spectrom. Ion Processes* **23**, 29–35 (1977).
25. Johnson, G. E. & Laskin, J. Soft landing of mass-selected gold clusters: Influence of ion and ligand on charge retention and reactivity. *Int. J. Mass Spectrom.* **377**, 205–213 (2015).
26. Johnson, G. E., Hu, Q. & Laskin, J. Soft landing of complex molecules on surfaces. *Annu. Rev. Anal. Chem.* **4**, 83–104 (2011).
27. Johnson, G. E., Gunaratne, D. & Laskin, J. Soft- and reactive landing of ions onto surfaces: concepts and applications. *Mass Spectrom. Rev.* **35**, 439–479 (2016).
28. Rauschenbach, S. *et al.* Electrospray ion beam deposition of clusters and biomolecules. *Small* **2**, 540–547 (2006).
29. Badu-Tawiah, A. K., Wu, C. & Cooks, R. G. Ambient ion soft landing. *Anal. Chem.* **83**, 2648–2654 (2011).
30. Verbeck, G., Hoffmann, W. & Walton, B. Soft-landing preparative mass spectrometry. *Analyst* **137**, 4393–4407 (2012).
31. Gologan, B., Green, J. R., Alvarez, J., Laskin, J. & Graham Cooks, R. Ion/surface reactions and ion soft-landing. *Phys. Chem. Chem. Phys.* **7**, 1490–1500 (2005).
32. Li, A. *et al.* Using ambient ion beams to write nanostructured patterns for surface enhanced raman spectroscopy. *Angew. Chem. Int. Ed.* **53**, 12528–12531 (2014).
33. Krásný, L. *et al.* In-situ enrichment of phosphopeptides on MALDI plates modified by ambient ion landing. *J. Mass Spectrom.* **47**, 1294–1302 (2012).
34. Evans, C. *et al.* Surface modification and patterning using low-energy ion beams: Si – O bond formation at the vacuum/adsorbate interface. *Anal. Chem.* **74**, 317–323 (2002).
35. Pepi, F. *et al.* Chemically modified multiwalled carbon nanotubes electrodes with ferrocene derivatives through reactive landing. *J. Phys. Chem. C* **115**, 4863–4871 (2011).
36. Mazzei, F., Favero, G., Frasconi, M., Tata, A. & Pepi, F. Electron-transfer kinetics of microperoxidase-11 covalently immobilised onto the surface of multi-walled carbon nanotubes by reactive landing of mass-selected ions. *Chemistry* **15**, 7359–7367 (2009).
37. Dubey, G. *et al.* Chemical modification of graphene via hyperthermal molecular reaction. *J. Am. Chem. Soc.* **136**, 13482–13485 (2014).
38. He, Q. *et al.* In Situ bioconjugation and ambient surface modification using reactive charged droplets. *Anal. Chem.* **87**, 3144–3148 (2015).
39. Gunaratne, K. D. D. *et al.* Design and performance of a high-flux electrospray ionization source for ion soft landing. *Analyst* **140**, 2957–2963 (2015).
40. Jaworek, A. Electrospray droplet sources for thin film deposition. *J. Mater. Sci.* **42**, 266–297 (2007).
41. Jaworek, A. & Sobczyk, A. T. Electrospraying route to nanotechnology: an overview. *J. Electrostat.* **66**, 197–219 (2008).
42. Morozov, V. N. & Morozova, T. Y. Electrospray deposition as a method to fabricate functionally active protein films. *Anal. Chem.* **71**, 1415–1420 (1999).
43. Morozov, V. N. & Morozova, T. Y. Electrospray deposition as a method for mass fabrication of mono- and multicomponent microarrays of biological and biologically active substances. *Analytical Chemistry* **71**, 3110–3117 (1999).
44. James, N. O. S. *et al.* Electrospray deposition of carbon nanotubes in vacuum. *Nanotechnology* **18**, 035707 (2007).
45. Saywell, A. *et al.* Electrospray deposition of C60 on a hydrogen-bonded supramolecular network. *J. Phys. Chem. C* **112**, 7706–7709 (2008).
46. Chaparro, A. M., Gallardo, B., Folgado, M. A., Martín, A. J. & Daza, L. PEMFC electrode preparation by electrospray: optimization of catalyst load and ionomer content. *Catal. Today* **143**, 237–241 (2009).
47. Beidaghi, M. & Gogotsi, Y. Capacitive energy storage in micro-scale devices: recent advances in design and fabrication of micro-supercapacitors. *Energy Environ. Sci.* **7**, 867–884 (2014).
48. Ervin, M. H., Miller, B. S., Hanrahan, B., Mailly, B. & Palacios, T. A comparison of single-wall carbon nanotube electrochemical capacitor electrode fabrication methods. *Electrochim. Acta* **65**, 37–43 (2012).
49. Gunaratne, K. D. D. *et al.* Controlling the charge state and redox properties of supported polyoxometalates via soft landing of mass-selected ions. *J. Phys. Chem. C* **118**, 27611–27622 (2014).
50. Conway, B. E. Transition from "supercapacitor" to "battery" behavior in electrochemical energy storage. *J. Electrochem. Soc.* **138**, 1539–1539 (1991).
51. Kowsari, E. *High-Performance Supercapacitors Based on Ionic Liquids and a Graphene Nanostructure, Ionic Liquids - Current State of the Art* (InTech, 2015).
52. Armand, M., Endres, F., MacFarlane, D. R., Ohno, H. & Scrosati, B. Ionic-liquid materials for the electrochemical challenges of the future. *Nat Mater* **8**, 621–629 (2009).
53. Bard, A. J. & Faulkner, L. R. *Electrochemical Methods: Fundamentals and Applications*, 2nd edn (Wiley, 2000).
54. Yang, X. *et al.* A high-performance graphene oxide-doped ion gel as gel polymer electrolyte for all-solid-state supercapacitor applications. *Adv. Funct. Mater.* **23**, 3353–3360 (2013).
55. Long, X.-L. *et al.* Adsorption of ammonia on activated carbon from aqueous solutions. *Environ. Prog.* **27**, 225–233 (2008).
56. Moradi, O. & Zare, K. Adsorption of ammonium ion by multi-walled carbon nanotube: kinetics and thermodynamic studies. *Fullerenes Nanotubes Carbon Nanostruct.* **21**, 449–459 (2012).
57. De Vos, N., Maton, C. & Stevens, C. V. Electrochemical stability of ionic liquids: general influences and degradation mechanisms. *ChemElectroChem* **1**, 1258–1270 (2014).
58. Coronado, E. & Gómez-García, C. J. Polyoxometalate-based molecular materials. *Chem. Rev.* **98**, 273–296 (1998).
59. Bellitto, C. *et al.* BEDT-TTF salts with.alpha.-Keggin poly(oxometallates): electrical, magnetic, and optical properties of (BEDT-TTF)8[PMo12O40] and (BEDT-TTF)8[SiW12O40] and X-ray crystal structure of (BEDT-TTF)8

[PMo12O40].cntdot.{(CH3CN.cntdot.H2O)2}. *Chem. Mater.* **7,** 1475–1484 (1995).

60. Moulder, J. F. & Chastain, J. *Handbook of X-ray photoelectron spectroscopy: a reference book of standard spectra for identification and interpretation of XPS data.* 6th edn (Physical Electronics Division, Perkin-Elmer Corp, 1992).

61. Taylor, A. in *Practical Surface Analysis* 2nd edn., Vol. I (eds Briggs, D. & Seah, M. P.) (John Wiley, New York, 1990).

62. Quincy, R. B., Houalla, M., Proctor, A. & Hercules, D. M. Distribution of molybdenum oxidation states in reduced molybdenum/titania catalysts: correlation with benzene hydrogenation activity. *J. Phys. Chem.* **94,** 1520–1526 (1990).

63. Nakayama, M., Ii, T. & Ogura, K. Spectrosopic and paramagnetic properties of molybdenum oxyhydroxide films electrochemically formed from the Keggin-type (PMo12O40)3 − complex. *J. Mater. Res.* **18,** 2509–2514 (2003).

64. Spevack, P. A. & McIntyre, N. S. A Raman and XPS investigation of supported molybdenum oxide thin films. 2. Reactions with hydrogen sulfide. *J. Phys. Chem.* **97,** 11031–11036 (1993).

65. Belanger, D. & Laperriere, G. Electrochromic molybdenum trioxide thin film preparation and characterization. *Chem. Mater.* **2,** 484–486 (1990).

66. Anwar, M., Hogarth, C. A. & Bulpett, R. Effect of substrate temperature and film thickness on the surface structure of some thin amorphous films of MoO3 studied by X-ray photoelectron spectroscopy (ESCA). *J. Mater. Sci.* **24,** 3087–3090 (1989).

67. Black, L., Garbev, K., Beuchle, G., Stemmermann, P. & Schild, D. X-ray photoelectron spectroscopic investigation of nanocrystalline calcium silicate hydrates synthesised by reactive milling. *Cement Concrete Res.* **36,** 1023–1031 (2006).

68. Li, Z., Gao, L. & Zheng, S. SEM, XPS, and FTIR studies of MoO3 dispersion on mesoporous silicate MCM-41 by calcination. *Mater. Lett.* **57,** 4605–4610 (2003).

69. Vasilopoulou, M., Douvas, A. M., Palilis, L. C., Kennou, S. & Argitis, P. Old metal oxide clusters in new applications: spontaneous reduction of Keggin and Dawson polyoxometalate layers by a metallic electrode for Improving efficiency in organic optoelectronics. *J Am. Chem. Soc.* **137,** 6844–6856 (2015).

70. Giannuzzi, L. A. & Stevie, F. A. A review of focused ion beam milling techniques for TEM specimen preparation. *Micron* **30,** 197–204 (1999).

Acknowledgements

This work was supported by the U.S. Department of Energy's (DOE) Office of Basic Energy Sciences, Division of Chemical Sciences, Geosciences & Biosciences and performed in EMSL, a national scientific user facility located at Pacific Northwest National Laboratory (PNNL). TEM work was supported by the Joint Center for Energy Storage Research (JCESR), an Energy Innovation Hub. PNNL is operated by Battelle for DOE. We thank Sigracet SGL carbon GmbH for providing the CNT substrates used in this study.

Author contributions

V.P., D.D.G., G.E.J. and J.L. conceived and designed the experiments. V.P. performed all the device fabrication and electrochemical experiments. B.L.M., J.J.D., D.C.J., N.D.B. performed STEM measurements; B.L.M. and V.P. analysed the STEM data. M.H.E. performed XPS measurements; V.P. and M.H.E. analysed the XPS data. B.W. performed SEM measurements; V.P. and B.W. analysed the SEM data. V.P., G.E.J. and J.L. co-wrote the paper. All authors contributed to the interpretation of the results and assisted in writing the paper.

Additional information

24

Li(V$_{0.5}$Ti$_{0.5}$)S$_2$ as a 1 V lithium intercalation electrode

Steve J. Clark[1,*], Da Wang[1,*], A. Robert Armstrong[2] & Peter G. Bruce[1]

Graphite, the dominant anode in rechargeable lithium batteries, operates at $\sim$0.1 V versus Li$^+$/Li and can result in lithium plating on the graphite surface, raising safety concerns. Titanates, for example, Li$_4$Ti$_5$O$_{12}$, intercalate lithium at $\sim$1.6 V versus Li$^+$/Li, avoiding problematic lithium plating at the expense of reduced cell voltage. There is interest in 1 V anodes, as this voltage is sufficiently high to avoid lithium plating while not significantly reducing cell potential. The sulfides, LiVS$_2$ and LiTiS$_2$, have been investigated as possible 1 V intercalation electrodes but suffer from capacity fading, large 1st cycle irreversible capacity or polarization. Here we report that the 50/50 solid solution, Li$_{1+x}$(V$_{0.5}$Ti$_{0.5}$)S$_2$, delivers a reversible capacity to store charge of 220 mAhg^{-1} (at 0.9 V), 99% of theoretical, at a rate of C/2, retaining 205 mAhg^{-1} at C-rate (92% of theoretical). Rate capability is excellent with 200 mAhg^{-1} at 3C. C-rate is discharge in 1h. Polarization is low, 100 mV at C/2. To the best of our knowledge, the properties/performances of Li(V$_{0.5}$Ti$_{0.5}$)S$_2$ exceed all previous 1 V electrodes.

[1] Departments of Materials and Chemistry, University of Oxford, Parks Road, Oxford OX1 3PH, UK. [2] School of Chemistry and EastChem, University of St Andrews, North Haugh, Fife, Saint Andrews KY16 9ST, UK. * These authors contributed equally to this work. Correspondence and requests for materials should be addressed to P.G.B. (email: peter.bruce@materials.ox.ac.uk).

In order to avoid the potential danger of Li plating[1,2] on the widely used graphite anodes[3–5] in Li-ion batteries, which intercalate Li at 0.1 V versus Li^+/Li, important efforts have been made to identify anodes operating at 1 V (refs 6–9), including organic intercalation compounds[10] and conversion reactions[11,12]. In the former case, the low density inherent in molecular intercalation electrodes (typically $\sim 1.6\,g\,cm^{-3}$) (ref. 13) leads to low volumetric capacities. Volumetric energy density is an important parameter for future Li-ion batteries. The majority of conversion reactions exhibit large polarization (often in excess of 1 V), and voltages above 1 V on Li removal (discharge when used as an anode) are observed frequently[7,14]. In contrast, inorganic intercalation electrodes operate by a well-established mechanism, without the drastic changes that accompany conversion reactions, without involving nanoparticles and with higher densities than organic intercalation compounds.

Li can be intercalated into the layered compound $LiVO_2$ (cubic close packed structure) but at the very low voltage of <0.1 V (refs 15–17). In contrast, Li intercalation into the early transition metal sulfides, $LiMS_2$, M = V, Ti, occurs in the region of 1 V $(M^{2+/3+})$ and these materials have been explored previously, in pioneering studies, as possible anodes for Li-ion batteries[18–20].

However, previous work on $LiVS_2$ has shown that following Li intercalation, the subsequent Li extraction (corresponding to discharge when the $LiVS_2$ is used as an anode in a Li-ion battery) occurs at 1.3 V even at low rates and increases further with rate[18,20]. $LiVS_2$ is therefore not a true 1 V anode. Li intercalation into $LiTiS_2$ occurs at a lower voltage of 0.5 V, and exhibits very poor capacity retention on cycling, the capacity reducing to only $120\,mAhg^{-1}$ after just five cycles. $LiTiS_2$ also exhibits a massive 1st cycle irreversible loss of capacity of over $180\,mAhg^{-1}$ (ref. 18). We investigate the complete solid solution range $Li(V_{1-x}Ti_x)S_2$ and find that the 50/50 composition $Li(V_{0.5}Ti_{0.5})S_2$ delivers a significant improvement in properties/performance over other compositions, including the end members. A total of 50% replacement of V by Ti is sufficient to lower the potential (by lowering the Fermi level in the d-states) even on deintercalation (charge) below 1 V (0.9 V), but not to the extent of $LiTiS_2$ (where the consequence is massive irreversible capacity loss). The irreversible loss of capacity for $Li(V_{0.5}Ti_{0.5})S_2$ on the 1st cycle is only $42\,mAhg^{-1}$ at C/2 (compared with $180\,mAhg^{-1}$ for $LiTiS_2$). Furthermore, the polarization on the first cycle is half that of $LiVS_2$ at the same rate. $Li(V_{0.5}Ti_{0.5})S_2$ exhibits a reversible capacity of $220\,mAhg^{-1}$ at C/2 (99% of the theoretical capacity) dropping only marginally to $200\,mAhg^{-1}$ at 3C. The point of the paper is not to claim that the $Li(V_{0.5}Ti_{0.5})S_2$ material reported here is a commercially viable anode, with all practical problems solved. Rather it is to show that $LiMS_2$ based electrodes are not limited to the relatively poor performance (drastic capacity

fading, large 1st cycle irreversible capacity and large polarization) reported previously and that sulfide based electrodes may merit further exploration in the future.

Results

$Li(V_{1-x}Ti_x)S_2$ solid solution. The range of compositions across the $Li(V_{1-x}Ti_x)S_2$ solid solution were synthesized as described in the Methods section, where the electrode fabrication and characterization methods are also described. The powder X-ray diffraction patterns are shown in Fig. 1a and correspond to the $LiMS_2$ structure, composed of hexagonal close packed S^{2-} with alternate sheets of octahedral sites occupied by Li and M cations, Fig. 1b. A comparison of powder X-ray patterns for $Li(V_{0.5}Ti_{0.5})S_2$ and the blend of 50 wt% $LiVS_2$ and 50 wt% $LiTiS_2$ is shown in Supplementary Fig. 1, confirming that the solid solution has been successfully synthesized. Li intercalation into this structure would require Li^+-occupying tetrahedral sites that share faces with Li in octahedral sites, and is therefore energetically unfavourable. Instead, upon intercalation, all the Li^+ ions in the octahedral sites in $LiMS_2$ are displaced to the neighboring tetrahedral sites in the alkali metal layers, as there are twice as many tetrahedral as octahedral sites the intercalating Li can be incorporated up to a theoretical maximum of Li_2MS_2 (ref. 18).

$Li(V_{0.5}Ti_{0.5})S_2$ was found to be the best performing of all the $Li(V_{1-x}Ti_x)S_2$ materials. Materials with more vanadium displayed increased polarization and deintercalation takes place above 1 V, comparisons with $Li(V_{0.6}Ti_{0.4})S_2$ and $LiVS_2$ are presented in Fig. 2a,b. Increasing the titanium content beyond 50% results in a decrease in capacity and increase in irreversible capacity on the 1st cycle, as is demonstrated by the $Li(V_{0.4}Ti_{0.6})S_2$ and $LiTiS_2$ data in Fig. 2c,d.

The electrochemical properties of $Li(V_{0.5}Ti_{0.5})S_2$. The charge–discharge curves for $Li(V_{0.5}Ti_{0.5})S_2$ are shown in Fig. 3. The capacity on cycling is $220\,mAhg^{-1}$ (2nd cycle) with the plateau on deintercalation (corresponding to the potential of ;the anode on discharge in a Li-ion cell) being located at 0.9 V. The polarization (separation between discharge and charge plateaus) is small, ~ 100 mV (rate = C/2). The load curves are invariant on subsequent cycling, save for a slow reduction in capacity. Concerning the first cycle, there is an initial capacity of $20\,mAhg^{-1}$ at ~ 2.1 V, associated with the $M^{4+/3+}$ redox couple and consistent with a small degree of lithium deficiency in the as-prepared material, $Li_{0.92}(V_{0.5}Ti_{0.5})S_2$, see refinement of structure, Supplementary Table 1. The main difference between the 1st and subsequent cycles is an additional, irreversible, capacity of $\sim 40\,mAhg^{-1}$ on cycle 1, Fig. 3, and is associated with reduction

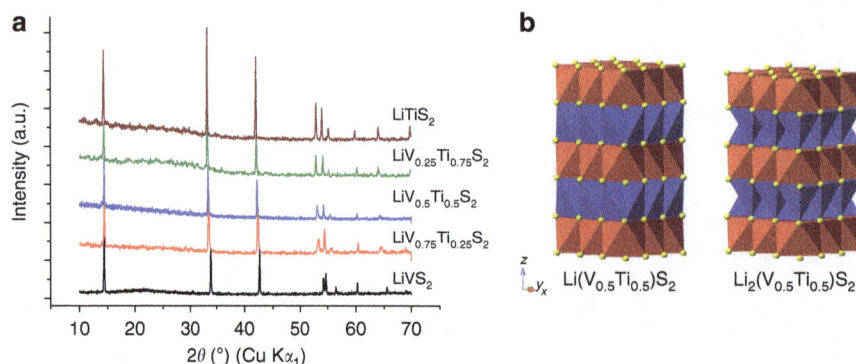

Figure 1 | Structures of $Li(V_{1-x}Ti_x)S_2$ and $Li_2(V_{0.5}Ti_{0.5})S_2$. (**a**) Powder X-ray diffraction patterns of as-prepared $Li(V_{1-x}Ti_x)S_2$. (**b**) Crystal structures of $Li(V_{0.5}Ti_{0.5})S_2$ and $Li_2(V_{0.5}Ti_{0.5})S_2$. Red octahedra—MS_6, blue polyhedra—LiS_x and yellow spheres—sulfur.

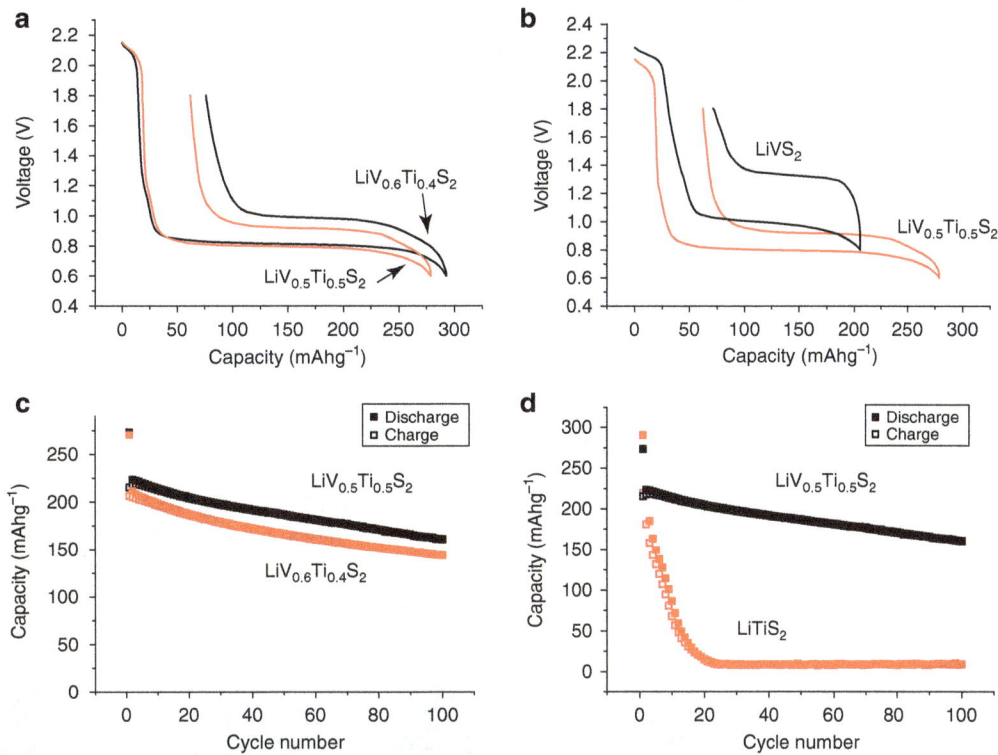

Figure 2 | Comparisons between Li($V_{0.5}Ti_{0.5}$)S_2 and other members of the Li($V_{1-x}Ti_x$)S_2 solid solution. (a,b) show the variation of potential with state-of-charge on the 1st cycle for Li($V_{0.5}Ti_{0.5}$)S_2 and Li($V_{0.6}Ti_{0.4}$)S_2, and LiVS$_2$ respectively. **(c,d)** show the variation of capacity with cycle number for Li($V_{0.5}Ti_{0.5}$)S_2 and Li($V_{0.4}Ti_{0.6}$)S_2, and LiTiS$_2$ respectively, Closed squares correspond to intercalation and open squares to deintercalation. All data were collected at a rate of 100 mAg^{-1}.

of the electrolyte. Electrolyte reduction/solid electrolyte interphase (SEI) layer formation has been observed previously for LiVS$_2$ and LiTiS$_2$ (ref. 18) and is discussed later for Li($V_{0.5}Ti_{0.5}$) S$_2$.

The capacity as a function of cycle number for several different rates is shown in Fig. 4. The capacity on cycling is well maintained with increasing rate, for example, dropping by only 6% on increasing the rate from C/2 to 3C at cycle 10. The capacities on intercalation/de-intercalation during the first few cycles are affected by the irreversible capacity (on cycle 1) and at higher rates by an increase in capacity on cycling, something not infrequently seen in intercalation electrodes at high rates and often related to changes in the composite electrode structure within the first few cycles[21,22]. After the irreversible capacity loss on cycle 1, the charge/discharge efficiency improves rapidly over the first few cycles at all rates, Fig. 4. Overall the capacity retention on cycling after the first few cycles corresponds to a loss of ~0.6 mAhg^{-1} per cycle. The origin of the capacity fading on cycling lies in the SEI layer formation and is discussed below.

The mechanism of intercalation into Li($V_{0.5}Ti_{0.5}$)S_2. Given the change in properties/performance exhibited by the Li($V_{0.5}Ti_{0.5}$)S_2 solid solution compared with the stoichiometric LiVS$_2$ and LiTiS$_2$ end members, it is important to ascertain if lithium intercalation into Li($V_{0.5}Ti_{0.5}$)S_2 operates by the same mechanism. The results below show that the improved properties of Li($V_{0.5}Ti_{0.5}$)S_2, are obtained despite a similar two-phase intercalation process and all three materials forming a SEI layer due to electrolyte degradation. Considering the structure of Li($V_{0.5}Ti_{0.5}$)S_2 and its evolution on intercalation/deintercalation and cycling, joint refinements were carried out on the as-prepared material using the powder X-ray diffraction data in Supplementary Fig. 2 and the neutron diffraction data in Supplementary Fig. 3. Refined parameters are

given in Supplementary Table 1. The refined composition, Li$_{0.92}$($V_{0.5}Ti_{0.5}$)S_2 is in good agreement with the capacity associated with the intercalation at 2.1 V (0.08 Li corresponds to 18 mAhg^{-1}). There is no evidence of site exchange between Li and V/Ti, that is, the structure remains layered, and no evidence of V/Ti long-range order. The V/Ti ratio refined to 1:1 within errors. The variation of structure with Li insertion/removal is shown in Fig. 5. The diffraction patterns were collected at the points on the load curve shown in Fig. 3. Initially, on intercalation Li fills the vacant octahedral sites in the Li layers of the as-prepared Li$_{0.92}$($V_{0.5}Ti_{0.5}$)S_2, accounting for the first 18–20 mAhg^{-1} (between A and B in Fig. 3), this results in an expansion in a and c, from $a = 3.4247(1)$ Å and $c = 6.1582(2)$ Å to $a = 3.4377(4)$ Å and $c = 6.1591(2)$ Å. There is no change in the diffraction patterns between points B and C. Thereafter, Li intercalation is a 2-phase process, in accord with the plateau in Fig. 3, in which the as-prepared phase, Li($V_{0.5}Ti_{0.5}$)S_2 is replaced continuously by Li$_2$($V_{0.5}Ti_{0.5}$)S_2. On lithium extraction the 2-phase process is reversed, associated with the plateau on charge, as seen at point I in Fig. 3. Two phase refinements at points D–G and at I are well described by a mixture of the two end members with the compositions Li($V_{0.5}Ti_{0.5}$)S_2 and Li$_2$($V_{0.5}Ti_{0.5}$)S_2 and the patterns at A, B, C, H and J by single phases of the two end members. A cyclic voltammogram collected on a 3-electrode cell with the Li($V_{0.5}Ti_{0.5}$)S_2 as the working electrode is shown in Supplementary Fig. 4 and exhibits one oxidation peak around 1 V and one reduction peak at 0.7 V, its shape is consistent with a 2-phase process, in accord with the powder X-ray diffraction results[23]. The above results show that the mechanism of Li intercalation into Li($V_{0.5}Ti_{0.5}$)S_2 is essentially similar to that of LiVS$_2$ (ref. 18).

The refined a and c lattice parameters for Li$_1$($V_{0.5}Ti_{0.5}$)S_2 and Li$_2$($V_{0.5}Ti_{0.5}$)S_2 are respectively $a = 3.4377(4)$ Å, $c = 6.1591(2)$ Å

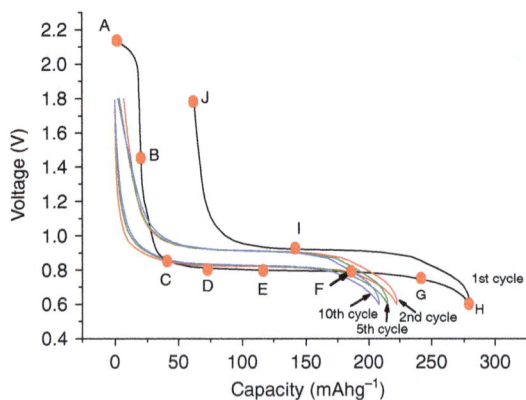

Figure 3 | Variation of potential (versus Li$^+$/Li) with state of charge for Li(V$_{0.5}$Ti$_{0.5}$)S$_2$. Rate 100 mAg^{-1}. Letters correspond to various states of charge on the 1st cycle, see also Fig. 5.

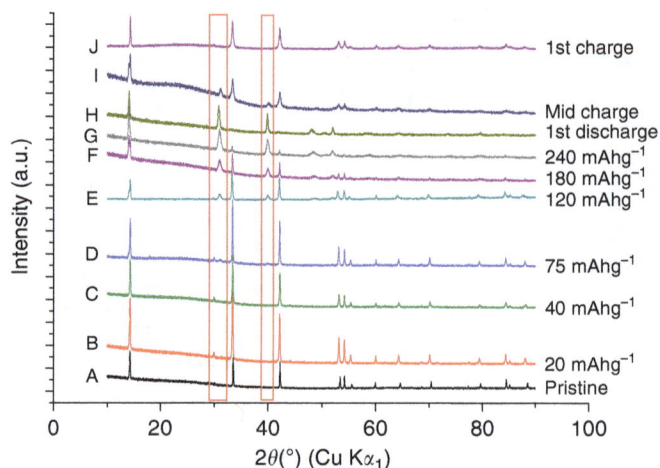

Figure 5 | Powder X-ray diffraction patterns of Li(V$_{0.5}$Ti$_{0.5}$)S$_2$ at various states of intercalation and deintercalation. Letters correspond to points on the 1st cycle in Fig. 3. Highlighted regions show the presence of Li$_2$(V$_{0.5}$Ti$_{0.5}$)S$_2$ phase.

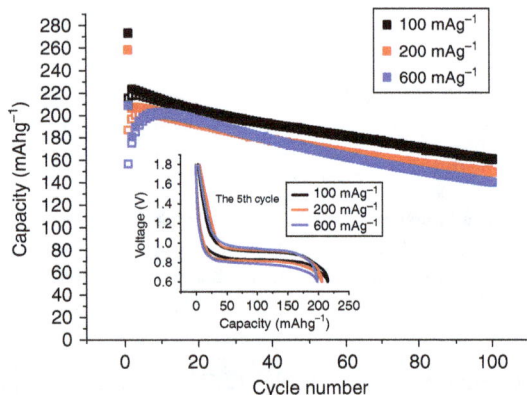

Figure 4 | Variation of capacity with cycle number for Li(V$_{0.5}$Ti$_{0.5}$)S$_2$. Closed squares correspond to intercalation and open squares to deintercalation. Inset shows load curves as a function of rate on the 5th cycle.

Figure 6 | SEM images. (**a**) Pressed pristine Li(V$_{0.5}$Ti$_{0.5}$)S$_2$ electrode and (**b**) Li(V$_{0.5}$Ti$_{0.5}$)S$_2$ electrode after 5 cycles. (**c**) Pristine Li(V$_{0.5}$Ti$_{0.5}$)S$_2$ electrode and (**d**) Li(V$_{0.5}$Ti$_{0.5}$)S$_2$ electrode after 1st discharge (intercalation) showing SEI layer formation on the surface of the electrode. Scale bar, (**a,b**) 10 μm, (**c,d**) 1 μm.

and $a = 3.7894(7)$ Å, $c = 6.229(2)$ Å. The volume expansion on intercalation is 23% yet the phase transformation is facile. The expansion along a is associated with the elongation of the (Ti/V)-S bond as formally (Ti/V)$^{3+}$ is reduced to (Ti/V)$^{2+}$. As the d-bands in sulfides are relatively wide compared with oxides, the electrons are well delocalized across the (Ti/V)-S$_2$ slabs. The c-axis expansion requires energy to increase the separation of the slabs. The small c-axis expansion, only 1.5%, may be responsible for the facile nature of the 2-phase intercalation reaction that exhibits a small polarization and excellent reversibility. Powder X-ray diffraction (PXRD) data collected on cycling, Supplementary Fig. 5 shows that the 2-phase intercalation/de-intercalation mechanism continues to operate. There is no change in lattice parameters with cycling. SEM images of electrodes were collected before and after cycling, Fig. 6a,b. The average particle size of Li(V$_{0.5}$Ti$_{0.5}$)S$_2$ is 20 μm and remains so on cycling. All of these results indicate that the material is stable on cycling.

SEI layer formation and capacity fade. Electrolyte reduction and SEI layer formation have been reported for LiVS$_2$ and LiTiS$_2$, where it was inferred only from the electrochemistry[18]. Of the ~40 mAhg^{-1} of irreversible capacity on the first cycle for Li(V$_{0.5}$Ti$_{0.5}$)S$_2$, Fig. 3, the 20 mAhg^{-1} of capacity between B and C may be assigned to electrolyte reduction, as there is no change

in the lattice parameters. Beyond C, the Rietveld refinements of the 2-phase mixtures of Li(V$_{0.5}$Ti$_{0.5}$)S$_2$ and Li$_2$(V$_{0.5}$Ti$_{0.5}$)S$_2$ along the plateau provide the ratios of the two phases, which in turn gives the amount of lithium and hence charge that has been inserted into the structure. These values are presented in Table 1 along with the values for the total charge passed minus the 20 mAhg^{-1} at the beginning of discharge associated with the M$^{4+/3+}$ redox couple. There is almost no difference in these values across the plateau, indicating that no significant electrolyte reduction occurs between C and G in Fig. 3. The remainder of the electrolyte reduction occurs beyond point G, in accord with the slow downturn in the potential. Extending the cut-off potential to lower values increases this electrolyte reduction and hence irreversible capacity significantly, which is why lower voltage anodes are less attractive. The amount of true intercalation capacity derived from the refinement of the PXRD data is consistent with the theoretical capacity for Li(V$_{0.5}$Ti$_{0.5}$)S$_2$ (222 mAhg^{-1}) and the observed capacity on the 2nd cycle,

Table 1 | Comparison between capacity observed electrochemically and intercalated capacity derived from X-ray refinements.

Point	Capacity from load curve minus the $M^{4+/3+}$ couple (mAhg^{-1})	Capacity of Li intercalated into Li(V$_{0.5}$Ti$_{0.5}$)S$_2$ from X-ray analysis (mAhg^{-1})	Difference (mAhg^{-1})
B	0	0	0
C	20	0	20
D	50	28	22
E	95	72	23
F	155	133	22
G	215	195	20
H	258	216	42

Letters correspond to the points on the 1st cycle in Fig. 3.

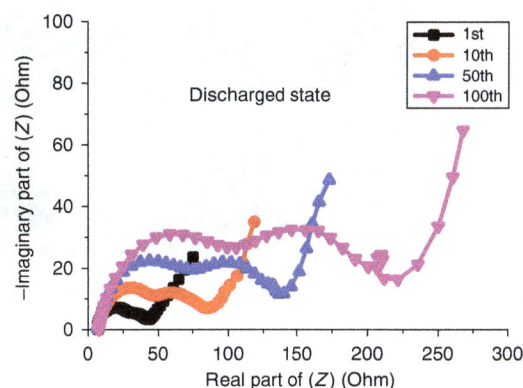

Figure 7 | AC impedance spectra. AC impedance of the Li(V$_{0.5}$Ti$_{0.5}$)S$_2$/electrolyte interface collected at the end of the 1st, 10th,50th and 100th cycles using 3-electrode cells.

Fig. 3. This demonstrates that the additional (irreversible) capacity on cycle 1 is due primarily to electrolyte reduction.

It is important to recall that the composite electrodes contain 10% super S carbon, and this will account for some of the observed electrolyte reduction (irreversible capacity). Cells were constructed with the same loading of Li(V$_{0.5}$Ti$_{0.5}$)S$_2$ per unit area as in Fig. 3, but without super S carbon, the comparison between the two cells is shown in Supplementary Fig. 6. The load curve without super S shows virtually no capacity between 0.93 and 1.3 V, suggesting that the ~ 12 mAhg^{-1} loss observed between those voltages for the electrode containing Super S is associated with electrolyte reduction on the carbon surface. The capacity between 0.93 V and point C, 8 mAhg^{-1}, could be due to reduction of the electrolyte on the surface of Li(V$_{0.5}$Ti$_{0.5}$)S$_2$ as it is present in the electrode with and without carbon. It might also arise from a very small amount of intercalation, corresponding to $x = 0.03$, too small to be seen in the X-ray data. The irreversible capacity loss occurring at the lower voltage beyond the plateau, <0.7 V, appears also due to a combination of carbon and Li(V$_{0.5}$Ti$_{0.5}$)S$_2$, since it increases when super S is present. Overall, Supplementary Fig. 6 shows that of the irreversible capacity observed in Fig. 3, only 25 mAhg^{-1} arises from electrolyte reduction on the Li(V$_{0.5}$Ti$_{0.5}$)S$_2$ surface that is, 11% of the reversible capacity, the rest is due to reduction on the surface of the super S.

To explore the SEI layer formation as a function of cycling, alternating current (AC) impedance spectroscopy and SEM (scanning electron microscopy) images were collected from cycles 1 to 100, Figs 7 and 8. The AC impedance data, from 3-electrode cells, show two semicircles, as is typical for the electrode/electrolyte interface[24]. The interfacial impedance grows continuously with cycle number indicative of a growing SEI layer. This was confirmed by examination of the SEM images, Fig. 8, where the SEI layer is seen to grow on the electrode particles from several nanometers at cycle 10 up to $>1\,\mu$m at cycle 100. The growing SEI layer is consistent with the capacity fading observed on cycling. Confirmation that the capacity fading is not intrinsic to the material but is due to the SEI layer growth was provided by the powder X-ray diffraction patterns at the 25th cycle, Supplementary Fig. 5, which showed the crystal structure is preserved and the 2-phase reaction remains on cycling. We have used standard LP30 electrolyte, but this is optimized for graphite and not for sulfide electrodes operating near 1 V. A detailed investigation of alternative electrolytes, electrolyte additives and of surface coatings on the sulfide may be able to reduce further the irreversible capacity on the 1st cycle even below the 11% associated with reduction on the Li(V$_{0.5}$Ti$_{0.5}$)S$_2$ itself, and to reduce significantly the capacity fading on cycling.

Li(V$_{0.5}$Ti$_{0.5}$)S$_2$ in a full lithium-ion cell. Li(V$_{0.5}$Ti$_{0.5}$)S$_2$ was incorporated as the anode in a full lithium-ion cell with a LiCoO$_2$ cathode. The cell was constructed such that the overall capacity was anode limited, in order to show the performance of the latter, see Methods section. First charge capacity, corresponding to insertion of lithium into Li(V$_{0.5}$Ti$_{0.5}$)S$_2$, exhibits a larger irreversible capacity than was observed when the counter electrode was lithium metal (~ 75 mAhg^{-1} as opposed to 42 mAhg^{-1}), see Fig. 9. In full cells the anode potential can reach lower values on charge than is the case for a lithium counter electrode. If this happens there will be a higher irreversible capacity, as observed here. The overall capacity on the first charge in Fig. 9 is 310 mAhg^{-1} compared with 278 mAhg^{-1} for the cell with a Li counter electrode, the difference is the same as the additional irreversible capacity (75–42 mAhg^{-1}), consistent with the additional irreversible capacity in Fig. 9. Lowering the voltage cutoff below 3.4 V seen in Fig. 9 leads to a loss of reversible capacity on the first and subsequent cycles. The discharge capacity is ~ 215 mAhg^{-1}, just below the theoretical capacity; this reduces to just over 180 mAhg^{-1} by the 10th cycle, see Supplementary Fig. 7. The purpose of this experiment was simply to show that a full cell can be constructed and cycled using the new anode.

Sulfides are used in a variety of applications ranging from pigments[25] to solar cells[26], they are not exotic materials. Although the sulfides here are synthesized in sealed tubes for convenience in the lab, this is not ubiquitous for sulfides, other methods can be used[27,28]. V is not the lowest cost element but several V based electrodes are under investigation and have been reported, for example, V$_2$O$_5$, Na$_x$VO$_2$, LiVO$_2$, and of course V redox-flow batteries are in use and continue to be studied[15,29–31]. The materials we report showed no change in their powder X-ray diffraction patterns after 12 h exposure to ambient air, Supplementary Fig. 8. After 15 h a small peak begins to appear at 11° in 2θ, Supplementary Fig. 8 and grows with time, showing that the sulfides are air sensitive as expected. The widely used LiFePO$_4$ cathode material is also not stable in air and is packed in sealed containers after synthesis for shipping. A similar approach could be taken with Li(V$_{0.5}$Ti$_{0.5}$)S$_2$. We suggest that the relatively slow rate of reaction in air is likely due to the formation of a thin oxide layer that slows decomposition.

Discussion

An intercalation anode, Li(V$_{0.5}$Ti$_{0.5}$)S$_2$, is reported that operates at 0.9 V versus Li$^+$/Li (56% lower voltage than Li$_4$Ti$_5$O$_{12}$

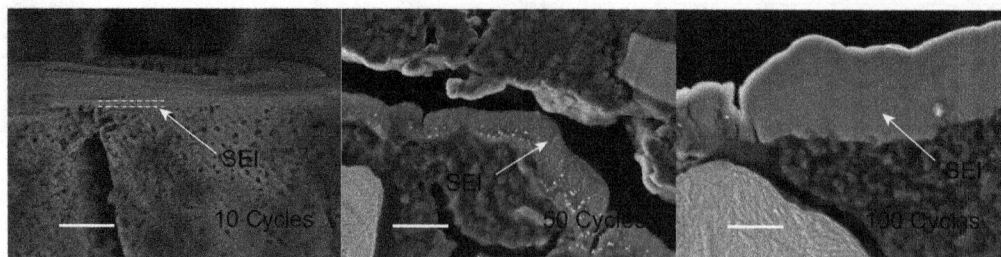

Figure 8 | Cross-sectional SEM images. Cross-sectional SEM images of the $Li(V_{0.5}Ti_{0.5})S_2$ electrode collected at the end of the 10th, 50th and 100th cycles. Scale bar, 1 μm.

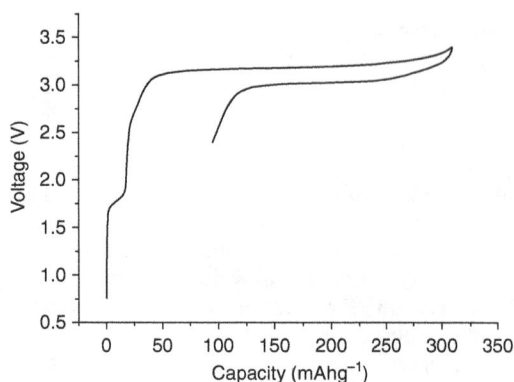

Figure 9 | Variation of potential with state of charge for the $Li(V_{0.5}Ti_{0.5})S_2/LiCoO_2$ cell. Rate 100 mAg^{-1} of anode, cells were cycled between 3.4 V and 2.4 V.

refs 32–34), with low polarization, 100 mV, a capacity of 220 mAhg^{-1} at C/2 on cycling, corresponding to 99% of theoretical capacity, good rate capability, 205 mAhg^{-1} at 1C and 200 mAhg^{-1} at 3C. The volumetric capacity is 740 mAhcm^{-3}, comparable to the reversible volumetric capacity of graphite (600–700 mAh cm^{-3})[35], but at a potential (0.9 V) that avoids Li plating. As such, this material delivers significantly better properties/performance than the previously studied end-members, $LiVS_2$ (the voltage of which is > 1 V versus Li^+/Li) and $LiTiS_2$ (which exhibits a very large irreversible capacity loss on the 1st cycle of 180 mAhg^{-1}). Electrolyte reduction on $Li(V_{0.5}Ti_{0.5})S_2$ during the first cycle accounts for only 11% of the first cycle capacity. Performance is better than previously reported 1 V anodes, to the best of our knowledge. Future work should focus on reducing the 1st cycle irreversible capacity while forming a robust SEI layer minimizing further electrolyte reduction, by examining lowering the amount of conductive additive in the electrode and investigating electrolyte additives or seeking alternative electrolytes to those optimized for graphite anodes.

Methods

Synthesis. $Li(V_{0.5}Ti_{0.5})S_2$ was synthesized from Li_2S (Aldrich, 99.9%), Ti powder (Aldrich, 99.98%), V powder (Aldrich, 99.5%) and S powder (Aldrich, 99.98%). Appropriate ratios of starting materials were mixed and ground together in an Ar filled MBraun glovebox. The mixture was then placed in a graphite crucible, which was in turn placed in a quartz tube. The quartz tube was subsequently sealed under vacuum. The reactants were heated at 1,050 °C for 72 h before being quenched to room temperature.

Electrochemical measurements. Composite electrodes were fabricated using the active material, super S carbon and Kynar Flex 2801 (a co-polymer based on polyvinylidene difluoride) binder in a mass ratio of 80:10:10. The mixture was then cast on copper foil using tetrahydrofuran as the solvent. The loading of active material was ~6–8 mg cm^{-2}. Electrochemical cells consisting of a $Li(V_{0.5}Ti_{0.5})S_2$ composite electrode, a lithium metal counter electrode and a glass microfiber GF/F

separator (WhatmanTM) saturated with electrolyte, a 1 M solution of $LiPF_6$ in ethylene carbonate–dimethyl carbonate 1:1 ((v/v) (BASF)), were constructed. Li-ion cells were constructed similarly with composite $LiCoO_2$ electrodes ($LiCoO_2$, Super S and Kynar Flex 2801 in a mass ratio of 80:10:10) replacing the lithium metal counter. Celgard monolayer polypropylene separator was used in addition to the glass microfiber separator. All handling was carried out in an Ar filled MBraun glovebox. Electrochemical measurements were conducted using a Maccor series 4,200 battery tester. Cyclic voltammetry and AC impedance were conducted on 3-electrode cells using a VMP3 electrochemical workstation (Biologic).

Structural analysis. Powder X-ray diffraction patterns were obtained using a Stoe STADI/P diffractometer employing CuKα$_1$ radiation operating in transmission mode with the samples sealed in 0.2 mm capillaries, expect for the experiments on air sensitivity, which were carried out in air. Time-of-flight powder neutron diffraction data were collected on the POLARIS high-intensity; medium resolution instrument at ISIS, Rutherford Appleton Laboratory (UK) with samples sealed in 2 mm quartz capillaries. The structures were refined by the Rietveld method using TOPAS Academic[36]. Samples with different Li amounts for X-ray diffraction were prepared electrochemically. After cycling, cells were transferred to an argon-filled glove box before opening and active material removed. The electrodes were then rinsed with a small amount of dry dimethyl carbonate to remove residual electrolyte. They were left under dynamic vacuum overnight to ensure all solvent had evaporated before measurements were collected. SEM studies were carried out using a Carl Zeiss Merlin instrument. Electrodes for cross-sectional imaging were rotary-etched in a Gatan Precision Etching Coating System (PECS 682).

References

1. Smart, M. C. & Ratnakumar, B. V. Effects of electrolyte composition on lithium plating in lithium-ion cells. *J. Electrochem. Soc.* **158,** A379–A389 (2011).
2. Petzl, M., Kasper, M. & Danzer, M. A. Lithium plating in a commercial lithium-ion battery—a low-temperature aging study. *J. Power Sources* **275,** 799–807 (2015).
3. Armand, M. & Tarascon, J. M. Building better batteries. *Nature* **451,** 652–657 (2008).
4. Etacheri, V., Marom, R., Elazari, R., Salitra, G. & Aurbach, D. Challenges in the development of advanced Li-ion batteries: A review. *Energy Environ. Sci.* **4,** 3243–3262 (2011).
5. Noel, M. & Suryanarayanan, V. Role of carbon host lattices in li-ion intercalation/de-intercalation processes. *J. Power Sources* **111,** 193–209 (2002).
6. Oumellal, Y., Rougier, A., Nazri, G. A., Tarascon, J. M. & Aymard, L. Metal hyrides for lithium-ion batteries. *Nat. Mater.* **7,** 916–921 (2008).
7. Goriparti, S. *et al.* Review on recent progress of nanostructured anode materials for li-ion batteries. *J. Power Sources* **257,** 421–443 (2014).
8. Gao, H., Liu, C. L., Liu, Y., Liu, Z. H. & Dong, W. S. MoO$_2$-loaded porous carbon hollow spheres as anode materials for lithium-ion batteries. *Mater. Chem. Phys.* **147,** 218–224 (2014).
9. Goodenough, J. B. & Kim, Y. Challenges for rechargeable Li batteries. *Chem. Mater.* **22,** 587–603 (2010).
10. Armand, M. *et al.* Conjugated dicarboxylate anodes for Li-ion batteries. *Nat. Mater.* **8,** 120–125 (2009).
11. Cabana, J., Monconduit, L., Larcher, D. & Palacin, M. R. Beyond intercalation-based Li-ion batteries: The state of the art and challenges of electrode materials reacting through conversion reactions. *Adv. Mater.* **22,** E170–E192 (2010).
12. Zhang, L., Wu, H. B. & Lou, X. W. Iron-oxide-based advanced anode materials for lithium ion batteries. *Adv. Energy Mater.* **4,** 1300958 (2014).
13. Kaduk, J. A. Terephthalate salts: salts of monopositive cations. *Acta Crystallogr. B-Struct. Sci.* **56,** 474–485 (2000).
14. Park, G. D. & Kang, Y. C. Superior lithium-ion storage properties of mesoporous CuO-reduced graphene oxide composite powder prepared by a two-step spray-drying process. *Chem. Eur. J.* **21,** 9179–9184 (2015).
15. Armstrong, A. R., Lyness, C., Panchmatia, P. M., Islam, M. S. & Bruce, P. G. The lithium intercalation process in the low-voltage lithium battery anode $Li_{1+x}V_{1-x}O_2$. *Nat. Mater.* **10,** 223–229 (2011).

16. Choi, N. S., Kim, J. S., Yin, R. Z. & Kim, S. S. Electrochemical properties of lithium vanadium oxide as an anode material for lithium-ion battery. *Mater. Chem. Phys.* **116,** 603–606 (2009).

17. Pourpoint, F. *et al.* New insights into the crystal and electronic structures of $Li_{1+x}V_{1-x}O_2$ from solid state NMR, pair distribution function analyses, and first principles calculations. *Chem. Mater.* **24,** 2880–2893 (2012).

18. Kim, Y., Park, K. S., Song, S. H., Han, J. T. & Goodenough, J. B. Access to M^{3+}/M^{2+} redox couples in layered $LiMS_2$ sulfides ($M = $ Ti, V, Cr) as anodes for Li-ion battery. *J. Electrochem. Soc.* **156,** A703–A708 (2009).

19. Murphy, D. W. & Carides, J. N. Low voltage behavior of lithium/metal dichalcogenide topochemical cells. *J. Electrochem. Soc.* **126,** 349–351 (1979).

20. Gupta, A., Mullins, C. B. & Goodenough, J. B. Electrochemical probings of $Li_{1+x}VS_2$. *Electrochim. Acta* **78,** 430–433 (2012).

21. Wei, X. *et al.* Electrochemical performance of high-capacity nanostructured $Li[Li_{0.2}Mn_{0.54}Ni_{0.13}Co_{0.13}]O_2$ cathode material for lithium ion battery by hydrothermal method. *Electrochim. Acta* **107,** 549–554 (2013).

22. Prosini, P. P., Zane, D. & Pasquali, M. Improved electrochemical performance of a $LiFePO_4$-based composite cathode. *Electrochim. Acta* **46,** 3517–3523 (2001).

23. Robertson, A. D., Tukamoto, H. & Irvine, J. T. S. $Li_{1+x}Fe_{1-3x}Ti_{1+2x}O_4$ ($0.0 \le x \le 0.33$) based spinels: possible negative electrode materials for future Li-ion batteries. *J. Electrochem. Soc.* **146,** 3958–3962 (1999).

24. Zhang, S. S., Xu, K. & Jow, T. R. EIS study on the formation of solid electrolyte interface in Li-ion battery. *Electrochim. Acta* **51,** 1636–1640 (2006).

25. Anglos, D. *et al.* Laser-induced fluorescence in artwork diagnostics: an application in pigment analysis. *Appl. Spectrosc.* **50,** 1331–1334 (1996).

26. Guo, Q. *et al.* Fabrication of 7.2% efficient CZTSSe solar cells using CZTS nanocrystals. *J. Am. Chem. Soc.* **132,** 17384–17386 (2010).

27. Murphy, D. W., Cros, C., Disalvo, F. J. & Waszczak, J. V. Preparation and properties of Li_xVS_2. *Inorg. Chem.* **16,** 3027–3031 (1977).

28. Vanlaar, B. & Ijdo, D. J. W. Preparation, crystal structure, and magnetic structure of $LiCrS_2$ and $LiVS_2$. *J. Solid State Chem.* **3,** 590–595 (1971).

29. Li, X. F., Zhang, H. M., Mai, Z. S., Zhang, H. Z. & Vankelecom, I. Ion exchange membranes for vanadium redox flow battery (VRB) applications. *Energy Environ. Sci.* **4,** 1147–1160 (2011).

30. Liu, Q. *et al.* Graphene-modified nanostructured vanadium pentoxide hybrids with extraordinary electrochemical performance for Li-ion batteries. *Nat. Commun.* **6,** 6127 (2015).

31. Guignard, M. *et al.* P2-Na_xVO_2 system as electrodes for batteries and electron-correlated materials. *Nat. Mater.* **12,** 74–80 (2013).

32. Ferg, E., Gummow, R. J., Dekock, A. & Thackeray, M. M. Spinel anodes for lithium-ion batteries. *J. Electrochem. Soc.* **141,** L147–L150 (1994).

33. Colbow, K. M., Dahn, J. R. & Haering, R. R. Structure and electrochemistry of the spinel oxides $LiTi_2O_4$ and $Li_{4/3}Ti_{5/3}O_4$. *J. Power Sources* **26,** 397–402 (1989).

34. Yang, Z. G. *et al.* Nanostructures and lithium electrochemical reactivity of lithium titanites and titanium oxides: a review. *J. Power Sources* **192,** 588–598 (2009).

35. Sivakkumar, S. R., Nerkar, J. Y. & Pandolfo, A. G. Rate capability of graphite materials as negative electrodes in lithium-ion capacitors. *Electrochim. Acta* **55,** 3330–3335 (2010).

36. Coelho, A. A. Whole-profile structure solution from powder diffraction data using simulated annealing. *J. Appl. Crystallogr.* **33,** 899–908 (2000).

Acknowledgements

P.G.B. is indebted to the EPSRC including the SUPERGEN program for financial support.

Author contributions

S.J.C. and D.W. carried out experiments. A.R.A. carried out structural refinements. P.G.B. wrote the manuscript. All authors contributed to the discussion and interpretation of the results.

Additional information

25

Charge localization in a diamine cation provides a test of energy functionals and self-interaction correction

Xinxin Cheng[1,†], Yao Zhang[1], Elvar Jónsson[2], Hannes Jónsson[1,2,3] & Peter M. Weber[1]

Density functional theory (DFT) is widely applied in calculations of molecules and materials. Yet, it suffers from a well-known over-emphasis on charge delocalization arising from self-interaction error that destabilizes localized states. Here, using the symmetric diamine *N,N'*-dimethylpiperazine as a model, we have experimentally determined the relative energy of a state with positive charge localized on one of the two nitrogen atoms, and a state with positive charge delocalized over both nitrogen atoms. The charge-localized state was found to be 0.33 (0.04) eV higher in energy than the charge-delocalized state. This provides an important test of theoretical approaches to electronic structure calculations. Calculations with all DFT functionals commonly used today, including hybrid functionals with exact exchange, fail to predict a stable charge-localized state. However, the application of an explicit self-interaction correction to a semi-local functional identifies both states and gives relative energy in excellent agreement with both experiment and CCSD(T) calculations.

[1] Department of Chemistry, Brown University, 324 Brook Street, Providence, Rhode Island 02912, USA. [2] COMP, Department of Applied Physics, Aalto University, FIN-00076 Espoo, Finland. [3] Faculty of Physical Sciences, VR-III, University of Iceland, 107 Reykjavík, Iceland. † Present address: Max Planck Institute for the Structure and Dynamics of Matter (MPSD), Luruper Chaussee 149, 22761 Hamburg, Germany. Correspondence and requests for materials should be addressed to P.M.W. (email: peter_weber@brown.edu).

Charge transfer (CT) between two or more charge centres and charge delocalization across separate parts of a molecular system are of great importance in many areas of chemistry as they often define molecular structure and reactivity[1–3]. Their relevance further extends to biological polymers such as DNA and to light-harvesting processes using artificial photosynthesis[3–7]. Localized electronic states and separation of charges are, furthermore, an important aspect of semiconductor devices and solar cells[8–12]. Yet, the calculation of charge localization and charge delocalization is challenging, especially for computational methods that are applicable to extended systems.

The commonly used Kohn–Sham density functional theory (DFT) functionals are known to suffer from bias towards delocalized states because of self-interaction error[13–15]. Hartree–Fock (HF), on the other hand, is overly biased towards localized states. Hybrid functionals that mix the two approaches are widely used in all branches of chemistry and are increasingly used in condensed matter calculations, but the balance between localized and delocalized states depends strongly on the proportions used in the mixing. This mixing ratio is sometimes treated as an empirical, system-dependent parameter. A more fundamental, parameter-free approach is, however, needed to accurately predict the delicate balance between localized and delocalized electronic states, especially in extended systems. Over 30 years ago, an explicit self-interaction correction approach was proposed by Perdew and Zunger (PZ-SIC)[16]. Although it was applied early on in studies of the electronic structure of molecules (see, for example, ref. 17), it has not become a commonly used approach. A variational, self-consistent implementation of this PZ-SIC using complex valued orbitals has recently been applied successfully in studies of molecules and solids, including Rydberg excited states[18–20]. A discussion of the self-interaction error and our implementation of PZ-SIC can be found in Supplementary Note 1. We show here that although all the commonly used DFT functionals, including hybrid functionals, fail to produce the localized charge state of the system studied, calculations using PZ-SIC give results in excellent agreement with the experimentally determined relative energy of a localized and delocalized electronic state.

To achieve this comparison, experimental benchmarks for charge-localized and charge-delocalized states are required. This is challenging because in most cases electronic systems exhibit just one particular charge distribution, making comparison between states with different charge distributions difficult. In a recent study, however, we have found that in N,N'-dimethylpiperazine (DMP), both a charge-localized state and a charge-delocalized state can be observed[21]. DMP has previously served as a prototype for exploring electron lone pair interactions and CT between nitrogen atoms[22–24]. In its cationic state, the positive charge can be localized on one of the nitrogen atoms (DMP-L$^+$), or delocalized over the two equivalent nitrogen atoms (DMP-D$^+$). The charge-localized ion DMP-L$^+$, however, has not been observed until the recent ultrafast time-resolved experiment[21]. It is possible to identify the different charge states because they have distinct spectral and temporal signatures. Upon optical excitation, the charge-localized state is initially generated. A subsequent CT then leads to the charge-delocalized state. This discovery of two states with distinct charge distributions has now laid the foundation for an experimental approach for measuring their relative energy. The approach is based on photoionization from Rydberg states, whose binding energies have been found to be remarkably dependent on the nuclear arrangements and the charge distribution of the molecular ion core. Further, the Rydberg electron-binding energy is independent of a molecule's internal energy, making it ideally suited for the exploration of high-energy charge states that are populated when the

molecules are highly energized[25–28]. Because the Rydberg electron has a small effect on the bonding and molecular structure, the configuration of the molecular ion cores of Rydberg states closely resemble those of the cationic states. The binding energies of the Rydberg states, therefore, yield information about the charge-localized and charge-delocalized cationic states.

The relative energy of a charge-localized state and a charge-delocalized state of a given molecule has not been determined previously, as far as we know, even though it is a critically important parameter for calibrating theoretical approaches. In the current study, the relative energy was determined using a newly devised experimental approach that takes advantage of an equilibrium established on a picosecond time scale upon excitation to the Rydberg states. Calculations using conventional DFT, self-interaction-corrected DFT and *ab initio* methods were tested against the experimental measurements. We aim to evaluate the capability of each method to properly describe the charge localization and delocalization in the ground electronic state of the molecular cation.

Results

Experimental determination of the relative energy. To measure the energy of the cationic state, we first experimentally determined the relative energy of 3sD and 3sL Rydberg states with the charge-delocalized ion core DMP-D$^+$ and the charge-localized ion core DMP-L$^+$, respectively, by measuring the equilibrium composition of the Rydberg states as a function of the excitation energy. As shown in Supplementary Fig. 1, because the binding energy, that is, the energy difference between the cationic state and the Rydberg state, is measured in the experiment, the relative energy of the cationic states is known once the relative energy of the 3s Rydberg states is known. DMP was excited from its ground state to the 3p or the 3s Rydberg state using pump photons with wavelengths tuned in the range of 193.0–240.8 nm. The probe photon monitored the time-dependent dynamics by ionizing the Rydberg-excited molecules. The kinetic energy of the ejected photoelectrons was measured to determine the binding energy of the Rydberg states. The 3p state, which was reached with the shorter wavelengths, decayed by internal conversion into 3s within several hundred femtoseconds. Details of the experimental setup and parameters are given in Supplementary Note 2. As we have shown previously, upon optical excitation to the localized charge state, an equilibrium between the localized and delocalized states is reached after several picoseconds[21].

The complete set of the time-resolved photoelectron spectra at several pump wavelengths are given in Supplementary Fig. 2. Because the energy is conserved in internal conversion, pump photons of different wavelengths deposit different amounts of energy into the vibrational manifolds. For pump wavelengths between 193.0 and 240.8 nm, the molecule has effective vibrational temperature between 565 and 980 K after relaxation, assuming the energy is distributed across all vibrational modes. The detailed calculation of the effective temperature is discussed in the Supplementary Note 6.

Two 3s peaks (Fig. 1), located at 2.70 (0.03) eV and at 2.81 (0.04) eV, are assigned to 3sD and 3sL, respectively[21]. The 3sL dominates at short delay times (Fig. 1a,b), whereas 3sD dominates at longer delay times (Fig. 1a,c,d), indicating an early population in 3sL and a lower energy for 3sD. As the temperature decreases from 980 to 565 K with the pump wavelength increasing from 193.0 nm to 240.8 nm, the intensity of 3sL decreases almost to the baseline in the spectra with the molecules at equilibrium (Fig. 1d)

Because the change of the Gibbs free energy (ΔG) scales linearly with the natural logarithm of the equilibrium constant

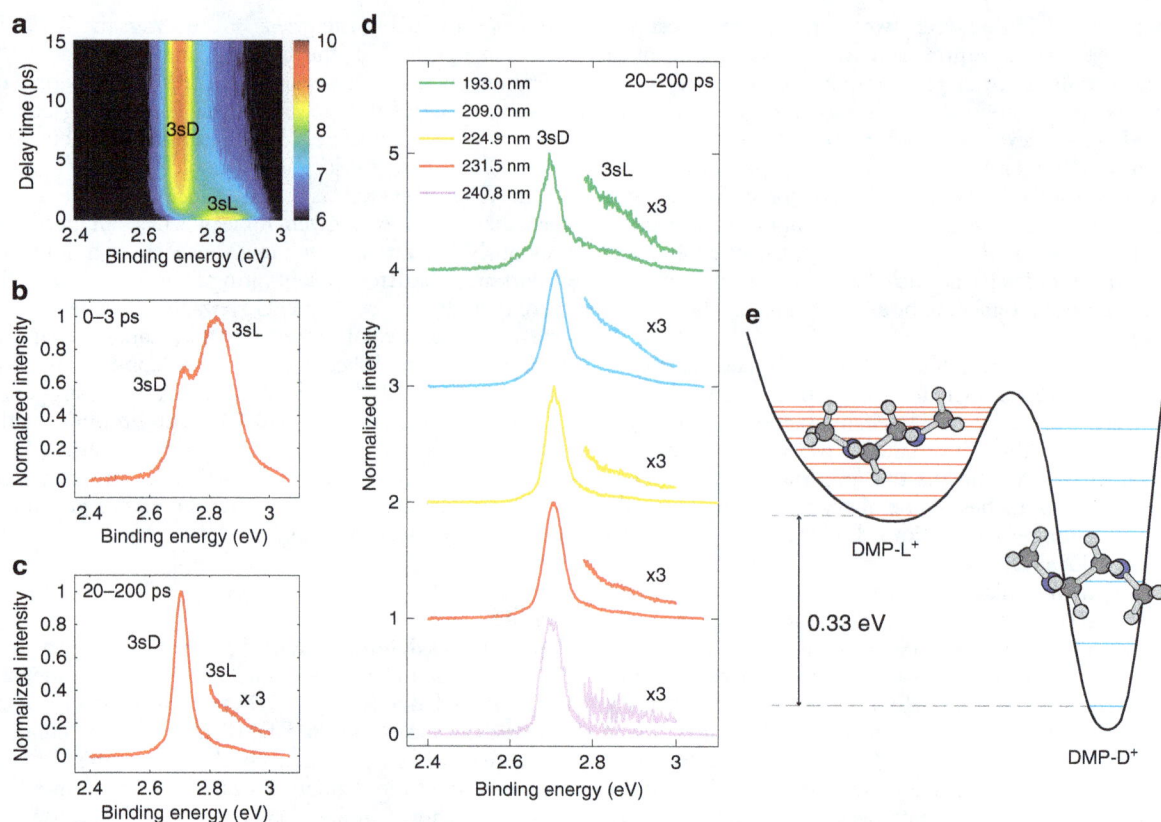

Figure 1 | Photoelectron spectra of DMP. (**a**) The time-resolved spectrum of DMP with 231.5 nm pump photon. The colour bar represents the logarithmic intensity scale. (**b**) The 0–3 ps time-integrated spectrum of **a**. (**c**) The 20–200 ps time-integrated spectrum of **a**. (**d**) The 20–200 ps time-integrated spectra of DMP at five selected pump wavelengths. The relative populations of the charge-localized (3sL) and the charge-delocalized (3sD) states can be determined as a function of temperature from this data, thus providing an estimate of the relative energy of the two states, which turns out to be 0.33 eV. (**e**) A schematic cut of the potential energy surface for DMP$^+$. The red and blue lines illustrate the vibrational states of DMP-L$^+$ and DMP-D$^+$, respectively.

(K), the enthalpy (ΔH) and entropy (ΔS) change in the transition can be determined from the logarithm of the equilibrium constant as a function of inverse temperature:

$$\ln K = -\Delta H/RT + \Delta S/R \qquad (1)$$

where R is the gas constant and T is the temperature.

The spectra at each point in time were fitted using two Lorentzians with variable peak centres to derive the equilibrium constants in the 3s states, as described in Supplementary Notes 4 and 5. Details of the fits are shown in Supplementary Notes 4 and 5, Supplementary Figs 3 and 4 and Supplementary Tables 1 and 2. The results are shown in Fig. 2. The logarithm of the equilibrium constant is indeed found to depend approximately linearly on the estimated reciprocal temperature. A fit using equation (1) gives -20.9 (3.7) kJ mol^{-1} or -0.22 (0.04) eV and -17.7 (4.6) J K^{-1} mol^{-1} for ΔH and ΔS of the transition from 3sL to 3sD. Because the binding energy difference between 3sL and 3sD is 0.11 (0.01) eV, the energy of the DMP-L$^+$ ion is determined to be 0.33 (0.04) eV higher than that of the DMP-D$^+$ ion. A schematic cut through the energy surface, deduced from these measurements, is shown in Fig. 1e.

Test of theoretical methods. The ability of various theoretical approaches to describe the charge localized and delocalized states can now be assessed by comparison with these experimental results. First, calculations were carried out to determine the optimal molecular geometries of the two states. The DMP-L$^+$ and DMP-D$^+$ ion structures were optimized with the Gaussian 09 (refs 29,30) and NWChem software[31] at various levels of

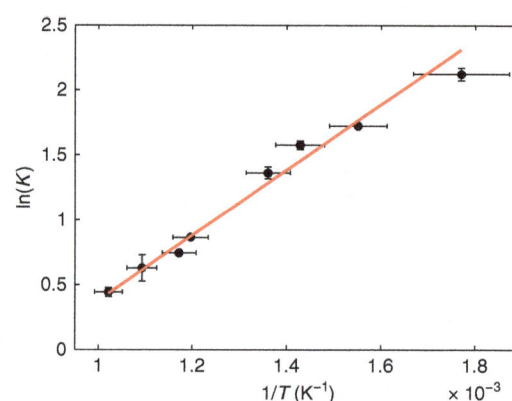

Figure 2 | Temperature dependence of the equilibrium constants. Measured values of the equilibrium constant for the 3sL to 3sD states of the DMP molecule are shown as a function of reciprocal temperature estimated from the photon energy. The red line shows a linear best fit providing an estimate of the energy and entropy difference between the two states.

theory including HF, MP2 (Møller–Plesset perturbation theory, truncated at the second order), DFT with all the commonly used functionals (complete list available in Supplementary Note 2), and CCSD (coupled cluster method with single and double excitations). Unless specified otherwise, the aug-cc-pVDZ basis set was used in the Rydberg state calculations and the cc-pVTZ basis set in the

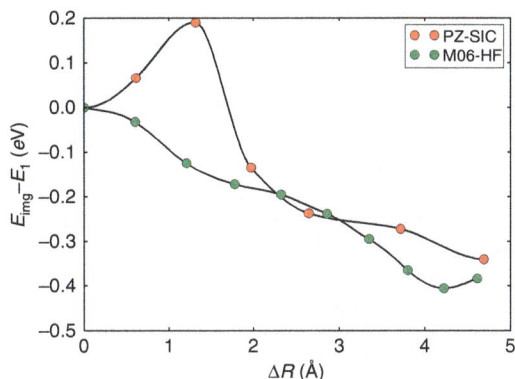

Figure 3 | Calculated minimum energy path between the localized and delocalized state of the DMP cation. The energy of images, E_{img}, in the nudged elastic band calculations is given with respect to the energy of the localized state, E_1, as a function of the accumulated displacement of the atoms, ΔR. The red dots show results of a PZ-SIC calculation where a barrier of 0.2 eV separates the metastable, localized state from the delocalized state. The green dots show results of calculations using the M06-HF functional where MP2 optimized structures are used for the end points. In the M06-HF calculations, the energy barrier is not present and a structure optimization starting from the localized state converges on the delocalized state. Similar results were obtained for all other commonly used DFT functionals.

Table 1 | RE of the DMP-L$^+$ and DMP-D$^+$ states obtained using various computational methods.

Method	RE (eV)
HF	− 0.53
MP2	0.81
DFT	
B3LYP	—[a]
M06	—[a]
M06-2X	—[a]
M06-HF	—[a]
PBE0	—[a]
BHandHLYP	0.19
PZ-SIC	0.34
CCSD[b]	0.23
MP2_CCSD(T)-SP	0.39
CCSD_CCSD(T)-SP	0.38
Experiment	0.33 (0.04)

CCSD, coupled cluster method with single and double excitations; DFT, density functional theory; DMP, *N,N*′-dimethylpiperazine; HF, Hartree-Fock; PZ-SIC, Perdew and Zunger self-interaction correction; RE, relative energy.
Single-point energy calculations were carried out with the CCSD(T) method, in one case using a structure obtained with MP2 and in the other case using a structure obtained with CCSD. Zero point energy correction has not been applied but an estimate based on ground vibrational states would reduce the calculated relative energy by 0.07 eV, see Supplementary Note 3. As the molecule is at high temperature, the full, ground vibrational state zero point correction is an overestimate.
[a]No value shown because the DMP-L$^+$ state was not stable at this level of theory.
[b]The CCSD optimizations were carried out with the aug-cc-pVDZ basis set because calculations with the cc-pVTZ were found to be too demanding for our computational resources.

cation calculations[32,33]. The cation structures were also optimized using PZ-SIC with the GPAW software[34–36], where a real space grid over a cubic simulation cell of 20 Å side length and 0.13 Å mesh was used. The PZ-SIC was applied to the PBE semi-local functional. Although in the neutral molecule each nitrogen atom assumes a pyramidal structure with the methyl group in equatorial position, the nitrogen becomes pseudo-axial and pseudo-planar in the cation. The Cartesian coordinates of selected optimized structures are listed in Supplementary Tables 4 and 5.

DFT calculations with any one of the available hybrid functionals implemented in the Gaussian 09 software failed to provide a stable localized state, except for the BHandHLYP functional that has previously been found to describe CT interactions well[37] while giving generally poor results for other molecular properties such as total energy (and therefore not commonly used)[38]. When DFT calculations were started from MP2 or HF-optimized structures for the localized state, the minimization of the energy resulted in a conversion to the charge delocalized state. Most surprisingly, the M06-HF functional, which contains 100% HF exchange and is therefore widely deemed to be particularly appropriate for CT[39–41], also fails to localize the charge in this case. However, the PZ-SIC calculation gives a stable localized state with similar structure as that obtained from MP2 and CCSD. The bond lengths are typically 0.02–0.04 Å shorter than those obtained from MP2 and CCSD. The minimum energy path between the two states calculated using the nudged elastic band method[42] and the PZ-SIC as well as the M06-HF functional is shown in Fig. 3. An energy barrier of 0.2 eV for the transition from the localized to the delocalized state is obtained with PZ-SIC, whereas no barrier is obtained in the M06-HF calculations.

Table 1 lists the calculated energy difference between the optimized DMP-L$^+$ and DMP-D$^+$ structures obtained using HF, MP2, CCSD, DFT with selected functionals and PZ-SIC. CCSD(T) calculations were carried out to obtain the single-point energy for each one of these structures. Although

calculations using HF, MP2 and CCSD produce both the DMP-L$^+$ and the DMP-D$^+$ states, the relative energy of these states is poorly estimated, especially by HF, which gives lower energy for the localized state than the delocalized state. Single-point CCSD(T) calculations using the MP2 or CCSD geometries give a relative energy in good agreement with the experimental measurements. The PZ-SIC calculation also yields a relative energy that is close to the experimental results, see Table 1 and Supplementary Table 3. Supplementary Note 7 also reports a satisfactory agreement of the experimental and computed entropy differences.

The close agreement between the relative energy obtained from the high-level CCSD(T) calculations and the experimental results confirms the validity of the interpretation of the experimental data. To further cement the correspondence of experimental and computational results, we have calculated the Rydberg electron-binding energy with the optimized DMP-L$^+$ and DMP-D$^+$ structures using the equation of motion CCSD. For comparison, the binding energy was also calculated using PZ-SIC, which has previously been shown to give good estimates of Rydberg binding energy of both molecules and molecular clusters[18,27,28,43]. The total energy of the Rydberg excited states using PZ-SIC was obtained using the delta self-consistent field method[44] and the binding energy obtained by subtracting the total energy of the excited state from that of the ion. As listed in Table 2, the calculated binding energy is in good agreement with the experimentally measured values for both the DMP-L$^+$ and DMP-D$^+$ structures, supporting the assignment of the observed spectroscopic features.

The Rydberg orbitals and the associated spin densities are shown in Fig. 4. The 3sL Rydberg orbital (Fig. 4a) anchors on the planar nitrogen, whereas the 3sD Rydberg orbital (Fig. 4b) is centred symmetrically between the two nitrogen atoms. Both orbitals are extended and comprise the whole molecule, as expected. The spin densities (shown in Fig. 3c,d), which were generated by subtracting the spin-down density from the spin-up density, illustrate the charge distributions of the localized and

Table 2 | Calculated Rydberg binding energy (in eV) of 3sD and 3sL states for MP2 optimized DMP-L$^+$ and DMP-D$^+$ structures using the EOM-CCSD and PZ-SIC methods.

Method	3sD	3sL
PZ-SIC	2.71	2.87
EOM-CCSD	2.65	2.72
Experiment	2.70 (0.03)	2.81 (0.04)

EOM-CCSD, equation of motion coupled cluster method with single and double excitations; PZ-SIC, Perdew and Zunger self-interaction correction.

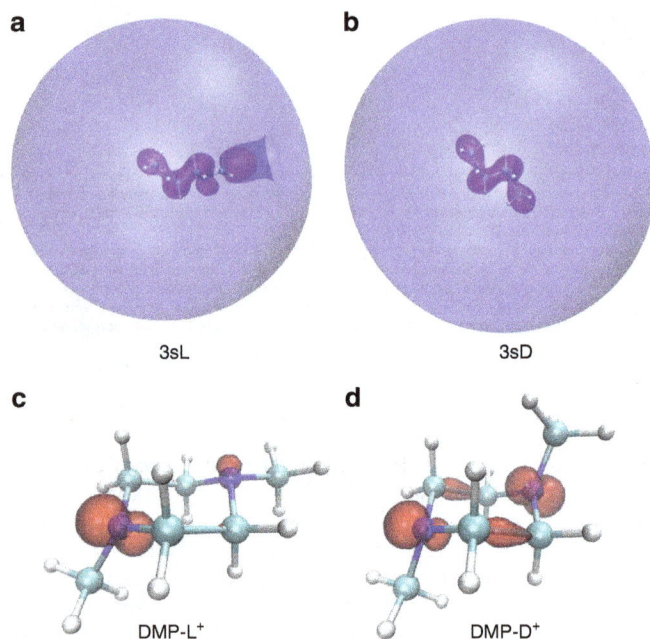

3sL

3sD

DMP-L$^+$

DMP-D$^+$

Figure 4 | The Rydberg orbitals and the associated spin densities.
(**a,b**) Calculated 3sL and 3sD Rydberg orbitals, respectively, rendered at 0.001 Å$^{-3/2}$ isovalues. (**c,d**) Calculated spin density of the DMP-L$^+$ and DMP-D$^+$ ion, respectively, at isovalue of 0.2 electron per Å^{-3}.

delocalized cations. The charge is localized on the planar nitrogen in the DMP-L$^+$ (Fig. 3c) but delocalized between the two nitrogen atoms in DMP-D$^+$ (Fig. 3d). Intriguingly, there is also net spin density between the two intermediate C-atoms, indicating a through-bond-interaction[22] in the charge-delocalized ion.

Discussion

The present study advances experimentally the state-of-the-art in exploring charge localization and delocalization in systems with multiple charge centres. The wavelength-dependent Rydberg electron-binding energy spectroscopy enables us to experimentally determine the energy and entropy change of the CT reaction. It is the experimental tool of choice to probe the CT process in the presence of large vibrational energy, as is needed to observe the higher-energy state before CT. Indeed, previous time-resolved spectroscopic studies on DMP radical cations prepared by a photo-induced electron transfer technique only observed the DMP-D$^+$ ion[23,24] possibly because of the low temperature in the system. Together with the binding energy information obtained from the spectra, the present experiment creates an important benchmark to evaluate various theoretical approaches.

The MP2 and CCSD methods are found to give good estimates of the molecular structure as the single-point CCSD(T) calculation gives good agreement with the experimental binding energies of the 3sL and 3sD states as well as the relative energy of the two cation states. The computational effort of these methods, however, scales unfavourably with system size and they are thus limited to small systems. The computational effort of DFT calculations increases slower with size as N^3, where N is the number of electrons, and is the only viable approach for many problems involving large molecules and condensed phase systems. Conventional DFT functionals are, however, found here not to predict the metastable localized state of DMP. The explicit inclusion of the self-interaction correction, as proposed by Perdew and Zunger, implemented in a variational and self-consistent way with complex-valued orbitals, can remedy these shortcomings of the DFT approach. The PZ-SIC calculations give similar results for the binding energies of the 3sL and 3sD states as the equation of motion CCSD calculations and for the relative energies of the DMP-L$^+$ and DMP-D$^+$ states as the CCSD(T) calculations. The computational effort in this approach is larger than conventional DFT but still scales with size as N^3. These results are expected to guide future improvements to energy functionals describing electronic systems.

References

1. Gaillard, E. R. & Whitten, D. G. Photoinduced electron transfer bond fragmentations. *Acc. Chem. Res.* **29**, 292–297 (1996).
2. Newton, M. D. Quantum chemical probes of electron-transfer kinetics - the nature of donor-acceptor interactions. *Chem. Rev.* **91**, 767–792 (1991).
3. Zewail, A. H. Femtochemistry: atomic-scale dynamics of the chemical bond using ultrafast lasers - (Nobel lecture). *Angew. Chem. Int. Ed. Engl.* **39**, 2587–2631 (2000).
4. Moser, C. C., Keske, J. M., Warncke, K., Farid, R. S. & Dutton, P. L. Nature of biological electron transfer. *Nature* **355**, 796–802 (1992).
5. Sanii, L. & Schuster, G. B. Long-distance charge transport in DNA: Sequence-dependent radical cation injection efficiency. *J. Am. Chem. Soc.* **122**, 11545–11546 (2000).
6. Barnett, R. N., Cleveland, C. L., Joy, A., Landman, U. & Schuster, G. B. Charge migration in DNA: ion-gated transport. *Science* **294**, 567–571 (2001).
7. Frischmann, P. D., Mahata, K. & Wurthner, F. Powering the future of molecular artificial photosynthesis with light-harvesting metallosupramolecular dye assemblies. *Chem. Soc. Rev.* **42**, 1847–1870 (2013).
8. Ge, N. H. *et al.* Femtosecond dynamics of electron localization at interfaces. *Science* **279**, 202–205 (1998).
9. Yeh, A. T., Shank, C. V. & McCusker, J. K. Ultrafast electron localization dynamics following photo-induced charge transfer. *Science* **289**, 935–938 (2000).
10. Bakulin, A. A. *et al.* The role of driving energy and delocalized States for charge separation in organic semiconductors. *Science* **335**, 1340–1344 (2012).
11. Falke, S. M. *et al.* Coherent ultrafast charge transfer in an organic photovoltaic blend. *Science* **344**, 1001–1005 (2014).
12. Najafi, E., Scarborough, T. D., Tang, J. & Zewail, A. Ultrafast dynamics. Four-dimensional imaging of carrier interface dynamics in p-n junctions. *Science* **347**, 164–167 (2015).
13. Cohen, A. J., Mori-Sanchez, P. & Yang, W. T. Insights into current limitations of density functional theory. *Science* **321**, 792–794 (2008).
14. Jónsson, H. Simulation of surface processes. *Proc. Natl Acad. Sci. USA* **108**, 944–949 (2011).
15. Baruah, T. & Pederson, M. R. Density functional study on a light-harvesting carotenoid-porphyrin-C$_{60}$ molecular triad. *J. Chem. Phys.* **125**, 164706 (2006).
16. Perdew, J. P. & Zunger, A. Self-interaction correction to density-functional approximations for many-electron systems. *Phys. Rev. B* **23**, 5048–5079 (1981).
17. Pederson, M. R., Heaton, R. A. & Lin, C. C. Local-density Hartree-Fock theory of electronic states of molecules with self-interaction correction. *J. Chem. Phys.* **80**, 1972–1975 (1984).
18. Gudmundsdóttir, H., Zhang, Y., Weber, P. M. & Jónsson, H. Self-interaction corrected density functional calculations of molecular Rydberg states. *J. Chem. Phys.* **139**, 194102 (2013).
19. Lehtola, S. & Jónsson, H. Variational, Self-consistent implementation of the Perdew-Zunger self-interaction correction with complex optimal orbitals. *J. Chem. Theory Comput.* **10**, 5324–5337 (2014); *J. Chem. Theory Comput.* **11**, 5052–5053 (2015).
20. Gudmundsdóttir, H., Jónsson, E. Ö. & Jónsson, H. Calculations of Al dopant in α-quartz using a variational implementation of the Perdew-Zunger self-interaction correction. *N. J. Phys.* **17**, 083006 (2015).

21. Deb, S., Cheng, X. & Weber, P. M. Structural dynamics and charge transfer in electronically excited N,N′-dimethylpiperazine. *J. Phys. Chem. Lett.* **4**, 2780–2784 (2013).
22. Hoffmann, R. Interaction of orbitals through space and through bonds. *Acc. Chem. Res.* **4**, 1–9 (1971).
23. Brouwer, A. M., Langkilde, F. W., Bajdor, K. & Wilbrandt, R. Through-bond interaction in the radical cation of N,N-dimethylpiperazine. Resonance Raman spectroscopy and quantum chemical calculations. *Chem. Phys. Lett.* **225**, 386–390 (1994).
24. Brouwer, A. M. *et al.* Radical cation of N,N-dimethylpiperazine: dramatic structural effects of orbital interactions through bonds. *J. Am. Chem. Soc.* **120**, 3748–3757 (1998).
25. Gosselin, J. L. & Weber, P. M. Rydberg fingerprint spectroscopy: a new spectroscopic tool with local and global structural sensitivity. *J. Phys. Chem. A* **109**, 4899–4904 (2005).
26. Kuthirummal, N. & Weber, P. M. Rydberg states: sensitive probes of molecular structure. *Chem. Phys. Lett.* **378**, 647–653 (2003).
27. Cheng, X. *et al.* Ultrafast structural dynamics in Rydberg excited N,N,N′,N′-tetramethylethylenediamine: conformation dependent electron lone pair interaction and charge delocalization. *Chem. Sci.* **5**, 4394–4403 (2014).
28. Cheng, X., Zhang, Y., Gao, Y., Jónsson, H. & Weber, P. M. Ultrafast structural pathway of charge transfer in N,N,N′,N′-tetramethylethylenediamine. *J. Phys. Chem. A* **119**, 2813–2818 (2015).
29. Frisch, M. J. *et al. Gaussian 09, Revision* **C.01** (Gaussian, Inc., 2009).
30. Frisch, M. J. *et al. Gaussian 09, Revision* **D.01** (Gaussian, Inc., 2009).
31. Valiev, M. *et al.* NWChem: a comprehensive and scalable open-source solution for large scale molecular simulations. *Comput. Phys. Commun.* **181**, 1477–1489 (2010).
32. Dunning, T. H. Gaussian basis sets for use in correlated molecular calculations. 1. The atoms boron through neon and hydrogen. *J. Chem. Phys.* **90**, 1007–1023 (1989).
33. Kendall, R. A., Dunning, T. H. & Harrison, R. J. Electron affinities of the firstrow atoms revisited. Systematic basis sets and wave functions. *J. Chem. Phys.* **96**, 6796–6806 (1992).
34. Mortensen, J. J., Hansen, L. B. & Jacobsen, K. W. Real-space grid implementation of the projector augmented wave method. *Phys. Rev. B* **71**, 035109 (2005).
35. Enkovaara, J. *et al.* Electronic structure calculations with GPAW: a real-space implementation of the projector augmented wave method. *J. Phys. Condens. Matter* **22**, 253202 (2010).
36. Valdes, A. *et al.* Solar hydrogen production with semiconductor metal oxides: new directions in experiment and theory. *Phys. Chem. Chem. Phys.* **14**, 49–70 (2012).
37. Zhao, Y. & Truhlar, D. G. Benchmark databases for nonbonded interactions and their use to test density functional theory. *J. Chem. Theory Comput.* **1**, 415–432 (2005).
38. Magyar, R. J. & Tretiak, S. Dependence of spurious charge-transfer excited states on orbital exchange in TDDFT: large molecules and clusters. *J. Chem. Theory Comput.* **3**, 976–987 (2007).
39. Zhao, Y. & Truhlar, D. G. Density functional for spectroscopy: no long-range self-interaction error, good performance for Rydberg and charge-transfer states, and better performance on average than B3LYP for ground states. *J. Phys. Chem. A* **110**, 13126–13130 (2006).
40. Zhao, Y. & Truhlar, D. G. The M06 suite of density functionals for main group thermochemistry, thermochemical kinetics, noncovalent interactions, excited states, and transition elements: two new functionals and systematic testing of four M06-class functionals and 12 other functionals. *Theor. Chem. Acc.* **120**, 215–241 (2008).
41. Zhao, Y. & Truhlar, D. G. Density functionals with broad applicability in chemistry. *Acc. Chem. Res.* **41**, 157–167 (2008).
42. Jónsson, H., Mills, G. & Jacobsen, K. W . in *Classical and Quantum Dynamics in Condensed Phase Simulations* (eds Berne, B. J., Ciccotti, G. & Coker, D. F.) 385–404 (World Scientific, 1998).
43. Gudmundsdóttir, H., Zhang, Y., Weber, P. M. & Jónsson, H. Self-interaction corrected density functional calculations of Rydberg states of molecular clusters: N,N-dimethylisopropylamine. *J. Chem. Phys.* **141**, 234308 (2014).
44. Gavnholt, J., Olsen, T., Engelund, M. & Schiotz, J. Delta self-consistent field method to obtain potential energy surfaces of excited molecules on surfaces. *Phys. Rev. B* **78**, 075441 (2008).

Acknowledgements

This project was supported by the National Science Foundation (Grant Number CBET-1336105), by the DTRA (Grant Number HDTRA1-14-1-0008), by the Academy of Finland through its COMP Center of excellence and FiDiPro grants (no. 263294 and 278260) and by the Icelandic Research Fund. The simulation and energy calculations of the molecular structures were conducted using computational resources and services at the Center for Computation and Visualization, Brown University, and at the Nordic High Performance Computer in Iceland.

Author contributions

P.M.W. and H.J. conceived and designed the research. X.C. did the spectroscopic experiments, analysed the data and ran the MP2, CCSD, CCSD(T) and conventional DFT calculations. Y.Z. carried out the PZ-SIC simulations to calculate the ionic and Rydberg states. E.J. performed PZ-SIC calculations to optimize the ion structures and minimum energy path calculation. X.C., P.M.W. and H.J. wrote the paper, with feedbacks from all co-authors.

Additional information

26

High-efficiency electrochemical thermal energy harvester using carbon nanotube aerogel sheet electrodes

Hyeongwook Im[1,*], Taewoo Kim[1,*], Hyelynn Song[1], Jongho Choi[1], Jae Sung Park[2], Raquel Ovalle-Robles[3], Hee Doo Yang[4], Kenneth D. Kihm[5], Ray H. Baughman[6], Hong H. Lee[7], Tae June Kang[8] & Yong Hyup Kim[1,9]

Conversion of low-grade waste heat into electricity is an important energy harvesting strategy. However, abundant heat from these low-grade thermal streams cannot be harvested readily because of the absence of efficient, inexpensive devices that can convert the waste heat into electricity. Here we fabricate carbon nanotube aerogel-based thermo-electrochemical cells, which are potentially low-cost and relatively high-efficiency materials for this application. When normalized to the cell cross-sectional area, a maximum power output of $6.6\,W\,m^{-2}$ is obtained for a $51\,^\circ C$ inter-electrode temperature difference, with a Carnot-relative efficiency of 3.95%. The importance of electrode purity, engineered porosity and catalytic surfaces in enhancing the thermocell performance is demonstrated.

[1] School of Mechanical and Aerospace Engineering, Seoul National University, Seoul 151-742, South Korea. [2] Institute of Advanced Machinery and Design, Seoul National University, Seoul 151-742, South Korea. [3] Nano-Science & Technology Center, Lintec of America, Inc., Richardson, Texas 75081, USA. [4] Department of NanoMechatronics Engineering, College of Nanoscience and Nanotechnology, Pusan National University, Busan 609-735, South Korea. [5] Department of Mechanical, Aerospace and Biomedical Engineering, University of Tennessee, Knoxville, Tennessee 37996, USA. [6] Alan G. MacDiarmid NanoTech Institute, University of Texas at Dallas, Richardson, Texas 75080, USA. [7] School of Chemical and Biological Engineering, Seoul National University, Seoul 151-744, South Korea. [8] Department of Mechanical Engineering, INHA University, Incheon 22212, South Korea. [9] Institute of Advanced Aerospace Technology, Seoul National University, Seoul 151-742, South Korea. * These authors contributed equally to this work. Correspondence and requests for materials should be addressed to T.J.K. (email: tjkang@inha.ac.kr) or to Y.H.K. (email: yongkim@snu.ac.kr).

Harvesting energy from waste heat has received much attention due to the world's growing energy problem[1–4]. Critical needs for harnessing waste heat are to improve the efficiency of thermal energy harvesters and decrease their cost[5]. Solid-state thermoelectric devices have been long investigated for the direct conversion of thermal energy to electrical energy, and many exciting advances have been made[6,7]. However, device performance relative to cost has so far limited application for waste heat recovery[8]. Thermal electrochemical energy harvesters[9] might have major advantages, as suggested in a very preliminary way by previous comparisons of Wh/dollar of solar and electrochemical thermocells[10]; however, they presently have no commercial applications because of their low energy conversion efficiencies and low areal output power. The goal of the present work is to increase the obtained energy conversion efficiencies and areal output power of thermocells to the point where they outperform thermoelectrics in energy output per device cost during device lifetime for low-grade thermal energy harvesting.

To increase the energy conversion efficiency of thermocells, carbon nanomaterials have been introduced as cell electrodes to take advantage of the fast redox processes, high thermal and electrical conductivities, and high gravimetric surface areas that these materials can provide[11]. Hu et al.[10] reported an energy conversion efficiency as high as 1.4%, relative to Carnot cycle efficiency, when carbon multi-walled nanotube (MWNT) buckypaper was used for thermocell electrodes. This efficiency was raised to 2.6% by introducing a carbon single-walled nanotube (SWNT)/reduced graphene oxide (rGO) composite electrode[12]. Improved mass transport due to enhanced porosity of the optimized SWNT/rGO composite was found responsible for the efficiency enhancement. Despite recent advances in thermocell technology, a significant efficiency increase is required for thermocells to become commercially attractive, considering that the Carnot-relative efficiency has to be 2–5% for commercial viability[13]. While efficiency is undoubtedly a key factor, another important quantity is the specific power the device can generate.

Here we exploit planar and cylindrically wound carbon nanotube (CNT) aerogel sheets as thermocell electrodes and devise various additional ways to optimize Carnot efficiency. The deployed optimization strategies to improve thermocell performance involve the use of CNT aerogel sheets as electrodes, removal of low activity carbonaceous impurities that limit electron transfer kinetics, decoration of CNT sheets with catalytic platinum nanoparticles, mechanical compression of nanotube sheets to tune conductivity and porosity, and the utilization of a cylindrical cell geometry. The output power density generated by a described cylindrical thermocell reaches 6.6 W m^{-2} for a 51 °C inter-electrode temperature difference, which corresponds to a Carnot-relative efficiency of 3.95% (that is, 3.95% of the maximum energy conversion efficiency possible for a heat engine operating between two given temperatures).

Results

CNT aerogel sheets as high-performance electrodes. A generic aqueous electrolyte thermocell, which utilizes the Fe(CN)$_6^{4-}$/Fe(CN)$_6^{3-}$ redox couple and K$^+$ as the counter ions, is schematically illustrated in the inset of Fig. 1a. An inter-electrode temperature difference causes a difference in the redox potentials of the electrolyte at the electrodes. This thermally generated potential difference drives electrons in the external circuit and ions in the electrolyte, thereby enabling electrical power to be generated. Continuous operation of the thermocell requires transport of the reaction products formed at one electrode to the other electrode. If either electrode is not furnished with the redox

molecules needed for electron generation or consumption, then power production will cease.

The thermoelectric potential in a thermocell is generated by the temperature dependence of the free energy difference between reactant and product of a reaction taking place at the electrolyte–electrode interface[9,13]. Minor effects on the cell potential arise from the thermal diffusion (Soret effect)[13–16] and the transport entropy of ions[17]. However, for most systems of interest, the electrode potentials dominate and the minor effects on potential can be neglected for practical purposes[13,16].

A planar-type thermocell, shown in Fig. 1a, was used to investigate the potential of CNT aerogel sheets as high-performance electrodes. For preparation of the thermocell electrode, CNT aerogel sheet was drawn from a CNT forest and laid on a rectangular tungsten frame that is connected to a motor. Then, the motor was rotated at 10 r.p.m. to simultaneously draw the sheet from the forest and warp it onto the frame (Fig. 1b). The thickness and area of the CNT electrode can be easily controlled by the number of motor rotations and the frame size, respectively. A 100-μm-thick CNT sheet electrode with an area of 1.0 × 1.0 cm^2 was used to evaluate thermocell performance. All thermocells and all electrochemical impedance measurements (to determine equivalent series resistance (ESR) and charge transfer resistance (R_{ct})) used as electrolyte a 0.4 M aqueous solution of potassium ferro/ferricyanide. Cyclic voltammetry (CV) measurements used an aqueous solution of 10 mM K$_3$Fe(CN)$_6$ and 0.1 M KCl.

Fast transport of the redox mediator ions into electrodes is required to obtain high-areal power generation from CNT thermocells. Randomly oriented CNT sheets, such as CNT buckypaper, may impede ion transport into electrode depths, due to the high tortuosity of the pore structure. On the other hand, the well-aligned CNTs in the aerogel sheets might result in faster ion transport deep within the electrodes, as schematically illustrated in Fig. 1c. The effectiveness of ion transport in CNT aerogel sheets was characterized by measuring the mass transport coefficient using an electrolytic flow cell (see Methods and Supplementary Fig. 1). Exploiting the limiting-current method, the mass transfer coefficient (k_c) was estimated from the dependence of limiting current on electrolyte concentration by using the following equation[18,19]:

$$k_c = \frac{i_L}{nFAC_\infty} \qquad (1)$$

where i_L is the limiting current, n is the number of moles of electrons transferred, F is the faradaic constant, A is the electrode area and C_∞ is the bulk species concentration.

Various aqueous 1:2 concentrations of Fe(CN)$_6^{3-}$/Fe(CN)$_6^{4-}$ redox couple in 0.5 M NaOH were used for the experiments, by varying the Fe(CN)$_6^{3-}$ concentration from 20 to 80 mM. The concentration of Fe(CN)$_6^{4-}$ was twice the Fe(CN)$_6^{3-}$ concentration for each experiment to ensure a limiting reaction rate at the cell cathode (that is, Fe(CN)$_6^{3-}$ + e$^-$ → Fe(CN)$_6^{4-}$). The electrolyte stored in a glass container was circulated through the cell by a peristaltic pump. The flow rate was kept low at 6.6 × 10^{-6} m^3 s^{-1} for allowing laminar flow (Reynolds number, Re ∼ 650, Supplementary Note 1 and Supplementary Table 1) in the cell.

The limiting-current method is based on driving an electrochemical reaction to the maximum possible reaction rate, which is limited by the mass transport of redox ions. The reaction rate limit is indicated by a current plateau on a polarization curve plot. Polarization curves of the thermocells based on CNT buckypaper and CNT aerogel sheet electrodes are shown in Fig. 1d,e, respectively. A linear relationship between limiting current and reactant concentration is evident from the insets of the figure panels. Using equation (1), the mass transfer coefficient of CNT

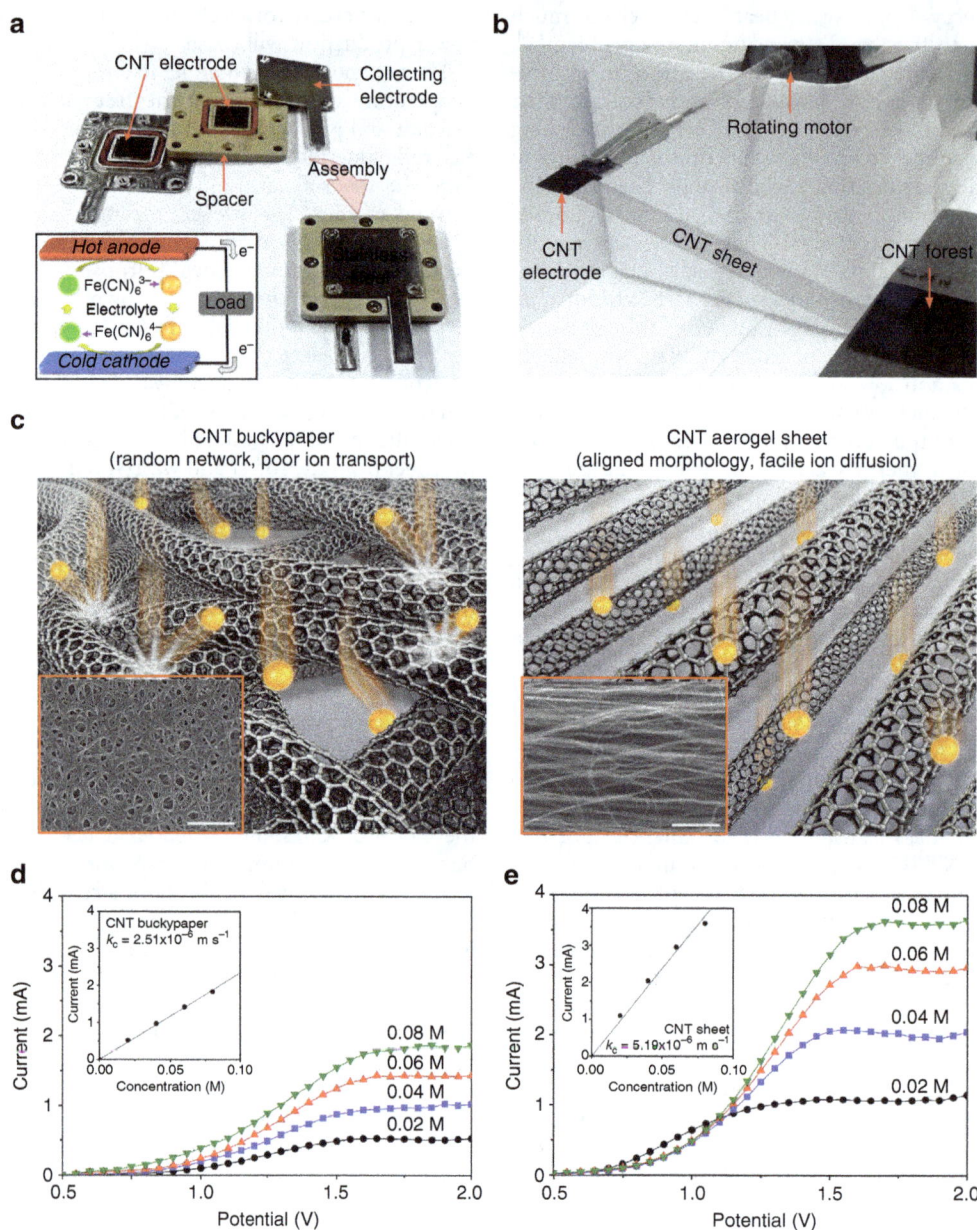

Figure 1 | Carbon nanotube aerogel sheets as high-performance electrodes. (**a**) Photographs of cell components and their assembly into a planar thermocell (inset: schematic drawing of thermocell operation). (**b**) An apparatus for continuously drawing a carbon nanotube (CNT) aerogel sheet from a CNT forest, and wrapping it around a metal frame to form an electrode for a planar thermocell. (**c**) Illustrations and SEM micrographs (insets) comparing CNT buckypaper and CNT aerogel electrodes and the relationship of these morphologies to ion transport (scale bars in the insets, 1-μm). MWNT bundling is not shown and only MWNT outer walls are pictured. Polarization curves for (**d**) CNT buckypaper and (**e**) CNT aerogel electrodes. The insets show the dependence of limiting current on ferrocyanide concentration.

aerogel sheet (5.19×10^{-6} m s^{-1}) is estimated to be twice that of CNT buckypaper (2.51×10^{-6} m s^{-1}). The transfer coefficient of CNT aerogel sheet electrode approaches the theoretically limiting mass transfer coefficient (5.76×10^{-6} m s^{-1}, see Supplementary Note 1), which corresponds to the unobstructed transport of reactants to a flat plate. These electrochemical results and the SEM micrographs suggest that the low tortuosity of the pore structure in the CNT aerogel, compared with that of CNT buckypaper, results in faster ion diffusion to deep within the electrode and corresponding higher limiting currents at a given redox concentration.

Optimization of thermocell performance. It is well-known that impurities, such as carbonaceous byproducts, are introduced during CNT synthesis[20]. A coating of these impurities on CNTs

could restrict charge exchange with redox ions in the electrolyte, thereby degrading cell performance[21]. Various efforts have been made to remove undesirable amorphous carbon from CNT surfaces, for the purpose of improving performance for diverse applications of CNTs[21,22]. We take advantage of the difference in oxidation rate in air between CNT and byproducts[23] to purify forest-drawn CNT sheets, which are ultimately used as electrodes.

To find the optimum heating schedule, the oxidation temperature was first varied in the range of 300–400 °C, while the oxidation time was held constant at 5 min. The purity of the CNT sheets was characterized using Raman spectroscopy for 514 nm excitation. The effect of oxidation temperature on the Raman intensity ratio of D band to G band (I_D/I_G) is shown in Fig. 2a, where the D band peak corresponds to defective carbon (like in amorphous carbon or carbon in defect sites) and the

G band arises from ordered sp^2 carbons[24]. The minimum ratio I_D/I_G occurs at anneal temperatures between 325 and 350 °C, signifying that higher anneal temperatures should not be used for annealing times as short as 5 min. For further optimization of CNT aerogel sheet quality, the anneal temperature was fixed at 340 °C and the heating time was varied between 0 and 60 min to further minimize the I_D/I_G ratio (Fig. 2b). The Raman spectra for an as-drawn sheet and for the optimally annealed sheet are shown in the figure inset. The merit of this method for removing CNT impurities was confirmed by high-resolution transmission electron microscopy (HR-TEM) images of MWNTs in the annealed sheets. The non-annealed MWNT (Fig. 2c) is covered with a carbonaceous coating, which is reduced by the 5-min anneal at 340 °C (Fig. 2d), and eliminated by a 15-min anneal at 340 °C (Fig. 2e), which corresponds to the minimum obtained ratio of I_D/I_G in Fig. 2b.

Thermocell performance was strongly affected by the annealing time at 340 °C, as shown in Fig. 3a,b. In this experiment, the hot plate temperature was maintained at 25 °C and the cold plate temperature was at 5 °C. While this temperature difference (ΔT) of 20 °C was applied between heating and cooling plates, the actual ΔT between the electrodes, which was calculated using the open-circuit voltage and the observed thermo-electrochemical Seebeck coefficient ($1.43\,\mathrm{mV\,K^{-1}}$)[16], was smaller (17.5 °C). This difference is caused by thermal resistances (and corresponding temperature drops) at the interfaces between hot and cold electrodes and the respective heating and cooling plates. The curves of cell voltage versus areal current density and the corresponding power density curves obtained for various thermal oxidation times (shown in Fig. 3a,b) reveal that a lower I_D/I_G ratio, indicating that a cleaner CNT sheet, yields a higher maximum output power. The maximized areal power density (P_{MAX}), normalized to the square of the inter-electrode

temperature difference (ΔT^2), is shown as a function of thermal oxidation time at 340 °C in Fig. 3c. A 186% increase (from 0.07 to $0.13\,\mathrm{mW\,m^{-2}\,K^{-2}}$) in the $P_{MAX}/\Delta T^2$ is shown to result from removing carbonaceous impurities from the surfaces of CNTs and CNT bundles. This optimum thermal oxidation condition (15 min at 340 °C) was used for further experiments.

To explain this major improvement in performance, we used electrochemical impedance measurements to characterize the three primary internal resistances of the thermocell (that is, the activation, ohmic and mass transport resistances)[16,25]. The activation resistance is the loss incurred in overcoming the activation barrier associated with reactions at the electrodes. The ohmic resistance is mainly due to the series resistances of electrode and electrolyte and the mass transport resistance is associated with the kinetics of ion diffusion and convection in the thermocell. The changes in these internal resistances with respect to heating time during sheet oxidation at 340 °C are shown in the inset of Fig. 3c. This figure shows that a major reduction in the activation resistance is realized by removal of carbonaceous impurities, while the decrease in the combined ohmic and mass transport resistances is much smaller. This result indicates that amorphous carbon covering the surface of CNT hinders electrochemical reaction, resulting in performance degradation.

The fact that the electrode is an aerogel implies that increased contact between CNTs (and correspondingly decreased inter-electrode electrical and thermal resistances) could be realized by compressing the electrode. However, the correspondingly increased density can restrict ion transport into the interior of the electrode (see Supplementary Fig. 2). To evaluate these opposing effects, mechanical compression of a planar electrode was accomplished by pressing CNT sheets (wrapped on the tungsten frame used for electrode fabrication), using two silicon wafers that were coated with gold films (which provide low

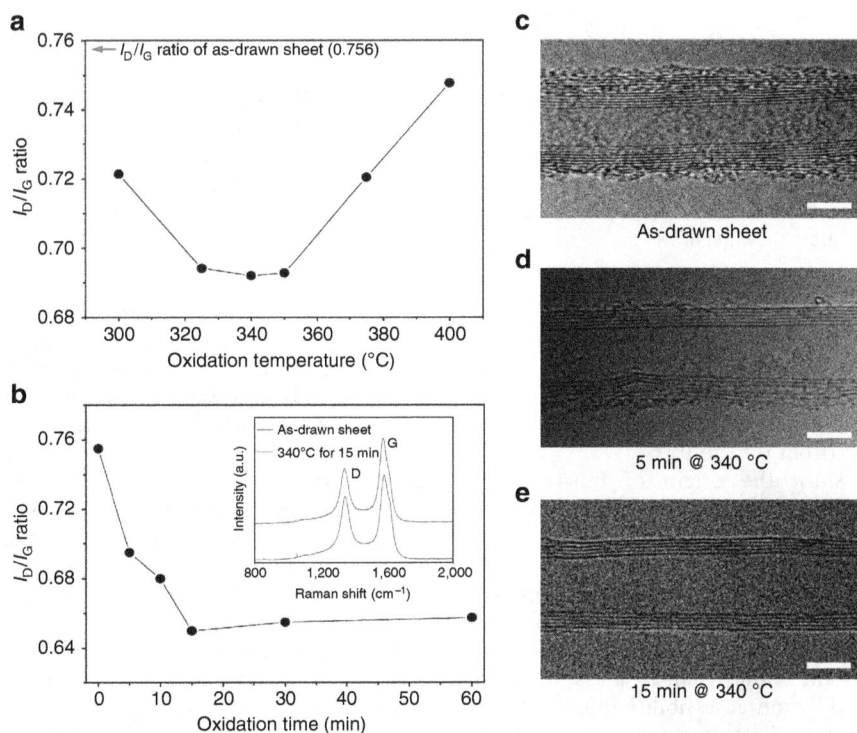

Figure 2 | The effects of thermal oxidation temperature and time. (a) The dependence of Raman I_D/I_G ratio on oxidation temperature for a 5-min anneal of CNT aerogel sheets in ambient air. (b) The dependence of the I_D/I_G ratio for CNT aerogel sheets on oxidation time in ambient air at 340 °C. The inset shows the Raman spectra for an as-drawn CNT sheet and for a CNT sheet that has been thermally oxidized in air for 15 min at 340 °C. (c-e) High-resolution transmission electron microscope (HR-TEM) images of (c) as-drawn, (d) 5-min oxidized, and (e) 15-min oxidized CNT aerogel sheets (all scale bars, 5-nm).

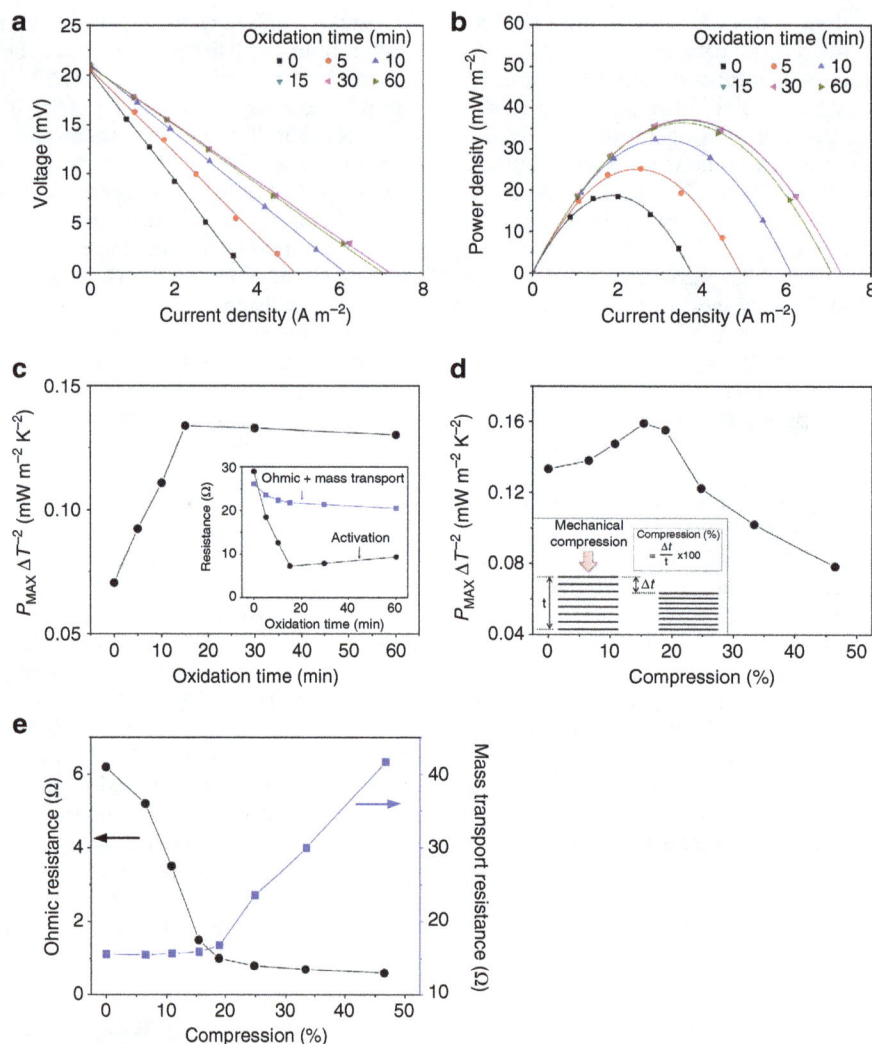

Figure 3 | The effect on thermocell performance of oxidation and compression of the CNT aerogel. (**a**) Cell voltage versus current density and (**b**) cell power density versus current density for samples having different thermal oxidation times. (**c**) Maximum power density normalized to the inter-electrode temperature difference ($P_{MAX}/\Delta T^2$) as a function of thermal oxidation time. (Inset: the dependence of activation resistance and the sum of ohmic and mass transport resistance on thermal oxidation time). (**d**) $P_{MAX}/\Delta T^2$ generated by the thermocell as a function of the per cent mechanical compression of cell electrodes. The inset illustrates the mechanical compression of a planar CNT aerogel electrode. (**e**) The dependence of ohmic and mass transport resistance on the compressive strain shown in **d**.

adhesion to the CNT sheet stack). The maximum obtained power density, normalized to ΔT^2, is plotted in Fig. 3d as a function of the per cent compression in the sheet thickness direction. These results show that $P_{MAX}/\Delta T^2$ is maximized for ∼15% sheet compression, and that this compression provides a 23% increase in this performance metric (from 0.13 to 0.16 mW m^{-2} K^{-2}).

The data in Fig. 3e show the origin of this effect of sheet compression. The mass transport resistance increases with increasing level of compression, whereas the ohmic resistance decreases with increasing compression. Ohmic resistance decreases with increasing compression, because a shorter electron pathway to the external circuit results from an increased number of contacts between CNTs and a decreased electrode thickness. On the other hand, mass transport resistance, typically affected by ion accessibility into electrodes, is increased by compression, indicating sluggish ion diffusion within the compressed CNT network (Supplementary Fig. 2). Therefore, $P_{MAX}/\Delta T^2$ is maximized at an intermediate compression (15%), where mass transport resistance is not much affected by compression, but ohmic resistance is dramatically decreased.

To further improve electrode performance, Pt nanoparticles were deposited on the thermally oxidized CNT aerogel sheet electrodes by chemical reduction of a platinum salt solution (see Methods and Supplementary Fig. 3). This Pt nanoparticle deposition increases the electrode surface area for the thermocell reaction, and the catalytic activity of platinum reduces the charge transfer resistance[26]. The HR-TEM image of Fig. 4a and the statistical size analysis of deposited Pt nanoparticles in Supplementary Fig. 3b show that platinum nanoparticles with an average diameter of 2.4 nm are uniformly deposited on the surfaces of individual CNTs and nanotube bundles to provide 86 wt% of Pt in the electrodes.

CV measurements were used to obtain the changes in electroactive surface area (ESA)[27] and redox potential difference between oxidation and reduction that Pt deposition provided. The CV curves of Fig. 4b indicate an increase in the Faradaic peak current for the thermally oxidized sheet electrode and Pt-decorated thermally oxidized sheet electrode (Pt-sheet) compared with that for the as-drawn sheet electrode. Moreover, the Pt-sheet has the highest faradaic current. This increased current can be attributed to the increased ESA, which according

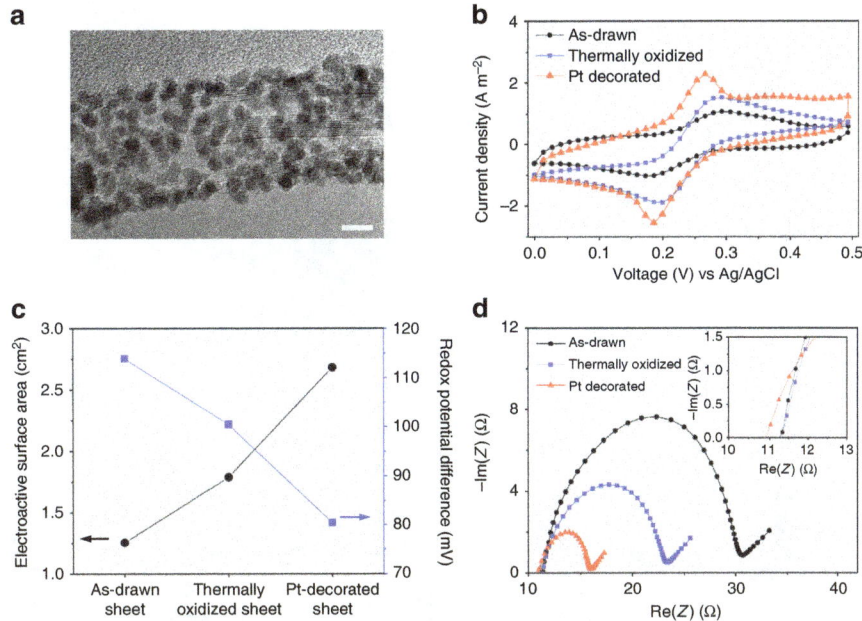

Figure 4 | Pt nanoparticle deposition for improving electrode performance. (**a**) HR-TEM image of individual CNT decorated by Pt nanoparticles. Scale bar, 5 nm. (**b**) Cyclic voltammograms obtained at a 100 mV s^{-1} scan rate using as-drawn, thermally oxidized, and thermally oxidized and Pt-deposited CNT electrodes. (**c**) Electroactive surface area and redox potential difference for the above various CNT electrodes. (**d**) Nyquist impedance plots for various electrodes. The inset shows a close-up of the high frequency region of the curves.

to the Randles–Sevcik equation[28] is given by:

$$I_p = 2.69 \times 10^5 \cdot \text{ESA} \cdot D^{1/2} \cdot n^{3/2} \cdot \upsilon^{1/2} \cdot C \qquad (2)$$

where I_p is the faradaic peak current, D is the diffusion coefficient, n is the number of electrons transferred during the redox reaction, υ is the potential scan rate, and C is the concentration of probe molecule.

The redox potential differences and ESAs determined by CV are shown in Fig. 4c for electrodes that are as-drawn, thermally oxidized and platinum-deposited after chemical cleaning by thermal oxidation. This figure shows that the cleaned CNT electrode has a higher ESA and a smaller redox potential difference than the as-drawn CNT electrode, which led to the improved performance shown in Fig. 3c. The largest ESA and the lowest potential difference resulted (Fig. 4c) when Pt nanoparticles were deposited onto the cleaned surface of CNTs. These results indicate that the Pt-decorated sheet electrodes provide the highest performance in thermocells, because they yield the largest effective surface area and the greatest reduction in the potential difference between peaks for oxidation and reduction. Electrochemical impedance spectroscopy analysis was also performed to support the performance improvement. The ESR (the intercept of the curve with the x-axis of the Nyquist plot[29] in the inset of Fig. 4d) is slightly reduced after Pt decoration. The Nyquist plot in Fig. 4d shows that the charge transfer resistance (the diameter of the semicircle)[30] is lower for the thermally oxidized sheet than for the as-drawn sheet and that it is lowest for the Pt-decorated, thermally oxidized sheet. These observations from electrochemical impedance spectroscopy measurements agree with the CV results.

Cylinder-type CNT thermocells and their performance. Building on our results for planar electrodes, we further optimized thermocell performance by using the cylindrical electrode configuration of Fig. 5a (while deploying the same 0.4 M ferro/ferricyanide aqueous electrolyte as for the flat thermocells).

The cylindrical electrodes pictured in this figure were fabricated by first using the apparatus of Fig. 5b to wind a forest-drawn CNT sheet onto a 300-μm diameter tungsten wire (which facilitates current collection) until a 3.0- to 3.5-mm outer scroll diameter was obtained. The SEM images of Fig. 5c show the realized highly uniform structure of the electrode sidewall and the orientation of the CNTs around the electrode circumference. Using the results of Fig. 3d on the optimal degree of electrode compression to produce densification, a cleaned CNT sheet electrode (wound to a 3.5 mm diameter) was compressed to 3.0 mm diameter by using two grooved templates (shown in Supplementary Fig. 4). Following deposition of Pt nanoparticles by chemical reduction of a platinum salt solution, two resulting cylindrical CNT sheet electrodes were inserted into a 3 mm inner diameter, cylindrical glass tube to form the thermocell of Fig. 5a.

The performance of the above thermocells containing differently prepared cylindrical CNT aerogel electrodes is presented in Fig. 5d,e for $\Delta T \approx 51\,°\text{C}$. The cylindrically configured as-drawn sheet electrode generates a P_{MAX} of 2.0 W m^{-2}. These results show that removal of carbonaceous impurities by the described optimized thermal oxidation increases the power output to 3.7 W m^{-2}, which is further increased for the cylindrical electrode thermocell to 6.0 W m^{-2} when Pt nanoparticles are deposited within the cylindrical electrodes. Finally, the maximum power density was further enhanced to 6.6 W m^{-2} when the cleaned, Pt-sheet electrode was laterally compressed. This output corresponds to an energy conversion efficiency of 3.95%, relative to the theoretically limiting Carnot cycle efficiency, as shown in Fig. 5f.

The energy conversion efficiency (η) of the thermocells was calculated using

$$\eta = \frac{(1/4)V_{\text{OC}} \cdot I_{\text{SC}}}{A_c \cdot \kappa(\Delta T/d)} \qquad (3)$$

where V_{OC} is the open-circuit voltage, I_{SC} is the short-circuit current, A_c is the cross-sectional area of the cell, κ is the thermal conductivity of the electrolyte, ΔT is the temperature difference

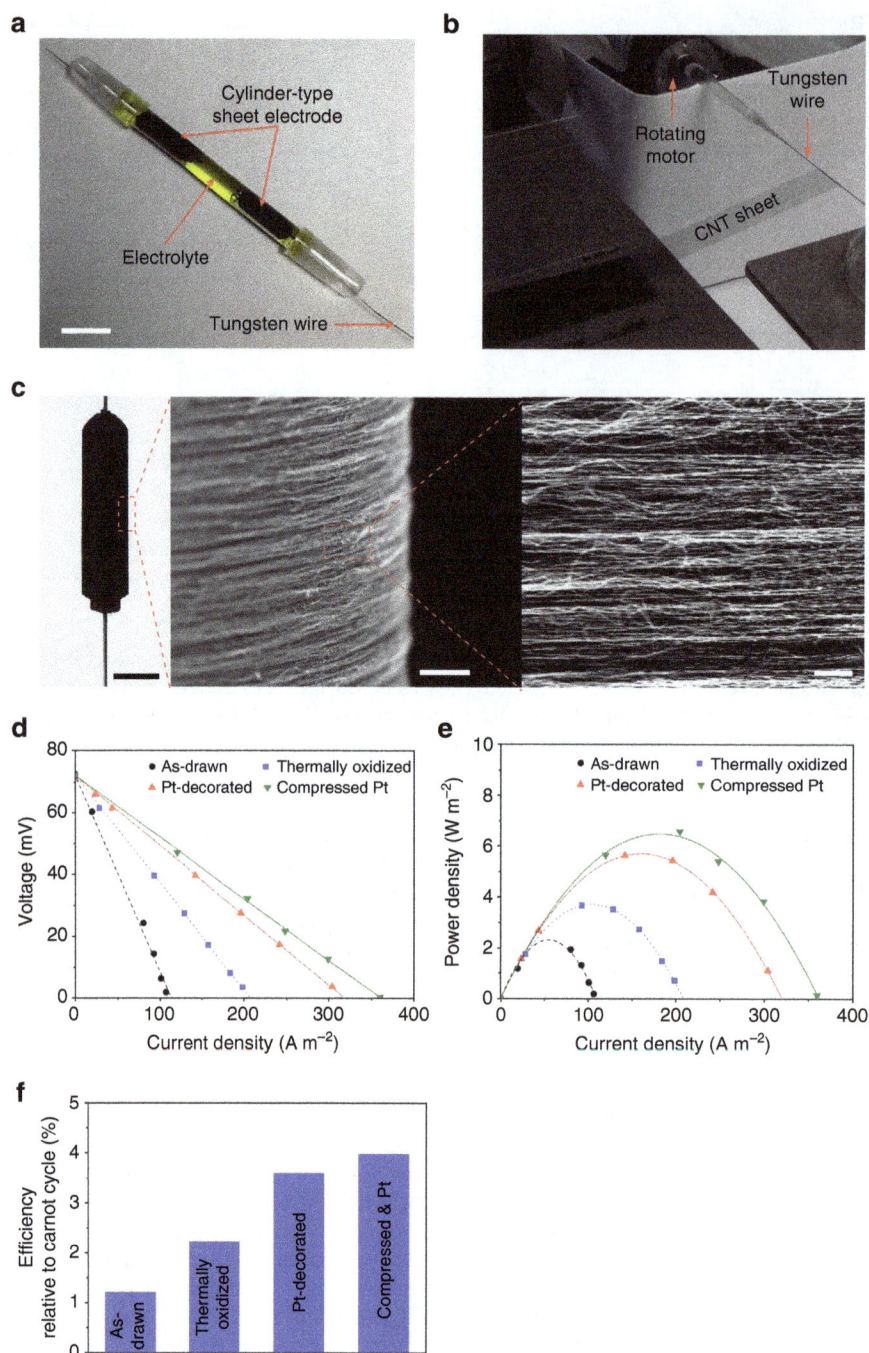

Figure 5 | Fabrication of cylinder-type CNT thermocells and their performance for $\Delta T \approx 51\,°C$. (a) Photograph of an assembled thermocell (scale bar, 2 cm) and (b) apparatus for fabrication of cylindrical CNT thermocell electrodes. (c) Photograph and SEM images of a cylindrical CNT electrode. Scale bars, 3 mm (left of panel), 30 μm (middle of panel) and 1 μm (right of panel). (d) Cell voltage and (e) power density versus current density for variously treated cell electrodes. (f) Energy conversion efficiency relative to Carnot cycle efficiency for various CNT electrodes in the cylindrical cell configuration. All sheet samples (except the as-drawn) were purified by thermal oxidation before assembly into cylindrical electrodes.

between the electrodes and d is the electrode separation distance (see Supplementary Note 2).

The open-circuit voltage was $V_{OC} = 72\,mV$ and the ΔT thereby calculated (from this voltage and the Seebeck coefficient of $1.43\,mV\,K^{-1}$, Supplementary Note 3, and Supplementary Fig. 5b) was $51.4\,°C$ for all electrode types in the cylindrical thermocell configuration. Thermal conduction dominates the heat transfer through the electrolyte between the hot and cold electrodes (see Supplementary Note 4, Supplementary Fig. 6 and Supplementary Table 2). The cell area and the thermal

conductivity[31] are $7.1 \times 10^{-6}\,m^2$ and $0.57\,W\,m^{-1}\,K^{-1}$, respectively. The electrode separation distance is 2.5 cm, corresponding to the closest approach distance between the hot and cold electrodes shown in Fig. 5a. The short-circuit current obtained from as-drawn, thermally oxidized, thermally oxidized and Pt-decorated; and thermally oxidized, compressed and Pt-decorated sheets in the cylindrical cell configuration are 0.8 mA, 1.5 mA, 2.3 mA and 2.6 mA, respectively. Using equation (3), η values of 0.17%, 0.32%, 0.51% and 0.56% are attained through the use of cylindrical thermocell configuration for these cylindrical thermocells from differently processed CNT aerogel sheets, respectively.

The energy conversion efficiency (η_r), relative to the Carnot efficiency limit of a heat engine, is

$$\eta_r = \frac{\eta}{(\Delta T / T_H)} \qquad (4)$$

where T_H is the hot temperature. We calculated the Carnot-relative energy conversion efficiency using equations (3) and (4) (See Supplementary Table 3 for parameters used in the equations), the measured $\Delta T = 51.4\,°C$, and an estimated T_H for the hot side electrode temperature of $90.7\,°C$ (estimated by considering symmetric geometry of cylindrical cell and electrodes arrangement, and ΔT). The resulting calculated Carnot-relative efficiency was $\eta_r = 3.95\%$ for the optimized cylindrical thermocell (based on cylindrical electrodes made of thermally purified, optimally compressed, Pt-decorated CNT sheets). This thermocell efficiency is substantially higher than the previously reported record Carnot-relative efficiency for a thermocell, which is 2.63% (ref. 12).

Discussion

The ability to simultaneously obtain a record Carnot-relative efficiency (3.95%) and a record areal current density ($6.6\,W\,m^{-2}$ for a temperature difference of only $51.4\,°C$) bodes well for eventual practical deployment of thermo-electrochemical cells for harvesting low-grade heat as electrical energy. This work demonstrates the importance of electrode purity, engineered porosity and catalytic surfaces on thermocell performance and the ways that each predominately affects thermocell parameters. It is doubtful that thermal electrochemical cells will ever match the efficiencies obtained for costly thermoelectrics. However, potentially low cost (using future low-cost catalyst) and convenient deployability (like the demonstrated wrapping of flexible CNT sheet thermocells about hot pipes[10]) might eventually lead to their importance in the menagerie of means for practically harvesting thermal energy.

Methods

Mass transfer coefficients of CNT electrodes. The mass transport characteristics of CNT buckypaper and CNT aerogel sheet electrodes were investigated using the electrolytic flow cell of Supplementary Fig. 1. CNT buckypaper was prepared by vacuum filtration of MWNTs (CM-95, Hanwha Nanotech) suspended in anhydrous N,N-dimethylformamide (N,N-DMF, Sigma-Aldrich) onto a membrane filter (Millipore PTFE filter, 0.2 µm pore size, 47 mm diameter), washing with deionized (DI) water and methanol, drying in vacuum and removal of the formed sheet from the filter. CNT aerogel sheet electrodes were prepared by sheet draw from a CNT forest. The electrolytic flow cell shown in Supplementary Fig. 1 consists of two parallel collecting electrodes that are separated by 10 mm. For both the CNT buckypaper and CNT aerogel sheet evaluations, 100-µm-thick CNT electrodes having an identical area of $1.0 \times 1.0\,cm^2$ are attached on the centre of both collecting electrodes using carbon paste. Other exposed areas of the collecting electrodes were covered with 100-µm-thick polyethylene terephthalate film. Various aqueous 1:2 concentrations of $Fe(CN)_6^{3-} / Fe(CN)_6^{4-}$ redox couple in 0.5 M NaOH were used for the experiments by varying the $Fe(CN)_6^{3-}$ concentration from 20 to 80 mM. The concentration of $Fe(CN)_6^{4-}$ was twice as high as the $Fe(CN)_6^{3-}$ concentration for each experiment to ensure a limiting reaction rate at the cell cathode (that is, $Fe(CN)_6^{3-} + e^- \rightarrow Fe(CN)_6^{4-}$). The electrolyte stored in a glass container was circulated through the cell by a peristaltic pump (Longer pump, BT100-2J), and the flow rate was kept low at $6.6 \times 10^{-6}\,m^3\,s^{-1}$ for allowing laminar flow in a cell channel. A power supply (Keithley, 2400 Source-meter) was used to drive electrochemical reactions.

Preparation and post-treatment of CNT aerogel sheet. Vertically aligned MWNT arrays were grown on an iron-catalyst-coated silicon (Si) substrate by chemical vapour deposition of acetylene gas[32]. The diameter and height of MWNTs are $\sim 10\,nm$ and $\sim 200\,µm$, respectively. CNT sheet was continuously drawn from a sidewall of the MWNT forests using a dry spinning process[32,33] and wrapped around a rectangular tungsten frame for planar-type electrodes and tungsten wire for cylinder-type electrodes, using a connected rotating motor at 10 r.p.m. The planar electrode has a thickness of 100 µm and area of $1.0 \times 1.0\,cm^2$

and the cylinder electrodes had a diameter and length of 3 mm and 2.0 cm, respectively. Thermal annealing of the CNT sheets to remove carbonaceous impurities was performed in ambient atmosphere using a halogen-lamp-heated quartz tube furnace, which had a ramp time to target temperatures (from 300 to 400 °C) of 1 min. For deposition of Pt nanoparticles by chemical reduction of a platinum salt solution[34], 3.75 mg of K_2PtCl_4 (Aldrich) was dissolved in 50 ml of diluted ethylene glycol solution (3:2 by volume ethylene glycol:DI water). Afterwards, the described thermally oxidized CNT sheet electrodes were immersed in the platinum salt solution and Pt nanoparticle deposition was permitted for 3 h at 110 °C with weak stirring to prevent mechanical damage to CNT sheet. After the reaction, the pH of the solution was decreased to 2 using HCl and the electrode was washed with DI water several times to remove excess ethylene glycol. To avoid liquid-based aerogel densification during liquid evaporation, the electrodes were not dried before use in the thermocells.

Thermocell testing and materials characterization. The electrolyte used in all tests was a 0.4 M aqueous solution of $Fe(CN)_6^{4-} / Fe(CN)_6^{3-}$. The ionic conductivity of the electrolyte at different temperatures was measured using a conductivity meter (Metter Toledo S-230), as shown in Supplementary Fig. 7. For evaluation of planar thermocells, the cell (Fig. 1a) was placed between two fluid-heated plates that were connected to hot and cold thermostatic baths to provide $\pm 0.1\,°C$ control of plate temperatures. For cylindrical thermocell testing, the glass wall surrounding each cell electrode (which are separated by 2.5 cm) was wrapped and bonded with a TYGON tube, through which cooling fluid or heating fluid was passed. The inter-electrode direction of the thermocell was oriented horizontally during measurements. While the hot and cold fluid temperatures were 100 °C and 30 °C, respectively, thermal resistances between the heating and cooling sources and the respective electrodes reduced the inter-electrode temperature difference from 70 °C to $\Delta T = 51.4\,°C$. This ΔT of $51.4\,°C$, obtained by dividing the measured V_{OC} by the thermo-electrochemical Seebeck coefficient, would correspond to a temperature difference between the cold and hot electrodes averaged over the radial shape of CNT electrode. A voltage-current meter (Keithley, 2000 Multimeter) was used for characterizing cell voltage versus cell current for different external resistive loads, and thereby determining power output. The glass tube having 3 mm inside diameter was used for the cylinder thermocell experiments (Fig. 5a). Raman spectra as a function of the thermal oxidation times and the temperature used for removing carbonaceous impurities were recorded for 514 nm excitation using a Renishaw inVia Raman Microscope. Sample structure was further characterized by field-emission scanning electron microscopy (FE-SEM, Hitachi-S4800) and transmission electron microscopy (JEM-300F). Electrochemical impedance measurements were conducted in the frequency range between 10 kHz and 50 mHz using a commercial instrument (Zahner, IM6ex). CV (using a Digi-Ivy, DY2100 instrument) used 10 mM $K_3Fe(CN)_6$ with 0.1 M KCl as the supporting electrolyte in aqueous solution and a scan rate of $100\,mV\,s^{-1}$. Platinum and Ag/AgCl electrodes were used as counter and reference electrodes, respectively, for the CV measurements.

References

1. Lan, Y., Minnich, A. J., Chen, G. & Ren, Z. Enhancement of thermoelectric figure-of-merit by a bulk nanostructuring approach. *Adv. Funct. Mater.* **20**, 357–376 (2010).
2. Biswas, K. *et al.* High-performance bulk thermoelectrics with all-scale hierarchical architectures. *Nature* **489**, 414–418 (2012).
3. Zhao, D. Waste thermal energy harvesting from a convection-driven Rijke–Zhao thermo-acoustic-piezo system. *Energy Convers. Manage.* **66**, 87–97 (2013).
4. Hochbaum, A. I. *et al.* Enhanced thermoelectric performance of rough silicon nanowires. *Nature* **451**, 163–167 (2008).
5. Hall, P. J. & Bain, E. J. Energy-storage technologies and electricity generation. *Energy Policy* **36**, 4352–4355 (2008).
6. Kraemer, D. *et al.* High-performance flat-panel solar thermoelectric generators with high thermal concentration. *Nat. Mater.* **10**, 532–538 (2011).
7. Hewitt, C. A. *et al.* Multilayered carbon nanotube/polymer composite based thermoelectric fabrics. *Nano Lett.* **12**, 1307–1310 (2012).
8. Vining, C. B. An inconvenient truth about thermoelectrics. *Nat. Mater.* **8**, 83–85 (2009).
9. Kuzminskii, Y. V., Zasukha, V. A. & Kuzminskaya, G. Y. Thermoelectric effects in electrochemical systems. Nonconventional thermogalvanic cells. *J. Power Sources* **52**, 231–242 (1994).
10. Hu, R. *et al.* Harvesting waste thermal energy using a carbon-nanotube-based thermo-electrochemical cell. *Nano Lett.* **10**, 838–846 (2010).
11. Nugent, J. M., Santhanam, K. S. V., Rubio, A. & Ajayan, P. M. Fast electron transfer kinetics on multiwalled carbon nanotube microbundle electrodes. *Nano Lett.* **1**, 87–91 (2001).
12. Romano, M. S. *et al.* Carbon nanotube—reduced graphene oxide composites for thermal energy harvesting applications. *Adv. Mater.* **25**, 6602–6606 (2013).
13. Quickenden, T. I. & Mua, Y. A review of power generation in aqueous thermogalvanic cells. *J. Electrochem. Soc.* **142**, 3985–3994 (1995).
14. Eastman, E. D. Thermodynamics of non-isothermal systems. *J. Am. Chem. Soc.* **48**, 1482–1493 (1926).

15. Eastman, E. D. Theory of the Soret effect. *J. Am. Chem. Soc.* **50,** 283–291 (1928).
16. Kang, T. J. *et al.* Electrical power from nanotube and graphene electrochemical thermal energy harvesters. *Adv. Funct. Mater.* **22,** 477–489 (2012).
17. deBethune, A. J., Licht, T. S. & Swendeman, N. The temperature coefficients of electrode potentials: the isothermal and thermal coefficients—the standard ionic entropy of electrochemical transport of the hydrogen ion. *J. Electrochem. Soc.* **106,** 616–625 (1959).
18. Wragg, A. A. & Leontaritis, A. A. Local mass transfer and current distribution in baffled and unbaffled parallel plate electrochemical reactors. *Chem. Eng. J.* **66,** 1–10 (1997).
19. Cañizares, P., García-Gómez, J., Fernández de Marcos, I., Rodrigo, M. A. & Lobato, J. Measurement of mass-transfer coefficients by an electrochemical technique. *J. Chem. Educ.* **83,** 1204 (2006).
20. Moon, J.-M. *et al.* High-yield purification process of singlewalled carbon nanotubes. *J. Phys. Chem. B* **105,** 5677–5681 (2001).
21. Fang, H.-T. *et al.* Purification of single-wall carbon nanotubes by electrochemical oxidation. *Chem. Mater.* **16,** 5744–5750 (2004).
22. Rinzler, A. G. *et al.* Large-scale purification of single-wall carbon nanotubes: Process, product, and characterization. *Appl. Phys. A* **67,** 29–37 (1998).
23. Dementev, N., Osswald, S., Gogotsi, Y. & Borguet, E. Purification of carbon nanotubes by dynamic oxidation in air. *J. Mater. Chem.* **19,** 7904–7908 (2009).
24. Dresselhaus, M. S., Dresselhaus, G., Jorio, A., Souza Filho, A. G. & Saito, R. Raman spectroscopy on isolated single wall carbon nanotubes. *Carbon* **40,** 2043–2061 (2002).
25. Im, H. *et al.* Flexible thermocells for utilization of body heat. *Nano Res.* **7,** 1–10 (2014).
26. Rubi, J. M. & Kjelstrup, S. Mesoscopic nonequilibrium thermodynamics gives the same thermodynamic basis to butler — volmer and nernst equations. *J. Phys. Chem. B* **107,** 13471–13477 (2003).
27. Pacios, M., del Valle, M., Bartroli, J. & Esplandiu, M. J. Electrochemical behavior of rigid carbon nanotube composite electrodes. *J. Electroanal. Chem.* **619–620,** 117–124 (2008).
28. Papakonstantinou, P. *et al.* Fundamental electrochemical properties of carbon nanotube electrodes. *Fuller. Nanotubes Carbon Nanostruct.* **13,** 91–108 (2005).
29. Park, S.-M. & Yoo, J.-S. Peer reviewed: electrochemical impedance spectroscopy for better electrochemical measurements. *Anal. Chem.* **75,** 455 A–461 A (2003).
30. Randles, J. E. B. Kinetics of rapid electrode reactions. *Discuss. Faraday Soc.* **1,** 11–19 (1947).
31. Romano, M. *et al.* Novel carbon materials for thermal energy harvesting *J. Therm. Anal. Calorim.* **109,** 1229–1235 (2012).
32. Zhang, M. *et al.* Strong, transparent, multifunctional, carbon nanotube sheets. *Science* **309,** 1215–1219 (2005).
33. Im, H. *et al.* Enhancement of heating performance of carbon nanotube sheet with granular metal. *ACS Appl. Mater. Interfaces* **4,** 2338–2342 (2012).
34. Jining, X., Nanyan, Z. & Vijay, K. V. Functionalized carbon nanotubes in platinum decoration. *Smart Mater. Struct.* **15,** S5 (2006).

Acknowledgements

This research was supported in Korea by the National Research Foundation of Korea (Grants 2009-0083512, 2014R1A2A1A05007760 and 2014R1A1A4A01008768). Financial support at the University of Texas at Dallas was from Air Force grants FA9550-13-C-0004, FA9550-15-1-0089 and Robert A. Welch Foundation grant AT-0029.

Author contributions

H.I. and T.K. contributed to experiment design, measurements, data analysis and manuscript preparation. H.S., J.C., J.S.P. and H.D.Y contributed to experimental measurements and data analysis. R.O.-R. made and characterized nanotube samples. K.D.K., R.H.B., H.H.L., T.J.K. and Y.H.K. contributed to planning experiments, data analysis and manuscript preparation.

Additional information

27

A method for controlling the synthesis of stable twisted two-dimensional conjugated molecules

Yongjun Li[1], Zhiyu Jia[1], Shengqiang Xiao[2], Huibiao Liu[1] & Yuliang Li[1]

Thermodynamic stabilization (π-electron delocalization through effective conjugation) and kinetic stabilization (blocking the most-reactive sites) are important considerations when designing stable polycyclic aromatic hydrocarbons displaying tunable optoelectronic properties. Here, we demonstrate an efficient method for preparing a series of stable two-dimensional (2D) twisted dibenzoterrylene-acenes. We investigated their electronic structures and geometries in the ground state through various experiments assisted by calculations using density functional theory. We find that the length of the acene has a clear effect on the photophysical, electrochemical, and magnetic properties. These molecules exhibit tunable ground-state structures, in which a stable open-shell quintet tetraradical can be transferred to triplet diradicals. Such compounds are promising candidates for use in nonlinear optics, field effect transistors and organic spintronics; furthermore, they may enable broader applications of 2D small organic molecules in high-performance electronic and optical devices.

[1] CAS Key Laboratory of Organic Solids, Beijing National Laboratory for Molecular Science (BNLMS), Institute of Chemistry, Chinese Academy of Sciences, Beijing 100190, China. [2] State Key Laboratory of Advanced Technology for Materials Synthesis and Processing, Wuhan University of Technology, Wuhan 430070, China. Correspondence and requests for materials should be addressed to Yongjun Li (email: liyj@iccas.ac.cn) or to Yuliang Li (email: ylli@iccas.ac.cn).

In the last two decades, methodologies for the controlled synthesis of large π-extended polycyclic aromatic hydrocarbons (PAHs) have been established using bottom-up principles[1–11]. The properties of PAHs depend heavily on their degrees of π-extension, shapes, widths and edge topologies[1,3,4,12–14]. Many linear acenes, including tetracene, pentacene and related derivatives, have been synthesized to produce highly desirable electronic properties, including remarkable charge-carrier mobilities[15–19]. Open-shell PAHs, which feature one or more π-electrons not tightly paired into the bonding molecular orbital in the ground state, are particularly interestingly structures that have properties very different from those of typical PAHs having closed-shell electronic structures[20–23]. The unique electronic structures of two-dimensional (2D) open-shell PAHs can impart attractive electronic, optical and magnetic properties for applications in materials science[24–26].

Acenes become increasingly reactive as the number of rings increases, with the central ring being the most reactive[27,28]. As a result, the central rings in these molecules are susceptible to oxidation, photodegeneration and Diels–Alder reactions[29,30]. Open-shell molecules are particularly vulnerable to degradation reactions[22,23]; therefore, instability remains a key obstacle affecting their practical applications. In most cases, thermodynamic stabilization (π-electron delocalization through effective conjugation) and kinetic stabilization (blocking of the most-reactive sites) are both necessary to obtain stable materials[18,31]. One strategy is to annulate the aromatic rings onto two neighbouring rings, creating 2D acene analogues[15]. Changing from the linear character of a condensed array (peri-condensed) to an angular geometry (cata-condensed) results in at least two sextets of π-electrons, increasing the stability relative to that of a linear analogue having only one sextet[32]. Tuning a planar structure to a nonplanar structure through twisting or saddling can also enhance the kinetic stability as a result of steric blocking[21,33].

2D acene analogues that represent various fragments of graphene have received significant attention because of their electronic properties[6,9,10,15]. Molecules of this class have large, planar π-surfaces allowing high-intermolecular surface overlap[34–36]. The packing of such molecules is dictated by multiple interactions that can effectively increase the dimensionality of the electronic structure, leading to enhanced transport properties[37,38]. Many 2D planar PAHs[11,32,34,36,39–41] and contorted polycyclic aromatics[5,42–44] have been synthesized and characterized recently.

In this paper, we report the transformation of a one-dimensional (1D) single conjugated dibenzoterrylene through the attachment of conjugated acene 'arms' to generate a 2D molecular structure. High dimensionality within the single molecule was assured by hybridizing dibenzoterrylene with acenes of different conjugation lengths; stability was provided by benzoannulating the active rings of acenes and inducing contortions from the cove or fjord regions. This protocol is applicable to construct various 2D twisted nanographene-like structures in a rapid manner.

Results

Design of target compounds. Figure 1a showed the conceptual approaches toward twisted 2D acenes. We positioned electron-donating/withdrawing groups (for example, CH_3, OMe, CN, F) to alter the optical absorbance and the energy levels of the highest occupied molecular orbital (HOMO) and lowest unoccupied molecular orbital (LUMO) of the tetracene-dibenzoterrylene hybrid, in addition to tuning the molecular packing in the solid state (Fig. 1b). We introduced phenanthrenyl groups to further increase the steric constraint, and naphthyl and perylenyl groups to elongate the acenes and allow annulation of benzene rings. Using this approach, 2D twisted acenes varying from closed-shell to open-shell structures were available through extension of the acenes vertical to the benzoterrylene and benzoannulation at the zigzag edge. Compound **4** can be considered (Fig. 1c) as a hybrid of two pentacene-annulated-tribenzo-octacenes, in which two phenalenyl radicals (5,10 positions of octacene) are annulated with a central tetracene unit, or having evolved from a dibenzo-pentacene with the reactive 5′,7′-positions of the pentacene connected and protected together. In both cases, one more Clar sextet was formed on resonance to tetraradical state, which also predicted the stability of the tetraradical.

Synthesis of target compounds. Figure 2 displays the synthesis of **1**, starting from the bromination of the amine **5** (ref. 45) with Br_2 followed by conversion of the amino group to the iodide through diazotization. Compound **7** was coupled with 2,7-bis(4,4,5,5-tetramethyl-1,3,2-dioxaborolan-2-yl)pyrene through Suzuki-Miyaura cross-coupling to give the key intermediate **8**. This compounds was coupled with a series of $RB(OH)_2$ derivatives **9** to provide the precursors **10a–g** for Scholl-type oxidative cyclode-hydrogenation, in which $FeCl_3$ was used as both a Lewis acid and an oxidant. *tert*-Butyl groups were introduced to increase the solubility of the final product. A phenanthrenyl group was used to further increase the steric constraint, leading to precursor **11**; naphthyl and perylenyl groups were introduced to elongate the acenes and introduce the annulation benzene rings, providing the precursors **12** and **13**, respectively.

For **10e**, C–C bonds were formed *ortho* and *para* to the *meta* fluorine atom, leading to a mixture where the fluorine atoms were located at the R_1 and R_3 positions, as supported in the X-ray structure (see below). For **11**, the Scholl reaction at the K-region (convex armchair edge) of the phenanthrenyl group led to **2** in 57%, which is too crowded at the fjord regions to allow two benzene rings to get sufficiently close to form a further C–C bond[46]. π-Conjugation extended compound **3** was obtained from **12** in 78% yield. The oxidative cyclohydrogenation of **13** is time-dependent. When the reaction time was <2 h, we obtained some incompletely cyclized products, which were difficult to separate from the target **4**; when the reaction time was longer than 12 h we obtained some byproducts with similar molecular weight as **4** but with different absorption spectra. The black compound **4** is well-soluble in $C_2H_2Cl_4$ and *o*-dichlorobenzene (dark-green solution), partially soluble in CH_2Cl_2 and $CHCl_3$, we could purify it (in 51% yield) through column chromatography on deactivated silica gel with CH_2Cl_2/MeOH (5/1). We characterized all of the intermediates and final products, except for **4**, using ^1H and ^{13}C nuclear magnetic resonance (NMR) spectroscopy and high-resolution mass spectrometry (MS); we characterized compound **4** only through high-performance liquid chromatography, high-resolution MS and elemental analysis.

Conformations characterization of compounds 1–3. At first, we performed density functional theory (DFT) optimization to gain insight into the molecular conformations of these 2D acenes. The nonplanar conformations of these compounds arose from severe steric strain between the two C–H bonds at the cove or fjord regions, as can be seen clearly in the model of **3** in Fig. 3a. DFT calculations revealed that **3** has five possible conformations (Fig. 3a); the hexacene units can take on conformations that are twisted or anti-folded with the central naphthalene unit, while the outer naphthalene rings in the hexacenes can be contorted out of the plane defined by the dibenzoterrylene core in the *cis* or *trans* direction: cAA-**3**, tAA-**3**, AT-**3**, cTT-**3** and tTT-**3**. Compound **1**

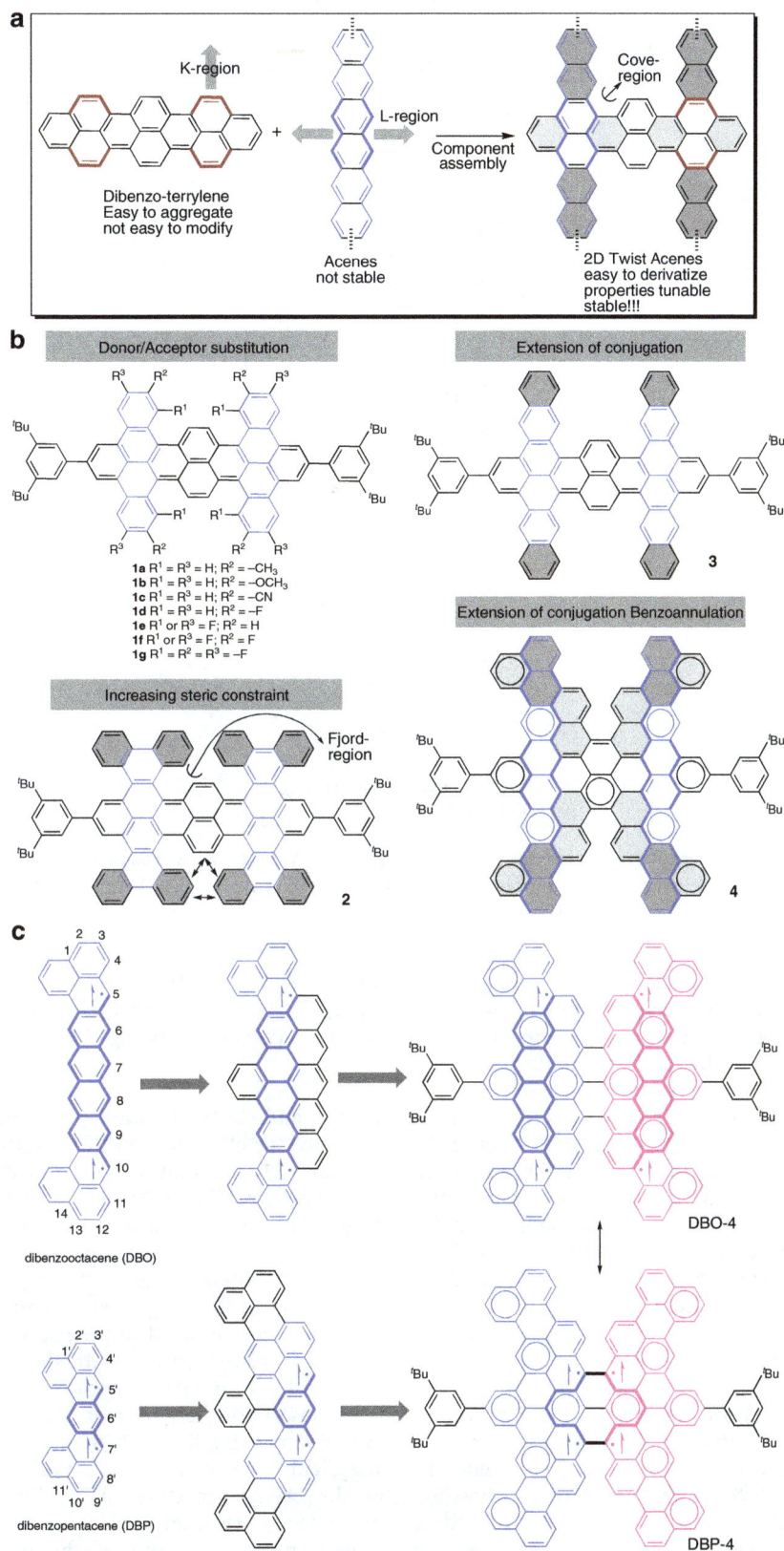

Figure 1 | Conceptual approaches toward twisted 2D acenes. (**a**) π-Components of 2D acenes. (**b**) Molecular structures of 2D acenes with increasing steric constraint (**2**), extension of conjugation (**3**) and with both extension of conjugation and benzoannulation at the L-region (zigzag edge; **4**). (**c**) Chemical structural evolution of **4**: (top) Benzoannulation of octacene to generate phenalenyl radicals annulated with a central tetracene and further pentacene-annulation on the other side to provide more stable diradicals, two of which connect in the central area; (bottom) evolution from a dibenzopentacene with benzo- and naphtheno-annulation, with the reactive 5′,7′-positions of the pentacene connected and protected together.

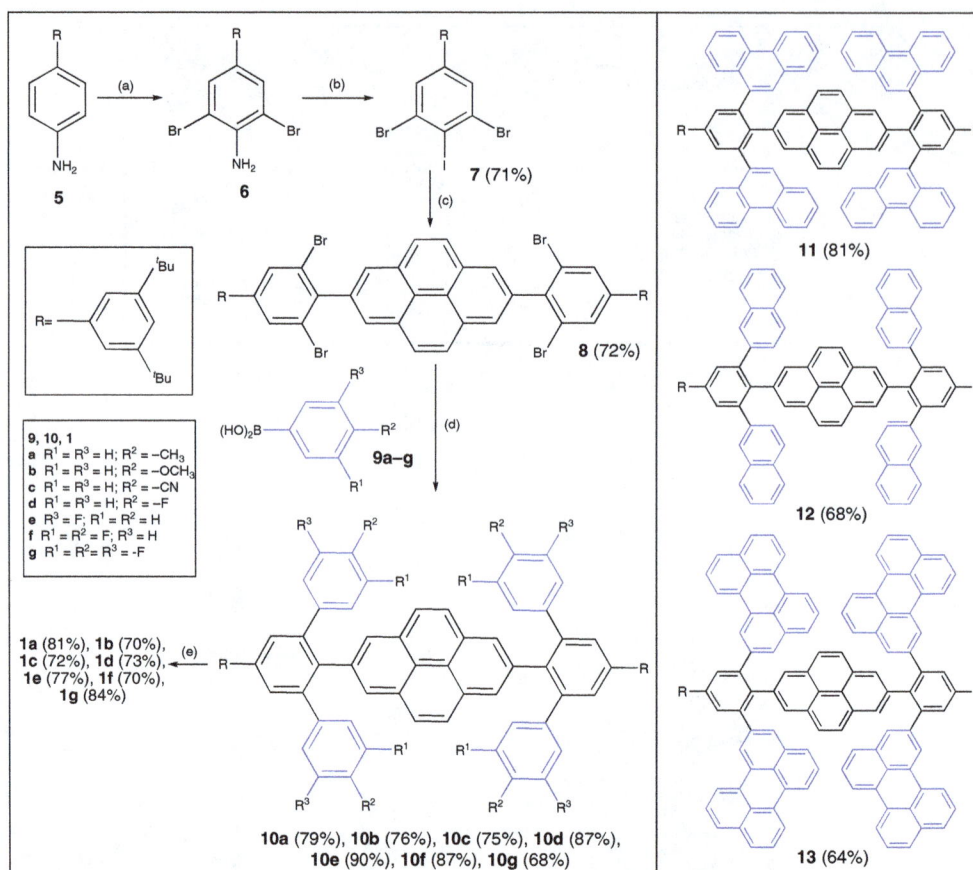

Figure 2 | Synthesis of the tetracene-dibenzoterrylene hybrids 1a–f and the precursors 11–13 for generation of 2–4. (a) Br$_2$, DCM/MeOH; (b) NaNO$_2$, AcOH/H$_2$SO$_4$, KI/I$_2$; (c) 2,7-bis(4,4,5,5-tetramethyl-1,3,2-dioxaborolan-2-yl)pyrene, Pd(PPh$_3$)$_4$, Na$_2$CO$_3$, THF/H$_2$O; (d) Pd(PPh$_3$)$_4$, Na$_2$CO$_3$, toluene/EtOH/H$_2$O; (e) FeCl$_3$, CH$_3$NO$_2$, CH$_2$Cl$_2$, 4 h.

existed in conformations analogous to those of **3** (Supplementary Fig. 69; Fig. 3b displays the most stable conformation of **1a**). In all of these cases, the C$_2$-symmetric *c*AA structures were calculated to be more stable than the other conformers by 3.03–24.65 kJ mol^{-1} (Supplementary Table 1).

We grew single crystals through diffusion of acetone into a solution of **1e** in CHCl$_3$ (Supplementary Table 2). Figure 3d displays the crystal structure of **1e**. The fluorinated outer aromatic rings are contorted, in the same direction, out of the plane defined by the dibenzoterrylene core, resulting in a saddle-like conformation, consistent with the calculated gas-phase structure (*c*AA-**1e**). The degree of twisting (θ), measured as the dihedral angle between the central naphthalene and the fluorinated outer benzene/terminal benzo group, were in the range of 32.2–37.3° (Fig. 3e). Molecules of **1e** were stacked in columns along the *c*-axis (Fig. 3f). Within a columnar stack, two saddle molecules with curvature in the same direction formed a π-dimer with effective π–π surface overlap; this π-dimer then stacked with another π-dimer of opposite molecular orientation.

For compound **2**, we predicted a similar set of five possible conformations (Supplementary Fig. 70), with *c*TT-**2** calculated as the most stable because of additional steric strain between the phenanthrenyl moieties (Fig. 3c). Fortunately, this zero-dipole-moment conformer *c*TT-**2** could be separated through column chromatography. Through slow diffusion of acetone into a solution of *c*TT-**2** in CHCl$_3$, were obtained crystals for X-ray analysis, although they diffracted only weakly (Supplementary Fig. 71). A partial solution revealed that the phenanthrenyl units were twisted with respect to the central naphthalene unit in the *cis* direction. Although frozen in a single conformation as a result of

crystal packing constraints in the solid state, these 2D acenes are highly flexible in solution and change their shape constantly on heating (Supplementary Fig. 72).

Optical and electrochemical properties of compounds 1–3. We measured the ultraviolet–visible absorption spectra of compounds **1–3** in CH$_2$Cl$_2$ solution (Fig. 4a–d, Supplementary Fig. 73, Supplementary Table 3). Absorption bands appeared in the ranges 300–400 nm and 400–600 nm with molar absorptivities of $>50,000 \, M^{-1}$. The substituents and lengths of the acenes appeared to influence the π-stacking of these 2D acenes. For **1**, three peaks in the range 400–600 nm, attributed to $0\rightarrow0$, $0\rightarrow1$ and $0\rightarrow2$ vibrational absorptions, were good handles for observation of its aggregation. With the increasing of concentration, the intensity of the $0\rightarrow0$ absorption decreased; in particular, the $0\rightarrow0$ absorption became a shoulder for the CN-substituted **1c**, even disappearing for the trifluorine-substituted **1g**. These results suggested the permanent formation of H-aggregated π-dimers, as in the crystal structure of **1e** (Fig. 3d) and the molecular modelling of **1g** (Fig. 3g), in solution. On increasing the concentration, the intensity of the $0\rightarrow0$ vibrational absorption decreased further (Supplementary Fig. 74), with the $0\rightarrow2$ vibrational absorptions becoming the main absorption bands in the spectra of spin-coated thin film of **1c–f**, consistent with further H-aggregation of their π-dimers. In the thin films of **1a** and **1b**, in addition to H-aggregation we also observed some J-aggregates, as indicated by the red-shifting of new bands. Such J-aggregates were also found for compounds **2** and **3** with bulky acenes, with red-shifting of the absorption maxima (λ_{max}).

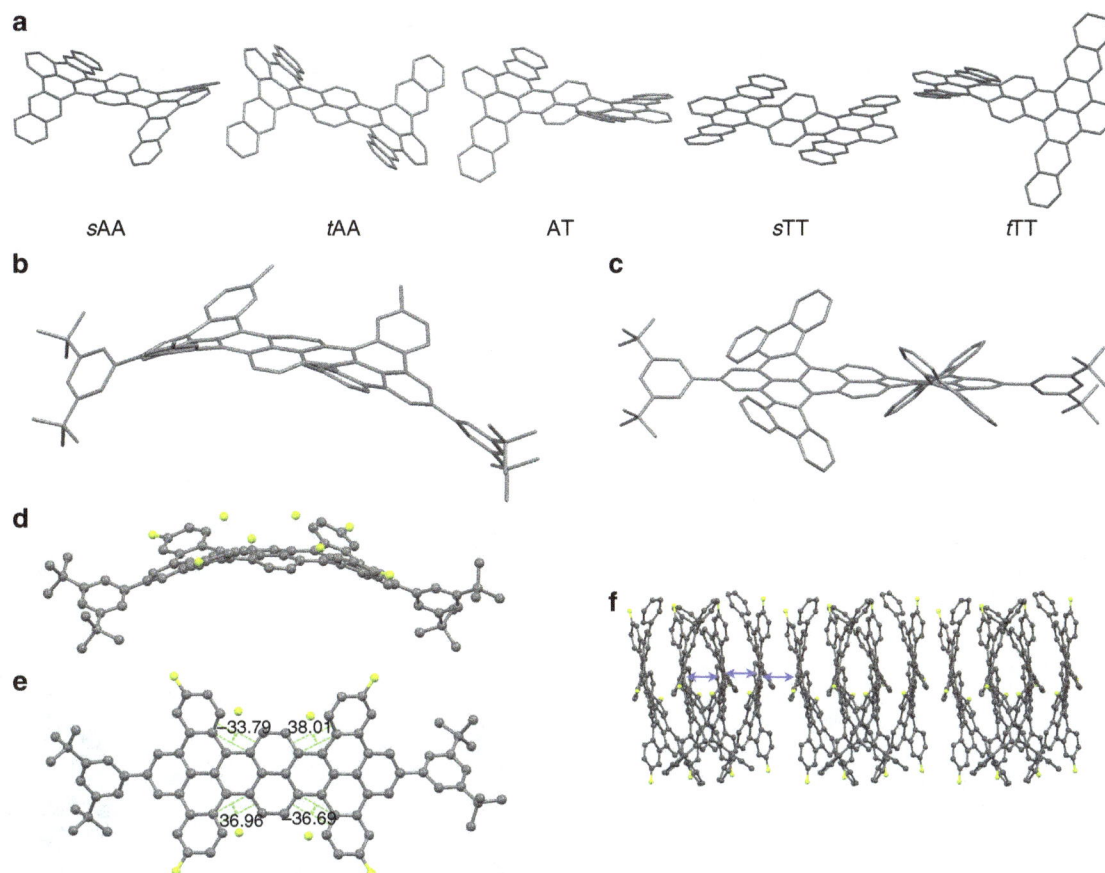

Figure 3 | Conformational flexibility of the 2D acenes. (a) Possible molecular conformations of **3** and **(b,c)** the most stable conformations of **(b) 1a** and **(c) 2**, calculated at the B3LYP/6-31G **(d)** level of theory. cAA-: anti-folded and anti-folded in the *cis* direction; tAA-: anti-folded and anti-folded in the *trans* direction; AT-: anti-folded and twisted; cTT-: twisted and twisted in the *cis* direction; tTT-: twisted and twisted in the *trans* direction. **(d)** Side and **(e)** top views of the π-backbone of **1e** from crystal structure analysis; **(f)** molecular packing of **1e**, viewed along the c-axis of the unit cell; the hydrogen atoms and *tert*-butyl groups have been omitted for clarity; carbon and fluorine atoms are coloured grey and green, respectively.

The photoluminescence features of these compounds in dilute solution (10^{-3} mmol l^{-1}) were complementary of their absorption spectra; quantum yields were in the range 0.52–0.89 (Supplementary Table 3). We observed large bathochromic emission features in the emission spectra of these compounds on concentrating or going from solution to the solid state, consistent with strong aggregation. For the tetracene-dibenzoterrylene **1**, we observed a bathochromic emission shift from 108 to 190 nm; the more bulky compounds **2** and **3** featured shifts of only 24 and 71 nm, respectively.

We used cyclic voltammetry (CV) and differential pulse voltammetry to examine the electrochemical behaviour of **1–3** in CH$_2$Cl$_2$ containing 0.1 M TBAPF$_6$ as the supporting electrolyte (Supplementary Fig. 75, Supplementary Table 3). The CV traces of **1a–c**, **1e**, **1f** and **2** each featured two well-defined, reversible oxidation potentials. The cyclic voltammogram of **3** was more remarkable, displaying three oxidation potentials for four electrons at +0.25 V (**1e**), +0.51 V (**1e**) and +0.72 V (**2e**), all as fully reversible waves (Fig. 4e). Compound **1d** exhibited a similar oxidation behaviour, with signals at +0.42 V (**1e**), +0.63 V (**1e**) and +0.87 V (**2e**) (versus Fc/Fc$^+$), implying multi-charge storage ability[47] for both **1d** and **3**. Compared with **1a**, the electron-donating MeO groups decreased the oxidation potentials by 70 mV, while the electron-withdrawing groups increase the oxidation potentials more evidently; for example, CN groups shifted the oxidation potentials by ~350 mV. The oxidation peaks shifted to higher potentials on increasing the

number of fluorine atoms (50–70, 250 and 500 mV for one, two and three fluorine substituents, respectively). Extension of the conjugation of the acene units also decreased the oxidation potentials; for example, to 0.25 V for **3** with its six conjugated benzene rings and to 0.21 V for **2** with its eight conjugated benzene rings. Consistent with these findings, **2** and **3** appear to be susceptible to slow oxidative degradation in solution; no such degradation appeared in the solutions of the other compounds. All of these compounds exhibited two reduction potentials, with the effects of the substituents on the reduction potentials occurring in the same direction as those on the oxidation potentials, but to a lesser extent. Accordingly, the band gaps of these species were influenced by the substituents and the conjugation length, with electron acceptors decreasing the HOMO energy level and, thereby, increasing the band gap (see Fig. 4f, especially for the trifluorinated derivative).

We performed the DFT calculations at the B3LYP/6-31G(d) level to examine the electronic structures of the dibenzoterrylene-acene hybrids. The calculated frontier orbitals indicated that they maintained the characteristics of their individual components. The HOMO and LUMO orbitals of **1a** are mainly localized in the central pyrene unit (Supplementary Table 4). We observed nodal planes perpendicular to the molecular skeleton in the HOMOs and LUMOs of these compounds, with the electron densities distributed symmetrically on the two sides of the nodal plane. The HOMO-1 and LUMO+1 orbitals of **1a** are comparable with the HOMO-2 and LUMO+2 orbitals, respectively, of

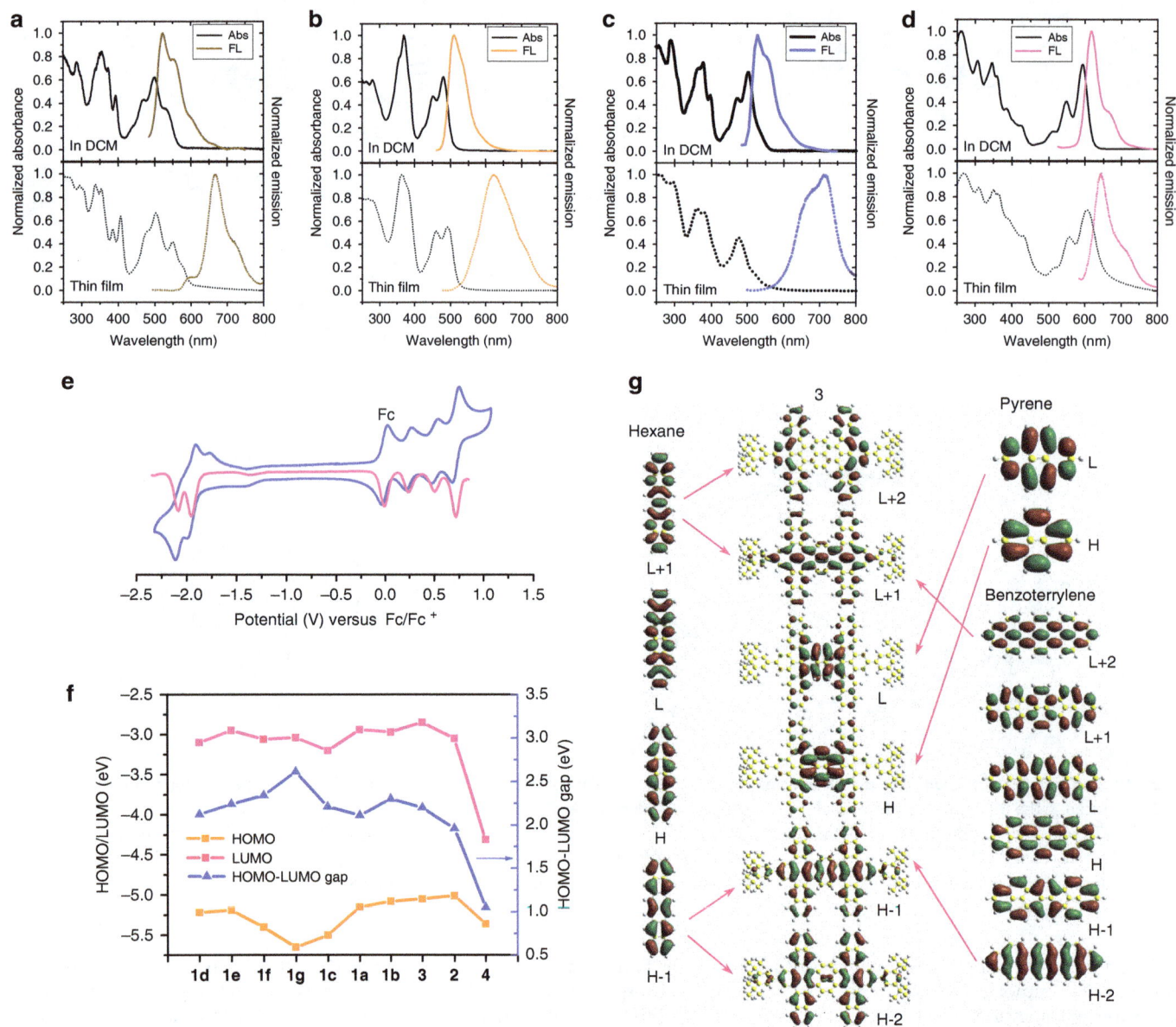

Figure 4 | Optical, electrochemical behaviour and orbital analysis of compounds 1–3. Absorbance (black) and emission (coloured) spectra of (**a**) **1a**, (**b**) **1g**, (**c**) **1c** and (**d**) **2** recorded in CH_2Cl_2 solution (solid lines) and in the solid state (spin-cast from CH_2Cl_2 solutions; dotted lines). (**e**) CV (blue) and DPV(red) traces of **3** in CH_2Cl_2 (0.1 mM) at room temperature. Scan rate: 100 mV s^{-1}; working electrode: glassy carbon; reference electrode: Ag wire; electrolyte: $TBAPF_6$. Fc/Fc$^+$: ferrocene/ferrocenium. (**f**) HOMO and LUMO energies and band gaps of **1–4**. (**g**) Frontier orbitals analysis of **3** calculated at the B3LYP/6-31G (**d**) level.

dibenzoterrylene. The other compounds had similarly appearing orbitals, except for **1f**, where the strongly electron-withdrawing CN group affected the polarization of its orbitals (Supplementary Table 4). Interestingly, extension of the acenes vertical to the benzoterrylene, as in **3**, led to more independent characteristics for the pyrene, benzoterrylene and acene subcomponents (Fig. 4g). Increasing the steric constraint of these acenes led to compounds exhibiting the properties of the central benzoterrylene (for example, for **2** in Supplementary Fig. 76). Time-dependent DFT calculations revealed that the absorption behaviour of these compounds was the result of a mixture of pyrene-, dibenzoterrylne- and acene-like orbitals (Supplementary Fig. 77, Supplementary Tables 5–11). For **1**, the main band in the range 400–600 nm, corresponding to the HOMO–LUMO transition, according to the calculation, was caused by a transition from a pyrene-like HOMO to a pyrene-like LUMO. The second band, near 350 nm, due to the HOMO − 1 → LUMO + 1 transition, was

related to the transition between the dibenzoterrylene-like HOMO-2 and LUMO + 2. The small sharp peak near 400 nm arose from the transition from a pyrene-like HOMO to a dibenzoterrylene-like LUMO + 2. We observed similar transition behaviour for **3**, although the band in the 300–400 nm region was red-shifted by ∼50 nm. Consistent with the molecular orbital profiles, **2** displayed a dibenzoterrylene-like HOMO → LUMO transition for the main band near 600 nm, with a transition between the dibenzoterrylene-like HOMO-2 and LUMO + 2 for the band near 350 nm. We also observed dibenzoterrylene-like HOMO-2 → LUMO and HOMO → LUMO + 2 transitions for the weak band at 344 nm.

Optical and magnetic properties of open-shell compound 4. Dark-green compound **4** provides an absorption spectrum (Fig. 5b) displaying bands with maxima at 834, 675 and 585 nm

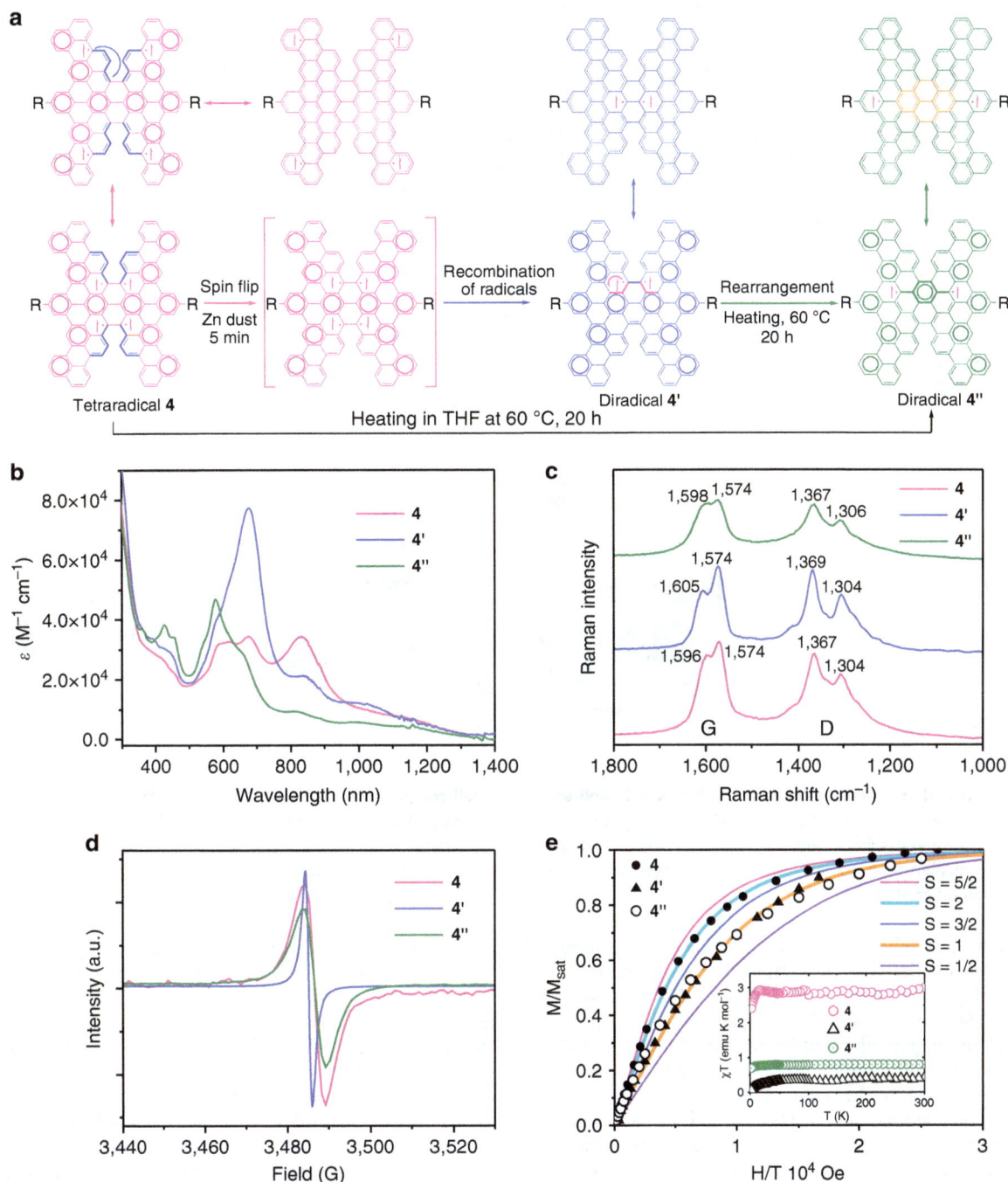

Figure 5 | Spectroscopic characterization of compound 4. (**a**) Transformation of tetraradical **4** to diradicals **4′** or **4″** mediated with spin catalysis by Zinc or trace radicals in THF, and the resonance structures of **4**, **4′** and **4″**. (**b**) UV–vis–near-infrared spectra of **4**, **4′** and **4″** in CH_2Cl_2. (**c**) Raman spectra of solid **4**, **4′** and **4″**. (**d**) ESR spectra of chloroform solutions of **4**, **4′** and **4″** recorded at room temperature. (**e**) The magnetization of **4**, **4′** and **4″** are plotted as M/M_{sat} versus H/T, with Brillouin functions for $S = 5/2$, 2, 3/2,1 and 1/2. The fitting parameters for **4**, **4′** and **4″** are $S = 2.08$ (1.9 K), 1.02 (3 K), 1.14 (1.9 K) and the corresponding $M_{sat} = 0.8$, 0.37, 0.88 μB, respectively. Inset plots: the magnetic susceptibility of **4** (2,000 Oe), **4′** (10,000 Oe) and **4″** (2,000 Oe) are plotted as χT versus T.

with a tail into the near-infrared region (ca. 1,200 nm). We calculated an optically measured band gap of $\sim$1.03 eV. The CV trace of **4** (Supplementary Fig. 75) in 1,2-dichlorobenzene (0.1 M tetrabutylammonium hexafluorophosphate) displayed a one-step irreversible oxidation and a one-step reduction. The HOMO and LUMO energies, calculated through CV, were -5.31 and -4.26 eV, respectively, corresponding to a band gap of 1.05 eV, very close to that determined *in silico* (1.1 eV). In addition, we observed almost no fluorescence from **4** in solution and in the solid state.

Raman spectroscopy is a very useful tool for characterizing benzene-type aromatic radicals[48], because characteristic benzene vibrational Raman bands exist near 1,600 cm^{-1} that are very sensitive to the electronic configuration within the six-membered benzene ring, either 'benzoquinoidal' or 'benzoaromatic.' The Raman spectrum of **4** featured two sets of peaks, much like the D and G bands of graphene (Fig. 5c). The peak near 1,367 cm^{-1} corresponds to the breathing vibrations of sp^2-hybridized carbon domains in aromatic rings (that is, the D band); we assign the peak near 1,596 cm^{-1} to the first-order scattering of the E$_{2g}$

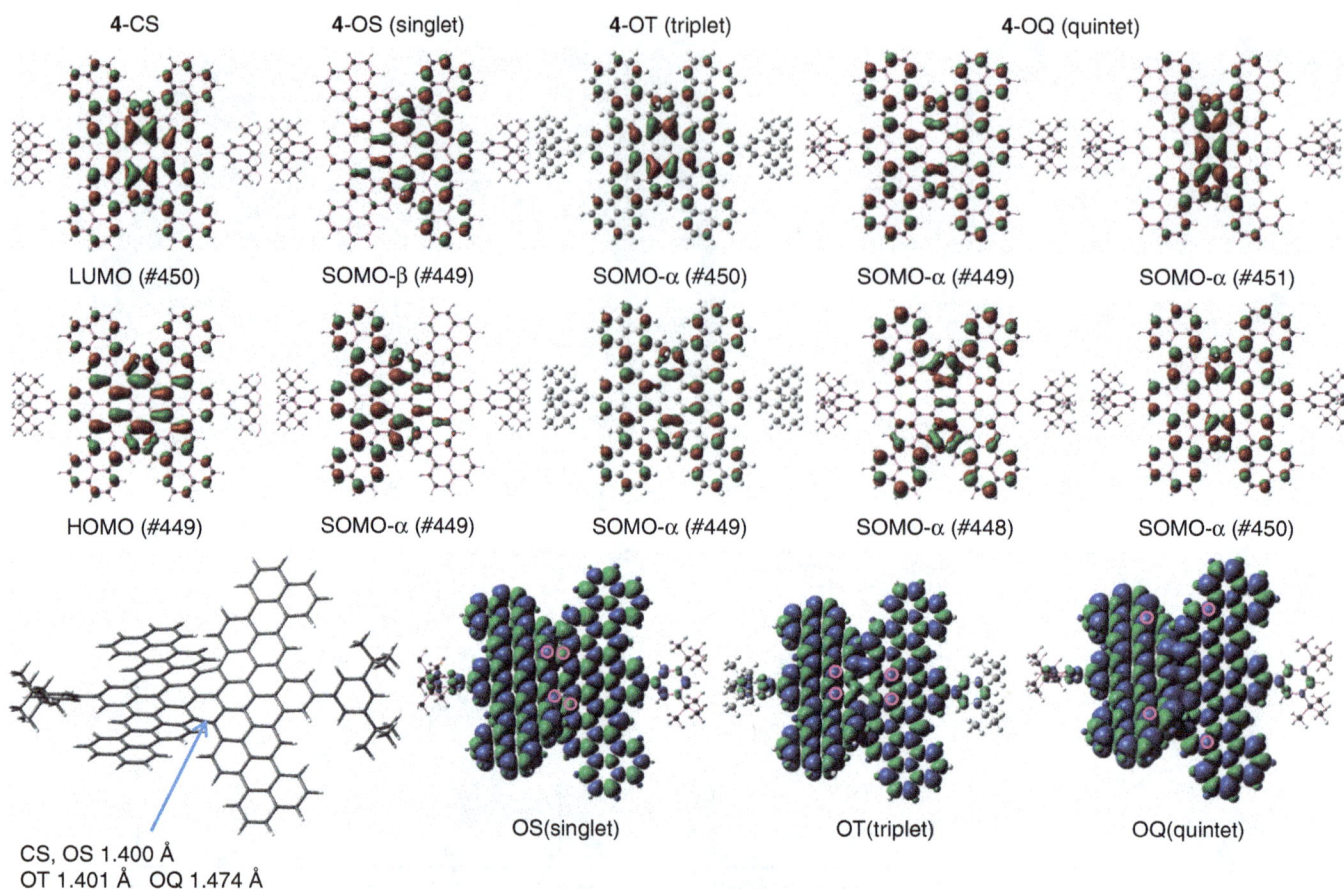

4-CS	4-OS (singlet)	4-OT (triplet)	4-OQ (quintet)	
LUMO (#450)	SOMO-β (#449)	SOMO-α (#450)	SOMO-α (#449)	SOMO-α (#451)
HOMO (#449)	SOMO-α (#449)	SOMO-α (#449)	SOMO-α (#448)	SOMO-α (#450)

CS, OS 1.400 Å
OT 1.401 Å   OQ 1.474 Å

OS(singlet)   OT(triplet)   OQ(quintet)

Figure 6 | Orbitals and spin-density distributions for closed-shell and open-shell compound 4. Calculated (UCAM-B3LYP) HOMOs and LUMOs for **4**-CS, SOMOs and spin-density distributions for **4** in the singlet (**4**-OS, S = 0), triplet (**4**-OT, S = 1) and quintet state (**4**-OQ, S = 2). Blue and green surfaces represent α- and β-spin densities, respectively. The red marks indicate the atoms exhibiting highest spin-density. The central bond connecting the two pentacene-annulated-tribenzo-octacene units in quintet-state pocesses the single-bond charateristic, which indicates the less conjugation between the two units than other states.

mode observed for in-phase stretching vibration of sp^2-hybridized carbon domains in aromatic rings (that is, the G band)[49]. The downfield-shifting of the D and G bands (from 1,367 to 1,304 cm^{-1} and from 1,596 to 1,574 cm^{-1}, respectively) indicated the coexistence of 'benzoquinoidal' and 'benzoaromatic' benzene rings.

The NMR spectrum of **4** in $C_2D_2Cl_4$ did not feature any signals at room temperature, nor after cooling to $-50\,°C$, suggesting the presence of a considerable paramagnetic species. In addition, the solution of **4** provided a featureless broad ESR signal (Fig. 5d), resulting from the long-distance spin − spin dipole interaction within the molecules and the extended spin-delocalization[48,50,51]. These data indicate that **4** exists as an open-shell multi-radical in the ground state.

We performed SQUID measurements for **4** in the powder form as a function of magnetic field ($H = 0 - 5 \times 10^4$ Oe and $T = 1.9$ K) and temperature ($T = 2 - 300$ K at $H = 2,000$ Oe). The value of S = 2.08, determined from the curvature of the Brillouin plots, indicate the quintet (S = 2) ground state for tetraradical **4** (Fig. 5d). The inset of Fig. 5d presents a plot of $\chi_M T(T)$ for **4**. The constant value of χT (2.84 emu K mol^{-1}), as evidenced by the flatness of the χT versus T plots in the T = 30 − 300 K range, indicates that there is no significant change in the thermal population of spin states up to 300 K. After removing the temperature independent contribution, the numerical fit of the data by the Currie-Weiss model give Currie constant of C = 2.84 emu K mol^{-1} and Weiss constant θ_W of − 0.38 K (Supplementary Fig. 78), which indicated weak intermolecular antiferromagnetic interaction below 20 K.

Interestingly, solid **4** displayed high stability, and there was no obvious decomposition when the solid was stored under ambient air and light conditions for months. However, mixing the tetraradical **4** with Zinc dust can generate another radical **4′**, and heating the solution of **4** or **4′** in THF at 60 °C for about 20 h will transform both of them to another species **4″** completely. **4′** and **4″** showed no molecular weight change, while the strongest absorption bands of **4′** and **4″** were shifted to 678 and 575 nm, respectively, with the absorption band at 834 nm became weak (Fig. 5b and Supplementary Fig. 81). Similar Raman spectra (Fig. 5c) and magnetic field dependent magnetic signal changes (Fig. 5e) indicate unequivocally that both of **4′** and **4″** are triplet biradicals at ground state (Supplementary Figs 79 and 80). The values of $\chi T \approx 1$ emu K mol^{-1} were found for **4′** and **4″**, after values of χT (0.77, 0.36 emu K mol^{-1}) are corrected by spin concentration ($M_{sat} = 0.88$, 0.37 μB). The transformation from tetraradical to diradicals is probably due to the fact that one pair of electrons in the tetraradical formed one covalent bond by Zinc surface catalysed spin flip[52,53] (Fig. 5a), the left two electrons delocalized on the two pentacene-annulated-tribenzo-octacene units, respectively, with parallel spin electron patterns at ground state. The obtained diradical **4′** can be rearranged to a more stable diradical **4″** on heating. Transformation directly from **4** to **4″** in hot THF processed through spin catalysis[52,53] by the trace radicals in THF, followed by the thermal rearrangement, without the observation of the intermediates. Sharpening ESR signal for **4′** (Fig. 5d), indicates that the spin − spin dipole interaction distance within the molecules of **4′** is shorter than that of **4** and **4″**. The transformation processes and the possible resonance structures of **4**, **4′** and **4″** are summarized in Fig. 5a, with the supports from the following DFT optimization of the tetraradical and diradicals. Tetraradical **4** can be considered as four phenalenyl radicals annulated by twisted H-shaped acene; intermediate diradical **4′** featured two close-contacted radical pairs like two over-lapped phenalenyl radicals; **4″** can be viewed as two phenalenyl radicals fused by central pyrene, with benzene rings surrounded. The detailed spin catalysis mediated transformation mechanism needs further investigation.

Although 1D PAHs (n-acenes) featured the antiferromagnetic ground state due to the zigzag-shaped boundaries, which cause π-electrons to localize and form spin orders at the edges, the longer acene ground states are polyradical in nature[54,55]. 2D PAHs (periacenes and circumacenes) develops a strong multiradical character with increasing zigzag chain length[56,57]. Our experimental results indicated tetraradical character for the twisted 2D molecule **4**. We performed DFT calculations at the UCAM-B3LYP level of theory to investigate the ground state of the tetraradical **4** (Fig. 6). Because of steric effects, the molecule adopted a twisted planar structure. The structure **4**-CS features extended π-electron delocalization from the pyrene core to the four annulated perylene units, with the central pyrene unit possessing the largest HOMO and LUMO coefficients (Fig. 6), suggesting that it would be the most reactive site. For the broken-symmetry singlet state **4**-OS, we observed disjointed singly occupied molecular orbitals (SOMOs), SOMO-α and SOMO-β, with orbital coefficients mainly localized at the terminal pentacene-annulated-tribenzo-octacene units. The spin-density distribution reveals delocalization with the central pyrene units having the largest density, especially the K-region carbon atoms (also the 5′, 7′ positions of dibenzopentacene) (Supplementary Fig. 82, Supplementary Tables 12–14); this situation is different from that found in bisphenalenyls, where the spin-density is delocalized mainly at the terminal phenalenyl units[25], with antiparallel spin electron patterns typical for this broken-symmetry singlet state. This S = 0 state can be considered as the **DBP-4** form in Fig. 1c, which also corresponds to an S = 1 state when the spin electrons parallel to each other (diradical **4′** in Fig. 5a). For the high-spin quintet state (**4**-OQ, S = 2), we observed a large spin-density at the 5,10 positions of the octacene core—positions that are blocked by the annulated pentacene, which is the **DBO-4** form (Fig. 1c, also the tetraradical **4**). However, for the S = 1 state (**4**-OT, also the diradical **4″** in Fig. 5a), the spins are delocalized on the pentacene-annulated-tribenzo-octacene units, with more distribution on the 7, 8 positions of the octacene core, well-protected by surrounding benzene rings. From the SOMO-α and SOMO-β profiles and the spin-distributions of the singlet **4**-OS and high-spin triplet **4**-OT and quintet **4**-OQ, we ascribe the good chemical stability of the tetraradical toward oxidation and dimerization/oligomerization to its thermodynamic stabilization[33], resulting from both delocalization and kinetic blocking through benzoannulation.

Discussion

We have developed an efficient method for the preparation of a series of stable 2D twisted dibenzoterrylene-acenes. We have investigated their electronic structures and geometries in the ground state using various experimental techniques, assisted by DFT calculations. The photophysical, electrochemical, and magnetic properties of these compounds were dependent on the length of the acene. The ground-state structures in this series of molecules were tunable, with **1–3** being closed-shell hydrocarbons and **4** being an open-shell singlet tetraradical. Compound **4** is the first stable twisted 2D hydrocarbon tetraradical obtained

by hybridizing dibenzoterrylene with acene, which can be transferred to more thermodynamic stable triplet diradical. The high stability of the tetraradical and diradical arose from thermodynamic stabilization, due to delocalization and kinetic blocking through benzoanullation. We believe that 2D twisted conjugated molecules with various width and edge structures could also be synthesized in a similar approach. With the installation of different functional groups such as heterocycles on the periphery of such 2D twisted conjugated molecules, specific functionalities on them could be obtained. Combination this approach with polymerization could give great promise for the synthesis of chemically precise, multiple-dimensional polymers. Moreover, the unique optical, electronic, and magnetic properties of these extended 2D acenes suggest that they might be promising candidates for use in nonlinear optics, field effect transistors, and organic spintronics.

Methods

Materials. Most of the chemical reagents were purchased from Alfa Aesar or Aldrich Chemicals and were utilized as received unless indicated otherwise. All solvents were purified using standard procedures. Column chromatography was performed on silica gel (size 200–300 mesh).

Synthesis procedures of compounds 1–4. For details of the synthetic procedures, see Supplementary Methods. For NMR and high-resolution mass spectra of compounds in this manuscript, see Supplementary Figs 1–68.

Sample characterization. ^{1}H and ^{13}C NMR spectra were recorded on a Bruker AVANCE 400 or Bruker AVANCE III 500WB instrument, at a constant temperature of 25 °C. Chemical shifts are reported in parts per million from low-to-high field and referenced to TMS. Matrix-assisted laser desorption/ionization Fourier transform ion cyclotron resonance (MALDI-FT-ICR-MS) MS were performed on a Bruker Solarix 9.4T FT-ICR-MS mass spectrometer. EI mass spectrometric measurements were performed on a SHIMADZU GCMS-QP2010 puls Spectrometer. Elemental analyses were recorded on a Carlo-Erba-1106 instrument. Electronic absorption spectra were measured on a JASCO V-579 spectrophotometer. Raman spectra were taken on a NT-MDT NTEGRA spectra Raman spectroscopy SPM system. CV experiments were performed using an electrochemical three-electrode configuration, glassy carbon electrode as a working electrode and Ag/AgCl as a reference electrode. All experiments were performed in CH_2Cl_2 with 0.1 M of nBu_4NPF_6 as a supporting electrolyte, the ferrocene/ferrocenium ion (Fc/Fc$^+$) as inter-reference. The $E_{1/2}$ values were determined as $1/2(E_{pa} + E_{pc})$, where E_{pa} and E_{pc} are the anodic and cathodic peak potentials, respectively.

X-ray diffraction data analysis. Crystals of **1e** and **2** were grown by slow diffusing acetone to the solutions in $CHCl_3$ solution. Single crystal X-ray diffraction data were collected on a Agilent SuperNova (Dual, Cu at zero, Atlas) diffractometer with Cu Kα radiation ($\lambda = 1.54184$ Å) from micro-focus sealed X-ray tube. Intensities were corrected for absorption effects using the multi-scan technique[58]. The structures were solved by direct methods and refined by a full matrix least squares technique based on F^2 using SHELXT program[59]. The refinement details: Refinement of F^2 against ALL reflections. The weighted R-factor wR and goodness of fit S are based on F^2, conventional R-factors R are based on F, with F set to zero for negative F^2. The threshold expression of $F^2 > 2$sigma (F^2) is used only for calculating R-factors (gt) and so on, and is not relevant to the choice of reflections for refinement. R-factors based on F^2 are statistically about twice as large as those based on F, and R- factors based on ALL data will be even larger. For X-ray data see Supplementary Data 1 and 2.

Attempts to refine peaks of residual electron density as solvents led to 12 $CHCl_3$ and 6 acetone and some additional peaks of residual electron density which can't be refined successfully. The data were corrected for disordered electron density through use of the SQUEEZE procedure[60] as implemented in PLATON[61]. A total solvent-accessible void volume of 2,179.1 A^3 with a total electron count of 584.3 was found in the unit cell in four voids.

The structure solution of crystal **2** (Cell: 19.2306(4) Å, 19.4633(5) Å, 58.6850(14) Å, 90°, orthorhombic P c c n, Z = 8) shows two independent half molecules, one of which is disordered with its symmetry equivalent. It is possible that the cell chosen for the current refinement in represents a subcell and that reflections of a larger supercell were too weak to be observed. The current result represents a preliminary connectivity only. We have grown the crystals of this structure several times, the crystals are thin plate, and we tried to collect better data, however, the data reported here is the best we can obtain. The six-membered rings in the phenanthrenyl groups and terminal benzenes with t-butyl groups were flattened with AFIX 66 and DFIX. As agreed by the reviewer, although the data is poor, it is perfectly fine for the assumptions that there is a twist associated with the molecule as predicted by the DFT calculation. We present with caution some geometrical parameters: phenanthrenyl units twisted with the central naphthene unit in the *cis* direction.

ESR spectra were obtained with a Bruker E500-10/12 spectrometer. No additional hyperfine coupling was detected at 0.1 G, thus experiments were performed at 1.0 or 2 G for improved signal intensities. The sweep width was 83.89 s, time constant was 40.96 ms and the microwave power was 10.1 mW (corresponding to attenuation = 13 dB, which was sufficiently high to avoid power saturation).

Magnetic measurements were performed on a Quantum Design MPMS-XL5 SQUID magnetometer. Small amounts of sample material (between 30 and 40 mg) were put into KLF containers and brought into measuring position using a straw. The variation of the magnetization as a function of temperature at fixed external magnetic field ($T = 2$–300 K at $H = 2{,}000$ Oe) and as a function of the external field at fixed temperature ($H = 0$–5×10^4 Oe and $T = 1.9$ K) was studied. The powder of **4** was sealed in a plastic tube. The signal of sample holder and plastic tube was deducted by measuring the sample holder and plastic tube under same conditions. A correction for the diamagnetism of the sample and the sample container was applied before calculating the susceptibilities from the magnetization data. For all other samples of solid di- and tetraradicals, the correction for diamagnetism was based on high-temperature extrapolation of the χ versus $1/T$ plots, that is, a suitable numerical factor (M_{dia}) was added to the magnetization (M), until the χT versus T plot becomes flat in the high-temperature range.

Theoretical calculations of target compounds 1–4. For compound **1**, **2** and **3**, DFT calculations were performed using the Gaussian 03 program[62], Geometries were optimized in the gas-phase using the B3LYP functional and 6–31 g (d) basis set on all atoms. Theoretical calculations of compound **4** were carried out by using the Gaussian 09 suite of programs[63]. The initial geometry optimization of **4** was performed with the UCAM-B3LYP level of theory on the singlet state and the Handy and co-workers'[66] long-range corrected version of B3LYP 6–31G*, and all electron basis sets were used for all atoms[64–66]. The resulting DFT solution (singlet 'closed-shell': zero-spin-density on all atoms) was further tested for its stability with the STABLE = OPT keyword[67]. A spin symmetry broken DFT solution was found with lower energy. Then the Guess = Read keyword was used to perform the optimization at the UCAM-B3LYP level (singlet open-shell). For DFT data see Supplementary Data 3–6.

References

1. Fujii, S. & Enoki, T. Nanographene and graphene edges: electronic structure and nanofabrication. *Acc. Chem. Res.* **46**, 2202–2210 (2013).
2. Yan, X., Cui, X. & Li, L.-S. Synthesis of large, stable colloidal graphene quantum dots with tunable size. *J. Am. Chem. Soc.* **132**, 5944–5945 (2010).
3. Chen, L., Hernandez, Y., Feng, X. & Muellen, K. From nanographene and graphene nanoribbons to graphene sheets: chemical synthesis. *Angew. Chem. Int. Ed.* **51**, 7640–7654 (2012).
4. Kawasumi, K. *et al.* A grossly warped nanographene and the consequences of multiple odd-membered-ring defects. *Nat. Chem.* **5**, 739–744 (2013).
5. Zhang, Q. *et al.* Facile bottom-up synthesis of coronene-based 3-fold symmetrical and highly substituted nanographenes from simple aromatics. *J. Am. Chem. Soc.* **136**, 5057–5064 (2014).
6. Narita, A. *et al.* Synthesis of structurally well-defined and liquid-phase-processable graphene nanoribbons. *Nat. Chem.* **6**, 126–132 (2014).
7. Cai, J. *et al.* Atomically precise bottom-up fabrication of graphene nanoribbons. *Nature* **466**, 470–473 (2010).
8. Treier, M. *et al.* Surface-assisted cyclodehydrogenation provides a synthetic route towards easily processable and chemically tailored nanographenes. *Nat. Chem.* **3**, 61–67 (2011).
9. Vo, T. H. *et al.* Large-scale solution synthesis of narrow graphene nanoribbons. *Nat. Commun.* **5**, 3189 (2014).
10. Ozaki, K. *et al.* One-shot K-region-selective annulative pi-extension for nanographene synthesis and functionalization. *Nat. Commun.* **6**, 6251 (2015).
11. Schuler, B. *et al.* From perylene to a 22-ring aromatic hydrocarbon in one-pot. *Angew. Chem. Int. Ed.* **53**, 9004–9006 (2014).
12. Kastler, M. *et al.* From armchair to zigzag peripheries in nanographenes. *J. Am. Chem. Soc.* **128**, 9526–9534 (2006).
13. Wang, Z. H. *et al.* Graphitic molecules with partial 'Zig/Zag' periphery. *J. Am. Chem. Soc.* **126**, 7794–7795 (2004).
14. Ritter, K. A. & Lyding, J. W. The influence of edge structure on the electronic properties of graphene quantum dots and nanoribbons. *Nat. Mater.* **8**, 235–242 (2009).
15. Zhang, L. *et al.* Unconventional, chemically stable, and soluble two-dimensional angular polycyclic aromatic hydrocarbons: from molecular design to device applications. *Acc. Chem. Res.* **48**, 500–509 (2015).
16. Anthony, J. E. The larger acenes: versatile organic semiconductors. *Angew. Chem. Int. Ed.* **47**, 452–483 (2008).
17. Chun, D., Cheng, Y. & Wudl, F. The most stable and fully characterized functionalized heptacene. *Angew. Chem. Int. Ed.* **47**, 8380–8385 (2008).
18. Purushothaman, B. *et al.* Synthesis and structural characterization of crystalline nonacenes. *Angew. Chem. Int. Ed.* **50**, 7013–7017 (2011).
19. Watanabe, M. *et al.* The synthesis, crystal structure and charge-transport properties of hexacene. *Nat. Chem.* **4**, 574–578 (2012).
20. Lambert, C. Towards polycyclic aromatic hydrocarbons with a singlet open-shell ground state. *Angew. Chem. Int. Ed.* **50**, 1756–1758 (2011).
21. Morita, Y., Suzuki, S., Sato, K. & Takui, T. Synthetic organic spin chemistry for structurally well-defined open-shell graphene fragments. *Nat. Chem.* **3**, 197–204 (2011).
22. Sun, Z., Ye, Q., Chi, C. & Wu, J. Low band gap polycyclic hydrocarbons: from closed-shell near infrared dyes and semiconductors to open-shell radicals. *Chem. Soc. Rev.* **41**, 7857–7889 (2012).
23. Abe, M. Diradicals. *Chem. Rev.* **113**, 7011–7088 (2013).
24. Kamada, K. *et al.* Strong two-photon absorption of singlet diradical hydrocarbons. *Angew. Chem. Int. Ed.* **46**, 3544–3546 (2007).
25. Kubo, T. *et al.* Synthesis, intermolecular interaction, and semiconductive behavior of a delocalized singlet biradical hydrocarbon. *Angew. Chem. Int. Ed.* **44**, 6564–6568 (2005).
26. Morita, Y. *et al.* Organic tailored batteries materials using stable open-shell molecules with degenerate frontier orbitals. *Nat. Mater.* **10**, 947–951 (2011).
27. Anthony, J. E. Functionalized acenes and heteroacenes for organic electronics. *Chem. Rev.* **106**, 5028–5048 (2006).
28. Zade, S. S. & Bendikov, M. Heptacene and beyond: the longest characterized acenes. *Angew. Chem. Int. Ed.* **49**, 4012–4015 (2010).
29. Einholz, R. & Bettinger, H. F. Heptacene: increased persistence of a $4n + 2$ π-electron polycyclic aromatic hydrocarbon by oxidation to the $4n$ π-electron dication. *Angew. Chem. Int. Ed.* **52**, 9818–9820 (2013).
30. Zade, S. S. *et al.* Products and Mechanism of acene dimerization. a computational study. *J. Am. Chem. Soc.* **133**, 10803–10816 (2011).
31. Kaur, I. *et al.* Design, synthesis, and characterization of a persistent nonacene derivative. *J. Am. Chem. Soc.* **132**, 1261–1263 (2010).
32. Pérez, D., Peña, D. & Guitián, E. Aryne cycloaddition reactions in the synthesis of large polycyclic aromatic compounds. *Eur. J. Org. Chem.* **2013**, 5981–6013 (2013).
33. Rath, H. *et al.* A stable organic radical delocalized on a highly twisted pi system formed upon palladium metalation of a mobius aromatic hexaphyrin. *Angew. Chem. Int. Ed.* **49**, 1489–1491 (2010).
34. Wu, J. S. *et al.* Controlled self-assembly of hexa-peri-hexabenzocoronenes in solution. *J. Am. Chem. Soc.* **126**, 11311–11321 (2004).
35. Pisula, W., Feng, X. & Muellen, K. Tuning the columnar organization of discotic polycyclic aromatic hydrocarbons. *Adv. Mater.* **22**, 3634–3649 (2010).
36. Chen, L. *et al.* Hexathienocoronenes: synthesis and self-organization. *J. Am. Chem. Soc.* **134**, 17869–17872 (2012).
37. Diez-Perez, I. *et al.* Gate-controlled electron transport in coronenes as a bottom-up approach towards graphene transistors. *Nat. Commun.* **1**, 31 (2010).
38. Yamamoto, Y. *et al.* Photoconductive coaxial nanotubes of molecularly connected electron donor and acceptor layers. *Science* **314**, 1761–1764 (2006).
39. Feng, X. *et al.* Controlled self-assembly of C-3-symmetric hexa-peri-hexabenzocoronenes with alternating hydrophilic and hydrophobic substituents in solution, in the bulk, and on a surface. *J. Am. Chem. Soc.* **131**, 4439–4448 (2009).
40. Hill, J. P. *et al.* Self-assembled hexa-peri-hexabenzocoronene graphitic nanotube. *Science* **304**, 1481–1483 (2004).
41. Kastler, M. *et al.* Influence of alkyl substituents on the solution- and surface-organization of hexa-peri-hexabenzocoronenes. *J. Am. Chem. Soc.* **127**, 4286–4296 (2005).
42. Xiao, S. *et al.* Controlled doping in thin-film transistors of large contorted aromatic compounds. *Angew. Chem. Int. Ed.* **52**, 4558–4562 (2013).
43. Cohen, Y. S. *et al.* Enforced one-dimensional photoconductivity in core-cladding hexabenzocoronenes. *Nano Lett.* **6**, 2838–2841 (2006).
44. Kang, S. J. *et al.* A supramolecular complex in small-molecule solar cells based on contorted aromatic molecules. *Angew. Chem. Int. Ed.* **51**, 8594–8597 (2012).
45. Marsden, J. A. & Haley, M. M. Carbon networks based on dehydrobenzoannulenes. 5. extension of two-dimensional conjugation in graphdiyne nanoarchitectures. *J. Org. Chem.* **70**, 10213–10226 (2005).
46. Luo, J., Xu, X., Mao, R. & Miao, Q. Curved polycyclic aromatic molecules that are pi-isoelectronic to hexabenzocoronene. *J. Am. Chem. Soc.* **134**, 13796–13803 (2012).
47. Echegoyen, L. & Herranz, M. A. Fullerene Electrochemistry. *Fullerenes: From Synthesis To Optoelectronic Properties* **4**, 267–293 (2002).
48. Li, Y. *et al.* Kinetically blocked stable heptazethrene and octazethrene: closed-shell or open-shell in the ground state? *J. Am. Chem. Soc.* **134**, 14913–14922 (2012).

49. Tuinstra, F. & Koenig, J. L. Raman spectrum of graphite. *J. Chem. Phys.* **53**, 1126 (1970).
50. Shimizu, A. *et al.* Alternating covalent bonding interactions in a one-dimensional chain of a phenalenyl-based singlet biradical molecule having Kekulé structures. *J. Am. Chem. Soc.* **132**, 14421–14428 (2010).
51. Sun, Z., Huang, K. & Wu, J. Soluble and stable heptazethrenebis(dicarboximide) with a singlet open-shell ground state. *J. Am. Chem. Soc.* **133**, 11896–11899 (2011).
52. Buchachenko, A. L. & Berdinsky, V. L. Electron spin catalysis. *Chem. Rev.* **102**, 603–612 (2002).
53. Khavryuchenko, O. V., Khavryuchenko, V. D. & Su, D. S. Spin catalysts: a quantum trigger for chemical reactions. *Chinese J. Catal.* **36**, 1656–1661 (2015).
54. Jiang, D.-E., Sumpter, B. G. & Dai, S. First principles study of magnetism in nanographenes. *J. Chem. Phys.* **127**, 124703 (2007).
55. Hachmann, J., Dorando, J. J., Avilés, M. & Chan, G. K.-L. The radical character of the acenes: a density matrix renormalization group study. *J. Chem. Phys.* **127**, 134309 (2007).
56. Jiang, D. E. & Dai, S. Electronic ground state of higher acenes. *J. Phys. Chem. A* **112**, 332–335 (2008).
57. Plasser, F. *et al.* The Multiradical character of one- and two-dimensional graphene nanoribbons. *Angew. Chem. Int. Ed.* **52**, 2581–2584 (2013).
58. CrysAlisPro 1.171.38.41; Agilent Technologies Ltd.: Yarton, Oxfordshire, UK, 2015.
59. Sheldrick, G. M. SHELXT- integrated space-group and crystal-structure determination. *Acta Cryst.* **A71**, 3–8 (2015).
60. van der Sluis, P. & Spek, A. L. BYPASS: an effective method for the refinement of crystal structures containing disordered solvent regions. *Acta Cryst A* **46**, 194–201 (1990).
61. Spek, A. L. *PLATON—a multipurpose crystallographic tool. Acta Cryst.* Vol. A 46, C34 (Utrecht University 1990).
62. Frisch, M. J. *et al. Gaussian 03, Revision B* (Gaussian, Inc., 2003).
63. Frisch, M. J. *et al. Gaussian09, Revision E.01* (Gaussian, Inc., 2009).
64. Lee, C., Yang, W. & Parr, R. G. Development of the Colle-Salvetti correlation-energy formula into a functional of the electron density. *Phys. Rev. B: Condens. Matter Mater. Phys.* **37**, 785–789 (1988).
65. Becke, A. D. Density-functional thermochemistry. III. The role of exact exchange. *J. Chem. Phys.* **98**, 5648–5652 (1993).
66. Yanai, T., Tew, D. & Handy, N. A new hybrid exchange − correlation functional using the coulomb-attenuating method (CAM-B3LYP). *Chem. Phys. Lett.* **393**, 51–57 (2004).
67. Seeger, R. Pople, self-consistent molecular orbital methods. XVIII. constraints and stability in Hartree-Fock theory. *J. Chem. Phys.* **66**, 3045–3050 (1977).

Acknowledgements

This work was supported by the National Natural Science Foundation of China (21290190, 21322301, 91227113), NSFC-DFG joint fund (21261130581), the National Basic Research 973 Program of China (2012CB932901), and the 'Strategic Priority Research Program' of the Chinese Academy of Sciences (XDA09020302, XDB12010300).

Author contributions

Yongjun Li and Yuliang Li conceived and designed the project and co-wrote the manuscript, with assistance from the other authors. Yongjun Li and Z.J. prepared the materials, Yongjun Li performed the computation, the ESR and SQUID data were analysed by S.X. and H.L., respectively. All authors discussed the results and contributed to manuscript preparation.

Additional information

28

Seawater usable for production and consumption of hydrogen peroxide as a solar fuel

Kentaro Mase[1], Masaki Yoneda[1], Yusuke Yamada[2] & Shunichi Fukuzumi[1,3,4]

Hydrogen peroxide (H_2O_2) in water has been proposed as a promising solar fuel instead of gaseous hydrogen because of advantages on easy storage and high energy density, being used as a fuel of a one-compartment H_2O_2 fuel cell for producing electricity on demand with emitting only dioxygen (O_2) and water. It is highly desired to utilize the most earth-abundant seawater instead of precious pure water for the practical use of H_2O_2 as a solar fuel. Here we have achieved efficient photocatalytic production of H_2O_2 from the most earth-abundant seawater instead of precious pure water and O_2 in a two-compartment photoelectrochemical cell using WO_3 as a photocatalyst for water oxidation and a cobalt complex supported on a glassy-carbon substrate for the selective two-electron reduction of O_2. The concentration of H_2O_2 produced in seawater reached 48 mM, which was high enough to operate an H_2O_2 fuel cell.

[1] Department of Material and Life Science, Graduate School of Engineering, Osaka University, ALCA and SENTAN, Japan Science and Technology Agency (JST), Suita, Osaka 565-0871, Japan. [2] Department of Applied Chemistry and Bioengineering, Graduate School of Engineering, Osaka City University, Osaka 558-8585, Japan. [3] Faculty of Science and Technology, Meijo University, ALCA and SENTAN, Japan Science and Technology Agency, JST, Shiogamaguchi, Tenpaku, Nagoya, Aichi 468-8502, Japan. [4] Department of Chemistry and Nano Science, Ewha Womans University, Seoul 120-750, Korea. Correspondence and requests for materials should be addressed to S.F. (email: fukuzumi@chem.eng.osaka-u.ac.jp).

Utilization of solar energy as a primary energy source has been strongly demanded to reduce emissions of harmful and/or greenhouse gases produced by burning fossil fuels. However, large fluctuation of solar energy depending on the length of the daytime is a serious problem[1,2]. To utilize solar energy in the night time, solar energy should be stored in the form of chemical energy and used as a fuel to produce electricity. In this context, H_2 has been regarded as the most promising candidate, because H_2 can be produced by photocatalytic water splitting and used as a fuel of H_2 fuel cells to generate electricity with a high efficiency without emission of harmful chemicals. However, the low solar energy conversion efficiency of H_2 production and the storage problem of gaseous H_2 have precluded the practical use of H_2 as a solar fuel[3]. In contrast to gaseous H_2, H_2O_2 can be produced as an aqueous solution from water and O_2 in the air by the combination of the photocatalytic two-electron reduction of O_2 and the catalytic four-electron oxidation of water[4,5]. H_2O_2 can be used as a fuel of an H_2O_2 fuel cell to generate electricity with emission of water and oxygen[5–10]. The energy density of aqueous H_2O_2 (60%) is $3.0\,MJ\,l^{-1}$ ($2.1\,MJ\,kg^{-1}$), which is comparable to the value ($2.8\,MJ\,l^{-1}$, $3.5\,MJ\,kg^{-1}$) of compressed hydrogen (35 MPa). However, the photocatalytic production of H_2O_2 from water and O_2 has yet to be combined with the consumption of the produced H_2O_2 in an H_2O_2 fuel cell because of the insufficient photocatalytic activity[4,5]. In order to realize the production of H_2O_2 and its use in an H_2O_2 fuel cell, a breakthrough is definitely required to improve the photocatalytic efficiency for H_2O_2 production.

We report herein efficient photocatalytic production of H_2O_2, which has been made possible by using the most earth-abundant resource, that is, seawater instead of pure water for the photocatalytic oxidation with a semiconductor and the catalytic two-electron reduction of O_2 with a cobalt chlorin complex supported on a glassy carbon substrate in a two-compartment photoelectrochemical cell under simulated solar illumination without an external bias potential. The H_2O_2 produced in seawater was used directly to generate electricity with the open-circuit voltage of 0.78 V and the maximum power density of $1.6\,mW\,cm^{-2}$ using an H_2O_2 fuel cell, and the solar-to-electricity conversion efficiency of the total system is estimated to be ca 0.28%.

Results

Performance of photoanode and cathode. As a semiconductor photocatalyst for the water oxidation, tungsten oxide (WO_3), which has a narrow band gap suitable for visible light ($<460\,nm$) absorption, was employed[11–14]. In WO_3, hole (h^+) generated in the valence band (VB) is positive enough to oxidize water with the long lifetime (equation (1))[15]. The use of seawater instead of pure water has also enabled us to combine the photocatalytic production of H_2O_2 from seawater and O_2 and its use in an H_2O_2 fuel cell. To perform selective reduction of O_2, a cobalt chlorin

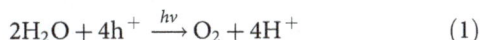

$$2H_2O + 4h^+ \xrightarrow{h\nu} O_2 + 4H^+ \qquad (1)$$

complex ($Co^{II}(Ch)$), which has been proved to function as a catalyst for the efficient and selective two-electron reduction of O_2 under homogeneous conditions (equation (2)), has been employed[16]. The overall photocatalytic reaction is given by equation (3), thus, H_2O_2 can be produced by the two-electron reduction of O_2 by water as an electron donor.

$$O_2 + 2H^+ + 2e^- \rightarrow H_2O_2 \qquad (2)$$

$$2H_2O + O_2 \xrightarrow{h\nu} 2H_2O_2 \qquad (3)$$

Mesoporous WO_3 (m-WO_3) prepared by a literature method was deposited on a fluorine-doped tin oxide (FTO) as a

photoanode (m-WO_3/FTO) and $Co^{II}(Ch)$ was adsorbed on a carbon paper (denoted as CP) as a cathode ($Co^{II}(Ch)$/CP; see Supplementary Information for details). Photocatalytic production of H_2O_2 was performed by using a two-compartment photoelectrochemical cell with the m-WO_3/FTO photoanode and the $Co^{II}(Ch)$/CP cathode, which were immersed in Ar-saturated and in O_2-saturated aqueous $HClO_4$ solutions (pH 1.3), respectively. These two electrodes were connected to each other by a conducting wire as an external circuit. The cathode and anode cells were separated by a Nafion membrane to prevent the decomposition of H_2O_2 produced in the cathode cell. Overall schematic diagram is shown in Fig. 1.

To evaluate the selectivity to the two-electron reduction of O_2 with $Co^{II}(Ch)$/CP, the number of transferred electrons during the catalytic reaction was estimated by performing the rotating ring disc electrode (RRDE) technique with a glassy carbon disk electrode modified with a $Co^{II}(Ch)$ adsorbed on multi-walled carbon nanotubes (MWCNTs; see Supplementary Information for details). From the ratio of the observed disk current and the ring current at various rotating rates, the average number of transferred electrons was determined to be 2.7 (70% selectivity; Supplementary Fig. 1). $Co^{II}(Ch)$ has previously been reported to catalyse the production of H_2O_2 with nearly 100% selectivity in PhCN under homogeneous conditions[16]. The decrease in the selectivity of two-electron reduction of O_2 may be attributed to the partial formation of μ-1,2-peroxo $Co^{III}(Ch)$ dimer, which has been reported to act as a reactive intermediate in the four-electron reduction of O_2 (ref. 17).

Photocatalytic production of H_2O_2. The simulated 1 sun illumination of m-WO_3/FTO in the anode cell afforded the efficient photocatalytic production of H_2O_2 in the cathode cell in the two-compartment photoelectrochemical configuration without an external bias potential. The time courses of photocatalytic H_2O_2 production are shown in Fig. 2. Very little amount of H_2O_2 was obtained in the absence of $Co^{II}(Ch)$ on CP electrode, indicating that $Co^{II}(Ch)$ adsorbed on CP efficiently catalyses the two-electron reduction of O_2 to produce H_2O_2 before the charge recombination of photoexcited electron in conduction band and h^+ in VB of WO_3. The rate of photocatalytic production of H_2O_2 in seawater was markedly enhanced compared with that in pure water. After the illumination for 24 h, the amount of produced H_2O_2 in seawater reached ca 48 mM, which is much larger than the value (2 mM) using a one-compartment system reported previously[4]. No structural change of m-WO_3/FTO electrode after

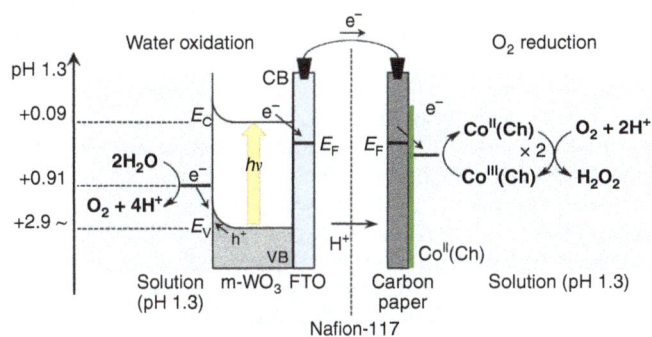

Figure 1 | Overall scheme of photocatalytic production of H_2O_2.
Photocatalytic production of H_2O_2 from water and O_2 using m-WO_3/FTO photoanode and $Co^{II}(Ch)$/CP cathode in water or seawater under simulated 1 sun (AM 1.5G) illumination.

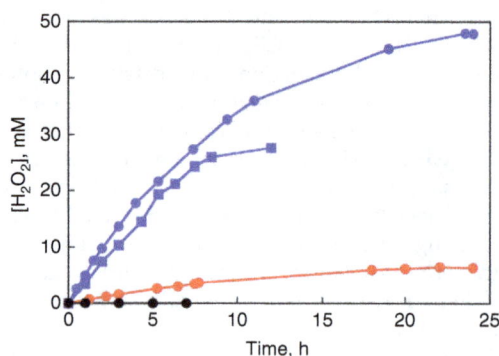

Figure 2 | Photocatalytic production of H₂O₂ in the two-compartment photoelectrochemical cell. Time courses of H_2O_2 production with m-WO₃/FTO photoanode and $Co^{II}(Ch)/CP$ cathode in pH 1.3 water (red circle), in pH 1.3 seawater (blue circle) and in an NaCl aqueous solution (pH 1.3; blue square) under simulated 1 sun (AM 1.5G) illumination. Time course of H_2O_2 production in the absence of $Co^{II}(Ch)$ on carbon paper under simulated 1 sun (AM 1.5G) illumination in pH 1.3 water is shown as black circle.

the photocatalytic reaction was confirmed by the powder X-ray diffraction measurements (Supplementary Fig. 2). The similar enhancement on photocatalytic activity was observed in an NaCl solution, in which the concentration of Cl^- was the same as that of Cl^- in seawater. The enhancement effect of Cl^- on the photocatalytic activity for water oxidation can be interpreted by the following Cl^--assisted mechanism[18–21]. First, the oxidation of Cl^- by photogenerated hole to form chlorine (Cl_2) occurs before the oxidation of water as given by equation (4)[18]. Cl_2 is in the disproportionation equilibrium with hypochlorous acid (HClO), as given by equation (5), where the population of Cl_2 and HClO varies depending on the pH the solution and Cl_2 is a major component under an acidic solution below pH 3 (refs 18,19). Second, HClO is decomposed to O_2 and Cl^- under solar irradiation, as given by equation (6)[20]. Thus, the overall reaction of the water oxidation assisted by Cl^- is given by equation (1). Indeed, the ultraviolet–visible absorption spectrum

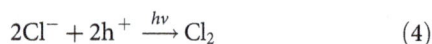

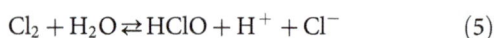

$$2Cl^- + 2h^+ \xrightarrow{h\nu} Cl_2 \tag{4}$$

$$Cl_2 + H_2O \rightleftarrows HClO + H^+ + Cl^- \tag{5}$$

$$2HClO \xrightarrow{h\nu} O_2 + 2H^+ + 2Cl^- \tag{6}$$

of the anode cell solution after the photocatalytic reaction for 24 h exhibited the absorption band around at 231 nm, which is identical to the spectrum of a standard HClO/Cl₂ solution below pH 3 (Supplementary Fig. 3)[19]. The amount of O_2 evolved in seawater in anode cell after 1 h (12.7 μmol) was more than three times larger than that in water (3.7 μmol) as shown in Supplementary Fig. 4. Thus, the enhancement of photocatalytic production of H_2O_2 in seawater (Fig. 2) results from the photocatalytic oxidation of Cl^- in seawater.

The effects of Cl^- on the catalytic performance of m-WO₃/FTO and $Co^{II}(Ch)/CP$ were also investigated by using a photoelectrochemical cell with three electrode configuration. The current–potential (I–V) curves of m-WO₃/FTO under simulated 1 sun illumination and dark are shown in Fig. 3a. The onset of photocurrent for the oxidation of water was observed at 0.2 V (versus saturated calomel electrode (SCE)) in water (pH 1.3), which corresponds to the 110 mV of overpotential with respect to the value of flat band potential of WO₃ (0.09 V versus SCE at pH

1.3)[22]. This overpotential is required as a driving force for the migration of the photogenerated electron from conduction band of WO₃ to FTO electrode. When the photocatalysis measurements of m-WO₃/FTO were performed under the same conditions in seawater instead of pure water, about four times larger photocurrent at 0.3 V (versus SCE) than that in water together with the negative shift of onset potential from 0.2 V (versus SCE) to 0.1 V (versus SCE) was observed (Fig. 3a). In addition, the stability of a photocurrent obtained by applying 0.3 V (versus SCE) was also improved as shown in Supplementary Fig. 5a. These improvements in the stability as well as the photocurrent in seawater can be explained by the efficient quenching of the photogenerated hole in VB by the oxidation of Cl^-, as described above. These are consistent with the enhancement of photocatalytic production of H_2O_2 in seawater (Fig. 2). The Faradaic efficiencies in the photoelectrochemical water oxidation in water and in seawater were determined to be 77% and 93%, respectively, from the simultaneous measurements of O_2 evolution (Supplementary Fig. 5b). The effect of Cl^- on the electrochemical property of $Co^{II}(Ch)/CP$ was also studied by measuring cyclic voltammograms using the three-electrode electrochemical cell. The addition of tetra-n-butylammonium chloride (0.1 M) to a N₂-saturated PhCN solution containing $Co^{II}(Ch)$ (1 mM) and tetra-n-butylammonium hexafluorophosphate (TBAPF₆; 0.1 M) resulted in the large negative shift of the redox potential for $[Co^{III}(Ch)]^+/Co^{II}(Ch)$ from 0.37 to 0.01 V (versus SCE; Supplementary Fig. 6a,b). In addition, no catalytic current was obtained for the O_2 reduction in the presence of Cl^- in an O_2-saturated PhCN solution containing HClO₄ (10 mM) as a proton source (Supplementary Fig. 6c), suggesting that Cl^- inhibits electron transfer from $Co^{II}(Ch)$ to O_2 because of the strong axial coordination of Cl^- to the reaction centre of $Co^{II}(Ch)$ to form 5- or 6-coordinated inactive species. In contrast, the catalytic current for the O_2 reduction with $Co^{II}(Ch)/CP$ measured in seawater (pH 1.3) appeared at ca 0.34 V (versus SCE), which is virtually the same onset potential and catalytic current measured in water (pH 1.3), as shown in Fig. 3b. This result indicates that the coordination of Cl^- to $Co^{II}(Ch)$ adsorbed on the electrode surface in an aqueous solution is negligibly weak as compared with that in PhCN[23]. Hence, the enhancement of photocatalytic production of H_2O_2 is mainly derived from the acceleration of water oxidation at the photoanode. The predicted operating current of the two-compartment photoelectrochemical cell was defined from the intersection of cyclic voltammograms of $Co^{II}(Ch)/CP$ and I–V curves of the m-WO₃/FTO photoanode, giving a value of 0.5 mA at 0.32 V (versus SCE) in water and 1.3 mA at 0.29 V (versus SCE) in seawater, respectively (Fig. 3c).

The effect of illumination intensity on the photocatalytic production of H_2O_2 was examined as shown in Supplementary Fig. 7a, where the produced amount of H_2O_2 increased in proportion to the intensity of the illumination. The solar energy conversion efficiency for the photocatalytic production of H_2O_2 in seawater was determined to be 0.55% under simulated 1 sun illumination. The best solar energy conversion efficiency was determined to be 0.94% when illumination intensity was reduced to 0.1 sun (Supplementary Fig. 7b). This efficiency exceeds that of swichgrass (0.2%), which has been considered as a promising crop for biomass fuel[24], and also the value in the one-compartment cell (0.25%)[4]. A much higher solar-to-hydrogen efficiency of 12.3% has recently been achieved by perovskite photovoltaics-based electrolysis[2]. However, the storage of hydrogen has still been a quite difficult issue, because hydrogen is a gas having a low volumetric energy density. In this contrast, in our system, the produced hydrogen peroxide in seawater can be used directly as a fuel in an H_2O_2 fuel cell.

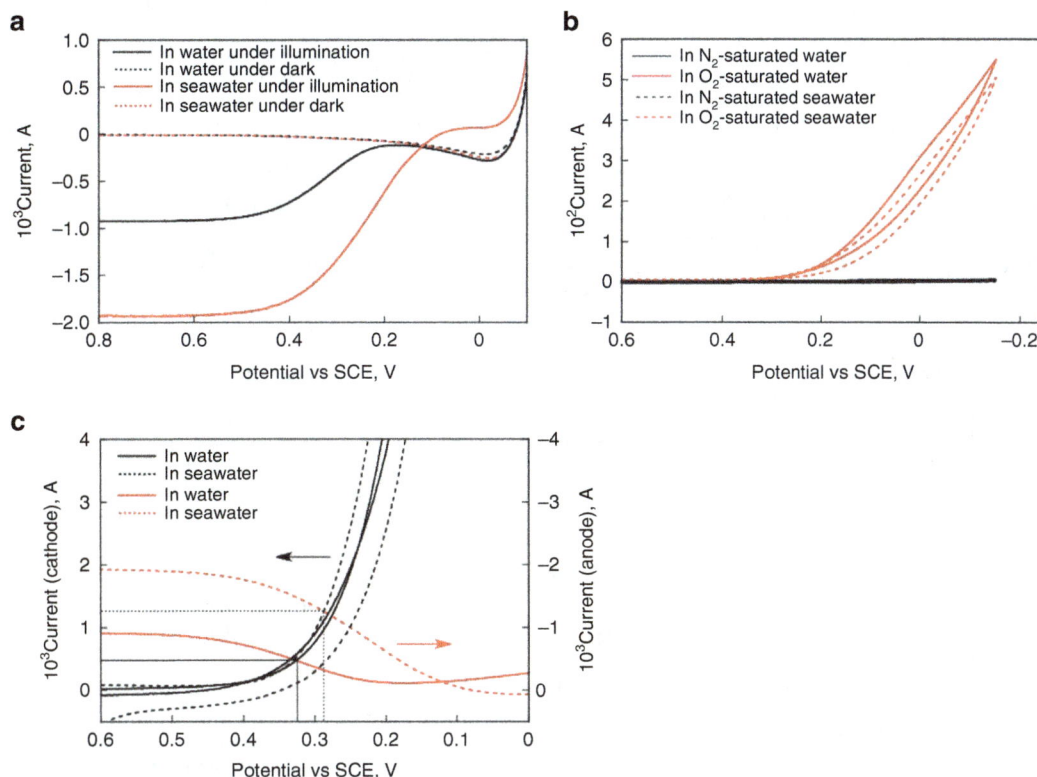

Figure 3 | Photoelectrochemical performance of m-WO$_3$/FTO and electrochemical performance of Co^{II}(Ch)/CP. (**a**) *I–V* curves of m-WO$_3$/FTO photoanode in pH 1.3 water (black solid) and in pH 1.3 seawater (red solid) under simulated 1 sun (AM 1.5G) illumination. *I–V* curves under dark are shown as dashed lines with the same colour definition. Sweep rate: 10 mV s^{-1}. (**b**) Cyclic voltammograms of Co^{II}(Ch)/CP in a N$_2$-saturated pH 1.3 water (black solid) and an O$_2$-saturated pH 1.3 water (red solid). The dashed lines show the cyclic voltammograms of Co^{II}(Ch)/CP recorded in pH 1.3 seawater. Sweep rate: 20 mV s^{-1}. (**c**) Cyclic voltammograms of Co^{II}(Ch)/CP in O$_2$-saturated pH 1.3 solutions (black) and *I–V* curves of m-WO$_3$/FTO photoanode in pH 1.3 solutions (red) under simulated 1 sun (AM 1.5G) illumination.

H$_2$O$_2$ fuel cell. Finally, the chemical energy of H$_2$O$_2$ produced by the photocatalytic reaction was converted to electrical energy through a H$_2$O$_2$ fuel cell composed of a polynuclear cyanide complexes, Fe^{II}$_3$[Co^{III}(CN)$_6$]$_2$, modified carbon cloth cathode and a nickel mesh anode in a one-compartment cell[10]. The reaction solution (seawater, pH 1.3) containing ca 48 mM of H$_2$O$_2$ in cathode was transferred to the H$_2$O$_2$ fuel cell. The obtained potential and power density depending on the current density are shown in Fig. 4. The cell has the open-circuit potential and the maximum power density of 0.78 V and 1.6 mW cm^{-2}, respectively. These values agree with those obtained from the H$_2$O$_2$ fuel cell using an aqueous solution (pH 1.0) containing authentic H$_2$O$_2$ (50 mM) and NaCl (1.0 M) as a supporting electrolyte (Supplementary Fig. 8). In addition, the energy conversion efficiency of the H$_2$O$_2$ fuel cell was determined to be ca 50% by the measurement of output energy as electrical energy versus consumed chemical energy H$_2$O$_2$ gas, which is comparable to the efficiency of an H$_2$ fuel cell (Supplementary Figs 9 and 10). Thus, the solar-to-electricity conversion efficiency of the total system is estimated to be ca 0.28% (0.55 × 50%), which is still much lower in contrast to the conventional solar-to-electricity device such as photovoltaic cells. However, there are still many things to do in the one-compartment H$_2$O$_2$ fuel cells including better anode materials to improve the performance, because the one compartment cell without membrane and use of an aqueous solution of H$_2$O$_2$ have significant advantages as compared with H$_2$ fuel cells. The production of chemical energy utilizing solar energy and its conversion to electrical energy based on H$_2$O$_2$ in seawater can provide practical solution to the construction of an ideal energy-sustainable society using seawater, which is the most earth-abundant resource.

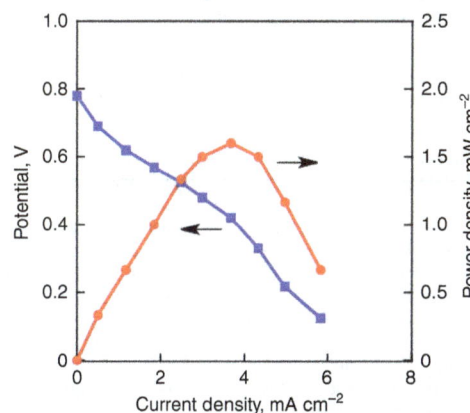

Figure 4 | Generation of electrical energy in the one-compartment H$_2$O$_2$ fuel cell. *I–V* (blue) and *I–P* (red) curves of the one-compartment H$_2$O$_2$ fuel cell with a Ni mesh anode and Fe^{II}$_3$[Co^{III}(CN)$_6$]$_2$/carbon cloth cathode in the reaction solution containing H$_2$O$_2$ (47.9 mM) produced by photocatalytic reaction in seawater as shown in Fig. 2 (blue circle).

Methods

Materials. Chemicals were purchased from commercial sources and used without further purification, unless otherwise noted. Benzonitrile (PhCN) used for spectroscopic and electrochemical measurements was distilled over phosphorus pentoxide before use[25]. Potassium hexacyanoferrate(III) (K$_3$[Fe(CN)$_6$]) and acetylacetone ($\geqq$99%) were purchased from Wako Pure Chemical Industries Ltd., Tungsten hexachloride (WCl$_6$, $\geqq$95%) was purchased from Nacalai Tesque. Scandium(III) nitrate tetrahydrate was purchased from Mitsuwa Chemicals. Potassium hexacyanocobaltate (K$_3$[Co^{III}(CN)$_6$], $\geqq$99.9%) was supplied by Stream Chemicals. Red sea salt was supplied by Red Sea. Oxo[5,10,15,20-tetra

(4-pyridyl)porphinato]titanium(IV) ([TiO(tpyp)]) was purchased from Tokyo Chemical Industry Co., Ltd. (TCI). Pluronic P-123, Triton X-100, Nafion perfluorinated ion exchange resin solution, Nafion perfluorinated membrane (Nafion 117) were received from Aldrich Chemical Co. CP electrode (EC-TP1-060T produced by Toray Industry Inc.) was obtained from Toyo Co. Glass slides coated with FTO (transmittance, 83.6%) were supplied by Aldrich Chemicals Co. and cut by Asahi Glass Co., Ltd. Tetra-n-butylammonium hexafluorophosphate (TBAPF$_6$) purchased from Wako Pure Chemical Industries, Ltd. was twice recrystallized from ethanol and dried in vacuo before use. Purified water was provided by a Millipore Milli-Q water purification system (Millipore, Direct-Q 3 UV) with an electronic conductance of 18.2 MΩ cm. Cobalt chlorin complex [Co^{II}(Ch)], mesoporous WO$_3$ (m-WO$_3$) and polynuclear cyanide complex (Fe$^{II}_3$[Co^{III}(CN)$_6$]$_2$) were synthesized according to the literature procedure (vide infra).

Preparation of Co^{II}(Ch)/CP electrode. Co^{II}(Ch)/CP electrode was prepared by an MeCN solution (1 ml) of Co^{II}(Ch) (0.3 mM), MWCNT (0.63 mg) and 5% Nafion (12 μl). For each experiment, the mixture was sonicated for 20 min and then a 50 μl of the mixture was applied on the both side of surface of a CP with a 3.0 cm^2 area by drop-casting and allowed to evaporate to afford a film containing a MWCNT loading of 50 μg cm^{-2} and a catalyst loading of 30 nmol.

Preparation of m-WO$_3$/FTO electrode. First, FTO glasses were cleaned before use by immersing into an MeOH/HCl (1/1 (v/v)) solution for 30 min and washed by purified water. The resulting FTO glasses were hydroxylated in H$_2$SO$_4$ for 2 h and then boiled in purified water for 30 min with subsequent drying under N$_2$ (ref. 26). m-WO$_3$/FTO electrode was prepared by a solution consisting of 1 ml of water containing 50 mg of m-WO$_3$ and acetyl acetone (30 μl) and 1 drop of Triton X-100. The mixture was sonicated for 5 min and then a 50 μl drop was applied on the surface of FTO electrode with a 2.5 cm^2 area and allowed to evaporate to afford a thin film. The resulting electrode was annealed to form crystalline m-WO$_3$ at 400 °C with ramping rate of 2 °C min^{-1} (held at 400 °C for 2 h) under air to remove surfactant species. The combustion of residual surfactant was confirmed by thermal gravimetric-differential thermal analysis (TG-DTA) of m-WO$_3$ dispersion prepared above (Supplementary Fig. 11). The exothermic current peak along with weight loss observed at around 180 °C was disappeared after the annealing at 400 °C. The morphology of obtained m-WO$_3$/FTO was observed by scanning electron microscope (SEM), as shown in Supplementary Fig. 12. The mesoporous structure of m-WO$_3$ was confirmed by N$_2$ adsorption–desorption isotherm measurements (Supplementary Fig. 13). The measurements performed at 77 K revealed a type IV isotherm^{27}, clearly indicating the presence of mesopores. The Brunauer–Emmett–Teller surface area of m-WO$_3$ was as high as 21 m^2 g^{-1} (Supplementary Fig. 13a). The size of the mesopores was determined to be 8 nm by Barrett–Joyner–Halenda plot (Supplementary Fig. 13b).

Preparation of seawater. The seawater was prepared by dissolving 33.4 g of red sea salt in 1 l of water to form a solution containing ca 550 mM of NaCl.

Characterization of m-WO$_3$/FTO. TG/DTA data were performed on an SII TG/DTA 7,200 instrument. A sample (ca 10 mg) was loaded into an Al pan and heated from 25 °C to 600 °C with a ramping rate of 2 °C min^{-1} under N$_2$. A certain amount of γ-Al$_2$O$_3$ was used as a reference for DTA measurements. Nitrogen-adsorption/desorption measurements were performed at 77 K on a Belsorp-mini (BEL Japan, Inc.) within a relative pressure range from 0.01 to 101.3 kPa. A sample mass of ca 100 mg used for adsorption analysis was pretreated at 150 °C for 2 h under vacuum conditions and kept in N$_2$ atmosphere until N$_2$-adsorption measurements. The resulting sample was exposed to a mixed gas of He and N$_2$ with a programmed ratio and adsorbed amount of N$_2$ was calculated from the change of pressure in a cell after reaching equilibrium (at least 5 min). Powder X-ray diffraction patterns were recorded on a Rigaku MiniFlex 600. Incident X-ray radiation was produced by a Cu X-ray tube, operating at 40 kV and 15 mA with Cu Kα radiation (λ = 1.54 Å). The scan rate was 1° min^{-1} from 2θ = 10–70°. SEM images of particles were observed by a FE-SEM (JSM-6320F or JSM-6701F) operating at 10 kV.

Electrochemical measurements. Cyclic voltammetry measurements were performed on an ALS 630B electrochemical analyser. Effects of Cl$^-$ on Co^{II}(Ch) was investigated in a N$_2$- or O$_2$-saturated PhCN solution containing 0.10 M TBAPF$_6$ as a supporting electrolyte at 298 K using a conventional three-electrode cell with a glassy carbon (GC) working electrode (surface area of 0.3 mm^2) and a platinum wire (Pt) as the counter electrode. The GC working electrode was routinely polished with polishing alumina suspension and rinsed with acetone before use. The potentials were measured with respect to the Ag/AgNO$_3$ (1.0 × 10^{-2} M) reference electrode. All potentials (versus Ag/AgNO$_3$) were converted to values versus SCE by adding 0.29 V (ref. 28). Redox potentials were determined using the relation $E_{1/2} = (E_{pa} + E_{pc})/2$.

Electrochemical performance of Co^{II}(Ch) deposited on CP electrode for the catalytic O$_2$ reduction was evaluated in a N$_2$- or O$_2$-saturated aqueous HClO$_4$

(pH 1.3) solution containing NaClO$_4$ (0.1 M) as a supporting electrolyte and in N$_2$- or O$_2$-saturated seawater containing HClO$_4$ (pH 1.3) and NaClO$_4$ (0.1 M) at 298 K using a conventional three-electrode cell consisting of Co^{II}(Ch)/CP as a working electrode and a platinum coil as the counter electrode. All the photoelectrochemical and electrochemical measurements in aqueous solutions were conducted using a reference SCE and all results in this work are presented against the SCE. The conversion of potentials versus SCE to versus normal hydrogen electrode (NHE) was performed according to the following equation (7).

$$E(\text{vs SCE at measured pH}) = E(\text{vs NHE at pH 0}) - 0.241\,\text{V} - 0.059\,\text{V} \times \text{pH} \quad (7)$$

The superior performance of Co^{II}(Ch)/CP in an aqueous solution was confirmed by the comparison with cobalt octaethylporphyrin (Co^{II}(OEP)), which is commonly used as an electrocatalyst for the O$_2$ reduction, modified CP (Co^{II}(OEP)/CP), as shown in Supplementary Fig. 14.

The RRDE measurements were carried out using a BAS RRDE-3A rotator linked to an ALS 730D electrochemical analyser. A three-electrode cell (100 ml) was employed with the RRDE consisting of a platinum ring (Pt) electrode and a GC disk electrode, platinum coil (Pt) as a counter electrode and SCE as a reference electrode. The voltammograms were measured in an O$_2$-saturated aqueous HClO$_4$ solution (pH 1.3) containing NaClO$_4$ (0.1 M) at 5 mV s^{-1} with various rotating rates (100, 300, 600, 900, 1,200, 1,500, 2,000, 2,500, 3,000, 3,500, 4,000 and 4,500 r.p.m.). A RRDE for the investigation of transferred electrons during catalytic O$_2$ reduction with Co^{II}(Ch)/CP was performed by the modification of GC disk electrode with a thin film of Co^{II}(Ch). The thin film was prepared by a solution consisting of MeCN (1 ml) containing Co^{II}(Ch) (0.3 mM), MWCNT (1.26 mg) and 5% Nafion (12 μl). For each experiment, the mixture was sonicated for 20 min and then a 10-μl drop was applied on the surface of a polished GC disk electrode and allowed to evaporate to afford a thin film containing a MWCNT loading of 100 μg cm^{-2} and a catalyst loading of 3 nmol.

The number of transferred electrons (n) is determined by following equation $n = 4I_D/(I_D + I_R/N)$, where I_D is the faradic current at the disk electrode, I_R is the faradic current at the ring electrode, and N is the collection efficiency of the RRDE. The N value is measured using an aqueous solution of K$_3$[Fe^{III}(CN)$_6$] (2 mM) as a standard one-electron redox couple ([Fe^{III}(CN)$_6$]$^{3-}$/[Fe^{II}(CN)$_6$]$^{4-}$) in the presence of KNO$_3$ (0.5 M) and is determined to be $N = 0.37$ when the GC disk electrode of RRDE is loaded with the same amount of MWCNT (100 μg cm^{-2}) as used above (Supplementary Fig. 15).

Photoelectrochemical measurements. Photoelectrochemical measurements were performed in a home-made quartz cell (light path length = 1 cm) composed of the as prepared m-WO$_3$/FTO electrode, a platinum coil counter electrode, and a SCE reference electrode in an Ar-saturated aqueous solution (8 ml) containing HClO$_4$ (pH 1.3) and 0.1 M of NaClO$_4$ at 298 K (Supplementary Fig. 16a). The photoanode was illuminated from the back side of FTO electrode (FTO/electrolyte interface) with a solar simulator (HAL-320, Asahi Spectra Co., Ltd.), where the light intensity was adjusted at 100 mW cm^{-2} (AM1.5G) at the sample position by a 1SUN checker (CS-20, Asahi Spectra Co., Ltd.). The Faradaic efficiency for O$_2$ evolution was determined by following equation (8), where F denotes Faradaic constant (9.65 × 10^4 C mol^{-1}).

The Faradaic efficiency for O$_2$ evolution (%)

$$= \frac{[\text{Amount of evolved O}_2,\ \text{mol}]}{\text{Total charge passed}/4 \times F,\ \text{mol}} \times 100 \quad (8)$$

Photocatalytic production of H$_2$O$_2$. Photocatalytic production of H$_2$O$_2$ was performed in a quartz anode cell (light path length = 1 cm) connected with a pyrex cathode cell through a Nafion membrane (Supplementary Fig. 16b). The anode cell consists of the as prepared m-WO$_3$/FTO photoanode for the water oxidation in an Ar-saturated aqueous solution (8 ml) containing HClO$_4$ (pH 1.3) and 0.1 M of NaClO$_4$. The cathode cell is composed of the as prepared Co^{II}(Ch)/CP cathode for the O$_2$ reduction in an O$_2$-saturated aqueous solution (10 ml) containing HClO$_4$ (pH 1.3) and 0.1 M of NaClO$_4$ at 298 K. These two electrodes were connected each other with alligator clips and copper wire as an external circuit. The photoanode was illuminated from the back side of the FTO electrode with the solar simulator (HAL-320, Asahi Spectra Co., Ltd.), where the light intensity was adjusted at 100 mW cm^{-2} (AM1.5G) at the sample position by the 1SUN checker (CS-20, Asahi Spectra Co., Ltd.). The anode and cathode solution was saturated by continuous bubbling with argon and oxygen gas for 30 min, respectively, before the photocatalytic reaction. The O$_2$ bubbling was continued during the photocatalytic reaction. The cathode cell was kept in dark to prevent the decomposition of produced H$_2$O$_2$ by ultraviolet-light irradiation during photocatalytic reaction. The amount of produced H$_2$O$_2$ was determined by spectroscopic titration with an acidic solution of [TiO(tpypH$_4$)]$^{4+}$ complex (Ti-TPyP reagent)^{29}. The Ti-TPyP reagent was prepared by dissolving 3.40 mg of the [TiO(tpyp)] complex in 100 ml of hydrochloric acid (50 mM). A small portion of the reaction solution was sampled and diluted with water depending on the concentration of produced H$_2$O$_2$. To 0.25 ml of 4.8 M HClO$_4$ and 0.25 ml of the Ti-TPyP reagent, a diluted sample was added. The mixed solution was then allowed to stand for 5 min at room temperature. This sample solution was diluted to 2.5 ml with water and used for the spectroscopic measurement. The absorbance at $\lambda = 434$ nm was measured by using a Hewlett Packard 8453 diode array spectrophotometer. A blank solution was prepared in a similar manner by adding distilled water instead of the sample solution to Ti-TPyP reagent in the same volume with its absorbance designated as A_B. The difference in absorbance was determined as follows: $\Delta A_{434} = A_B - A_S$. Based on ΔA_{434} and the volume of the solution, the amount of H_2O_2 was determined (Supplementary Fig. 17).

Detection of O_2. The concentration of O_2 in the anolyte was monitored during both photocatalytic production of H_2O_2 and photoelectrochemical measurements (*vide supra*) by using a fluorescence-based oxygen sensor (FOXY Fiber Optic Oxygen Sensor, Ocean Optics). The O_2-sensing needle probe was installed in a gas-tight quartz anode cell filled with 8 ml of a solution, which left 7 ml of a headspace, through a rubber septum on the end of the cell. The solution and headspace were purged with argon gas for 30 min before measurements. Two-point calibration of the O_2 sensor was performed against solutions (air, 20.9% O_2, and Ar, 0% O_2) used in each measurement. The amount of O_2 leaked in the anode cell during measurements was determined under dark and subtracted from the data obtained under illumination. The amount of dissolved O_2 in solutions was recorded as mole %. The total amount O_2 evolved in the anode cell was determined using Henry's Law and converted, using the ideal gas law, into μmol.

Solar-to-H_2O_2 energy conversion efficiency. Measurement of solar energy conversion efficiency of the photocatalytic production of H_2O_2 was carried out in a quartz anode cell (light path length = 1 cm) connected with a pyrex cathode cell through a Nafion membrane as used in photocatalytic production of H_2O_2 as described above. The photoanodes were illuminated from back side of the FTO electrode with the solar simulator (HAL-320, Asahi Spectra Co., Ltd.), where the light intensity was adjusted at 10–100 mW cm^{-2} (AM1.5G) at the sample position by the 1SUN checker (CS-20, Asahi Spectra Co., Ltd.). The amount of produced H_2O_2 was determined by the titration with the Ti-TPyP reagent (*vide supra*). The solar energy conversion efficiency was determined by following equations (3 and 9), where output energy as H_2O_2 was calculated by

Solar Energy Conversion Efficiency (%)

$$= \frac{[\text{Output energy as } H_2O_2]}{[\text{Energy density of incident solar light}] \times [\text{Irradiation area}]} \times 100$$
$$= \frac{[\text{Enthalpy change of equation}(\Delta H)] \times [\text{Produced amount of } H_2O_2]}{[\text{Energy density of incident solar light}] \times [\text{Irradiation area}]} \times 100$$

(9)

the multiplication of enthalpy change ($\Delta H = 98.3$ kJ mol^{-1}) and the produced amount of H_2O_2 (the concentration of $H_2O_2 \times$ volume of cathode solution (10 ml)). Energy density of incident solar light was adjusted at 10–100 mW cm^{-2} (0.1–1 SUN, Air Mass 1.5 (AM1.5)) at the sample position for whole irradiation area (2.5 cm^2) by the 1 SUN checker (CS-20, Asahi Spectra Co., Ltd.) at room temperature.

Spectroscopic measurements. Ultraviolet–visible spectroscopy was carried out on a Hewlett Packard 8453 diode array spectrophotometer at room temperature using quartz cell (light path length = 1.0 cm).

H_2O_2 fuel cell. $Fe^{II}_3[Co^{III}(CN)_6]_2$ was mounted onto a carbon cloth by drop-casting or by spraying a dispersion of $Fe^{II}_3[Co^{III}(CN)_6]_2$ in isopropanol with an airbrush (TAMIYA Spray-work HG). An aqueous solution of Nafion (0.2 wt.%) was used to protect the film of $Fe^{II}_3[Co^{III}(CN)_6]_2$ on a carbon cloth. A Ni mesh (150 mesh) and $Fe^{II}_3[Co^{III}(CN)_6]_2$ that was mounted onto a carbon cloth were immersed in the solution of H_2O_2. The performance tests were conducted in a one-compartment cell with the reaction solution containing H_2O_2 produced by the photocatalytic reaction transferred from the cathode cell of the two-compartment cell system. The current and power values normalized by the geometric surface area of an electrode were recorded on an ALS 630B electrochemical analyser and KFM 2005 FC impedance meter at 25 °C. The performance tests in solutions containing various concentrations of standard H_2O_2, $HClO_4$ (pH 1) and NaCl (1.0 M) were performed for the control experiment (Supplementary Fig. 8).

Energy conversion efficiency of H_2O_2 fuel cell. $Fe^{II}_3[Co^{III}(CN)_6]_2$/carbon cloth and Ni mesh electrodes were prepared as noted above. Each electrode was connected with Pt wire and protected by PP (polypropylene) sheet to avoid electrical short circuit. The performance tests were conducted in a well-sealed one-compartment cell with a rubber septum (Supplementary Fig. 9). The reaction solution containing H_2O_2 (0.3 M), NaCl (1.0 M) and $Sc(NO_3)_3 \bullet 4H_2O$ (0.1 M)[10] and the headspace (6.5 ml) of the one-compartment cell were purged separately with argon gas for 30 min before measurements. After the argon-saturated reaction solution was transferred to the one-compartment cell using gas-tight syringe, cell voltage, applying constant current of 3.3 mA, was recorded on a KFM 2005 FC impedance meter at 25 °C. The amount of evolved O_2 gas in the headspace of one-compartment cell was quantified by a Shimadzu GC-17A gas chromatograph

(Ar carrier, a capillary column with molecular sieves (Agilent Technologies, 19095PMS0, 30 m × 0.53 mm) at 313 K) equipped with a thermal conductivity detector. The energy conversion efficiency of H_2O_2 fuel cell was determined by following equations (3) and (10), where consumed chemical energy as H_2O_2 was calculated by the multiplication of enthalpy change ($\Delta H = -98.3$ kJ mol^{-1}) and twice of the produced amount of O_2 (Supplementary Fig. 10).

Energy Conversion Efficiency of H_2O_2 Fuel Cell (%)

$$= \frac{[\text{Output energy as electrical energy}]}{[\text{Consumed chemical energy as } H_2O_2]} \times 100$$
$$= \frac{[\text{Cell voltage}] \times [\text{Current}] \times [\text{Reaction time}]}{[\text{Enthalpy change of equation}(\Delta H)] \times [\text{Produced amount of } O_2] \times 2} \times 100$$

(10)

References

1. Faunce, T. A. *et al.* Energy and environment policy case for a global project on artificial photosynthesis. *Energy Environ. Sci.* **6**, 695–698 (2013).
2. Luo, J. *et al.* Water photolysis at 12.3% efficiency via perovskite photovoltaics and Earth-abundant catalysts. *Science* **345**, 1593–1596 (2014).
3. Maeda, K., Higashi, M., Lu, D., Abe, R. & Domen, K. Efficient nonsacrificial water splitting through two-step photoexcitation by visible light using a modified oxynitride as a hydrogen evolution photocatalyst. *J. Am. Chem. Soc.* **132**, 5858–5868 (2010).
4. Kato, S., Jung, J., Suenobu, T. & Fukuzumi, S. Production of hydrogen peroxide as a sustainable solar fuel from water and dioxygen. *Energy Environ. Sci.* **6**, 3756–3764 (2013).
5. Shiraishi, Y. *et al.* Sunlight-driven hydrogen peroxide from water and molecular oxygen by metal-free photocatalysts. *Angew. Chem. Int. Ed.* **53**, 13454–13459 (2014).
6. Fukuzumi, S., Yamada, Y. & Karlin, K. D. Hydrogen peroxide as a sustainable energy carrier: electrocatalytic production of hydrogen peroxide and the fuel cell. *Electrochim. Acta* **82**, 493–511 (2012).
7. Yamada, Y., Yoneda, M. & Fukuzumi, S. A robust one-compartment fuel cell with a polynuclear cyanide complex as a cathode for utilizing H_2O_2 as a sustainable fuel at ambient conditions. *Chem. Eur. J.* **19**, 11733–11741 (2013).
8. Yamada, Y., Yoneda, M. & Fukuzumi, S. High power density of one-compartment H_2O_2 fuel cells using pyrazine-bridged $Fe[M^C(CN)_4]$ ($M^C = Pt^{2+}$ and Pd^{2+}) complexes as the cathode. *Inorg. Chem.* **53**, 1272–1274 (2014).
9. Shaegh, S. A. M., Ehteshami, S. M. M., Chan, S. H., Nguyen, N.-T. & Tan, S. N. Membraneless hydrogen peroxide micro semi-fuel cell for portable applications. *RSC Adv.* **4**, 37284–37287 (2014).
10. Yamada, Y., Yoneda, M. & Fukuzumi, S. High and robust performance of H_2O_2 fuel cells in the presence of scandium ion. *Energy Environ. Sci.* **8**, 1698–1701 (2015).
11. Hisatomi, T., Kubota, J. & Domen, K. Recent advances in semiconductors for photocatalytic and photoelectrochemical water splitting. *Chem. Soc. Rev.* **43**, 7520–7535 (2014).
12. Kudo, A. & Miseki, Y. Heterogeneous photocatalyst materials for water splitting. *Chem. Soc. Rev.* **38**, 253–278 (2009).
13. Mi, Q., Coridan, R. H., Brunschwig, B. S., Gray, H. B. & Nathan, S. L. Photoelectrochemical oxidation of anions by WO_3 in aqueous and nonaqueous electrolytes. *Energy Environ. Sci.* **6**, 2646–2653 (2013).
14. Hill, J. C. & Choi, K.-S. Effect of Electrolytes on the Selectivity and Stability of n-type WO_3. *J. Phys. Chem. C* **116**, 7612–7620 (2012).
15. Pesci, F. M., Cowan, A. J., Alexander, B. D., Durrant, J. R. & Klug, D. R. Charge carrier dynamics on mesoporous WO_3 during water splitting. *J. Phys. Chem. Lett.* **2**, 1900–1903 (2011).
16. Mase, K., Ohkubo, K. & Fukuzumi, S. Efficient two-electron reduction of dioxygen to hydrogen peroxide with one-electron reductants with a small overpotential catalyzed by a cobalt chlorin complex. *J. Am. Chem. Soc.* **135**, 2800–2808 (2013).
17. Fukuzumi, S. *et al.* Catalytic four-electron reduction of O_2 via rate-determining proton-coupled electron transfer to a dinuclear cobalt-μ-1,2-peroxo complex.. *J. Am. Chem. Soc.* **134**, 9906–9909 (2012).
18. Chen, Z., Concepcion, J. J., Song, N. & Meyer, T. J. Chloride-assisted catalytic water oxidation. *Chem. Commun.* **50**, 8053–8056 (2014).
19. Nakagawara, S. *et al.* Spectroscopic characterization and the pH dependence of bactericidal activity of the aqueous chlorine solution. *Anal. Sci.* **14**, 691–698 (1998).
20. Huang, L. *et al.* Cl^- making overall water splitting possible on TiO_2-based photocatalysts. *Catal. Sci. Technol.* **4**, 2913–2918 (2014).
21. Miseki, Y. & Sayama, K. High-efficiency water oxidation and energy storage utilizing various reversible redox mediators under visible light over surface-modified WO_3. *RSC Adv.* **4**, 8308–8316 (2014).
22. Saito, R., Miseki, Y. & Sayama, K. Highly efficient photoelectrochemical water splitting using a thin film photoanode of $BiVO_4/SnO_2/WO_3$ multi-composite in a carbonate electrolyte. *Chem. Commun.* **48**, 3833–3835 (2012).

23. Kadish, K. M. *et al.* Electrochemistry, spectroelectrochemistry, chloride binding, and O_2 catalytic reactions of free-base porphyrin-cobalt corrole dyads. *Inorg. Chem.* **44,** 6744–6754 (2005).

24. Melis, A. Solar energy conversion efficiencies in photosynthesis: minimizing the chlorophyll antennae to maximize efficiency. *Plant Sci.* **177,** 272–280 (2009).

25. Armarego, W. L. F. & Chai, C. (eds) *Purification of Laboratory Chemicals 7th edn* (Butterworth-Heinemann, 2013)

26. Krasnoslobodtsev, A. V. & Smirnov, S. N. Effect of water on silanization of silica by trimethoxysilanes. *Langmuir* **18,** 3181–3184 (2002).

27. Rouquerol, F., Rouquerol, J. & Sing, K. in *Adsorption by Powder & Porous Solid* (Academic, 1999).

28. Izutsu, K. in *Electrochemistry in Nonaqueous Solutions* (Wiley-VCH GmbH & Co. KGaA, 2009).

29. Matsubara, C., Kawamoto, N. & Takamura, K. Oxo[5, 10, 15, 20-tetra(4-pyridyl) porphyrinato]titanium(IV): an ultra-high sensitivity spectrophotometric reagent for hydrogen peroxide. *Analyst* **117,** 1781–1784 (1992).

Acknowledgements

This work was supported by an Advanced Low Carbon Technology Research and Development and SENTAN projects from Japan Science Technology Agency to S.F. and Grants-in-Aid (Nos. 24350069 and 25600025 to Y.Y.) for Scientific Research from Japan Society for the Promotion of Science (JSPS). K.M. gratefully acknowledge support from JSPS by Grant-in-Aid for JSPS fellowship for young scientists (No. 25●727).

Author contributions

S.F. conceived and designed the experiments. K.M., M.Y. and Y.Y. performed the experiments and analysed the data. S. F., K.M., M.Y. and Y.Y. co-wrote the paper.

Additional information

Permissions

The contributors of this book come from diverse backgrounds, making this book a truly international effort. This book will bring forth new frontiers with its revolutionizing research information and detailed analysis of the nascent developments around the world.

We would like to thank all the contributing authors for lending their expertise to make the book truly unique. They have played a crucial role in the development of this book. Without their invaluable contributions this book wouldn't have been possible. They have made vital efforts to compile up to date information on the varied aspects of this subject to make this book a valuable addition to the collection of many professionals and students.

This book was conceptualized with the vision of imparting up-to-date information and advanced data in this field. To ensure the same, a matchless editorial board was set up. Every individual on the board went through rigorous rounds of assessment to prove their worth. After which they invested a large part of their time researching and compiling the most relevant data for our readers.

The editorial board has been involved in producing this book since its inception. They have spent rigorous hours researching and exploring the diverse topics which have resulted in the successful publishing of this book. They have passed on their knowledge of decades through this book. To expedite this challenging task, the publisher supported the team at every step. A small team of assistant editors was also appointed to further simplify the editing procedure and attain best results for the readers.

Apart from the editorial board, the designing team has also invested a significant amount of their time in understanding the subject and creating the most relevant covers. They scrutinized every image to scout for the most suitable representation of the subject and create an appropriate cover for the book.

The publishing team has been an ardent support to the editorial, designing and production team. Their endless efforts to recruit the best for this project, has resulted in the accomplishment of this book. They are a veteran in the field of academics and their pool of knowledge is as vast as their experience in printing. Their expertise and guidance has proved useful at every step. Their uncompromising quality standards have made this book an exceptional effort. Their encouragement from time to time has been an inspiration for everyone.

The publisher and the editorial board hope that this book will prove to be a valuable piece of knowledge for researchers, students, practitioners and scholars across the globe.

List of Contributors

Xiao Liang, Connor Hart, Quan Pang and Linda F. Nazar
Department of Chemistry, University of Waterloo, 200 University Avenue West, Waterloo, Ontario, Canada N2L 3G1

Arnd Garsuch and Thomas Weiss
BASF SE, 67056 Ludwigshafen

Tatjana Sentjabrskaja and Stefan U. Egelhaaf
Condensed Matter Physics Laboratory, Heinrich Heine University, 40225 Du¨sseldorf, Germany

Emanuela Zaccarelli, Cristiano De Michele and Francesco Sciortino
CNR-ISC, Universitàdi Roma 'La Sapienza', Piazzale A. Moro
2Roma 00185, Italy
Dipartimento di Fisica, Universitàdi Roma 'La Sapienza', Piazzale A. Moro 2, Roma 00185, Italy

Piero Tartaglia
Dipartimento di Fisica, Universita` di Roma 'La Sapienza', Piazzale A. Moro 2, Roma 00185, Italy

Thomas Voigtmann
Institut für Materialphysik imWeltraum, Deutsches Zentrum für Luft- und Raumfahrt (DLR), 51170 Köln, Germany
Heinrich Heine University, Universitätsstra_e 1, 40225 Düsseldorf, Germany

Marco Laurati
Condensed Matter Physics Laboratory, Heinrich Heine University, 40225 Düsseldorf, Germany
División de Ciencias e Ingeniería, Universidad de Guanajuato, Loma del Bosque 103, Leo´n 37150, Mexico

Guosheng Li, Xiaochuan Lu, Jin Y. Kim, Kerry D. Meinhardt, Hee Jung Chang, Nathan L. Canfield and Vincent L. Sprenkle
Electrochemical Materials and Systems Group, Energy Processes and Materials Division, Pacific Northwest National Laboratory, Richland, 99352 Washington

Uta Hejral, Patrick Müller and Andreas Stierle
Deutsches Elektronen-Synchrotron (DESY), NanoLab, Notkestrasse 85, D-22607 Hamburg, Germany
Universität Hamburg, Fachbereich Physik Jungiusstra_e 9, 20355 Hamburg, Germany
Universität Siegen, Fachbereich Physik, Walter-Flex-Stra_e 3, 57072 Siegen, Germany

Olivier Balmes
MAX IV Laboratory, Fotongatan 2, 225 94 Lund, Sweden
ESRF - The European Synchrotron, Radiation Facility, 71 Avenue des Martyrs, 38043 Grenoble, France

Diego Pontoni
ESRF - The European Synchrotron, Radiation Facility, 71 Avenue des Martyrs, 38043 Grenoble, France
Takanari Ouchi, Brian L. Spatocco and Donald R. Sadoway
Department of Materials Science and Engineering, Massachusetts Institute of Technology 77 Massachusetts Avenue, Cambridge, Massachusetts
02139-4307, USA

Hojong Kim
Department of Materials Science and Engineering, The Pennsylvania State University, 320 Forest Resources Laboratory, University Park, Pennsylvania 16802-4705, USA

Lukasz Piatkowski and Esther Gellings
ICFO—Institut de Ciencies Fotoniques, The Barcelona Institute of Science and Technology, 08860 Castelldefels (Barcelona), Spain

Niek F. van Hulst
ICREA—Institució Catalana de Recerca i Estudis Avançats, 08010 Barcelona, Spain

Ji-Sen Li
Jiangsu Key Laboratory of Biofunctional Materials, College of Chemistry and Materials Science, Nanjing Normal University, Nanjing 210023, China
Key Laboratory of Inorganic Chemistry in Universities of Shandong, Department of Chemistry and Chemical Engineering, Jining University, Qufu, Shandong 273155, China

Chun-Hui Liu, Shun-Li Li, Long-Zhang Dong, Zhi-Hui Dai, Ya-Fei Li and Ya-Qian Lan
Jiangsu Key Laboratory of Biofunctional Materials, College of Chemistry and Materials Science, Nanjing Normal University, Nanjing 210023, China

Yu-Guang Wang
Key Laboratory of Inorganic Chemistry in Universities of Shandong, Department of Chemistry and Chemical Engineering, Jining University, Qufu, Shandong 273155, China

Jakub Tymoczko and Wolfgang Schuhmann
Center for Electrochemical Sciences — CES, Ruhr-Universität Bochum, Universita'tsstrasse 150, D-44780 Bochum, Germany
Analytische Chemie — Elektroanalytik & Sensorik, Ruhr-Universität Bochum, Universita'tsstrasse 150, D-44780 Bochum, Germany

Federico Calle-Vallejo
Leiden Institute of Chemistry, Leiden University, PO-Box 9502, 2300 RA Leiden, The Netherlands

Aliaksandr S. Bandarenka
Center for Electrochemical Sciences — CES, Ruhr-Universität Bochum, Universitäts strasse 150, D-44780 Bochum, Germany
Physik-Department ECS, Technische Universita't München, James-Franck-Strasse 1, 85748 Garching, Germany
Nanosystems Initiative Munich (NIM), Schellingstrassee 4, 80799 Munich, Germany

Thierry Stoecklin and Philippe Halvick
Universitéde Bordeaux, Institut des Sciences Moléculaires, UMR 5255 CNRS, 33405 Talence, France

Mohamed Achref Gannouni and Majdi Hochlaf
Université Paris-Est, Laboratoire Modélisation et Simulation Multi Echelle, MSME UMR 8208 CNRS, 5 bd Descartes, 77454 Marne-la-Vallée, France

Eric R. Hudson
Department of Physics and Astronomy, University of California, 475 Portola Plaza, Los Angeles, California 90095, USA

Amirmehdi Saedi and Marcel J. Rost
Huygens-Kamerlingh Onnes Laboratory, Leiden University, Niels Bohrweg 2, Leiden 2333 CA, The Netherlands. w Present address: ARCNL, Science Park 102, Amsterdam 1098 XG, The Netherlands

Lu Huo, Ian Davis, Shingo Esaki and Aimin Liu
Department of Chemistry, Georgia State University, Atlanta, Georgia 30303, USA
Molecular Basis of Disease Area of Focus Program, Georgia State
University, Atlanta, Georgia 30303, USA

Fange Liu
Department of Chemistry, Georgia State University, Atlanta, Georgia 30303, USA

Babak Andi
Photon Sciences Directorate, Brookhaven National Laboratory, Upton, New York 11973, USA

Hiroaki Iwaki and Yoshie Hasegawa
Department of Life Science and Biotechnology and ORDIST, Kansai University, Suita, Osaka 564-8680, Japan

Allen M. Orville
Photon Sciences Directorate, Brookhaven National Laboratory, Upton, New York 11973, USA
Biosciences Department, Brookhaven National Laboratory, Upton, New York 11973, USA

Lin-Tai Da, Fátima Pardo-Avila and Lu Zhang
Department of Chemistry, School of Science and Institute for Advance Study, Hong Kong University of Science and Technology, Clear Water Bay, Kowloon,Hong Kong

Liang Xu and Dong Wang
Department of Cellular and Molecular Medicine, School of Medicine; Skaggs School of Pharmacy and Pharmaceutical Sciences, University of California San Diego, La Jolla, California 92093, USA

Daniel-Adriano Silva
Department of Chemistry, School of Science and Institute for Advance Study, Hong Kong University of Science and Technology, Clear Water Bay, Kowloon, Hong Kong
Department of Biochemistry, University of Washington, Seattle, Washington 98195, USA

Xin Gao
King Abdullah University of Science and Technology, Computational Bioscience Research Center, Computer, Electrical and Mathematical Sciences and Engineering Division, Thuwal 23955-6900, Saudi Arabia

Xuhui Huang
Department of Chemistry, School of Science and Institute for Advance Study, Hong Kong University of Science and Technology, Clear Water Bay, Kowloon, Hong Kong
Division of Biomedical Engineering, School of Science and Institute for Advance Study, Hong Kong University of Science and Technology, ClearWater Bay, Kowloon, Hong Kong
Center of Systems Biology and Human Health, School of Science and Institute for Advance Study, Hong Kong University of Science and Technology, Clear Water Bay, Kowloon, Hong Kong

Mei-Ling Wu
Key Laboratory of Molecular Nanostructure and Nanotechnology, Institute of Chemistry, Chinese Academy of Sciences (CAS), Beijing 100190, China
Beijing National Laboratory for Molecular Sciences, Beijing 100190, China
University of CAS, Beijing 100049, China

Dong Wang and Li-Jun Wan
Key Laboratory of Molecular Nanostructure and Nanotechnology, Institute of Chemistry, Chinese Academy of Sciences (CAS), Beijing 100190, China
Beijing National Laboratory for Molecular Sciences, Beijing 100190, China

Calvin Boerigter, Robert Campana, Matthew Morabito and Suljo Linic
Department of Chemical Engineering, University of Michigan, Arbor, Michigan 48109, USA

Borja Cirera, Francisco Martínez-Peña, Alberto Martin-Jimenez, Ana M. Pizarro and David Ecija
IMDEA Nanoscience, c/Faraday 9, Cantoblanco, 28049 Madrid, Spain

Nelson Giménez-Agullo
Institute of Chemical Research of Catalonia, Barcelona Institute of Science and Technology, Avinguda Països Catalans 16, Tarragona 43007, Spain

Jonas Björk
Department of Physics, Chemistry and Biology, IFM, Linko¨ping University, 58183 Linköping, Sweden

Jonathan Rodriguez-Fernandez
Departamento de Física de la Materia Condensada, Universidad Autónoma de Madrid, c/Francisco Tomásy Valiente 7, Cantoblanco, 28049 Madrid, Spain

Roberto Otero
IMDEA Nanoscience, c/Faraday 9, Cantoblanco, 28049 Madrid, Spain
Departamento de Física de la Materia Condensada, Universidad Autónoma de Madrid, c/Francisco Toma´s y Valiente 7, Cantoblanco, 28049 Madrid, Spain

JoséM. Gallego
Instituto de Ciencia de Materiales de Madrid, c/ Sor Juana Inés de la Cruz 3, Cantoblanco, 28049 Madrid, Spain

Pablo Ballester and JoséR. Galan-Mascaros
Institute of Chemical Research of Catalonia, Barcelona Institute of Science and Technology, Avinguda Països Catalans 16, Tarragona 43007, Spain
Catalan Institution for Research and Advanced Studies, Passeig Lluis Companys 23, Barcelona 08010, Spain

Jinsoo Kim, Hyeokjun Park, Won Mo Seong, Hee-Dae Lim and Youngjoon Bae
Department of Materials Science and Engineering, Research Institute of Advanced Materials, Seoul National University, 1 Gwanak-ro, Gwanak-gu, Seoul 151-742, Republic of Korea

Byungju Lee, Haegyeom Kim and Kisuk Kang
Department of Materials Science and Engineering, Research Institute of Advanced Materials, Seoul National University, 1 Gwanak-ro, Gwanak-gu, Seoul 151-742, Republic of Korea
Center for Nanoparticle Research at Institute for Basic Science, Seoul National University, 1 Gwanak-ro, Gwanak-gu, Seoul 151-742, Republic of Korea

Won Keun Kim and Kyoung Han Ryu
Environment and Energy Research Team, Division of Automotive Research and Development, Hyundai Motor Company, 37 Cheoldobangmulgwan-ro, Uiwang, Gyeonggi-do 437-815, Republic of Korea

Sarah Holliday, Raja Shahid Ashraf, Andrew Wadsworth, Derya Baran, Christian B. Nielsen, Ching-Hong Tan, Stoichko D. Dimitrov and James R. Durrant
Department of Chemistry and Centre for Plastic Electronics, Imperial College London, London SW7 2AZ, UK

Amber Yousaf
Department of Physics, Government College University, Lahore 54000, Pakistan

Zhengrong Shang and Alberto Salleo
Department of Materials Science and Engineering, Stanford University, 476 Lomita Mall, Stanford, California 94305, USA

Nicola Gasparini and Christoph J. Brabec
Institute of Materials for Electronics and Energy Technology (I-MEET), Friedrich-Alexander-University Erlangen-Nuremberg, 91058 Erlangen, Germany

Maha Alamoudi and Frédéric Laquai
King Abdullah University of Science and Technology (KAUST), Solar and Photovoltaics Engineering Research Center (SPERC), Thuwal 23955-6900, Saudi Arabia

Iain McCulloch
Department of Chemistry and Centre for Plastic Electronics, Imperial College London, London SW7 2AZ, UK
King Abdullah University of Science and Technology (KAUST), Solar and Photovoltaics Engineering Research Center (SPERC), Thuwal 23955-6900, Saudi Arabia

William D. Richards and Yan Wang
Department of Materials Science and Engineering, Massachusetts Institute of Technology, Cambridge, Massachusetts 02139, USA

Tomoyuki Tsujimura, Ichiro Uechi and Naoki Suzuki
Samsung R&D Institute Japan, Minoh Semba Center Building 13F, Semba Nishi 2-1-11, Minoh, Osaka 562-0036, Japan

Lincoln J. Miara
Samsung Advanced Institute of Technology—USA, 255 Main Street, Suite 702, Cambridge, Massachusetts 02142, USA

Jae Chul Kim
Department of Materials Science and Engineering, Massachusetts Institute of Technology, Cambridge, Massachusetts 02139, USA
Materials Sciences Division, Lawrence Berkeley National Laboratory, Berkeley, California 94720, USA

Shyue Ping Ong
Department of NanoEngineering, University of California San Diego, La Jolla, California 92093-0448, USA

Gerbrand Ceder
Department of Materials Science and Engineering, Massachusetts Institute of Technology, Cambridge, Massachusetts 02139, USA
Materials Sciences Division, Lawrence Berkeley National Laboratory, Berkeley, California 94720, USA
Department of Materials Science and Engineering, University of California Berkeley, Berkeley, California 94720, USA

Liang Chen, Hezhu Shao, Xufeng Zhou, Guoqiang Liu, Jun Jiang and Zhaoping Liu
Ningbo Institute of Materials Technology and Engineering, Chinese Academy of Sciences, Ningbo 315201, China

Yudan Yin, Lin Niu and Dehai Liang
Beijing National Laboratory for Molecular Sciences, MOE Key Laboratory of Polymer Chemistry and Physics, College of Chemistry and Molecular Engineering, Peking University, Beijing 100871, China

Xiaocui Zhu and Meiping Zhao
Beijing National Laboratory for Molecular Sciences, MOE Key Laboratory of Bioorganic Chemistry and Molecular Engineering, College of Chemistry and Molecular Engineering, Peking University, Beijing 100871, China

Zexin Zhang
Center for Soft Condensed Matter Physics and Interdisciplinary Research, Soochow University, Suzhou 215006, China

Stephen Mann
Centre for Protolife Research, School of Chemistry, University of Bristol, Bristol BS8 1TS, UK

Johannes W. Cremer and Ruth Signorell
Department of Chemistry and Applied Biosciences, Laboratory of Physical Chemistry, ETH Zurich, Vladimir-Prelog-Weg 2, CH-8093 Zurich, Switzerland

Klemens M. Thaler and Christoph Haisch
Laboratory for Applied Laser Spectroscopy, Chair of Analytical Chemistry, Technical University of Munich, Marchioninistrasse 17, D-81377 Munich, Germany

Deli Wang, Sufen Liu and Jie Wang
Key laboratory of Material Chemistry for Energy Conversion and Storage (Huazhong University of Science and Technology), Ministry of Education, Hubei Key Laboratory of Material Chemistry and Service Failure, School of Chemistry and Chemical Engineering, Huazhong University of Science and Technology, Wuhan 430074, China

Ruoqian Lin and Huolin L. Xin
Center for Functional Nanomaterials, Brookhaven National Laboratory, Upton, New York 11973, USA

Masahiro Kawasaki
JEOL USA, Inc., Peabody, Massachusetts 01960, USA
Eric Rus, Katharine E. Silberstein, Michael A. Lowe and Héctor D. Abruña
Department of Chemistry and Chemical Biology, Cornell University, Ithaca, New York 14853, USA

Feng Lin
Energy Storage and Distributed Resources Division, Lawrence Berkeley National Laboratory, Berkeley, California 94720, USA

Dennis Nordlund
Stanford Synchrotron Radiation Lightsource, SLAC National Accelerator Laboratory, Menlo Park, California 94025, USA

David A. Muller
School of Applied and Engineering Physics, Cornell University, Ithaca, New York 14853, USA
Kavli Institute at Cornell for Nanoscale Science, Cornell University, Ithaca, New York 14853, USA

Venkateshkumar Prabhakaran, B. Layla Mehdi, K. Don D. Gunaratne, Grant E. Johnson and Julia Laskin
Physical Sciences Division, Pacific Northwest National Laboratory, PO Box 999, MSIN K8-88, Richland, Washington 99352, USA

Jeffrey J. Ditto and David C. Johnson
Department of Chemistry, University of Oregon, Eugene, Oregon 97403, USA

Mark H. Engelhard and Bingbing Wang
Environmental Molecular Sciences Laboratory, Pacific Northwest National Laboratory, Richland, Washington 99352, USA. w Present address: State Key Laboratory of Marine and Environmental Science and College of Ocean and Earth Sciences, Xiamen University, Xiamen 361102, China

Steve J. Clark, Da Wang and Peter G. Bruce
Departments of Materials and Chemistry, University of Oxford, Parks Road, Oxford OX1 3PH, UK

A. Robert Armstrong
School of Chemistry and EastChem, University of StAndrews, North Haugh, Fife, Saint Andrews KY16 9ST, UK

Xinxin Cheng, Yao Zhang and Peter M. Weber
Department of Chemistry, Brown University, 324 Brook Street, Providence, Rhode Island 02912, USA

Elvar Jónsson
COMP, Department of Applied Physics, Aalto University, FIN-00076 Espoo, Finland

Hannes Jónsson
Department of Chemistry, Brown University, 324 Brook Street, Providence, Rhode Island 02912, USA COMP, Department of Applied Physics, Aalto University, FIN-00076 Espoo, Finland Faculty of Physical Sciences, VR-III, University of Iceland, 107 Reykjavı ́k, Iceland. w Present address: Max Planck
Institute for the Structure and Dynamics of Matter (MPSD), Luruper Chaussee 149, 22761 Hamburg, Germany

Hyeongwook Im, Taewoo Kim, Hyelynn Song and Jongho Choi
School of Mechanical and Aerospace Engineering, Seoul National University, Seoul 151-742, South Korea

Jae Sung Park
Institute of Advanced Machinery and Design, Seoul National University, Seoul 151-742, South Korea

Raquel Ovalle-Robles
Nano-Science & Technology Center, Lintec of America, Inc., Richardson, Texas 75081, USA

Hee Doo Yang
Department of NanoMechatronics Engineering, College of Nanoscience and Nanotechnology, Pusan National University, Busan 609-735, South Korea

Kenneth D. Kihm
Department of Mechanical, Aerospace and Biomedical Engineering, University of Tennessee, Knoxville, Tennessee 37996, USA

Ray H. Baughman
Alan G. MacDiarmid NanoTech Institute, University of Texas at Dallas, Richardson, Texas 75080, USA

Hong H. Lee
School of Chemical and Biological Engineering, Seoul National University, Seoul 151-744, South Korea

Tae June Kang
Department of Mechanical Engineering, INHA University, Incheon 22212, South Korea

Yong Hyup Kim
School of Mechanical and Aerospace Engineering, Seoul National University, Seoul 151-742, South Korea Institute of Advanced Aerospace Technology, Seoul National University, Seoul 151-742, South Korea

Yongjun Li, Zhiyu Jia, Huibiao Liu and Yuliang Li
CAS Key Laboratory of Organic Solids, Beijing National Laboratory for Molecular Science (BNLMS), Institute of Chemistry, Chinese Academy of Sciences, Beijing 100190, China

Shengqiang Xiao
State Key Laboratory of Advanced Technology for Materials Synthesis and Processing,Wuhan University of Technology, Wuhan 430070, China

Kentaro Mase and Masaki Yoneda
Department of Material and Life Science, Graduate School of Engineering, Osaka University, ALCA and SENTAN, Japan Science and Technology Agency (JST), Suita, Osaka 565-0871, Japan

Yusuke Yamada
Department of Applied Chemistry and Bioengineering, Graduate School of Engineering, Osaka City University, Osaka
558-8585, Japan

Shunichi Fukuzumi
Department of Material and Life Science, Graduate School of Engineering, Osaka University, ALCA and SENTAN, Japan Science and Technology Agency (JST), Suita, Osaka 565-0871, Japan Faculty of Science and Technology, Meijo University, ALCA and SENTAN, Japan Science and Technology Agency, JST, Shiogamaguchi, Tenpaku, Nagoya, Aichi 468-8502, Japan Department of Chemistry and Nano Science, Ewha Womans University, Seoul

Index